Functional groups and their IUPAC numenclature suffixes (Table 1.1).

Structure	Condensed structure	Name	Suffix
* $-\overset{\overset{O}{\|}}{C}-OH$	$-COOH$ or $-CO_2H$	**Carboxylic acid**	-oic acid (-carboxylic acid)
$-\overset{\overset{O}{\|}}{\underset{\underset{O}{\|}}{S}}-OH$	$-SO_3H$	**Sulfonic acid**	-sulfonic acid
* $-\overset{\overset{O}{\|}}{C}-O-$	$-COO-$ or $-CO_2-$	**Ester**	-oate (-carboxylate)
* $-\overset{\overset{O}{\|}}{C}-Cl$	$-COCl$	**Acid chloride**	-oyl chloride
* $-\overset{\overset{O}{\|}}{C}-NH_2$	$-CONH_2$	**Amide**	-amide (-carboxamide)
* $-C\equiv N$	$-CN$	**Nitrile**	-nitrile (-carbonitrile)
* $-\overset{\overset{O}{\|}}{C}-H$	$-CHO$	**Aldehyde**	-al (-carbaldehyde)
$-\overset{\overset{O}{\|}}{C}-$	$-CO-$	**Ketone**	-one
$-OH$	$-OH$	**Alcohol or Phenol**	-ol
$-SH$	$-SH$	**Thiol**	-thiol
$-\overset{H}{\underset{H}{N}}$	$-NH_2$	**Amine**	-amine
$-C=C-$	$-C=C-$	**Alkene**	-ene
$-C\equiv C-$	$-C\equiv C-$	**Alkyne**	-yne
$-O-$	$-O-$	**Ether**	ether
(benzene ring structure)	(benzene ring)	**Benzene**	benzene

Increasing priority →

ORGANIC CHEMISTRY

ORGANIC CHEMISTRY

Second Edition

Thomas N. Sorrell
The University of North Carolina at Chapel Hill

University Science Books
Sausalito, California

University Science Books
www.uscibooks.com

Developmental editor: Mary Castellion
Production manager: Ann Knight
Manuscript editor: Jeannette Stiefel
Text design: Caroline McGowan
Cover design: George Kelvin
Compositor: Publishers' Design and Production Services, Inc.
Illustrator: Lineworks
Manufacturer: VonHoffmann Corporation

This book is printed on acid-free paper.

University Science Books, Sausalito, California
©1999, 2006
All rights reserved. First edition 1999. Second edition 2006.

ISBN 10: 1-891389-38-6
ISBN 13: 978-1-891389-38-2

Library of Congress Cataloging-in-Publication Data

Sorrell, Thomas N.
    Organic Chemistry / Thomas N. Sorrell.—2nd ed.
        p. cm.
    Includes bibliographical reference and index.
    ISBN 1-891389-38-6 (alk. paper)
       1. Chemistry, Organic—Textbooks.   I. Title.

QD251.3.S67 2005
547—dc22   2005052854

Printed in the United States of America
10 9 8 7 6 5 4 3 2 1

This book is dedicated to Courtney—
a loving daughter and burgeoning research chemist,
who perceives the beauty of organic chemistry as
readily as she sees the splendors of the Carolina coast
and the majesty of the Rocky Mountains.

CONTENTS OVERVIEW

CONTENTS

Chapter 8   ELIMINATION REACTIONS OF ALKYL HALIDES, ALCOHOLS, AND RELATED COMPOUNDS *239*

Chapter 9   ADDITION REACTIONS OF ALKENES AND ALKYNES *267*

Chapter 10   ADDITION REACTIONS OF CONJUGATED DIENES *303*

Chapter 11   OXIDATION AND REDUCTION REACTIONS *331*

The manner of teaching organic chemistry has changed somewhat since my days as a student in the early 1970s. Most notably, organic chemistry textbooks offer more and better descriptions of topics in related fields, such as biochemistry and materials science; the internet allows one to search for information about specific topics; and computer software is readily available for modeling chemical structures and reactions. The overall level of sophistication has also risen for the presentation of traditional themes such as stereochemistry, bonding, reaction mechanisms, spectroscopy, and synthesis.

In spite of these changes, however, the mastery of organic chemistry as a course of study still requires a sound knowledge of the principles of molecular structure and chemical reactivity, which are topics introduced in most general chemistry courses. With such a background, a student studying organic chemistry begins to focus on a more limited set of atomic building blocks, particularly of carbon and its elemental neighbors. And while the study of a smaller portion of the periodic table might be expected to be easily manageable, understanding organic chemistry can still seem overwhelming because of the diverse ways that this handful of elements can combine and interact. To learn organic chemistry, one must grasp the recurring patterns that correlate the presented facts.

Toward that end, this textbook organizes and discusses the patterns of chemical reactivity, which constitutes the majority of the subject matter, by combining information about the structures of functional groups (the reactive portions of a molecule) with the reaction mechanisms (pathways of chemical reactions) that these functional groups undergo. This approach differs from the one presented in many other texts, which describe *every* type of reaction that can occur for a given functional group; each approach has its advantages and disadvantages. The one I have utilized here evolved from my objective to integrate discussions about biochemical processes with the types of reactions that are carried out in chemistry laboratories. With the use of two points of reference—structures *and* mechanisms—the similarities that associate biochemical and synthetic reactions can be appreciated more easily.

ORGANIZATION OF THE TEXT

A sound knowledge of structures, bonding, and stereochemistry—the three-dimensional arrangements and shapes of molecules—is required in order to understand the patterns of chemical reactivity. For that reason, this book begins in Chapters 1–4 with a detailed treatment of molecular structures. After Chapter 5, which describes some general aspects of chemical reactions, specific transformations are presented in the next several chapters.

Spectroscopic methods are then covered in Chapters 13 and 14 (and used in subsequent chapters), with the emphasis in the first of these two chapters on nuclear magnetic resonance (NMR) spectroscopy. Chapter 14 introduces other analytical methods and integrates a number of techniques for the elucidation of molecular structures.

The topic of chemical synthesis constitutes a sizable portion of Chapters 15 and 16, with the latter focusing on enantioselective reactions. As with the spectroscopic methods presented in the previous two chapters, synthetic methods are used throughout the remainder of the text.

Chapter 17 summarizes the structures and reactions of aromatic compounds, and then the next several chapters present information about the reactions associated with the carbon–oxygen double bond, a structural feature of organic molecules that is pervasive in biochemical systems. Chapters 18–24 pay particular attention to chemical reactions that occur in nature.

The final four chapters cover specific topics that make use of the basic structures and reactions that have already been presented. The closing two chapters describe the chemistry of amino and nucleic acids, which establish the background for subsequent studies in biology, biochemistry, and molecular biology.

EXERCISES

Rodger Griffin, Jr., who taught me organic chemistry, was fond of saying that organic chemistry was best learned by solving problems and not by reading chapter-by-chapter as one would do for a history or philosophy course. This text has many exercises incorporated within the text as well as at the end of each chapter. Furthermore, this second edition has included many more examples (solved exercises) within the chapters so that students can see successful approaches to problem solving. The accompanying *Solutions to Exercises* book has the answer to every exercise; in many instances, the approach needed to work toward the answer is included along with the factual solution.

"Organic Chemistry" at most colleges and universities carries the unfortunate status of being among the more difficult and demanding courses offered. With its position in the curriculum as the prerequisite for many upper level chemistry, biology, and pharmacy courses, its negative repute is unrivaled. No wonder many students refer to this two-semester sequence as the "premed weed-out courses!"

Your experience does not have to be that way, however. Below are listed several suggestions I make at the beginning of each semester aimed at maximizing my students' success in this subject.

- *Go to class and participate.*

  I have written as clearly about organic chemistry as I can, but let's be honest— if you could learn this stuff on your own by reading the book and working exercises, then you're a whiz kid and going to class probably won't matter. If you're a typical student, however, verbal explanations *will* help, and that's one thing that class is about. It's also about seeing how problems are solved and how others look at them. Students who genuinely participate usually learn a lot. They also impress their professors, which can't hurt.

- *Be prepared.*

  Read the sections in the book that will be covered in class on a given day and think about how the material fits with what you know or have covered recently. Much of the material that you will learn in your organic chemistry courses is conceptually new—organic chemistry is *not* a repeat of general chemistry. If you have an idea beforehand about the key points of the topics being discussed, you will learn the material faster. On the other hand, don't spend hours and hours reading and taking notes—organic chemistry is not an English or history course.

- *Study (even a little) every day.*

  Easier said than done: I had many teachers tell me this when I was a student, and I myself have said the same thing to countless students. Does anyone listen? Maybe for about a week at the beginning of the semester. Then the pressures of life and your other courses get in the way, and you're back to cramming the night before an exam. I learned this lesson by experience: If you want to do well, studying one hour each day of a week is at least twice as good as studying seven hours once each week.

- *Work the assigned exercises.*

  Understanding organic chemistry is about seeing and learning the recurring *patterns* that correlate the many facts being presented. Those who see and learn these patterns—among structures, chemical reactions, molecular properties— usually do well. Those who don't (or can't) learn these patterns will struggle. Your brain may be wired in ways that make it difficult for you to see the underlying design of organic chemistry. There's not much that can be done about this, but it doesn't mean you will fail—it just means that you may have to work harder than others in order to succeed. The best way to learn these patterns is to work problems. And then work more problems. And then work even more problems.

  The answers to many of the study exercises may not be obvious when you first read the problem. Do not give up quickly and do not consult the *Solutions to Exercises* book without making a *determined* effort to solve the exercise on your own. Many people can read an answer and understand it. Don't be fooled into thinking that you can solve problems because you understand the answers when you see them. Write the solution to each exercise on paper. Even if the answer seems obvious, writing it will help you remember and learn. Also, do several problems before you look at their answers. If you look up the solution as you do each exercise, you may catch a glimpse of the answer to the next problem. The result is that you are not really working the next one.

- *Make flash cards.*

  In addition to doing the study exercises, some students find it useful to make "flash cards," especially to learn the many chemical reactions that you will have to commit to memory. A set of cards with important information makes it easy to review the course material while you're waiting for a movie to begin, sitting between classes, riding the bus, or any time during the day when you may have five or ten minutes to study. Organic chemistry is a cumulative subject. At the end of the semester, you will need to know material presented on the first day of the course just as much as you will need to know it for the first exam. A set of cards will be invaluable for review. NOTE WELL: Using flash cards prepared by someone else has limited benefit. Much of their advantage comes from having to think about structures and chemical reactions enough to prepare the cards.

- *Before an exam, get a good night of sleep.*

  Many of you will be expected to solve problems on exams that you have not seen before. If you have studied regularly, then being alert and relaxed is more important than last minute cramming. If you haven't studied regularly, you're probably in trouble whether you're rested or not. So you might as well be rested.

- *Get help early.*

  The sooner you realize that you're in trouble, the better. Go talk to your professor, a teaching assistant, or a tutor. Or form a study group. Some students find that other students can help them sort out confusing facts as well as anyone, and an added benefit is a measure of accountability to your peers.

I am always interested to hear from students about what works and what doesn't. If you find factual errors or discussions that are confusing, please send me an email to let me know (sorrell@unc.edu). Such comments are important to keep this text evolving so that it is as useful as possible to its readers—you, the students. Sometimes the most innocent remark can make me understand where a student has lost the thread of thought that winds through the presentation of a particular topic, and it is that type of input that teaches me and helps me to teach others better.

As noted in the Preface to the first edition of this book, I became dissatisfied with many organic chemistry texts when I started teaching biochemistry. The same students who had done well in organic chemistry could not seem to remember even simple reactions related to the ones being covered in the biochemistry course. Yet I remembered, for example, covering the structures and reactions of carbohydrates, amino acids, and nucleic acids in the previous courses. I concluded that the text we were using had failed to *associate* the material shared by the two disciplines because the topics related to biochemistry were segregated within their own chapters and students had compartmentalized the material they had learned, mentally labeling the information as "organic chemistry" or "biochemistry."

In writing this textbook, and especially this second edition, my goal has been to integrate the information about many of the fundamental biochemical reactions with the corresponding transformations that are carried out in organic chemistry laboratories. Many who teach the biochemistry courses will give scant attention to the details of the chemical reactions that constitute metabolic processes, which I think is useful information to be learned by the students who hope to pursue a career in the health professions and who constitute the majority of those taking organic chemistry. The organic chemistry courses provide the best (and perhaps only) forum to make students aware of such details. I believe that this awareness can be facilitated by illustrating the parallels between biochemical processes and the simpler chemical reactions that are being discussed. Even if students cannot remember specific facts later, they will likely recall that such associations exist; and they will be more likely later to dig out their organic chemistry texts to review these topics when they want to understand a particular biochemical process at the molecular level. As most organic chemistry instructors know, the key for understanding chemical reactions resides in comprehending their mechanisms. Therefore, reaction mechanisms comprise the major organizational theme of this text, and their details are presented in order to illustrate and underscore the similarities between synthetic and biochemical processes.

To develop these connections about chemical reactions, however, one cannot gloss over the facts about molecular structures, stereochemistry, and thermodynamics. Therefore, these topics are developed in the first five chapters of this book, and I trust they are presented with sufficient depth that students who intend to become organic chemists will not be shortchanged. Toward that same end of serving the future chemists in the organic chemistry classes, I have also included presentations of synthetic reactions that appear frequently in the current literature (e.g., the Misunobu, Swern, and Suzuki reactions) as well as topics such an enantioselective synthesis and molecular recognition.

NEW TO THIS EDITION

Based on the feedback given to me by hundreds of students as well as from the critiques of the dedicated reviewers listed in the Acknowledgments, several of whom taught their courses using the first edition of this text, I have made changes that are meant to make the material better organized and easier to understand. I estimate that one-quarter to one-third of the text has been completely rewritten. A substantial change in the content comprises the inclusion of more examples (solved exercises).

As is often true of second editions, a number of the more specialized topics (and chemical reactions) that appeared in the first edition have been left out. Most notably absent is the first chapter of the previous edition, which provided a historical perspective of organic chemistry. The current first chapter—on basic structures and nomenclature—still stands as the organizational pillar of structural chemistry for the book, but it has been modified by incorporating the strategies for naming compounds (which was Appendix A in the first edition) and eliminating the discussions of some types of groups that are better saved until later in the book (esters, in particular).

The first chapter that concerns a specific type of chemical reaction (nucleophilic substitution) has been expanded and divided into two chapters in this edition (Chapters 6 and 7). In this way, alkyl halides and alcohol substitution reactions are treated separately, and some sulfur-containing molecules are now included in the chapter on alcohols.

The oxidation reactions of alcohols have been removed from the chapter on elimination reactions (first edition), and a separate chapter on reduction and oxidation reactions has been created here (Chapter 11). The information in this chapter also includes discussions about the reduction and oxidation reactions of alkenes, which were covered previously in a chapter about alkene addition reactions. Also, the chemistry of dienes, including the Diels–Alder reaction, have been collected in a chapter separate from the one devoted to the addition reactions of simple alkenes.

The order of topics in the chapters that present spectroscopic methods has been reversed from that in the previous edition, so nuclear magnetic resonance spectroscopy is now covered first. The subsequent spectroscopy chapter integrates the use of several techniques, including elemental analysis, for the elucidation of molecular structures.

The chapter that introduces synthetic methods has been largely preserved from the first edition, but it is followed directly by the chapter on enantioselective synthesis (previously, these two chapters were separated by the spectroscopy chapters). The discussion of enantioselective reactions has been completely rewritten, and its emphasis has been changed to encourage students to think about designing enantioselective syntheses without having to memorize a lot of details about specific reagents and conditions.

The topic of aromatic compounds (benzene and its derivatives) has been moved (from Chapter 12 in the first edition to Chapter 17 in the second) with the material on polycyclic arenes combined with an abbreviated discussion of heterocycles in Chapter 24. The presentations about diazonium compounds and nucleophilic aromatic substitution reactions, which were in the chapter on nitrogen-containing compounds (Chapter 23, first edition), have been incorporated into Chapter 17 (second edition).

The chapters describing carbonyl compounds have been kept largely intact. The most significant change is the division of Chapter 17 (first edition) about aldehydes, ketones, and carbohydrates into two chapters in the current edition. The division has been made according to the reaction mechanisms involved (nucleophilic addition versus nucleophilic addition—substitution), not according to the functional groups that are undergoing the reactions.

The chapter on nitrogen-containing compounds (Chapter 23 in the first edition) has been parceled in this edition among several chapters, as appropriate. In contrast, the discussions of polymer chemistry, which were interspersed throughout the book in the first edition, have been collected to form Chapter 26 in this edition. Some basic presentations of polymerization processes have been left in the earlier chapters to show the mechanistic similarities to other fundamental reactions, but many of the details are developed in this later chapter.

The final two chapters on amino and nucleic acids, including the topic of molecular recognition, are similar in style and content to those in the first edition.

ACKNOWLEDGMENTS

For its creation, a good textbook requires participation by many people besides the author, and I am profoundly grateful to the following individuals for the work they did to make this second edition a reality.

With their comments and questions while using the first edition of this text, the students that I have taught during the past several years contributed invaluably to the development and writing of this second edition.

My daughter Courtney, who is now a graduate student in chemistry at Georgia Tech, was especially helpful during the early and middle stages of the writing process as she helped with the typing, both of the text and *Solutions to Exercises*. Her work at Synthematix (a local startup company that has since been acquired by Symyx Technologies, Inc.), gave her the experience to offer suggestions that helped organize several topics better.

My professional peers offered much in the way of their technical comments, corrections, and support. At the University of North Carolina at Chapel Hill, my colleagues Maurice Brookhart, Joseph DeSimone, Richard Hiskey, Paul Kropp, and Gary Pielak shared their valuable expertise on specific topics that appeared in both editions. Paul Kropp suggested some especially constructive changes for this edition as a result of using the book in the Honors courses at UNC. At other institutions, the following persons read all or parts of the manuscripts and provided detailed critiques and suggestions that helped make the text more readable, more accurate, and more pedagogical.

SECOND EDITION

David Bergbreiter	*Texas A&M University*	Neil Marsh	*University of Michigan*
René Boeré	*University of Lethbridge*	William le Noble	*SUNY Stonybrook*
Andrew Bororvik	*University of Kansas*	Jennifer Radkiewicz	*Old Dominion University*
David Bundle	*University of Alberta*	Christian Rojas	*Barnard College*
Marjorie Caserio	*UC, San Diego*	Dalibor Sames	*Columbia University*
Chuck Doubleday	*Columbia University*	Martin Semmelhack	*Princeton University*
Andrew French	*Albion College*	J. William Suggs	*Brown University*
Warren Giering	*Boston University*	Marcus Thomsen	*Franklin & Marshall College*
John Hubbard	*Marshall University*	F. Dean Toste	*UC, Berkeley*
Steven Kass	*University of Minnesota*	Nick Turro	*Columbia University*
Susan King	*UC, Irvine*	Nanine Van Draanen	*California Polytechnic State University*
Michael Zagorski	*Case Western Reserve University*		

FIRST EDITION

Tom Bond	*UC, San Diego*	Dagmar Ringe	*Brandeis University*
Mark Burk	*Duke University*	Charles Rose	*University of Nevada*
Keith Buszek	*Kansas State University*	John Swenson	*Ohio State University*
Barry Carpenter	*Cornell University*	K. Barry Sharpless	*Scripps Institute*
Sherin Halfon	*UC, San Francisco*	Bill Suggs	*Brown University*
Bob Hanson	*St. Olaf College*	Audrey Miller	*University of Connecticut*
F. Chris Pigge	*University of Missouri at St. Louis*	Carol Dempster	*Long Island Research Institute*

During and after the writing process, a large amount of work is required to produce a finished book from hundreds of files containing the text and figures. As a publisher, Bruce Armbruster is among the best at orchestrating this transformation, and along with Kathy Armbruster and Jane Ellis, this team at University Science Books did their usual excellent job. Mary Castellion, herself an author, was an especially effective collaborator during the writing of this edition in helping me to edit the previous text and to present more succinctly and accurately some of the difficult concepts. Jeannette Stiefel did a masterful job with the copy edit, while Ann Knight coordinated the production.

The accompanying *Solutions to Exercises* was originally fashioned with significant help from Julius Beau Lucks, a former undergraduate student who was a Churchill Scholar after graduation and is currently a graduate student in chemistry at Harvard University. Beau worked every exercise in the first edition and checked their solutions. Subsequent groups of students have found and corrected some errors in those solutions; the second edition of the *Solutions to Exercises* text was copyedited and checked by Christine Cleveland, a student in my 2002–03 course who will be attending the UNC–CH School of Medicine beginning in the 2005 fall semester.

I sincerely thank all of you for the efforts and support you have contributed to this creative process.

Thomas Sorrell
Chapel Hill

THE STRUCTURES OF ORGANIC MOLECULES

The organic art of Hans Hoffman. Organic gardening. Frank Lloyd Wright's organic style of architecture. Organic wines. The adjective *organic* became increasingly popular during the latter half of the twentieth century because of its association with the notion of life or naturalness regarding its subject. Similarly, the term *organic chemistry* was used during the early nineteenth century to describe the study of substances related to living, or once-living, organisms, a concept that grew out of the prevailing belief of that day in vitalism, the life-giving force of the universe. Those substances such as rocks and minerals that had no obvious relationship to living things were referred to as *inorganic*.

With passage of time, the differentiation between living and nonliving substances became less clear as newly discovered chemical reactions could be used to interconvert materials. Consequently, scientists came to care more about whether chemical compounds had high or low melting points, were soluble or not in water, or were liquids, solids, or gasses. Using these criteria, chemists realized that molecules containing carbon share many similar properties. Gradually, **organic chemistry** came to be defined as the chemistry of substances containing carbon, the definition we currently recognize and use.

The field of organic chemistry is extensive, encompassing structures and reactions of literally millions of molecules. It is an important area for study and research for many reasons, two of which warrant mention.

1. The structures of organic molecules are well understood, so new types of compounds are readily prepared by rational approaches. *This ability to make new substances with predictable structures and properties* is one aspect that sets organic chemistry apart from other branches of chemistry, physics, and biology.

2. The essential processes of biology are chemical reactions of organic molecules. Studying biochemistry requires one to understand organic chemistry, a primary reason why many students—perhaps you—have chosen to study the subject. Learning organic chemistry is a natural prerequisite for entry into many medical, dental, pharmacy, and graduate schools.

This chapter presents some central concepts about the structures of organic molecules and ways to depict them. In addition, basic rules of nomenclature are presented so that you are able to communicate with others about these substances. When you have finished this chapter, you should be able to draw structural formulas, and you should be familiar with the general strategies required to interpret the names of many of the molecules that you will encounter in this course.

1.1   STRUCTURAL COMPONENTS OF ORGANIC MOLECULES

1.1a   A CARBON ATOM FORMS FOUR BONDS TO NEIGHBORING ATOMS IN STABLE MOLECULES

If you are unfamiliar with the electronic structures of atoms and how bonds are formed, you may be puzzled about the diversity of structures you see in this chapter. Chemists in the nineteenth century knew little about bonding, yet they performed and predicted the products of reactions readily. You can therefore assume that much can be learned without a detailed knowledge of bonding models, which are presented in the next chapter (Chapter 2).

For now, it is sufficient to know that *a carbon atom forms four bonds in a stable, isolable compound.* This pattern exists because the shell of electrons around the nucleus of the carbon atom (and the other second-period elements N, O, and F) will be filled, hence stable, when eight electrons are present. Each bond has two electrons, so four bonds provide eight electrons. The bonds may be single, double, or triple, as illustrated in Figure 1.1.

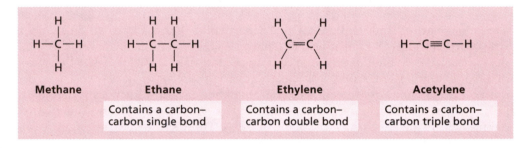

Figure 1.1
Representative structures of carbon-containing substances.

In a stable organic molecule, nitrogen usually forms *three* bonds to neighboring atoms, and oxygen normally forms *two* bonds. Except in rare instances, hydrogen and the halogens [fluorine (F), chlorine (Cl), bromine (Br), and iodine (I)] form one bond to another atom. Figure 1.2 illustrates these trends.

Nitrogen:
normally forms three bonds

Ammonia   Aminomethane   Acetonitrile

Oxygen:
normally forms two bonds

Water   Methanol   Formaldehyde

Figure 1.2
Representative structures of substances with elements in addition to carbon and hydrogen. Notice that each carbon atom forms four bonds and each hydrogen atom forms one.

Halogen atom (F, Cl, Br, I):
normally forms one bond

Chloromethane   Bromoethene

1.1b   A FUNCTIONAL GROUP IS A REACTIVE CENTER THAT CONTAINS HETEROATOMS OR A MULTIPLE CARBON–CARBON BOND

Given the general bonding properties of atoms that constitute a variety of molecules, a survey of organic compounds would reveal features that are common among them. Four principal characteristics would emerge:

1. All contain carbon, and bonds usually exist between two or more carbon atoms.

2. Most contain hydrogen, often, but not always bonded to carbon.

3. Many contain elements besides C and H. These are called **heteroatoms,** and the most common are oxygen, nitrogen, sulfur, phosphorus, silicon, and the halogens. *Heteroatoms* can form bonds with each other or with carbon. The bonds may be single, double, or triple.

4. Most contain a **functional group,** which is defined as any group of atoms other than those with only carbon–carbon and carbon–hydrogen single bonds. *Functional groups* are normally the reactive portions of a molecule, and they range in complexity from the carbon–carbon double bond to groups with multiply bonded heteroatoms.

For the sake of understanding the structures and reactions of organic compounds, you will need to learn the identities of common functional groups. Many, but not all, of the functional groups you will encounter in this text are listed and illustrated in Table 1.1. The condensed structures and the use of suffixes will be discussed shortly.

EXERCISE 1.1

For each of the following compounds, name each functional group that is present.

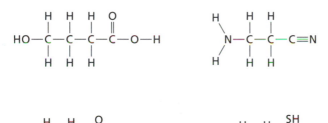

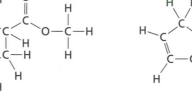

1.2   STRUCTURAL FORMULAS AND CONDENSED STRUCTURES

1.2a   CONDENSED STRUCTURES ELIMINATE THE NEED TO SHOW EVERY ATOM AND BOND IN A MOLECULE

Drawing every bond between pairs of atoms, as in the structural formulas already shown, becomes tedious and cumbersome, especially in complex molecules. Fortunately, easier ways suffice to represent structures of organic compounds. A common way to simplify a structure is to express groups of atoms that are bonded together as a single entity, generating a **condensed structure.**

Table 1.1 Functional groups and their IUPAC nomenclature suffixes.

Structure[a]	Condensed structure	Name	Suffix[b]
* —C(=O)—OH	—COOH or —CO₂H	Carboxylic acid	-oic acid (-carboxylic acid)
—S(=O)(=O)—OH	—SO₃H	Sulfonic acid	-sulfonic acid
* —C(=O)—O—	—COO— or —CO₂—	Ester	-oate (-carboxylate)
* —C(=O)—Cl	—COCl	Acid chloride	-oyl chloride
* —C(=O)—NH₂	—CONH₂	Amide	-amide (-carboxamide)
* —C≡N	—CN	Nitrile	-nitrile (-carbonitrile)
* —C(=O)—H	—CHO	Aldehyde	-al (-carbaldehyde)
—C(=O)—	—CO—	Ketone	-one
—OH	—OH	Alcohol or Phenol	-ol
—SH	—SH	Thiol	-thiol
—N(H)(H)	—NH₂	Amine	-amine
—C=C—	—C=C—	Alkene	-ene
—C≡C—	—C≡C—	Alkyne	-yne
—O—	—O—	Ether	ether
(benzene ring structure)	(benzene ring structure)	Benzene	benzene[c]

Increasing priority

[a]Functional groups with an * must appear at the end of a carbon chain or be attached to a ring.
[b]The suffixes in parentheses are used in names of cyclic compounds.
[c]The benzene ring is a functional group, but its ring is also the carbon skeleton in many aromatic compounds as explained in Section 1.3a.

As an illustration, consider the following structure:

$$\begin{array}{c} \quad\;\; H \;\; H \;\; H \;\; O \\ \quad\;\; | \;\;\;\; | \;\;\;\; | \;\;\;\; \| \\ H-C-C-C-C-O-H \\ \quad\;\; | \;\;\;\; | \;\;\;\; | \\ \quad\;\; H \;\; H \;\; H \end{array}$$

All of the atoms have the normal number of bonds (C, four; O, two; and H, one). Starting with the fragments that have only carbon and hydrogen atoms, which are called **hydrocarbon** fragments, we condense them to *methyl* (CH_3), *methylene* (CH_2), and *methine* (CH) groups, which are shown in Table 1.2.

This first level of simplification gives the following condensed structures:

$$\begin{array}{cc} \qquad\qquad\quad O & \qquad\qquad\quad O \\ \qquad\qquad\quad \| & \qquad\qquad\quad \| \\ CH_3-CH_2-CH_2-C-O-H \quad\text{or}\quad & CH_3CH_2CH_2-C-O-H \end{array}$$

Notice that we can show the bonds between the hydrocarbon fragments or we can leave them out. Because we know each hydrogen atom forms only one bond, it is accepted that the hydrogen atoms are attached only to the carbon atom on their immediate left.

The second level of condensed structures puts functional groups into their condensed formats. These structures are listed in Table 1.1. For this molecule, therefore, we would draw one of the following condensed structures:

$$CH_3CH_2CH_2COOH \qquad\text{or}\qquad CH_3CH_2CH_2CO_2H$$

Table 1.2   Hydrocarbon fragments.

Name	Structure	Condensed form
Methyl	$-\overset{\displaystyle H}{\underset{\displaystyle H}{C}}-H$	$-CH_3$
Methylene	$-\overset{\displaystyle H}{\underset{\displaystyle H}{C}}-$	$-CH_2-$
Methine	$-\overset{\displaystyle |}{\underset{\displaystyle |}{C}}-H$	$-\overset{|}{\underset{|}{C}}H$

EXAMPLE 1.1

Represent the following structural formula in a condensed format:

$$\begin{array}{c} H \qquad\; H \;\; Cl \;\; H \\ \backslash \qquad\;\; | \;\;\;\; | \;\;\;\; | \\ \;\; N-C-C-C-H \\ / \qquad\;\; | \;\;\;\; | \;\;\;\; | \\ H \qquad\; H \;\; H \;\; H \end{array}$$

First, convert the hydrocarbon units to their condensed representations.

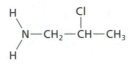

Next, express the functional group with its condensed notation. Any of the following would be acceptable:

$$H_2N-CH_2-CHCl-CH_3 \qquad H_2N-CH_2-\overset{\overset{\textstyle Cl}{|}}{CH}-CH_3 \qquad H_2NCH_2CHClCH_3$$

Notice that we write H_2N rather than NH_2 when the amino group appears on the left. The same holds true for alcohols (HO, not OH), carboxylic acids, (HOOC, not COOH), aldehydes (OHC, not HOC or CHO), and thiols (HS). We can also write H_3C for a methyl group, but CH_3 is preferred.

EXERCISE 1.2

Represent each of the compounds in Exercise 1.1 in a condensed format. For cyclic compounds, connect the condensed hydrocarbon units with bonds to form the appropriate size ring.

EXERCISE 1.3

Draw structural formulas for each of the following compounds, showing every bond:

a. CH_3CH_2CHO

b.
$$\begin{array}{c} H_2C-CH_2 \\ | \qquad \qquad CHOH \\ H_2C-CH_2 \end{array}$$

c. $HSCH_2CH_2OCH_3$

d. $CH_3CHBrCH_2COOCH_3$

1.2b   HYDROCARBON FRAGMENTS MAY BE REPRESENTED BY ANGLED LINES AND POLYGONS

The propensity of carbon to form four bonds, coupled with the ubiquity of hydrogen in organic compounds, has led to an even more abbreviated notation than the one shown in Section 1.2a. Using lines for the carbon–carbon bonds and ignoring all carbon–hydrogen bonds, we use an intersection of lines to define the position of each carbon atom. Heteroatoms and hydrogen atoms attached to heteroatoms are written.

As an example, consider the following molecule:

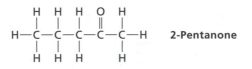

2-Pentanone

We can draw a variety of condensed structures according to the procedure outlined in Section 1.2a, and several are shown in the following scheme. Make note that the orientation of the structure on the page is not crucial, a point that will become clearer in subsequent chapters when you learn more about three-dimensional representations.

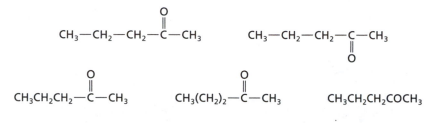

$$CH_3CH_2CH_2\text{—}\overset{\overset{\textstyle O}{\|}}{C}\text{—}CH_3 \qquad CH_3(CH_2)_2\text{—}\overset{\overset{\textstyle O}{\|}}{C}\text{—}CH_3 \qquad CH_3CH_2CH_2COCH_3$$

The orientation of a structure on the page is not important; all of the structures and formulas shown here are equivalent.

To condense the structures further, we omit the atom labels (C and H) and draw lines for each carbon–carbon bond. The meeting point of two lines, which we draw to occur at an angle, represents the position of a carbon atom. Any heteroatoms (in this example, oxygen) are always shown. We know that each carbon atom forms four bonds, so we can mentally include the requisite number of hydrogen atoms at each position to complete the structure. Again, the orientation of the structures shown below is arbitrary.

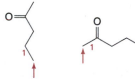   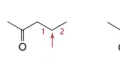

At the end of the chain, only one bond to the carbon atom is drawn; the three hydrogen atoms are not drawn.

Only two bonds to the indicated carbon atoms are shown. The two hydrogen atoms on each carbon atom are not shown.

All four bonds of this carbon atom are drawn. A double bond counts as two bonds.

At the beginning, you may want to place a dot at each carbon atom position. This measure will help you keep track of the carbon atoms while you learn to convert between line and condensed structures.

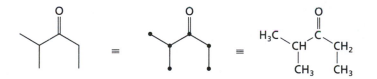

There are instances in which it is helpful to include some of the carbon or hydrogen atom labels. A specific case occurs when a functional group appears at the end of the chain.

Hydrogen atom shown.

Condensed functional group notation used; other carbon atom positions indicated by angles.

To represent compounds with a ring, we draw a regular polygon in which each vertex of the ring represents a carbon atom position. Functional groups attached to a ring are usually written out.

EXAMPLE 1.2

Draw the line structure for the condensed representation given here:

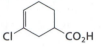

First, indicate each carbon atom position, including the bonds between carbon atoms, leaving the heteroatoms and functional groups in place (below, a.).

Finish by expanding the carboxylic acid group to show all atoms, and then add zero, one, two or three hydrogen atoms to each carbon atom to give each a total of four bonds (below, b.).

a.   b.

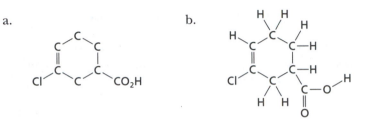

EXERCISE 1.4

Draw the structural formulas for each of the following condensed representations:

a.   b.

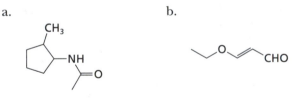

EXERCISE 1.5

Draw a condensed structure for each of the following compounds:

a.   b.

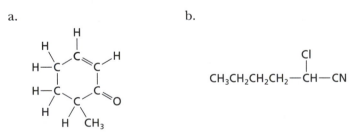

1.3 SYSTEMATIC NOMENCLATURE: IUPAC NAMES

Nomenclature is not the most interesting topic in any new subject you study, but to understand what others are talking about, it is crucial to gain knowledge of the basic concepts. When you were learning to talk, your parents made sounds that you tried to imitate, even though it took months to construct words and sentences yourself. Likewise, if you have studied a foreign language, you probably spent the first part of the course learning to imitate your instructor. In either case, it was more important at the beginning for you to understand what was said than to make statements yourself. If your parents said, Don't touch the stove, that command was crucial to your safety. It is unlikely that you had to warn your parents about getting burned.

So it is in your study of organic chemistry: It is initially more important for you to understand the names that appear in this book or are used by your instructor. For that reason, the emphasis in this chapter will be on *interpretation of names*. As you master this "new language", you will learn much about the structures of organic molecules. This knowledge will later be applied to assimilate information about chemical reactions when you encounter them. The process of naming molecules will follow naturally from learning how to interpret names.

When learning nomenclature, you can easily become overwhelmed by hundreds of detailed rules. Focus instead on the basic pattern that underlies most of the compound names that you will encounter in this book. Because systematic names share the same formal construction, which is presented in Section 1.3a, you can easily learn how to interpret many names within a short period of time. Some of the more complex types of molecules as well as more detailed rules will be presented in later chapters.

1.3a   THE SYSTEMATIC NAME OF AN ORGANIC MOLECULE IS BASED ON THE NAME OF THE LONGEST CARBON CHAIN OR ON THE SIZE OF A RING

In 1949, the International Union of Pure and Applied Chemistry (IUPAC) formulated rules for naming organic compounds. These **IUPAC rules** are most commonly used today. Insofar as possible, this book will use IUPAC rules and names.

Before the IUPAC rules were accepted, *trivial* or *common names* were widely employed, and many of these were incorporated into the IUPAC system. Because trivial names are still used (and are common in the older literature), you will need to learn many of them. This situation is analogous to that for languages, which can be either formal or colloquial. In written documents, the formal system is followed more strictly, but in speaking, colloquial words are used because they are often easier to say.

The IUPAC name of an organic compound can be divided and allocated among four fields, a process illustrated in the chart below and discussed in subsequent sections.

- The *first* field includes the names and positions of **substituents**.
- The *second* field contains the **compound root** word.
- The *third* field is what we call the **multiple-bond index**.
- The *fourth* field defines the **principal functional group**.

First field	Second field	Third field	Fourth field
Substituents	Compound root	Mulitple-bond index	Principal Functional group
Examples:			
4-Chloro-2-pentanone:			
4-Chloro-	pent	an	2-one
2,3-Dimethyl-2-hexene:			
2,3-Dimethyl-	hex	2-ene	
3,3-Diphenylcyclobutanol:			
3,3-Diphenyl	cyclobut	an	ol
2-Chloro-5-nitrophenol			
2-Chloro-5-nitro	phenol		
3-Hydroxybutanal			
3-Hydroxy	but	an	al

Table 1.3   Structures and names of alkanes with 1–20 carbon atoms.

Alkane name	Root word	Structure	Alkane name	Root word	Structure
Methane	Meth-	CH_4	Undecane	Undec-	$CH_3-(CH_2)_9-CH_3$
Ethane	Eth-	CH_3-CH_3	Dodecane	Dodec-	$CH_3-(CH_2)_{10}-CH_3$
Propane	Prop-	$CH_3-CH_2-CH_3$	Tridecane	Tridec-	$CH_3-(CH_2)_{11}-CH_3$
Butane	But-	$CH_3-CH_2-CH_2-CH_3$	Tetradecane	Tetradec-	$CH_3-(CH_2)_{12}-CH_3$
Pentane	Pent-	$CH_3-(CH_2)_3-CH_3$	Pentadecane	Pentadec-	$CH_3-(CH_2)_{13}-CH_3$
Hexane	Hex-	$CH_3-(CH_2)_4-CH_3$	Hexadecane	Hexadec-	$CH_3-(CH_2)_{14}-CH_3$
Heptane	Hept-	$CH_3-(CH_2)_5-CH_3$	Heptadecane	Heptadec-	$CH_3-(CH_2)_{15}-CH_3$
Octane	Oct-	$CH_3-(CH_2)_6-CH_3$	Octadecane	Octadec-	$CH_3-(CH_2)_{16}-CH_3$
Nonane	Non-	$CH_3-(CH_2)_7-CH_3$	Nonadecane	Nonadec-	$CH_3-(CH_2)_{17}-CH_3$
Decane	Dec-	$CH_3-(CH_2)_8-CH_3$	Eicosane	Eicos-	$CH_3-(CH_2)_{18}-CH_3$

Start by identifying the root word, which derives from the longest continuous carbon chain in the case of **aliphatic compounds.** The term *aliphatic* originates from a Greek word for "fat", the source of many hydrocarbons isolated in the 1800s. Today, we use the word *aliphatic* to refer to those organic compounds that have chains of carbon atoms. These chains can be straight or branched, but they lack rings. The root word in the name of an aliphatic compound is based on the identity of the corresponding **alkanes,** hydrocarbon molecules with only single bonds between atoms. The names and structures of the 20 simplest alkanes—those with unbranched chains—are shown in Table 1.3.

The root words for compounds with rings are only slightly more complicated. **Alicyclic** (*ali*phatic *cyclic*) **compounds,** which have rings and a preponderance of single bonds, use the root word of the alkane with the same number of carbon atoms, but the prefix "cyclo" is inserted just in front of this root, as illustrated in Figure 1.3.

Benzene is a cyclic compound that is the simplest member of the family of substances known as **aromatic compounds** or **arenes.** Their distinguishing feature is the presence of apparently alternating single and double bonds within a six-membered ring. Figure 1.4a shows several ways to draw benzene.

Many aromatic compounds are known by the common names given in Figure 1.4. The IUPAC system incorporated many of these common names, so "toluene", "phenol",

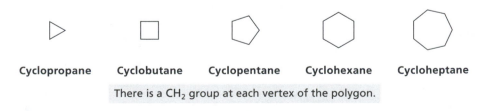

| Cyclopropane | Cyclobutane | Cyclopentane | Cyclohexane | Cycloheptane |

There is a CH_2 group at each vertex of the polygon.

Figure 1.3
Structures and names of cycloalkanes with three-to-seven carbon atoms.

a.

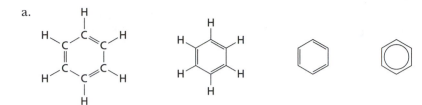

Benzene, C_6H_6, is the simplest of the "aromatic compounds." All of the structures shown above are equivalent. For the two structures at the right, a hydrogen atom is attached at each vertex of the hexagon. The significance of the circle within the ring will be discussed in more detail in Chapters 2 and 17.

b.

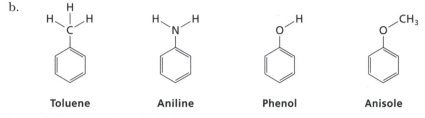

| Toluene | Aniline | Phenol | Anisole |

Just as for benzene, there is a hydrogen atom at each vertex of the hexagon that does not have another atom already attached.

Figure 1.4
The structures of benzene (a.) and of aniline, anisole, phenol, and toluene (b.).

and so on are considered systematic names. The names of other arenes will be covered later, but for now concentrate on knowing the names and structures shown in Figure 1.4b and on recognizing the root word "benz", used for many of benzene's derivatives such as *benz*aldehyde (C_6H_5CHO), *benz*oic acid (C_6H_5COOH), and *benz*onitrile (C_6H_5CN).

1.3b   THE MULTIPLE-BOND INDEX SPECIFIES THE NUMBER AND TYPE OF MULTIPLE BONDS BETWEEN CARBON ATOMS

In an IUPAC name, the *multiple-bond index* follows directly after the compound root and indicates whether double or triple bonds are present. If no carbon–carbon multiple bond is present, then the suffix "ane" (or "an") follows the root.

The suffix "ene" refers to the presence of a double bond between two carbon atoms (C=C), and "yne" refers to a triple bond between carbon atoms (C≡C). A compound that has a carbon–carbon double bond is called an **alkene,** as a general class, and one with a triple bond is an **alkyne.** If the molecule has two double bonds, the word "diene" follows the compound root; for three double bonds, "triene"; and so on. Table 1.4 matches these suffixes with specific examples.

Table 1.4   Suffixes for the multiple-bond index illustrated for compounds with four carbon atoms (root = but—).

Suffix	Meaning		Example
-ane	No C–C double or triple bonds	Butane	$CH_3-CH_2-CH_2-CH_3$
-ene	One C–C double bond	1-Butene	$\overset{1}{C}H_2=\overset{2}{C}H-\overset{3}{C}H_2-\overset{4}{C}H_3$
-yne	One C–C triple bond	1-Butyne	$H-\overset{1}{C}\equiv\overset{2}{C}-\overset{3}{C}H_2-\overset{4}{C}H_3$
-diene	Two double bonds	1,3-Butadiene	$\overset{1}{C}H_2=\overset{2}{C}H-\overset{3}{C}H=\overset{4}{C}H_2$

A multiple bond can appear anywhere within a chain of carbon atoms, so a numeral is often included in the name to specify at which carbon atom the double or triple bond begins. Figure 1.5 shows some examples. A molecule with no other functional group is numbered from the end that gives the lower possible number to the position of the multiple bond. For a diene or triene, position numbers are given for each double bond. In a cycloalkene or cycloalkyne, the double or triple bond is assumed to start at C1, unless otherwise specified. (Incidentally, carbon atom positions are described using one of the following equivalent phrases: "at the 2-position", "at C2", or "at the number 2 carbon atom". When the numerals are subscripts [e.g., C_2H_5], they refer to the total numbers of atoms.)

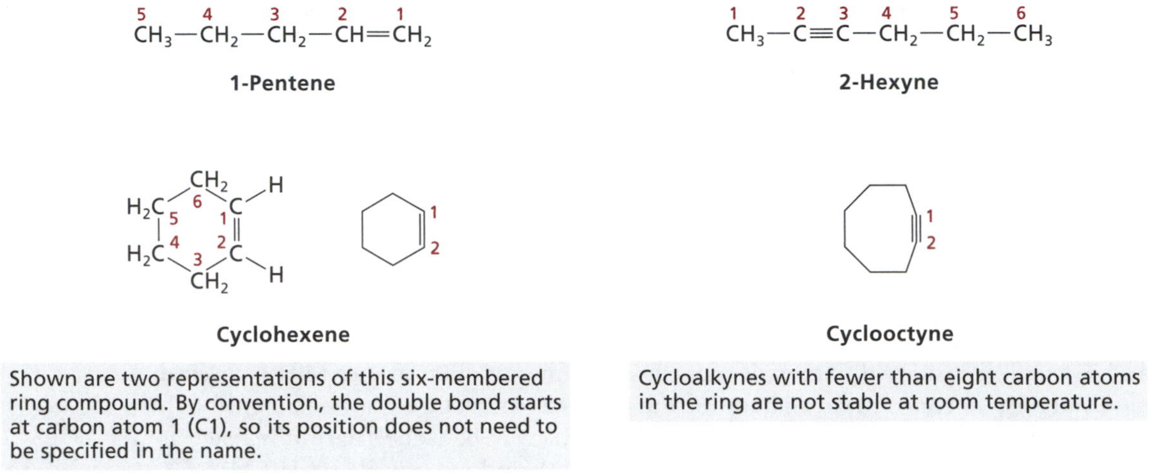

$$\overset{5}{CH_3}-\overset{4}{CH_2}-\overset{3}{CH_2}-\overset{2}{CH}=\overset{1}{CH_2}$$

1-Pentene

$$\overset{1}{CH_3}-\overset{2}{C}\equiv\overset{3}{C}-\overset{4}{CH_2}-\overset{5}{CH_2}-\overset{6}{CH_3}$$

2-Hexyne

Cyclohexene

Shown are two representations of this six-membered ring compound. By convention, the double bond starts at carbon atom 1 (C1), so its position does not need to be specified in the name.

Cyclooctyne

Cycloalkynes with fewer than eight carbon atoms in the ring are not stable at room temperature.

Figure 1.5
Examples of numbering in compounds with carbon–carbon double and triple bonds.

EXERCISE 1.6

Based on the structures in Tables 1.3 and 1.4 and Figure 1.5, draw structural formulas and condensed structures for each of the following alkenes and alkynes:

a. 3-Hexene   b. 4-Octyne   c. 1-Butene-3-yne   d. Cyclobutene

---

1.3c   THE IDENTITY OF THE PRINCIPAL FUNCTIONAL GROUP APPEARS AT THE END OF THE COMPOUND'S NAME

In the IUPAC system, the identity of the principal functional group appears at the end of the name. Common suffixes for the functional groups were given in Table 1.1. Other functional groups may be present in the molecule as well, and these will be specified as substituents. What constitutes the "principal" group is based on a priority ranking, and Table 1.1 is organized so that the highest priority group is at the top, and the lowest is at the bottom. (Even though benzene is listed at the bottom of Table 1.1, its presence in a molecule is reflected by the compound root word.)

Some functional groups, by their very nature, *must* be at the end of the carbon chain because the carbon atom within the group already has three bonds to hydrogen or heteroatoms. Examples include aldehydes, carboxylic acids, and nitriles (an asterisk next to each name in Table 1.1 identifies these groups). Incidentally, the carbon atom at the end of a chain, whether part of a functional group or not, is called the *terminal* carbon atom.

Some functional groups can be located at practically any position of a chain, so a numeral is added in those cases to specify its position, as was the case for multiple bonds. Ketones and alcohols are the most common functional groups in this category. This numeral immediately precedes the functional group suffix if a multiple bond is present; otherwise it appears in front of the root word. The following examples illustrate names that have been separated into their parts (root, multiple-bond index, and principal functional group).

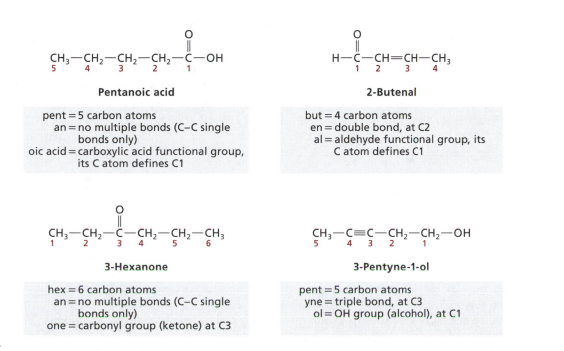

Pentanoic acid

pent = 5 carbon atoms
an = no multiple bonds (C–C single
bonds only)
oic acid = carboxylic acid functional group,
its C atom defines C1

2-Butenal

but = 4 carbon atoms
en = double bond, at C2
al = aldehyde functional group, its
C atom defines C1

3-Hexanone

hex = 6 carbon atoms
an = no multiple bonds (C–C single
bonds only)
one = carbonyl group (ketone) at C3

3-Pentyne-1-ol

pent = 5 carbon atoms
yne = triple bond, at C3
ol = OH group (alcohol), at C1

When a functional group that is required to be at the end of chain is attached to a ring, then a different suffix is required (Table 1.1) because the ring's root word does not include the carbon atom of the functional group. Common groups in this category include the aldehyde (suffix = carbaldehyde) and carboxylic acid (suffix = carboxylic acid) groups. The following structures illustrate how these suffixes are treated. Notice that the carbon atom *in the ring* at which this functional group is attached is designated C1.

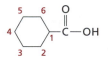

Cyclohexanecarboxylic acid

cyclohex = a ring of 6 carbon atoms
ane = no multiple bonds (C–C single
bonds only)
carboxylic acid = carboxylic acid functional group
attached to the ring at C1

2-Cyclopentenecarbaldehyde

cyclopent = a ring of 5 carbon atoms
ene = double bond, at C2
carbaldehyde = aldehyde functional group
attached to the ring at C1

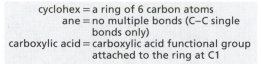

EXERCISE 1.7

What is the principal functional group in each of the following compounds? Draw the full and condensed structures of this generalized functional group, ignoring the compound root and substituents.

Example: 3-Chloropentanoic acid.

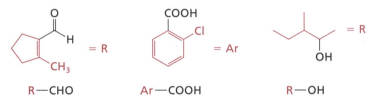

Suffix = -oic acid = carboxylic acid = —COOH

a. 2,2-Dimethyl-3-hexanone   b. 3-Methoxybenzaldehyde   c. 2-Methyl-2-butanol

A practical and common way to designate the part of the molecule other than its functional group is by use of **R,** which may be thought of as "the **R**emainder of the molecule". Similarly, **Ar** is used to designate an aromatic ring, usually a derivative of benzene. These symbols are used when you want to focus attention on a particular functional group, and the identity of the remaining structure is not crucial. Most of the time, R and Ar are used to represent portions of the molecule that contain only carbon and hydrogen, but other substituents (Section 1.3d) are sometimes also included, depending on the context.

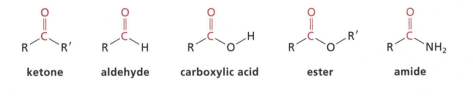

R—CHO   Ar—COOH   R—OH

This introduction to functional groups would be incomplete without specific reference to the **carbonyl group,** which is the structural unit with a carbon–oxygen double bond, C=O. ("Carbonyl" is pronounced carbon-EEL.) Looking at Table 1.1, you will see several functional groups that contain the carbonyl group. The ketone functional group consists of a carbonyl group attached to two carbon-containing fragments. The aldehyde functional group has a carbonyl group attached to a carbon-containing fragment and a hydrogen atom. Several important functional groups are shown below in abbreviated format with the carbonyl group highlighted in color. Notice the use of R and R', which means that the two "remainder" groups may not be the same.

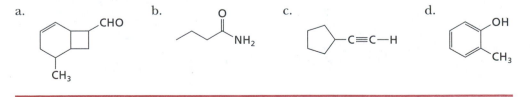

ketone   aldehyde   carboxylic acid   ester   amide

EXERCISE 1.8

Draw the following structures in the more general form using the label R or Ar in combination with the principal functional group.

a.   b.   c.   d.

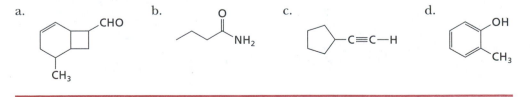

1.3d THE IDENTITIES AND POSITIONS OF SUBSTITUENTS ARE SPECIFIED IN THE FIRST FIELD OF A COMPOUND'S NAME

Substituents are atoms or groups that appear in place of hydrogen atoms attached to the carbon skeleton. The identities of the substituents constitute the first field of an IUPAC name, and their positions of attachment are specified by numerals. The carbon chain is numbered so that the functional group with the highest priority (Table 1.1) has the lowest number. If a substituent is attached to a heteroatom instead of the carbon skeleton, its placement is denoted by the italicized *N, O,* or *S,* for nitrogen, oxygen, or sulfur, respectively.

Common examples of substituents are given in Table 1.5. When more than one of the same type of substituent is present, the name of the atom or group follows a prefix for that number of items: *di* = two, *tri* = three, *tetra* = four, and so on. (When the substituents are more complex than those listed in Table 1.5, you will instead see and use the prefixes bis, tris, tetrakis, etc. Meanings are the same, e.g., di = bis = two.) Furthermore, there must be a corresponding numeral for each atom or group. Thus,

2, 2-Dichloro . . . (correct)

but not   **2-Dichloro . . .** (incorrect: not enough specifying numerals)

or   **2,2-Chloro . . .** (incorrect: prefix specifying two chlorine atoms is missing)

Exceptions to the numeral rule occur when there is no chance for ambiguity, for example, when there is only one carbon atom that can be substituted by the specified group. Then, numerals are not needed, but a prefix is still required. For example,

Difluoromethane
$$F-\overset{\overset{\displaystyle H}{|}}{\underset{\underset{\displaystyle H}{|}}{C}}-F$$

EXERCISE 1.9

Correct any of the following names that are inconsistent.

a. 2,3,4-Hydroxyhexanal   b. 2,2,4,4-Tetrachloropentane   c. Triiodomethane

---

Table 1.5   A summary of prefixes for common substituents.

Substituent	Prefix	Substituent	Prefix
—R	Alkyl- (see text)	—F	Fluoro-
—OR	Alkoxy- (see text)	$-\overset{\overset{\displaystyle O}{\parallel}}{C}\diagdown_{H}$	Formyl-
$-\overset{\overset{\displaystyle O}{\parallel}}{C}\diagdown_{CH_3}$	Acetyl-	—OH	Hydroxy-
—NH$_2$	Amino-	—I	Iodo-
—Br	Bromo-	—NO$_2$	Nitro-
—COOH	Carboxy-	—SH	Mercapto-
—Cl	Chloro-	=O	Oxo-
—C≡N	Cyano-	—O—⬡	Phenoxy-

The following examples illustrate how the IUPAC format is applied to interpreting compound names.

EXAMPLE 1.3

Draw the structural formula and condensed structure of 5,5-dichloro-3-hexanone.

First, identify the compound root and draw the carbon skeleton. *Root* = hex (six carbon atoms).

$$C—C—C—C—C—C$$

Next, translate the multiple-bond index. The suffix that directly follows the root word is "an" meaning that there are no carbon–carbon double or triple bonds. The suffix is "one", which means that the compound is a ketone, and the numeral "3" indicates that the carbon–oxygen double bond of the ketone is at C3:

$$\underset{1}{C}—\underset{2}{C}—\underset{3}{\overset{\overset{\displaystyle O}{\|}}{C}}—\underset{4}{C}—\underset{5}{C}—\underset{6}{C}$$

Next, we place the substituents, in this case chlorine atoms, at the indicated positions:

$$\underset{1}{C}—\underset{2}{C}—\underset{3}{\overset{\overset{\displaystyle O}{\|}}{C}}—\underset{4}{C}—\underset{5}{\underset{\underset{\displaystyle Cl}{|}}{\overset{\overset{\displaystyle Cl}{|}}{C}}}—\underset{6}{C}$$

Finish by adding hydrogen atoms to give each carbon atom a total of four bonds.

$$\underset{\underset{\displaystyle H}{|}}{\overset{\overset{\displaystyle H}{|}}{H—C}}—\underset{\underset{\displaystyle H}{|}}{\overset{\overset{\displaystyle H}{|}}{C}}—\overset{\overset{\displaystyle O}{\|}}{C}—\underset{\underset{\displaystyle H}{|}}{\overset{\overset{\displaystyle H}{|}}{C}}—\underset{\underset{\displaystyle Cl}{|}}{\overset{\overset{\displaystyle Cl}{|}}{C}}—\underset{\underset{\displaystyle H}{|}}{\overset{\overset{\displaystyle H}{|}}{C}}—H \quad \equiv \quad CH_3—CH_2—\overset{\overset{\displaystyle O}{\|}}{C}—CH_2—\underset{\underset{\displaystyle Cl}{|}}{\overset{\overset{\displaystyle Cl}{|}}{C}}—CH_3$$

EXAMPLE 1.4

Draw the structural formula and condensed structure of 3-hydroxybutanoic acid.

The compound root is "but" (four carbon atoms), and the multiple-bond index is "an", which means that there are no carbon–carbon double or triple bonds. The suffix is "oic acid", so the compound is a carboxylic acid. The carboxylic acid functional group has to be at the end of the chain, and any principal functional group at the end defines the 1-position with its carbon atom.

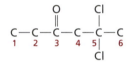

Next, we place the substituent, in this case the hydroxy group (OH), at the 3-position:

$$\underset{4}{C}—\underset{3}{\underset{\underset{\displaystyle OH}{|}}{C}}—\underset{2}{C}—\underset{1}{\overset{\overset{\displaystyle O}{\|}}{C}}—OH$$

The last step is to add hydrogen atoms to give each carbon atom a total of four bonds.

$$H-\overset{\overset{\displaystyle H}{|}}{C}-\overset{\overset{\displaystyle OH}{|}}{\underset{\underset{\displaystyle H}{|}}{C}}-\overset{\overset{\displaystyle H}{|}}{\underset{\underset{\displaystyle H}{|}}{C}}-\overset{\overset{\displaystyle O}{\|}}{C}-OH \quad \equiv \quad CH_3-\overset{\overset{\displaystyle OH}{|}}{CH}-CH_2-\overset{\overset{\displaystyle O}{\|}}{C}-OH$$

EXERCISE 1.10

Draw the structural formula for each of the following compounds:

a. 4-Fluorobutanal   b. 3-Mercapto-2-pentanol   c. Trichloroethanenitrile

---

EXAMPLE 1.5

Draw the structural formula and condensed structure of *N*-chloro-2-propenamide.

The compound root is "prop" (three carbon atoms), and the multiple-bond index is "en", which means that the compound has a carbon–carbon double bond, which starts at C2.

$$C-C-C \qquad \underset{3}{C}=\underset{2}{C}-\underset{1}{C}$$

The suffix is "amide", so the compound is an amide (also called a carboxamide), which has a carbonyl group attached to a carbon-containing fragment and a nitrogen atom. The amide functional group must be at the end of the chain, and its carbon atom defines C1. Next, place the substituents. A chlorine atom is attached to the nitrogen atom, as indicated by the letter *N* at the beginning of the name. A numeral would appear if the chlorine atom were attached to one of the carbon atoms in the chain.

$$\underset{3}{C}=\underset{2}{C}-\underset{1}{\overset{\overset{\displaystyle O}{\|}}{C}}-N \qquad \underset{3}{C}=\underset{2}{C}-\underset{1}{\overset{\overset{\displaystyle O}{\|}}{C}}-N-Cl$$

The last step is to add hydrogen atoms to give each carbon atom a total of four bonds and to the nitrogen atom to give it three bonds.

$$H-\overset{\overset{\displaystyle H}{|}}{C}=\overset{\overset{\displaystyle H}{|}}{C}-\overset{\overset{\displaystyle O}{\|}}{C}-\underset{\underset{\displaystyle H}{|}}{N}-Cl \quad \equiv \quad H_2C=CH-\overset{\overset{\displaystyle O}{\|}}{C}-NH-Cl$$

EXERCISE 1.11

Draw the structural formula for each of the following compounds:

a. 3-Oxopentanoic acid   b. 2-Nitropentanal   c. 4-Hydroxy-2-hexyne

---

1.3e   THE PLACEMENT OF SUBSTITUENTS IN CYCLIC COMPOUNDS DEPENDS ON DEFINING A STARTING POINT FOR NUMBERING

In an aliphatic compound, numbering starts at the end of the chain that gives the principal functional group the lower possible number. Numbering is done in such a way that the functional group, multiple bonds, and substituents have the lowest numerical values possible. For example,

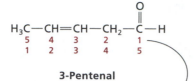

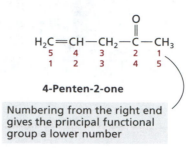

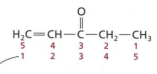

3-Pentenal

Numbering from the right end gives the principal functional group a lower number

4-Penten-2-one

Numbering from the right end gives the principal functional group a lower number

1-Penten-3-one

The principal functional group is at the same position whether the chain is numbered left to right or right to left; numbering from the left end gives the double bond a lower number, however.

A ring, whether it is alicyclic or aromatic, has no "end". Therefore, numbering starts with the position at which the principal functional group or a substituent is attached and proceeds around the ring. The direction of numbering is the one with the lower number at the first point of difference. Placement of the other groups is defined relative to the starting point, as illustrated by the examples in Figure 1.6.

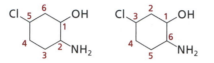

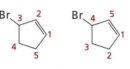

2-Amino-5-chlorocyclohexanol

The principal functional group (alcohol) defines C1. Numbering is clockwise because 1,2,5 is lower than 1,3,6.

3-Bromocyclopentene

The presence of the double bond defines C1. Numbering is counterclockwise because carbon atoms in multiple bonds are numbered consecutively.

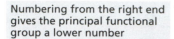

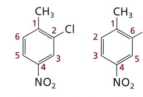

1-Bromo-4-chlorocycloheptane

When only two substituents are present and there is no principal functional group, the substituent that appears first alphabetically defines C1.

2-Chloro-4-nitrotoluene

The methyl group of toluene, which is the compound root, defines C1. Numbering is clockwise because 1,2,4 is lower than 1,4,6.

Figure 1.6
Examples of numbering patterns in cyclic compounds.

EXERCISE 1.12

Draw the structure for each of the following compounds:

a. 2-Bromobenzoic acid   b. 3-Fluoro-2-cyclohexenone   c. 3-Chloro-4-nitrophenol

---

1.4   CONSTITUTIONAL ISOMERS AND HYDROCARBON SUBSTITUENTS

1.4a   STRUCTURES OF CARBON-CONTAINING SUBSTITUENTS ARE DIVERSE

Until now, substituents in the examples have mainly been groups containing heteroatoms. To deal with substituents that contain carbon, you also have to begin to learn about isomerism. **Isomers** exist whenever a molecular formula can be repre-

Skeletal isomers	CH₃—CH₂—CH₂—CH₃ **Butane: C₄H₁₀**	CH₃ over CH₃—CH—CH₃ **2-Methylpropane: C₄H₁₀**
Positional isomers	CH₃—CH₂—CH₂—OH **1-Propanol: C₃H₆O**	OH over CH₃—CH—CH₃ **2-Propanol: C₃H₆O**
Functional isomers	CH₃—O—CH₃ **Dimethyl ether: C₂H₆O**	CH₃—CH₂—OH **Ethanol: C₂H₆O**

Figure 1.7
Types of constitutional isomers

sented by different arrangements of the constituent atoms, and Figure 1.7 summarizes three types of **constitutional isomers,** which are isomers having different connectivities between neighboring atoms. The three categories of constitutional isomers are **skeletal, positional,** and **functional isomers.**

Related to the concept of isomerism is the classification of carbon atoms *with four single bonds* as primary, secondary, tertiary, and quaternary. A carbon atom attached to only one other carbon atom is a **primary carbon atom,** which we designate 1°. A carbon atom attached to two other carbon atoms is **secondary** (2°); to three other carbon atoms is **tertiary** (3°); and to four carbon atoms, **quaternary** (4°).

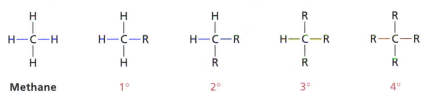

Methane   1°   2°   3°   4°

A carbon atom with four single bonds that is attached to *no other carbon atom* is designated methyl, methylene, or methine according to the number of attached hydrogen atoms (three, two, and one, respectively).

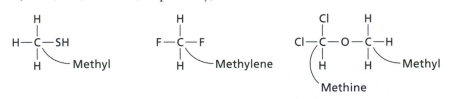

Notice that *within a carbon* chain, a CH₃ group can be designated as either methyl or 1°, depending on the context. Likewise, a CH₂ group can be labeled either methylene or 2°; and a CH group can be identified as either methine or 3°.

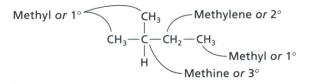

The terms primary, secondary, and tertiary can be applied to other substances, too. For amines and amides, these words are used to indicate how many carbon-containing groups are attached to the *nitrogen* atom.

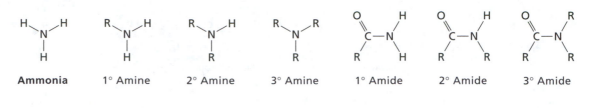

| Ammonia | 1° Amine | 2° Amine | 3° Amine | 1° Amide | 2° Amide | 3° Amide |

EXERCISE 1.13

For the carbon atoms with four single bonds in the following structures, classify each as 1°, 2°, 3°, or 4°. First, expand each structure to show all of the hydrogen atoms.

a.   b.

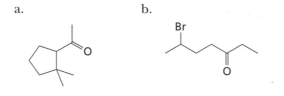

1.4b   THE NAMES OF CARBON-CONTAINING SUBSTITUENTS ARE DERIVED FROM THE NAMES OF THE PARENT HYDROCARBONS

The simplest carbon-containing substituents are **alkyl groups,** formed by removing a hydrogen atom from an alkane. For example, removing a hydrogen atom from methane creates the methyl (often abbreviated as Me) group. Removing a hydrogen atom from ethane generates the ethyl (abbreviated as Et) group. Taking a hydrogen atom from the end of any alkane chain yields a "straight-chain" alkyl group.

CH_4 ⟶ CH_3- $CH_3-CH_2-CH_3$ ⟶ $CH_3-CH_2-CH_2-$
Methane   **Methyl**   **Propane**   **Propyl**

CH_3-CH_3 ⟶ CH_3-CH_2- $CH_3CH_2CH_2CH_2CH_2CH_3$ ⟶ $CH_3CH_2CH_2CH_2CH_2CH_2-$
Ethane   **Ethyl**   **Hexane**   **Hexyl**

If an alkane has three carbon atoms or more, then isomeric alkyl groups can be made. Taking a hydrogen atom from the C1 of propane gives the propyl (Pr) group. Removing a hydrogen atom from C2 yields the 2-propyl or isopropyl (iPr) group. The IUPAC system includes both propyl and isopropyl. Notice that the prefix "iso" is not italicized.

$CH_3-CH_2-CH_2-$

Propyl

$CH_3-\overset{\overset{\displaystyle H}{|}}{\underset{|}{C}}-CH_3$

**2-Propyl
(or Isopropyl)**

Among substances with four carbon atoms are examples of alkane isomers: butane and isobutane (IUPAC name: 2-methylpropane). Isobutane is an example of an alkane with a branched-carbon chain. While alkanes are normally named according to IUPAC rules, terms like "isobutane" are still commonly used.

$CH_3-CH_2-CH_2-CH_3$

Butane

$CH_3-\overset{\overset{\displaystyle CH_3}{|}}{\underset{\underset{\displaystyle H}{|}}{C}}-CH_3$

**Isobutane
(2-Methylpropane)**

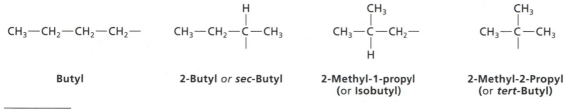

Figure 1.8
Structures of the alkyl groups derived from the butane isomers.

The result of this carryover is that alkyl groups generated from alkanes with four carbon atoms are frequently called butyl, isobutyl (the isomer derived from removing a hydrogen atom from the primary carbon atom of isobutane), *sec*-butyl (*sec* stands for "secondary" to indicate removal of a hydrogen atom from the 2° carbon atom of butane), and *tert*-butyl (*t*-Bu; *tert* stands for "tertiary"). Figure 1.8 shows the structures of these isomeric alkyl groups. Notice that he prefixes "*sec*-" and "*tert*-" are italicized.

EXAMPLE 1.6

Draw the structural formula of 3-ethyl-2-hexanone.

The compound root is "hex" (six carbon atoms), the multiple-bond index is "an", which means that there are no carbon–carbon double or triple bonds, and the suffix is "one." The compound is a ketone, and the carbonyl group is at C2.

$$C-C-C-C-\overset{\overset{\displaystyle O}{\|}}{C}-C$$
$$654321$$

We place the substituent, an ethyl group, at C3. The structure is completed by adding hydrogen atoms until each carbon atom has a total of four bonds.

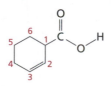

$$H_3C-CH_2-CH_2-\overset{\overset{\displaystyle H}{|}}{\underset{\underset{\displaystyle H_2C-CH_3}{|}}{C}}-\overset{\overset{\displaystyle O}{\|}}{C}-CH_3$$

EXAMPLE 1.7

Draw the structure of 3-*tert*-butyl-2-cyclohexenecarboxylic acid.

The compound root is "cyclohex" (six carbon atoms in a ring), and the multiple-bond index is "en", which means a carbon–carbon double bond is present, starting at C2. The suffix is "-carboxylic acid," so the compound is a carboxylic acid, with the –COOH group attached to the ring. Its point of attachment defines C1 of the ring.

At C3, we place the substituent, a *tert*-butyl group. The structure is completed by adding hydrogen atoms until each carbon atom has a total of four bonds.

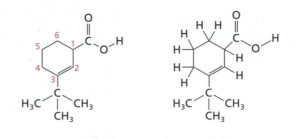

EXERCISE 1.14

Draw the structural formula for each of the following compounds containing alkyl group substituents:

a. Isobutylbenzene   b. 3-*tert*-Butylhexanol   c. 3-Ethylcyclopentanone

If an alkyl group has substituents itself, its name is set apart from the root name of the parent compound by enclosing it in parentheses. We deduce *its* identity by finding *its* root name and *its* substituents, just as we do for the parent compound. The following example illustrates this procedure.

EXAMPLE 1.8

Draw the structure of 3-(2- chloroethyl)-2-heptanol.

The compound root is "hept" (seven carbon atoms), the multiple-bond index is "an", which means no carbon–carbon double or triple bonds, and the suffix is "ol." The compound is an alcohol, a functional group that can be attached to any carbon atom; the "2" in front of the root word tells us the OH group is at C2. At position 3, we attach the alkyl substituent, "X" (= 2- chloroethyl).

$$\underset{7\quad6\quad5\quad4\quad3\quad\overset{|}{2}\quad1}{C-C-C-C-C-\overset{\overset{OH}{|}}{C}-C}\qquad\underset{7\quad6\quad5\quad4\quad3\quad\overset{|}{2}\quad1}{C-C-C-C-\overset{\overset{X}{|}}{C}-\overset{\overset{OH}{|}}{C}-C}$$

Now, interpret the name of "X". Its root is "eth" (2 carbon atoms), and it is attached to the main chain at its C1 by convention. A chlorine atom is attached at C2 of this side chain.

$$C-C-C-C-\overset{\overset{2}{\overset{|}{C}}}{\underset{\underset{OH}{|}}{\overset{\overset{1}{\overset{|}{C}}}{C}}}-C-C \qquad C-C-C-C-\overset{\overset{2}{\overset{|}{C}}-Cl}{\underset{\underset{OH}{|}}{\overset{\overset{1}{\overset{|}{C}}}{C}}}-C-C$$

Finally, we add hydrogen atoms to every carbon atom to give each a total of four bonds.

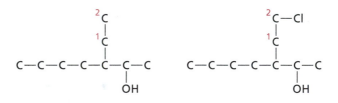

EXERCISE 1.15

Draw the structure of 2-(1,1-dimethylpropyl)hexanoic acid.

Related to the alkyl groups are **alkoxy groups,** which comprise an alkyl group attached to the main compound framework through an oxygen atom. The names of these substituents are generated by inserting "oxy" after the alkyl root name.

CH_3—	Methyl	CH_3—O—	Methoxy
CH_3CH_2—	Ethyl	CH_3CH_2—O—	Ethoxy
$(CH_3)_2CH$—	Isopropyl	$(CH_3)_2CH$—O—	Isopropoxy

EXAMPLE 1.9

Draw the structure of 3-methoxy-4-pentenal

The compound root is "pent" (5 carbon atoms), the multiple-bond index is "en", which means a carbon–carbon double is present starting at C4, and the suffix is "al." The compound is an aldehyde, a functional group that must be at the end of the chain.

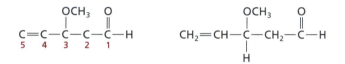

A methoxy group (methyl group + oxygen atom) is attached at C3.

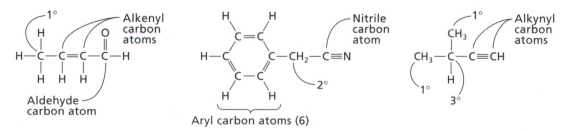

1.4c   NAMES OF UNSATURATED SUBSTITUENTS ARE DERIVED FROM THE NAMES OF THE CORRESPONDING ALKENES, ALKYNES, AND ARENES

Carbon-containing substituents are not limited to those with only four single bonds to the carbon atoms: multiple bonds can also be present. First, realize that a carbon atom that forms a multiple bond cannot be designated as 1°, 2°, 3°, or 4°. Instead, as shown below, these carbon atoms are identified according to the functional group of which they are part. A carbon atom in a carbon–carbon double bond is termed an **alkenyl** (or **vinyl**) **carbon atom;** in a triple bond, an **alkynyl carbon atom;** and in a benzene ring, an **aryl carbon atom.** The following examples illustrate the use of these terms, along with the designations for the carbon atoms that have only single bonds.

The simplest substituents with multiple bonds have names incorporated into the IUPAC system from common names: vinyl, allyl, phenyl, and benzyl, which are summarized in Figure 1.9. These are treated in a name in the same fashion as alkyl groups.

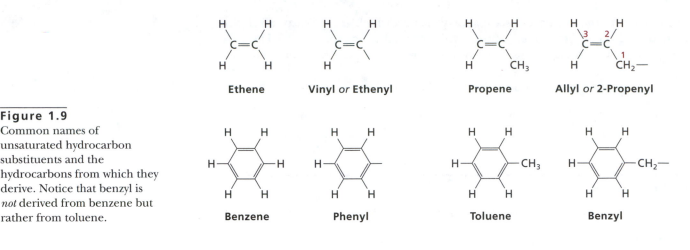

Figure 1.9
Common names of unsaturated hydrocarbon substituents and the hydrocarbons from which they derive. Notice that benzyl is *not* derived from benzene but rather from toluene.

EXAMPLE 1.10

Draw the structure of 3-vinylcyclohexanone.

The compound root is "cyclohex" (six carbon atoms), and the multiple-bond index is "an", which means no carbon–carbon multiple bonds. The principal functional group is a ketone (one), which in a ring defines C1.

At position 3, we attach the substituent, which is the vinyl group. We finish by adding hydrogen atoms to give every carbon atom a total of four bonds.

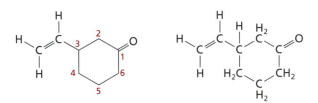

EXAMPLE 1.11

Draw the structure of 3-phenylcycloheptene.

The compound root is "cyclohept" (seven carbon atoms in a ring), the multiple-bond index is "ene", which means a carbon–carbon double is present, starting at C1.

A phenyl (Ph) group is attached at C3. Hydrogen atoms are added to give each carbon atom four bonds.

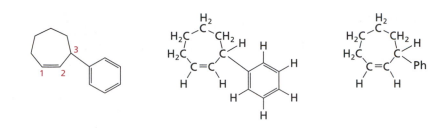

EXERCISE 1.16

Draw the structural formula of 3-allyl-5-chlorobenzoic acid.

1.5 NAMING ORGANIC COMPOUNDS

Having seen how IUPAC names can be interpreted by translating each portion of a name, you are now in a position to write the name of a compound given its structure. The steps in the naming process follow:

1. Identify the *principal* functional group, which is defined by the priority listing given in Table 1.1, and choose the suffix that will appear at the end of the name.

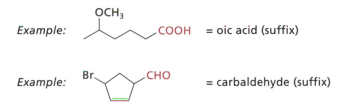

2. The longest carbon chain *that also contains the principal functional group* is then identified, and its name becomes the compound's root. If the compound is cyclic, then the ring *connected* to the principal functional group is chosen as the root.

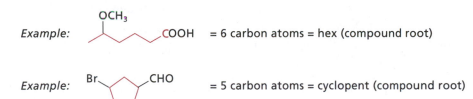

3. The chain or ring is numbered in such a way that the principal functional group is attached at (or is part of) the carbon atom with the lowest possible number. In a cyclic compound, the next priority comprises the positions of multiple bonds.

Numbering can be a problem when no high priority group is present. In such a case, the numbering scheme chosen is the one with the lower number at the first point of difference. In the following example, 1, 2, 4 < 1, 3, 4 (2 lower than 3) and 1, 2, 4 < 1, 2, 5 (4 lower than 5) so the name is 2-bromo-1-chloro-4-methylcyclohexane.

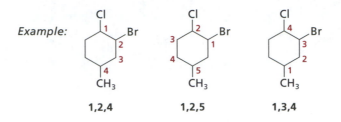

1,2,4	**1,2,5**	**1,3,4**

4. Once the order of numbering is defined, then the designation of the appropriate multiple-bond index is inserted between the root word and the functional group suffix. If necessary, a numeral is included to indicate at which carbon atom the double or triple bond begins.

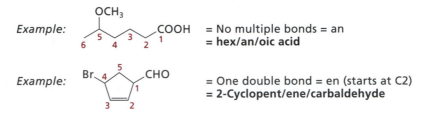

Example: = No multiple bonds = an
     = **hex/an/oic acid**

Example: = One double bond = en (starts at C2)
     = **2-Cyclopent/ene/carbaldehyde**

5. Finally, the substituents are specified by appending the appropriate prefixes and numerals to the name.

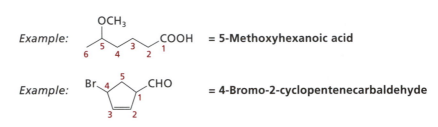

Example: = **5-Methoxyhexanoic acid**

Example: = **4-Bromo-2-cyclopentenecarbaldehyde**

When more than one substituent is present, their names are arranged in alphabetical order, ignoring any italicized words such as *sec-* or *tert-* as well as prefixes having to do with how many of a given substituent are present.

Example: *tert-*Butyl comes before chloro because butyl comes before chloro alphabetically.

Example: Ethyl comes before dimethyl because ethyl comes before methyl alphabetically (i.e., the "di" prefix is ignored).

The following examples illustrate the procedure used to name organic molecules.

EXAMPLE 1.12

Give an acceptable systematic name for the following compound:

This compound has the ketone functional group, so the suffix is "one." The longest carbon chain has six carbon atoms, so the root is "hex". There are no carbon–carbon double or triple bonds, so the multiple-bond index is "an." The name so far is hex/an/one = hexanone.

The position of the ketone functional group must be specified by a numeral. With respect to numbering, rule 3 states that the chain is numbered to give the functional group the lowest possible number. Therefore, numbering begins at the right end of this structure instead of at the left end.

Once the numbering is done, the position of the methyl group is established, which is C4. The name is therefore **4-methyl-2-hexanone.**

EXAMPLE 1.13

Give an acceptable systematic name for the following compound:

This compound has a nitrile functional group, so the name ends in "nitrile." The longest carbon chain, *including the nitrile carbon atom,* has seven carbon atoms, so the root is "hept". There are no carbon–carbon double or triple bonds, so the multiple-bond index is "ane". The name so far is hept/ane/nitrile = heptanenitrile.

The principal functional group is one that must be at the end of the chain (Table 1.1), so numbering begins at the right end of the structure.

$$CH_3-CH_2-\underset{Br}{CH}-CH_2-\underset{CH_3}{CH}-CH_2-CN$$
$$\quad 7 \qquad 6 \qquad 5 \qquad 4 \qquad 3 \qquad 2 \qquad 1$$

The positions of attachment for the bromine atom and the methyl group are subsequently established, and *bromo* precedes *methyl* in the name because of alphabetical order, even though the methyl group is attached to a carbon atom with a lower number. The name is **5-bromo-3-methylheptanenitrile.**

EXAMPLE 1.14

Give an acceptable systematic name for the compound shown here.

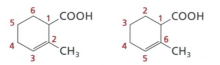

This compound has the carboxylic acid as its principal functional group, which is attached to a six-membered ring. The root of "cyclohex" must be compounded with the suffix "carboxylic acid". There is a carbon–carbon double bond, so the multiple-bond index is "ene". The name so far is cyclohexenecarboxylic acid.

The point of attachment of the principal functional group defines where the numbering begins, and we number in the clockwise direction so as to give the position of the double bond the lower possible number (2 instead of 5, which would be its position if we numbered counterclockwise).

The methyl group is attached at C2, so the name is **2-methyl-2-cyclohexenecarboxylic acid.**

EXAMPLE 1.15

Give an acceptable systematic name for the compound shown here.

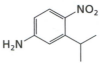

This compound has both amine and nitro functional groups. The nitro group is always listed as a substituent, so the amino group is the principal functional group. Aromatic compounds with an amino group are named as derivatives of *aniline* (Fig. 1.4), and the point of attachment of the amino group defines the C1 position of the ring.

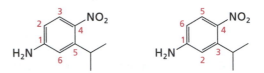

Numbering is counterclockwise because 3,4– is lower than 4,5– at the first point of difference. Isopropyl comes before nitro because of alphabetical order, so the name of this compound is **3-isopropyl-4-nitroaniline.**

EXAMPLE 1.16

Give an acceptable systematic name for the compound shown here.

This compound has no principal functional group or multiple bond. The longest carbon chain is 6 carbon atoms, which is hexane. There are two substituents and chloro will precede methyl in the name because of alphabetical order. The chain can be numbered from either end, however.

3-Chloro-5-methylhexane   **4-Chloro-2-methylhexane**

Because 2,4 is lower than 3,5 at the first point of difference, the correct name is **4-chloro-2-methylhexane.**

EXERCISE 1.17

Give an acceptable systematic name for each of the following compounds:

a.   b.   c.

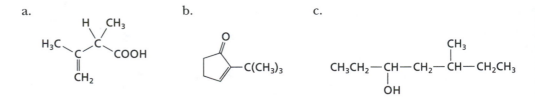

Section 1.1 Structural components of organic molecules

- Carbon always forms four bonds to other atoms in stable molecules. Nitrogen normally forms three bonds, oxygen forms two bonds, and hydrogen and the halogens (F, Cl, Br, and I) form one bond. These trends are summarized in Figures 1.1 and 1.2.

- Multiple bonds—double and triple—count as two and three bonds, respectively.

- Organic compounds comprise carbon atoms (always), hydrogen atoms (often), and heteroatoms (sometimes). Heteroatoms are elements besides C and H, and their presence often creates a functional group, the reactive portion of a molecule. Common functional groups are listed in Table 1.1.

Section 1.2 Structural formulas and condensed structures

- Condensed formulas eliminate the need to show every atom and bond. Methyl (CH_3), methylene (CH_2), and methine (CH) groups are condensed hydrocarbon units.

- Carbon atom positions can be represented by angled meetings of lines or as the vertices of a regular polygon.

- The carbon–oxygen double bond, called the carbonyl group, is a constituent of several functional groups including ketone, aldehyde, and carboxylic acid.

Section 1.3 Systematic nomenclature: IUPAC names

- IUPAC names are systematic and read as follows from left to right:

  a. The nature and positions of substituents (Section 1.3d).

  b. The root word (Section 1.3a).

  c. The nature and number of double or triple carbon–carbon bonds (Section 1.3b).

  d. The identity of the principal functional group (Section 1.3c).

- For aliphatic compounds, the root word is the longest carbon chain related in structure to the corresponding alkane, a molecule with the general formula $CH_3(CH_2)_nCH_3$. The chain is numbered to give the principal functional group the lowest possible number.

- For alicyclic compounds, the root word is based on the size of the ring.

- For aromatic compounds, the root word contains "benz", short for benzene, or a common name such as toluene, anisole, aniline, or phenol, among others.

- The abbreviation "R" is used to generalize the identity of a substituent, especially one that contains carbon and hydrogen atoms. The abbreviation "Ar" is likewise used as a shorthand notation for an aromatic ring.

- The positions of substituents are specified by numerals that precede the name of the substituent group. Multiple substituents of the same type are identified by a prefix (di-, tri-, tetra-, etc.) that tells how many such groups are present.

Section 1.4 Constitutional isomers and hydrocarbon substituents

- Isomers are compounds that have the same chemical formula but a different spatial arrangement of atoms. Figure 1.7 summarizes the types of constitutional isomers.

- A carbon atom that forms four single bonds is classified as primary (1°), secondary (2°), tertiary (3°), or quaternary (4°) according to the number of other carbon atoms attached. Carbon atoms with four single bonds can also be designated methyl, methylene, and methine (CH) groups according to the number of hydrogen atoms attached (3, 2, and 1, respectively).

- A nitrogen atom that forms three single bonds in amines and amides is classified as primary (1°), secondary (2°), or tertiary (3°) according to the number of carbon atoms attached.

- Removing the hydrogen atom from an alkane generates an alkyl group.

- An alkoxy group is a substituent in which an alkyl group is connected to the main chain or ring via an oxygen atom.

- Substituents with carbon–carbon double and triple bonds are formed from unsaturated hydrocarbons and are called alkenyl, alkynyl, and aryl groups.

Section 1.5   Naming organic compounds

- IUPAC names are generated by identifying the longest carbon chain or ring, the principal functional group that is present, the kind and number of carbon–carbon multiple bonds present, and the types and positions of substituents attached to the chain or ring.

KEY TERMS

Introduction
organic chemistry

Section 1.1b
heteroatom
functional group

Section 1.2a
condensed structure
hydrocarbon

Section 1.3a
IUPAC rules
aliphatic compounds
alkane
alicyclic
aromatic compound
arene

Section 1.3b
alkene
alkyne

Section 1.3c
R
Ar
carbonyl group

Section 1.4a
isomers
constitutional isomers
skeletal isomers
positional isomers
functional isomers
primary carbon atom
secondary carbon atom
tertiary carbon atom
quaternary carbon atom

Section 1.4b
alkyl group
alkoxy group

Section 1.4c
alkenyl carbon atom
vinyl carbon atom
alkynyl carbon atom
aryl carbon atom

ADDITIONAL EXERCISES

1.18. Draw structural formulas for the following alkanes, showing all of the atoms in the longest carbon chain:

a. The five isomers of C_6H_{14}.

b. The nine isomers of C_7H_{16}.

1.19. Draw a condensed structure for each answer in Exercise 1.18.

1.20. Show all of the hydrogen atoms in each of the following condensed representations of some naturally occurring organic compounds. You may use CH$_3$ to represent the methyl groups instead of showing all three hydrogen atoms with their bonds.

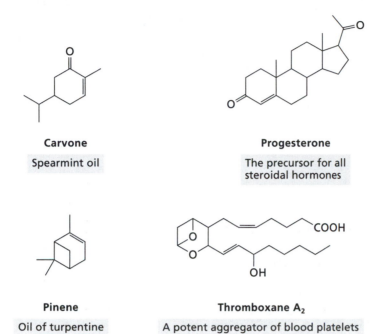

Carvone

Spearmint oil

Progesterone

The precursor for all
steroidal hormones

Pinene

Oil of turpentine

Thromboxane A$_2$

A potent aggregator of blood platelets

1.21. Each of the following molecules has a single functional group. Give its identity according to those listed in Table 1.1. Represent each molecule in the form that makes use of "R" along with the functional group (see Exercise 1.8).

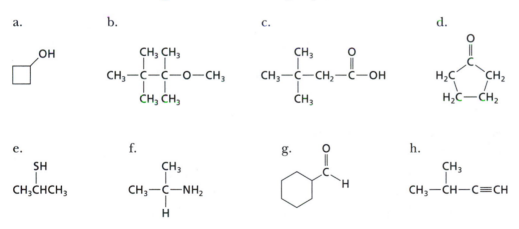

a.

b.

c.

d.

e.

f.

g.

h.

1.22. Draw two structures that exemplify each of the following types of compounds:

a. A ketone with the formula C$_5$H$_{10}$O.

b. A chloro ketone with four carbon atoms.

c. An aromatic amine.

d. An aldehyde with six carbon atoms.

e. A hydroxy aldehyde.

f. An alicyclic carboxylic acid.

1.23. For the compounds shown in Exercise 1.22, identify the carbon atoms as 1°, 2°, 3°, 4° (Section 1.4a) or alkenyl, alkynyl, aryl, or specific functional group (Section 1.4c).

1.24. Shown below are condensed representations of some organic compounds that have been used as drugs. Show all of the hydrogen atoms and identify the following functional groups: alcohol, aldehyde, amide, amine, carboxylic acid, ether, carboxylic acid ester, ketone, and thiol (some of these groups appear more than once; some do not appear). You may use CH_3 to represent the methyl groups instead of showing all three hydrogen atoms with their bonds.

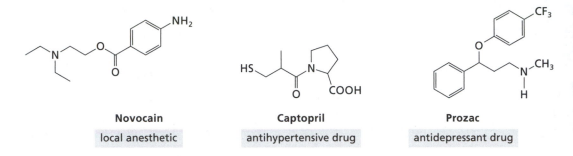

Novocain	**Captopril**	**Prozac**
local anesthetic	antihypertensive drug	antidepressant drug

1.25. For the compounds shown in Exercise 1.24, identify the carbon atoms as 1°, 2°, 3°, 4° (Section 1.4a) or alkenyl, alkynyl, aryl, or specific functional group (Section 1.4c).

1.26. The root word for an aliphatic compound is the longest carbon chain that also contains the principal functional group. Together with the other rules you have learned, explain why each of the following names is incorrect:

a. Methylheptane

b. 3-Propylhexane

c. 2,2-Dimethyl-3-ethylbutane

d. 2-Dimethylpentane

e. 2-Isopropyl-1-propanol

f. Dichloroheptane

1.27. Draw a structure for each of the seven isomers of C_6H_{10} that have a triple bond. You may use condensed structures. Name each compound.

1.28. Draw every alcohol with the molecular formula $C_5H_{12}O$. Label each carbon atom as 1°, 2°, 3°, or 4°. You may use condensed structures. Name each compound.

1.29. Draw a structural formula for each of the following alcohols:

a. 2-Phenylethanol

b. 1,3-Dibromo-2-pentanol

c. 3-Chloropropanol

d. 2-Methyl-3-buten-2-ol

e. 2,2,2-Trifluoroethanol

f. 2-Amino-2-methylbutanol

g. 2,3-Butadienol

h. 4-Hexynol

1.30. Draw a structural formula for each of the following carboxylic acids:

a. 2-Aminobenzoic acid

b. 2,2-Difluorobutanoic acid

c. 2,3-Dibromopropanoic acid

d. 4-Isobutylbenzoic acid

e. 3-Methoxycycloheptanecarboxylic acid

f. 3-Mercapto-4-hexenoic acid

g. 5-Hydroxy-3-heptenoic acid

h. 2,5-Dimethylbenzoic acid

1.31. Within each group of compounds, which of the following structures are identical and which are constitutional isomers?

a.

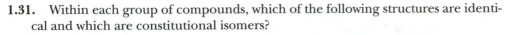

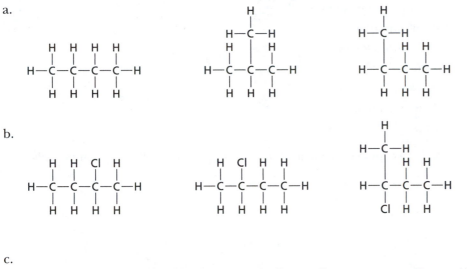

b.

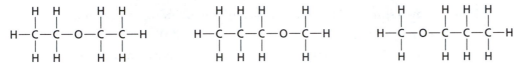

c.

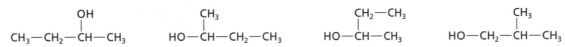

d.

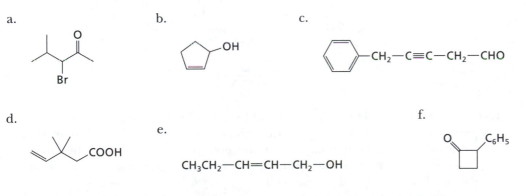

1.32. Give an acceptable systematic name for each of the following aliphatic or alicyclic compounds:

a.

b.

c.

d.

e.

f.

1.33. Draw a structural formula for each of the following amines and amides:

a. *N,N*-Dimethylaniline

b. 1,2-Diaminocyclohexane

c. 1-Amino-3-phenylbutane

d. 3-Hydroxycyclopentanecarboxamide

e. *N*-Methylbutanamide

f. 2,4-Dimethylaniline

1.34. Draw a structural formula for each of the following aldehydes and ketones:

  a. 3-Methyl-2-butanone

  b. 1-Chloro-3-hexene-2-one

  c. 3-Methoxypentanal

  d. 3-Isopropylcyclohexanecarbaldehyde

  e. 4-Bromobenzaldehyde

  f. 3-Ethoxy-2-hexanone

  g. 2-Cyclohexenone

  h. 3-*tert*-Butylcyclobutanone

1.35. Give an acceptable systematic name for each of the following compounds:

a.   b.   c.   d.

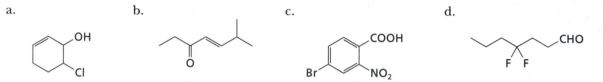

1.36. Draw a structural formula for each of the following compounds:

  a. 1-Chloro-2-hexyne   b. 1,3-Dicyanobenzene

  c. 6-Bromohexanoic acid   d. 2-Nitrobenzaldehyde

  e. 4-Nitrotoluene   f. 2-Allyl-6-chlorophenol

  g. 4-Pentynol   h. 3,5-Heptadienal

  i. 1-Decanethiol   j. 1-Nitropropane

  k. 2, 3, 4-Hexanetriol   l. 2-Butyl-3-chlorobenzonitrile

1.37. Give an acceptable systematic name for each of the following aromatic compounds:

a.   b.   c.   d.

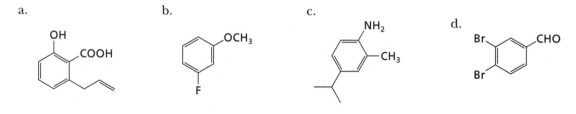

1.38. Draw a structural formula for each of the following compounds that have a complex substituent. Follow the procedure shown in Example 1.8: First interpret the name of the parent compound, then interpret the name of the substituent given in parentheses. By convention, the carbon atom of the substituent attached to the parent chain or ring is designated C1.

  a. 4-(1,1-Dimethylethyl)-4-octanol

  b. 3-Chloro-2-(1-hydroxyethyl)-6-nitrophenol

  c. 3-(2-Fluoro-2-propenyl)-5-hepten-2-one

  d. 4-(1-Methylethyl)-5-methyl-3-hexenal

BONDING IN ORGANIC MOLECULES

A prodigious amount of work on the reactions of organic compounds was accomplished long before chemists knew much about bonding. Structural formulas, line structures, and basic molecular shapes were understood before the twentieth century began, so it should be no surprise that simple concepts will go far in helping us to predict structures and reactions of molecules.

Nevertheless, it is also true that knowing how electrons are distributed within a molecule permits us to go beyond the reactions of the 1800s, and instead, to make predictions about transformations that have not yet been encountered. Therefore, it is crucial to build a solid base of knowledge about electronic structures of molecules and bonds between atoms.

The purpose of this chapter is to acquaint you with different models that are used to represent the arrangements of electrons in organic molecules. These models include Lewis structures as well as pictorial illustrations. After completing this chapter, you should be able to predict the shape of a simple organic molecule by looking at its structure. In addition, you should be able to anticipate how electrons are distributed. The latter is essential for making predictions about reactions of organic molecules.

2.1 LEWIS STRUCTURES

2.1a LEWIS STRUCTURES DESCRIBE THE DISTRIBUTION OF VALENCE ELECTRONS

In 1916, Gilbert N. Lewis published a theory of bonding for organic molecules based on the *octet rule,* which says that C, N, O, and F atoms attain stable configurations when they have eight electrons in their outer shell of orbitals. Despite the evolution of more comprehensive theories, **Lewis structures** are commonly used to depict the distribution of valence electrons in organic molecules. In compounds that contain heteroatoms, Lewis structures are used to account for unshared electron pairs (imprecisely called *lone pairs* in older texts) that are not involved in forming covalent bonds to other atoms. For that reason, you need to understand how to draw Lewis structures.

Because you are already familiar with line structures from Chapter 1, you need only add the nonbonding electrons to those in order to generate the corresponding Lewis structures. The following procedure outlines a systematic method for creating the Lewis structure of a molecule.

1. Draw the structural formula of the molecule, making use of the fundamental rule that there are four bonds to each carbon atom; three bonds to each nitrogen, phosphorus, or boron atom; two bonds to each oxygen or sulfur atom, and one bond to each halogen or hydrogen atom (Figures 1.1 and 1.2).

2. Tally the number of valence shell electrons that each element contributes to the molecule. The number of valence electrons for a given element is determined from its position in the periodic table [group 1 (IA) elements have one valence electron, group 2 (IIA) elements have 2 electrons; group 13 (IIIA) elements have 3 electrons, and so on (see Figure 2.1)].

3. Subtract the number of electrons in the covalent bonds of the structure you have already drawn (two electrons per bond) to calculate how many unshared electrons need to be added to the structure.

4. Distribute these remaining electrons in pairs so that each non-hydrogen atom is surrounded by eight electrons. Electrons in the bond between atoms are included in the electron count of *each* atom. Note, too, the following exceptions to the octet rule:

   • Hydrogen needs only two electrons to complete its valence shell, so it normally forms only one bond and does not have additional electrons associated with it.

   • Boron and aluminum have only six electrons surrounding them when they are bonded to three atoms.

   • Elements in the third period and higher can accommodate 10 or 12 electrons in their valence shell. Phosphorus and sulfur are the most common elements that may have more than 8 electrons surrounding them.

Group number	1	2	3	4	5	6	7	8	9	10	11	12	13	14	15	16	17	18
	(IA)	(IIA)	(IIIB)	(IVB)	(VB)	(VIB)	(VIIB)	(VIII)	(VIII)	(VIII)	(IB)	(IIB)	(IIIA)	(IVA)	(VA)	(VIA)	(VIIA)	(VIIIA)
Valence electrons	1	2											3	4	5	6	7	8
	H																	He
	Li	Be											B	C	N	O	F	Ne
	Na	Mg											Al	Si	P	S	Cl	Ar
	K	Ca	Sc	Ti	V	Cr	Mn	Fe	Co	Ni	Cu	Zn	Ga	Ge	As	Se	Br	Kr
	Rb	Sr	Y	Zr	Nb	Mo	Tc	Ru	Rh	Pd	Ag	Cd	In	Sn	Sb	Te	I	Xe
	Cs	Ba	La	Hf	Ta	W	Re	Os	Ir	Pt	Au	Hg	Ti	Pb	Bi	Po	At	Rn

Figure 2.1
The periodic table with the numbers of valence electrons for the main group elements.

EXAMPLE 2.1

Draw the Lewis structure for 2-aminobutane.

1. Draw the line structure according to the procedure outlined in Chapter 1. The structural formula of 2-aminobutane is shown here.

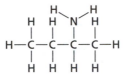

2. Tally the total number of valence electrons (= 32).

Valence electrons		
C	4×4e =	16
H	11×1e =	11
N	1×5e =	5
Total		32e

3. Subtract the number of electrons already present in the covalent bonds (2 electrons × 15 bonds = 30), so 32 (total) − 30 (bonded) = 2 unshared electrons need to be added to the structural formula.

4. Because each hydrogen atom already has two electrons (hence, a filled shell) and each carbon atom has eight electrons, the two remaining electrons are placed on the nitrogen atom, which lacks an octet. The Lewis structure of 2-aminobutane is shown below.

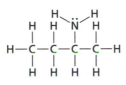

Recall from Chapter 1 that many functional groups contain atoms that are bonded to another atom by a double or triple bond. To give each atom an octet of electrons, we sometimes need to include double and triple bonds, resulting in the placement of four or six electrons *between* the two atoms joined by the multiple bond. As was the case for the electrons in single bonds, the electrons in multiple bonds are included in the electron count of *each* atom.

EXAMPLE 2.2

Draw the Lewis structure for 2-bromo-3-pentanone.

1. Draw the line structure according to the procedure outlined in Chapter 1. The structural formula of 2-bromo-3-pentanone is shown below.

2. Tally the total number of valence electrons (= 42).

Valence electrons		
C	5×4e =	20
H	9×1e =	9
O	1×6e =	6
Br	1×7e =	7
Total		42e

16 bonds × 2e = 32 electrons accounted for; 10 electrons unaccounted for

3. Subtract the number of electrons already present in the covalent bonds (2 electrons × 16 bonds = 32), so 42 (total) − 32 (bonded) = 10 unshared electrons need to be added to the structural formula.

4. Because the hydrogen and carbon atoms already have filled shells, the additional 10 electrons must be associated with the oxygen and bromine atoms, each of which lacks an octet. The Lewis structure of 2-bromo-3-pentanone is below. Notice that the octets for the oxygen and its attached carbon atom include the four electrons in the double bond.

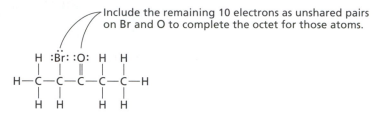

EXERCISE 2.1

Draw a Lewis structure for each of the following compounds:

a. $CH_3CH_2CH_2Cl$ c. CO_2 e. 3-Buten-2-one
b. 3-Fluoro-1-butanol   d. 2-Cyanopentane

2.1b   FORMAL CHARGES COMPENSATE FOR THE UNEQUAL DISTRIBUTION OF VALENCE ELECTRONS

In the procedure outlined in Section 2.1a, we simply distributed the electrons so that each atom attained an octet (with the noted exceptions). In any Lewis structure we draw, however, the valence electrons that each atom contributes to a molecule may be distributed on another atom in the molecule. When this situation occurs, some atoms acquire a **formal charge,** which is the difference between its number of valence electrons (as determined from the periodic table) and the number of electrons it either has in its unshared electron pairs or has contributed to covalent bonds. Each atom forming a covalent bond contributes one electron to that bond. The following formula is used to calculate the formal charge on an atom:

Formal charge = Number valence electrons − (Number bonds + Number nonbonding electrons)   (2.1)

Formal charges of +1 or −1 are indicated simply by writing a plus or minus sign next to the atom. If the formal charge is +2 or −2, then the numeral is also included. Nothing is written if the formal charge is zero.

In *most* stable organic molecules, formal charges do not occur unless nitrogen, sulfur, or phosphorus atoms are present. Examples of formal charges frequently encountered in organic compounds are summarized in Table 2.1.

Table 2.1   Common functional groups with formal charges on one or more atoms.

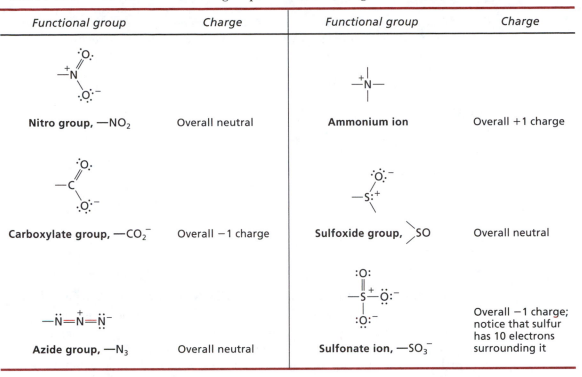

Functional group	Charge	Functional group	Charge
Nitro group, —NO₂	Overall neutral	Ammonium ion	Overall +1 charge
Carboxylate group, —CO₂⁻	Overall −1 charge	Sulfoxide group, >SO	Overall neutral
Azide group, —N₃	Overall neutral	Sulfonate ion, —SO₃⁻	Overall −1 charge; notice that sulfur has 10 electrons surrounding it

EXAMPLE 2.3

Calculate formal charges for the atoms in nitroethane.

First, draw the Lewis structure:

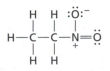

Each nonhydrogen atom has an octet of electrons.

Then, apply Eq. 2.1 to calculate the formal charge on each atom.

For the singly bonded O atom:
formal charge = $(6) - (1+6) = -1$

For the N atom:
formal charge = $(5) - (4 + 0) = +1$

For the doubly bonded O atom:
formal charge = $(6) - (2 + 4) = 0$

For each C atom:
formal charges = $(4) - (4 + 0) = 0$

Finish by including the charges on the complete structure. (Remember that nothing is written next to the atom if its formal charge equals zero.)

Recognize that a carbon atom does not carry a charge when it has four bonds. Because carbon atoms in stable organic molecules rarely have other than four bonds, carbon atoms normally do not have formal charges.

EXERCISE 2.2

For the following compounds, add the unshared electrons to the given structure to generate a Lewis structure, and then calculate the formal charge on each non-hydrogen atom:

a. 2-Nitropropane   b. Diazomethane   c. Dimethyl sulfone

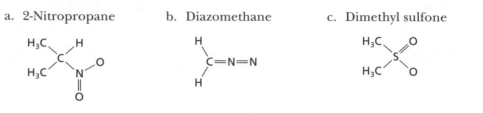

2.2   BOND PROPERTIES

2.2a   THE PAULING SCALE OF ELECTRONEGATIVITY QUALITATIVELY PREDICTS HOW ELECTRONS ARE DISTRIBUTED IN COVALENT BONDS

A Lewis structure—even with formal charges—does not reveal how the electron pair in a covalent bond is shared between the two atoms it connects. When two atoms are identical, as in molecular hydrogen, H–H, the electrons in the bond are attracted toward each nucleus to the same degree and so are shared equally. When the atoms bonded together are different, however, then the sharing of electrons is unequal, and the bond is a *polar* one. The property of an atom to attract electrons to itself is called **electronegativity,** and Linus Pauling was the first to devise a scale that gauges this attractive force. Figure 2.2 shows the periodic table with common electronegativity values for the main group elements. We can use these values to predict the polarity of any bond based on the assumption that an atom with a higher electronegativity value will attract a greater share of the bonded electrons toward *its* nucleus.

In the first two periods, the nonmetallic elements are more electronegative than carbon with the notable exceptions of hydrogen and boron. This electronegativity difference makes carbon susceptible to reaction with ions and molecules that contain heteroatoms, most of which are also more electronegative.

Group number   1   2   3   4   5   6   7   8   9   10   11   12   13   14   15   16   17   18

(IA) (IIA) (IIIB) (IVB) (VB) (VIB) (VIIB) (VIII) (VIII) (VIII) (IB) (IIB) (IIIA) (IVA) (VA) (VIA) (VIIA) (VIIIA)

H 2.1																	He
Li 1.0	Be 1.5											B 2.0	C 2.5	N 3.0	O 3.5	F 4.0	Ne
Na 0.9	Mg 1.2											Al 1.5	Si 1.8	P 2.1	S 2.5	Cl 3.0	Ar
K 0.8	Ca 1.0	Sc	Ti	V	Cr	Mn	Fe	Co	Ni	Cu	Zn	Ga 1.6	Ge 1.8	As 2.0	Se 2.4	Br 2.8	Kr
Rb 0.8	Sr 1.0	Y	Zr	Nb	Mo	Tc	Ru	Rh	Pd	Ag	Cd	In 1.7	Sn 1.8	Sb 1.9	Te 2.1	I 2.5	Xe
Cs 0.8	Ba 0.9	La	Hf	Ta	W	Re	Os	Ir	Pt	Au	Hg	Tl 1.8	Pb 1.9	Bi 1.9	Po 2.0	At 2.2	Rn

Figure 2.2
Electronegativity values for the main group elements.

A simple way to symbolize the polarity of a bond is to use the lower case Greek letter delta (δ) along with a plus or minus sign to indicate which atom is more electronegative and which is more electropositive. *The δ signifies a partial charge and should not be confused with formal charges.* Another way to indicate bond polarity is to use an arrow with a plus sign as the "tail" to indicate the direction in which the bond is polarized. The unequal electron distribution in a covalent bond gives rise to a **dipole** (two poles) because the bond has a positive and a negative end.

Two representations for the polarity of the carbonyl group.

Two representations for the polarity of the atoms in the alcohol group. Note that there are different dipoles for the two polar bonds.

Bond polarity is not solely the result of electronegativity differences between atoms. Another factor that contributes to the polarity of bonds is the *polarizability* of atoms. A **polarizable** atom is one in which the distribution of its electrons is readily distorted or deformed by outside influences (think of such an atom as a pliable, foam rubber sphere). Polarizability, which results from the weakening attraction between the nucleus and its outermost electrons as an atom becomes larger, is most important for elements such as Br and I. Polarizability of a large atom makes its bonds polar, even if the electronegativity values indicate otherwise. For example, the electronegativity values of carbon and iodine are identical (2.5), so you might assume that the C–I bond is not polar. Yet, CH_3I, reacts as if the carbon atom were partially positive and the iodine atom, partially negative. In general, the bonds between carbon and any element in groups 15–17 (VA)–(VIIA) are considered to be polar, with the carbon atom bearing the δ+ designation. For bonds between carbon and a metal atom, including those of the transition metals, the carbon atom is δ–.

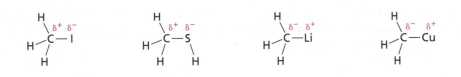

For the reactions of organic molecules, the "δ" notation will be used almost exclusively throughout this text to denote the polarity differences of bonds and atoms in a molecule. Figure 2.3 illustrates some functional groups, along with the expected polarities of their atoms. For some molecules, particularly those in which a hydrogen atom is attached to an oxygen or nitrogen atom, several δ symbols are needed to indicate the polarization of every atom in the functional group. Note that the polarities of the C–H bonds are largely ignored. The electronegativity difference between carbon and hydrogen is normally small, so hydrocarbon portions of a molecule are considered nonpolar.

Even though the use of δ+ and δ– is a practical way to denote atom polarities when writing chemical reactions, the arrow notation is more useful if you want to *compare* bond polarities or portray the polarity of the molecule as a whole. The length and direction of a polarity arrow is related to the magnitude of the electronegativity differences of the atoms as well as to the orientation of the bond in three dimensions.

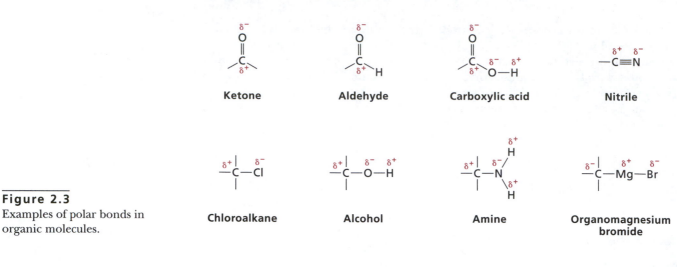

Figure 2.3

Examples of polar bonds in organic molecules.

Thus, to show that the C–Cl bond in chloromethane is more polar than the C–Br bond in bromomethane, we can draw an arrow for the C–Cl bond that is longer. To depict the polarity of the chloromethane molecule and not just its C–Cl bond, we add the polarity arrows for each bond. As shown below at the right, the arrow for the whole molecule is longer than for the isolated C–Cl bond because the polarities of the C–H bonds will contribute dipoles that enhance the C–Cl dipole.

Comparison of the C–X bond polarities in chloromethane (a) and bromomethane (b)

Depiction of the individual bond polarities in chloromethane (c) and for the whole molecule (d)

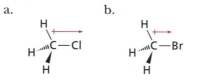

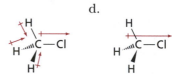

To specify the polarity of a molecule or group, you have to know its three-dimensional structure (to be discussed shortly), the magnitudes of the polarity differences for each bond, and rudimentary vector algebra.

EXERCISE 2.3

Using the δ+ and δ– formalism and ignoring the carbon–carbon and carbon–hydrogen bonds, indicate the polarity of the bonds in the following molecules:

a.   b.   c.   d.

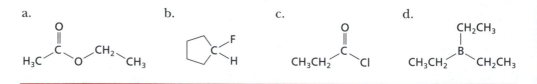

2.2b   BOND LENGTHS VARY AS A FUNCTION OF ATOM IDENTITY AND OF BOND ORDER

Many organic molecules contain single, double, and triple bonds between adjacent atoms. Experimentally, multiple bonds shorten the distance between any two atoms. A single bond is therefore longer than a double bond, which, in turn, is longer than a triple bond. Expected lengths for bonds in simple molecules are given in Table 2.2.

Within a group of the periodic table, bond lengths increase as the atomic radii of the elements increase. For example, a carbon–oxygen single bond is shorter than a carbon–sulfur single bond, and a carbon–chlorine bond is shorter that a carbon–iodine bond.

Table 2.2   Average bond lengths for atoms in organic molecules.

Bond type	Length (Å)	Bond type	Length (Å)
C—C	1.53	C—N	1.47
C=C	1.31	C=N	1.28
C≡C	1.18	C≡N	1.14
C—H	1.09	C—F	1.40
C—O	1.43	C—Cl	1.79
C=O	1.21	C—Br	1.97
C—S	1.82	C—I	2.16

2.3   RESONANCE STRUCTURES

2.3a   RESONANCE STRUCTURES ARE USED TO INDICATE THAT ELECTRONS ARE NOT LOCALIZED ON SPECIFIC ATOMS

When drawing Lewis structures, we make an implicit assumption, namely, that electrons are **localized,** which means that they occupy a specific region of space either between or on individual atoms. In certain cases, bonding in a molecule cannot be represented by a *single* Lewis structure. An example is nitroethane (cf. Example 2.3), for which we draw two equivalent Lewis structures, as illustrated in Figure 2.4. Note the presence of formal charges on the nitrogen and oxygen atoms in these structures.

Double-headed arrow denotes that these two Lewis structures are resonance structures.

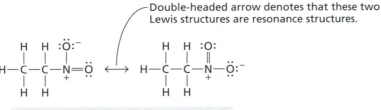

Each non-hydrogen atom has an octet of electrons: *neither structure actually exists.*

Figure 2.4
Possible Lewis structures for nitroethane, showing electrons localized on an atom or in a bond between atoms.

Each Lewis structure has a nitrogen–oxygen single bond, as well as a nitrogen–oxygen double bond. However, the experimentally determined nitrogen–oxygen bond lengths in this molecule are *identical*. Therefore, a *single* structure cannot be used to portray the distribution of electrons in this molecule—we must adopt a different formalism. We call such representations **resonance structures** (or *resonance forms*). They indicate the possible ways that electrons can be *localized* on or between the atoms.

It is crucial to understand that resonance forms are *imaginary* descriptions of electron density in a molecule; that is, *they do not actually exist as separate entities*. The "real" picture is called the **resonance hybrid,** a composite of all of the possible resonance structures. These hybrids frequently involve multiple bonds and nonbonding electrons that can be associated with more than one atom.

In molecules for which more than one structure can be drawn, we first draw all of the possible Lewis structures that depict what the bonding would look like if *only* a localized electron distribution were possible. We then relate those structures with double-headed arrows (↔) to indicate that the actual electron density picture is a combination of these imaginary forms. Notice that the word *resonance* is an adjective, so do not make it into a verb: The individual structures do *not* "resonate".

2.3b REDISTRIBUTING THE ELECTRONS IN A LEWIS STRUCTURE GENERATES OTHER RESONANCE FORMS

To generate resonance structures, follow the procedure outlined here:

1. Draw a single Lewis structure as you normally would, giving each atom eight electrons except for H (two electrons) and B and Al (six electrons, usually). Include formal charges if they are present.

2. Redraw the atom positions of the Lewis structure and connect the atoms via single bonds in the same fashion as in the original. Remember this primary rule: *Nuclei do not move*. Once you draw the structure, you may only change the way in which the electrons are distributed. This redistribution is done by moving an electron pair *to an adjacent position* (note that the formal charges often change during these movements). In this text, a curved arrow will be used to show movement of a pair of electrons (more about this topic in Chapter 5). Thus, allowed patterns of electron movement are as follows:

    • One bond of a double bond between two atoms can become an unshared pair on one of those atoms; both choices should be considered:

    • An unshared pair on an atom can form a double (or triple) bond; we draw the arrow from the electron pair on one atom to the next atom, which creates the multiple bond:

    • One bond of a double bond can shift to the adjacent position, forming a different double bond:

Any of the movements illustrated above may be associated with redistribution of the electrons on other atoms, too. For example, two pairs of electrons are required to move in the N_2O structure (below) at the left to maintain an octet at each atom in the one on the right.

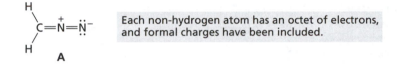

3. Calculate the formal charge on each atom. As already mentioned, these may change as electrons are redistributed within the molecule.

EXAMPLE 2.4

Draw resonance structures for diazomethane, CH_2N_2, the structure of which was shown in Exercise 2.2.

1. Draw a Lewis structure for the compound and include formal charges:

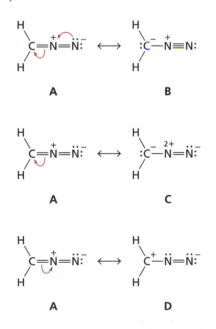

Each non-hydrogen atom has an octet of electrons, and formal charges have been included.

2. As many times as necessary, redraw the original Lewis structure and move the electrons in each structure on the left no farther than the adjacent position to generate another structure. Make certain that you use all of the electrons and calculate formal charges correctly.

After generating several structures, you must subsequently evaluate them in order to decide which are important to describe the distribution of electrons in the molecule. How to do this evaluation is described in Section 2.3c. The evaluation of these structures will be presented in Example 2.5.

2.3c   THE RELATIVE IMPORTANCE OF A RESONANCE STRUCTURE IS ASSESSED ON THE BASIS OF SEVERAL CRITERIA

Evaluating the resonance structures you have drawn is often the most difficult step of the entire process because a structure may contribute to the overall resonance hybrid in a limited way yet still be important. The best strategy is to generate several structures (say, between three and nine), and then eliminate the less important ones by evaluation, which makes use of the following guidelines:

- The most important resonance forms have an octet of electrons around each non-hydrogen atom (except B and Al). A second-period element may have *fewer*—but never more—than eight electrons.

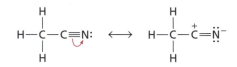

> Less important because carbon lacks an octet. This structure also has fewer bonds and more formal charges than the structure at the left.

- Important contributors have *more bonds* and/or *fewer formal charges* than other structures (no formal charges, if possible).

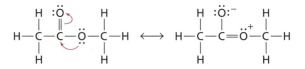

> Less important because of formal charges; each non-hydrogen atom has an octet.

- Structurally equivalent forms are especially important contributors because the resonance hybrid gains additional stabilization energy, a fact that can be demonstrated experimentally.

> Equal contributors; additional stabilization energy is acquired.

- When formal charges *are* present, the more important contributors have the negative charge on a more electronegative atom. For structures with two or more charges, opposite charges should be as close together as possible; like charges should be as far apart as possible.

> Less important because oxygen is more electronegative than nitrogen.

- Elements in period 3 and higher, especially sulfur and phosphorus, may have 10 or 12 electrons in their resonance forms.

- Representations with a substantially *higher* charge on one atom than in other structures are less important. Ignore structures in which the charge on an atom has changed by ±2 (or more) relative to another structure.

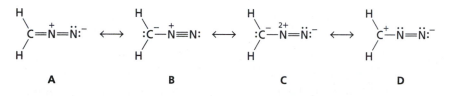

> Less important; N has a higher charge than in the other structures and adjacent atoms have the same charge.

EXAMPLE 2.5

Evaluate the four resonance forms for diazomethane that were generated in Example 2.4. Those structures, **A–D**, are reproduced below.

Forms **A** and **B** are important resonance structures for diazomethane because each non-hydrogen atom has an octet. We would rank form **A** as more significant because the negative charge is on the more electronegative atom (N instead of C, as it is in **B**). Structure **D** contributes less than either **A** or **B** because the carbon atom lacks an octet of electrons and has fewer bonds. We can eliminate structure **C** because one of the nitrogen atoms lacks an octet, it has the fewest number of bonds, and it has a greater number and magnitude of formal charges than the other structures.

EXAMPLE 2.6

Draw and evaluate resonance structures for methyl isocyanate, CH_3NCO.

1. Draw the Lewis structure:

2. As many times as necessary, redraw structure **A** and move the electrons in each structure on the left to an adjacent position to generate another structure. Make certain that you use all of the electrons and calculate formal charges correctly.

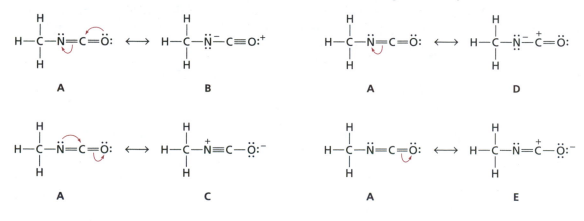

3. Evaluate the forms. Structure **A** is the best because each atom has an octet and there are no formal charges. Structures **B** and **C** are important because each atom has an octet. Structure **C** is more important than **B,** however, because the more electronegative oxygen atom bears the negative charge. Structures **D** and **E** are minor contributors because the carbon atom in each lacks an octet of electrons.

Resonance structures are important in describing the electronic structures of benzene and its derivatives. By shifting the three pairs of π electrons simultaneously, we can draw two equivalent and therefore equally important resonance structures for benzene.

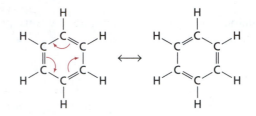

Experimentally, we find that the lengths of each of the carbon–carbon bonds in benzene are identical. If either *localized* structure were to exist, we would expect three of the bonds to be longer than the other three because of the difference in bond lengths for carbon–carbon single and double bonds (see Table 2.2). The actual distribution of the electrons in benzene is therefore best represented by the two structures (the resonance hybrid) shown above.

Resonance structures are often required to depict the electronic structures of ions. In these instances, the net charge on each resonance form must be the same. An example will illustrate this point.

EXAMPLE 2.7

Draw and evaluate resonance structures for the acetate ion, $CH_3CO_2^-$ (OAc⁻).

First, draw a Lewis structure for the compound, making certain to include the negative charge on the oxygen atom.

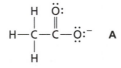

Next, redraw structure **A** and move the electrons to generate another structure. Make certain that you use all of the electrons and calculate formal charges correctly. Notice that the original structure has a negative charge, therefore *every structure drawn must also have an overall negative charge.*

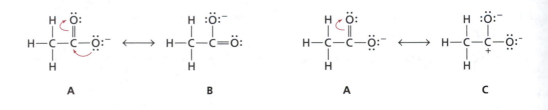

Finally, evaluate the forms. Structures **A** and **B** are equivalent and each non-hydrogen atom has eight electrons. Structure **C** is less important because one carbon atom lacks an octet. The resonance hybrid for the acetate ion is therefore represented as a composite of the two structures shown below:

$$ \begin{array}{ccc} \text{H} & :\!\text{O}\!: & \\ | & \| & \\ \text{H—C—C—}\ddot{\text{O}}\!:^- & \longleftrightarrow & \text{H—C—C=}\ddot{\text{O}} \\ | & & \\ \text{H} & & \end{array} $$

EXERCISE 2.4

Draw a complete Lewis structure and resonance forms for each of the following compounds. Remember that phosphorus can accommodate 10 or 12 electrons.

a.

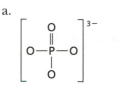

Phosphate ion

b.

H₃C—NNN

Azidomethane

2.4   THE FORMATION OF BONDS

2.4a   THE THEORY OF QUANTUM MECHANICS OPENED THE WAY FOR MATHEMATICAL DESCRIPTIONS OF BONDING INTERACTIONS

Although Lewis structures can be used for many applications in which we want to depict the bonds between atoms, we soon encounter difficulties if we begin to think about what a bond really is. Such problems arise because electrons do not move about the nuclei of atoms according to the laws of classical mechanics. A different concept is needed, and the theory of quantum mechanics, independently proposed in the mid-1920s by Werner Heisenberg, Paul Dirac, and Erwin Schrödinger, provided a way to describe electron motion mathematically using *wave equations*. The solutions of **wave equations,** called wavefunctions, are normally denoted by the Greek letter psi, Ψ. A more important quantity is Ψ^2, which relates to the probability of finding the electron at any particular point in space.

The plot of Ψ^2 in three dimensions generates the characteristic shapes of atomic orbitals, which are denoted with the letters *s*, *p*, *d*, and *f*. If you have studied chemistry before, you have undoubtedly seen these representations, some of which are illustrated in Figure 2.5. The *s* orbitals are spherical, but *p* orbitals are dumbbell shaped and are oriented along each of the three Cartesian axes. The mathematical signs of the two lobes of a *p* orbital are different because the orbital has a **node** at the nucleus, a point at which the probability of finding the electron is zero. Each orbital can have a maximum of two electrons.

2.4b   THE OVERLAP BETWEEN ATOMIC ORBITALS DEFINES A CHEMICAL BOND

Quantum mechanics can be simplified by making certain assumptions that lead to different models for describing chemical bonds. **Valence bond theory** defines bond formation as the result of overlap between atomic orbitals that occupy the same region of

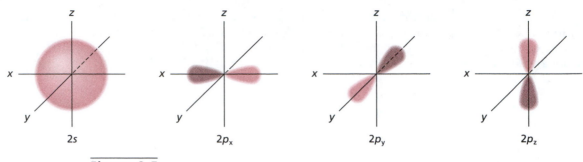

Figure 2.5
Representations of the s and p orbitals.

space. A bond is formed when an orbital with only a single electron overlaps with another singly occupied orbital. For example, Figure 2.6 shows the result of the overlap of the $1s$ orbitals of two hydrogen atoms, generating the hydrogen molecule (H_2). Hydrogen has the electron configuration $1s$, which shows that its s orbital has one electron. When overlap between orbitals lies along the line connecting the two nuclei, it is called a σ (**sigma**) **bond.**

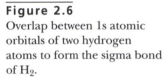

Figure 2.6
Overlap between 1s atomic orbitals of two hydrogen atoms to form the sigma bond of H_2.

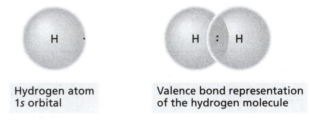

Hydrogen atom
1s orbital

Valence bond representation
of the hydrogen molecule

Similarly, a σ bond can also form between a p orbital and an s orbital because overlap occurs along the axis between the nuclei.

A specific example is the molecule HF, illustrated in Figure 2.7. Hydrogen has the electron configuration $1s$, and fluorine has the electron configuration $1s^2 2s^2 2p_x 2p_y^2 2p_z^2$; overlap takes place between the singly occupied $1s$ orbital of hydrogen and the $2p_x$ orbital of fluorine (x, y, and z are assigned arbitrarily).

Figure 2.7
Overlap of the atomic orbital of hydrogen with a p orbital of fluorine to generate the hydrogen fluoride molecule. Each p orbital comprises two lobes that have different mathematical signs (positive and negative), designated as shaded and unshaded.

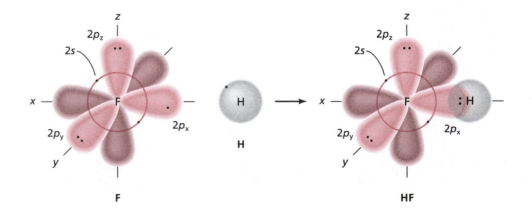

Now, consider bringing together a carbon atom ($1s^2 2s^2 2p_x 2p_y$) with hydrogen atoms to produce a hydrocarbon. Using the description based on overlap of atomic orbitals, we would predict that CH_2 is a stable molecule because only two p orbitals on carbon contain single electrons. Overlap between these p orbitals and the $1s$ orbitals of the hydrogen atoms would generate two bonds. Since the mid-1800s, however, chemists have known that carbon forms *four* bonds in stable molecules, so the simplest hydrocarbon is CH_4, methane. The four C–H bonds in methane extend to the corners of a tetrahedron. When bonding theories were developed during the 1900s, they had to account for the tetrahedral shape surrounding carbon atoms in simple hydrocarbons.

2.5   HYBRID ORBITALS AND SHAPES OF MOLECULES

2.5a   THE CONCEPT OF HYBRID ORBITALS EXPLAINS THE SHAPES OF ORGANIC MOLECULES

A straightforward way to rationalize the tetrahedral arrangement of bonds to carbon atoms in organic compounds is called the **valence-shell electron-pair repulsion (VSEPR) model.** The underlying concept of the VSEPR model is that electron pairs repel each other, whether they are involved in bonding or not. As a result, they will be oriented in directions that put them as far apart as possible. Table 2.3 summarizes the structures for an element surrounded by two, three, or four electron pairs. These electron-pair arrangements are manifestations of the hybridization models described on the next several pages.

Table 2.3   Arrangements of electron pairs about an atom A that minimize electrostatic repulsion.

Structure	Number of electron pairs	Arrangement of electron pairs	Hybridization	Idealized bond angle
:—A—:	2	Linear	sp	180°
:—A	3	Trigonal planar	sp^2	120°
A	4	Tetrahedral	sp^3	109.5°

When applying the VSEPR model to atoms in a molecule, you have to take into account the number and types of bonds and unshared electron pairs that are present. When a molecule contains atoms that only form single bonds and/or have unshared electron pairs, then we use the data from Table 2.3 directly. For example, the carbon and oxygen atoms in methanol, shown below, are predicted to be tetrahedral according to the VSEPR model because each atom is surrounded by four electron pairs.

Four single bonds = four electron pairs = tetrahedral

Two single bonds + two unshared pairs = four electron pairs = tetrahedral

To apply the data in Table 2.3 to atoms that form multiple bonds, we consider instead how many *regions* of electron density surround each atom. Thus, a carbon atom with a double and two single bonds has three regions, and its bond angles should be ~ 120°. For nearly all organic molecules, the number of regions equals the number of unshared pairs plus the number of sigma bonds. In other words, a multiple bond is considered the same as a single bond in the VSEPR model.

Three regions of electrons around the C atom

Three regions of electrons around the O atom

:N≡C⊕H

Two regions of electrons around the C atom

:N≡C—H

Two regions of electrons around the N atom

In 1931, Linus Pauling published the first paper in which mathematical models, utilizing quantum mechanics, explained the shapes of molecules by making use of a new concept, called **hybrid orbitals,** in which orbitals of similar energies can "mix". For methane, the simplest hydrocarbon, formation of bonds requires the use of the four sp^3 *hybrid orbitals* formed by a combination of one $2s$ and three $2p$ orbitals. The superscript associated with the hybrid designation indicates the number of each orbital type that has combined to create the hybrid. The number of hybrid orbitals always equals the number of atomic orbitals that combined.

Because a p orbital has a node at the nucleus, hybrid orbitals that have a contribution from a p orbital also have a node at the nucleus. We generally ignore the smaller lobe of the hybrid orbitals because it does not extend as far from the nucleus as the larger one, and so it is not involved in the bonding per se. The four sp^3 orbitals are oriented along lines that pass through the corners of a tetrahedron, as illustrated in Figure 2.8.

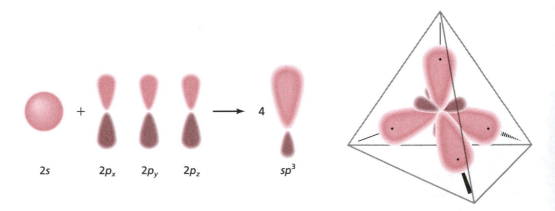

Figure 2.8
The conceptual combination of the atomic orbitals of carbon that generates sp^3 hybrid orbitals.

Overlap between the sp^3 hybrid orbitals of carbon and the atomic orbital of four hydrogen atoms generates the valence bond representation of methane, shown in Figure 2.9. Because the overlapping orbitals lie along the line connecting the carbon and hydrogen nuclei, these C–H bonds are σ bonds. From knowledge of Euclidian geometry, we calculate that the H–C–H angle should be 109.5°. This fact has been verified experimentally for the structures of many organic molecules.

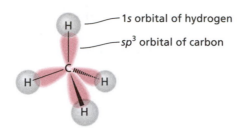

Figure 2.9
A valence bond representation of methane. All H–C–H angles are 109.5°. The smaller lobe of each hybrid orbital has been omitted for clarity.

EXERCISE 2.5

a. Draw a Lewis structure for water. b. Draw a valence bond representation of water assuming that the oxygen atom has sp^3 hybridization (the two pairs of unshared electrons are placed in two of the hybrid orbitals).

2.5b SOLID AND DASHED WEDGES AND LINES ARE USED TO REPRESENT MOLECULES IN THREE DIMENSIONS

Because a tetrahedron is a nonplanar object, its representation on a two-dimensional surface, such as a piece of paper, can create problems until you familiarize yourself with ways to illustrate molecules that have atoms with this geometry. A detailed presentation of three-dimensional shapes of organic compounds will be covered in the next two chapters (Chapters 3 and 4), but now is a good time to begin learning how to draw them.

For any nonplanar species, we use a solid wedge (►—) to represent a bond that is sticking out of the plane of the paper and a dashed wedge (▦—) or dashed line (-----) to depict a bond receding behind the plane of the page. Normal straight lines are used to represent bonds that are wholly in the plane. If you are looking at the side of a cyclic compound, a bold line (▬▬) is often used to illustrate the front edge of the plane. The bonds of the ring receding behind the page should be shown as dashed lines (▦— or -----), but more often, normal lines are employed. We rely on the fact that our eyes pick up the bold lines in front, automatically "reading" the other lines as farther back in the page.

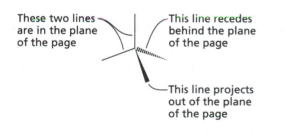

These two lines are in the plane of the page

This line recedes behind the plane of the page

This line projects out of the plane of the page

The side view of a six-membered ring

The figure on the right is technically correct, showing the three carbon atoms in back behind the plane of the page, but the figure on the left is more commonly used.

EXAMPLE 2.8

Draw the valence bond representation and the three-dimensional structure of ethane.

First, draw the Lewis structure, accounting for all of the electrons:

When only single bonds are present in the Lewis structure, we assume that each carbon atom has sp^3 hybridization. Thus, we can draw the carbon atoms, and then allow one sp^3 hybrid orbital from each to overlap, generating the C–C bond.

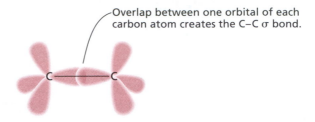

Overlap between one orbital of each carbon atom creates the C–C σ bond.

Four sp^3 orbitals on each carbon atom

Next, overlap the $1s$ orbital of each H atom with an sp^3 hybrid orbital on each carbon atom. The final picture shows that all of the bonds are σ bonds, and all of the angles (∠H–C–H, and ∠C–C–H) are ~ 109.5°.

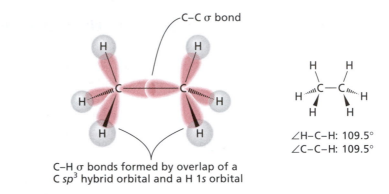

C–C σ bond

C–H σ bonds formed by overlap of a C sp^3 hybrid orbital and a H $1s$ orbital

∠H–C–H: 109.5°
∠C–C–H: 109.5°

EXERCISE 2.6

Draw the valence bond representation for propane, $CH_3CH_2CH_3$.

2.5c A CARBON ATOM WITH A DOUBLE AND TWO SINGLE BONDS IS MODELED WITH SP^2 HYBRIDIZATION OF CARBON

From experimental data, we know that ethylene, $CH_2=CH_2$, is a flat molecule. Therefore, the tetrahedral arrangement of orbitals for carbon atoms in methane, ethane, propane, and other alkanes cannot be used to portray the bonding in ethylene. Instead, we model the bonding in ethylene with use of two $2p$ orbitals of carbon, hybridized with its $2s$ orbital. This arrangement, termed sp^2 hybridization, leaves unmixed the orbital that we designate $2p_z$, as illustrated in Figure 2.10.

Figure 2.10
(a.) The conceptual combination of the atomic orbitals that generates sp^2 hybrid orbitals.
(b.) Orientation of the three hybrid orbitals and the unhybridized p_z orbital.

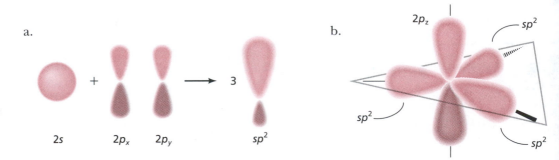

a.

$2s$ $2p_x$ $2p_y$ sp^2

b.

$2p_z$

sp^2
sp^2
sp^2

Figure 2.11 depicts the valence bond representation of ethylene, in which orbitals of two sp^2-hybridized carbon atoms overlap with the $1s$ orbital of four hydrogen atoms. This picture shows that two carbon–hydrogen bonds are formed at each carbon atom, and a carbon–carbon bond is created, too. The bond created by overlap between two sp^2 orbitals that lie along the line connecting the nuclei is a σ bond.

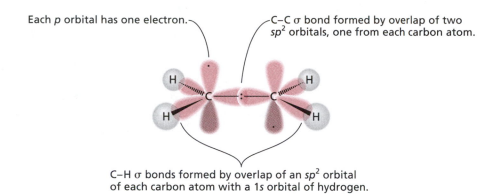

Each p orbital has one electron.

C–C σ bond formed by overlap of two sp^2 orbitals, one from each carbon atom.

C–H σ bonds formed by overlap of an sp^2 orbital of each carbon atom with a $1s$ orbital of hydrogen.

Figure 2.11
A valence bond representation of the σ bonds and carbon p orbitals in ethylene. Each p orbital has two lobes, one above and the other below the plane of the nuclei. They have different mathematical signs (positive and negative), indicated as shaded and unshaded.

The p orbital on each carbon atom is perpendicular to the line connecting the nuclei. The lobes of the p orbital lie above and below the plane defined by the six nuclei of the carbon and hydrogen atoms. The side-to-side overlap of two p orbitals, illustrated in Figure 2.12, creates a π **(pi) bond**, which is not coincident with the line connecting the carbon nuclei. Instead, a π bond has a node in the plane defined by the six atoms, just as a p orbital has a node at the nucleus of an atom.

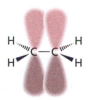

Pairing of the electrons from each of the p_z orbitals creates the π bond. Notice that overlap of the orbitals occurs both above and below the plane.

Figure 2.12
Representation of the π bond in ethylene created by overlapping p orbitals.

The complete valence bond representation for ethylene (Figure 2.13) shows that two bonds are formed between the carbon atoms, one a σ bond and one a π bond. The $\angle$H–C–H and $\angle$C–C–H angles in ethylene, and for all atoms with sp^2 hybridization, are 120°, and the nuclei of all six atoms lie in a plane.

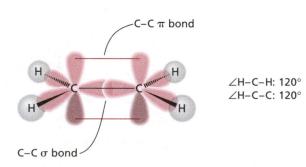

C–C π bond

$\angle$H–C–H: 120°
$\angle$H–C–C: 120°

C–C σ bond

Figure 2.13
A valence bond representation of ethylene, showing all of the σ and π bonds.

A π bond does not exist unless a σ bond is also formed, a fact that has important consequences for the reactivity of organic molecules. As you can see in the picture, a π bond constitutes a region of electron density that lies away from the line connecting the two nuclei. Therefore, electrons in a π bond are available to react with other nuclei and with electron-poor species. Sigma bonds are closer to the positively charged nuclei and are therefore less reactive. When a carbon–carbon double or triple bond is present in a molecule, the π bond reacts preferentially.

EXAMPLE 2.9

Draw the valence bond representation for propene, which has two sp^2-hybridized carbon atoms, and one sp^3-hybridized carbon atom.

First, draw the Lewis structure and assign the hybridization of each carbon atom. A carbon with four single bonds has sp^3 hybridization, and a carbon atom with a double and two single bonds has sp^2 hybridization (below, a.). Next, draw the carbon atoms with their hybrid orbitals, allowing one orbital from each to overlap with its neighbor, forming the C–C σ bonds (below, b.). (The hybridization of each orbital lobe is given in this example, but it is not necessary to label the orbitals every time.)

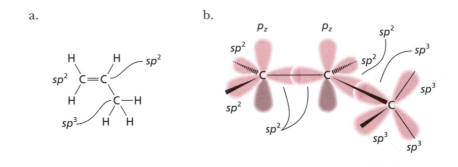

Include the hydrogen atoms by allowing the 1s orbital of each H to overlap with the remaining hybrid orbitals on the carbon atoms, and show the overlap between the carbon atoms for the π bond (remember that the overlap occurs both above and below the plane). Finally, indicate the bond angles (not all of them are shown in the figure).

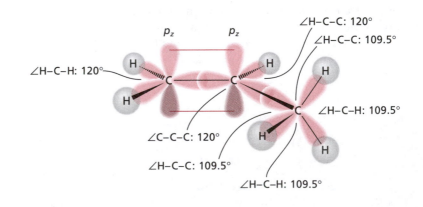

EXERCISE 2.7

Draw the valence bond representation for 1,3-butadiene, $H_2C=CH–CH=CH_2$, which has only sp^2-hybridized carbon atoms.

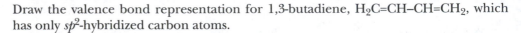

2.5d   A CARBON ATOM WITH A TRIPLE AND A SINGLE BOND IS MODELED WITH *SP* HYBRID ORBITALS

Acetylene (ethyne), the simplest hydrocarbon with a carbon–carbon triple bond, is a linear molecule; therefore, its pictorial representation must have a H–C–C angle of 180°. To produce an appropriate model for the bonding in acetylene, the orbitals needed to create the sigma framework have *sp* hybridization, which is a combination of the *s* orbital and a single *p* orbital of carbon. By creating *sp* hybrid orbitals, two *p* orbitals remain to overlap and establish two π bonds that form the triple bond. Figure 2.14 shows the orbital arrangement for a carbon atom with *sp* hybridization.

a.

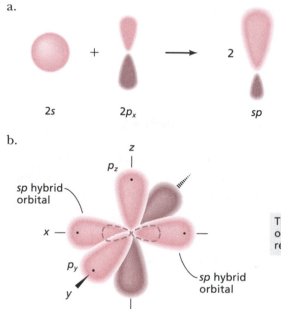

2s   2p_x   sp

b.

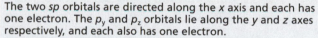

The two *sp* orbitals are directed along the *x* axis and each has one electron. The p_y and p_z orbitals lie along the *y* and *z* axes respectively, and each also has one electron.

Figure 2.14
(a.) The conceptual combination of the atomic orbitals that generates the *sp* hybrid orbitals. (b.) Orientation of the hybrid orbitals and the unhybridized *p* orbitals. The smaller lobe of each hybrid orbital is shown as a dashed shaded loop. Both lobes of each of the two *sp* hybrid orbitals lie along the *x* axis.

When two *sp*-hybridized carbon atoms are brought together with two hydrogen atoms, we generate the structure depicted in Figure 2.15. There is a σ bond between the two carbon atoms and between each carbon atom and a hydrogen atom. In addition, the two orthogonal *p* orbitals form π bonds.

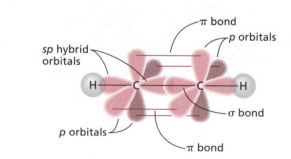

Figure 2.15
The valence bond representation of acetylene.

Notice that the two π bonds are oriented within perpendicular planes that intersect along the line connecting the carbon nuclei, as shown in the following scheme:

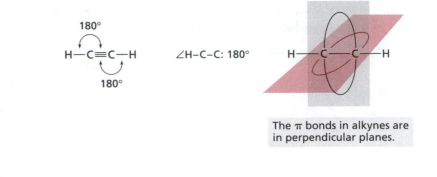

The π bonds in alkynes are in perpendicular planes.

EXERCISE 2.8

Draw the valence bond representation for 1-propyne, which has two *sp*-hybridized carbon atoms and an *sp*³-hybridized carbon atom.

2.5e   THE HYBRIDIZATION OF CARBON ATOMS IN REACTIVE INTERMEDIATES VARIES

Until now, we have looked only at structures of stable molecules. During many chemical reactions, however, intermediates are formed in which a carbon atom has fewer than four bonds. These species—**carbocations, carbanions, radicals,** and **carbenes**—are illustrated in Figure 2.16.

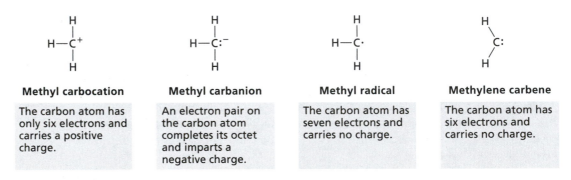

Methyl carbocation

The carbon atom has only six electrons and carries a positive charge.

Methyl carbanion

An electron pair on the carbon atom completes its octet and imparts a negative charge.

Methyl radical

The carbon atom has seven electrons and carries no charge.

Methylene carbene

The carbon atom has six electrons and carries no charge.

Figure 2.16
Carbon-containing species in which carbon forms fewer than four bonds.

The carbon atom in a *carbocation* forms three bonds and has six electrons instead of eight. A carbocation therefore adopts a planar geometry to minimize repulsive interactions between the three electron pairs. Like a carbon atom that forms a double and two single bonds, which also has three groups attached, the structure of a carbocation can be described with an *sp*²-hybridized carbon atom. Figure 2.17 shows the representation of the methyl carbocation, in which the three hydrogen atoms are in the plane and the *p* orbital extends above and below that plane.

Figure 2.17
The valence bond representation of the methyl carbocation.

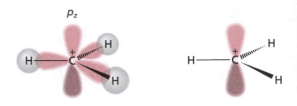

The carbon atom of a *carbanion*, which also has only three bonds, can be either sp^2 or sp^3 hybridized. In the absence of possible overlap with a neighboring p orbital (a point discussed shortly), a carbanion is tetrahedral because the unshared electron pair occupies space and repels the electrons in the three bonds. An electron pair actually occupies more volume than a carbon–hydrogen bond.

The methyl *radical* is best described using an sp^2-hybridized carbon atom, but the geometry of radical species larger than the methyl radical is tetrahedral, in which case sp^3 hybridization provides a better model.

The carbon atom of a *carbene* forms only two bonds but has six electrons (normally three electron pairs), so it has sp^2 hybridization (Figure 2.18).

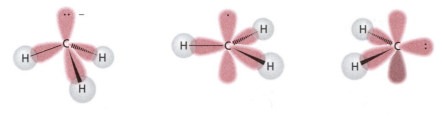

Methyl carbanion, :CH$_3^-$ Methyl radical, CH$_3$·   Methylene carbene, :CH$_2$

Figure 2.18
Valence bond structures for the methyl carbanion, the methyl radical, and methylene carbene.

2.5f   HYBRIDIZATION OF CARBON ATOMS: A SUMMARY

Having now looked at substances that contain only carbon and hydrogen, we summarize in Table 2.4 the types of hybridization expected for the carbon atoms in these species, which includes stable molecules as well as reactive intermediates.

Table 2.4   Summary of hybridization patterns.

Type of C	Type of species	Hybridization	No. of σ bonds	No. of π bonds	Approx bond angle
—C—	Alkane	sp^3	4	0	109.5°
C=	Alkene	sp^2	3	1	120°
—C≡	Alkyne	sp	2	2	180°
C$^+$—	Carbocation	sp^2	3	0	120°
—C·	Radical	sp^2 or sp^3	3	0	120° or 109.5°
—C:$^-$	Carbanion	sp^2 or sp^3	3	0	120° or 109.5°
C:	Carbene	sp^2	2	0	120°

2.6   BONDS WITH HETEROATOMS

2.6a   THE HYBRIDIZATION OF HETEROATOMS OFTEN FOLLOWS THE VSEPR MODEL

Because many interesting organic compounds also have heteroatoms, especially oxygen and nitrogen, we must have a way to depict their bonds, too. We can make use of the VSEPR model to deduce the type of hybridization that a heteroatom has, and the common patterns are shown in Figure 2.19. Unshared electron pairs are placed in the hybrid orbitals that are not used to form σ bonds.

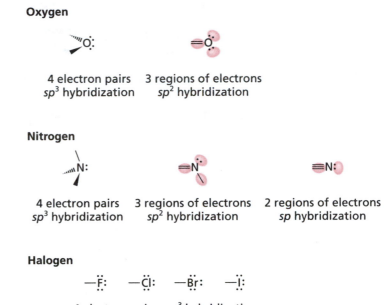

Figure 2.19
Hybridization modes for common heteroatoms found in organic molecules.

Figure 2.20a shows the valence bond representation of methanol, in which the carbon atom, with four σ bonds, is assigned sp^3 hybridization. The oxygen atom, with two bonds and two unshared pairs, also has sp^3 hybridization. The idealized bond angles around oxygen in methanol should be 109.5°, but the H–O–C angle is, in fact, slightly smaller than that value because the electron pairs in the hybrid orbitals tend to take up more room than the O–H bond, compressing the angle slightly.

The simplest organic molecule with a double bond between carbon and oxygen is formaldehyde, $H_2C=O$, the valence bond representation of which is shown in Figure 2.20b. The unshared electron pairs on the oxygen atom are in hybrid orbitals, leaving a single p orbital on both the carbon and oxygen atoms to overlap, creating the π bond.

a. Methanol

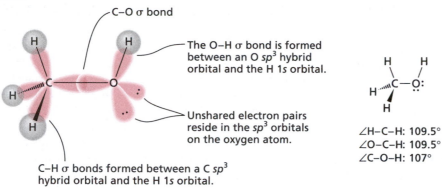

The O–H σ bond is formed between an O *sp³* hybrid orbital and the H 1*s* orbital.

Unshared electron pairs reside in the *sp³* orbitals on the oxygen atom.

C–H σ bonds formed between a C *sp³* hybrid orbital and the H 1*s* orbital.

∠H–C–H: 109.5°
∠O–C–H: 109.5°
∠C–O–H: 107°

b. Formaldehyde

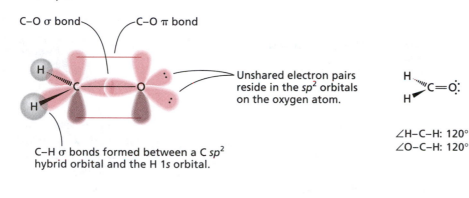

Unshared electron pairs reside in the *sp²* orbitals on the oxygen atom.

C–H σ bonds formed between a C *sp²* hybrid orbital and the H 1*s* orbital.

∠H–C–H: 120°
∠O–C–H: 120°

Figure 2.20
Valence bond representation of methanol, CH_3OH (a.) and formaldehyde, CH_2O (b.).

EXERCISE 2.9

Draw the valence bond representation for dimethyl ether, $H_3C–O–CH_3$.

Hydrogen cyanide represents the simplest compound that has a triple bond to a heteroatom, in this case nitrogen. We model it, as shown in Figure 2.21, by analogy with acetylene. The unshared electron pair on nitrogen is placed in the *sp* hybrid orbital on nitrogen, leaving two *p* orbitals to form π bonds and create a total of three bonds between the carbon and nitrogen atoms.

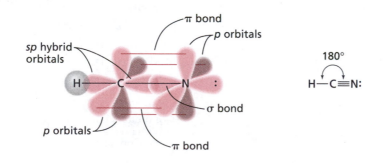

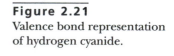

Figure 2.21
Valence bond representation of hydrogen cyanide.

EXERCISE 2.10

Draw the valence bond representation for acetonitrile, CH_3CN.

2.6b THE HYBRIDIZATION OF A HETEROATOM CAN DIFFER FROM ITS PREDICTED TYPE WHEN RESONANCE FORMS ARE IMPORTANT

Knowing which hybridization model applies to a given atom allows one to establish an atom's geometry—tetrahedral (sp^3), trigonal (sp^2), or linear (sp). The hybridization in turn provides information about the bond angles and spatial orientations of atoms and groups. Recognizing the hybridization of a heteroatom is often not as crucial as it is for carbon because a heteroatom normally forms fewer bonds than carbon. The hybridization of a halogen atom is least important because the halogens mainly form one bond.

For nitrogen and oxygen atoms, however, hybridization changes can affect their geometries significantly. The most obvious case that you will encounter is that of the amide functional group. Looking at the structure of dimethylformamide (DMF), shown below in (a.), you would most likely assign sp^3 hybridization to the nitrogen atom because it has four electron pairs surrounding it (three bonding and one nonbonding). You would therefore predict that the C–N–C bond angles would be ~ 109°.

a.   b.

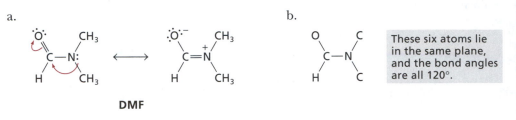

DMF

> These six atoms lie in the same plane, and the bond angles are all 120°.

The actual structure shows that the bond angles are closer to 120°, and the six atoms of the amide unit are coplanar, a geometry more consistent with sp^2 hybridization of nitrogen. An important resonance form of DMF, also shown above, fits with sp^2 hybridization: *The nitrogen atom in this resonance form is surrounded by three regions of electron density* (see Fig. 2.19). Remember the primary rule about resonance structures: *Nuclei cannot move* (Section 2.3b). Because changes in hybridization *will* lead to geometric changes—hence movement of nuclei—the hybridization of an atom must be the same in every structure whenever multiple resonance forms are involved. This situation is common when a heteroatom is bonded to a carbon atom with sp^2 hybridization.

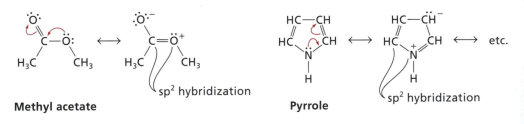

Methyl acetate   **Pyrrole**

When a heteroatom has sp^2 hybridization, its electron pair is placed into a *p* orbital instead of a hybrid orbital, as illustrated in Figure 2.22 for DMF.

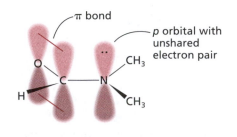

Figure 2.22

The valence bond representation of DMF showing the unshared pair of electrons in the *p* orbital that remains on the nitrogen atom after hybridization. The nitrogen, oxygen, and carbonyl carbon atoms all have sp^2 hybridization. The sigma bonds are shown as black lines.

Draw the valence bond representation for methyl acetate, the structure of which is shown above. Both oxygen atoms have sp^2 hybridization, as does the carbonyl carbon atom. The methyl carbon atoms have sp^3 hybridization.

2.7   DELOCALIZED π ELECTRON SYSTEMS

2.7a   BENZENE HAS DELOCALIZED ELECTRONS

Benzene, a hydrocarbon, cannot be adequately described by a single Lewis structure (Section 2.3c): All six carbon–carbon bonds in benzene have the same length, so a structure with alternating single and double bonds is incorrect, and at least two resonance structures are needed to represent its structure accurately.

Nevertheless, these two principal resonance forms of benzene show that every carbon atom forms a double bond to one of its neighbors; therefore, to draw a valence bond representation of benzene, we start with sp^2-hybridized carbon atoms. (From this point on, we will use a line to depict a σ bond, because orbitals that overlap to create σ bonds lie between the nuclei, just as the lines do.)

Putting six sp^2-hybridized carbon atoms together to form the benzene ring, we use the hybrid orbitals on each carbon atoms to construct the sigma framework, as illustrated in Figure 2.23. With sp^2-hybridized carbon atoms, benzene is a planar molecule, and every carbon atom has a p orbital perpendicular to the ring.

The p orbitals in benzene are evenly spaced as a result of the identical carbon–carbon bond lengths, so choosing which pairs of adjacent orbitals overlap to form π bonds is arbitrary. In fact, it cannot be done: Overlap occurs between each pair of adjacent orbitals. This type of overlap leads to interactions among all six p orbitals in benzene, so the electrons are delocalized, which means that they can move throughout the π system (Figure 2.24).

Whenever we draw a Lewis structure, we represent a bond with a line, which signifies that those electrons are *localized* between the atoms that are connected (Section 2.3a). Likewise, two dots next to an atom mean that an unshared pair of electrons is *localized* on that atom. But in compounds such as benzene, which have delocalized electrons, a

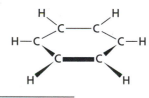

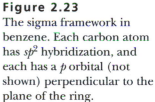

Figure 2.23
The sigma framework in benzene. Each carbon atom has sp^2 hybridization, and each has a p orbital (not shown) perpendicular to the plane of the ring.

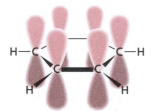

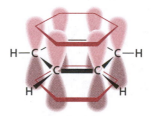

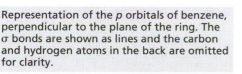

Representation of the p orbitals of benzene, perpendicular to the plane of the ring. The σ bonds are shown as lines and the carbon and hydrogen atoms in the back are omitted for clarity.

Because the p orbitals are equally spaced, overlap occurs between all six, allowing the electrons to be delocalized.

Figure 2.24
Orbital representation of the π bonds in benzene, showing electron delocalization.

single structure cannot accurately portray where the electrons are, so resonance forms are needed to show the different ways in which the electrons are distributed.

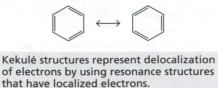

Kekulé structures represent delocalization of electrons by using resonance structures that have localized electrons.

Another way to represent the delocalized electrons is to use a hexagon with an inscribed circle, analogous to the valence bond picture in Figure 2.24.

Example 2.10 shows how delocalization can occur in an ionic compound, a carbocation.

EXAMPLE 2.10

Draw resonance structures and a valence bond representation for the benzyl cation, $C_6H_5\text{-}CH_2^+$.

First, draw a Lewis structure for the compound, including the positive charge on carbon:

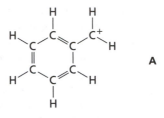

Next, draw several structures that have only single bonds between the atoms, and move electrons between the atom positions in **A**, making certain to account for all the valence electrons. Notice that the original structure, **A**, has an overall positive charge, and one carbon atom has only six electrons. Therefore every structure that you draw will have a net positive charge, and one carbon atom will have only six electrons. We cannot draw a resonance form that gives each non-hydrogen atom an octet.

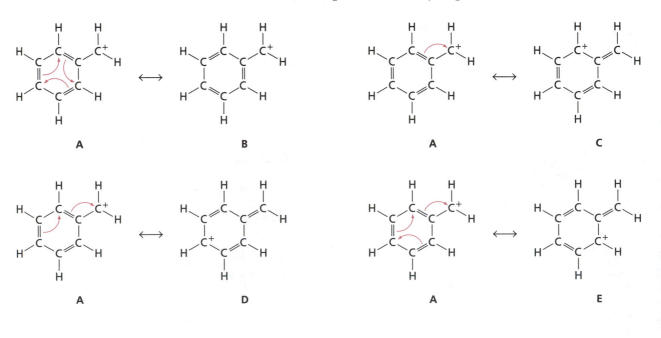

To draw the valence bond representation, recognize first that every carbon atom has sp^2 hybridization. The sigma framework is drawn between the carbon and hydrogen atoms (below, a.); then a p orbital is placed at each carbon atom position (below, b.). Overlap occurs among all seven carbon atoms, and the overall structure bears a positive charge (the lower portions of the orbitals in back have been omitted for clarity of presentation).

a.   b.

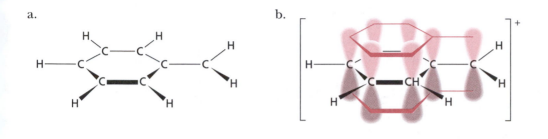

EXERCISE 2.12

Draw a valence bond representation for the acetate ion, CH_3COO^- (Example 2.7). Use lines to represent the σ bonds. The two carbon–oxygen bond lengths are equal.

2.7b LEWIS STRUCTURES CAN BE MODIFIED TO REPRESENT DELOCALIZATION

One way to relate orbital pictures with resonance structures is to modify Lewis structures to indicate delocalization. For benzene and its derivatives, this procedure simply requires drawing a circle within the ring. For the acetate ion, we use a dotted, curved line with a negative charge, as shown in Figure 2.25. This type of notation is actually valid only when the delocalized system is symmetrical (or nearly so, as in derivatives of benzene).

Benzene   **Toluene**   **Benzoic acid**   **Acetate ion**

Figure 2.25
Representations of delocalized electrons using line structures.

For a reactive intermediate such as a carbocation, radical, or carbanion (Section 2.5e), delocalization occurs whenever the bond-deficient carbon atom (i.e., the one with fewer than four bonds) is adjacent to a π system. In such cases each carbon atom is assigned sp^2 hybridization, just as heteroatom hybridization is modified when a heteroatom is bonded to an sp^2 hybridized carbon atom (Section 2.6b): minimizing the p-orbital contribution to the hybrid leaves a p orbital on the atom adjacent to the π bond, which makes it possible for the electrons to be delocalized. A modified Lewis structure used to represent delocalization in the allyl radical is shown in Figure 2.26.

a.

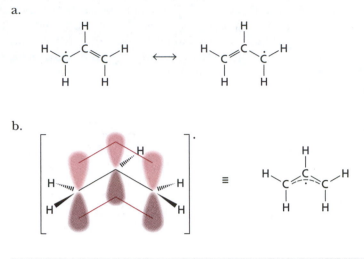

b.

Each carbon atom has *sp*^{2} hybridization so that a *p* orbital is available to permit delocalization of the electrons over the entire π system.

Figure 2.26
Representations of the allyl radical: resonance forms (a.); valence bond drawing (b. *left*), and modified Lewis structure indicating delocalization (b. *right*).

2.8   NONCOVALENT INTERACTIONS

2.8a   ELECTROSTATIC FORCES INFLUENCE THE INTERACTIONS BETWEEN ATOMS AND MOLECULES

For well over a century, structural organic chemistry has been dominated by theories aimed at understanding and explaining covalent bonding. In biological systems, covalent bonds are just as necessary to link atoms together, but structures are often influenced by *noncovalent interactions,* which at their core result from **electrostatic forces,** meaning that they occur between species with full or partial positive and negative charges. Electrostatic influences vary considerably according to the magnitude, duration, and physical separation of the charges on the species involved. Some important categories of electrostatic forces are summarized in Table 2.5.

Ion–ion, ion–dipole, and *dipole–dipole interactions* are intuitively easy to understand because the ions or molecules involved have established charges that result from the distribution of electrons (Section 2.1) or the polarity of the bonds (Section 2.2). If the interacting charges are opposite (as shown in Table 2.5), then the resulting interaction is attractive. If the charges are the same, then the ensuing interaction will be repulsive. As the magnitude of the charges on interacting species become smaller (i.e., δ– for a dipole instead of –1 for an ion), the electrostatic force diminishes more rapidly as a function of distance. Thus, doubling the distance between two oppositely charged ions decreases the attractive force between them by half, but doubling the distance between an ion and a molecule with a dipole decreases the attraction by a factor of 4, and between two molecules with dipoles by a factor of 8.

Dispersion forces, also called *London forces* to honor Fritz London, the chemical physicist who explained their origin and properties, are extremely weak interactions that occur between nonpolar molecules when they approach each other and momentarily induce a dipole in each other. The attractive forces that exist between hydrocarbons provide a classic example of dispersion forces. Because dispersion forces decrease with distance to the inverse sixth power, they are only important when the molecules are actually touching each other.

Table 2.5 Types of noncovalent interactions between molecules.

Type of Interaction	General form	Example	Distance (r) factor
Ion–ion	⊕ ⊖	$(CH_3)_4N^+$ $-\overset{O}{\underset{O}{\parallel}}C-R$	$1/r$
Ion–dipole	⊕ $(\delta^- \delta^+)$	$(CH_3)_4N^+$ $\overset{\delta^-}{:}O\overset{H\,\delta^+}{\underset{H\,\delta^+}{}}$	$1/r^2$
Dipole–dipole	$(\delta^- \delta^+)(\delta^- \delta^+)$	$\overset{\delta^-}{:}O\overset{H\,\delta^+}{\underset{H\,\delta^+}{}}$ $\overset{\delta^-}{:}O\overset{H\,\delta^+}{\underset{H\,\delta^+}{}}$	$1/r^3$
Dispersion	$\delta^+\ \delta^-$ induced		$1/r^6$
Hydrogen bond ^{a}	—D—H--- :A	$\overset{H}{\underset{}{\backslash}}O-H---\overset{H}{\underset{H}{:O}}$	Fixed length highly directional

^{a}Hydrogen-bond donor = D; hydrogen-bond acceptor = A.

2.8b HYDROGEN BONDS ARE DIRECTED ELECTROSTATIC INTERACTIONS

Among electrostatic forces, the **hydrogen bond** is a much different type because the atoms have to be aligned and the interacting groups must be separated by a specific distance. For organic molecules, a hydrogen bond forms when an O–H or N–H group is close enough to another heteroatom—usually O or N—that an attraction exists between the unshared electron pair on the heteroatom and the nucleus of the hydrogen atom. This attraction results from the polarity of the atoms, in which the heteroatom is δ– and the H is δ+. Table 2.6 shows typical hydrogen-bond lengths, as well as specific examples of hydrogen bonds that form between functional groups in organic molecules. The strongest hydrogen bonds exist when the hydrogen and two heteroatoms are oriented in a linear fashion, but weak hydrogen bonds may still form whenever the three atoms are proximal.

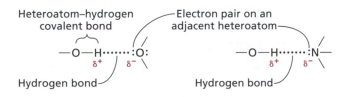

The species with the –D–H bond (D means donor in this context) is the **hydrogen-bond donor,** and the compound providing the electron pair is the **hydrogen-bond acceptor.** Any compound bearing an O–H or N–H group can form a hydrogen bond with other like molecules by intermolecular interactions because oxygen and nitrogen atoms have an unshared electron pair. (*Intermolecular* means between molecules; *intramolecular* means within the same molecule.) The converse is not true, however: a compound that has only a heteroatom cannot form hydrogen bonds to other molecules

Table 2.6   Examples of hydrogen bonds important in organic and biological molecules.

Bond	Distance (Å)	Example
O—H······O	2.70	
O—H······O⁻	2.63	
O—H······N	2.88	
N—H······O	3.04	
N—H······N	3.10	

*Distance refers to the separation between the two heteroatoms.

of itself if no D–H bond is available. For example, liquid acetone, CH_3–CO–CH_3, does not form hydrogen bonds between neighboring molecules because no O–H bonds are present.

EXERCISE 2.13

Illustrate, where appropriate, the hydrogen bonds that can exist in the following systems:

a. Pure ethanol
b. Pure water
c. A mixture of ethanol in water
d. Pure dimethyl ether
e. A mixture of dimethyl ether and water

2.8c   HYDROGEN BONDS INFLUENCE INTERMOLECULAR INTERACTIONS AND ARE CRUCIAL FOR STABILIZING STRUCTURES OF BIOMOLECULES

A hydrogen bond has only a fraction of the strength of a covalent bond, so to have much influence, large numbers of hydrogen bonds must normally be present before their effects become noticeable. Biological molecules such as proteins and nucleic acids satisfy this criterion. As a consequence, the effects of hydrogen-bond formation have long been of interest to biochemists. Organic chemists have become increasingly interested in hydrogen-bond formation as their attention has been drawn by the topic of molecular recognition. Now, many examples of hydrogen-bond formation in low molecular weight compounds are being exploited, a topic discussed in Chapter 28.

Apart from the importance of hydrogen bonds to stabilize structures of biomolecules, it has been long known that hydrogen bonds influence chemical structures and reactions in profound ways. Physical properties such as melting point and boiling point provide simple examples. Because the boiling points of alkanes are proportional to their molecular weights, a consequence of the effects of dispersion forces, we can use boiling point values as a measure of attractive forces between molecules.

Table 2.7 compares the boiling points of butane, methoxyethane, chloroethane, 1-propanol, and acetic acid. All have the same approximate molecular weight, yet their boiling points differ by > 100°C. Substances without an OH group boil within several degrees of each other because the forces holding the molecules together in solution are weak, being mostly dispersion or dipole–dipole forces. The compounds with OH groups—namely, ethanol and acetic acid—form hydrogen bonds between neighboring molecules. Additional heat has to be added to overcome the attractive forces so that the molecules can vaporize.

Table 2.7   Influence of hydrogen bonding on boiling points.

Compound	MW [a]	Boiling point (°C)
$CH_3{-}CH_2{-}CH_2{-}CH_3$	58	0
$CH_3{-}CH_2{-}O{-}CH_3$	60	7
$CH_3{-}CH_2{-}Cl$	64.5	12
$CH_3{-}CH_2{-}CH_2{-}OH$	60	97
$CH_3{-}\overset{\overset{\textstyle O}{\|\|}}{C}{-}O{-}H$	60	116

[a]Molecular weight = MW.

In the vapor phase, acetic acid actually exists as a dimer via formation of hydrogen bonds, as shown below. Its boiling point (bp = 116°C) is more similar to that of octane (bp = 125°C) than to butane (bp = 0°C). (FW = Formula weight.)

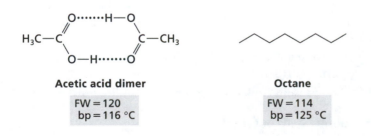

Acetic acid dimer   **Octane**

FW = 120   FW = 114
bp = 116 °C   bp = 125 °C

The effects of hydrogen bonds on chemical reactions, while more subtle than their influence on physical properties like boiling points, become significant for transformations carried out in solvents such as water or alcohols. Examples of solvent effects resulting from hydrogen bonding will be presented in Chapter 6, when we consider their influences on substitution reactions.

CHAPTER SUMMARY

Section 2.1 Lewis structures

- Lewis structures are structural formulas that also show the unshared electrons.
- The procedure used to create a Lewis structure is outlined in Section 2.1a.
- Some atoms in a Lewis structure carry a formal charge, which is the numeric difference between the number of valence electrons an isolated atom has and the number of electrons it has in the Lewis structure, either as unshared pairs or as a contribution to covalent bond formation.

Section 2.2 Bond properties

- Electronegativity is a measure of an atom's attraction for electrons; the difference in the electronegativity values for bonded atoms accounts for the polarity of bonds.
- Bond polarity is designated with the symbols δ^+ and δ^-, the latter indicating the more electronegative element of each pair of atoms bonded together.
- Bond lengths vary according to the size of atoms that constitute the bond, as well as the multiplicity (single, double, or triple) of the bond. Typical bond lengths are summarized in Table 2.2.

Section 2.3 Resonance structures

- Resonance structures show the different ways that electrons can be distributed among atoms in a molecule. Rules for drawing such structures are summarized in Section 2.3b. The criteria used to evaluate the relative importance of resonance forms are given in Section 2.3c.

Section 2.4 The formation of bonds

- A chemical bond is defined as the overlap between orbitals, which are mathematical functions that describe the probability of finding an electron in a particular region of space.
- A sigma (σ) bond is formed when overlap occurs between two orbitals that lie along the line connecting nuclei of adjacent atoms.

Section 2.5 Hybrid orbitals and shapes of molecules

- The geometry of the bonds around a given atom is related to the number of electron pairs surrounding that atom, a correlation known as the valence-shell electron-pair repulsion (VSEPR) model.
- Hybridization—the mixing of atomic orbitals—accounts for the observed shapes predicted by the VSEPR model. These data are summarized in Tables 2.3 and 2.4.
- Drawings that depict orbitals and their overlap to form bonds are referred to as valence bond representations.
- A pi (π) bond results from overlap between parallel p orbitals on adjacent atoms.

Section 2.6 Bonds with heteroatoms

- The hybridization of a heteroatom is assigned according to the VSEPR model making use of the number of regions of electron density when multiple bonds are present.
- Oxygen and nitrogen atoms often have sp^2-hybridization if the heteroatom is bonded to a carbon atom with sp^2 hybridization. This type of hybridization permits electron pairs on a heteroatom to interact with π bonds in the molecule.

Section 2.7   Delocalized π electron systems

- Benzene and its derivatives have a π system in which overlap occurs between adjacent p orbitals, allowing the electrons to be delocalized.
- Lewis structures can be modified to depict delocalized π electrons. The most common modification is used for benzene and its derivatives, in which the π system is represented by a circle circumscribed by a hexagon.

Section 2.8   Noncovalent interactions

- Several types of noncovalent interactions occur readily because of electrostatic forces between ions, polar molecules, and nonpolar molecules that are momentarily deformed to create dipoles. Electrostatic forces are generally much weaker than covalent bonds.
- Hydrogen bonds are noncovalent interactions formed by the electrostatic attraction between a heteroatom ($\delta-$) and hydrogen ($\delta+$) bonded to a second heteroatom.
- Hydrogen bonds can influence the structures and physical properties of molecules such as solubility or melting and boiling points.

KEY TERMS

Section 2.1a
Lewis structure

Section 2.1b
formal charge

Section 2.2a
electronegativity
dipole
polarizable

Section 2.3a
localized
resonance structure
resonance hybrid

Section 2.4a
wave function
node

Section 2.4b
valence bond theory
σ (sigma) bond

Section 2.5a
VSEPR model
hybrid orbital

Section 2.5c
π (pi) bond

Section 2.5e
carbocation
carbanion
radical
carbene

Section 2.7a
delocalized

Section 2.8a
electrostatic forces

Section 2.8b
hydrogen bond
hydrogen-bond donor
hydrogen-bond acceptor
intermolecular
intramolecular

ADDITIONAL EXERCISES

2.14. Draw a Lewis structure for each of the following molecules:

   a. 2-Methyl-3-pentanone
   b. 3-Bromopropanenitrile
   c. 3-Hydroxycyclopentanecarboxylic acid
   d. 3-Methoxybenzaldehyde

2.15. Based only on the VSEPR model, what is the hybridization of each non-hydrogen atom in the structures you drew as answers for Exercise 2.14?

2.16. Label each bond as either σ and π in the following structures:

a.

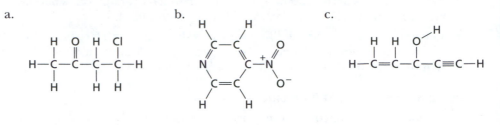

b.

c.

2.17. In the following comparison of bonds, indicate which bond is the shorter:

a. Nitrogen–oxygen bonds in

CH_3—O—N=O

b. Carbon–oxygen bonds in

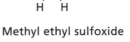

c. Carbon–nitrogen bonds in

H_3C—NH_2 or   H—C≡N

d. Carbon–halogen bonds in

CH_3CH_2—Br   or CH_3—I

2.18. Calculate formal charges for all of the non-hydrogen atoms in the following compounds. Draw a complete Lewis structure first.

a.

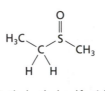

Nitrobenzene

b.

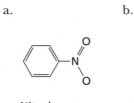

Methyl ethyl sulfoxide

c.

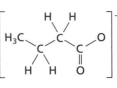

The anion derived from butanoic acid

2.19. Draw Lewis structures for sulfur trioxide, SO_3, and for the sulfite ion, SO_3^{2-}. Include formal charges, if any, on each atom. Draw resonance forms where necessary. Comment on their differences with respect to the distribution of electrons.

2.20. The following compounds were introduced in Exercise 1.24. Assign appropriate hybridization to the carbon atoms in each of these drug molecules (it might be helpful to consult your previous solutions to the Chapter 1 exercise to see the expanded structures).

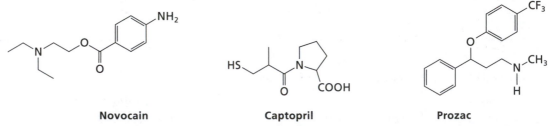

Novocain

local anesthetic

Captopril

antihypertensive drug

Prozac

antidepressant drug

2.21. Which of the following pairs of Lewis structures do not constitute resonance structures?

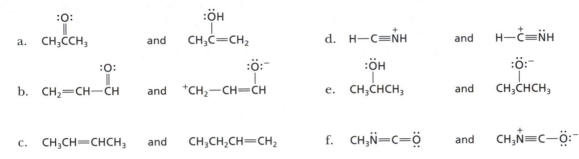

2.22. For each set of resonance structures, rank each in order of its relative importance to contribute to the resonance hybrid.

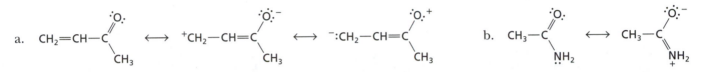

c. $CH_2=C=\ddot{O} \longleftrightarrow {}^-{:}CH_2-C\equiv O{:}^+ \longleftrightarrow {}^+CH_2-\ddot{\ddot{C}}=\ddot{O}$

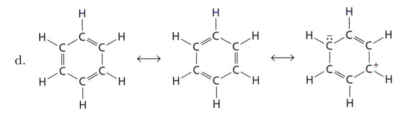

d.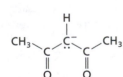

2.23. The boron atom in boron trifluoride is an exception to the octet rule because it has only six electrons.

a. Draw a Lewis structure for BF_3.

b. BF_3 reacts with diethyl ether, $C_2H_5-O-C_2H_5$, to form a substance called boron trifluoride etherate, which has the formula $BF_3[O(C_2H_5)_2]$. In this molecule, the oxygen atom of the ether group forms a bond with the boron atom. Draw a Lewis structure for this substance, including formal charges. Is it surprising that this substance is stable (it can be distilled at 126°C)?

2.24. Draw at least two resonance structures for each of the following species:

a.

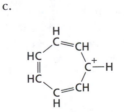

Anion of 2,4-pentadione

b.

CH_3-ONO

Methyl nitrite

c.

Cycloheptatrienyl cation

2.25. In Section 2.5e, you were introduced to carbenes, species in which the carbon atom has six electrons and is bonded to only two groups. The simplest carbene is :CH₂, methylene, and the nonbonded electrons can either be paired (singlet methylene) or unpaired (triplet methylene). Figure 2.18 showed the orbital picture for singlet methylene.

Draw the valence bond representation for triplet methylene and predict the hybridization of the carbon atom.

Singlet methylene   Triplet methylene

2.26. Draw a complete valence bond representation for each of the following compounds:

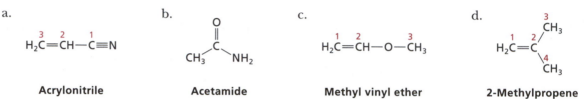

a.   b.   c.   d.

Acrylonitrile   Acetamide   Methyl vinyl ether   2-Methylpropene

2.27. For the corresponding structures shown in Exercise 2.26, give the expected bond angles for each set of atoms indicated.

    a. ∠C2—C1—N   ∠C1—C2—C3   ∠H—C3—H

    b. ∠C—C—N   ∠C—C—O   ∠N—C—O   ∠H—C—H

    c. ∠C2—O—C3   ∠C1—C2—O   ∠H—C1—H   ∠H—C3—O

    d. ∠C4—C2—C1   ∠C4—C2—C3   ∠H—C3—C2

2.28. Among the following compounds, the amide has the highest boiling point. Explain briefly:

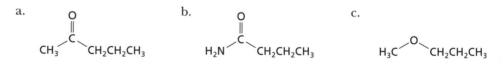

a.   b.   c.

2.29. Of the compounds shown in Exercise 2.28, which one would you expect to have the greatest solubility in water? Explain briefly.

2.30. The following two compounds react with each other to produce a hydrogen-bonded dimer. Include the unshared electrons on each heteroatom, then indicate the hydrogen bonds that are expected to form.

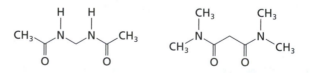

2.31. Of the following compounds, which can function as a hydrogen-bond donor to water in an aqueous solution? Which is a hydrogen-bond acceptor with water as the hydrogen bond donor in an aqueous solution? Which is a hydrogen-bond donor and acceptor with itself in a nonpolar solvent like hexane?

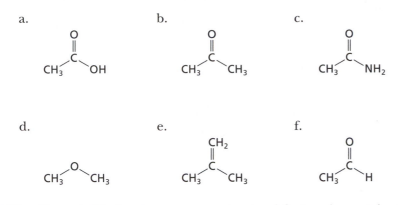

a.   b.   c.

d.   e.   f.

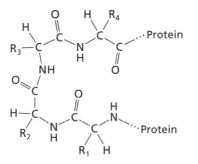

2.32. Illustrated below is a common structural feature in proteins called a reverse turn, which allows a protein to fold back on itself (the identity of the R groups is irrelevant for the purpose of the exercise). Hydrogen bonds are formed across the intervening space to stabilize a reverse turn. Include the unshared electrons on each heteroatom, and then indicate three such hydrogen bonds that could form.

THE CONFORMATIONS OF ORGANIC MOLECULES

Even a cursory look at the historical development of organic chemistry makes it clear that chemists knew a great deal about the three-dimensional shapes of molecules by the late 1800s, particularly with respect to the tetrahedral geometry of a carbon atom with four single bonds. The development of bonding theories has since added to our understanding of molecular structures, and we can often make accurate predictions about the shapes of new compounds. This predictive power has also led to the recent development of computer software to calculate structures of molecules. As yet, however, we can obtain only approximate structures for complex molecules such as nucleic acids and proteins, so interpreting those calculations still requires caution.

This chapter and Chapter 4 will explore the three-dimensional shapes of organic molecules, relating spatial orientations with the idealized geometries and representations covered in Chapter 2. Because the specific structures of biologically important substances influence their interactions with other molecules, it is important to understand these fundamental aspects of *stereochemistry*, the three-dimensional relationships between atoms in a molecule.

This chapter will focus on the conformations of molecules. A conformer represents the instantaneous orientation of the atoms in a molecule that results from rotation about carbon–carbon σ bonds. Under normal conditions, sufficient energy is present to interconvert the infinite numbers of conformers in which a molecule can exist. Many chemical transformations occur, however, only when two substituents are specifically oriented. Knowing which structural arrangements, or limiting conformations, are likely to predominate will help later to understand reaction chemistry.

3.1 CONFORMATIONS OF ACYCLIC COMPOUNDS

3.1a CONFORMERS ARE INTERCONVERTIBLE SPATIAL ORIENTATIONS OF THE SAME STRUCTURE PRODUCED BY ROTATION ABOUT σ BONDS

In discussions about the structures of molecules, we must first make a distinction between *isomerism* and *conformation*. Isomers are different compounds that have the same molecular formula, and they are usually stable toward interconversion at room temperature. Figure 1.7 summarized the types of constitutional isomers. Transforming one such isomer to another normally requires that at least one bond is broken and made.

Conformers, on the other hand, constitute arrangements of atoms that are readily converted to other spatial orientations by rotation about σ bonds. The specific orientation of a given conformer is termed its **conformation.** *Interconversion of different conformations does not require a bond to be broken.* (Some chemists use the term "conformational isomer" as a synonym for *conformer,* but this text will use the term conformer exclusively.)

To understand why different conformations exist, recall that a σ bond lies along the line connecting two nuclei, so there is no orientation in which overlap between the orbitals is compromised. Conformers exist because the relative positions of the substituents on different atoms, especially carbon, change as rotation about the bond axis occurs.

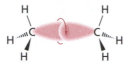

Rotation about the σ bond between two atoms occurs readily because overlap is not lost in any orientation.

3.1b   ETHANE HAS TWO LIMITING CONFORMATIONS, ECLIPSED AND STAGGERED

The simplest molecule that has discernable conformations is ethane, which has only one carbon–carbon bond about which rotation occurs. If we fix the position of one carbon and its hydrogen atoms in space, we can consider where the hydrogen atoms of the second carbon atom are, relative to the first ones. To see and to represent these conformations, you must first draw the molecule in a way that reveals its three-dimensional structure. One type of representation, shown in Figure 3.1a, is called a **sawhorse projection,** which is displayed both as a three-quarter view and a side view. The other [Fig. 3.1b] is called a **Newman projection,** which is the view looking along the carbon–carbon bond. In a Newman projection, you are looking end-on at a single carbon atom with its three attached atoms. The lines go to the center of the circle to indicate that these substituents are connected to the carbon atom in front. The other three bonds go only to the edge of the circle, indicating that they are attached to the carbon atom that cannot be seen from this orientation.

In the ethane molecule, the hydrogen atoms can assume an infinite number of orientations relative to each other, but only two unique limiting conformations exist, as shown in Figure 3.1. In one, the hydrogen atoms are **eclipsed** with one another, which means that each hydrogen atom is directly in front of another when you look along the carbon–carbon bond in the Newman projection. In the other limiting conformation,

a.

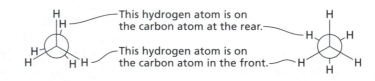

Sawhorse projection of ethane in the eclipsed conformation

Sawhorse projection of ethane in the staggered conformation

b.

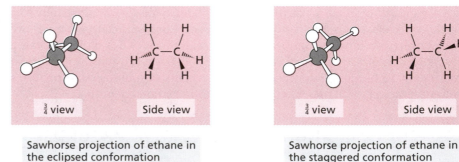

This hydrogen atom is on the carbon atom at the rear.

This hydrogen atom is on the carbon atom in the front.

Newman projection of ethane in the eclipsed conformation

Newman projection of ethane in the staggered conformation

Figure 3.1
Limiting conformations of ethane, showing ball-and-stick models and line structures in the sawhorse (a.) and Newman (b.) projections.

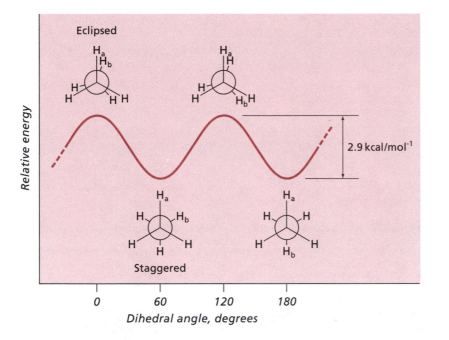

Figure 3.2
An energy diagram for the conformations of ethane as a function of torsional angle. All of the hydrogen atoms are equivalent. Two are arbitrarily labeled H_a and H_b to illustrate their relationship to each other as rotation occurs about the C–C bond. There are only two unique, *limiting* conformations: eclipsed and staggered.

the hydrogen atoms are **staggered,** which places them as far apart as possible. The staggered conformation is more stable than the eclipsed conformation.

Another way to evaluate conformations is to make a graph that plots energy versus the angle of rotation of one carbon atom relative to the other. For ethane, such a plot is shown in Figure 3.2. The starting point is the eclipsed conformation.

As we look along the carbon–carbon bond, we can envision the angle formed between two carbon–hydrogen bonds on adjacent atoms, a relationship known as the **torsional** or **dihedral angle.** In connection with the torsional angle, the energy value passes through a maximum or minimum point with every 60° of rotation. When the hydrogen atoms are staggered (60°, 180°, etc.), the energy value reaches its minimum; when the hydrogen atoms are eclipsed (0°, 120°, etc.), the energy value is at its maximum.

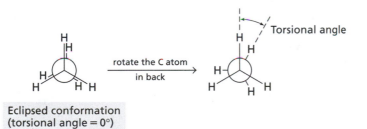

A simple rationalization for the lower energy value—and therefore greater stability–of the staggered conformation is that the hydrogen atoms are farther apart. As a result of these greater distances, repulsive electrostatic forces between the hydrogen atoms (and the C–H bonds) are smaller. Predictions about the relative stabilities of conformations based on this simple notion are usually correct although other factors are involved, too.

3.1c   CONFORMATIONS OF SUBSTITUTED ALIPHATIC COMPOUNDS ARE CALLED *SYN*-PERIPLANAR, *GAUCHE*, ANTICLINAL ECLIPSED, AND *ANTI*

Consider now a slightly more complex molecule—butane—in which one hydrogen atom on each carbon atom of ethane is replaced by a methyl group. For butane, several additional conformations correspond to energy minima and maxima caused by rotation about the central carbon–carbon single bond. These conformations are shown in Figure 3.3, using both the sawhorse and Newman projections. When the dihedral angle equals

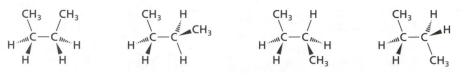

These sawhorse projections are seen from the side; notice that rotation about the C–C bond is occurring with the orientation of the carbon atom on the left held fixed.

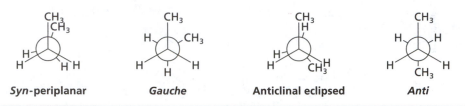

Syn-periplanar   **Gauche**   **Anticlinal eclipsed**   **Anti**

Each of these Newman projections corresponds to the conformation shown directly above in the sawhorse representation. The view is along the C–C bond from the left. Notice that the orientation of the carbon atom in front stays fixed while rotation occurs about the C–C bond.

Figure 3.3
Some conformations for butane created by rotation about the C2–C3 bond.

0°, both methyl groups are eclipsed, which is the highest energy form of butane and is called the *syn*-periplanar conformation. The electron density of the two methyl groups fills a volume of space that leads to substantial electrostatic repulsion between them. This repulsive interaction is called a **steric effect,** which has the connotation that the methyl groups act as "hard spheres" that cannot occupy the same space at the same time.

An energy diagram for the conformations of butane is shown in Figure 3.4.

EXERCISE 3.1

Make a model of butane and examine how closely the methyl groups approach each other in the *syn*-periplanar conformation. Draw the sawhorse projection of your model, including all of the hydrogen atoms.

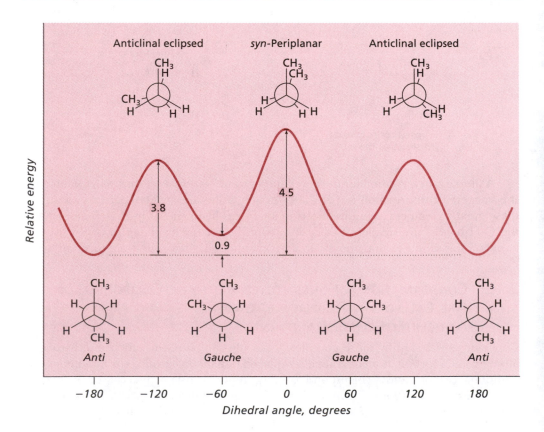

Figure 3.4
An energy diagram for the conformations about the C2–C3 bond in butane. The energy values are given in kilocalories per mole (kcal mol^{-1}). The least stable conformation, shown at the center, has the two methyl groups eclipsed.

As rotation about the C2–C3 bond in butane continues past the *syn*-periplanar conformation, the molecule passes through a minimum energy conformation in which the torsional angle between the methyl groups is 60°. This orientation, analogous to a staggered conformation in ethane, is called the **gauche conformation.**

As rotation proceeds, the molecule passes through another energy maximum as the methyl groups and hydrogen atoms become eclipsed with one another. This is the *anticlinal eclipsed conformation.* Further rotation puts the molecule in its most stable conformation in which the torsional angle between the methyl groups is 180°, and the methyl groups reach the point of greatest separation. This is called the **anti (antiperiplanar) conformation.** Rotation from 180° to 360° repeats the process of producing anticlinal-eclipsed, *gauche,* and finally *syn*-periplanar conformations.

The eclipsed hydrogen–hydrogen interaction in ethane is estimated to be ~ 1 kcal mol^{-1}. In butane, we find that creating a hydrogen–methyl *eclipsed* interaction costs ~ 1.4 kcal mol^{-1}; a methyl–methyl *eclipsed* interaction, ~ 2.5 kcal mol^{-1}; and a methyl–methyl *gauche* interaction, ~ 0.9 kcal mol^{-1}. These values can be used to estimate the energy barriers among different conformations of other alkanes.

As a molecule's structure increases in complexity, the number of possible conformations that exist at energy minima and maxima increases. To predict a compound's most stable conformation, you would have to look individually at each pair of adjacent hydrocarbon units. For example, the conformations illustrated in Figure 3.3 for butane considered only the bond between C2 and C3, each with its substituents. For the entire butane molecule, the conformations of the two terminal carbon–carbon bonds (C1–C2 and C3–C4) would have to be assessed, too. In general, the more stable conformations of alkanes have the largest groups staggered, but entropy effects (not discussed here) become more significant for longer chains, and the evaluation of overall conformations for larger molecules can be complicated.

To draw accurate conformations of the groups attached to a single bond in an aliphatic molecule is straightforward, and the following example shows how to represent and assess the relative stabilities of the possible conformations.

EXAMPLE 3.1

Draw the most stable conformation of 1-phenylpropane with respect to the C1–C2 bond, showing both Newman and sawhorse projections.

First, interpret the name of the compound and draw its structure, making certain to show the bond about which you will rotate to generate its conformations.

$$\overset{\overset{\text{H}\quad\text{H}}{|\;^2\;\;|\;^1}}{\underset{\underset{\text{H}\quad\text{H}}{|\quad|}}{^3CH_3-C-C-C_6H_5}}\qquad\textbf{1-Phenylpropane}$$

Next, arrange the carbon framework in the sawhorse and Newman projections. For simplicity, consider only the staggered conformations because you already know that eclipsed conformations will be less stable.

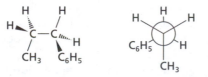

Choose a staggered conformation to begin because it will always be lower in energy than any of the eclipsed conformations.

Next, consider rotations about the C1–C2 bond, holding the position of C2 (the carbon atom on the left) fixed:

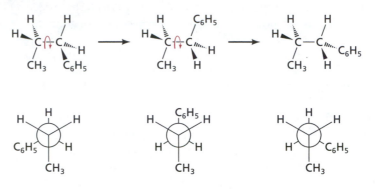

Finally, choose the structure that has the largest groups *anti* to one another. In this case, the structure in the middle has the methyl and phenyl groups *anti*. In the other two conformations, the methyl and phenyl groups are *gauche* with respect to each other.

EXERCISE 3.2

Draw the most stable conformation for each of the following compounds, showing both the Newman and sawhorse projections. For parts (a.) and (b.), draw an energy diagram like that in Figure 3.4 for the rotation process, and estimate the energy differences between limiting conformations.

a. Propane
   (C1–C2 bond)

b. 2-Methylpropane
   (C1–C2 bond)

c. 2,2-Dichlorobutane
   (C2–C3 bond)

The energy differences that exist between conformations in the aliphatic compounds presented so far are relatively small quantities. The thermal energy available to a molecule at room temperature is sufficient to permit rapid interconversion between all possible conformations. It is also true, however, that a larger percentage of the most stable conformation is present at any given time, while the percentages are insignificant for those conformations that correspond to energy maxima. As the temperature is lowered, the amount of energy available to cause rotation about the σ bond decreases, so it is possible to "freeze" certain conformations. At low temperatures, only those conformations that correspond to energy minima are present.

3.2   CONFORMATIONS OF CYCLIC COMPOUNDS

3.2a   THE BONDS IN RINGS HAVING THREE, FOUR, AND FIVE ATOMS ARE STRAINED

Cyclic hydrocarbons have different influences on their conformations because rotations cannot always occur to stagger the orientations between hydrogen atoms and other substituents. The bonds in small ring compounds—those with three or four atoms—experience **Baeyer strain,** a destabilizing energy effect that results from the deviation of the angles in these rings from the idealized 109.5° angle of a tetrahedron. For cyclopropane, the angles at every corner of its triangular structure would have to be 60°, which is very far from 109.5°. For the square ring of cyclobutane, the theoretical C–C–C angles would be 90°.

These theoretical angles are based on the assumption that the carbon–carbon σ bonds lie along the lines between the carbon atom nuclei (Section 2.4b). For cyclopropane especially, such an arrangement would have a prohibitively high energy. One model used to describe orbital overlap in cyclopropane considers the σ bonds to be

a.   b.

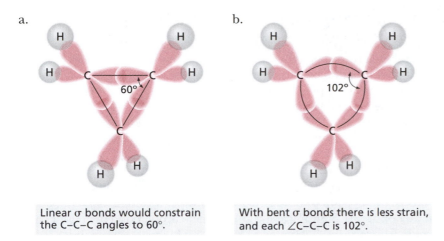

Linear σ bonds would constrain
the C–C–C angles to 60°.

With bent σ bonds there is less strain,
and each ∠C–C–C is 102°.

Figure 3.5
Structures showing sigma
bond formation in
cyclopropane. (a.) The
valence bond representation
of cyclopropane assuming the
σ bonds lie along the lines
connecting the carbon atom
nuclei. (b.) The valence bond
representation of cyclopropane
assuming the σ bonds are bent.

"bent", as illustrated in Figure 3.5. Even so, the angle at each vertex is ~ 102°, which is still strained compared with the tetrahedral ideal of 109.5°. The overlap between the hybrid orbitals in these bent bonds is also not as good as overlap between collinear orbitals. As a result, compounds that have a three-membered ring are more reactive than other cyclic compounds, as you will see later.

Cyclopentane also experiences strain in its bonds, but in a different way than either cyclopropane or cyclobutane does. The angles at each vertex in a regular pentagon are 108°, which is close to the value for a tetrahedral angle of 109.5°. As expected, cyclopentane is relatively free of Baeyer strain. In this nearly ideal geometry, however, hydrogen atoms on adjacent carbon atoms are eclipsed, as shown in Figure 3.6. This eclipsing geometry creates **torsional strain,** so-called because the conformational energy has its maximum value when the torsional angle is 0° (eclipsed).

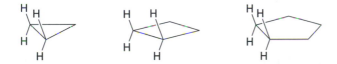

If the rings were flat, substituents on the ring would necessarily
be eclipsed, as shown for the hydrogen atoms on adjacent carbon
atoms of these 3-, 4-, and 5-membered rings.

Figure 3.6
Planar rings with three, four,
and five carbon atoms. Other
hydrogen atoms have been
omitted for clarity; add them
to the structures and confirm
that all are eclipsed.

To relieve either the Baeyer strain in the four-membered ring, or the torsional strain in the five-membered ring, one atom moves out of the plane, which decreases the number of eclipsing conformations between hydrogen atoms, as illustrated in Figure 3.7. Cyclopropane cannot undergo such a change because three points *define* a plane, so its strain relief comes in the form of bent bonds, as shown in Fig. 3.5.

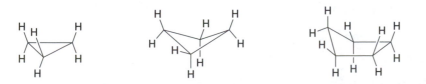

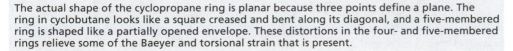

The actual shape of the cyclopropane ring is planar because three points define a plane. The
ring in cyclobutane looks like a square creased and bent along its diagonal, and a five-membered
ring is shaped like a partially opened envelope. These distortions in the four- and five-membered
rings relieve some of the Baeyer and torsional strain that is present.

Figure 3.7
Actual shapes of three-, four-,
and five-membered rings.

3.2b   ALIPHATIC SIX-MEMBERED RINGS ARE FREE OF STRAIN IN A CONFORMATION CALLED THE CHAIR FORM

The six-membered ring of cyclohexane is the only ring with fewer than 14 carbons that can adopt a conformation essentially free from strain.

Figure 3.8a shows the planar form of cyclohexane in which neighboring hydrogen atoms are eclipsed with one another. In this geometry, the C–C–C bond angles would have to be 120° instead of 109.5°. But if one carbon atom is moved to a position above the plane and one atom on the other side of the ring to a point below the plane, the positions of the hydrogen atoms shift enough that each is staggered with respect to its neighbors, and the C–C–C bond angles "relax" toward 109.5° (the actual angle is cyclohexane is close to 111° (Figure 3.8b). In this form, called the **chair conformation,** all of the carbon–carbon bonds are *gauche* with respect to bonds emanating from the adjacent carbon atoms. Sighting along any carbon-carbon bond in cyclohexane reveals these *gauche* relationships, illustrated in Figure 3.9 for two of the bonds.

a.   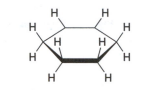   b.

Figure 3.8
The shape of cyclohexane as a planar molecule and in the chair form.

If cyclohexane were planar, the hydrogen atoms would have to be eclipsed...

...but by adopting this "chair conformation," the hydrogen atoms can all be staggered with respect to those on adjacent carbon atoms.

EXERCISE 3.3

Draw some of the other Newman projections for individual carbon–carbon bonds in cyclohexane, and verify that all of the carbon–carbon bonds adopt a *gauche* orientation. Use a model if necessary.

Three-dimensional models will help as you examine the shapes of cyclohexane and its derivatives. *It is difficult to stress too much the value of using models to learn this material.* To transfer the structural information from a model to paper, however, you must learn to draw the chair form of cyclohexane reliably, so that you can convey in two dimensions the relationships that exist between the substituents attached to the ring. As you will soon discover, the geometric arrangement of substituents has a significant influence on the conformations, stabilities, and reactivity of substances that have a six-membered ring.

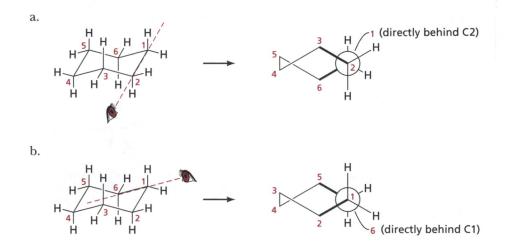

Figure 3.9
Newman projections for two bonds in the chair form of cyclohexane, illustrating some gauche relationships between bonds. (a.) Sighting along the C2–C1 bond. (b.) Sighting along the C1–C6 bond.

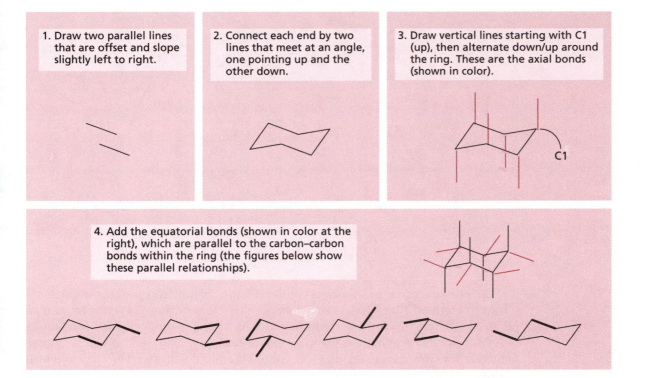

1. Draw two parallel lines that are offset and slope slightly left to right.

2. Connect each end by two lines that meet at an angle, one pointing up and the other down.

3. Draw vertical lines starting with C1 (up), then alternate down/up around the ring. These are the axial bonds (shown in color).

C1

4. Add the equatorial bonds (shown in color at the right), which are parallel to the carbon–carbon bonds within the ring (the figures below show these parallel relationships).

Figure 3.10
A scheme for drawing the chair form of cyclohexane.

Figure 3.10 shows how to make a perspective drawing of a chair form of cyclohexane. In the first step, draw two parallel lines of equal length, slightly offset from each other and sloped slightly down from left to right. Then connect the ends that are near each other with two V-shaped lines, one up and one down. Next, position what are called the **axial bonds** alternating up, down, up, down, up down around the ring, starting at the far right of the molecule. Finally, add the **equatorial bonds.** Notice that these lines are drawn parallel to the bond between the next two carbon atoms in the ring, as illustrated at the bottom of Figure 3.10. Compare your representations with those throughout this chapter, and you will soon be able to draw a six-membered ring reliably and without difficulty.

3.2c   CYCLOHEXANE CAN EXIST IN SEVERAL CONFORMATIONS THAT READILY INTERCONVERT

The chair conformation represents the lowest energy form in which cyclohexane can exist. But, as we saw for acyclic compounds, the carbon framework is flexible, so numerous other conformations exist, too. At room temperature, the cyclohexane ring undergoes a rapid **ring-flip,** in which one end moves down and the other end moves up, as illustrated in Figure 3.11. The other carbon atoms have to twist at the same time (best observed with a set of molecular models), and several changes occur as a result: *positions that were axial become equatorial, and positions that were equatorial become axial.* Notice, however, that positions that were *up*, relative to the mean plane of the ring, are still up; and positions that were down are still below the plane. *A ring-flip converts axial and equatorial substituents but does not change the relative orientations of up and down.*

Figure 3.11
A ring-flip between chair conformations in cyclohexane. This interconversion is an equilibrium process, which means the species on the left is converting to the one on the right, and vice versa. The double arrows ($\rightleftharpoons$) are used to indicate equilibrium processes.

This end flips up.

C1

C4

This end flips down.

Dashed = equatorial
Solid = axial

C4

C1

Dashed = axial
Solid = equatorial

Key:
—— Above ring
—— Below ring

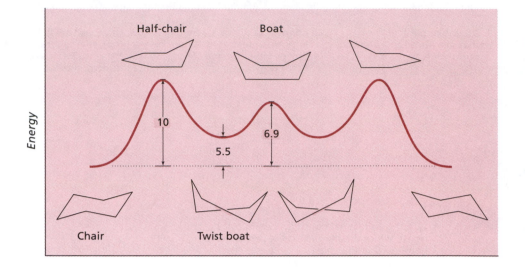

Figure 3.12
An energy diagram showing the relative stabilities for some conformations of cyclohexane. Numerical values are in kilocalories per mole (kcal mol⁻¹).

Cyclohexane can exist in several conformations, the relative energies of which are shown in Figure 3.12. As you saw for butane, there is sufficient energy at room temperature to interconvert these conformations. As a result, the ring-flip that equilibrates the two chair forms readily takes place in solution at 25°C. Even though several conformations exist for cyclohexane, we will focus only on the two limiting forms—chair and **boat conformations**—and these are shown in Figure 3.13.

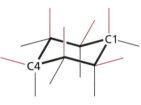

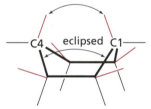

Figure 3.13
Limiting conformations of cyclohexane. The bonds shown in color are those that are above the plane of the ring.

The chair conformation has only *gauche* interactions between carbon–carbon bonds.

The boat conformation has several eclipsed interactions between carbon-carbon bonds, as well as a flagpole interaction when groups larger than H are bonded at C1 and C4.

EXAMPLE 3.2

Draw a Newman projection (use a model if necessary) looking along the C2–C3 bond from the right side of the boat conformation of cyclohexane illustrated in Figure 3.13 (also see Fig. 3.9). Identify the conformational relationships (eclipsed, *gauche*, or *anti*) between the bonds.

First, draw the conformational form, identify the bond to be examined, and add (and label) the substituents, which are hydrogen atoms in this molecule.

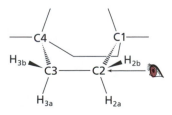

Next, draw the Newman projection, with C2 at the front and C3 behind. Using a model, you can see that the C2–C1 and C3–C4 bonds point to the right, H_{2a} and H_{3a} extend down, and H_{2b} and H_{3b} extend to the left and up.

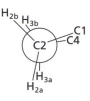

In the boat conformation, all of the bonds radiating from C2 are eclipsed with the bonds emanating from C3. The eclipsing interactions between the bonds on C2 and C3, as well as those on C5 and C6, make the boat conformation much less favorable than the chair form of cyclohexane.

EXERCISE 3.4

Draw a Newman projection looking along the C1–C2 bond from the top of the boat conformation of cyclohexane illustrated in Figure 3.13. Identify the conformational relationships (eclipsed, *gauche*, or *anti*) between the bonds on these two carbon atoms.

3.3   CONFORMATIONS OF SUBSTITUTED CYCLOHEXANES

3.3a   SUBSTITUTION OF A HYDROGEN ATOM IN CYCLOHEXANE STABILIZES CONFORMATIONS IN WHICH THE SUBSTITUENT IS EQUATORIAL

Cyclohexane undergoes a rapid ring-flip between two *equivalent* forms, which simply interchanges the axial and equatorial hydrogen atoms. As noted in Figure 3.11, this is an equilibrium process. Replacement of a hydrogen atom with a methyl group produces two different chair conformations, however, in which the methyl group is axial in one and equatorial in the other. These two conformations are no longer equivalent in energy, and the conformation with the equatorial methyl group is more stable. The two species are still involved in an equilibrium process, but we use double arrows of unequal length to indicate which species exists in a greater quantity.

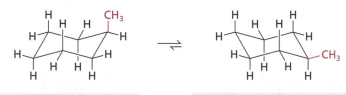

The methyl group is axial.   The methyl group is equatorial.

In fact, if you were to survey the results from experiments on hundreds of substituted six-membered ring compounds, you would find this same result: *Any group larger than hydrogen on a cyclohexane ring generally stabilizes the conformation in which that group is in the equatorial position.*

Gauche interactions between the methyl group and adjacent carbon–carbon bonds constitute one factor that destabilizes any conformation with an axial substituent. As shown in Figure 3.14, two *gauche* interactions exist when the methyl group occupies the axial position, but there are no such interactions when the methyl group is equatorial.

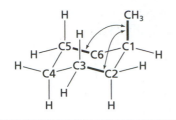

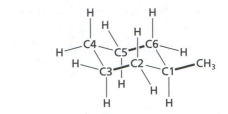

Arrows indicate *gauche* interactions between the ring C–C bonds and the C–CH₃ bond.

In this conformation, the C–C bonds in the ring are *anti* to the C–CH₃ bond.

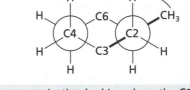

Newman projection looking along the C2–C1 and C4–C5 bonds, showing the *gauche* interaction between the C2–C3 bond and the axial methyl group.

Newman projection looking along the C2–C1 and C4–C5 bonds, showing the *anti* arrangement of the C2–C3 bond to the equatorial methyl group.

Figure 3.14
Chair conformations of methylcyclohexane illustrated with perspective drawings and Newman projections.

The second destabilizing influence comprises the **1,3-diaxial interactions,** which are caused by steric repulsions between an axial substituent and another axial substituent (including hydrogen atoms) located two carbon atoms away.

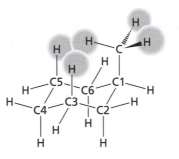

1,3-Diaxial interactions are caused by steric repulsion between substituents two carbon atoms away.

(To see this type of interaction better, you may find it helpful to build a model of methylcyclohexane to study its three-dimensional structure.) The larger a substituent is, the greater will be the repulsive forces between it and the hydrogen atoms at C3 and C5. The next section presents a way to quantify these relationships.

EXAMPLE 3.3

Draw the most stable conformer of phenylcyclohexane.

This exercise has an obvious solution, because there is only one substituent on the cyclohexane ring. Draw one chair conformer and then generate the other with a ring-flip. The conformer with the larger substituent in the equatorial position is the more stable.

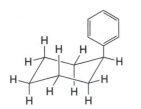

| The phenyl group is axial. | The phenyl group is equatorial. |

EXERCISE 3.5

Two of the possible conformations of methylcyclopentane are shown below. Which do you expect to be more stable? Cyclopentane derivatives do not have significant 1,3-diaxial interactions, so the *gauche* interactions mainly influence their conformational stabilities. Identify all of the *gauche* interactions that are present in each conformation.

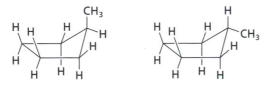

3.3b   THE PERCENTAGE OF EACH CHAIR CONFORMER CAN BE CALCULATED FROM STANDARD FREE ENERGY DIFFERENCES

If you have studied thermodynamics in a previous course, then you may recall that the difference in energy between reactants and products for a chemical reaction is called the Gibbs free energy change, ΔG. This quantity indicates whether or not a particular process will occur—if $\Delta G < 0$, the products are energetically more stable than the reactants and the reaction can proceed from left to right as written. A more useful quantity, because it is constant for a particular set of reactants, is the *standard* free energy change, $\Delta G°$. The degree symbol (°) designates that the reaction takes place under "standard conditions", which means that the reactants and products are in their most stable states at 25°C and that they have an initial solution concentration of 1 M or a gas pressure of 1 atmosphere (atm).

We can use the same formalisms to quantify the equilibrium between two conformers A $\rightleftharpoons$ B. If $\Delta G° < 0$, then production of B is favored; and if $\Delta G° > 0$, then A predominates. The concentrations of A and B, which are indicated by [A] and [B], are related to the equilibrium constant, K_{eq}, by the expression

$$K_{eq} = \frac{[B]}{[A]} \tag{3.1}$$

The correlation of $\Delta G°$ with K_{eq} is given by the equation

$$\Delta G° = -RT \ln K_{eq} \tag{3.2}$$

where $R = 1.986$ cal $\cdot$ mol^{-1} $\cdot$ K^{-1} and T is the temperature in Kelvin (K). By rearranging Eq. 3.2, we derive the following expression for K_{eq}:

$$K_{eq} = e^{-(\Delta G° / RT)} \tag{3.3}$$

At 298 K, when $\Delta G°$ is measured in kilocalories per mole

$$K_{eq} = e^{-1.69\Delta G°} \tag{3.4}$$

Even small differences in the free energy lead to nearly complete conversion of **A** to **B**. The percentage of component **B** in such a mixture can be calculated from the expression

$$\%\mathbf{B} = 100[K_{eq}/(1 + K_{eq})] \qquad (3.5)$$

A compilation of these values is presented in Table 3.1.

Table 3.1 Correlation among values for $\Delta G°$, K_{eq}, and %**B** for the equilibrium $\mathbf{A} \rightleftharpoons \mathbf{B}$ at 298 K.

$\Delta G°$ (kcal mol^{-1})	K_{eq}	%B
+1.0	0.18	15
+0.5	0.43	30
0.0	1.0	50
−1.0	5.4	84
−2.0	29	97
−3.0	159	99.4

EXAMPLE 3.4

Calculate the values of K_{eq} and %**B** at 298 K for an equilibrium reaction $\mathbf{A} \rightleftharpoons \mathbf{B}$ that has $\Delta G° = -0.5$ kcal mol^{-1}.

To calculate the value of K_{eq} we substitute the given value of $\Delta G°$ into Eq. 3.4 and solve the expression

$$K_{eq} = e^{-1.69(-0.5)} = e^{0.845} = 2.33$$

To calculate the percentage of **B** in the equilibrium mixture, we use Eq. 3.5

$$\%\mathbf{B} = 100[K_{eq} / (1 + K_{eq})] = 100(2.33/3.33) = 70.0\%$$

EXERCISE 3.6

Calculate the values of K_{eq} and percentage of **B** at 298 K for an equilibrium reaction $\mathbf{A} \rightleftharpoons \mathbf{B}$ that has the following $\Delta G°$ values.

a. +5.0 kcal mol^{-1} b. −5.0 kcal mol^{-1}

For a monosubstituted cyclohexane derivative, the identity of the substituent determines the energy difference between the axial and equatorial conformations, and these differences are summarized in Table 3.2. Using values for the free energy change between the two forms (illustrated in the figure at right of Table 3.2), we can readily calculate the percentage of each chair conformer that is present at any given temperature. For some substituents, such as fluorine, the ratio of X_{axial} and $X_{equatorial}$ does not deviate far from the 50:50 ratio of cyclohexane itself. For a larger group, such as methyl, only a small portion of the conformer exists with the substituent in the axial position.

Table 3.2   Energy differences ($\Delta G°$) for equilibration of monosubstituted cyclohexanes, $C_6H_{11}X$.

Substituent, X	$\Delta G°$ (equatorial – axial) (kcal mol^{-1})
–H	0.0
–F	–0.2
–CN	–0.2
–Cl	–0.5
–Br	–0.5
–C≡CH	–0.5
–OCH$_3$	–0.6
–OH	–1.0
–COOH	–1.4
–CH$_3$	–1.7
–CH=CH$_2$	–1.7
–CH$_2$CH$_3$	–1.8
–CH(CH$_3$)$_2$	–2.1
–C$_6$H$_5$	–2.9
–C(CH$_3$)$_3$	–5.4

(Less stable conformations: half-chair, boat, twist-boat)

Energy

$\Delta G°$

EXERCISE 3.7

From the given $\Delta G°$ values, calculate the values of K_{eq} and the percentage of equatorial isomer present at 298 K for each monosubstituted cyclohexane in Table 3.2. Refer also to Eq. 3.4 and Eq. 3.5.

3.3c   THE *TERT*-BUTYL GROUP HAS A LARGE CONFORMATIONAL BIAS FOR THE EQUATORIAL POSITION

The possible chair conformations for ethylcyclohexane are like those for methylcyclohexane, and even their $\Delta G°$ values are similar (Table 3.2). When the ethyl group is axial, it can adopt a conformation that orients its methyl group away from the ring, as illustrated in Figure 3.15. As a result, there is little increase in the 1,3-diaxial interactions

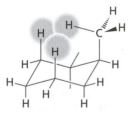

Methylcyclohexane

As already seen, the hydrogen atom on the methyl group experiences a steric interaction with the hydrogen atoms at positions 3 and 5.

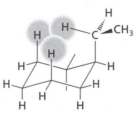

Ethylcyclohexane

The "extra" methyl group can be oriented so that steric interactions are not much greater than those in methylcyclohexane.

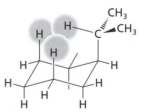

Isopropylcyclohexane

As in the ethyl derivative, the isopropyl group can be oriented so that the 1,3-diaxial interactions are not much more severe than those in methylcyclohexane.

Figure 3.15
Chair conformations of methyl-, ethyl-, and isopropylcyclohexane with the substituent in the axial position.

relative to those of methylcyclohexane. Likewise, isopropylcyclohexane is also not much different from methylcyclohexane. It, too, can adopt a conformation that minimizes the interaction of both methyl groups with the hydrogen atoms at the 3- and 5-positions of the ring. For all three compounds at room temperature, a small percentage of the molecules exist with the alkyl group in the axial position.

tert-Butylcyclohexane, shown in Figure 3.16, is a different case altogether. The addition of the third methyl group to the alkyl substituent requires that there will always be a large steric interaction with the hydrogen atoms at the other axial positions when the *tert*-butyl group is axial. Therefore, a *tert*-butyl group will almost always be equatorial (cf. Exercise 3.7).

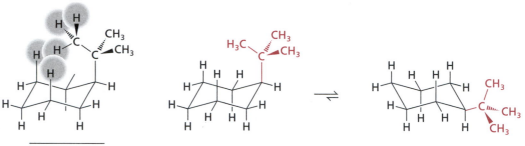

Figure 3.16
The chair conformation of *tert*-butylcyclohexane with the *tert*-butyl group in the axial position (left) to illustrate the extremely unfavorable 1,3-diaxial interactions. The conformation with an equatorial *tert*-butyl group is greatly preferred.

3.3d DISUBSTITUTED CYCLOHEXANES CAN EXIST AS ISOMERS AS WELL AS CONFORMERS

Having looked at conformers for monosubstituted cyclohexane derivatives, we now turn our attention to disubstituted compounds, beginning with 1,2-dimethylcyclohexane. This simple disubstituted derivative can exist in two isomeric forms. When the hydrogen atoms are on the same side of the ring, the compound is called the **cis isomer;** when the hydrogen atoms are on opposite sides of the ring, it is the **trans isomer.** In the names of disubstituted cycloalkanes, the prefix *cis* or *trans* (italicized) is placed in the front field to indicate that two groups are on the same or opposite sides of the ring. Thus, the names of the compounds of this type shown below are *cis*-1,2-dimethylcyclohexane and *trans*-1,2-dimethylcyclohexane, respectively.

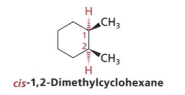

cis-1,2-Dimethylcyclohexane

trans-1,2-Dimethylcyclohexane

| The hydrogen atoms are on same side of the plane of the ring (below); therefore both methyl groups are above the plane of the page. | The hydrogen atoms are on opposite sides of the plane of the ring (one above and one below); thereofore, one methyl group is up and one is down. |

Notice that in these "flat" representations (you are looking directly down onto the ring), "up" and "down" have to be indicated by wedges and dashed lines, respectively. Bonds shown by wedge lines are above the plane of the paper, and those with dashed lines are behind the page (Section 2.5b). Realize that these line structures do not indicate anything about the conformation of the six-membered ring. A perspective chair

cis-1,2-Dimethylcyclohexane

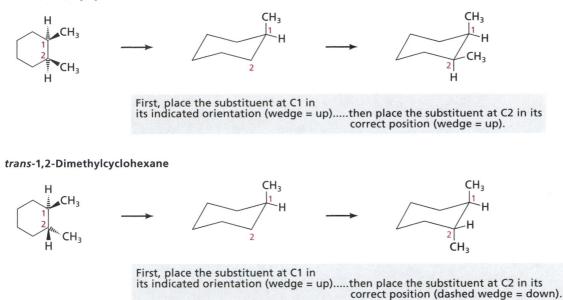

First, place the substituent at C1 in
its indicated orientation (wedge = up).....then place the substituent at C2 in its
correct position (wedge = up).

trans-1,2-Dimethylcyclohexane

First, place the substituent at C1 in
its indicated orientation (wedge = up).....then place the substituent at C2 in its
correct position (dashed wedge = down).

Figure 3.17
Drawing the chair conformations of the 1,2-dimethylcyclohexane isomers.

form will be used to render that. Figure 3.17 shows how to ensure that you draw the correct isomer, which must be done before you consider possible conformations.

Once the correct isomer has been drawn, you can consider its possible chair conformations and answer the question, "Which is the more stable conformer?" For *cis*-1,2-
dimethylcyclohexane, one methyl group is axial and the other is equatorial. A ring-flip
converts the axial one to an equatorial one, but it also converts the equatorial one to
axial. Therefore, the two chair conformers (Figure 3.18) have equal energy values. In
each conformer, one of the two substituents must be axial. Also, for each conformer, the
1,3-diaxial interactions will be identical. In one instance, they are above the ring; in the
other, below. That fact does not change the magnitude of these repulsions, however.

On the other hand, *trans*-1,2-dimethylcyclohexane has two energetically unequal
chair conformations. In the one depicted at the left side of Figure 3.19, both substituents are axial; a ring-flip puts each substituent equatorial.

Even though both methyl groups are equatorial in the more stable conformation
of this compound, they are still on opposite sides of the ring. One is up and one is
down. Remember that a ring-flip does not convert different isomeric forms—cis is still
cis, trans is still trans. A ring-flip converts only axial to equatorial and vice versa.

For the two chair conformers of *trans*-1,2-dimethylcyclohexane, there is a large energy difference. The diequatorial conformer is more stable than the diaxial conformer
because it has no 1,3-diaxial methyl–hydrogen interactions. In the conformer with two
axial methyl groups, there are numerous 1,3-diaxial interactions, both above and below
the ring.

In this conformation, one methyl group
is axial, and one is equatorial. After a ring-flip...one methyl group is still axial,
and the other is equatorial.

Figure 3.18
The two chair conformations
of *cis*-1,2-dimethylcyclohexane
related by a ring-flip.

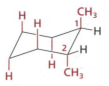

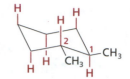

In the diaxial conformation, there are numerous 1,3-diaxial interactions between the methyl groups and the axial hydrogen atoms on the ring.

In the diequatorial conformation, there are 1,3-diaxial interactions only between the axial hydrogen atoms on the ring. This conformation is much more stable because the bulkier groups are in equatorial positions. Notice that this is still the *trans* isomer—one methyl group is up, the other is down.

Figure 3.19
Chair conformations of *trans*-1,2-dimethylcyclohexane.

Two isomers exist for 1,3-dimethylcyclohexane as well. The isomer with methyl groups on the same side of the ring is *cis*-1,3-dimethylcyclohexane. Both groups must be either axial or equatorial, as shown in Figure 3.20. The preferred conformation as you would expect, places both methyl groups in equatorial positions to minimize the 1,3-diaxial interactions.

The other isomer, *trans*-1,3-dimethylcyclohexane, has the methyl groups on opposite sides of the ring. This is the trans isomer, for which one methyl group must be axial and the other equatorial. The two conformations have identical energies, just as you saw for *cis*-1,2-dimethylcyclohexane.

The isomers and conformers for 1,4-dimethylcyclohexane are analogous to those for 1,2-dimethylcyclohexane in terms of the relationships of axial and equatorial substituents for the cis and trans isomers. Thus, *cis*-1,4-dimethylcyclohexane has one substituent axial and the other equatorial in either chair conformation. For the trans isomer, both substituents are either axial or equatorial.

A summary of the relationships for dimethylcyclohexanes is given in Table 3.3.

EXERCISE 3.8

Draw the chair conformations for *cis*- and *trans*-1,4-dimethylcyclohexane and indicate which conformer is more stable for each of the isomers.

---

cis-1,3-Dimethylcyclohexane   **Severe 1,3-diaxial interaction**

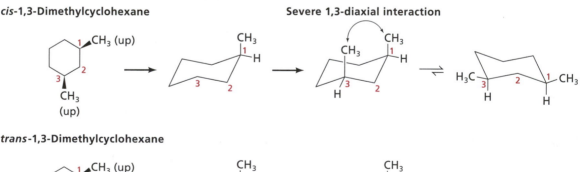

trans-1,3-Dimethylcyclohexane

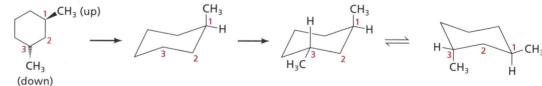

Figure 3.20
The chair conformations of *cis*- and *trans*-1,3-dimethylcyclohexane. The 1,3-diaxial interaction between the methyl group in the *cis* isomer is more destabilizing than one between a methyl group and a hydrogen atom. This repulsive force makes the diaxial conformer much less stable than the diequatorial conformer that is formed by a ring-flip.

Table 3.3   A summary of isomer and conformer relationships for the chair forms of 1,n-di-X-substituted cyclohexanes.

Isomer	Orientation of substituent at		Orientation of substituent after ring-flip at		
	C1	Cn	C1	Cn	
cis-1,2-	Axial	Equatorial	Equatorial	Axial	a
trans-1,2-	Axial	Axial	Equatorial	Equatorial	b
cis-1,3-	Axial	Axial	Equatorial	Equatorial	b
trans-1,3-	Axial	Equatorial	Equatorial	Axial	a
cis-1,4-	Axial	Equatorial	Equatorial	Axial	a
trans-1,4-	Axial	Axial	Equatorial	Equatorial	b

[a] Conformers of equal energy.
[b] More stable chair conformer.

3.3e   THE FAVORED CONFORMATION OF A DISUBSTITUTED CYCLOHEXANE WITH UNLIKE SUBSTITUENTS HAS THE LARGER GROUP IN THE EQUATORIAL POSITION

Six-membered rings that have more than two substituents or those in which the two substituents are not of equal size are more complicated to analyze than the simple dimethylcyclohexane derivatives described in Section 3.3d. For compounds with more than two substituents, you must make a judgment about the relative bulk of each group and how it will affect the conformation. Alkyl groups are relatively easy to evaluate because they are more likely to be found in the equatorial position. For example, the two possible conformers of *cis*-1-bromo-2-methylcyclohexane are shown in Figure 3.21. The conformation with an equatorial methyl group is more stable.

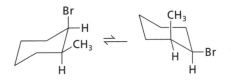

For this cis isomer, the bromine atom is axial in one conformation, and the methyl group is axial in the other. The conformation that places the methyl group in the equatorial position is more stable.

Figure 3.21
Chair conformations for *cis*-1-bromo-2-methylcyclohexane.

To predict the relative stabilities of substituted cyclohexane derivatives, you can make use of the data presented in Table 3.2 and subtract the $\Delta G°$ value for the axial substituent from the $\Delta G°$ value for the equatorial group:

- If $\Delta G° < 0$, then the conformation drawn first is favored.
- If $\Delta G° > 0$, then the conformation after performing a ring-flip is favored.
- If $\Delta G° = 0$, then neither conformation is favored.

Examples:

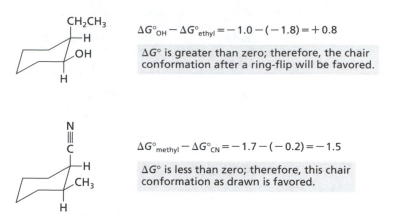

$\Delta G°_{OH} - \Delta G°_{ethyl} = -1.0 - (-1.8) = +0.8$

$\Delta G°$ is greater than zero; therefore, the chair conformation after a ring-flip will be favored.

$\Delta G°_{methyl} - \Delta G°_{CN} = -1.7 - (-0.2) = -1.5$

$\Delta G°$ is less than zero; therefore, this chair conformation as drawn is favored.

Notice that this type of analysis is needed only for the *cis*-1,2-, *trans*-1,3-, and *cis*-1,4-isomers. Chair conformations for the other isomers are more stable when *both* groups are equatorial (cf. Table 3.3).

EXAMPLE 3.5

Draw the more stable chair conformation of *trans*-3-chlorocyclohexanecarboxylic acid.

First, draw the correct isomer of the compound, then draw the chair conformation, as shown below:

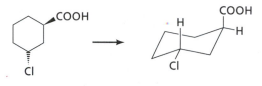

trans-3-Chlorocyclohexanecarboxylic acid

Then, calculate the energy difference using the equation, $\Delta G° = \Delta G°_{(eq\ substituent)} - \Delta G°_{(ax\ substituent)}$. For this molecule, as drawn, the chlorine atom is equatorial (eq) and the acid group is axial (ax). Therefore, $\Delta G° = \Delta G°_{(Cl)} - \Delta G°_{(COOH)} = (-0.5) - (-1.4) = +0.9$. Because $\Delta G° > 0$, the conformation formed after a ring flip is the more stable.

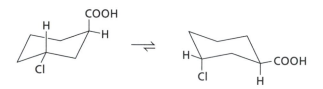

This chair form is favored.

EXERCISE 3.9

Draw the more stable chair conformation for the following molecules:

a. *trans*-1-Bromo-3-fluorocyclohexane
b. *cis*-2-Ethylcyclohexanecarboxylic acid
c. *trans*-4-Chlorocyclohexanol
d. *cis*-1-*tert*-Butyl-3-methylcyclohexane
e. *cis*-1-Isopropyl-4-phenylcyclohexane
f. *trans*-1,2-Dimethoxycyclohexane

3.4 CONFORMATIONS OF OTHER CYCLIC COMPOUNDS

3.4a SOME COMPOUNDS WITH A SIX-MEMBERED RING ARE LOCKED INTO A SINGLE CONFORMATION

As mentioned in Section 3.3c, the *tert*-butyl group must adopt an equatorial orientation, so it can be used to hold a cyclohexane ring in a specified conformation. Such derivatives are said to be *conformationally biased*. As an example, consider *cis*-1-*tert*-butyl-4-methylcyclohexane, in which the *tert*-butyl group must be equatorial, fixing the methyl group in the axial position:

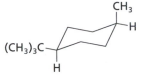

> This derivative is "conformationally biased" because of the propensity for the *tert*-butyl group to occupy the equatorial position.

The other chair conformation is much higher in energy because of the unfavorable steric interactions that would result if the *tert*-butyl were axial, so its quantity is negligible.

Another way to study conformationally biased cyclohexane derivatives is to prepare derivatives of *trans*-decalin. Decalin is the common name for decahydronaphthalene. The rings in the trans isomer cannot flip because that would require one ring to span adjacent, axial positions in the other.

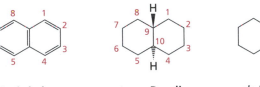

Naphthalene   *trans*-Decalin   *cis*-Decalin

Therefore, the orientation of substituents is fixed, and the compound is said to be conformationally locked. As illustrated in Figure 3.22, however, *cis*-decalin is not subject to this constraint.

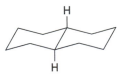

trans-Decalin

> The other conformation cannot be formed because the second six membered ring cannot span from one axial position to the other.

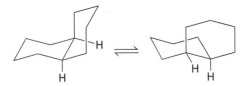

cis-Decalin

> Here, a ring-flip can occur.

Figure 3.22
Chair–chair conformations of *trans*- and *cis*-decalin.

EXERCISE 3.10

a. Draw the chair–chair conformation for each of the two isomers of 2-methyl-*trans*-decalin. b. Draw the more stable chair–chair conformation for each of the two isomers of 2-methyl-*cis*-decalin.

3.4b THE CONFORMATIONS OF COMPOUNDS WITH LARGE OR MULTIPLE RINGS CAN BE COMPLICATED

When rings have seven or more carbon atoms, some of the conformations that exist are similar to those observed for cyclohexane. For example, cycloheptane has a chairlike conformation (Figure 3.23), but the "twist–chair" form is actually most stable. In rings larger than six atoms, chair conformations necessarily create some eclipsed hydrogen–hydrogen interactions. When those higher energy interactions are removed through the twisting of certain bonds, the chairlike conformation is distorted.

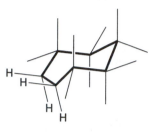

Figure 3.23
Some conformations of cycloheptane

Cycloheptane
The "chair" form has eclipsed hydrogen–hydrogen interactions (C–C ring bonds are shown in bold).

Cycloheptane
The "twist-chair" conformation is the most stable one for cycloheptane.

EXERCISE 3.11

For the idealized "boat" conformation of cycloheptane shown below, identify the eclipsed and *gauche* interactions.

With even larger rings, **transannular** (across-ring) **interactions** exist and may even predominate, as illustrated for cyclodecane in the left side of Figure 3.24.

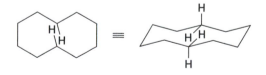

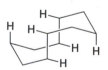

Solid-state structure for cyclodecane in the boat–chair conformation

Figure 3.24
Some conformations of cyclodecane

If cyclodecane adopted this conformation, the hydrogen atoms at positions 1 and 6 would overlap, creating transannular steric interference.

There are still several through-space H–H interactions that destabilize this structure relative to that of cyclohexane.

3.4c COMPOUNDS WITH TWO RINGS ADOPT STABLE CONFORMATIONS OF THE INDIVIDUAL COMPONENTS BUT MAY BE CONSTRAINED

We close this chapter by looking at the conformations of compounds with two rings, which can share one, two, or three atoms between them.

- Substances in which *one atom* is shared between two rings are called **spirocyclic compounds.** Their conformations are not much different from those adopted by monocyclic systems, but you have to pay attention to the conformations of *both* rings. The nomenclature for this type of compound starts with the prefix spiro, followed by a pair of brackets containing two numerals that specify the number of carbon atoms in each chain connecting the common carbon atom, followed by the root word for the *total number* of carbon atoms in both rings. The numerals in brackets are arranged in ascending order and are separated by a period. Numbering begins in the smaller ring at the carbon atom adjacent to the spiro carbon atom and proceeds around the ring and into the other.

Spiro[5.5]undecane

Each six-membered ring is in its stable chair form.

Spiro[2.4]heptane

The five-membered ring is in the stable envelope form.

- **Fused bicyclic compounds** are those in which two rings share two atoms, or a common *bond.* They are ubiquitous in nature, comprising many examples among substances known as alkaloids and terpenes, which include steroids. The conformational analysis of these compounds depends in large part on the flexibility, or perhaps the lack thereof, in the ring system. The simplest fused system with six-membered rings is decalin, discussed in Section 3.4a. Both cis and trans isomers exist. The trans isomer, although conformationally fixed, is the more stable of the two because it has fewer *gauche* interactions than does the cis isomer. The same analysis can be applied to more complex ring systems such as those in steroids, which have four fused rings.

Decalin
Bicyclo[4.4.0]decane

Bicyclo[4.1.0]heptane

Bicyclo[3.1.0]hexane

Bicyclo[2.2.0]hexane

- **Bridged bicyclic compounds** have two rings that share three or more atoms, or a common *face.*

Compounds of this type are named starting with the prefix bicyclo, followed by a pair of brackets containing the numerals that specify the number of carbon atoms in each chain connecting the common carbon atoms, which are called the **bridgehead carbon atoms,** followed by the root word for the *total number* of carbon atoms in the rings. The numerals within the brackets are arranged in decreasing order and are separated by periods. Numbering begins at a bridgehead carbon atom, proceeds around the larger bridging group, then includes the next shorter bridge, and finally the shortest bridging chain. Two common bridged bicyclic compounds are bicyclo[2.2.1]heptane, also called norbornane, and bicyclo[2.2.2]octane.

**Bicyclo[2.2.1]heptane
(norbornane)**   **Bicyclo[2.2.2]octane**

Key:

• = Bridgehead carbon atoms

Bicyclic compounds are often conformationally locked, and as depicted below for *cis*-2,3-dimethylbicyclo[2.2.2]octane, substituents must sometimes adopt positions that are eclipsed because there is no conformational change that can relieve that strain.

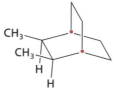

Even though the methyl groups are eclipsed, they cannot move because of the rigidity of the bridged bicyclic ring system.

Simple fused ring bicyclic compounds are named by the method used for bridged bicyclo compounds. For example, decalin is named bicyclo[4.4.0] decane. The inclusion of a "0" within the brackets tells you that the molecule has a fused ring system instead of a bridged bicyclic one.

EXERCISE 3.12

Draw structures for each of the following compounds. Identify the bridgehead carbon atoms if they are present.

a. spiro[3.5]nonane   b. bicyclo[3.2.1]octane   c. bicyclo[3.3.0]octane

CHAPTER SUMMARY

Section 3.1   Conformations of acyclic compounds

• Conformers of molecules exist because of rotation about the axis of a sigma bond.

• Ethane has two limiting conformations called eclipsed and staggered, which correspond to an energy maximum or minimum related to the dihedral angle between hydrogen atoms on the adjacent carbon atoms.

• Conformations of aliphatic compounds are commonly depicted by Newman or sawhorse projections.

• A Newman projection shows the orientation between substituents on adjacent atoms as one looks along the bond axis.

• A sawhorse projection shows the orientation between substituents on adjacent atoms with a perspective drawing.

• Aliphatic compounds larger than ethane have additional conformations called *syn*-periplanar, *gauche*, anticlinal eclipsed, and antiperiplanar (*anti*), all of which are related to the dihedral angle between substituents on adjacent carbon atoms.

Section 3.2   Conformations of cyclic compounds

- Rings with three or four atoms are strained because their bond angles are compressed relative to those associated with the hybridization of the carbon atoms.
- A three-membered ring is planar, a four-membered ring looks like a square creased and bent along its diagonal, and a five-membered ring is shaped like a partially opened envelope.
- A cyclohexane ring can adopt a shape like a chair or a boat, and these are the limiting conformations of a compound with a six-membered ring. Other conformations of cyclohexane include the half-chair and twist–boat forms.
- The hydrogen atoms in the chair form of cyclohexane are attached in orientations called axial (perpendicular to the plane of the ring) and equatorial (parallel to the sides of the ring).
- In cyclohexane, a ring flip converts axial hydrogen atoms to equatorial ones and vice versa.

Section 3.3   Conformations of substituted cyclohexanes

- A monosubstituted cyclohexane ring has two possible chair conformations in equilibrium with one another, and the more stable of these two forms has the substituent in an equatorial orientation.
- Standard free energy differences can be used to calculate the percentages of the chair forms in which the substituent is axial or equatorial.
- A *tert*-butyl group attached to the chair form of cyclohexane cannot be axial.
- A cycloalkane with two or more substituents can exist as cis and trans isomers.
- Cyclohexane derivatives that have more than one substituent exist in chair forms that have the larger group in an equatorial position.

Section 3.4   Conformations of other cyclic compounds

- For cyclic compounds larger than about eight carbon atoms, transannular effects can occur between groups across the ring from one another.
- Compounds with two adjacent rings are called spirocyclic, fused bicyclic, or bridged bicyclic according to the number of atoms shared by the rings.

KEY TERMS

Section 3.1a
conformer
conformation

Section 3.1b
sawhorse projection
Newman projection
eclipsed conformation
staggered conformation
torsional angle
dihedral angle

Section 3.1c
syn-periplanar
    conformation
steric effect
gauche conformation
anti conformation

Section 3.2a
Baeyer strain
torsional strain

Section 3.2b
chair conformation
axial bonds
equatorial bonds

Section 3.2c
ring-flip
boat conformation

Section 3.3a
1,3-diaxial interaction

Section 3.3d
cis isomer
trans isomer

Section 3.4b
transannular interactions

Section 3.4c
spirocyclic compound
fused bicyclic compound
bridged bicyclic compound
bridgehead carbon atom

ADDITIONAL EXERCISES

3.13. Draw an eclipsed, *gauche*, and *anti* conformation for each of the following compounds and indicate which is the most stable. Show both the Newman and sawhorse projections.

    a. 1-Chloropropane (C1–C2 bond)

    b. 2-Iodobutane (C2–C3 bond)

3.14. The ratio between the *anti* and *gauche* conformations of 1,2-dibromoethane is 89:11 at 298 K. Calculate the value of $\Delta G°$ for this process.

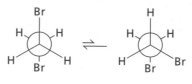

Ratio: 89 : 11

3.15. Draw the least and most stable conformations for each of the following compounds with respect to the bond indicated. Show both Newman and sawhorse projections.

    a. Propanal (C2–C3).

    b. 1-Pentyne (C3–C4 bond).

3.16. For each given name, draw a structure for each of the following compounds. Ignore conformations for this exercise.

    a. *cis*-1-Bromo-2-methylcyclopentane   b. 2,2-Difluorocyclohexanone

    c. 2-Cyclohexenol   d. *cis*-3-Chlorocyclobutanol

    e. *trans*-4-*tert*-Butylcyclohexanecarboxylic acid

3.17. For each set of structures, indicate which pairs of compounds are cis/trans isomers, which pairs are constitutional isomers, which pairs are conformers, and which pairs are identical.

    a.

    b.

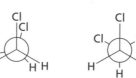

    c.

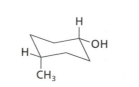

3.18. Draw the structure of the two chair conformations for each of the following compounds, and identify the more stable conformer:

a. *cis*-1-Bromo-3-ethylcyclohexane

b. *cis*-4-Hydroxycyclohexanecarboxylic acid

c. *trans*-2-Isopropylcyclohexanol

d. *trans*-3-Chloro-1-bromocyclohexane

3.19. The energy difference between the two chair conformations of cyanocyclohexane is small, suggesting that *gauche* interactions and 1,3-diaxial repulsions are slight. Draw each chair conformation and explain why these interactions are much less for a cyano group than are those same interactions experienced by a methyl group.

3.20. When three different carbon atoms of cyclohexane bear substituents, the conformer with two groups in the equatorial positions is usually the more stable. Draw the preferred conformation for each of the following compounds:

a.   b.   c.   d.

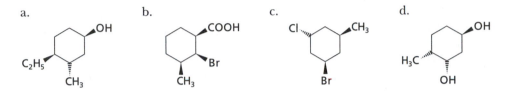

3.21. Draw the structure for each of the following compounds, then draw the conformer that you expect to be most stable:

a. Spiro[3.5]nonane   b. *trans*-Bicyclo[3.3.0]octane

c. Spiro[2.4]heptane   d. *trans*-2,2-Dibromobicyclo[4.4.0]decane

e. 2-Methylspiro[3.3]heptane   f. Bicyclo[4.3.1]decane

3.22. Draw all of the isomers of dimethylcyclobutane (there are five). For four of the isomers, there are two conformations that can interconvert, changing axial substituents to equatorial and vice versa. Which is the more stable conformer for each isomer?

3.23. Some members of a class of natural products called terpenes, which are hydrocarbons with 10 carbon atoms derived from isoprene (2-methyl-1,3-butadiene), have a bridged bicyclic carbon framework. For each of the following compounds, give the name of the parent bridged bicyclic hydrocarbon (i.e., ignore the methyl and isopropyl substituents in each structure). Draw the structure of isoprene.

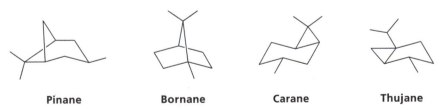

Pinane   **Bornane**   **Carane**   **Thujane**

3.24. Menthol (peppermint oil) is a terpene natural product that is monocyclic. Draw its two chair conformations and indicate which is more stable.

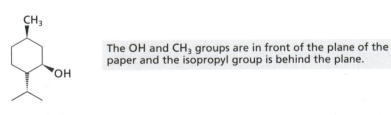

The OH and CH₃ groups are in front of the plane of the paper and the isopropyl group is behind the plane.

Menthol

3.25.   Draw the four possible boat conformations of methylcyclohexane. Which conformer is the most stable? Why?

3.26.   Hydrindane, or hexahydroindane, is an analogue of decalin. It is a fused bicyclic compound that has a six- and a five-membered ring. Draw the cis and trans isomers and any conformer of those in which the six-membered ring adopts a chair form.

Hydrindane

3.27.   *trans*-Decalin is more stable than the cis isomer. Identify the 1,3-diaxial interactions in *trans*- and *cis*-decalin and rationalize the greater stability of the trans isomer.

3.28.   Kemp's triacid, a compound described in Chapter 28, is the common name for the isomer of 1,3,5-trimethyl-1,3,5-cyclohexanetricarboxylic acid that has all of the methyl groups on the same side of the ring. Which conformation is the most stable? Why?

3.29.   Organophosphorus derivatives of inositol are important constituents of cell membranes. Inositol is 1,2,3,4,5,6-hexahydroxycyclohexane, and it exists in several isomeric forms depending on whether the OH groups are up or down. Draw the most stable isomer of inositol in its most stable conformation.

THE STEREOCHEMISTRY OF ORGANIC MOLECULES

The previous chapter (Chapter 3) described the conformations of molecules that interconvert by rotation about carbon–carbon σ bonds. It also described the cis and trans isomers of cycloalkanes, which exist when substituents are attached to the same or opposite sides of a ring.

This chapter will focus attention on *stereoisomers,* molecules in which the constituent atoms have the same connectivities, but differ in their overall shape. In studying reactions of organic molecules, especially those involved in biological processes, it is essential to understand the consequences of the specific three-dimensional arrangement of atoms, how to distinguish between different stereoisomers, and how to represent a three-dimensional shape on a two-dimensional surface. *The use of molecular models is strongly encouraged for your study,* at least until you become familiar enough with drawings to manipulate structures mentally or by drawing them. Even then, the use of models can provide insights about interactions within a molecule.

This chapter begins with a description of alkene isomers. The π bond of an alkene prevents rotation about the carbon–carbon σ bond, which means that substituents can have different orientations within the alkene plane. The prefixes *cis, trans,* (*E*), and (*Z*) are used to denote these relative arrangements, and you will learn how to interpret names with these stereochemical descriptors. The focus of the chapter then shifts to the stereochemistry of saturated compounds, beginning with those that have an asymmetric carbon atom, which is one with four different groups attached. Such molecules exist as stereoisomers because there are two ways to spatially orient the four substituents.

4.1 GEOMETRIC ISOMERS OF ALKENES

4.1a ISOMERS OF ALKENES WITH TWO HYDROGEN ATOMS ARE DESIGNATED CIS AND TRANS

Constitutional isomers differ in the connectivity patterns among their atoms (Section 1.4a). **Stereoisomers**, in which the connectivities between the atoms in each isomer are

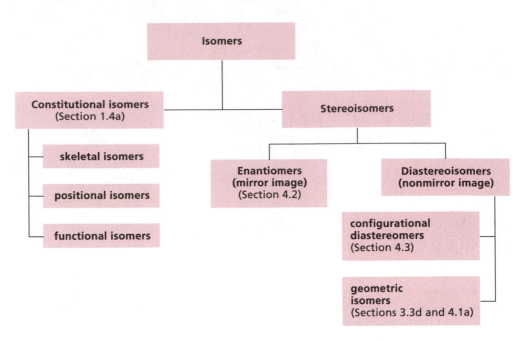

Figure 4.1
A diagram summarizing the relationships among types of isomers.

the same, differ in the arrangement of their atoms in three dimensions. A full summary of isomer types is presented in Figure 4.1.

Cis and trans isomers of disubstituted cycloalkanes, which were discussed in Section 3.3d, represent a type of stereoisomer called **geometric isomers.**

 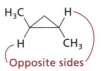

cis-1,2-Dimethylcyclopropane   trans-1,2-Dimethylcyclopropane

The terms cis and trans are also used for alkenes that have two hydrogen atoms, one at each end of the carbon–carbon double bond. The alkene molecule with the hydrogen atoms on the same side of the line connecting the carbon nuclei is the cis isomer, and the one with hydrogen atoms on the opposite sides of the line connecting the carbon nuclei is the trans isomer. In naming an alkene, the prefix *cis-* or *trans-* (italicized) is added to the name. Notice that an alkene with two hydrogen atoms at the *same* end of the double bond does not exist as geometric isomers.

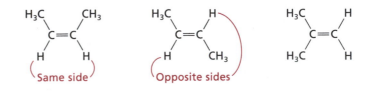

EXERCISE 4.1

Draw the structural formula that corresponds to each of the following names:

a. *cis*-3-Octene   b. *trans*-1,1-Dichloro-2-pentene   c. *trans*-2-Hexenal

4.1b   ISOMERS OF ALKENES WITH TWO OR MORE SUBSTITUENTS ARE IDENTIFIED BY THE PREFIX (*E*) OR (*Z*)

An alkene with less than two hydrogen atoms may also exist in isomeric forms, but the prefixes cis and trans have no meaning. Instead, the spatial relationships among the groups are denoted by the italicized letter *E* (from the German *entgegen,* meaning across) or *Z* (from the German *zusammen,* or together) and are based on *the priorities of the substituents at each end of the double bond.* If the higher priority substituents at each end are on the same side of the double bond, the compound is the (*Z*) isomer. If the higher priority substituents at each end are on opposite sides of the double bond, then the compound is the (*E*) isomer. Notice that a disubstituted alkene can also be labeled as *E* or *Z* rather than as trans or cis. In an IUPAC name, the (*E*) and (*Z*) descriptors are included as prefixes placed within parentheses, italicized, and connected by a hyphen to the rest of the name.

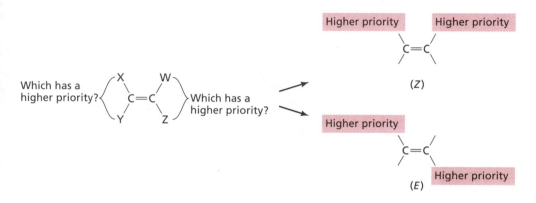

To determine the priorities of the substituents, you apply a set of rules called the **Cahn–Ingold–Prelog convention,** which are based on the magnitude of the atomic number of each atom: *A higher atomic number means a higher priority.* An unshared electron pair, which is sometimes encountered when the double bond is between carbon and a heteroatom, has the lowest priority. Among *atoms,* hydrogen has the lowest priority.

The rules are summarized as follows:

1. Look at the two atoms that are attached directly to each carbon atom connected by a double bond. Considering each end of the double bond *separately,* assign a higher priority to the element with the *higher atomic number.*

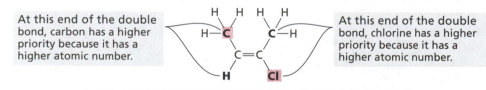

At this end of the double bond, carbon has a higher priority because it has a higher atomic number.

At this end of the double bond, chlorine has a higher priority because it has a higher atomic number.

Because the higher priority groups are on opposite sides of the double bond axis, this compound is (*E*)-2-Chloro-2-butene.

2. If the attached atom is an isotope of the same element, the one with higher mass has the higher priority. For example, D (deuterium) has a higher priority than H (hydrogen).

3. If the two atoms bonded to one end of the double bond are the same, look at the next "shell" of atoms connected to each of the identical ones, and list them in order of decreasing atomic number. The group with an element having a *higher atomic number at the first point of difference* has the higher priority.

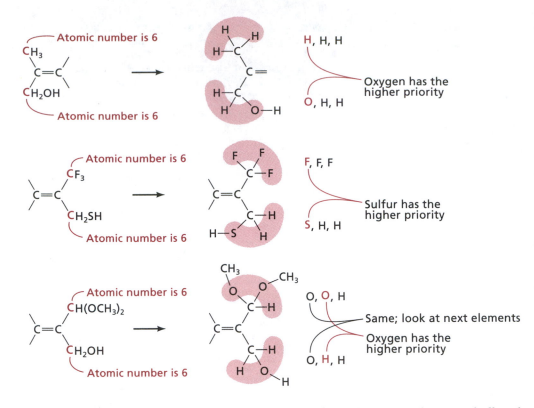

If all three atoms in the second shells are identical, then go to the next shell and repeat the process until a difference is found.

4. If multiple bonds are present, these are assigned "single-bond equivalents". Each atom in the double bond is instead connected by single bonds to two copies of the other atom. Replication of the element at each end of the multiple bond is depicted in Figure 4.2.

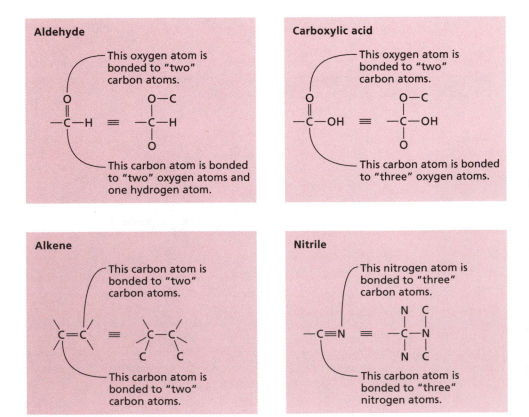

Figure 4.2
Some single-bond equivalents for functional groups with multiple bonds.

EXAMPLE 4.1

Assign (*E*) or (*Z*) to the stereochemistry of the double bond in the isomer of 4-chloro-3-methyl-2-butenoic acid shown below.

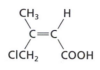

The first step is to assign priorities to the substituents attached to the carbon atom in the 2-position. The first atoms of each substituent are H and C, respectively; so the carboxylic acid group has the higher priority.

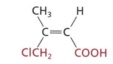

Next, assign priorities to the substituents attached to the carbon atom at the 3-position. Here, the first atom is the same, namely, C, so it is necessary to look at the next shell of atoms. On one side, the atoms are H, H, H; on the other, they are Cl, H, H. The higher atomic number takes precedence, so the chloromethyl group has the higher priority.

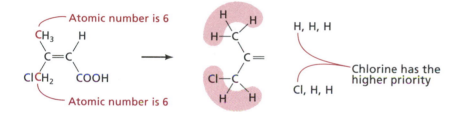

Finally, answer the question: Is the higher priority group at each end on the same side or opposite sides of the double bond? In this case, they are on the same side, so this is the (*Z*) isomer.

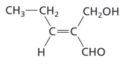

(*Z*)-4-Chloro-3-methyl-2-butenoic acid

EXAMPLE 4.2

Assign (*E*) or (*Z*) to the stereochemistry of 2-(hydroxymethyl)-2-pentenal, shown below.

CH₃—CH₂   CH₂OH
    \\   /
     C = C
    /   \\
   H   CHO

First, assign priorities to the substituents on the carbon atom at the 3-position because the difference between the attached atoms is obvious (C vs. H), so the ethyl group is easily identified as having the higher priority:

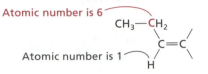

The ethyl group has the higher priority

Next, assign priorities to the substituents on the carbon atom in the 2-position. The first atom of each substituent is the same (C), so it is necessary to look at the next shell of atoms. However, first write out the single-bond equivalents for the aldehyde group. This step is required whenever the substituent has a multiple bond.

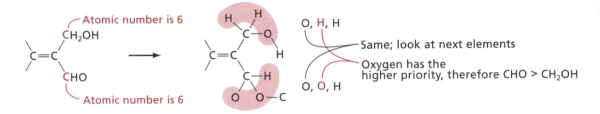

Thus, the higher priority groups are on opposite sides of the C=C bond, so this is the (*E*)-isomer.

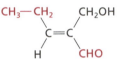

(*E*)-2-(Hydroxymethyl)-2-pentenal

EXERCISE 4.2

Assign (*E*) or (*Z*) for the geometry of the double bond in each of the following compounds:

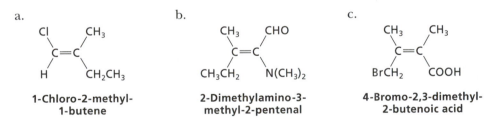

a.

1-Chloro-2-methyl-
1-butene

b.

2-Dimethylamino-3-
methyl-2-pentenal

c.

4-Bromo-2,3-dimethyl-
2-butenoic acid

EXERCISE 4.3

Draw the structural formula for each of the following compounds, including its stereochemistry:

a. (*Z*)-3-Heptene
b. (*Z*)-3-Methoxy-2-octenal

c. (*E*)-1,3-Dichloro-2-methyl-2-hexene
d. (*E*)-3,4-Dibromo-3-heptene

4.2   CHIRALITY AND ENANTIOMERS

4.2a   COMPOUNDS THAT HAVE A CARBON ATOM WITH FOUR DIFFERENT SUBSTITUENTS CAN EXIST AS STEREOISOMERS CALLED ENANTIOMERS

The stereoisomers described so far comprise substituted cycloalkanes and alkenes, which are stereoisomers by virtue of geometric relationships created by the presence of rings and π bonds, respectively. However, stereoisomers of unconstrained acyclic compounds may exist if the carbon framework and its substituents can be oriented differently in three dimensions.

First, consider how to tell whether or not two molecules are identical. If presented with the two structures shown below, you would first look to see if the atoms are connected in the same ways.

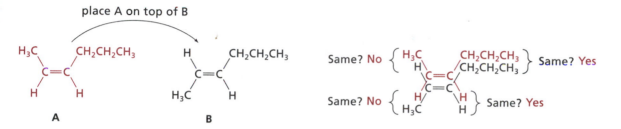

Any change in the connectivity tells you immediately that these compounds are constitutional isomers (Fig. 4.1). If the bond connections are the same (as in fact they are), you might then construct a model of each and see if the models are superimposable, meaning that one fits exactly on top of the other. If they *are* superimposable, then you know the molecules are identical. If they are not, then they must be *stereoisomers*.

With alkenes, you can probably superimpose them mentally. Consider the following two isomers of 2-hexene. Are these two compounds superimposable? If you move structure **A** onto structure **B**, you would see that they are not superimposable because all of the atoms or groups do not align.

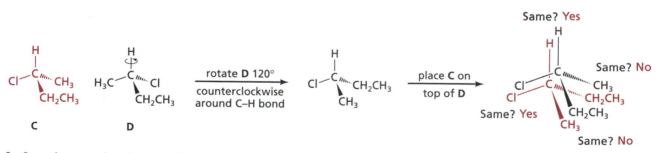

One is cis and the other is trans. You already know that these molecules comprise a set of geometric isomers, which represent one type of stereoisomer.

Now consider the two compounds **C** and **D** shown below. The bond connections are the same, therefore the molecules are either identical, or else they are stereoisomers. There are no double bonds or rings, so they cannot be geometric isomers. Are they superimposable? If you make a model of each, and then try to align them, you would find that they are *not* superimposable.

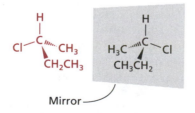

In fact, these molecules are mirror images.

Compounds that are nonsuperimposable mirror images are called **enantiomers,** and they are said to be **chiral.** The opposite of chiral is **achiral.** Your hands represent the classic example of chirality: *They are mirror images and they are not superimposable.* For this reason, chiral substances, even molecules, are sometimes described as having "handedness".

Every object, hence every molecule, has a mirror image. To deduce whether a molecule is chiral, you have only to ascertain whether it and its mirror image are superimposable. If they are, then the molecules are identical. If not, the molecules are enantiomers.

A more general question you might ask is this: Is there a structural feature of a molecule that *always* renders mirror images nonsuperimposable? The answer is yes: Any molecule in which *only one* of its carbon atoms has four different groups attached will always be chiral. A carbon atom with four different substituents is called a **chiral, asymmetric,** or **stereogenic carbon atom.** We often refer to a chiral carbon atom as a *stereocenter.* The three-dimensional arrangement of the groups attached to a chiral carbon atom is called its **configuration.** A summary of definitions for many of the terms used to describe stereoisomers is given in Table 4.1.

Molecules may possess more than one chiral carbon atom. In these cases, the whole molecule itself may or may not be chiral. Additionally, there are other types of chiral molecules that do not contain chiral carbon atoms. These types of molecules will be discussed later.

EXERCISE 4.4

Which of the following pairs of molecules are identical? Which are stereoisomers?

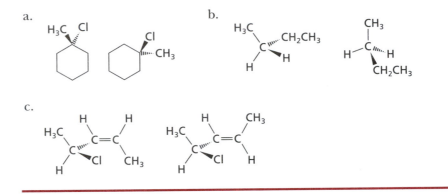

a.

b.

c.

4.2b   A CHIRAL SUBSTANCE ROTATES PLANE-POLARIZED LIGHT

Enantiomeric compounds, because they differ only by the spatial orientations of their atoms, have identical physical properties, as you might expect for compounds with the same types of bonds. They also have identical chemical properties when they react with *achiral* substances. It is possible to differentiate between enantiomers, however, and one way is to probe how each interacts with another *chiral* molecule. For example, migration of two enantiomers on a chiral stationary phase during chromatography occurs at different rates. Another method to differentiate enantiomers is by the way they rotate **plane-polarized light.** In fact, referring to a chiral molecule as **optically active,** a term still in use today, relates to this property.

In 1815, the French physicist Biot demonstrated that certain organic compounds rotate plane-polarized light, in a manner represented by Figure 4.3. What he termed "optical activity" was proposed to be an inherent property of such molecules. Polarized light is made by passing ordinary light through a filter that allows only those light waves aligned with the filter to pass. This is the same principle that is used in polarized

Table 4.1   Definitions of terms used to discuss stereoisomers and their properties.

Absolute configuration	The arrangement of groups about a stereogenic carbon atom, normally indicated by its (R) or (S) configuration.
Achiral	Used to describe a substance that is superimposable on its mirror image.
Asymmetric carbon atom	See chiral carbon atom.
Chiral	The property of a substance that it is not superimposable on its mirror image; also used to refer to a compound that is optically active.
Chiral carbon atom	A carbon atom that has four different groups attached to it—also called a "asymmetric carbon atom" or a "stereogenic carbon atom."
Chiral center	Any tetrahedral atom, often carbon, with four different substituents.
Configuration	The three-dimensional arrangement of a chiral center with its substituents.
Dextrorotatory	The property of a compound that causes it to rotate plane-polarized light in the clockwise direction; abbreviated by the use of a lower case, italicized d or by the symbol "(+)" in front of the compound's name.
Diastereomers	Any set of steroisomers that are not enantiomers; the singular form is used to refer to any one of such a set of compounds.
Enantiomers	Any set of isomers with the same atom connectivities that are nonsuper-imposable mirror images; the singular form refers to either one of the two mirror-image isomers.
Fischer projection	A two-dimensional representation of chiral substances that makes use of vertical and horizontal lines to specify the positions in space of substituents attached to an asymmetric carbon atom.
Levorotatory	The property of a compound that causes it to rotate plane-polarized light in the counterclockwise direction; abbreviated by the use of a lower case, italicized ℓ or by the symbol "(−)" in front of compound's name.
Meso compound	An isomer that has chiral centers, but an internal mirror plane that makes it achiral.
Mirror plane	An imaginary plane between two molecules, or dividing a molecule, through which the two molecules or groups are related as mirror images.
Optically active	The ability of a substance to rotate plane-polarized light.
(R, S) configuration	The absolute configuration assigned by applying the Cahn–Ingold–Prelog priority rules.
Racemate	A 50:50 mixture of enantiomers; equivalent terms are racemic mixture, racemic modification, d, ℓ-mixture, (±)-mixture, D, L-mixture, or (R), (S)-mixture; a racemate does not rotate plane-polarized light even though it contains chiral substances.
Racemic	An adjective used to describe a 50:50 mixture of enantiomers.
Racemic mixture	See racemate.
Racemization	A process wherby the absolute configuration of a chiral center is converted to its mirror-image configuration, leading eventually to a 50:50 mixture of enantiomers.
Scalemic	An adjective used to describe a mixture of enantiometric compounds that is enriched in the concentration of one of the enantiomers (not a universally accepted term).
Stereocenter	See chiral center.
Stereogenic carbon atom	See chiral carbon atom.
Stereoisomers	Compounds in which the atoms have the same connectivities but that have different shapes in three dimensions.

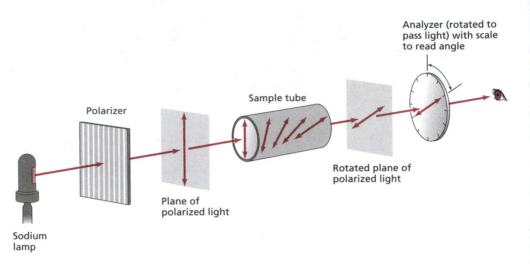

Figure 4.3
Rotation of plane-polarized light by a solution containing a chiral molecule.

sunglasses. If the polarized light then passes through a solution containing a chiral compound, it will be rotated by a certain amount. Enantiomers differ only in the *direction* that they rotate the plane-polarized light: The absolute value of the magnitude is identical, as long as the temperature, concentration, solvent, and wavelength of light are the same.

When a portion of one enantiomer is combined with an equal amount of the other, a **racemic mixture** or **racemate** is formed. A solution containing a racemic mixture does not rotate plane-polarized light because the two compounds rotate the light equally in opposite directions, canceling any effect. Even though enantiomers each have identical physical properties, their 50:50 mixture—the racemate—often displays different physical properties, such as melting point and solubility, compared with the properties of either pure enantiomer. Any characteristic that is influenced by the forces between molecules in the solid state can be altered by the presence of a molecule's mirror image.

To illustrate this effect, try folding your hands (which are nonsuperimposable mirror images) together, and then try folding your left hand with someone else's left hand. The mirror image hands fit together in a more spherical orientation, whereas the two identical hands form a stacking arrangement.

Although optical activity was known as early as 1815, the existence of enantiomeric compounds was not recognized until the 1840s. Working with salts of tartaric acid, a by-product of fermentation, Louis Pasteur manually separated crystals with mirror-image crystal faces into two piles. Dissolving the same amount of compound from each pile in water, he found that the solutions differed only by the property that they rotate plane-polarized light in opposite directions. Thereafter, and until the three-dimensional structures of enantiomers were understood, each of a pair of enantiomers was identified by the direction in which it rotated plane-polarized light.

An example that is simpler than tartaric acid is 2-butanol because it contains a single chiral carbon atom. The isomer that rotates plane-polarized light in the clockwise direction is called "(+)-2-butanol", and its enantiomer, "(−)-2-butanol". The letters *d* (for **dextrorotatory,** meaning "right rotating") and *l* (for **levorotatory,** meaning "left rotating") are used to convey the same information, so the names *d*-2-butanol and (+)-2-butanol or *l*-2-butanol and (−)-2-butanol are synonymous. The racemic mixture is represented by *d,l*-2-butanol or (±)-2-butanol.

Being able to specify the direction that each isomer rotates plane-polarized light does not tell us the actual arrangement of the atoms in three dimensions. All that we know is that the two molecules must have mirror-image configurations. By carrying out chemical reactions in which we convert 2-butanol to other substances, we could relate the configuration of any compound to one of the two isomers of 2-butanol, eventually building an inventory of compounds that have the same configuration. This process— except that glyceraldehyde was employed as the reference compound—was actually used for decades to specify the configurations of atoms in chiral molecules.

EXERCISE 4.5

Draw the structural formula for the enantiomer of each of the following compounds:

a.   b.   c.

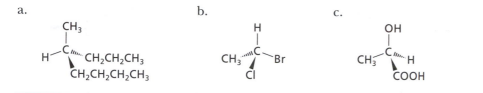

4.2c   ABSOLUTE CONFIGURATIONS WERE EVENTUALLY ASSIGNED ON THE BASIS OF A COMPOUND'S STRUCTURAL SIMILARITY TO *d*-GLYCERALDEHYDE

As knowledge about chemical structures and reactivity grew, and as more chiral substances were isolated and characterized, it became apparent that many optically active compounds could be interconverted by known reactions. If the stereochemistry stayed the same in these processes, the reaction was said to occur by *retention of configuration*. If the stereochemistry changed, the reaction proceeded by *inversion of configuration*. The only matter to settle was finding an appropriate reference compound for which the absolute configuration could be specified and to which other compounds of known configuration could be related.

Emil Fischer, one of the most prominent organic chemists in the late 1800s, was instrumental in correlating configurations of organic structures by their reactions. Most of his work, for which he received the 1902 Nobel Prize in chemistry, was focused on the chemistry of sugars, many of which are chiral substances. In 1891, Fischer proposed the use of *d*-glyceraldehyde as a reference material for specifying the configurations of stereocenters, and he *arbitrarily* chose the configuration shown below.

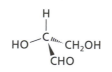

d-Glyceraldehyde (later, D-Glyceraldehyde)

Realize that he had a 50:50 chance of choosing the correct absolute configuration for this compound. Fortunately, when the actual structure of the compound now known as D-glyceraldehyde was determined by X-ray crystallography in 1951, the assignment proved to be correct. Therefore, the absolute configuration of all compounds that had been related to glyceraldehyde by reaction chemistry during the previous decades had been assigned correctly.

Fischer's early system for assigning absolute configurations employed the designations *d* and *l*, even though these descriptors were already in use to denote the directions that a compound rotates plane-polarized light. Later D and L were adopted instead. These absolute configurations do not necessarily correlate with the sign of optical rotation for a particular compound, however. In other words, D does not mean the same as *d,* which in the present day refers only to the direction of rotation of plane-polarized light by the compound.

As chemists began to recognize the broader significance of chirality, a realization that evolved from the symbiotic advances within the fields of chemistry and biology, it became clear that a method was needed by which to specify the configurations of stereocenters.

EXERCISE 4.6

Draw the structure of L-glyceraldehyde, the enantiomer of D-glyceraldehyde.

4.2d   THE CAHN–INGOLD–PRELOG CONVENTION IS NOW USED TO SPECIFY THE ABSOLUTE CONFIGURATION OF A CHIRAL CARBON ATOM

The method now used to specify the absolute configuration of a chiral center is the Cahn–Ingold–Prelog convention, the same set of priority rules you learned in Section 4.1b to assign the (*E*) or (*Z*) configuration of alkenes. Specifying the absolute configuration of a carbon atom follows the procedure illustrated in Figure 4.4.

1. Assign a priority to each of the four substituents attached to a carbon atom.

2. Turn the chiral carbon atom so that the priority 4 group is pointing toward the back.

3. Determine the relative orientation of the substituents, going from priority 1 to 2 to 3. If their arrangement is clockwise, the chiral center has the (*R*) configuration (from the Latin *rectus,* meaning right). If the arrangement is counterclockwise, the chiral center has the (*S*) configuration (from the Latin for *sinister,* meaning left).

As with the descriptors D and L, the absolute configurations (*R*) and (*S*) have nothing to do with the (+) and (−) labels, or the *d* and *l* designations, which refer to the directions that a stereoisomer rotates plane-polarized light,. Figure 4.5 illustrates, as an example, the relationships between (*S*)-(−)-2-chloropropanoic acid and (*S*)-(+)-2-chloro-1-propanol. These compounds have the same absolute configuration but rotate plane-polarized light in opposite directions.

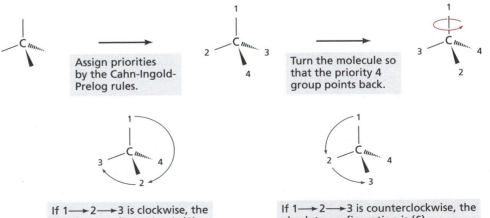

Figure 4.4
The procedure for specifying the absolute configuration of a chiral carbon atom.

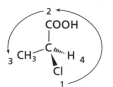

(S)-(−)-2-Chloropropanoic acid

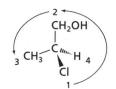

(S)-(+)-2-Chloro-1-propanol

(−) indicates rotation of plane-polarized light in the counterclockwise direction.

After priorities are assigned to the substituents and the structure is arranged so that the lowest priority group points away from you, the direction from priority 1⟶2⟶3 is counterclockwise.

(+) indicates rotation of plane-polarized light in the clockwise direction.

After priorities are assigned to the substituents and the structure is arranged so that the lowest priority group points away from you, the direction from priority 1⟶2⟶3 is counterclockwise.

Figure 4.5

Compounds that illustrate the absence of correlation between absolute configuration and the direction of rotation of plane-polarized light.

To specify the absolute configuration of a chiral carbon atom in an IUPAC name, we add the prefix (R) or (S) to the name. When the molecule has only one chiral atom the letter R or S is placed in parentheses at the beginning of the name. When there is more than one chiral center, a numeral precedes each letter R or S, and these are placed in parentheses in front of the name.

(R)-2-Chlorobutane   **(2S,3R)-3-Bromo-2-butanol**

EXAMPLE 4.3

Assign the absolute configuration to the chiral carbon atom in the enantiomer of 1-chloro-2-butanol shown below.

First, identify the chiral carbon atom, and then assign priorities to each group.

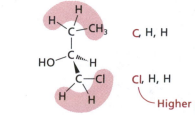   

C, H, H

Cl, H, H

└ **Higher priority**

The four atoms attached to the chiral carbon atom are O, C, C, and H. The O has the highest atomic number and H has the lowest, so these are priority groups 1 and 4, respectively. The two groups with C have to be expanded.

After expanding the chloromethyl and ethyl groups to see the next shell of atoms, we see that the higher atomic number of Cl makes the chloromethyl group priority 2 and the ethyl group priority 3.

Because the priority 4 group is pointing back (dashed line), the molecule is in the correct orientation to assign (R) or (S), so determine whether the priority sequence 1 → 2 → 3 is clockwise or counterclockwise.

In this molecule, the sequence in going from priority 1 to 2 to 3 is counterclockwise, so the configuration of the central carbon atom is (S). This molecule is named (S)-1-chloro-2-butanol.

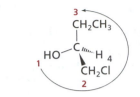

For a situation in which you are asked to assign the configuration of a stereocenter, Example 4.3 illustrates the best-case scenario: *The priority 4 group is already oriented toward the rear.* Therefore, once you have assigned priorities to the substituents of the chiral carbon atom, determining its configuration is straightforward.

Suppose, however, that the priority 4 group, as drawn, does not point away from you. There are at least three options you have to reorient the substituents so that the priority 4 group points away from you.

- VISUALIZE a rotation of the molecule in your mind, and then draw the reoriented structure. In this example, the process is fairly straightforward, requiring a simple rotation around one bond. In other cases, however, the required movements may be more complex. For people who have trouble picturing three-dimensional objects, visualization is the most difficult option.

 in your mind, visualize rotating around the vertical C–C bond (in this example) until the H atom points toward the rear, and then draw this reoriented structure →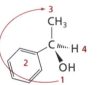

With the priority 4 group toward the rear, the sequence in going from priority 1 to 2 to 3 is clockwise, so the configuration of the central carbon atom is (R).

- MAKE A MODEL. This option requires the most work, but it is relatively fool-proof, and you can subsequently move the molecule into any other orientation that you may need. *To overstate the value of making models to study stereochemical relationships is difficult.* Realize that you do not have to include every atom of the molecule in your model if you only want to assign the configuration of a chiral center. Instead, use single spheres to represent each of the four groups. Once you have the model in hand—literally—you can turn it any way you need. To assign the absolute configuration, this means turning the model so that the priority 4 group points away from you.

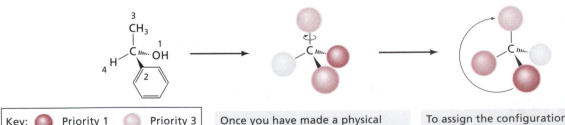

Key: ● Priority 1   ● Priority 3
      ● Priority 2   ○ Priority 4

Once you have made a physical model, you can hold it in your hand and turn it any way you want.

To assign the configuration of the chiral carbon atom, turn the model so that the priority 4 group points away. The sequence in going from priority 1 to 2 to 3 is clockwise, so the configuration of the central carbon atom is (R).

- INTERCHANGE TWO GROUPS—WITH CAUTION! The following statement conveys an essential fact about configurations: *whenever you interchange any two groups attached to a chiral atom, you invert its configuration.* To exploit this option to assign configurations, switch the position of the group pointing away with that of the priority 4 group. It is helpful to use a designation other than (R) or (S) to remind yourself that you have switched the groups. For example, (S^I) has been used here to signify that the priority sequence $1 \rightarrow 2 \rightarrow 3$ is counterclockwise but that two groups have been interchanged (I). An assignment of (S^I) means that the original configuration was (R). Likewise, (R^I) means that the original configuration was (S).

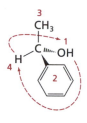

interchange the priority 4 group with the group that points away

(in this example, the priority 1 group)

The direction of priority 1 to 2 to 3 is counterclockwise = (S^I) (the superscript I means that two groups were first interchanged). (S^I) = (R)

> This molecule now has the correct orientation to assign the configuration of the chiral carbon atom, so look at the relationship between priority groups 1, 2, and 3 (clockwise or counterclockwise).

EXAMPLE 4.4

Assign the absolute configuration to the chiral carbon atom in the enantiomer of 2-methylbutanoic acid shown here:

First, identify the chiral carbon atom, and then assign priorities to each group. The carbon-containing groups have to be expanded, and the carboxylic acid group has to be further expanded to include its single-bond equivalent structure (Fig. 4.2).

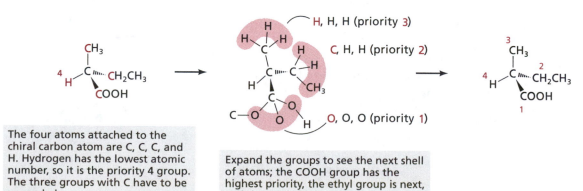

The four atoms attached to the chiral carbon atom are C, C, C, and H. Hydrogen has the lowest atomic number, so it is the priority 4 group. The three groups with C have to be expanded.

Expand the groups to see the next shell of atoms; the COOH group has the highest priority, the ethyl group is next, and the methyl group is 3.

Because the priority 4 group is not directed toward the rear, we have to reorient the molecule. Using the interchange option, we switch the priority 2 and 4 groups so that the H will point toward the back.

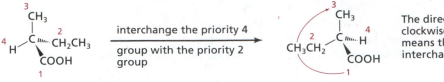

The priority order $1 \rightarrow 2 \rightarrow 3$ is clockwise, but we have switched two groups, so the assignment is (R^I), which means that the original configuration was (S). This molecule is named (S)-2-methylbutanoic acid.

EXAMPLE 4.5

Draw the structural formula of (S)-4-bromo-*cis*-2-pentene.

Interpret the name in the same way you learned in Chapter 1, ignoring the stereochemistry of the chiral carbon atom, at first. The structural formula is shown here:

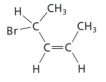

4-Bromo-*cis*-2-pentene

Next, arrange the chiral center (C4) so that the three highest priority groups are disposed in a counterclockwise fashion when the priority 4 group points toward the rear. Notice that you can simply draw the molecule with the hydrogen atom pointing behind the plane, and then add the two remaining groups so that the (S) configuration is obtained.

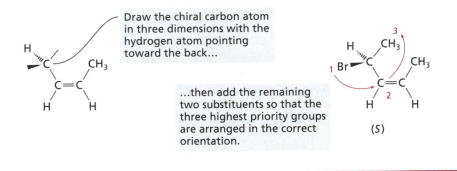

Draw the chiral carbon atom in three dimensions with the hydrogen atom pointing toward the back...

...then add the remaining two substituents so that the three highest priority groups are arranged in the correct orientation.

EXERCISE 4.7

Assign the absolute configuration of the chiral carbon atom in each of the compounds shown below.

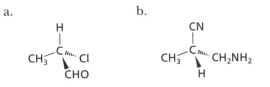

a.   b.

EXERCISE 4.8

Draw the three-dimensional structure for each of the following compounds:

a. (R)-2-Bromopentanoic acid   b. (S)-3-Hexanol   c. (R)-3-Chloro-2-butanone

4.2e   BREAK THE RING TO ASSIGN THE CONFIGURATION OF A STEREOCENTER IN A CYCLIC COMPOUND

Applying the priority rules to a carbon atom in an acyclic compound is straightforward in most cases because it is easy to see that four groups are different. For example, is the three-dimensional carbon atom of the following compound chiral?

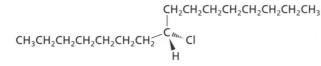

The answer is yes. If asked how you arrived at that conclusion, you would probably say that you looked at the four substituents, counted the number of carbon atoms in the two long chains to find out if they were the same length or not, and concluded that no two groups were the same.

Now consider the molecule shown below in which the carbon atom with the asterisk is thought to be chiral. Does it have four different groups attached?

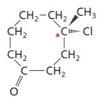

The answer again is yes, but when two of the groups are composed of atoms in a ring, some people have trouble seeing whether four *different* groups are attached. To solve this type of problem, draw a line through the potentially chiral carbon atom that also divides the ring into two equally sized portions. (For a ring with an even number of carbon atoms, this division will bisect the atom in the ring directly opposite the carbon atom being evaluated.) Then look at the four groups and decide if any two are the same. In this example, the division is done in the manner illustrated below, and one of the carbon atoms is split in half:

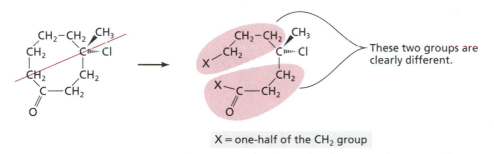

EXAMPLE 4.6

Assign the absolute configuration of the alcohol carbon atom in the structure below.

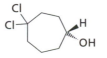

The first task is to determine the identities of the four groups attached to the chiral carbon atom and assign a priority to each. This step is accomplished by drawing a line across the ring from the putative asymmetric carbon atom, splitting the ring. In this case, one of the C–C bonds is broken.

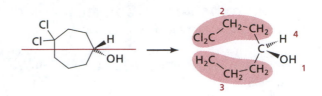

The compound in this example is already oriented so that the priority 4 group is toward the rear, so the counterclockwise sequence of $1 \to 2 \to 3$ leads to the assignment of the (S) configuration. This compound is (S)-4,4-dichlorocycloheptanol.

EXERCISE 4.9

Assign the configuration of the asymmetric carbon atom in each of the compounds shown below.

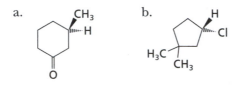

4.2f   ANY TETRAHEDRAL CENTER WITH FOUR DIFFERENT GROUPS CAN BE CHIRAL

The property of chirality, which is important in naturally occurring substances such as carbohydrates and amino acids, is not limited to single carbon atoms. Any atom with four different groups attached to it has the potential to be chiral, as long as the molecule is not superimposable on its mirror image. Nitrogen often has three different groups and an electron pair, so it is chiral. Nitrogen atoms, however, usually undergo inversion too quickly for the two mirror-image isomers to be *isolated*. This process, termed *pyramidal inversion* has been likened to an umbrella being blown inside out on a windy day. Inversion occurs via rehybridization of the nitrogen atom.

On the other hand, quaternary ammonium ions that have four different groups attached to the nitrogen atom can be isolated as enantiomerically pure substances.

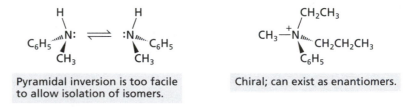

Sulfur and phosphorus compounds, even with electron pairs, can often be isolated in chiral form. These atoms do not undergo pyramidal inversion as rapidly as nitrogen does. Recall that the unshared pair of electrons is the *lowest priority group* (Section 4.1b).

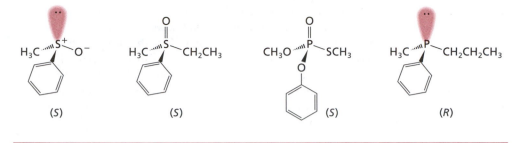

(S)   (S)   (S)   (R)

EXERCISE 4.10

Assign the absolute configuration to each of the chiral atoms in the following compounds. The phenyl group is C_6H_5 and the pentafluorophenyl group is C_6F_5.

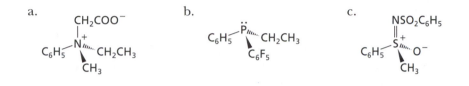

a.   b.   c.

4.3   DIASTEREOMERS

4.3a   MOLECULES WITH TWO OR MORE CHIRAL CENTERS CAN EXIST AS STEREOISOMERS CALLED DIASTEREOMERS

Up to now, you have mainly encountered compounds with a single stereogenic center. Suppose a molecule has two chiral carbon atoms. As an example, consider (2R,3S)-2-bromo-3-methylpentane, shown in Figure 4.6a. If the configuration of both carbon atoms is inverted, the new compound is the enantiomer of the original. Enantiomers of compounds with more than one stereocenter can be identified readily *because the configuration of every chiral center in the molecule changes.*

Now look at the molecule (2R,3R)-2-bromo-3-methylpentane in Figure 4.6c. This compound is *not* the same as the one above it. Nor is it the mirror image of that one. Remember, a molecule only has one mirror image, within which the absolute configuration of every asymmetric carbon is inverted—carbon atoms that are (R) become (S) and those that are (S) become (R). Molecules in which at least one of the chiral carbon

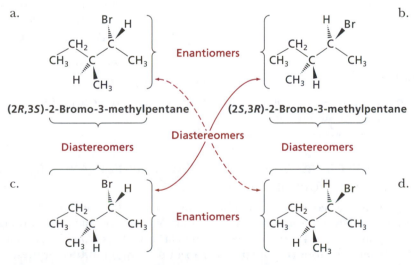

Figure 4.6
The stereoisomers of 2-bromo-3-methylpentane and their stereochemical relationships.

atoms has retained its configuration compared with the other are called **diastereomers** or *diastereoisomers. Diastereomers are stereoisomers that are not enantiomers.*

The maximum number of stereoisomers for a molecule with n asymmetric carbon atoms is 2^n, but there may be fewer, as you will see shortly. 2-Bromo-3-methylpentane has two stereocenters, so it exists as any one of four stereoisomers (all are shown in Fig. 4.6 with their assigned configurations). A compound with three asymmetric carbon atoms can exist as 2^3, or 8, different isomers. Notice that among the many stereoisomers, there exist *pairs* of enantiomers.

EXAMPLE 4.7

Draw all of the stereoisomers for 4-chloro-2-pentanol, and identify the relationships among the stereoisomers.

First, interpret the name by drawing a structural formula without regard to stereochemistry, and then represent the molecule with a perspective drawing (if you place the hydrogen atoms pointing back, it will be easier to assign the configurations initially). Assign the stereochemistry at the two chiral centers.

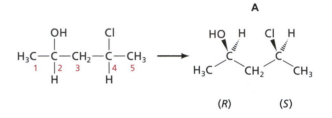

By interchanging two groups on each of the chiral centers, one carbon atom at a time, generate all of the other possible isomers and write their configurations.

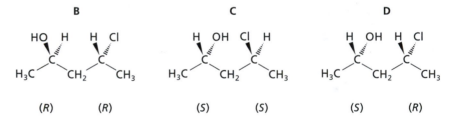

Finally, pick the enantiomeric pairs—those in which every configuration is inverted. In this example, the enantiomers are **A** (*2R, 4S*) and **D** (*2S, 4R*) and **B** (*2R, 4R*) and **C** (*2S, 4S*). The diastereomeric relationships are **A** and **B**, **A** and **C**, **B** and **D**, and **C** and **D**.

EXERCISE 4.11

Draw all of the stereoisomers of 3-bromo-2-hydroxybutanedioic acid. Identify the relationships among the stereoisomers.

4.3b   STEREOISOMERS WITH AN INTERNAL MIRROR PLANE ARE CALLED MESO COMPOUNDS AND ARE NOT CHIRAL

A molecule with n stereocenters can exist as a maximum of 2^n stereoisomers. Consider the compound 2,3-dibromobutane. A maximum of four possible stereoisomers exist for this compound ($2^2 = 4$). By numbering the carbon atoms from the left end of the molecule, we generate four combinations of absolute configurations for these isomers, which are indicated below each structure in Figure 4.7.

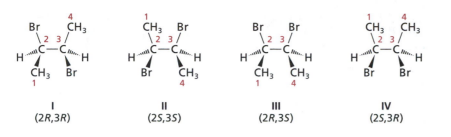

Figure 4.7
Possible isomers of 2,3-dibromobutane.

However, *if we simply number the carbon atoms from the right end,* as illustrated below, the chiral centers in compound **IV** have the same configurations as those in **III**. In fact, compounds **III** and **IV** are superimposable (make a model of each as they appear in Fig. 4.7 and verify that they are superimposable). Why are two of the isomers the same? If you consider different conformations for **III** (or **IV**), you will notice than in the eclipsed conformation (below, b.), *the molecule has an internal mirror plane.*

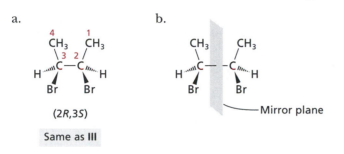

In a sense, this molecule is like a racemic mixture within the same molecule. *One stereocenter is the mirror image of the other.*

Compounds that have such an internal mirror plane in *any* of its possible conformations are called **meso compounds,** and they are not optically active. The overall molecule is achiral, even though it contains chiral carbon atoms (there are four different groups attached to both C2 and C3). The existence of meso compounds decreases the number of possible stereoisomers that can exist for a particular substance. In this instance, we calculated that there should be four stereoisomers, yet only three actually exist. Two of the possible isomers are *identical.*

EXAMPLE 4.8

Draw the isomers of 1,2-cyclopentanediol. Which are meso compounds?

First, interpret the name of the compound by drawing its structural formula (below, at the left); then draw the perspective forms and assign the configuration of each asymmetric carbon atom.

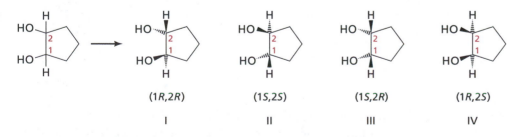

Next, look closely at any isomer that has an equal number of carbon atoms with opposite configurations (*R* and *S*) and determine if that molecule has an internal mirror plane. If yes, then that isomer has the prefix *meso–*. In this example, the isomers shown as (1*S*, 2*R*), **III,** and (1*R*, 2*S*), **IV,** have an internal mirror plane, and they are identical compounds. Thus, 1,2-pentanediol exists as three unique stereoisomers:

I (1*R*,2*R*)-1,2-Cyclopentanediol

II (1*S*,2*S*)-1,2-Cyclopentanediol

III *meso*-1,2-Cyclopentanediol

EXERCISE 4.12

Which of the following compounds have a stereoisomer that is a meso compound?

a. 2,4-Dichloropentane   b. 1,3-Dimethylcyclohexane   c. 2,3-Dibromopentane

---

4.3c   CONFORMATIONAL STEREOISOMERS CAN EXIST WHEN ROTATION ABOUT SINGLE BONDS IS RESTRICTED

The notion that meso compounds have an internal mirror plane in an eclipsed conformation (Section 4.3b) raises the question of conformational effects in molecules that have stereocenters. For example, consider the chair structure of *cis*-1,2-dimethylcyclohexane, which exists as two equivalent forms related by a ring-flip (Fig. 3.18). Notice that the configurations of the stereocenters stay the same during the ring-flip, as expected since no bonds are broken during this conformational change.

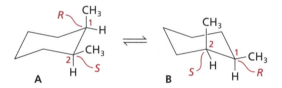

However, conformers **A** and **B** are actually mirror images and they are not superimposable, as you can see by performing the manipulations shown below (if necessary, make models of **A** and **B** to see these relationships).

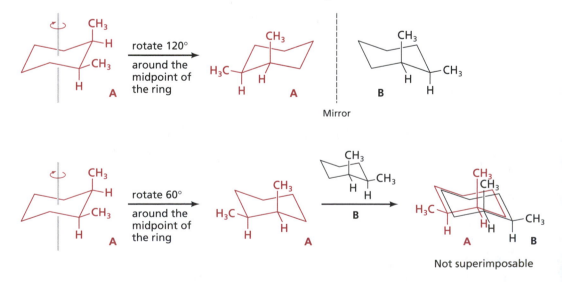

Does this mean that **A** and **B** are enantiomers? If these conformers could be *isolated,* then yes, they would exist as enantiomers. Remember, however, that there is sufficient energy under normal conditions to rapidly interconvert the chair conformers of cyclohexane (Section 3.3b, Table 3.1).

Because of this facile equilibration, the average structure for this molecule is planar—as it is for most cycloalkanes—and the planar form is the one that you examine to evaluate chirality. For *cis*-1,2-dimethylcyclohexane, this planar form has an internal mirror plane, which means that it is meso.

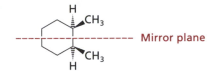

meso-**1,2-Dimethylcyclohexane**

Conformational stereoisomers *can* exist if the rate of interconversion can be slowed sufficiently to permit their isolation. Technically, cooling a solution of certain cyclohexane derivatives can lead to "freezing" of the different chair forms. Separating such mixtures into their stereoisomer forms is usually not practical (or even possible).

One class of molecule that exists as isolable conformational stereoisomers at room temperature comprises derivatives of biphenyl (phenylbenzene) that have large groups at the positions on the rings adjacent to the carbon–carbon single bond as shown below in (a.). Molecules of this type can be isolated as enantiomers, although heating will cause racemization if enough heat is supplied to overcome the energy barrier and bring about rotation. Molecules like the mirror images below (b.), in which there is restricted rotation, are called *atropisomers*.

a.

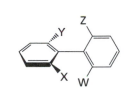

At least three of the four indicated positions have to be substituted with a group larger than H.

b.

Enantiomers of 6,6′-Dimethyl-2′-nitro-2-biphenylcarboxylic acid

4.4   FISCHER PROJECTIONS

Up to now, we have made perspective drawings to represent the three-dimensional structures of organic compounds. A **Fischer projection** presents a chiral molecule as a flat structure, and it is most commonly used to represent the structures of carbohydrates.

A Fischer projection, shown in Figure 4.8a, uses the intersection of two perpendicular lines to define the position of a carbon atom. The substituents are then placed at the ends of the lines. The corresponding perspective form is illustrated in Figure 4.8b. The two substituents on the *horizontal* line of the Fischer projection (a and c) represent groups in *front* of the plane of the paper. The two substituents on the *vertical* line (b and d) lie *behind* this plane.

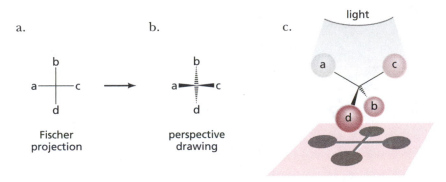

Figure 4.8
Representations of chiral carbon atoms with a Fischer projection (a.), compared with a perspective drawing (b.). Part (c.) is an illustration that shows the genesis of the Fischer projection, which is like the shadow cast by a molecule when lighted from above.

a.

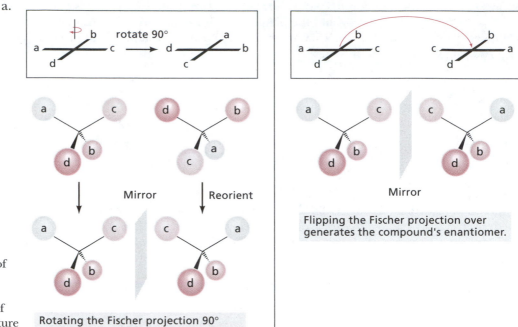

Rotating the Fischer projection 90° generates the compound's enantiomer.

b.

Flipping the Fischer projection over generates the compound's enantiomer.

Figure 4.9

Forbidden manipulations of Fischer projections that generate the compound's enantiomer. (a.) rotation of 90°; (b.) flipping the structure over.

Because the three-dimensional structure of a molecule is represented as a flat form in a Fischer projection, it cannot be moved about in the same way that a perspective drawing can. The structure cannot be rotated by 90° (or −90°) because that manipulation interchanges the bonds coming out of the page with those receding behind the page, which inverts the absolute configuration of the stereocenter. Flipping the structure front-to-back is also forbidden because that movement also interchanges the groups in front with those behind the plane of the page. Both of these operations are illustrated in Figure 4.9, and you may want to make models to confirm these results.

Allowed manipulations of a Fischer projection include rotation in the plane by 180°, and interchanging *three* of the groups, as illustrated in Figure 4.10.

a.

b.

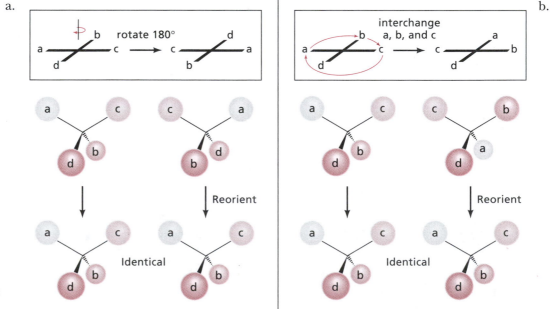

Figure 4.10

Allowed manipulations of Fischer projections. (a.) Rotation of 180°; (b.) interchanging three of the groups.

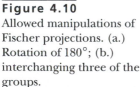

To assign the absolute configuration of a carbon atom in a Fischer projection, you can proceed as before by first assigning priorities to each of the substituents. Because the groups in the vertical positions are pointing back, you need only to move the priority 4 substituent to one of those two positions by interchanging three groups. Example 4.9 illustrates the process by which the configuration of a chiral center is determined from a Fischer projection.

EXAMPLE 4.9

Assign (R) or (S) to the chiral center in the given structure of 3-hydroxybutanal.

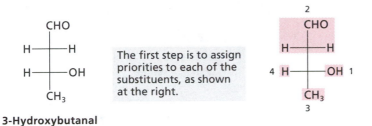

The first step is to assign priorities to each of the substituents, as shown at the right.

3-Hydroxybutanal

Then, interchange three groups so that the lowest priority group is in one of the vertical positions, which is behind the plane of the paper.

Interchange three groups (1, 3, and 4) to place 4 in the vertical position, which is toward the rear.

Now, look at the relationship of groups with priorities 1, 2, and 3. These are positioned in a clockwise relationship, so the configuration is (R). A perspective drawing is included for comparison.

Clockwise = (R)

EXERCISE 4.13

Assign (R) and (S) configurations to the chiral carbon atom(s) in the following compounds shown as their Fischer projections:

a.   b.

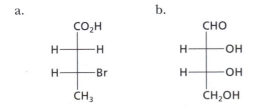

4.5   REACTION STEREOCHEMISTRY: A PREVIEW

Some chemical reactions begin with chiral substances, while in other transformations, only achiral starting materials are used. Chemical reactions can also yield products in which at least one carbon atom is chiral. In those situations, the most important thing

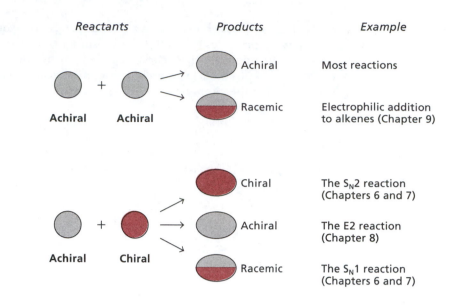

Figure 4.11
A summary preview of
reaction stereochemistry.

to remember is this: When two *achiral* molecules react, they will either generate achiral products or, if the product has chiral carbon atoms, a racemic mixture. (If more than one new stereocenter is produced from achiral reactants, you may obtain several racemic pairs.) *You cannot make one enantiomer of a product preferentially when starting only with achiral molecules.* Figure 4.11 provides a summary preview of reaction stereochemistry, a topic that will be more fully developed in subsequent chapters.

CHAPTER SUMMARY

Section 4.1   Geometric isomers of alkenes

- Stereoisomers are compounds in which the atoms are connected identically but that differ in their overall three-dimensional shapes. Figure 4.12 is a flow chart to assess whether two compounds are conformers, isomers, or identical.

- An alkene with only one hydrogen atom at each end of a double bond exists as isomers, which are designated cis- or trans- [or, alternatively, (*E*) or (*Z*)].

- An alkene that has three or four non-hydrogen substituents attached to the double-bonded carbon atoms may exist as isomers, designated (*E*) or (*Z*).

- Assignment of the (*E*) or (*Z*) configuration depends on the mutual relationship between the higher priority substituent at each end of the double bond. If these groups are on the same side of the bond axis, the alkene carries the (*Z*) designation. If the groups are on opposite sides of the bond axis, the alkene is designated (*E*).

- The priority of substituents is assigned according to the Cahn–Ingold–Prelog convention, which is based on the atomic number of atoms in the substituent.

Section 4.2   Chirality and enantiomers

- Stereoisomers that are nonsuperimposable mirror images are called enantiomers.

- Enantiomers are chiral.

- A carbon atom that has four different groups attached is called an asymmetric, chiral, or stereogenic carbon atom.

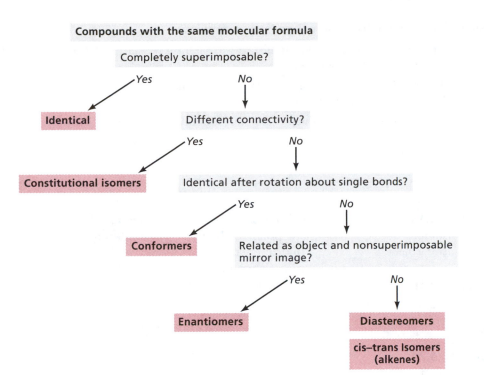

Figure 4.12
A flow chart to assess whether two components are conformers, isomers, or identical.

- The three-dimensional arrangement of groups bonded to an asymmetric carbon atom is called its configuration.
- The absolute configuration of an asymmetric carbon atom is indicated by the letter (*R*) or (*S*,) which is based on the Cahn–Ingold–Prelog priority system (higher atomic number). The process to assign the configuration of a chiral atom is given in Figure 4.4.
- A chiral substance is said to be optically active, which means that it rotates plane-polarized light.
- A 50:50 mixture of enantiomers is called a racemic mixture or racemate. A mixture of enantiomers that is not 50:50 is called a scalemic mixture.
- Any tetrahedral atom with four different groups can be chiral.

Section 4.3   Diastereomers

- Molecules with more than one chiral atom that are neither superimposable nor mirror images are called diasteroisomers or diastereomers.
- A molecule with chiral carbon atoms that has an internal mirror plane in at least one conformation is called a meso compound. Meso compounds do not rotate plane-polarized light.
- Compounds with restricted rotation about a carbon–carbon single bond can be isolated as stereoisomers. Biphenyl derivatives are the most common examples of conformational stereoisomers.

Section 4.4   Fischer projections

- A Fischer projection represents the structure of a substance with chiral centers on a two-dimensional surface with use of intersecting vertical and horizontal lines. At each intersection of a vertical and horizontal line is a carbon atom.
- The substituents attached at the horizontal positions are in front of the plane of the paper; the substituents at the vertical positions recede behind the plane.

Section 4.5   Reaction stereochemistry: a preview

- The products of a reaction that start with achiral reactants and reagents will either be achiral or racemic.
- The products of a reaction that starts with both achiral and chiral reactants may be chiral, achiral, or racemic.
- Figure 4.11 summarizes the possible stereochemical outcome of chemical reactions.

KEY TERMS

Section 4.1a
stereoisomer
geometric isomer

Section 4.1b
Cahn–Ingold–Prelog
    convention

Section 4.2a
enantiomer
chiral
achiral
asymmetric carbon atom
stereogenic carbon atom
chiral carbon atom
configuration

Section 4.2b
plane-polarized light
    optically active
    racemic mixture
racemate
dextrorotatory
levorotatory

Section 4.3a
diastereomer

Section 4.3b
meso compound

Section 4.4
Fischer projection

ADDITIONAL EXERCISES

4.14. Each of the following names is flawed because of a stereochemical descriptor that is either missing or incorrect. Provide a corrected name for each substance and draw its structure.

 a. *trans*-2-Chloro-3-methyl-2-pentene   b. 3,4-Dimethylcyclohexene

 c. 3-Methylcyclopentene   d. 3-Hexene

4.15. Draw all of the geometric isomers and assign complete names using the (*E*) and (*Z*) descriptors, where appropriate, for each of the following alkenes:

 a. 1,3-Pentadiene   b. 3-Methyl-2,4-hexadiene

 c. 2,3-Dimethyl-2-butene   d. 3-Ethyl-4-octene

4.16. Draw a structure for each of the following compounds:

 a. (*Z*)-3-Bromo-2-hexene   b. (3*Z*,6*E*)-1,3,6-Octatriene

 c. (*Z*)-1-Methylcyclononene   d. (*E*)-2-Methoxy-2-pentene

4.17. Label each of the following as (*Z*), (*E*), or neither.

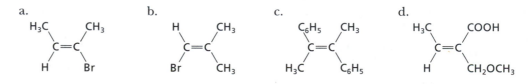

4.18. Using the scheme summarized in Figure 4.12, classify each of the following pairs of substances as identical, conformer, or isomers. If isomers, specify which type.

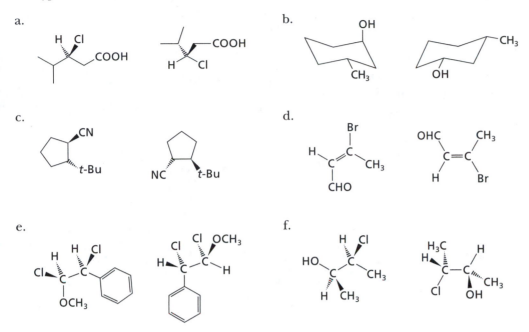

4.19. Arrange each of the following sets of substituents in decreasing order of priority according to the Cahn–Ingold–Prelog convention.

  a. –Cl, –H, –NH$_2$, –OCH$_3$

  b. –CN, –OCH$_3$, –COOH, –CH$_2$CH$_3$

  c. –OH, –CH$_2$CH$_3$, –CHO, –Br

  d. –CH$_3$, –C≡CH, –CN, –COOCH$_3$

4.20. Draw and name any geometric isomers that can exist for the following compounds:

  a. 2-Heptene   b. 1-Chlorocyclobutene   c. 1,3-Butadiene

  d. 3,4-Dimethyl-3-hexene   e. 3-Phenyl-2-pentene

4.21. Label each of the following as (Z), (E), or neither:

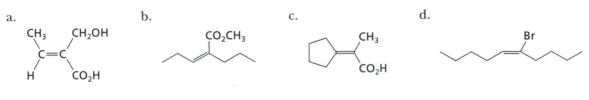

4.22. Which of the following objects are chiral?

  a. A golf club   b. A pair of scissors   c. A baseball glove

  d. A corkscrew   e. A telephone   f. A pencil

  g. A basketball   h. A hammer   i. A spiral staircase

4.23. Aldrich Chemical Co. sells tiglic aldehyde as *trans*-2-methyl-2-butenal. Provide a correct IUPAC name for this substance, the structure of which is shown below.

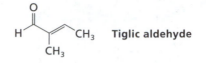

4.24. A key intermediate in the metabolism of many carboxylic acids is commonly referred to in biochemistry texts as "*cis*-aconitate". What is the correct name of this substance with regards to its stereochemistry?

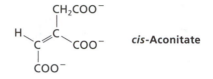

4.25. For each of the following compounds, assign (*R*) or (*S*) to each stereogenic center:

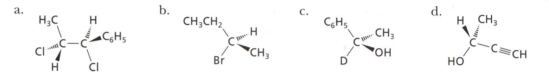

a.   b.   c.   d.

4.26. Draw structures for each of the following compounds:

  a. (3*S*,4*S*)-4-Methyl-3-hexanol

  b. An optically active isomer of 1,2-dimethylcyclopentane

  c. (1*R*, 3*S*)-3-Methylcyclohexanol

  d. The meso isomer of 1,3-dichlorocyclopentane

4.27. Draw all of the stereoisomers that can exist for each of the following compounds. Identify the relationships between the stereoisomers (enantiomers or diastereomers).

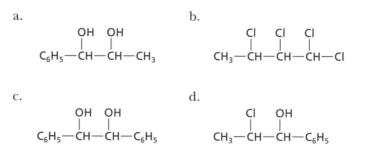

a.   b.

c.   d.

4.28. For each of the following compounds, assign (*R*) or (*S*) to each stereogenic center:

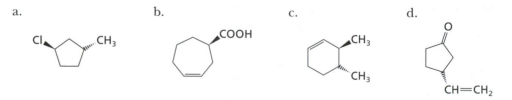

a.   b.   c.   d.

4.29. In Exercise 1.28, you were asked to draw every alcohol with the molecular formula $C_5H_{12}O$. Which of these can exist as stereoisomers? For each stereoisomer, assign (*R*) or (*S*) to each stereogenic center.

4.30. Convert the following perspective formulas to Fischer projections. By convention, the carboxylic acid or aldehyde group is placed at the top of the structure and the longest carbon chain runs vertically.

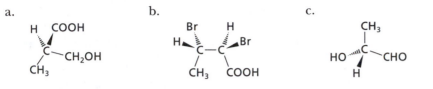

a.   b.   c.

4.31. Draw all of the stereoisomers of 2,3,4-trichloropentane.

a. Label each structure as chiral or meso.

b. In each structure, assign the absolute configuration of C2 and C4 as (*R*) or (*S*).

c. Is C3 chiral in any of these stereoisomers? Explain.

4.32. Convert the following Fischer projections to perspective formulas:

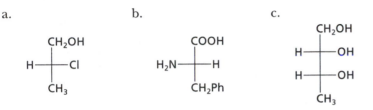

a.   b.   c.

4.33. Draw all of the stereoisomers that exist for 2,4-dimethylcyclobutanecarboxylic acid. Identify the meso compounds, if present, and assign the absolute configuration of each chiral center.

4.34. For each of the following structures, label each chiral carbon atom as (*R*) or (*S*) and identify any meso compounds:

a.   b.   c.

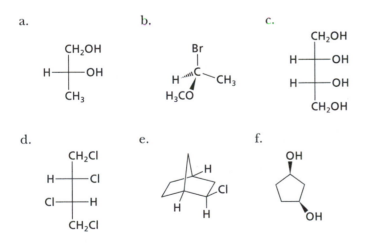

d.   e.   f.

4.35. Draw all of the stereoisomers that can exist for each of the following compounds. Identify the relationships between the stereoisomers (enantiomers or diastereomers).

a.

b.

c.

CH_3—$CHBr$—$CHBr$—CH_3 C_5H_6—$CHOH$—CH_2—$CHOH$—C_6H_5

$\underset{\underset{OH}{|}}{\overset{\overset{OH}{|}}{CH_3CH_2CHCHCH_2CH_3}}$

4.36. For the following bioactive compounds, assign (*R*) or (*S*) to each stereogenic center:

a. Brefeldin A, an antiviral agent

b. Imipenim, an antibacterial

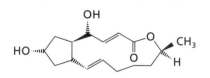

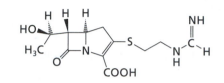

CHAPTER

5

CHEMICAL REACTIONS AND MECHANISMS

5.1 GENERAL ASPECTS OF REACTIONS

5.2 ACID–BASE REACTIONS

5.3 REACTION MECHANISMS

5.4 REACTION COORDINATE DIAGRAMS

5.5 REACTIONS IN BIOCHEMICAL SYSTEMS

CHAPTER SUMMARY

Previous chapters have acquainted you with the molecular structures of carbon-containing compounds, yet much of organic chemistry is devoted to the study of their chemical reactions. This chapter provides an overview of the types of reactions that you will encounter in this course, and its purpose is to familiarize you with the formalisms needed to describe the molecular interactions that convert reactants into products.

The cornerstone of chemical reactivity is the acid–base reaction. The importance of this seemingly simple process would be difficult to overstate. By understanding the nature and reactions of acids and bases, you will begin to organize vast amounts of information needed to make predictions about reactions that you have never seen before.

An important perspective on chemical reactions is also gained by considering the energies of reactants, intermediates, and products. This viewpoint confers an understanding about whether a reaction will proceed and at what rate it will go. The topic of energy in chemical reactions leads naturally to the subject of catalysis, a concept with relevance to industrial as well as biochemical processes. An overview of chemical reactions that occur in biological systems, particularly those involved in metabolism, rounds out this chapter. One goal of this text is to relate reactions done in the laboratory with those that occur in living systems.

5.1 GENERAL ASPECTS OF REACTIONS

5.1a A REACTION IS PRESENTED AS AN EQUATION THAT DEPICTS THE STRUCTURES OF THE REACTANTS AND PRODUCTS

In presenting reactions, we write an equation showing **reactants** on the left side going to **products** on the right side, with a horizontal arrow indicating the direction in which the reaction proceeds. Many organic reactions are equilibria, so *they are reversible,* a condition shown by a double arrow: ⇌. A species that accelerates a reaction but is not changed at the conclusion of the reaction is a **catalyst.** Of the following equations, the third equation illustrates a reaction that is catalytic with respect to aluminum chloride.

137

Reactant(s)   Product(s)

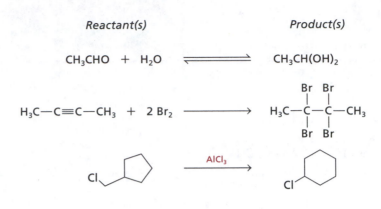

When we want to focus attention on a particular organic reactant, we often write the equation with some of the reactants or **reagents** on the arrow. Reagents are reactants of secondary importance—they can either be organic or inorganic compounds, which contain elements other than carbon. A reactant written on the arrow can still show up in the product, or it may be a catalyst. When the reagents are written on the arrow, the equation is NOT necessarily balanced except for the principal reactant and product.

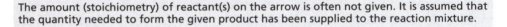

The amount (stoichiometry) of reactant(s) on the arrow is often not given. It is assumed that the quantity needed to form the given product has been supplied to the reaction mixture.

Solvent molecules, temperature, and other aspects of the reaction conditions are also included with the arrow. For example, if ultraviolet light (UV) is used, the notation hv is written. If the reaction is heated, the symbol Δ is placed with the arrow.

hv The reaction is irradiated with UV or visible (vis) light.

Δ The reaction is heated.

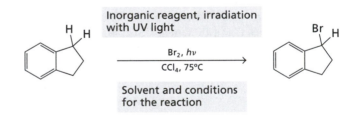

When sequential operations are carried out with a given compound, common practice is to *list* the reagents with the arrow rather than to show each reaction separately. For example, the following equation shows that two independent steps are being done. As is often the case in sequential operations, the reagents are not compatible with each other, so they cannot be mixed in a single reaction flask.

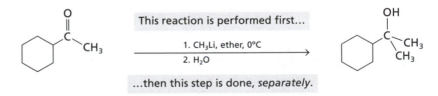

The meaning of the numerals associated with the reagents is that you perform Step 1, and then wait until that reaction is complete. Next, Step 2 is carried out and allowed to proceed until that reaction is done. The use of numbered steps can also mean that the product of the first step is actually isolated and purified before it is subjected to the next reaction. If no numbers are used with the reagents, you can assume that all of the reagents are added to the reaction flask at the same time. *It is important to learn whether a transformation requires separate steps or if a combination of reagents is used.*

Another piece of information that is often included with an equation is the yield of product, which is expressed as a percentage listed at the far right side. This number is simply the number of moles of product actually obtained, divided by the number of moles that can theoretically be produced, multiplied by 100.

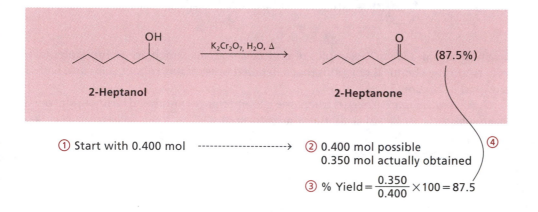

① Start with 0.400 mol - - - - - - - - - - - - - - - →   ② 0.400 mol possible
                                                              0.350 mol actually obtained   ④

③ % Yield $= \dfrac{0.350}{0.400} \times 100 = 87.5$

The percent yield is supposed to give an indication about how well a reaction works, but yields are variable and depend on the skill of the investigator as well as on the inherent efficiency of the transformation shown. Reported yields of product greater than 70 or 80% usually mean that the reaction works well and creates few side products. When reported yields are lower, you will probably have to be concerned with separation of unwanted by-products. When a reaction does not work (i.e., the percent yield = 0%), the letters "N.R.," meaning *No Reaction*, are written in place of the product structure. A "quantitative yield" means 100%. Ideally, reaction yields should be reported as a range, which more accurately indicates the reaction's efficiency.

Finally about reactions, you may encounter transformations for which the actual reagents are not listed. Instead, the name of a particular reaction is given. For example, the Swern reaction (or Swern oxidation), illustrated below, proceeds in two steps using a mixture of oxalyl chloride ($ClCOCOCl$) and dimethyl sulfoxide (CH_3SOCH_3 or DMSO) at low temperature, followed by treatment with triethylamine (Section 11.46b).

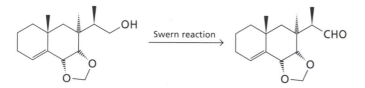

Rather than list the reagents, a chemist will sometimes provide only the name of the transformation, and other chemists generally know what is meant. There is an entire lore of *name reactions* in organic chemistry, and you will undoubtedly learn many of the common ones.

5.1b REACTIONS CAN BE CLASSIFIED BY SEVEN BASIC TYPES

Chemists generally agree about four characteristic types of reactions that organic molecules undergo: **proton-transfer, substitution, addition,** and **elimination. Rearrangements** are sometimes put into their own category, as are **oxidation** and **reduction** reactions. Certainly, all organic reactions can be classified as one of these seven types, which is what we do in this text.

The proton-transfer reaction occurs when H^+ is either lost or gained by a molecule.

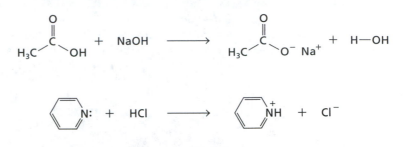

Reactions of this type are examples of *acid–base reactions,* which are so fundamental to understanding chemical reactions that a detailed presentation about acids and bases is provided in Section 5.2.

A *substitution reaction* occurs when one group replaces another. During this process, a sigma bond is broken and another one is formed.

Substitution of OH by Br

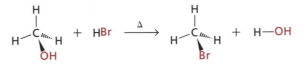

Substitution of H by Br

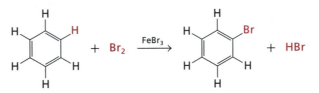

EXERCISE 5.1

Given that the preceding equation is balanced, what is the function of $FeBr_3$?

An *addition reaction* occurs when a substance adds to a π bond, whether that bond is between two carbon atoms or between a carbon atom and a heteroatom.

Addition to a C=C bond

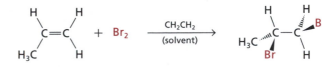

Addition to a C=O bond

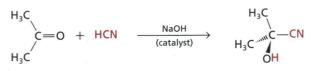

A third characteristic process for organic compounds is the *elimination reaction,* which occurs in the reverse sense of an addition reaction, creating a new π bond by removing atoms or groups from an organic molecule:

Elimination of H and Cl

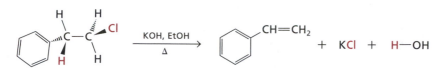

Like addition and elimination, *oxidation* and *reduction* reactions of organic compounds are also the reverse of each other. If you have studied chemistry before, you probably learned that a "redox reaction" is one in which one reactant is oxidized while another one is reduced. In organic chemistry, the same holds true, but the oxidizing or reducing agent is very often an inorganic substance, and we classify a process as an oxidation or reduction *based on what happens to the organic reactant. Oxidation* occurs when hydrogen atoms are removed from or heteroatoms are incorporated into an organic molecule. *Reduction* takes place when hydrogen atoms are added to or heteroatoms are removed from an organic molecule.

Oxidation of an alcohol to form an aldehyde

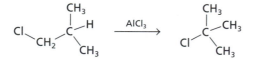

Reduction of an alkene to form an alkane

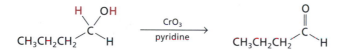

A *rearrangement,* as the name implies, occurs when the carbon skeleton of the original molecule changes to another.

EXAMPLE 5.1

Classify the following reaction by one or more of the designations just described:

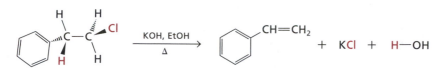

This transformation of an alcohol to a ketone creates a C=O π bond by removing two hydrogen atoms; therefore, this is an elimination reaction. We can also classify the transformation as an oxidation reaction because two hydrogen atoms are removed from the starting compound.

EXERCISE 5.2

Classify each of the following reactions by one or more of the designations just described:

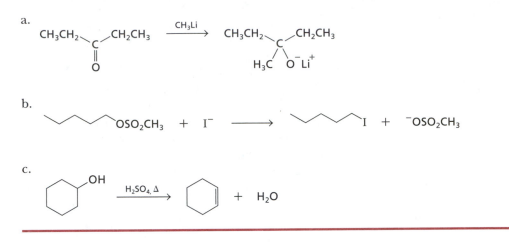

a.

b.

c.

5.1c   BONDS CAN BE BROKEN AND FORMED BY POLAR, RADICAL, AND PERICYCLIC PROCESSES

Besides the classification scheme that makes use of reaction types, organic transformations can be categorized by the manner in which bonds are broken or made. In an *ionic* or **polar reaction,** which is by far the most common type, the bond to be broken is polarized so that one atom assumes a positive, or partial positive, charge and the other assumes a negative, or partial negative, charge. These bond polarities result from the electronegativity or polarizability of atoms that constitute the molecule (Section 2.2a). In a polar reaction, bonds undergo **heterolysis,** meaning that both electrons in the bond stay with *one* of the two atoms that are initially bonded together, so the bond breaks in an unsymmetric fashion.

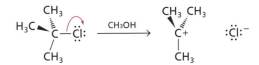

In a heterolytic process, the two electrons in the bond go with one of the atoms, leaving a species with a positive charge. In this case, a carbocation is formed.

When the carbon atom wholly loses its electrons to a more electronegative element, it forms a reactive intermediate called a *carbocation* (Section 2.5e), which is the normal course for heterolysis of a carbon–heteroatom bond. *Carbocations are normally only short-lived intermediates in a chemical reaction.* If the bond between carbon and a more *electropositive* element (usually H) undergoes heterolysis, then the carbon atom retains the electron pair and acquires a negative charge, forming a *carbanion.*

In **radical reactions** bonds undergo **homolysis,** meaning that one electron in a bond moves to one atom, and the other electron in the same bond moves to the other atom. Bonds are broken (or made) in a symmetric fashion, and odd-electron species are formed. Radical reactions will be discussed in Chapter 12.

In a **pericyclic reaction,** electron movement occurs in a concerted process, which means that all of the bonds are made and broken at the same time. No intermediate radicals, carbocations, or carbanions are formed. Pericyclic reactions will be described in Section 10.4a.

5.2   ACID–BASE REACTIONS

5.2a   ACIDS ARE PROTON DONORS

Polar reactions, which are the most common processes in organic and biological chemistry, often begin with the reaction between an acid and a base. According to the **Brønsted–Lowry** definition, an acid is a substance that donates a proton, and a base is a substance that accepts a proton.

In a general way, we write an acid–base reaction as follows:

$$\text{HA} + \text{B} \rightleftharpoons \text{A}^- + \text{HB}^+ \qquad (5.1)$$

Acid   *Base*   *Conjugate base of HA*   *Conjugate acid of B*

In Eq. 5.1, HA is an acid, B is a base, A$^-$ is called the **conjugate base** of *acid* HA, and HB$^+$ is called the **conjugate acid** of *base* B. As described in Section 5.2b, *an acid–base equilibrium lies in the direction of the weaker acid,* so it is important to know the identity of a base's conjugate acid and the acid's conjugate base.

For an acid–base reaction written in the form of Eq. 5.1, an acid and a base appear on each side of the arrows. The equilibrium constant, K_{eq}, for Reaction 5.1 is defined as follows:

$$K_{eq} = \frac{[\text{A}^-][\text{HB}^+]}{[\text{HA}][\text{B}]} \qquad (5.2)$$

When water is the solvent and plays the role of base, its concentration is assumed to be constant. Equation 5.1 simplifies to the following:

$$\text{HA}_{(aq)} + \text{H}_2\text{O} \rightleftharpoons \text{A}^-_{(aq)} + \text{H}_3\text{O}^+_{(aq)} \qquad (5.3)$$

We can now define the quantities K_a and pK_a:

$$K_a = \frac{[\text{A}^-_{(aq)}][\text{H}_3\text{O}^+_{(aq)}]}{[\text{HA}_{(aq)}]} \qquad (5.4)$$

$$pK_a = -\log K_a \qquad (5.5)$$

The **pK_a value** is a measure of how strong an acid is. It is a convenient quantity because values are normally positive, and we do not have to use exponents. Realize that in aqueous solution, any substance more acidic than H_3O^+ transfers its proton completely to the water to produce the hydronium ion, and any species more basic than OH$^-$ removes a proton from water to generate the hydroxide ion. Therefore, measurements of pK_a values below about −2 and above 16 have to be done in a solvent other than water. For our purposes, we make the assumption (which is convenient but not always accurate) that pK_a values measured in other solvents are equal to those in water.

Two essential facts about acid strength are,

1. The stronger the acid, the lower its pK_a value.
2. The conjugate base of a strong acid is a weak base, and the conjugate base of a weak acid is a strong base.

Because it is necessary to know pK_a values in order to judge the strength of an acid or its conjugate base—and to make these data easily accessible—a table of pK_a values is printed on the inside of the front cover of this book. A more manageable compilation is given in Table 5.1, in which pK_a values have been rounded to multiples of 5, which makes them easier for many people to remember. While certain applications require that you know pK_a values more accurately than the values in this table, the majority of acid–base reactions in this text are readily grasped if you know these approximate numbers.

Table 5.1   The pK_a values for types of compounds frequently encountered in organic chemistry.

pK_a	Compound types
−10	Mineral acids (H_2SO_4, HI, HBr, HCl); sulfonic acids (RSO_3H)
0	H_3O^+, H_3PO_4
5	Carboxylic acids ($RCOOH$), HF, aromatic thiols ($ArSH$), HN_3
10	Weak inorganic acids (H_2S, HCN, NH_4^+), amine salts (RNH_3^+), phenols ($ArOH$), thiols (RSH), aromatic amides ($ArCONH_2$)
15	H_2O, alcohols (ROH), amides ($RCONH_2$)
20	Ketones (RCH_2COR')
25	Esters (RCH_2COOR'), alkynes ($RC{\equiv}CH$), nitriles (RCH_2CN)
30	Anilines ($ArNH_2$)
40	Ammonia (NH_3), amines (RNH_2), benzylic C–H ($ArCH_2X$)
45	Arenes (ArH) and alkenes ($RCH{=}CH_2$)
50	Alkanes (RH)

5.2b   ACID–BASE REACTIONS ARE EQUILIBRIA THAT FAVOR FORMATION OF THE WEAKER ACID

An acid–base reaction occurs so rapidly that its equilibrium is affected only by the relative energies of the reactants and products. Let us reconsider Eq. 5.1, in which acid HA reacts with base B to generate their conjugate pairs; for an acid–base reaction, there is an acid and base on each side of the arrows.

$$\underset{\text{Acid}}{HA} \; + \; \underset{\text{Base}}{B} \; \rightleftharpoons \; \underset{\substack{\text{Conjugate} \\ \text{base of HA}}}{A^-} \; + \; \underset{\substack{\text{Conjugate} \\ \text{acid of B}}}{HB^+} \tag{5.1}$$

The equilibrium constant, K_{eq}, for this reaction is given by Eq. 5.2. We can separate the overall process (5.1) into distinct steps as follows:

$$HA \rightleftharpoons A^- + H^+ \tag{5.6}$$

$$B + H^+ \rightleftharpoons HB^+ \tag{5.7}$$

The equilibrium constants for Reactions 5.6 and 5.7 are:

$$K_{eq}(HA) = \frac{[A^-][H^+]}{[HA]} \quad \text{which is simply } K_a \text{ for HA} \tag{5.8}$$

and

$$K_{eq}(HB^+) = \frac{[HB^+]}{[H^+][B]} \quad \text{which is } \frac{1}{K_a} \text{ for HB}^+ \tag{5.9}$$

Thus, for the acid–base Reaction 5.1,

$$K_{eq} = \{K_{eq}(HA)\}\{K_{eq}(HB^+)\} = \frac{K_a \text{ for HA}}{K_a \text{ for HB}^+} \tag{5.10}$$

If HA is a stronger acid than HB^+, then $K_{a(HA)} > K_{a(HB+)}$, so the equilibrium constant K_{eq} for the overall reaction will have a value > 1. If you learned about equilibrium constants before, you will recall that $K_{eq} > 1$ means that a reaction will proceed from left to right as written. In this example, this means that it will proceed *toward the weaker acid.* But if HA is the weaker acid, then solving Eq. 5.10 will give $K_{eq} < 1$. Thus, Reaction 5.1 will proceed from right to left, which is *also toward the weaker acid.* The conclusion is simple: *Acid–base reactions proceed to generate the weaker acid.*

A specific example will serve to illustrate how to predict in which direction an acid–base equilibrium lies.

EXAMPLE 5.2

In which direction does the equilibrium lie for the reaction of acetic acid and cyanide ion, producing acetate ion and hydrogen cyanide?

$$CH_3COOH + CN^- \rightleftharpoons CH_3COO^- + HCN$$

To predict the position of the equilibrium, we first identify which species are the acids and which are the bases. In this example, one reactant is acetic acid, a neutral compound with a proton attached to oxygen, a common structural feature of an acidic substance. The other reactant, cyanide ion, has a negative charge, which is often an important feature of a base. When they react, an acid and a base will produce the corresponding *conjugate base* and *conjugate acid,* so an acid and base exist on each side of the equation. These species should differ only by the presence or absence of a proton.

CH_3COOH	+	CN^-	$\rightleftharpoons$	CH_3COO^-	+	HCN
Acid		Base		Conjugate base		Conjugate acid

Next, we look up pK_a values for each of the two acids in Table 5.1. Acetic acid has a pK_a value of ~ 5 ($pK_a = 4.75$) and HCN has a pK_a value of ~ 10 ($pK_a = 9.2$). Knowing that an acid–base equilibrium lies in the direction of the weaker acid (larger numeric value), we expect this equilibrium to lie toward the right.

EXERCISE 5.3

Draw the structure for the conjugate base of the following acids:

a. CH_3CH_2COOH b. HN_3 c. $(CH_3)_3CSH$ d. C_6H_5OH e. NH_4^+

EXERCISE 5.4

In which direction does the equilibrium lie for each of the following reactions? Make use of the data in Table 5.1 or in the table on the inside front cover of the book.

a.

$$CH_3CO_2H + Cl^- \rightleftharpoons CH_3CO_2^- + HCl$$

b.

$$CH_3OH + CN^- \rightleftharpoons CH_3O^- + HCN$$

c.

d.

$$NH_4^+ + CH_3CH_2S^- \rightleftharpoons NH_3 + CH_3CH_2SH$$

5.2c   THE ACIDITY OF A PROTON IS INFLUENCED BY THE LOCAL ENVIRONMENT

Many factors influence the strength of an acid, but these effects ultimately reflect the stability of its conjugate base. *The more stable the conjugate base, the less attraction it has for a proton, so the stronger the acid.* A base's stability depends on *solvation energy,* the energy needed to dissolve and keep the base soluble. Solvation energy in turn is influenced by three main structural and electronic features of the base: size, electronegativity, and electron delocalization (resonance).

- *Size.* If you consider the pK_a values of the hydrogen halides, which decrease from HF to HI, you will see that this trend parallels the increasing radii of the halide ions, which are the conjugate bases of these acids. (Recall from general chemistry that atomic radii increase from top to bottom within a group of the periodic table.) As the radius of an atom increases, so does its volume, which means that the negative charge is not as concentrated in a large ion as in a smaller one. Because its volume is greater and its charge is more highly dispersed, a larger conjugate base can interact with solvent molecules to a greater degree and gain stability.

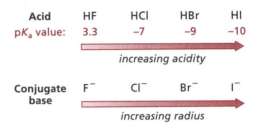

Acid	HF	HCl	HBr	HI
pK_a value:	3.3	–7	–9	–10

increasing acidity

Conjugate base	F⁻	Cl⁻	Br⁻	I⁻

increasing radius

- *Electronegativity.* In any period of elements, atomic size decreases from left to right, but these changes in atomic radii are much smaller than for those within a group. Solvation energies are therefore not affected much by size differences within a period. On the other hand, electronegativity values *increase* from left to right across a period. The greater an atom's electronegativity, the more strongly electrons are attracted to its nucleus, and the more stable it will be as an anion.

Acid	CH_4	NH_3	H_2O	HF
pK_a value:	48	38	15.7	3.3

increasing acidity

Conjugate base	CH_3^-	NH_2^-	OH^-	F^-

increasing electronegativity of the non–hydrogen atom

For a given atom—carbon, in the following example—electronegativity increases with increasing *s* character of hybrid orbitals. Electrons in an *s* orbital have a higher probability of being nearer the nucleus, so they are more strongly attracted to the nucleus. Electrons in an *sp* orbital are therefore closer to the nucleus, on average, than those in an sp^2 or sp^3 hybrid orbital because of the greater *s* orbital contribution. With its higher electronegativity, an *sp*-hybridized carbon atom bears a negative charge better, so its conjugate base derivative is more stable.

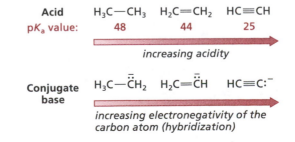

Acid	$H_3C—CH_3$	$H_2C=CH_2$	$HC≡CH$
pK_a value:	48	44	25

increasing acidity

Conjugate base	$H_3C—\ddot{C}H_2$	$H_2C=\ddot{C}H$	$HC≡C:^-$

increasing electronegativity of the carbon atom (hybridization)

- *Electron delocalization.* When electrons can be delocalized over more than one atom, the influences are classified as **resonance effects.** Resonance stabilization provides a simple way to rationalize why acetic acid is more acidic than ethanol. The negative charge of the anion generated by loss of the proton can be delocalized to the carbonyl oxygen atom as well, which disperses the charge and makes the conjugate base more stable.

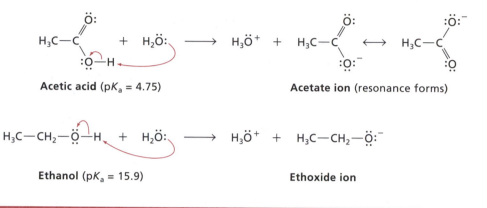

Acetic acid (pK_a = 4.75)   **Acetate ion** (resonance forms)

Ethanol (pK_a = 15.9)   **Ethoxide ion**

EXAMPLE 5.3

Circle the most acidic proton in each compound and indicate which is a stronger acid.

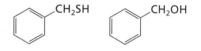

First, expand each line structure to reveal every proton type. Each of these substances has five aromatic protons; two benzylic, aliphatic protons; and a proton attached to a heteroatom. In most cases, protons attached to heteroatoms are more acidic than those attached to carbon. Second, assign an approximate pK_a value to each proton type. These data are found in Table 5.1 or in the table on the inside front cover of this text.

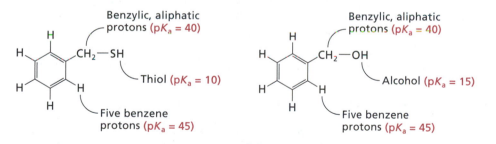

The proton with the lowest pK_a value is the most acidic, so we circle the SH and OH protons in the two compounds, respectively. Of the pair of compounds, the thiol is the more acidic.

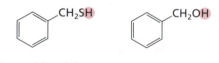

More acidic of the pair

EXERCISE 5.5

For each pair of compounds, circle the most acidic proton in each molecule, and then indicate which is the stronger acid of the two.

a.   b.   c.

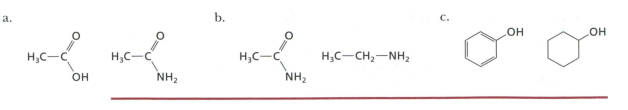

5.2d ACID STRENGTH IS INFLUENCED BY INDUCTIVE EFFECTS

Carboxylic acids are the strongest acids among compounds that contain only C, H, and O. Acetic acid provides the reference mark for a typical carboxylic acid, and it has a pK_a value of 4.75. The pK_a value of chloroacetic acid is less than that for acetic acid.

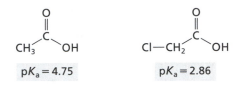

The variation in acidity among structurally similar compounds like these can be explained by differences in the electronegativity values of the *substituents*. These electronegativity differences manifest themselves by donating or withdrawing electrons through the bonds between atoms, an influence known as an **inductive effect.** In the comparison of acetic and chloracetic acids, the argument goes like this: chlorine, being electronegative, renders the carbon atom adjacent to the carbonyl group partially positive.

In turn, this withdraws electron density from the carbonyl carbon atom and the oxygen atom bearing the proton. The effect is to weaken the O–H bond, making the proton more acidic.

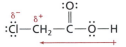

Thus, this inductive effect of the chlorine atom makes chloroacetic acid more acidic than acetic acid. In general, an electron-withdrawing substituent near the COOH group increases the acidity of acetic acid. The more electronegative a substituent, the stronger the acid. Conversely, an electron-donating substituent makes the acid less acidic than acetic acid. Alkyl groups are the most common substituents that donate electrons (Section 6.2b). Table 5.2 lists some derivatives of acetic acid and their corre-

Table 5.2   The pK_a values of acetic acid and derivatives.

Carboxylic acid	pK_a	Carboxylic acid	pK_a	Carboxylic acid	pK_a
CH_3—COOH	4.75	Cl—CH_2—COOH	2.85	Br—CH_2—COOH	2.90
F—CH_2—COOH	2.59	Cl_2CH—COOH	1.48	I—CH_2—COOH	3.12
CH_3—CH_2—COOH	4.87	Cl_3C—COOH	0.70		

Table 5.3   The pK_a values of butanoic acid and its derivatives.

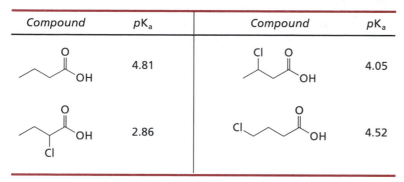

Compound	pK$_a$	Compound	pK$_a$
	4.81		4.05
	2.86		4.52

sponding pK_a values, illustrating the effect of substituents on the carbon atom adjacent to the carboxylic acid group.

As you might expect, more than one electron-withdrawing group increases the magnitude of a substituent's effect, so dichloroacetic acid is stronger than chloroacetic acid, and trichloroacetic acid is the strongest acid of this series.

Inductive effects depend on the distance between atoms, too. As a chlorine atom is moved farther from the carboxylic acid group, its electron-withdrawing effect diminishes rapidly. The series of butanoic acids that have a chlorine atom attached to C2, C3, and C4 shows this influence, and the corresponding pK_a values are listed in Table 5.3.

EXAMPLE 5.4

Among the following compounds, circle the strongest acid and put a rectangular box around the weakest acid:

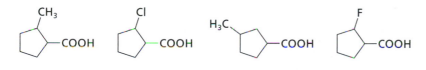

To begin, chose a reference compound with the same carbon skeleton as the compounds to be evaluated: cyclopentanecarboxylic acid has no substituent, so it is the logical choice.

Alkyl groups are electron donating, so derivatives with a methyl group will be less acidic than the reference compound. The 2-methyl compound places the methyl group closer to the carboxylic acid group than the 3-methyl derivative does, so its inductive effect will be greater. The chlorine and fluorine atoms are electron withdrawing, so the 2-chloro and 2-fluoro compounds will be stronger than the reference compound. Fluorine is more electronegative than chlorine, so its inductive influence will be greater.

The solution is therefore as follows:

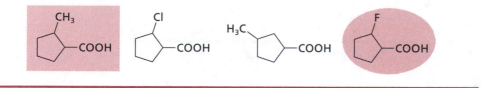

EXERCISE 5.6

For each set of compounds circle the strongest acid and put a rectangle around the weakest.

a.

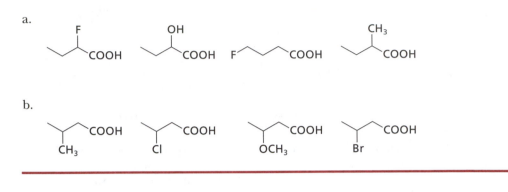

b.

5.2e   BASES OFTEN CONTAIN A NITROGEN ATOM

A base is a substance that accepts a proton, according to the Brønsted–Lowry definition, and several common ones are listed in Table 5.4, along with the corresponding conjugate acids and their associated pK_a values. Remember that the conjugate base of a strong acid is a weak base, and the conjugate base of a weak acid is a strong base. If you know the pK_a value for an acid, then you also know the relative strength of its conjugate base.

Amines are derivatives of ammonia, and they are considered organic bases. As such, they are relatively weak; but they are readily soluble in organic solvents, so they find widespread use. When a strong base is required with which to carry out a reaction, then sodium hydroxide, NaOH; sodium methoxide, Na^+ CH_3O^- or ethoxide, Na^+ $CH_3CH_2O^-$; potassium *tert*-butoxide, K^+ $(CH_3)_3CO^-$, sodium amide, Na^+ NH_2^-, or lithium diisopropylamide, LDA, Li^+ $[CH(CH_3)_2]_2N^-$ is chosen as a reagent.

Table 5.4   Substances commonly used as bases in organic reactions

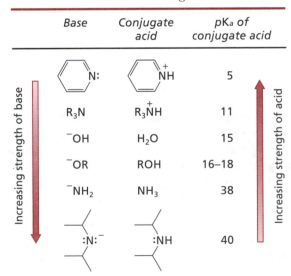

5.2f   THE BASICITY OF AN AMINE IS INFLUENCED BY INDUCTIVE AND RESONANCE EFFECTS

As noted in Section 1.4a, we classify amines according to the number of carbon atoms attached to the nitrogen atom of ammonia.

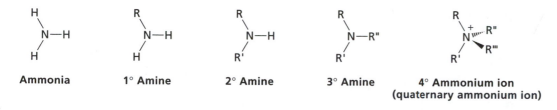

Simple aliphatic amines are named by adding the name of the alkyl group, with a prefix, if needed, to the suffix amine. Primary amines can also be named with the prefix amino and the alkane or cycloalkane root. Cyclic amines often have common names; in the IUPAC system, the prefix aza along with the cycloalkane root indicates that one of the carbon atoms (and attached hydrogen atoms) has been replaced by a nitrogen atom.

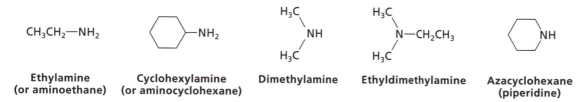

Ammonia is the conjugate base of the ammonium ion, which has a pK_a value of ~ 9. We can quantify the basicity of an amine in a manner similar to that used for acidity, defining a pK_b value according to Eq. 5.13.

$$RNH_2 + H_2O \rightleftharpoons RNH_3 + OH^- \tag{5.11}$$

$$K_b = \frac{[RNH_3^+][OH^-]}{[RNH_2]} \tag{5.12}$$

$$pK_b = -\log K_b \tag{5.13}$$

If we use ammonia as a reference point for the basicity of nitrogen-containing compounds, we see from Table 5.5 that an alkyl group makes the nitrogen atom of a primary amine more basic than that of ammonia.

Table 5.5   The pK_b values of ammonia and aliphatic amines and pK_a values of their conjugate acids.

Amine	pK_b	Ammonium Ion	pK_a
NH_3	4.75	$\overset{+}{N}H_4$	9.25
CH_3NH_2	3.35	$CH_3\overset{+}{N}H_3$	10.65
$(CH_3)_2NH$	3.27	$(CH_3)_2\overset{+}{N}H_2$	10.73
$(CH_3)_3N$	4.22	$(CH_3)_3\overset{+}{N}H$	9.78
$CH_3CH_2NH_2$	3.29	$CH_3CH_2\overset{+}{N}H_3$	10.71
$(CH_3CH_2)_2NH$	3.00	$(CH_3CH_2)_2\overset{+}{N}H_2$	11.00
$(CH_3CH_2)_3N$	3.25	$(CH_3CH_2)_3\overset{+}{N}H$	10.75
$CH_2CH_2CH_2CH_2NH_2$	3.31	$CH_3CH_2CH_2CH_2\overset{+}{N}H_3$	10.69

This effect results from the electron-donating influence of alkyl groups. As two and then three alkyl groups are substituted for H on the nitrogen atom, the basicity of the amine first increases, then decreases. This pattern is a result of solvation of the conjugate acid.

Instead of referring to pK_b values, organic chemists normally use the pK_a value of the corresponding *ammonium ion.* In aqueous solution, pK_a and pK_b are related by the following expression:

$$pK_a + pK_b = 14 \qquad \qquad (5.14)$$

We make the assumption that this relationship also holds for nonaqueous solvents. Table 5.5 lists the corresponding pK_a values for the conjugate acids of ammonia and aliphatic amines. Although potentially confusing, a pK_a value is normally given when discussing the basicity of the amine, *with the understanding that the number actually refers to the* pK_a *value of the corresponding ammonium ion, the amine's conjugate acid.*

Compared with acyclic amines, cyclic analogues have a higher basicity, which results from less steric hindrance toward protonation of the nitrogen atom within a ring. (Remember, the pK_a values listed refer to the amine's conjugate acid.)

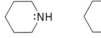

Diethylamine $pK_a = 11.00$ **Piperidine** $pK_a = 12.20$

Hybridization of a nitrogen atom also affects base strength because of electronegativity effects (Section 5.2c). Increasing the *s* character of a hybrid orbital will decrease the molecule's basicity. Because an sp^2-hybridized orbital is less basic that an sp^3-hybridized one, pyridine is a weaker base than piperidine. Conversely, the conjugate acid of pyridine is more acidic (lower pK_a value) than the conjugate acid of piperidine.

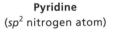

Piperidine $pK_a = 12.20$ **Pyridine** $pK_a = 5.3$
(sp³ nitrogen atom)   *(sp² nitrogen atom)*

The basicity of aniline derivatives is influenced by resonance effects in addition to the inductive effects. The electron pair on the nitrogen atom interacts with the π electrons of the benzene ring, and delocalization into the ring makes this electron pair less available to react with a proton. In fact, triphenylamine is not at all basic with regard to protonation by water (Table 5.6).

Attaching substituents to the aromatic ring also influences the basicity of the amino group in derivatives of aniline. For example, a nitro group at C4 decreases basicity because the electron pair on the aniline nitrogen atom is delocalized with the benzene ring and nitro group, so it is less available to react with a proton.

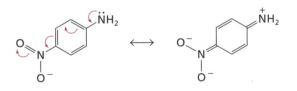

Other substituents can exert inductive effects, too, increasing or decreasing the electron density in the ring and on the nitrogen atom of the amino group.

Table 5.6   The pK_a values of the conjugate acids of aniline and related compounds.

Compound	Conjugate acid	pK$_a$ value	Compound	Conjugate acid	pK$_a$ value
aniline $\ddot{N}H_2$	aniline $\overset{+}{N}H_3$	4.62	H_3C—aniline $\ddot{N}H_2$	H_3C—aniline $\overset{+}{N}H_3$	5.08
diphenylamine $\ddot{N}H$	diphenylamine $\overset{+}{N}H_2$	0.80	O_2N—aniline $\ddot{N}H_2$	O_2N—aniline $\overset{+}{N}H_3$	1.00
triphenylamine $\ddot{N}$	triphenylamine $\overset{+}{N}H$	< –2			

EXERCISE 5.7

p-Anisidine (4-methoxyaniline) is more basic than aniline. Draw the important reso-nance structures that rationalize this greater basicity of *p*-anisidine.

5.2g   ACIDS AND BASES CAN ALSO BE CLASSIFIED BY THEIR ABILITY TO ACCEPT OR DONATE ELECTRON PAIRS

The Brønsted–Lowry definition of proton donors and acceptors is sufficient for many applications. For some reactions, however, especially among *intermediates* of certain re-actions, protons are not involved. For these situations, the concepts of **Lewis acids** and **Lewis bases** are more appropriate.

A *Lewis acid* is a substance that is an *electron-pair acceptor*, and a Lewis base is a sub-stance that is an *electron-pair donor*. Note that the Lewis concept of an acid encompasses the Brønsted–Lowry definition because a proton, which has no electrons, is the proto-type of an electron-pair acceptor. Similarly, substances that attract protons (bases by the Brønsted definition) have at least one unshared pair of electrons, so they satisfy the Lewis definition as well.

A Lewis acid is a species that lacks an octet of electrons and includes such com-pounds as boron trifluoride, BF_3, and aluminum chloride, $AlCl_3$. Other substances like tin tetrachloride, $SnCl_2$, and iron(III) chloride, $FeCl_3$, accept an electron pair to fill a vacant *d* orbital. All of these substances react to form a new bond with molecules or ions that have an unshared pair of electrons.

A specific example of a Lewis acid–base reaction is the interaction of BF_3 with di-ethyl ether. The product of this reaction is a substance known as boron trifluoride etherate, a distillable liquid. Ether is a liquid itself, but boron trifluoride is a gas. The combination is quite stable and, when dissolved in another solvent, can dissociate to form a small amount of BF_3 and diethyl ether. Many Lewis acids are stabilized by com-plexation with a Lewis base, yet they dissociate to form the reactive Lewis acid in solu-tion. This situation is not unlike that of a substance like hydrogen chloride, which ionizes in water to form solvated protons and chloride ions.

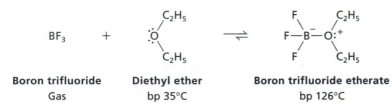

Boron trifluoride	Diethyl ether	Boron trifluoride etherate
Gas	bp 35°C	bp 126°C

Any molecule with an atom that has an unshared pair of electrons, particularly oxygen and nitrogen, is a Lewis base. It can and will react with substances that seek an electron pair to complete their valence shells, which are Lewis acids.

EXAMPLE 5.5

For the following Lewis acid–Lewis base reaction, draw the structural formula of the product:

$$BH_3 + (CH_3)_2S \longrightarrow$$

First, draw the Lewis structures for each of the reactants. A Lewis base reacts with a Lewis acid by donating a pair of electrons to the atom with a less-than filled shell. Boron, a group 13 (IIIA) element, is one of the elements that violates the octet rule: when it forms three bonds, it has only six electrons, which makes it a Lewis acid. After a bond has been formed between the electron pair donor atom, S, and the electron pair acceptor atom, B, we include formal charges on atoms in the product.

EXERCISE 5.8

For each of the following Lewis acid–Lewis base reactions, draw the structural formula of the product. Reaction (b) is an example in which a chlorine atom forms two bonds.

a. $BF_3 + OH^- \longrightarrow$ b. $AlCl_3 + CH_3Cl \longrightarrow$

5.3   REACTION MECHANISMS

5.3a   THE MECHANISM OF AN ORGANIC REACTION IS A RATIONALIZATION OF THE ELECTRON MOVEMENT INVOLVED IN MAKING AND BREAKING COVALENT BONDS

Formally, the **mechanism** of a chemical reaction is defined as the set of molecular events that results in the observed conversion of reactants to products. For many applications, this pathway is indicated by a series of equations called "elementary steps", the slowest of which is the **rate-determining step.**

For reactions in organic chemistry, the term "mechanism" refers to how we rationalize the movement of electrons during the conversion of reactants to products: *A mechanism is depicted by use of curved arrows that show electron movement.* For example, in the following substitution reaction, a bromide ion reacts with chloromethane, replacing chlorine.

A mechanism for a substitution reaction

One curved arrow shows that the electrons of the bromide ion react with the carbon atom to form a new C–Br bond. The other curved arrow shows that the electrons in the C–Cl bond associate with Cl, which produces the chloride ion. The most important principle to remember when showing electron movement with use of curved arrows is this:

In a polar reaction, the tail of the arrow starts
at an electron pair and ends at a nucleus.

5.3b   THE PARTICIPANTS IN A POLAR REACTION ARE CALLED THE NUCLEOPHILE AND THE ELECTROPHILE

In a *polar reaction* (Section 5.1c), which is the most common type in organic chemistry, re-actants carry either a positive or a negative charge, or they develop a charge as the reac-tion progresses. In the case of a positively charged species, one of the atoms is normally electron deficient, which means it lacks an octet of electrons in its valence shell. As a re-sult, this atom actively "seeks" electrons. Such a species is an **electrophile** and is said to be *electrophilic* (electron loving). A carbocation certainly fills this role, as does a proton, H^+. Some atoms, like boron and aluminum, lack eight valence electrons yet carry no charge. They are also electrophilic. Some common electrophiles are listed in Table 5.7.

tert-Butyl carbocation

The central carbon atom has
only six e$^-$ and is electrophilic.

Trimethylborane

The boron atom has only six e$^-$
and is electrophilic.

Species that have an octet of electrons around each atom sometimes carry a nega-tive charge. These species are called **nucleophiles** and are *nucleophilic* (nucleus loving). Even a neutral (noncharged) molecule with an electron pair or a π bond is nucleophilic because it possesses a region with a high density of electrons. A nucleophile reacts with an electrophile, providing the basis for a majority of reactions that you will learn. There-fore, understanding which species in organic reactions are nucleophilic or electrophilic is extremely important. Common nucleophiles are also listed in Table 5.7.

Recognize that many substances listed as electrophiles in Table 5.7 have both elec-trophilic and nucleophilic centers as a result of having an atom with unshared electron pairs or because it has polar bonds. For example, the following species all contain a heteroatom with unshared electron pairs. In certain reactions, these groups can react as nucleophiles, especially the ones that contain oxygen. Also, molecules such as alco-hols can have more than one electrophilic atom (C and H). Molecules that are overall neutral often have both nucleophilic and electrophilic atoms.

Sulfur has an unshared pair of
electrons, but the +1 formal charge
makes the sulfur atom electrophilic.

On the other hand, species that carry a −1 charge are normally considered as nucle-ophiles, and those with a +1 charge or that lack an octet of electrons are electrophiles, even if bonds in the species are polar.

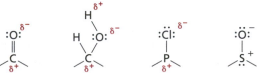

A charge of −1 makes each of these ions a nucleophile.
The carbon atom in cyanide ion carries a formal charge
of −1, so it is the more nucleophilic atom.

An overall charge of +1 or an atom lacking an octet
makes each of these species an electrophile.

Table 5.7   Some electrophiles and nucleophiles commonly encountered in chemical reactions.

Substances with electrophilic centers			Substances with nucleophilic centers		
General form	Example	Species	General form	Example	Species
H_3O^+	H_2SO_4 in H_2O	Proton (acid)	$:\ddot{C}l:^-$		Chloride ion
			X^- — $H\ddot{O}:^-$		Hydroxide ion
$\overset{\delta^+}{\underset{}{C}}$—$\overset{\delta^-}{X}$	CH_3—Br	Alkyl halide	$CH_3\ddot{O}:^-$		Methoxide ion
	CH_3—OH	Alcohol	$^-:CN:$		Cyanide ion
$\overset{\delta^-}{:\ddot{O}}$ ‖ C $\overset{}{\underset{\delta^+}{}}$	CH_3—CHO	Aldehyde	$\overset{\delta^-}{\underset{}{C}}$—$\overset{\delta^+}{M}$	CH_3—Li	Organolithium compound
	CH_3—CO—CH_3	Ketone			
	CH_3—CO—Cl	Acid chloride		CH_3—MgBr	Grignard reagent
	CH_3—CO—OCH_3	Ester			
$\underset{}{B}$	BH_3	Borane	$\underset{:}{N}$	$:N(CH_3)_3$	Amine
	BF_3	Boron trifluoride			
$:\ddot{O}:^-$ $\underset{}{S^+}$	Cl—SO—Cl	Thionyl chloride	$\underset{:}{P}$	$:P(CH_3)_3$	Phosphine
			$C{=}C$	$H_2C{=}CH_2$	π Bond
$\overset{\delta^-}{X}$ $\underset{\delta^+}{P}$	PCl_3	Phosphorus trichloride	—C≡C—	HC≡CH	π Bond

---

EXAMPLE 5.6

Predict whether the following species are nucleophilic or electrophilic (or both) by comparison with the properties of the species shown in Table 5.7:

a. $:\ddot{B}r^+$ b. H_2S

---

a. Species that bear an overall charge of +1, even if unshared electron pairs are present, react mainly as electrophiles. In this species, the fact that the Br atom has only six electrons makes it more likely that it seeks to add an electron pair, which makes it an electrophile.

b. First, draw a full Lewis structure for this molecule. Look to see if unshared electron pairs or formal charges are present, and indicate the relative electronegativity effects (δ+ or δ−), especially if the molecule is neutral.

The sulfur atom bears a partial negative charge because its bonds to hydrogen are polar, and it has two unshared electron pairs. The hydrogen atoms each have a partial positive charge because of the polar bonds.

This molecule will likely have both electrophilic and nucleophilic properties, and its reactions will depend on what species it reacts with.

EXERCISE 5.9

Predict whether the following species are nucleophilic or electrophilic (or both) by comparison with the properties of the species shown in Table 5.7:

5.3c   Electrophiles and Nucleophiles (Lewis Acids and Lewis Bases) Are Integral to Chemical Reactions

As already noted (Section 5.3a), we depict mechanisms of polar reactions by showing electron movement with curved arrows: The tail of the arrow starts at an electron pair, which is most commonly either a π bond or an unshared pair on a heteroatom—in short, a *nucleophile*. The head of the arrow points to an atom—an *electrophilic* center, which is deficient in electrons. *The species from which the reaction is initiated is always an electron pair.*

Consider the mechanism of the reaction between 2,3-dimethyl-2-butene and hydrogen bromide, which yields 2-bromo-2,3-dimethylbutane as the product (an addition reaction, see Section 5.1b).

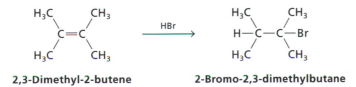

2,3-Dimethyl-2-butene   **2-Bromo-2,3-dimethylbutane**

The electrophile in this reaction is the proton of HBr, and the nucleophile is the π bond of the alkene (Table 5.7). According to formalisms for depicting movement of electrons, the tail of the arrow starts at the π bond, and the head of the arrow is directed at the proton.

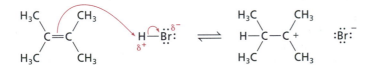

Several points to bear in mind:

- When a new bond forms, one or more bonds will likely be broken. In this example, formation of the bond between the carbon and hydrogen atoms requires that the bond between H and Br is broken, otherwise the hydrogen atom will have more than two electrons. Likewise, when the new C–H bond forms, the carbon–carbon π bond breaks, otherwise one of the carbon atoms will have more than an octet of electrons.

- When a bond breaks, the electrons in that bond often move onto an atom, creating a new unshared pair. Bromine acquires the pair in this example to maintain its octet. It is helpful, at least initially, to show all of the unshared pairs on every atom of the reactants and products to make sure that the octet rule is not violated.

- When a neutral molecule reacts with a positively charged species, a cation is formed as an intermediate. For an organic molecule, that often means that a carbocation is formed. *Make certain that charges on one side of the horizontal reaction arrow (or equilibrium arrows) balance with the charges on the other side.*

After movement of electrons to form the intermediate carbocation, as shown in the last equation, a new electrophile and a new nucleophile are formed. A second step then takes place, and electron movement occurs in the same fashion: The nucleophile reacts with the electrophile, so we draw the arrow from an electron pair on the bromide ion to the positively charged carbon atom.

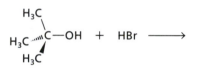

The feature common to both steps just shown is that the *nucleophile* (a Lewis base) reacts with the *electrophile* (a Lewis acid), producing a new bond. Although a chemical transformation may comprise several steps, individual processes can often be conceptualized as acid–base reactions. Because an acid is an electrophile and a base is a nucleophile, you want to be able to identify whether molecules, ions, and intermediates are either electrophiles or nucleophiles.

EXAMPLE 5.7

Identify the nucleophilic and electrophilic center(s) in each reactant of the following reaction. Using curved arrows, propose a possible *first* step for the reaction.

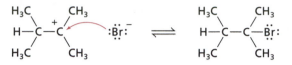

First, identify the nucleophilic and electrophilic sites in each reactant. It is helpful to add the unshared electrons to the expanded structures, as shown at left, because their presence defines the nucleophilic centers. A proton attached to a heteroatom is the most common and easily identified electrophilic site. For these reactants, the oxygen and bromine atoms are nucleophiles, and the proton attached to each is a possible electrophile. (The carbon atom attached to the oxygen atom is also electrophilic but protons are often more potent electrophiles, as you will learn.)

What we definitely know about the first step of this reaction is that it will occur by movement of electrons from a nucleophile to an electrophile. We can conceive of at least two possibilities with the assumption that the protons are the most likely electrophiles: (a) the oxygen atom reacts with the proton of HBr, or (b) the bromine atom reacts with the proton on the OH group.

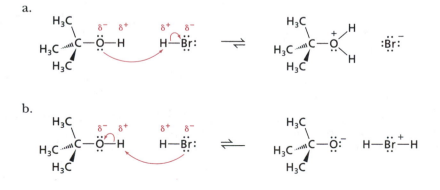

[In this example, you were not asked to identify the more likely possibility, but eventually you will have to make such evaluations. Pathway (a.) is the more likely of the two because it creates an acid that is *weaker* than HBr, whereas pathway (b.) creates an acid that is *stronger* than the starting alcohol. Recall from Section 5.2b that an acid–base reaction is an equilibrium that favors formation of the weaker acid. The equilibrium in pathway (a.) lies toward the right, and the equilibrium of pathway (b.) lies far to the left. This example shows why it is important to know in which direction acid–base equilibria lie.]

EXERCISE 5.10

Identify the nucleophilic and electrophilic center(s) in each reactant of the following reactions. Using curved arrows, propose a possible *first* step in each reaction.

a.   b.

$$H_2C \overset{CH_2}{\underset{CH_2}{\Big\langle}} \overset{CH_2}{\underset{}{\Big\rangle}} C{=}O \ + \ {}^-CN \longrightarrow$$

$$\overset{O}{\underset{}{\overset{\|}{H_3C-C-CH_3}}} \ + \ HCl \longrightarrow$$

5.4   REACTION COORDINATE DIAGRAMS

5.4a   THE ENERGY PROFILE OF A REACTION PROVIDES ADDITIONAL INSIGHTS ABOUT THE FEASIBILITY OF A REACTION AND ITS RATE

The use of curved arrows to represent the mechanism of a reaction provides a way to rationalize what changes take place as reactants are converted to products. "Electron pushing", as this exercise is sometimes called, is done with the assumption that the reaction that has been written *will* occur, and at a reasonable rate. But how do we know if a reaction will take place or how fast it will be? Understanding a transformation on the basis of its energy profile allows you to answer these two questions.

As an example, consider the acid–base reaction between hydrogen sulfide, H_2S, and cyanide ion, CN^-.

$$H_2S \ + \ CN^- \ \rightleftharpoons \ HS^- \ + \ HCN$$

pK_a=7.0   pK_a=9.2

We already know something about this reaction based on information presented in Section 5.2b: The equilibrium lies to the side of the weaker acid, in this case toward HCN.

We can propose a mechanism for this reaction because we know the nucleophile, cyanide ion, will react with the electrophile, the proton attached to sulfur in H_2S.

$$H\ddot{S}-H \qquad {}^-{:}C{\equiv}N{:} \longrightarrow H\ddot{S}{:}^- \ + \ H-C{\equiv}N{:}$$

In physical terms, the course of this reaction can be visualized by thinking of the molecules as three-dimensional objects. We know that molecules are in motion because they have kinetic energy provided by the heat of their surroundings. As a result of their motion, they collide with one another and with other substances like the solvent molecules or the walls of the flask. In our illustration, cyanide ion must collide with the H_2S molecule in the proper orientation for proton transfer to occur, and a favorable approach would look like that shown at the right. During stages **A–C**, the molecules are coming together and in stages **E–G**, they are moving apart. Species **D** is called the **transition state,** which will be discussed shortly.

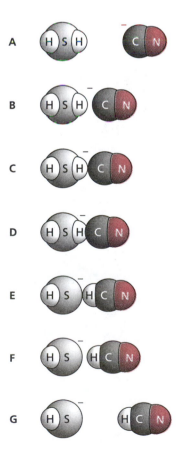

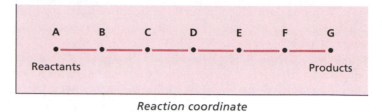

Reaction coordinate

Figure 5.1
The reaction coordinate for $H_2S + CN^- \rightleftharpoons HS^- + HCN$.

If we could film a movie of the reaction between these two molecules as it proceeds, then species **A** to **G** amount to a series of snapshots taken at various stages between where reactants exist (**A**) and where products have formed (**G**). If we now plot these points on a graph, as in Figure 5.1, we label the horizontal axis as the **reaction coordinate,** which we define as the progression of changes in the molecular structures as reactants are converted to products.

The second variable, energy, is plotted on the vertical axis of an energy profile and shows how energy varies with progression along the reaction coordinate. If you consider the illustrated collision between molecules of cyanide ion and hydrogen sulfide, you will realize that *even when they have the correct orientation to react,* electrons surrounding the carbon and sulfur atoms will tend to repel each other because of their like charges. Additionally, the magnitude of this repulsion increases as the molecules get closer together; so, for example, the repulsive forces will be greater for **D** than for **B.**

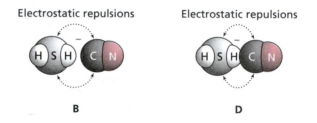

At some point along the reaction coordinate, the repulsive energy reaches a maximum, a point called the *transition state.* When this point is reached, the molecules must move away from each other to form products, or they may move apart to regenerate reactants. Figure 5.2 plots the reaction coordinate versus energy for the reaction between H_2S and, CN^-.

Even though the transition state represents the maximum energy that must be overcome so that a particular process will occur, there are sometimes even higher energy pathways that exist for the collision between reactants. For example, two molecules can collide without having the proper orientation to react. Because they simply bounce apart after the collision, they are not on the same reaction coordinate that leads to a viable reaction, so we can disregard them.

The changes in energy between some of the states shown in Figure 5.2 are important quantities. The energy difference between that of reactants and the transition state is the **free energy of activation,** $\Delta G^{\ddagger}$, and the magnitude of this quantity determines how fast a reaction proceeds. The double dagger is a symbol that is always associated with the transition state, to differentiate it from other energy changes. A small value for $\Delta G^{\ddagger}$ means that the energy barrier between reactants and products is low, so repulsive forces are small. In those instances, the reaction proceeds rapidly because there is sufficient energy available to counteract repulsive forces. Many acid–base reactions, like the one exemplified above, have a small free energy of activation value. A large value for $\Delta G^{\ddagger}$ means that the barrier is high, and the reaction will be slow. In those cases, energy, usually in the form of heat, must be supplied to overcome unfavorable interactions along the reaction pathway. The free energy of activation for a typical organic reaction is 5–30 kcal mol^{-1}.

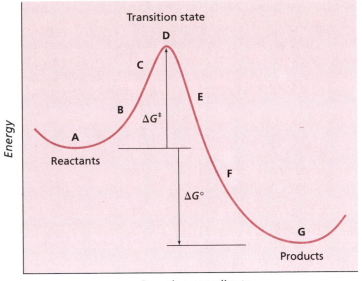

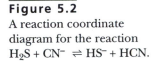

Figure 5.2
A reaction coordinate diagram for the reaction $H_2S + CN^- \rightleftharpoons HS^- + HCN$.

The difference in energy between that of reactants and that of products when standard conditions are employed is called the standard Gibbs free energy change, $\Delta G°$, a quantity that was mentioned briefly in Section 3.3b. If the products have a *lower* energy than the reactants do, as is the case represented by Figure 5.2, then $\Delta G°$ is negative. Such a reaction proceeds in the direction written (left to right), and it is said to be **exergonic** or **product favored.** On the other hand, if the energy of the products is *higher* than that of the reactants, then $\Delta G°$ is positive, and the process is said to be **endergonic** or **reactant favored.** [The terms *exergonic* and *endergonic* refer to the free energy difference ($\Delta G°$) between the reactants and products. You may be familiar from general chemistry with the terms *exothermic* and *endothermic*, which refer to the *enthalpy* difference ($\Delta H°$) between reactants and products. Enthalpy is one contribution to the free energy, the other being the *entropy*, a term that refers to how ordered a system is.]

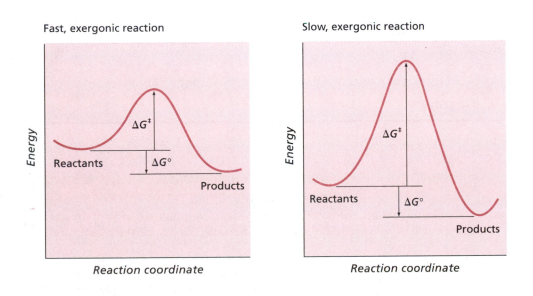

EXERCISE 5.11

Draw reaction coordinate diagrams for a slow, endergonic process and for a fast, endergonic process.

5.4b   FREE ENERGY IS RELATED TO THE EQUILIBRIUM CONSTANT

The correlation between the free energy change of a process and its equilibrium constant was mentioned already in Section 3.3b for the interconversion that takes place between cyclohexane conformers. For a general chemical reaction $aA + bB \rightleftharpoons cC + dD$, in which more than one reactant or product is involved, the concentrations of A, B, C, and D are related to the equilibrium constant, K_{eq}, by the expression

$$K_{eq} = \frac{[C]^c[D]^d}{[A]^a[B]^b} \qquad (5.15)$$

As we saw in Section 3.3b, $\Delta G°$ is related to K_{eq} by the equation

$$\Delta G° = -RT \ln K_{eq} \qquad (5.16)$$

where $R = 1.986 \text{ cal} \cdot \text{mol}^{-1} \cdot \text{K}^{-1}$ and T is the temperature in Kelvins. The sign of $\Delta G°$ indicates whether the reaction is product favored under standard conditions, producing C and D ($\Delta G° < 0$), or whether it favors production of the reactants A and B ($\Delta G° > 0$). From Eq. 5.16, you can see that when $\Delta G° < 0$, then $K_{eq} > 1$, and when $\Delta G° > 0$, then $K_{eq} < 1$.

If the initial concentrations of reactants are different from those that define standard conditions, the following expression is employed for calculating the free energy change:

$$\Delta G = \Delta G° + RT \ln Q \quad \text{where} \quad Q = \frac{[C]^c[D]^d}{[A]^a[B]^b} \qquad (5.17)$$

A consequence of this relationship is that the overall free energy, ΔG, can be altered by increasing the concentration of one or more of the reactants or products, which changes the magnitude of the second part of the expression on the right side of Eq. 5.17. Therefore, even though the equilibrium for a reaction might lie to the left under standard conditions ($\Delta G° > 0$), it can be shifted to the right ($\Delta G < 0$) by adding a large excess of one of the reactants A or B. This effect of concentration on the equilibrium is called **Le Chatelier's Principle.** Increasing the concentration of A or B makes Q smaller, so its logarithm also becomes negative and can cause ΔG to become negative.

EXERCISE 5.12

Calculate the value of K_{eq} for the following $\Delta G°$ values at 298 K:

a. +3.0 kcal mol^{-1} b. −3.0 kcal mol^{-1} c. −6.0 kcal mol^{-1}

5.4c   THE ENERGY PROFILE FOR A REACTION THAT PROCEEDS VIA FORMATION OF AN INTERMEDIATE HAS TWO TRANSITION STATES

Having looked at a one-step reaction between an acid and a base in Section 5.4a, we now consider a reaction that proceeds in two steps. As an illustration, we will use the example that was presented in Section 5.3c, the reaction between 2,3-dimethyl-2-butene and HBr.

Recall that the mechanism for this transformation comprises two steps, which we represent as follows:

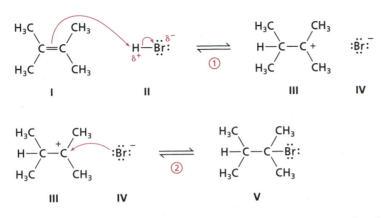

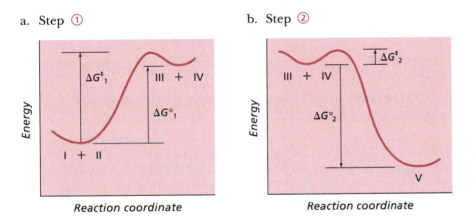

For Step 1, we know that one of the carbon atoms of the carbocation has only six electrons, so it must be a fairly unstable species. We would predict that this equilibrium lies toward the left, where both compounds—the alkene and HBr—are stable species. Therefore, $K_{eq} < 1$, and with use of Eq. 5.16, we can see that $\Delta G° > 0$. Step 1 is therefore an endergonic process, which we represent by the reaction coordinate diagram in Figure 5.3a. By contrast, the equilibrium of Step 2 must lie toward the right, toward the stable product. The value for $\Delta G°$ is < 0, and the reaction coordinate diagram can be represented as in Figure 5.3b.

a. Step ①

b. Step ②

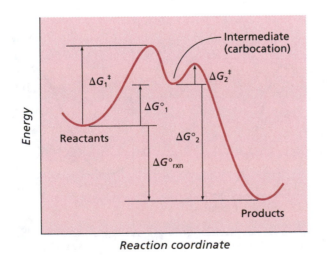

Figure 5.3
Energy profiles for the two steps in the reaction of 2,3-dimethyl-2-butene and HBr.

For the overall reaction **I + II ⇌ V,** we can represent the energy profile with the plot shown in Figure 5.4. Because this transformation proceeds by way of a discrete intermediate, namely, carbocation **III,** the highest energy portions of the curve display a local minimum that corresponds to this intermediate.

Figure 5.4
An energy profile for the overall reaction of 2,3-dimethyl-2-butene and HBr.

There is a large activation barrier for its formation ($\Delta G_1^{\ddagger}$), but only a small one for its further reaction ($\Delta G_2^{\ddagger}$). Notice that the free energy for this reaction, $\Delta G°_{rxn}$, is equal to the sum of $\Delta G°_1 + \Delta G°_2$ (Fig. 5.4).

EXERCISE 5.13

Without concerning yourself with the details of electron movement in the following reaction, draw a suitable energy-level diagram for this substitution process that proceeds via a free radical intermediate. The radical intermediate is a high-energy species (rxn is the abbreviation for reaction).

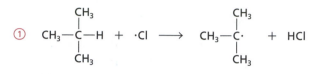

Starting materials   **Intermediate**

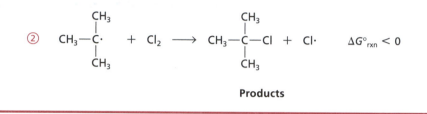

Products

5.4d CATALYSTS INCREASE REACTION RATES BY STABILIZING TRANSITION STATES

When a reaction proceeds slowly, its rate can be increased in some cases by adding a catalyst. The most common catalysts are acids, bases, or metal-containing compounds. In theory, a catalyst can be recovered at the end of the reaction, but experimentally this procedure is sometimes troublesome. As illustrated by the diagram in Figure 5.5a, a catalyst can operate by stabilizing the transition state of a reaction, which lowers the free energy of activation and increases the rate of reaction. In many catalyzed reactions, the catalyst may also alter the mechanism of reaction, changing its energy profile more dramatically, as illustrated in Figure 5.5b. The free energy of activation is lowered in these cases, too. In either case, $\Delta G°$, the free energy of reaction, *stays the same* for both the uncatalyzed and catalyzed routes.

Figure 5.5
Energy profiles for an uncatalyzed versus catalyzed reaction. (a.) A reaction in which the free energy of activation has been decreased by catalysis. (b.) A reaction in which the mechanism has been changed by catalysis; the free energy of activation has also been decreased. In both cases, the free energy of reaction, $\Delta G°$, remains the same for the uncatalyzed and catalyzed mechanisms.

a.
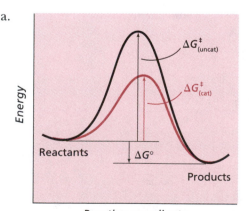

Reaction coordinate

b.
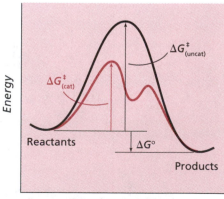

Reaction coordinate

Catalyzed reactions are crucial for the success of many biochemical reactions, where conditions of temperature and pH are restricted. Catalysis is also important for industrial processes as a way to increase the rates of reactions, so they occur under milder conditions, or to provide selectivity that cannot be obtained otherwise. Catalyzed reactions can also be used to prevent waste, both of substrates and reagents. This application is related to the concept of "atom economy" (Section 15.5a).

5.5   REACTIONS IN BIOLOGICAL SYSTEMS

5.5a   BIOCHEMICAL PROCESSES ARE CATALYZED ORGANIC REACTIONS

Fundamentally, reactions that occur in biological systems are organic reactions. Biomolecules contain carbon, hydrogen, nitrogen, oxygen, sulfur, and phosphorus atoms, so they certainly qualify as organic materials. In many instances, these molecules have greater mass and complexity than those normally encountered in an organic chemistry laboratory, but the fundamental reactions of the functional groups are the same. Therefore, everything that has been said about reactions of organic compounds applies to transformations that occur in biochemical systems:

- Functional groups react principally by movement of electrons from nucleophilic centers to electrophilic atoms.
- The acid–base reaction is a fundamental step that underlies reaction mechanisms.
- The free energy difference between starting materials and products governs whether reactions will be product favored or reactant favored.
- Catalysis occurs by decreasing the free energy of activation for the reaction.

In living systems, catalysts are vital to increase reaction rates because of the exacting conditions of temperature (20–37°C), solvent (H_2O), and pH (usually 6–8) that must be satisfied. Most commonly, catalysts for biological reactions are **enzymes,** most of which are proteins with three-dimensional structures that are able to bring organic molecules together and stabilize the transition state structure of the reaction. The question is, How does catalysis occur?

To fulfill one of the goals of this text, namely, to illustrate the parallels between biochemical transformations and the organic reactions performed in the laboratory, some rudimentary aspects of protein structures are presented here so that illustrations of important biochemical processes will make better sense later. Section 5.5b outlines some basic aspects of protein structures in order to provide a foundation for understanding how enzymes work.

5.5b   PROTEINS COMPRISE A SEQUENCE OF α-AMINO ACIDS FOLDED INTO A SPECIFIC THREE-DIMENSIONAL SHAPE

The building blocks for proteins are the **L-α-amino acids**, which have an amino and a carboxylic acid group attached to the same carbon atom, as represented by the following general structure:

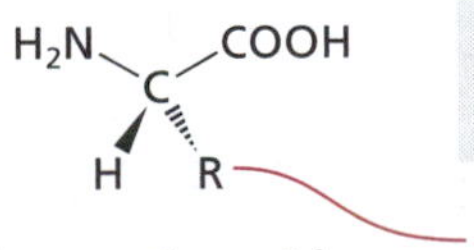

For the amino acids that constitute the majority of proteins, the configuration of the α-carbon atom is normally (*S*). An exception is cysteine, in which the higher priority of the sulfur atom in the side chain makes the configuration (*R*).

An α-amino acid

Side chain (varies in structure among the amino acids)

The Greek letter α is a way to designate the relative relationship between two groups in a molecule, especially when one of them is a carbonyl group. The α-position is the atom *adjacent* to the carbonyl group, β is the next position, and so on. Greek letters are often used when referring to a general class of molecules without having to specify details about the remainder of the structure. For example,

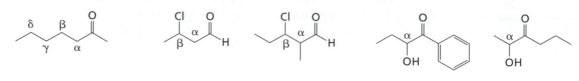

β-chloro aldehydes   **α-hydroxy ketones**

The absolute configuration of the α-amino acids used to construct the majority of proteins is normally specified using descriptors D and L instead of (R) and (S), mostly for historical reasons. We represent the general form of an L-amino acid by the following structure; its enantiomer is the D isomer.

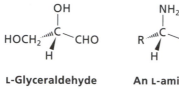 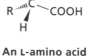 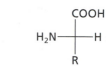

L-Glyceraldehyde   **An L-amino acid**   **A D-amino acid**   **An L-amino acid**
(Fischer projection)

About 20 different R groups are common to thousands of proteins that exist in Nature. Some specific examples are illustrated below (and in Table 5.8).

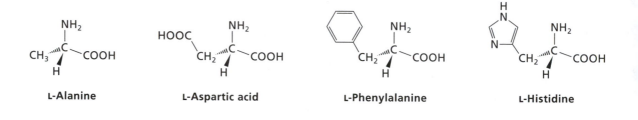

L-Alanine   **L-Aspartic acid**   **L-Phenylalanine**   **L-Histidine**

EXERCISE 5.14

Draw the structural formulas of D-aspartic acid and D-phenylalanine.

The amino and carboxylic acid groups of amino acids can react with each other to form an amide functional group, and this link between the two amino acids is called a **peptide bond.** A prefix is used to indicate how many amino acids constitute a polypeptide. For example, a *di*peptide (shown below) has two amino acids. (This word may be somewhat confusing because there is only one peptide bond.) Each amino acid is called a **residue,** so a dipeptide comprises two residues.

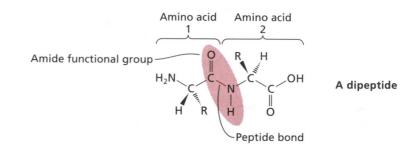

A dipeptide

A **polypeptide,** which means that many amino acids have been linked by peptide bonds, has an amino group at one end, which is called the N-**terminus,** and a carboxy group at the other end, which is the C-**terminus.** In general, molecules called polypeptides are smaller than those referred to as proteins.

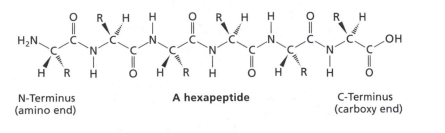

N-Terminus   **A hexapeptide**   C-Terminus
(amino end)   (carboxy end)

EXAMPLE 5.8

Draw the full structure of the dipeptide Asp-Glu, showing all of the atoms.

A peptide sequence is normally written as a string of the one- or three-letter abbreviations for the constituent amino acids. The first amino acid in the sequence (reading from left to right) is at the N-terminus (so it has the amino group, H_2N-), and the last one defines the C-terminus ($-COOH$). In between, amide groups link the amino acids together. The general structures of a dipeptide and a tripeptide, including the stereochemistry of the chiral carbon atoms, are as follows:

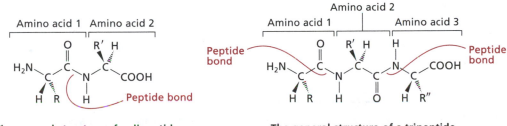

The general structure of a dipeptide   **The general structure of a tripeptide**

To draw the dipeptide Asp-Glu, we look up the structures of the side chains for Asp and Glu (see Table 5.8), respectively, and replace the R groups from left to right with those structures.

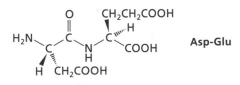

Asp-Glu

EXERCISE 5.15

Draw the full structure of the following tripeptides, showing all of the atoms:

a. His–Glu–Lys   b. Val–Cys–Phe

Proteins are polypeptides that fold into specific three-dimensional shapes, often with specific interior and exterior portions essential to their biological function. Myoglobin, for example, is the protein used to store O_2 in muscles. It is known as a "globular protein" because it folds into a ball-like shape, as illustrated in Figure 5.6. Here the interior and exterior are readily apparent; the interior part holds an iron-containing heme molecule, where O_2 is bound.

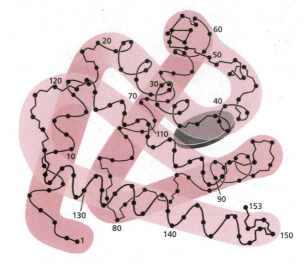

Figure 5.6
The three-dimensional structure of myoglobin, an oxygen-storage protein. The numbered dots represent the positions of the α-carbon atoms of the 153 amino acids that constitute the protein. The heme unit is shown as an ellipse.

Because proteins often function in an aqueous medium, the outside is normally *hydrophilic* (water-loving) and the interior is *hydrophobic* (water-fearing), but there are plenty of exceptions. The sequence of side chains of the constituent amino acids helps define the specific properties of a protein and dictates how the protein folds to place the residues on the outside or inside. Table 5.8 shows a selection of some common amino acids with their names, one- and three-letter abbreviations, and some characteristics. We will describe these specific characteristics further in Section 5.5c and in Chapter 27.

Table 5.8   Properties of the side chains for several amino acids present in proteins and enzymes.[a]

Side chain, R	Name	Abbreviations	Characteristics
$HOOC-CH_2-$	Aspartic acid	Asp, D	The carboxylic acid group is acidic.
$HS-CH_2-$	Cysteine	Cys, C	The thiol group is an excellent nucleophile.
$HOOC-CH_2CH_2-$	Glutamic acid	Glu, E	The carboxylic acid group is acidic.
(imidazole)$-CH_2-$	Histidine	His, H	The N atom is a good Lewis base; its protonated form is a weak acid.
$H_2\ddot{N}-CH_2CH_2CH_2CH_2-$	Lysine	Lys, K	The amine group is a good base and a good nucleophile.
(phenyl)$-CH_2-$	Phenylalanine	Phe, F	The benzene ring makes this a very hydrophobic group.
$HO-$(phenyl)$-CH_2-$	Tyrosine	Tyr, Y	The phenol proton is slightly acidic; the phenol ring can form a stabilized free radical.
H_3C $CH-$ H_3C	Valine	Val, V	The alkyl group is very hydrophobic.

[a]The general form of the amino acid:

$$R-\underset{\underset{NH_2}{|}}{\overset{\overset{H}{|}}{C}}-COOH$$

5.5c   ENZYMES OFTEN USE ACID AND BASE GROUPS TO CATALYZE REACTIONS

As noted in Section 5.5a, most enzymes that catalyze reactions in biochemical systems are proteins. An enzyme normally interacts with functional groups of the **substrate**—the molecule undergoing reaction—by forming hydrogen bonds or by exploiting dispersion and electrostatic forces (Table 2.5) to orient the reactant molecule and stabilize the transition state of the reaction. The **active site** of the enzyme—where the reaction actually takes place—often consists of two parts, the *binding site* and the *catalytic site*.

Consider, for example, lysozyme, an enzyme that many organisms use to catalyze the rupture of bacterial cell walls. The workings of an enzyme's active site are represented by drawings like Figure 5.7. The *screened color* in these drawings depicts the protein "backbone", which consists of the amide groups (peptide bonds) as well as the α-carbon atoms to which the side chains of the constituent amino acids are attached. Most of the individual atoms are not shown in these representations, except those required to bind the substrate or for catalysis. If the side chain of a particular amino acid is important, these side-chain atoms will be shown, but the rest of the amino acid will be symbolized by its three-letter abbreviation (and sometimes a number to signify its position in the protein starting from the N-terminus). Thus, Glu_{35}–CH_2–CH_2–$COOH$ signifies that residue number 35 is glutamic acid. Because this side chain contains the carboxylic acid functional group that is crucial for catalyzing the reaction, its atoms are included. *Drawings like this will be employed throughout this text to represent enzyme active sites and to conceptualize enzyme-catalyzed reactions.*

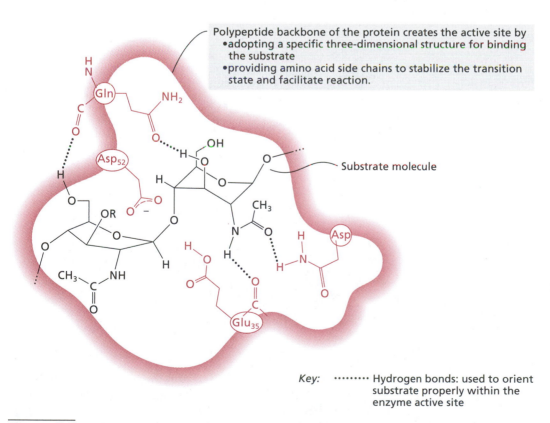

Figure 5.7

A representation of the active site in lysozyme that illustrates binding of the substrate molecule (shown in black type).

From studies by enzymologists on lysozyme, we know that several amino side chains provide hydrogen bonds to orient the substrate molecule near the catalytic amino acid residues, which are Glu$_{35}$ and Asp$_{52}$.

The following drawings depict the first step of the reaction being catalyzed, in which the oxygen atom linking two of the carbohydrate rings reacts with the proton (shown as highlighted in color) of the carboxylic acid side chain of Glu$_{35}$ (the hydrogen bonds that orient the substrate have been omitted from this picture for clarity, but they are still present). *This protonation of the oxygen atom is the same initial step that occurs during the analogous reaction in solution (below); only the identity of the acidic group is different.*

In the enzyme active site:

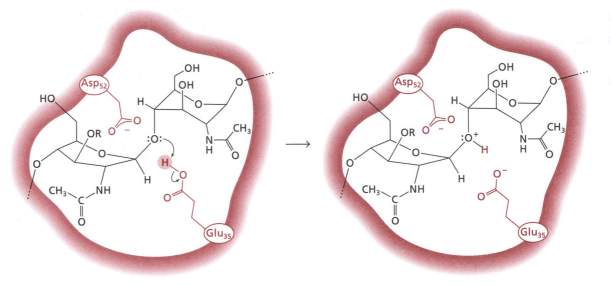

In solution:

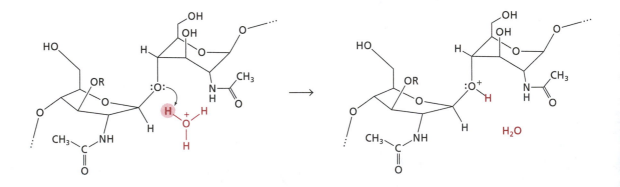

Rationalizations like this one can also be used to portray the subsequent steps in the reaction catalyzed by lysozyme, as well as those of many other enzymes and their reactions. A large part of research in enzymology seeks to understand, *at the molecular level,* how enzyme catalysis occurs. As you learn chemical reactions that are typical for organic molecules, you will recognize that as bonds are broken, the structures of reacting molecules change in predictable ways. These same patterns apply in biochemistry.

Several types of catalytic mechanisms occur in biological systems. The most important, and simplest, is acid–base catalysis, the type that occurs in lysozyme. Catalysis can also be promoted by metal ions, by orientation and proximity effects, by electrostatic influences, and by formation of covalently bonded intermediates. In every instance, *the enzyme brings the reactant molecules together and lowers the energy of the transition state.*

This text will focus on natural *enzyme*-catalyzed processes to illustrate fundamental organic reactions in biology. A particular emphasis will be on reactions that occur in the metabolism of carbohydrates, fats, and amino acids because in converting these fuels to energy, living organisms demonstrate the classic processes of proton transfer, substitution, addition, elimination, oxidation, and reduction. A goal of this text is to help you acquire the background to understand biochemical processes.

EXERCISE 5.16

Classify the following biochemical processes according to the fundamental reaction types of substitution, addition, or elimination. What molecules, atoms, or groups are lost, added, or substituted?

a.

$$CH_3-\underset{\underset{OH}{|}}{\overset{\overset{H}{|}}{C}}-CH_2-\overset{\overset{O}{\|}}{C}-S-ACP \longrightarrow CH_3-\overset{\overset{H}{|}}{C}=CH-\overset{\overset{O}{\|}}{C}-S-ACP$$

The third step during the biosynthesis of fats in mammals; ACP = acyl carrier protein

b.

$$CH_3-\overset{\overset{}{\underset{\underset{O}{\|}}{C}}}{C}-COO^- \longrightarrow CH_3-\overset{\overset{H}{|}}{\underset{\underset{OH}{|}}{C}}-COO^-$$

The last step during the metabolism of sugars by certain microorganisms

CHAPTER SUMMARY

Section 5.1 General aspects of reactions

- A chemical reaction is represented by an equation that shows reactants on the left side of an arrow pointing to products on the right.
- Sequential reactions are indicated by numbering the reagents listed on the arrow.
- The yield of a chemical reaction is the molar percentage of product that is isolated relative to the amount that could be produced.
- Reactions can be classified by the type of changes that functional groups undergo or by the manner in which bonds are made or broken.
- Classification by functional group changes includes proton transfer, substitution, addition, elimination, oxidation, reduction, and rearrangement.
- Classification according to the manner in which bonds are made or broken includes polar, radical, and pericyclic reactions.

Section 5.2 Acid–base reactions

- An acid is a substance that donates a proton (Brønsted–Lowry definition) or accepts an electron pair (Lewis definition).
- A base is a substance that accepts a proton or donates an electron pair.
- The negative logarithm of the equilibrium constant for proton dissociation from an acid in aqueous solution is defined as its pK_a value.
- The more positive an acid's pK_a value, the weaker the acid.
- An acid–base reaction is an equilibrium process that favors formation of the weaker acid.
- The species formed after dissociation of a proton from an acid is called its conjugate base.
- The conjugate base of a strong acid is a weak base and vice versa.
- The relative strength of an acid is influenced by stabilization of the conjugate base by size, electronegativity, and/or electron delocalization effects.

Section 5.3 Reaction mechanisms

- Curved arrows that represent the movement of electrons portray the mechanism of a reaction, which rationalizes how reactants are converted to products.
- In a polar reaction, a nucleophile reacts with an electrophile to create a bond.
- An electrophile is a Lewis acid. A nucleophile is a Lewis base.
- When electron movement in a polar reaction is shown with use of curved arrows, the tail of the arrow starts at the nucleophile and its head points to the electrophile.

Section 5.4 Reaction coordinate diagrams

- A reaction coordinate diagram correlates the progress of a chemical reaction with energy.
- The increase in energy as reactants begin to be converted to products is a barrier to the reaction. The value of this increase is called the free energy of activation.
- The local maximum on an energy curve is the transition state for that step.
- The energy difference between reactants and products is called the free energy of reaction, $\Delta G°$. If $\Delta G° < 0$, the reaction is exergonic and proceeds from left to right as written. If $\Delta G° > 0$, the reaction is endergonic and proceeds from right to left.
- A catalyst decreases the free energy of activation of a reaction by stabilizing the transition state, but it does not affect the free energy, $\Delta G°$, of the reaction.

Section 5.5 Reactions in biological systems

- Biological reactions are governed by the same principles as organic reactions.
- Amino acids are the building blocks of proteins and enzymes. Amino acids are connected in proteins via amide functional groups, called peptide bonds.
- A protein folds into a specific three-dimensional shape. An enzyme is a protein that catalyzes a chemical reaction, and folding creates an active site that can bind a substrate molecule (the reactant).
- An enzyme active site provides groups that can function in place of reagents that would be used in a laboratory setting to carry out a chemical reaction. Acidic and basic groups are the most common of such groups.
- An enzyme serves to lower the free energy of activation for a particular transformation.

KEY TERMS

Section 5.1a
reactants
products
catalyst
reagent

Section 5.1b
proton-transfer reaction
substitution reaction
addition reaction
elimination reaction
rearrangement reaction
oxidation reaction
reduction reaction

Section 5.1c
polar reaction
heterolysis
radical reaction
homolysis
pericyclic reaction

Section 5.2a
Brønsted–Lowry acid
Brønsted–Lowry base
conjugate base
conjugate acid
pK_a value

Section 5.2c
resonance effects

Section 5.2d
inductive effect

Section 5.2g
Lewis acid
Lewis base

Section 5.3a
mechanism
rate determining step

Section 5.3b
electrophile
nucleophile

Section 5.4a
transition state
reaction coordinate
free energy of activation
exergonic, product
    favored
endergonic, reactant
    favored

Section 5.4b
Le Chatelier's principle

Section 5.5a
enzyme

Section 5.5b
L-α-amino acids
peptide bond
residue
polypeptide
N-terminus
C-terminus

Section 5.5c
substrate
active site

ADDITIONAL EXERCISES

5.17. Classify each of the following reactions as substitution, addition, elimination, oxidation, reduction, or rearrangement:

a.

b.

c.

d.

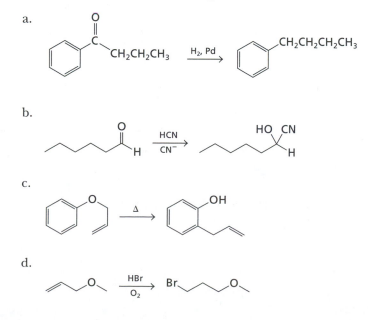

5.18. For the reactions shown in Exercise 5.17, what is the stoichiometry of each with respect to the different reactants? Which reactions are catalyzed? What are the catalysts?

5.19. Write an equation for each of the following processes. Put any inorganic reagents, solvents, and reaction conditions on the arrow. Classify each of these processes as substitution, elimination, addition, oxidation, reduction, or rearrangement.

    a. Hexanal reacts with $NaBH_4$ in 50% ethanol—water at 35°C to produce 1-hexanol.

    b. 3-Heptanone is treated with $LiAlH_4$ in ether at 0°C. After 2 h, the solution is treated with aqueous hydrochloric acid at room temperature to yield 3-heptanol.

    c. Benzenethiol and 1-iodopropane react at 25°C in DMSO solution containing triethylamine. The product is phenyl propyl sulfide, $CH_3CH_2CH_2$–S–C_6H_5.

    d. 1-Pentanol is heated with aqueous $K_2Cr_2O_7$ for 3 h to form pentanoic acid.

5.20. What is the percent yield for each of the following reactions?

    [grams ÷ formula weight = moles]

    a.

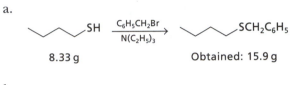

        8.33 g   Obtained: 15.9 g

    b.

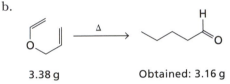

        3.38 g   Obtained: 3.16 g

5.21. Classify each of the following species as an electrophile, nucleophile, both, or neither:

    a. Cl^+ b. OH^- c. $(C_6H_5)_3P$ d. $Br·$ e. $(CH_3)_3Al$

    f. I^- g. $(CH_3)_2S$ h. CH_3^+ i. CH_3^- j. Li metal

5.22. Within each group, circle the molecule with the greater base strength, which is mainly influenced by inductive effects for each molecule. Briefly explain the reasons for your choices.

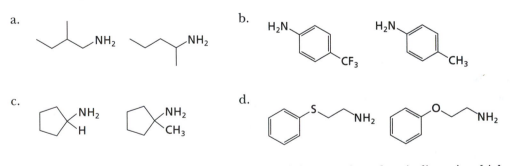

5.23. Identify the acid and base on each side of the equation, then indicate in which direction each equilibrium lies.

    a.

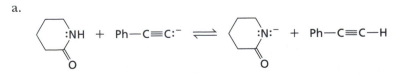

b.

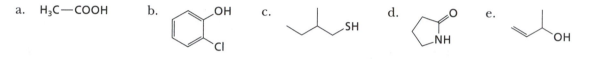

c.

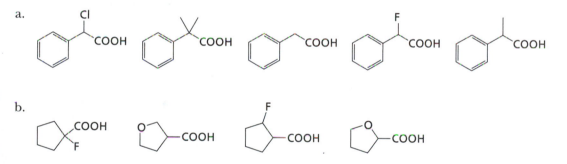

5.24. Draw a reaction coordinate diagram for a one-step endergonic process. Label it with respect to the reactants, products, transition state, free energy of activation, and standard free energy of reaction.

5.25. Repeat Exercise 5.24 for a two-step exergonic process in which the first step is slower than the second step.

5.26. Write an equation showing the acid–base reaction between each of the following molecules and aqueous sodium hydroxide, NaOH. In which direction does the equilibrium lie?

a. $H_3C-COOH$ b.   c.   d.   e.

5.27. Within each group, indicate which compound is the most acidic and which is the least acidic.

a.

b.

5.28. Write an equation showing the acid–base reaction between each of the following compounds and hydrochloric acid, HCl. In which direction does the equilibrium lie?

a. $CH_3CH_2NH_2$ b.   c.   d.

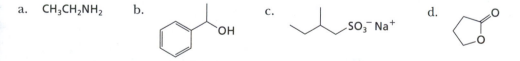

5.29. Repeat Exercise 5.28 with aqueous acetic acid in place of hydrochloric acid.

5.30. Consider the following molecules:

3-Pentanone   1,1,1-Trifluoroethane   3-Chlorobenzamide   4-Chlorobutanol

a. Draw a structural formula for each compound (Benzamide is $C_6H_5CONH_2$).

b. Assuming that you could add a base strong enough to remove the most acidic proton in each, identify which proton is most acidic, and give its approximate pK_a value (to the nearest multiple of 5 is sufficient—Table 5.1).

c. Which is the most acidic compound of the four? Which is the least acidic?

5.31.   Suppose a reaction has an equilibrium constant $K_{eq} = 1$. What is the value of $\Delta G°$ for this process? What will its reaction coordinate diagram look like?

5.32.   For aspartic and glutamic acids, draw the structure of the side chain in its conjugate base form.

5.33.   For each of the following reaction steps, suggest a possible *first* step (there may be more than one possibility for some) using arrows to illustrate the movement of electrons. Identify the electrophile and nucleophile in each reaction.

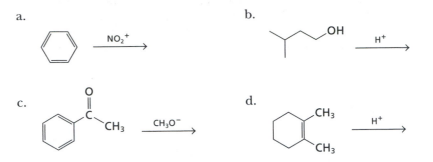

a.   b.

c.   d.

5.34.   The amino acids lysine and histidine normally function as bases in biological systems. Draw the structure of the side chain of each in its conjugate acid form. What is the approximate pK_a value for the conjugate acid form of the lysine side chain?

5.35.   For the amino acids shown in Table 5.8, identify those that have a functional group in their side chains that can function as (a) an electrophile, (b) a nucleophile, or (c) neither.

5.36.   Classify the following biological processes according to the fundamental reaction types of substitution, addition, or elimination. What molecules, atoms, or groups are lost, added, or substituted? Which of these reactions can be described as an oxidation or reduction process?

a.

The first step during the degradation of the amino acid proline.

b.

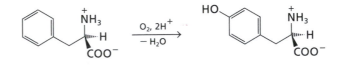

The last step during the biosynthesis of the amino acid tyrosine.

c.

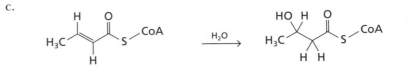

The second step during the metabolic cycle for degrading fats in animals.

SUBSTITUTION REACTIONS OF ALKYL HALIDES

The substitution reaction is conceptually one of the simplest in organic chemistry—*one group replaces another in the structure of the molecule.* One type of substitution reaction, the **nucleophilic substitution reaction,** occurs when a nucleophile replaces another group. A variety of substrates can be used, notably alkyl halides, alcohols, ethers, epoxides, and sulfonate esters. Even within this somewhat limited type of reaction, however, a variety of mechanisms is possible. This chapter focuses on the nucleophilic substitution reactions of alkyl halides, and the next chapter (Chapter 7) will describe the nucleophilic substitution reactions of alcohols and related molecules.

In presenting details about a single reaction type, this chapter will build on and exemplify the general characteristics of reactions that were described in Chapter 5. This chapter is the first one that introduces you to specific reactions, and your goals should be to learn the features of these reactions, and then to apply that knowledge to predict the products of transformations that you have not explicitly encountered. To help you assimilate all of the given information, a summary of reactions is presented at the end of this chapter, just before the Additional Exercises.

6.1 FUNDAMENTAL ASPECTS OF SUBSTITUTION REACTIONS

6.1a A NUCLEOPHILE REPLACES A LEAVING GROUP IN SUBSTITUTION REACTIONS OF ALKANE DERIVATIVES

Figure 6.1 shows four, typical, nucleophilic substitution reactions, which you might initially consider to represent a random collection of transformations. These reactions share three common features, however.

1. Each starting compound has a *heteroatom*—a halogen or oxygen atom (in OH and OSO_2CH_3)—attached to carbon. This heteroatom-containing group is called a **leaving group,** and its capacity for replacement varies considerably.

2. In every case, the leaving group renders its attached carbon atom electrophilic, making it susceptible to react with a nucleophile (Nuc). Note explicitly *that the leaving group is attached to an sp^3-hybridized carbon atom.*

3. A nucleophilic reagent is the other reactant.

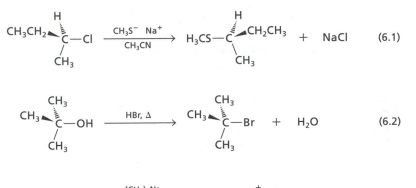

Figure 6.1
Representative examples of substitution reactions.

Simply put: *Alkyl halides, alcohols, and their derivatives are the common substrates that undergo nucleophilic substitution reactions.* Ethers and alkylsulfonium salts (R–$^+SR'_2$) react in a similar fashion, and the latter are especially important in biological systems, as you will learn in Chapter 7.

The strength of the nucleophile can vary substantially, and the temperature and nature of the solvent affects the overall reaction, too. We will examine each of these details in turn as we study the nucleophilic substitution reactions of alkyl halides.

EXERCISE 6.1

What is the nucleophile in each of the Reactions 6.1–6.4? (*Hint:* It is the group that replaces the leaving group.)

6.1b GOOD LEAVING GROUPS ARE WEAK BASES

The bond between a carbon atom and a heteroatom is a polar one, and the carbon atom normally bears a partial positive charge (section 2.2a), which means it is electrophilic. A nucleophile reacts readily with many types of electrophilic carbon atoms. (Recall from Section 5.3b that polar reactions occur *between nucleophiles and electrophiles.*)

$$\ddot{X}{:}^- \quad \overset{\delta^+}{C}{-}\overset{\delta^-}{\ddot{B}r}{:} \quad \longrightarrow \quad X{-}C \quad + \quad :\ddot{B}r:^-$$

The electron pair on the nucleophile, X^-, is attracted to the electrophilic carbon atom…	…leading to replacement of the leaving group Br^- by X^-. Bromide ion is a good leaving group because it is a weak base.

For a substitution to take place, however, the bond between the carbon atom and the heteroatom (the leaving group) has to break. This only happens when the *leaving group is a weak base.* The best leaving groups are therefore the halide ions, Cl^-, Br^-, and I^-, which are the conjugate bases of three of the strongest acids (HCl, HBr, and HI, respectively), and alkyl halides are among the best substrates for nucleophilic substitution reactions.

6.1c   NUCLEOPHILICITY DEPENDS ON FACTORS OF BASICITY AND POLARIZABILITY

What makes a good nucleophile? Any number of anions (ions with a negative charge) can serve as nucleophiles, and common examples include halide, hydroxide, azide, and thiolate ions. A variety of oxygen-containing anions derived from alcohols and phenols (alkoxide and phenolate ions) are nucleophiles, and certain carbanions also participate in nucleophilic substitution reactions. (Carbanion nucleophiles will be discussed in Chapters 15 and 22.) Table 6.1 lists some common reagents, as well as the actual nucleophiles, that are used in substitution reactions. Many neutral molecules with nitrogen, oxygen, sulfur, and phosphorus atoms are also nucleophilic; these are included at the right side of Table 6.1. In neutral molecules, the actual nucleophile is the heteroatom that is present.

Table 6.2 shows the expected products from typical nucleophilic substitution reactions that start with a primary or a tertiary alkyl bromide. In the reactions of a charged nucleophile with an alkyl bromide (entries 1–5), the inorganic product of the reaction

Table 6.1   Common reagents used in organic chemistry for substitution reactions.[a]

Reagent	Nucleophile	Reagent	Nucleophile	Nucleophilic molecules	
HCl	Cl^-	NaOPh[b]	PhO^-	H_2O	R_3N
HBr	Br^-	KOAc[c]	AcO^-	ROH	R_3P
HI	I^-	NaN_3	N_3^-	NH_3	H_2S
NaOH	OH^-	NaCN	CN^-	RNH_2	RSH
$NaOCH_3$	CH_3O^-	$NaSCH_3$	CH_3S^-	R_2NH	R_2S
$NaOCH_2CH_3$	$CH_3CH_2O^-$	$LiAlH_4$	H^-		
$KOC(CH_3)_3$	$(CH_3)_3CO^-$				

[a]In most cases, both potassium and sodium salts of anionic nucleophiles can be used.
[b]Phenyl, $C_6H_5- = Ph$.
[c]Acetate ion, $CH_3COO^- = AcO^-$.

Table 6.2   Representative nucleophilic substitution reactions.

	Substrate	Reagent	Products		Type of Molecule
(1)	$CH_3CH_2CH_2—Br$	+ $Na^+ \ I^-$	$\longrightarrow$ $CH_3CH_2CH_2—I$	+ $Na^+ \ Br^-$	Iodoalkane
(2)	$CH_3CH_2CH_2—Br$	+ $K^+ \ OH^-$	$\longrightarrow$ $CH_3CH_2CH_2—OH$	+ $K^+ \ Br^-$	Alcohol
(3)	$CH_3CH_2CH_2—Br$	+ $Na^+ \ N_3^-$	$\longrightarrow$ $CH_3CH_2CH_2—N_3$	+ $Na^+ \ Br^-$	Azidoalkane
(4)	$CH_3CH_2CH_2—Br$	+ $Na^+ \ CN^-$	$\longrightarrow$ $CH_3CH_2CH_2—CN$	+ $Na^+ \ Br^-$	Nitrile
(5)	$CH_3CH_2CH_2—Br$	+ $Li^+ \ CH_3S^-$	$\longrightarrow$ $CH_3CH_2CH_2—SCH_3$	+ $Li^+ \ Br^-$	Thioether
(6)	$CH_3CH_2CH_2—Br$	+ NH_3	$\longrightarrow$ $CH_3CH_2CH_2—\overset{+}{N}H_3$	Br^-	Amine salt
(7)	$CH_3CH_2CH_2—Br$	+ PPh_3	$\longrightarrow$ $CH_3CH_2CH_2—\overset{+}{P}Ph_3$	Br^-	Phosphonium salt
(8)	$(CH_3)_3C—Br$	+ H_2O	$\longrightarrow$ $(CH_3)_3C—OH$	+ HBr	Alcohol
(9)	$(CH_3)_3C—Br$	+ CH_3OH	$\longrightarrow$ $(CH_3)_3C—OCH_3$	+ HBr	Ether
(10)	$(CH_3)_3C—Br$	+ PhSH	$\longrightarrow$ $(CH_3)_3C—SPh$	+ HBr	Thioether

is a bromide salt. When the nucleophile is neutral (entries 6–10), the product may or may not carry a charge, depending on whether or not a proton is lost. An amine salt is a weak acid (entry 6), so it normally retains its protons unless a stronger base like NaOH is added after the reaction is complete. On the other hand, a protonated alcohol (the initial product of reaction between the alkyl bromide and water, entry 8) is a strong acid; so it loses a proton under the reaction conditions to generate the neutral alcohol and HBr as products.

The reagents listed in Tables 6.1 and 6.2 are not equally nucleophilic. In fact, the degree of nucleophilicity varies greatly and depends on several factors:

- For a given atom type or within a period of the periodic table, nucleophilicity parallels basicity; nucleophiles with a negative charge are usually better nucleophiles than their neutral analogues. For example,

$$C_2H_5O^- \quad > \quad OH^- \quad > \quad C_6H_5O^- \quad > \quad CH_3COO^- \quad > \quad H_2O$$
Ethoxide ion   Hydroxide ion   Phenolate ion   Acetate ion   Water

- For each group in the periodic table, nucleophilicity increases with increasing atomic size (polarizability of the electrons) so that $I^- > Br^- > Cl^- > F^-$ and $H_2S > H_2O$. This ordering depends on the solvent used, as you will see later. A smaller ion is often more highly solvated, making it is less nucleophilic.

- Steric effects can reduce the nucleophilicity of large species. For example, $(CH_3)_3CO^-$ (*tert*-butoxide ion) is more basic than $CH_3CH_2O^-$ (ethoxide ion), but *tert*-butoxide ion is a weaker nucleophile because its bulk prevents it from approaching an electrophilic center as readily.

EXAMPLE 6.1

Predict whether F^- or NH_2^- is the better nucleophile, and explain the reasons for your choice.

Both N and F are period 2 elements, so the nucleophilicities of F^- and NH_2^- should parallel their base strengths. Their conjugate acids are HF and NH_3, which have pK_a values of ~ 5 and 40, respectively (Table 5.1). Therefore, NH_2^- is the stronger base, which means it should be more nucleophilic than F^-.

EXERCISE 6.2

For each pair, predict which species is expected to be the better nucleophile, and explain the reasons for your choice.

a. SH^- or OH^- b. $P(CH_3)_3$ or $N(CH_3)_3$ c. NH_3 or H_2O

Another important point about nucleophiles: Even though nucleophilicity parallels base strength for many species, an ion or molecule can be a good nucleophile yet a weak base and vice versa. A nucleophile is always a Lewis base, so it is important to consider both properties—basicity and nucleophilicity—when looking at their reactions. Table 6.3 summarizes the properties of some common nucleophiles, which can vary depending on the specific reaction conditions.

6.1d THE LEAVING GROUP MUST BE ATTACHED TO AN SP³-HYBRIDIZED CARBON ATOM

For a nucleophilic substitution reaction to be successful, a nucleophile must be present, and the substrate molecule must have a good leaving group. A third required fea-

Table 6.3 A classification of nucleophiles according to their basicity and nucleophilicity properties.[a]

Weak Base (pK_a conjugate acid < 5)		Moderate Base (pK_a conjugate acid ~5–12)		Strong Base (pK_a conjugate acid > 15)	
Poor nucleophile	*Good nucleophile*	*Poor nucleophile*	*Good nucleophile*	*Poor nucleophile*	*Good nucleophile*
HSO_4^-	H_2O	$RCOO^-$	N_3^-	$(CH_3)_3CO^-$	OH^-
$ROSO_2^-$	ROH^c	R_3N	CN^-	LDA^d	RO^-
$H_2PO_4^-$	$RCOOH$		ArS^-		$H^{-\,e}$
F^-	H_2S		RS^-		NH_2^-
ROH^b	RSH		NH_3		
	R_3P		RNH_2		
	Cl^-		R_2NH		
	Br^-		ArO^-		
	I^-				

[a] Nucleophiles shown in color participate in S_N2 reactions (see Table 6.4).
[b] 2° and 3° alcohols.
[c] 1° alcohols.
[d] Lithium diisopropylamide = LDA.
[e] From $LiAlH_4$.

ture is that *the leaving group must be attached to an* sp³-*hybridized carbon atom;* that is, a carbon atom with four single bonds.

A carbon atom that has *sp³* hybridization can be classified by the number of other carbon atoms to which it is bonded (Section 1.4a). A primary carbon atom is attached to one other carbon atom, a secondary carbon atom to two other carbon atoms, and so on. Alkyl halides and alcohols can be classified likewise as methyl, primary (1°), secondary (2°), or tertiary (3°) according to the type of carbon atom to which the halogen atom or the OH group is attached.

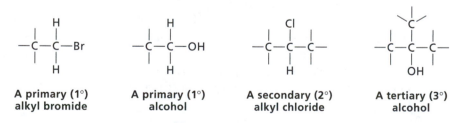

A primary (1°) alkyl bromide   A primary (1°) alcohol   A secondary (2°) alkyl chloride   A tertiary (3°) alcohol

A benzylic halide is one that has the halogen atom attached to an *sp³*-hybridized carbon atom adjacent to a benzene ring, and an allylic halide is one in which the halogen atom is bonded to an *sp³*-hybridized carbon atom *adjacent* to a double bond. Notice that allylic and benzylic halides and alcohols are also classified as 1°, 2°, or 3°.

A primary (1°) benzylic chloride   A primary (1°) benzylic alcohol   A primary (1°) allylic bromide   A tertiary (3°) allylic alcohol

When the halogen atom is attached to a carbon atom that has sp^2 hybridization, it is classified as vinyl or aryl. An aryl "alcohol" is a phenol, and a vinyl alcohol is unstable. *Vinyl and aryl halides and phenols do not undergo nucleophilic substitution under normal conditions.*

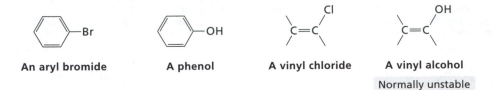

| An aryl bromide | A phenol | A vinyl chloride | A vinyl alcohol |
| | | | Normally unstable |

If we were to survey the substitution reactions of hundreds of different alkyl halides, we would find two limiting types of behavior. In one, the reaction rate is independent of the nucleophile concentration. In the other, the reaction rate increases upon raising the concentration of either the substrate or the nucleophile. By definition, the first mechanism is **unimolecular** (the reaction rate depends on the concentration of a single species), and the second one is **bimolecular** (the reaction rate is dependent on the concentration of two species).

The structure of the alkyl halide has a significant effect on which mechanism is observed:

- Tertiary alkyl and many secondary, benzylic, and allylic halides react via a process in which the nucleophile concentration is immaterial. This process is an **S_N1 reaction; S_N1** means *substitution, nucleophilic, unimolecular.*

$$R—X \ + \ Y^- \longrightarrow R—Y \ + \ X^- \quad \text{rate} = k[R—X]$$
R = 3°, 2°, benzylic, allylic

- Methyl, primary, and many secondary alkyl halides (including, in many cases, 1° benzylic and allylic halides) react with nucleophiles via a bimolecular process called an **S_N2 reaction; S_N2** stands for *substitution, nucleophilic, bimolecular.*

$$R—X \ + \ Y^- \longrightarrow R—Y \ + \ X^- \quad \text{rate} = k[R—X][Y^-]$$
R = methyl, 1°, 2°, and some benzylic, allylic

The molecularity of the reaction, however, is not the only parameter that varies according to the alkyl halide structure. The stereochemistry of each process also differs. Therefore, we will examine the detailed pathways of the S_N1 (Section 6.2) and the S_N2 (Section 6.3) mechanisms to see how structure and reactivity are related.

EXERCISE 6.3

Classify each of the following organobromides (1°, 2°, 3°, benzylic, etc.), and predict whether a substitution reaction will favor the S_N1, S_N2, both, or neither pathway:

a. $CH_3CH_2CH_2Br$

b.

c.

d. $CH_3—CH—CH_3$
             |
            Br

e.

6.2   THE S_N1 REACTION OF ALKYL HALIDES

6.2a   THE S_N1 REACTION IS OFTEN A SOLVOLYSIS PROCESS THAT PROCEEDS VIA A CARBOCATION INTERMEDIATE

If you studied chemical kinetics previously, you may recall that every reaction has a *rate-determining step* (*rds*), through which some or all of the reactants must pass as they are converted to products. The rate-determining step controls the overall reaction rate, just as the velocity of water flowing through a series of pipes is regulated by the diameter of the smallest one. Any step that occurs *after* the rds occurs at least at the same rate as the rds itself, so it has no effect on the reaction rate.

In a nucleophilic substitution reaction, two changes occur: The bond between the carbon atom and the leaving group is broken, and the bond between the carbon atom and the nucleophile is created. If the rate of a transformation is independent of the nucleophile concentration, as it is for the S_N1 reaction, *then the bond-forming process must occur after the rate-determining step,* which means that at least two separate steps are involved.

If we employ the curved-arrow formalism to depict electron movement during each stage of the reaction, as outlined in Section 5.3c, we can portray the S_N1 process for a generic bromoalkane that is consistent with the kinetic data. In the first step, the electron pair that is the carbon–bromide bond moves onto the more electronegative bromine atom, which dissociates as bromide ion and generates a carbocation intermediate.

In the first step, the leaving group (bromide ion) dissociates along with the electron pair.

This leaves the carbon atom with six electrons, so it adopts a planar structure.

In the next step, the electron pair of the nucleophile reacts with the electron-deficient carbon atom to yield the product. Because dissociation of the bromide ion in the first step is rate determining, the overall rate is independent of the nucleophile concentration, a fact that has been determined and confirmed experimentally.

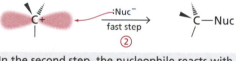

In the second step, the nucleophile reacts with the carbocation, generating a new bond.

The S_N1 mechanism predominates when three conditions are satisfied.

1. The substrate molecule can form a relatively stable carbocation. Tertiary, benzylic, and allylic halides form the most stable carbocations, and they are therefore the best substrates.

2. The nucleophile is a weak base, often the solvent itself. A reaction in which the solvent acts as the nucleophile is a **solvolysis reaction.**

3. The solvent is polar and *protic,* which means that is has a relatively acidic OH group (water, alcohols, and carboxylic acids). Because the S_N1 pathway produces a carbocation intermediate, polar solvents can stabilize this charged species through ion–dipole interactions (Table 2.5). Furthermore, the leaving group, a halide anion, is solvated (stabilized) by forming hydrogen bonds if the solvent is protic. These hydrogen bonds can form even before dissociation of the halide ion, and their formation assists in the bond-breaking step.

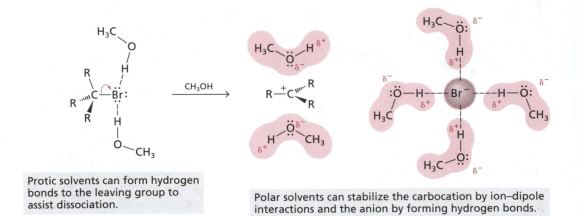

Protic solvents can form hydrogen bonds to the leaving group to assist dissociation.

Polar solvents can stabilize the carbocation by ion–dipole interactions and the anion by forming hydrogen bonds.

The nucleophile in a solvolysis reaction is uncharged, so the initial product formed by reaction of the carbocation with the solvent still carries a positive charge. In most instances, this initial product loses a proton to form a neutral product, as governed by the thermodynamics of the subsequent acid–base reaction. The following example illustrates the situation in which water is the nucleophile and a tertiary alkyl bromide is the substrate.

In the first step, bromide ion dissociates to form the intermediate carbocation.

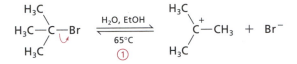

Then, an unshared pair of electrons on the oxygen atom of water reacts with the positively charged center to form an initial adduct, a protonated alcohol.

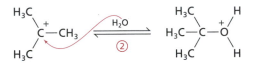

Finally, a proton is removed by the solvent, acting as a base, to generate the alcohol product. *The loss of a proton is actually a separate acid–base reaction that takes place after the substitution process is finished.*

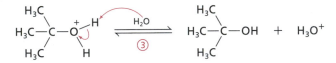

Note that each of the individual steps of the overall S_N1 reaction is an acid–base equilibrium. The carbocation is a Lewis acid in the first and second steps.

EXAMPLE 6.2

Propose a reasonable mechanism for the following solvolysis reaction, showing electron movement with curved arrows:

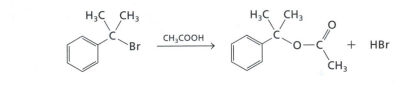

First, decide if the S$_N$1 mechanism is likely. The substrate has a bromide ion (good leaving group) attached to a 3° (and benzylic) carbon atom, and the solvent is a protic one, namely, acetic acid. Reaction via the S$_N$1 mechanism is highly probable.

The first step of the S$_N$1 mechanism is dissociation of the leaving group, so we show the electrons leaving with the bromide ion to form a carbocation. Start the arrow at the electron pair of the C–Br bond, and *make certain the charges balance on each side of the equation.*

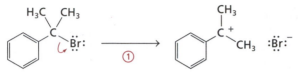

Next, the solvent acts as the nucleophile and intercepts the carbocation. We draw the arrow from electrons on one of the oxygen atoms to the carbocation center. Notice the positive charge on *each side of the equation.*

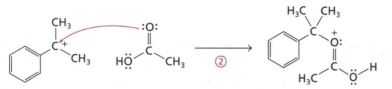

Because this is one of the first specific mechanisms that you have seen, a number of questions will undoubtedly arise. One may be, Why did the electrons on the double-bonded oxygen atom react, and not the electrons on the OH group of acetic acid? In complicated cases, you may have to draw all of the possible steps you can think of, and then choose which one looks most likely. This skill will come as you practice and learn the patterns that are common to many polar mechanisms.

In this case, we look at the possible resonance forms for acetic acid, which are shown below. The resonance form to the far right shows that the carbonyl oxygen atom has a negative charge (highlighted in color), so that oxygen atom is more basic, hence more nucleophilic, than the OH oxygen atom.

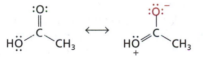

Returning to the mechanism of the exercise, the last step is an acid–base reaction in which the bromide ion removes a proton to generate the given products. Again notice that the charges balance: The left side of the equation has a plus and minus charge (overall charge = 0), and the right side of the equation has only neutral molecules (overall charge = 0).

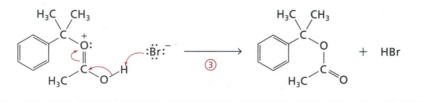

You may wonder why we chose to move the electrons as we did in Step (3) of Example 6.2. Do not forget the two main rules for drawing the curved arrows that represent electron movement in a mechanism (Section 5.3a):

- An arrow starts at a pair of electrons
- An arrow points to a nucleus

Most mechanistic problems will provide you with the structures of the reactant(s), product(s), and reagent(s). When proposing a mechanism, you cannot supply reagents besides the ones given. As you work through each step, be aware of the final products that are formed. Knowing which bonds have to form or break and which atoms have to be added or removed will often give clues as to how to move the electrons. For Step 3 of Example 6.2, this evaluation would be as follows:

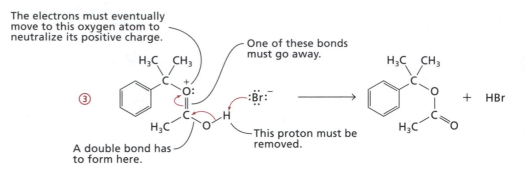

The electrons must eventually move to this oxygen atom to neutralize its positive charge.

One of these bonds must go away.

A double bond has to form here.

This proton must be removed.

EXERCISE 6.4

Propose a reasonable mechanism for the following solvolysis reaction, showing electron movement with curved arrows.

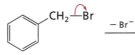

6.2b CARBOCATION FORMATION REQUIRES STABILIZATION BY DELOCALIZATION OR HYPERCONJUGATION

Dissociation of a halide ion from an alkyl halide is only one way by which a carbocation can be generated. With only six electrons on its carbon atom, however, a carbocation forms only when it can be stabilized. Electron delocalization (resonance) is one effect with which you are already familiar, and it accounts for the stability of benzylic and allylic carbocations (Section 2.7).

Benzylic halides

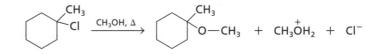

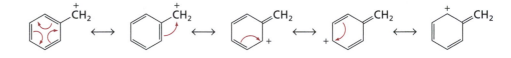

Allylic halides

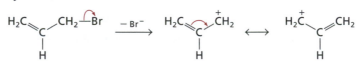

An oxygen atom (and other heteroatoms) also stabilizes a carbocation by resonance, as illustrated in the following scheme:

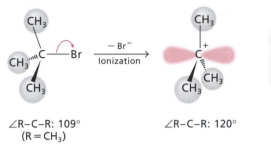

Saturated aliphatic carbocations lack π bonds and unshared electron pairs, so typical resonance forms do not exist. In the case of a 3° alkyl halide such as *tert*-butyl bromide, formation of a carbocation leads to relief of strain inherent in the tetrahedral geometry. By forming a planar species (Section 2.5e), the methyl groups move farther apart and relieve the effects of electron repulsion.

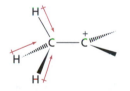

By losing Br⁻ and adopting a planar structure, the carbon center is less hindered because the bulky groups are farther apart.

∠R–C–R: 109°   ∠R–C–R: 120°
(R = CH$_3$)

In relieving these repulsive forces, which are destabilizing, a higher energy species is actually formed because the central carbon atom of a carbocation has only six electrons. Carbocation formation can occur only in those systems that can provide other stabilizing forces. In other words, relief from steric strain (the electron–electron repulsion between the substituents) is not enough to promote carbocation formation.

In saturated aliphatic systems, stabilization of a carbocation results from electron donation from alkyl group substituents. The methyl group, and to a lesser extent other alkyl groups, are considered to be electron donating because the additive effects (see Section 2.2a) of the carbon–hydrogen bond dipoles generate a dipole for the entire methyl group that is aligned with the sigma bond to the adjacent carbon atom. The negative end of this dipole points away from the methyl group; that is, the methyl group is donating electron density to the adjacent atom or group.

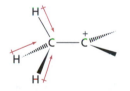
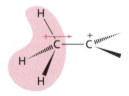

Individual bond dipoles for the carbon–hydrogen bonds.

The dipole for the methyl group resulting from the sum of the individual bond dipoles.

Stabilization of a carbocation also results from a phenomenon known as **hyperconjugation,** which is electron donation to the carbocation center that occurs by overlap between the vacant *p* orbital and an adjacent σ bond orbital, as shown below in (a.).

The sigma bond between carbon and hydrogen overlaps with the *p* orbital of the carbocation. Only one of the C–H bonds is aligned with the *p* orbital. The other two C–H bonds are staggered in their relationship to the *p* orbital.

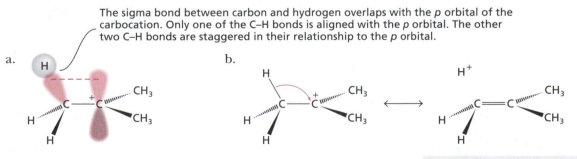

a.   b.

A "no-bond resonance form"

Using Lewis structures, we can represent hyperconjugation for the *tert*-butyl carbocation by drawing "no-bond resonance forms" as illustrated in the previous scheme in (b.). These unusual forms delocalize the positive charge onto a hydrogen atom of a neighboring methyl group.

The *tert*-butyl carbocation actually has three hyperconjugation contacts with which to stabilize the positive charge, as shown below in (a.). Generally, *the more highly substituted the carbocation, the more stable it is*. The *tert*-butyl carbocation has actually been isolated at low temperature and its structure determined using X-ray crystallographic methods. The ball-and-stick model of this structure constructed on the basis of the crystallographic data, shown below in (b.), reveals that the carbon atoms all lie in a plane.

a.   b.

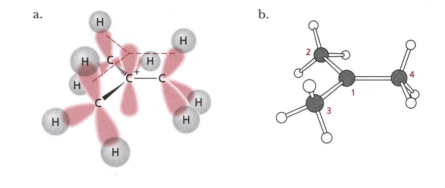

EXERCISE 6.5

Draw the three no-bond resonance forms for the *tert*-butyl carbocation that illustrate its stabilization by hyperconjugation.

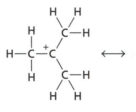

Primary carbocations do not exist because they are not sufficiently stabilized by hyperconjugation. A secondary carbocation can form in certain cases, depending on the amount of stabilization that it obtains from relieving steric congestion. Vinyl and phenyl carbocations are not stabilized as well as their aliphatic analogues because the effects of hyperconjugation are less significant, as illustrated below.

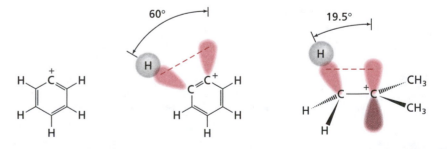

In the phenyl carbocation, the empty orbital is an sp^2-hybrid orbital that points away from the center of the ring. Overlap of this empty orbital with the adjacent C–H sigma bond is not as good as the overlap in an aliphatic carbocation because the orbitals are splayed farther apart.

By summarizing the foregoing discussion, we can arrange the different types of car-bocations on the basis of their stabilities as follows. An important consequence of this order is that *primary, vinyl, and aryl halides do not undergo S$_N$1 reactions.*

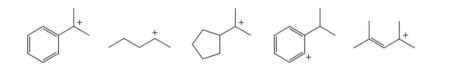

3° benzylic, allylic > 3°~2° benzylic, allylic > 2°~1° benzylic, allylic >> 1° > methyl > vinyl, phenyl

decreasing stability of the carbocation

These types of carbon atoms (in color) do not form carbocations in most cases.

EXERCISE 6.6

Arrange the following carbocations in order from least to most stable. First, expand the structures to show all of the carbon and hydrogen atoms, at least in the vicinity of the charge.

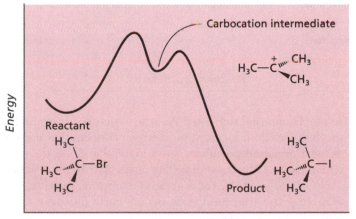

6.2c   THE HAMMOND POSTULATE PROVIDES INSIGHTS INTO THE TRANSITION STATES OF THE S$_N$1 REACTION

With an understanding of the ways by which a carbocation is stabilized, we return to the S$_N$1 reaction. In Chapter 5, you learned that electron movement is not the only way to convey information about reaction mechanisms; a reaction coordinate diagram (Section 5.4) sometimes provides additional insights. For the reaction between *tert*-butyl bromide and iodide ion, we can draw the reaction coordinate diagram shown in Figure 6.2 (this example uses iodide ion as the nucleophile so that we do not have to be concerned about the additional acid–base reaction that occur when water is the nucleophile).

Figure 6.2
A reaction coordinate diagram for the S$_N$1 reaction between $(CH_3)_3CBr$ and I^-.

As with any S$_N$1 process, there are two steps: In Step 1, the carbocation is formed by dissociation of bromide ion. In the second step, iodide ion reacts with the carboca-tion to form the product.

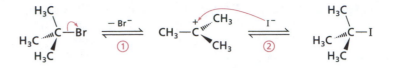

We can depict the reaction coordinate for each step as in Figure 6.3.

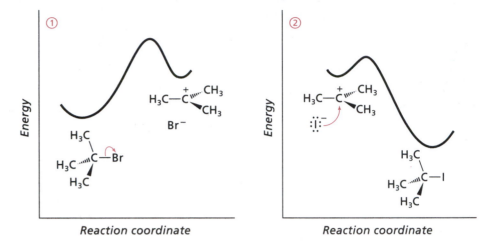

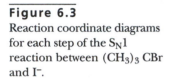

Figure 6.3
Reaction coordinate diagrams
for each step of the S_N1
reaction between $(CH_3)_3$ CBr
and I^-.

Each step has a transition state at higher energy than that of the carbocation intermediate. However, unlike a carbocation, which has a finite lifetime, the transition state of each step is transient, so it cannot be observed. For many reactions, we can only speculate as to its structure. In this instance, we can draw reasonable structures for points along the reaction coordinate of the first step, knowing that the bromide ion must eventually dissociate from the carbon atom. Thus, the bond must get longer and longer until it breaks. This exercise does not tell us which structure best represents the structure of the transition state, however.

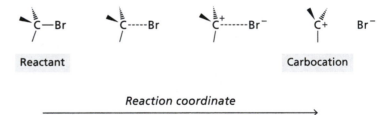

In 1955, George Hammond suggested that the structure of a transition state should resemble the species to which it is closest in energy, a concept that has subsequently become known as the **Hammond postulate.** For a general endergonic process, shown in the following scheme in (a.), the transition state occurs late in the reaction and its structure looks like that of the product. For a general exergonic process, shown in the following scheme in (b.), the transition state occurs early during the reaction and its structure resembles that of the reactant.

For the S_N1 reaction between *tert*-butyl bromide and iodide ion, the first step is endergonic, and the second step is exergonic. The Hammond postulate tells us that the first transition state will have a structure much like that of the carbocation, which is the product of the first step. Therefore, we expect that the C–Br bond will be long and

a. Endergonic reaction

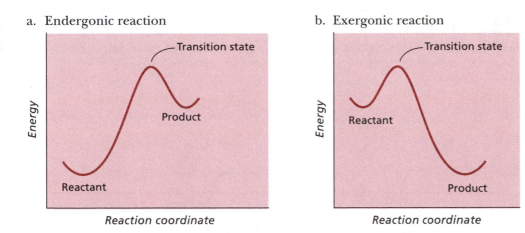

b. Exergonic reaction

weak in the transition state, and the carbon atom will bear most of the positive charge. The transition state of the second step will have a structure similar to its reactant, which is also the carbocation.

Let us now apply this insight to understand the relative rates of S$_N$1 reactions for *tert*-butyl bromide and isopropyl bromide. We already know that a tertiary carbocation is more stable than a secondary one, largely because of electron donation from the methyl groups and hyperconjugation (Section 6.2b). So the first steps in the S$_N$1 reactions of *tert*-butyl bromide and of isopropyl bromide can be represented by energy profiles that are related to the stabilities of the carbocations formed after dissociation of bromide ion, as pictured in Figure 6.4.

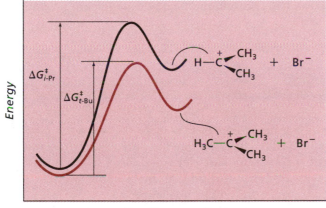

Figure 6.4
Energy profiles for the dissociation of bromide ion from $(CH_3)_3C–Br$ (black line) and from $(CH_3)_2CH–Br$ (colored line).

The Hammond postulate tells us that the first transition state in each of these S$_N$1 reactions will have a structure like that of the corresponding carbocation, so the pathway leading to the more stable carbocation must also have a lower free energy of activation. Therefore, S$_N$1 reactions of *tert*-butyl bromide will occur faster than those of isopropyl bromide because the free energy of activation is less. Applying this same type of analysis to other classes of organohalide substrates allows us to rank order the rates for the S$_N$1 reactions of various compounds, which are summarized as follows. Notice that this ranking is the same order as that for carbocation stability.

3° benzylic, allylic > 3° ~ 2° benzylic, allylic > 2° ~ 1° benzylic, allylic >> 1° > methyl > vinyl, phenyl

decreasing rate of reaction via the S$_N$1 mechanism   These types of organohalides (in color) do not undergo the S$_N$1 reaction.

EXERCISE 6.7

Within each set of compounds, predict which molecule would react the fastest and the slowest in an S_N1 reaction. Explain the reasons for your choices briefly.

a.   2-Bromoheptane   3-Bromo-3-methylhexane   2-Bromo-2-heptene

b.

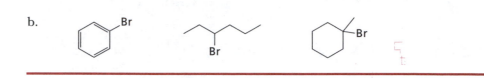

6.2d   A CARBOCATION CAN REARRANGE TO FORM A MORE STABLE CARBOCATION

The S_N1 reaction is not the only process in which a carbocation can participate. In fact, a carbocation can react in at least three ways:

1. *Reaction* with a Lewis base—this process is what occurs in the second step of the S_N1 reaction, as described in Section 6.2a.

2. *Elimination* of a proton from a neighboring carbon atom to form a new π bond. Elimination reactions will be discussed in Chapter 8.

3. *Rearrangement* to form a more stable carbocation.

The **Wagner–Meerwein rearrangement**—a rather imprecise term for many reactions that involve a carbocation intermediate—was first recognized by Frank Whitmore in 1932. A specific example, the reaction of 2-bromo-3-methylbutane with aluminum tribromide, illustrates an occasion in which rearrangement predominates.

2-Bromo-3-methylbutane   **2-Bromo-2-methylbutane**

The first step is a Lewis acid–base reaction between the alkyl bromide and aluminum tribromide. Recall that aluminum has only six electrons when bonded to three atoms, so many of its compounds are electron deficient, hence Lewis acids. Calculating formal charges for the adduct, we find that the bromine atom bears a positive charge, and the aluminum atom bears a negative charge.

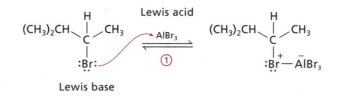

This species can ionize in one of two ways. The first is simply a reverse of the initial step; the second possibility involves loss of the tetrabromoaluminate ion to form a secondary carbocation.

A hydrogen atom *with its pair of electrons*—a hydride ion—subsequently migrates from the neighboring carbon atom to generate a more stable tertiary carbocation.

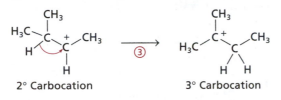

2° Carbocation   3° Carbocation

This carbocation in turn reacts with the tetrabromoaluminate ion, which is a nucleophile, to yield the rearranged product after dissociation of aluminum tribromide in another acid–base reaction.

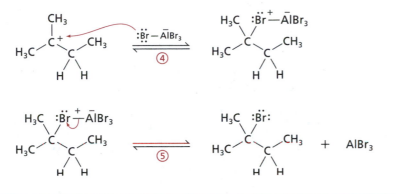

EXAMPLE 6.3

Draw the structures of the substitution products of the following S$_N$1 reaction. The products are formed by direct replacement of the leaving group as well as by rearrangement. Name each product.

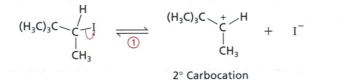

2-Iodo-3,3-dimethylbutane

In the first step, the iodide ion dissociates to produce a secondary carbocation.

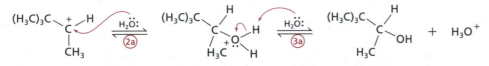

2° Carbocation

This carbocation can react with water to form the protonated alcohol, which undergoes an acid–base reaction to produce the secondary alcohol by loss of a proton. Even though both water and methanol (the solvents) can react as nucleophiles, water is a better nucleophile and often is present at a higher concentration than methanol, so it reacts preferentially. (If we wanted methanol to react, we would leave out water.) This product is named 3,3-dimethyl-2-butanol.

$(H_3C)_3C$ ... H₂Ö: (2a) ... H₂Ö: (3a) ... $+ H_3O^+$

3,3-dimethyl-2-butanol
(direct substitution product)

If a methyl group adjacent to the secondary carbocation center migrates, a more stable tertiary cation is formed. This example illustrates that groups other than hydride can change positions in order to generate a more stable carbocation.

Rearrangement step

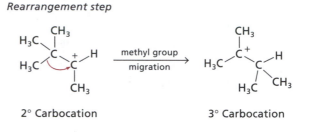

2° Carbocation   3° Carbocation

This 3° carbocation (a Lewis acid) subsequently reacts with water (a Lewis base) in the same way that the secondary carbocation did, producing the tertiary alcohol in this instance. The IUPAC name of the product is 2,3-dimethyl-2-butanol.

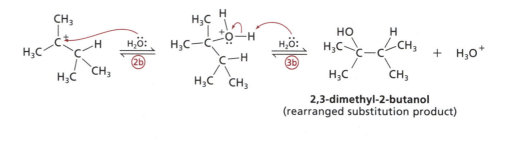

2,3-dimethyl-2-butanol
(rearranged substitution product)

A question that naturally comes up is, When does rearrangement occur? Most often, *a rearrangement process is recognized after the fact,* when an unexpected compound is identified as a product of a substitution reaction. If we observe the formation of a rearranged product, we conclude that a carbocation intermediate formed and rearranged. A priori, we can reasonably expect rearrangement any time that a carbocation is generated in which migration of a hydrogen atom or an alkyl group will produce a *more stable* carbocation. Thus, common instances are 2° → 3°, 2° → benzylic, and 3° → benzylic rearrangements. Because 1° carbocations are so high in energy, their formation, hence rearrangement, is less common. Some examples of carbocation rearrangements are illustrated in Figure 6.5.

EXERCISE 6.8

What are the substitution products of the following S_N1 reaction, formed by direct replacement of the leaving group as well as by rearrangement? Name each of the products.

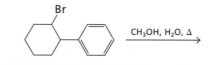

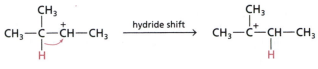

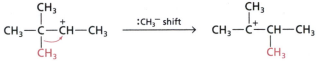

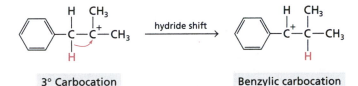

Figure 6.5
Examples of carbocation intermediates that are likely to rearrange.

6.2e THE S$_N$1 REACTION PRODUCES RACEMIC PRODUCTS FROM CHIRAL REACTANTS

If you carry out a nucleophilic substitution reaction under ideal S$_N$1 conditions starting with a chiral alkyl halide, you would find that the product is not optically active. The reason becomes apparent when you remember that the structure of the carbocation intermediate is planar, hence *achiral*. The nucleophile, iodide ion in the following example, can react with the carbocation from either side of the plane defined by the substituents. There is an equal chance for reaction at each side, so two chiral products are generated, in a ratio of 1:1. This is a racemic mixture, which is not optically active.

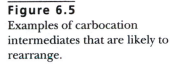

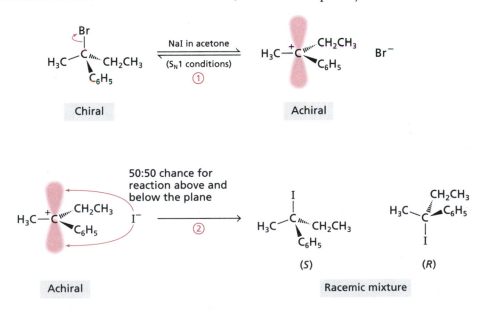

If the stereochemistry of a product is the same as that of a chiral starting material, we say that the reaction proceeds with **retention of configuration.** When the mirror-image product of a chiral starting material is formed, the reaction is said to proceed with **inversion of configuration.** If a racemic mixture is obtained from a chiral reactant, the reaction occurs with **racemization.** *The S_N1 reaction occurs with racemization.*

EXERCISE 6.9

Draw the structures of the expected substitution products of the following reaction (they are stereoisomers). Water is the nucleophile and Br^- is the leaving group. Clearly indicate the stereochemistry of each product, and assign the configuration of any stereocenters.

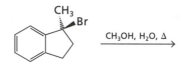

6.3   THE S_N2 REACTION OF ALKYL HALIDES

6.3a   THE S_N2 REACTION IS A CONCERTED PROCESS

Methyl, primary, and some secondary alkyl halides do not form stable carbocations, so their reactions with nucleophiles must proceed by a mechanism other than S_N1. The rates of these reactions vary with the concentration of both substrate and nucleophile, meaning that both reactants are involved in the rate-determining step. This type of process is an *S_N2 reaction.*

Nucleophilic substitution reactions are polar processes and involve both a nucleophile and an electrophile. An organic compound with a heteroatom has an electrophilic carbon atom, so a nucleophile is attracted to this carbon atom to initiate the substitution process. At some stage, the leaving group departs.

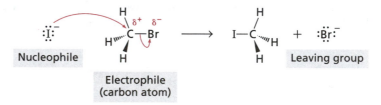

As the new bond forms between the electrophilic carbon atom and the nucleophile, the carbon atom begins to gain additional electrons. If the new bond formed *before* the leaving group departed, then the carbon atom would have 10 electrons, which is improbable for a second-row element like carbon. Therefore, the process illustrated below *cannot* represent the mechanism.

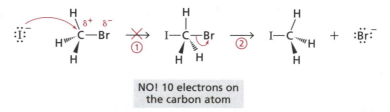

NO! 10 electrons on the carbon atom

If the leaving group dissociates first—before the new bond forms—then the mechanism would be S_N1, which is inconsistent with the kinetics results. *Therefore, the leaving group must depart at the same time that the new bond is formed.* When bond formation and bond breaking occur simultaneously, the process is **concerted.**

A reaction coordinate diagram for the S$_N$2 reaction is shown in Figure 6.6. Unlike the S$_N$1 mechanism, an S$_N$2 reaction does not generate a discrete intermediate. Instead, the transition state is a five-coordinate species. Normally, five bonds to carbon violates the octet rule, but in this transition state, bonds from carbon to the leaving group and nucleophile are only partial bonds. When we show a transition state structure in a mechanism, we place it within brackets with a double dagger superscript [ ]‡. The overall charge (if any) is also included as a superscript.

The involvement of a five-coordinate transition state in the S$_N$2 reaction explains the relative rates observed for these transformations. When smaller substituents are attached to the carbon atom bearing the leaving group, fewer electrostatic repulsion forces occur in the transition state, so the free energy of activation is lower. A methyl halide should therefore react the fastest among possible substrates in an S$_N$2 reaction because it is the least hindered.

methyl > 1° (including benzylic, allylic) > 2° (including benzylic, allylic) >> 3° (including benzylic, allylic) > vinyl, phenyl

decreasing rate of reaction via the S$_N$2 mechanism

These types of organohalides (in color) do not undergo the S$_N$2 reaction.

Tertiary alkyl halides do not react by an S$_N$2 mechanism because steric repulsions among three alkyl groups, a nucleophile, and the leaving group would make the energy of the transition state too great. The electrophilic carbon atom of a tertiary alkyl halide still attracts the nucleophile, but the nucleophile removes a hydrogen atom from the adjacent carbon atom, leading to elimination, as described in the Chapter 8.

Aryl and vinyl halides also do not react by the S$_N$2 mechanism, but for a different reason. For them, the leaving group is attached to an sp^2-hybridized carbon atom, so the carbon–leaving group bond is stronger and the carbon is not as electrophilic.

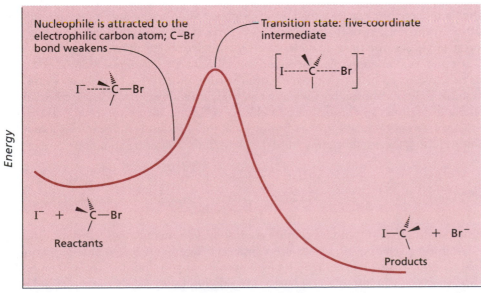

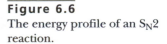

Figure 6.6
The energy profile of an S$_N$2 reaction.

6.3b   THE STEREOCHEMICAL COURSE OF AN S$_N$2 REACTION IS INVERSION OF CONFIGURATION

Nearly 100 years ago, Paul Walden demonstrated that (+)-malic acid (2-hydroxy-butanedioic acid) could be converted to either (+) or (−)-chlorosuccinic acid (2-chlorobutanedioic acid) with different reagents. Although the absolute configuration of each substance was not known at the time, it was clear that one of these processes occurred by inversion of configuration at the stereocenter, and the other by retention.

Not until 1923, and in the years following, did Joseph Kenyon and Henry Phillips carry out a series of investigations definitively showing that the stereochemistry of a chiral substrate is normally inverted during the course of an S$_N$2 reaction.

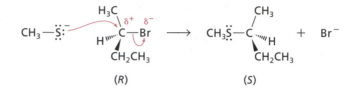

To commend Walden for his pioneering work on the stereochemical course of substitution reactions, we sometimes refer to the stereochemistry of an S$_N$2 reaction as **Walden inversion.**

EXERCISE 6.10

What is the major product for each of the following S$_N$2 reactions? In each instance, assign the absolute configuration (*R* or *S*) of the chiral center in reactant and product. The solvents for these reactions are DMSO and acetone, respectively. Identify the nucleophile in each reaction.

a.   b.

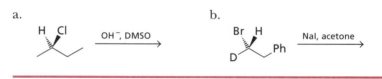

6.3c   SOLVENT EFFECTS HAVE A PROFOUND INFLUENCE ON S$_N$2 REACTIONS

When a nucleophile carries a negative charge, solvation of the anion can vary substantially with the polarity of the solvent. This degree of solvation influences the S$_N$2 reaction strongly because the rate depends on the nucleophile concentration.

Certain combinations of nucleophile and solvent are used almost without exception. For example, reactions with sodium iodide are nearly always done in acetone because NaI is soluble in it. In contrast, both sodium chloride and sodium bromide are essentially insoluble in acetone, so conversion of a chloro- or bromoalkane to an iodoalkane (the *Finkelstein reaction*) is readily driven by precipitation of sodium chloride (or sodium bromide), shifting the equilibrium toward the right in the following equation.

Sodium chloride and sodium bromide are insoluble in many other solvents, too.

Anionic nucleophiles, which are common reagents in S$_N$2 reactions, require a polar solvent to dissolve them. So let us suppose that you want to carry out an S$_N$2 reaction in which you intend to replace a bromide ion by an iodide ion. Solid NaI is first added to a solvent. If a moderately polar solvent is used, the substance may dissolve to a certain extent, but remain as an ion pair, which means that the ions remain associated with each other. This ion–ion attraction between the sodium cation and the iodide anion will mask the nucleophilicity of I⁻. If a highly polar solvent is used, then the individual ions will be separated, and interactions with the solvent will then become important.

$$\text{Na}^+\text{I}^-\text{(s)} \xrightarrow{\text{solvent}} \text{Na}^+\cdots\text{I}^-\text{(solvated)} \longrightarrow \text{Na}^+\text{(solvated)} \quad \text{I}^-\text{(solvated)}$$

Ion pair   Solvated ions

If the solvent is protic—methanol, for example—then the compound will dissolve well, but the nucleophile will engage in significant hydrogen-bond formation, as illustrated below.

Protic solvent

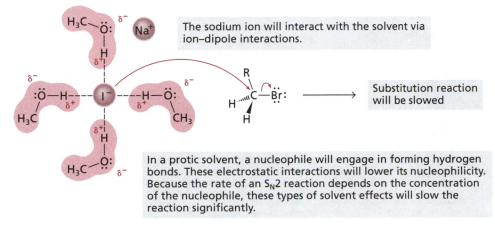

The sodium ion will interact with the solvent via ion–dipole interactions.

Substitution reaction will be slowed

In a protic solvent, a nucleophile will engage in forming hydrogen bonds. These electrostatic interactions will lower its nucleophilicity. Because the rate of an S$_N$2 reaction depends on the concentration of the nucleophile, these types of solvent effects will slow the reaction significantly.

The effect of forming hydrogen bonds will lower the nucleophilicity of the reactant (iodide ion in this case), which will impede the substitution reaction.

If the solvent is aprotic and polar, then the compound will dissolve well because the sodium ion will be solvated, and the nucleophile will be reactive because it is separated from its counterion. The iodide ion will also interact with the solvent via ion–dipole forces, but these are less significant than hydrogen bonds are. Furthermore, the positive end of the dipole is often shielded by substituents in the solvent molecule. In an aprotic solvent, therefore, the nucleophile will be highly reactive, and the substitution reaction will be facile.

Polar, aprotic solvent

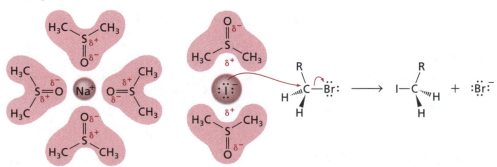

In a polar, aprotic solvent, the cation portion of the reagent will be stabilized by ion–dipole interactions with the solvent, but the nucleophile will not form hydrogen bonds with the solvent. The normal ion–ion attraction between the cation and anion will therefore be broken up, which will make the nucleophile seem as if it has a greater charge or higher concentration. This effect will enhance the rate of the substitution process.

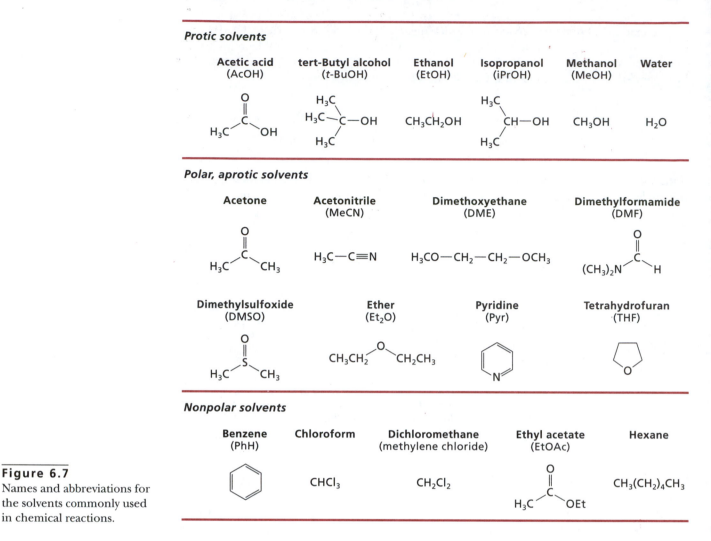

Figure 6.7
Names and abbreviations for the solvents commonly used in chemical reactions.

For this reason, many S_N2 reactions are carried out in **polar, aprotic solvents,** the structures of which are illustrated in Figure 6.7, along with protic and nonpolar solvents commonly used in chemical reactions. Tetrahydrofuran (THF) is an especially good solvent for lithium salts of nucleophiles.

6.3d AMMONIA AND AMINES REACT VIA THE S_N2 PATHWAY, BUT A MIXTURE OF PRODUCTS IS OFTEN OBTAINED

An S_N2 reaction commonly makes use of an anionic nucleophile and yields a neutral product. When ammonia or an amine is used as the nucleophile, however, the product is charged. Hydroxide ion is added in a second step to facilitate isolation of the neutral organic product.

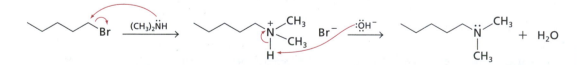

A complication, however, is that ammonia and amines are bases themselves, so the deprotonation step, shown above as a separate step, can occur even as the reaction proceeds.

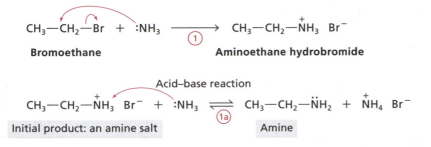

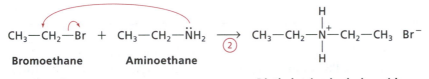

The product of the acid–base reaction (1a), aminoethane, is a *better* nucleophile than ammonia, because aminoethane is more basic (Section 5.2e). Therefore, it begins to compete with ammonia for reaction with the alkyl halide.

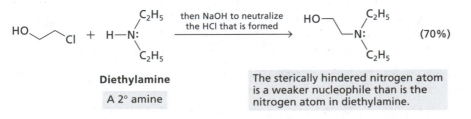

Diethylamine, which is the product of this second substitution step (after deprotonation by ammonia; step not shown) is even more basic than either aminoethane or ammonia (Section 5.2e). As its concentration increases, it *also* begins to compete for reaction with the alkyl halide. The result is a mixture of products, which is summarized in the following equation. Even under the conditions given, which use a large excess of ammonia, a substantial amount of diethylamine (and even some triethylamine) is obtained.

$$CH_3-CH_2-Br \quad \xrightarrow[\substack{\text{2. NaOH (to neutralize} \\ \text{the amine salt)}}]{\substack{\text{1. 16-fold excess of NH}_3 \\ \text{in EtOH}-H_2O}} \quad \begin{array}{ll} CH_3-CH_2-NH_2 & (34\%) \\ (CH_3-CH_2-)_2NH & (57\%) \\ (CH_3-CH_2-)_3N & (1\%) \end{array}$$

Bromoethane

The point is this: The preparation of a pure primary amine is difficult to accomplish using a substitution reaction between an alkyl halide and ammonia. The same is true for the reaction of a primary amine with an alkyl halide, which produces a mixture of secondary and tertiary amines. However, a secondary amine reacts to form mainly a tertiary amine. The latter is hindered enough that a further reaction is slow, and a tertiary amine is *less* basic (Section 5.2e), hence less nucleophilic, than a secondary amine (Table 6.3).

A tertiary amine, as the principal reactant, can function as a nucleophile to form a quaternary ammonium salt. In the majority of cases, this transformation is limited to the use of methyl iodide as the reagent, because methyl iodide is the only alkyl halide that reacts fast enough with the weakly nucleophilic 3° amine. In the product, the nitrogen atom in fully alkylated. It has no unshared electron pair and is no longer nucleophilic.

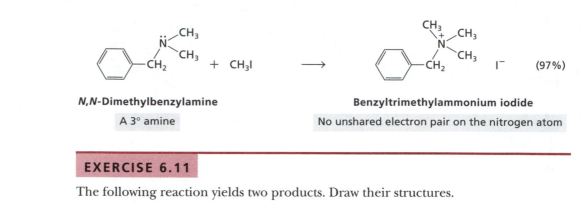

N,N-Dimethylbenzylamine

A 3° amine

Benzyltrimethylammonium iodide

No unshared electron pair on the nitrogen atom

EXERCISE 6.11

The following reaction yields two products. Draw their structures.

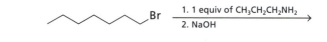

A practical substitution reaction that makes use of an amine nucleophile is the cyclization of a haloalkyl amine. As with most ring-forming processes, three- to six-membered rings can be made, but the four-membered ring is formed extremely slowly. Cyclization is favored by having a dilute solution of the aminoalkyl halide so that intermolecular substitution is disfavored. (Remember that the rate of an S_N2 reaction is proportional to the concentration of *each* reactant. In *dilute* solution, the rate of reaction between an amine group on one molecule and the halide end of a second molecule is therefore slow.)

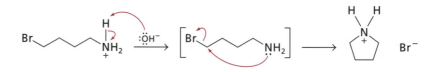

Ring size	Rate constant
3	0.036
4	0.0005
5	30.0
6	0.5
7	0.001

The haloalkylamine starting material is often prepared in its protonated (conjugate acid) form. Hydroxide ion is then added to deprotonate the nitrogen atom, unveiling the unshared pair of electrons. Ring formation involves an S_N2-like transition state and occurs with nucleophilic displacement of the halide ion by the nitrogen atom.

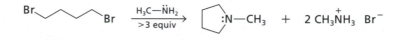

A dihaloalkane, when treated with a primary amine, also yields the corresponding cyclic product, especially if a five- or six-membered ring is being formed.

EXERCISE 6.12

Propose steps for the mechanism of the cyclization reaction directly above. Show electron movement using curved arrows. Number each step.

6.4   A COMPARISON OF THE S$_N$1 AND S$_N$2 MECHANISMS

In this chapter, you have learned that nucleophilic substitution reactions of alkyl halides occur readily. The features required by the reactants of these reactions include the presence of

- A good leaving group, which is a weak base (Cl$^-$, Br$^-$, or I$^-$ for a haloalkane).
- An sp^3-hybridized carbon atom bonded to the leaving group.
- A nucleophile (Lewis base).

You have also learned two common mechanisms: S$_N$1 and S$_N$2. These reaction pathways actually represent the *limiting* mechanisms that are possible; not all reactions fall conveniently into one of these two categories, however. Without collecting kinetic data to determine whether the rate law of a reaction is unimolecular or bimolecular, we take advantage of other information to decide which mechanism is operating.

The simplest criteria that we can use to decide on the mechanism of a substitution reaction are the structure of the alkyl halide and the type of nucleophile undergoing reaction. A summary of these data is given in Table 6.4.

Nucleophiles can be classified as poor or good and are further categorized by their basicities (Table 6.3). Nucleophiles that are poor do not react appreciably in either the S$_N$1 or S$_N$2 mechanism, so they are not included in Table 6.4. For weak base nucleophiles, both mechanisms are possible, but only the nucleophiles shown in color in the weak base category of Table 6.3 react with methyl and primary alkyl halides. Secondary and tertiary substrates can form carbocations, so they react with good, weak base nucleophiles via the S$_N$1 mechanism.

Table 6.4   A summary of the expected reactivity patterns for the nucleophilic substitution reactions of alkyl halides.

Substrate structure	Carbon atom classification	Weak base, good nucleophile[a]	Moderate base, good nucleophile[a]	Strong base, good nucleophile[a]
CH$_3$—X	Methyl	S$_N$2[b]	S$_N$2	S$_N$2
RCH$_2$—X	1°	S$_N$2[b]	S$_N$2	S$_N$2
R$_2$CH—X	2°	S$_N$1[c]	S$_N$2[d]	e
R$_3$C—X	3°	S$_N$1[c]	e	e
[benzylic structure]	1° benzylic 2° benzylic 3° benzylic	S$_N$1[c], S$_N$2[b] S$_N$1[c] S$_N$1[c]	S$_N$2 S$_N$2[d] e	S$_N$2[d] e e
[allylic structure]	1° allylic 2° allylic 3° allylic	S$_N$1[c], S$_N$2[b] S$_N$1[c] S$_N$1[c]	S$_N$2 S$_N$2[d] e	S$_N$2[d] e e
[aryl structure]	Aryl	No reaction	No reaction	No reaction
[alkenyl structure]	Alkenyl (vinyl)	No reaction	No reaction	e

[a] See Table 6.3.
[b] These results are for the nucleophiles in this category that are shown in color; otherwise, no reaction.
[c] These results are for all nucleophiles in this category except R$_3$P, which do not generally react with 3° alkyl halides.
[d] These substrates may also give elimination products (see Chapter 8).
[e] These substrates give only elimination products if they react (see Chapter 8).

With good moderate base nucleophiles, there are four possible results: substitution via the S$_N$2 mechanism, formation of a mixture of substitution and elimination products, elimination, or no reaction. Aryl and alkenyl substrates do not react at all. Tertiary substrates undergo elimination solely, while methyl and primary alkyl halides undergo substitution exclusively. Secondary substrates can give a mixture of products, especially if the nucleophile is a strong base.

Notice that secondary substrates can participate in S$_N$1, S$_N$2, *and* elimination processes. To decide whether the S$_N$1 or S$_N$2 mechanism predominates for a 2° substrate, we can sometimes consider another feature that differentiates the two mechanisms—the stereochemical outcome of the reaction when a chiral alkyl halide is employed as the reactant. These results are summarized in Figure 6.8.

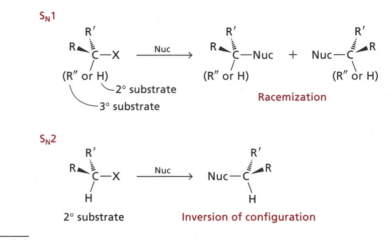

Figure 6.8
The reaction stereochemistry associated with the S$_N$1 and S$_N$2 mechanisms.

Notice that primary substrates are rarely chiral (except for molecules that have a deuterium instead of a hydrogen atom as in Exercise 6.10b), so only 2° and 3° alkyl halides have to be considered. Because 3° alkyl halides do not undergo the S$_N$2 reaction, the use of stereochemistry as a mechanistic probe is important only when dealing with 2° alkyl halides.

Although the stereochemical results are clear for the *ideal* S$_N$1 and S$_N$2 mechanisms, some substrates give mostly inversion of configuration even when the mechanism is S$_N$1. This result is a consequence of *ion pairing*, which means that the leaving group stays associated with the carbocation. If an ion pair is formed, then one side of the carbocation—the side the halide ion dissociated from—is shielded from approach of the nucleophile. In these cases, inversion of configuration occurs.

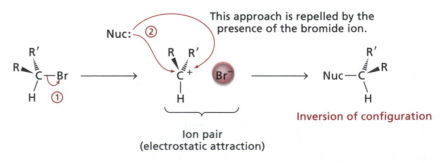

This result speaks to a point made at the beginning of this section, namely, that *the S$_N$1 and S$_N$2 pathways are limiting mechanisms*. Some reactions have characteristics of both mechanisms. Thus, the criteria used to classify reactions are not always unambiguous. For the most part, we will make use of the summary information in Table 6.4 and Figure 6.8 to decide whether a reaction occurs by the S_N1 or S_N2 mechanism.

EXAMPLE 6.4

Which nucleophilic substitution mechanism (S_N1 or S_N2) is expected to operate in each of the following reactions? Draw the structure of the major product, including stereochemistry.

a.

---

a. First, classify the type of substrate and the nucleophile. The alkyl chloride in this re-action is primary, and the nucleophile is a strong one. Furthermore, the solvent is a polar, aprotic one. The data in Table 6.4 indicates that this reaction should occur via the S_N2 mechanism. The starting materials are achiral, so we do not have to decide about reaction stereochemistry. The product is an azidoalkane (Table 6.2).

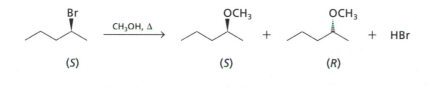

b. First, classify the type of substrate and the nucleophile. This alkyl bromide is sec-ondary, and the nucleophile is a weak base, neutral, and functions as the solvent. The data in Table 6.4 suggests that this reaction should occur via the S_N1 mechanism. The sub-strate is chiral, so racemization should occur. The product is an ether (Table 6.2).

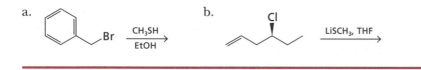

EXERCISE 6.13

Which nucleophilic substitution mechanism (S_N1 or S_N2) is expected to operate in each of the following reactions? Draw the structure of the major product, including stereochemistry.

a.   b.

Section 6.1   Fundamental aspects of substitution reactions

- Nucleophilic substitution occurs when a good leaving group attached to an sp^3-hybridized carbon atom is replaced by a nucleophile.
- A good leaving group is a weak base (the conjugate base of a strong acid).
- Nucleophilicity increases with base strength for a given atom type, or with in-creasing polarizability within a group of the periodic table. Typical properties of nucleophiles are summarized in Table 6.3.

- Bulky nucleophiles are less reactive than smaller ones because of steric effects.

- Two limiting mechanisms for nucleophilic substitution reactions are S_N1 (substitution, nucleophilic, unimolecular) and S_N2 (substitution, nucleophilic, bimolecular).

Section 6.2   The S_N1 reaction of alkyl halides

- The S_N1 reaction is favored in those cases involving substrate molecules that can form a relatively stable carbocation. Use of a weak nucleophile and a polar solvent also favor the S_N1 pathway.

- An S_N1 process, in which solvent also functions as the nucleophile, is called a solvolysis reaction.

- A carbocation is stabilized through the delocalization of electrons by resonance or hyperconjugation. Relief of steric crowding assists in the loss of a leaving group to form the positively-charged carbon atom.

- Hyperconjugation occurs when there is overlap between an adjacent σ-bond and the vacant p orbital of the carbocation.

- Relative rates for the reactions of substrates in an S_N1 process decrease in the order benzylic, allylic > 3° > 2° > 1° > methyl > vinyl, phenyl.

- The Hammond postulate correlates transition state energies with structures of reactants, products, and intermediates in a reaction. Relative rates for substrate reactivity in the S_N1 reaction correlate with relative stabilities of the resulting carbocations.

- In addition to formation of a bond by reaction with a nucleophile, a carbocation can also undergo elimination or rearrangement

- A rearrangement is likely to occur when a more stable carbocation can be formed by migration of a hydrogen atom or alkyl group from a carbon atom adjacent to the original carbocation center.

- The stereochemical consequences of a substitution reaction are retention, inversion, or racemization of configuration of the carbon atom bearing the leaving group.

- The typical S_N1 reaction proceeds by racemization of configuration.

Section 6.3   The S_N2 reaction of alkyl halides

- The S_N2 reaction is a concerted process that proceeds with formation of a bond to the nucleophile while the bond to the leaving group is being broken.

- The relative rates for substrate reactivity in an S_N2 process decrease in the order methyl > 1° > 2° > 3° > vinyl, phenyl.

- The S_N2 reaction is best carried out in a polar, aprotic solvent such as DMF, DMSO, acetone, or THF. A protic solvent can form hydrogen bonds with the nucleophile, which slows the reaction significantly.

- The typical S_N2 reaction proceeds by inversion of configuration, a process called Walden inversion.

- The preparation of amines by substitution is complicated by acid–base reactions between the starting nucleophile and the product, which is often a better nucleophile than the original.

Section 6.3   A comparison of the S_N1 and S_N2 mechanisms

- A correlation between substrate structure, nucleophile characteristics, and the types of mechanisms is given in Table 6.4.

Introduction
nucleophilic substitution

Section 6.1a
leaving group

Section 6.1d
unimolecular
bimolecular
S_N1 reaction
S_N2 reaction

Section 6.2a
rate-determining step
solvolysis reaction

Section 6.2b
hyperconjugation

Section 6.2c
Hammond postulate

Section 6.2d
Wagner–Meerwein
    rearrangement

Section 6.2e
retention of configuration
inversion of configuration
racemization

Section 6.3a
concerted

Section 6.3b
Walden inversion

Section 6.3c
polar aprotic solvent

Section 6.1c

Examples of typical nucleophilic substitution reactions: Table 6.2

Section 6.2a

The S_N1 reaction (solvolysis). The nucleophile is a weak base, usually a protic solvent such as water or alcohol. The order of reactivity of the organohalide is 3° benzylic, allylic > 3° ~ 2° benzylic, allylic > 2° ~ 1° benzylic, allylic > 1° > methyl > phenyl, vinyl. A chiral substrate undergoes racemization via the S_N1 mechanism.

$$-\overset{|}{\underset{|}{C}}-X \xrightarrow[\text{H—Sol}=H_2O, ROH, RCOOH]{\text{H—Sol}} -\overset{|}{\underset{|}{C}}-Sol \;+\; HX$$

X = Cl, Br, I

Section 6.2e

Carbocations rearrange to form more stable species; typical examples are shown in Figure 6.5.

Section 6.3a

The S_N2 reaction. The nucleophile is often a moderate base and usually carries a negative charge. The best solvents are polar and aprotic (no OH group). The order of reactivity of the organohalide is methyl > 1° (including benzylic and allylic) > 2° (including benzylic and allylic) > 3° (including benzylic and allylic) > phenyl, vinyl. A chiral substrate undergoes substitution with inversion of configuration.

$$-\overset{|}{\underset{|}{C}}-X \xrightarrow{\text{Nuc}^-} -\overset{|}{\underset{|}{C}}-Nuc \;+\; X^-$$

X = Cl, Br, I

Section 6.3d

The synthesis of amines from alkyl halides and ammonia or amines is complicated by the formation of multiple products.

$$R-X \xrightarrow{NH_3 \,(xs)} R-NH_2 \;+\; R_2NH \;+\; R_3N \;+\; R_4N^+X^-$$

X = Cl, Br, I
(xs) = excess

Product mixture; the quantity of each depends on the exact reaction conditions and reagents

6.14.   Draw structures for the reactant and the expected substitution product in each of the following reactions. Include stereochemistry where appropriate. If no reaction is expected, write N. R.

   a. (S)-CH$_3$CHBrCH$_2$CH$_2$CH$_3$ + NaCN (in DMF) →

   b. (R)-CH$_3$CHClCH(CH$_3$)$_2$ + NaI (in acetone) →

   c.   BrCH$_2$CH$_2$CH$_2$CH$_3$ + (CH$_3$)$_3$NH (in THF)

   d.   *cis*-4-Bromo-1-*tert*-butylcyclohexane + NaSCH$_3$ (in DMSO) →

6.15.   For each pair of reactants, which one do you expect to react faster in the indicated substitution reaction? Why? For each reaction, indicate whether the S$_N$1 or S$_N$2 pathway is the more likely.

   a.

   CH$_3$I + OH$^-$ $\xrightarrow{\text{DMSO}}$ CH$_3$OH + I$^-$

   CH$_3$I + OAc$^-$ $\xrightarrow{\text{DMSO}}$ CH$_3$OAc + I$^-$

   b.

   (CH$_3$)$_3$CBr + H$_2$O ⟶ (CH$_3$)$_3$COH + HBr

   (CH$_3$)$_2$CHBr + H$_2$O ⟶ (CH$_3$)$_2$CHOH + HBr

   c.

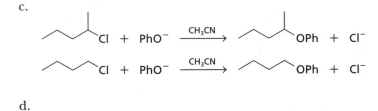

   d.

   ◇—Br + CN$^-$ $\xrightarrow{\text{DMF}}$ ◇—CN + Br$^-$

   ⬡—Br + CN$^-$ $\xrightarrow{\text{DMF}}$ ⬡—CN + Br$^-$

6.16.   Explain how and why the solvolysis of 2-bromo-3-methylbutane in aqueous methanol yields 2-methyl-2-butanol as the major product.

6.17. (R)-2-Bromooctane was stirred with sodium iodide in acetone at room temperature for 2 days. Write an equation for this reaction. Further experimentation showed the rate of the reaction was dependent on the concentrations of both the bromooctane and the sodium iodide, yet the product was a racemic mixture. Rationalize the stereochemical result.

6.18.   Draw a Lewis structure for the nitrate [NO$_3^-$], dimethylphosphate [(CH$_3$O)$_2$PO$_2^-$], and methylsulfate [CH$_3$OSO$_3^-$] ions. Explain why these species generally are poor nucleophiles in substitution reactions.

6.19. Silver(I) halides are among the least soluble compounds known. Explain how Ag(I) might facilitate solvolysis of a 2° alkyl chloride in alcohol as the solvent.

6.20. What is the expected major product for each of the following transformations? Indicate stereochemistry where appropriate and assign (R) and (S) to each stereocenter. In no reaction occurs, write N. R.

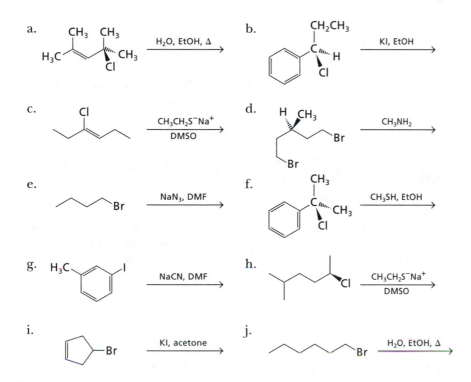

6.21. Compounds that have an aryloxy (ArO–) group at the α-position of a carboxylic acid are frequently used in agricultural treatments (see Section 5.5b if you do not remember what the "α" designation means).

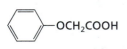

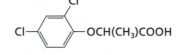

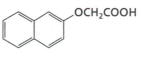

Phenoxyacetic acid   **2-(2′,4′-Dichlorophenoxy)propanoic acid**   **2-Naphthoxyacetic acid**

(fungicide)   (plant growth regulator and herbicide)   (plant hormone)

Such compounds are made by the reaction between a phenolate ion and an α-chloro carboxylic acid:

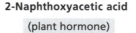

Draw the structures of the specific phenolate ion nucleophiles and α-chloro carboxylic acids needed to make the compounds shown above.

6.22. Indicate what reagent(s) and solvents can be used to carry out each of the following transformations:

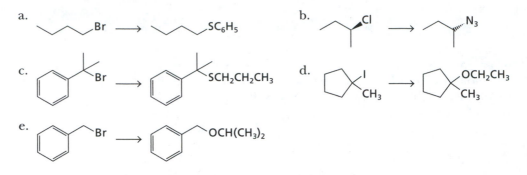

a.

b.

c.

d.

e.

6.23. For each of the following nucleophilic substitution reactions, identify which ones will probably occur and which ones will be unlikely or occur very slowly. Briefly explain your reasoning.

a. $CH_3CH_2Cl + I^- \rightarrow CH_3CH_2I + Cl^-$

b. $CH_3CH_2OH + Br^- \rightarrow CH_3CH_2Br + OH^-$

c. $CH_3CH_2Br + CN^- \rightarrow CH_3CH_2CN + Br^-$

d. $CH_3CH_2CN + SCH_3^- \rightarrow CH_3CH_2SCH_3 + CN^-$

6.24. A mechanism designated as S_N2' occurs whenever the π bond of an allylic substrate can participate in the substitution process. Show the electron movement for the following transformation that involves competing S_N2 and S_N2' reactions.

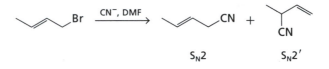

6.25. Allylic rearrangements like that shown in Exercise 6.24 occur *any time* an S_N1 process is involved. What is the difference in the involvement of electrons for the S_N2/S_N2' process compared with the S_N1 reaction?

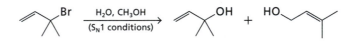

SUBSTITUTION REACTIONS OF ALCOHOLS AND RELATED COMPOUNDS

This chapter continues the presentation of nucleophilic substitution reactions begun in Chapter 6. The material here covers the reactions that are typical for alcohols, ethers, and their sulfur analogs, and it concludes with a description of nucleophilic substitution reactions that are important in biochemical processes. These examples validate the previously stated principle (Section 5.5a) that reactions in living systems follow the same conventions as those that take place in the laboratory.

7.1 SUBSTITUTION REACTIONS OF ALCOHOLS

7.1a A STRONG ACID PROTONATES THE ALCOHOL OH GROUP TO GENERATE A GOOD LEAVING GROUP

For the nucleophilic substitution reactions of alkyl halides, you learned that the substrate molecule has an sp^3-hybridized carbon atom bonded to a good leaving group, which is a weak base, and the other reactant is a nucleophile (Lewis base). You also learned that there are two common mechanisms: S_N1 and S_N2. These pathways differ in their stereochemical outcomes, the types of substrates and nucleophiles that normally participate, and the types of solvents that prove useful. An abbreviated summary is given below (also see Table 6.4).

Table 7.1   A summary of the S_N1 and S_N2 reactions of alkyl halides

Mechanism	Substrate structures	Type of nucleophile	Solvent	Reaction stereochemistry
S_N1	Benzylic, allylic, 3°, 2°	Weakly basic	Protic	Racemization
S_N2	Methyl, 1°, 2°	Strong	Polar, aprotic	Inversion of configuration

The combination of an alcohol and a nucleophile satisfies two of the three principal criteria needed to carry out a successful substitution reaction. The obstacle is the OH group, which is a poor leaving group because it is *not* a weak base.

OH⁻ is a poor leaving group because it is a strong base. This reaction does not occur.

There are several ways to convert the OH group into a better leaving group in order to perform a substitution reaction. One of the simplest methods uses a hydrohalic acid (HCl, HBr, or HI), which protonates the OH group while delivering a good nucleophile at the same time. For example, *tert*-butyl alcohol reacts with HBr as shown in the following reaction.

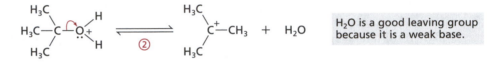

An acid-base reation generates the protonated alcohol.

With protonation, the leaving group is now the water molecule, a weak base and a good leaving group. Ionization subsequently occurs as in any S_N1 process.

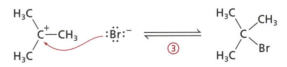

H_2O is a good leaving group because it is a weak base.

In the final step of the mechanism, the nucleophile released in the first step reacts with the carbocation to yield the substitution product.

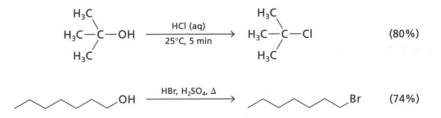

This reaction is quite limited, however. Remember that a nucleophile is also a Lewis base, so a mineral acid will protonate it unless it is an extremely weak base. The only nucleophiles that work well under these conditions are the halide ions—chloride, bromide, and iodide. Fluoride ion is generally a poor nucleophile because it is strongly solvated in most solvents, and HF is a weak acid ($pK_a = 3.3$). Primary alcohols do not react with HCl or HF; they are even sometimes unreactive toward HBr unless sulfuric acid is also added. The following examples are typical (aq = aqueous).

The mechanisms that apply to the reactions between alcohols and the hydrohalic acids are the same ones you learned for the reactions of alkyl halides. Thus, methyl and primary alcohols react by an S_N2 mechanism, and secondary, tertiary, and all types of allylic and benzylic substrates react by the S_N1 pathway. Phenols do not react under these conditions for the same reasons that aryl halides do not react in either the S_N1 and S_N2 reactions (Table 6.4).

Propose a mechanism for the conversion of 1-heptanol to 1-bromoheptane by treatment with HBr and H_2SO_4, as shown in the preceding equation. Sulfuric acid is included to increase the amount of protons for the first step of the reaction: protonation of the alcohol OH group.

7.1b   AN ALCOHOL CAN BE CONVERTED TO AN ALKYL SULFONATE ESTER, WHICH HAS A GOOD LEAVING GROUP, A SULFONATE ION

A more general way to convert the alcohol OH group to a good leaving group is to make the alkyl sulfonate ester derivative. **Alkyl sulfonates,** as they are also called, are prepared by treating an alcohol with p-toluenesulfonyl chloride ($CH_3C_6H_4SO_2Cl$, or TsCl) or methanesulfonyl chloride (CH_3SO_2Cl, or MsCl) and a base, which is usually pyridine, triethylamine, or hydroxide ion, as shown in the following examples:

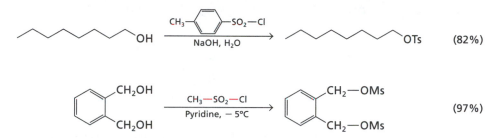

The mechanism for the formation of these alkyl sulfonate derivatives involves displacement of the chloride ion from the sulfonyl chloride reactant by the alcohol, which functions in Step 1 as a nucleophile. In the second step, an added base removes the proton to yield the neutral product.

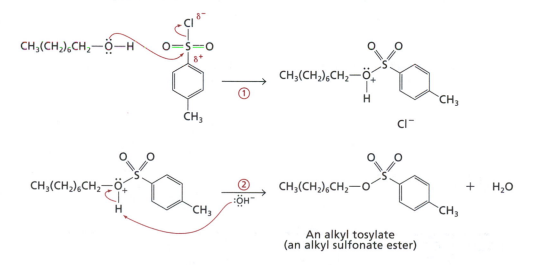

An alkyl tosylate
(an alkyl sulfonate ester)

When using an alkyl sulfonate for a substitution reaction, the actual leaving group is a sulfonate anion, which is the conjugate base of a sulfonic acid and not unlike sulfuric acid in its acidity ($pK_a < -5$). The common sulfonates are methanesulfonate (also called **mesylate,** abbreviated **–OMs**), trifluoromethanesulfonate (also called **triflate,** abbreviated **–OTf**), and p-toluenesulfonate (also called **tosylate,** abbreviated **–OTs**).

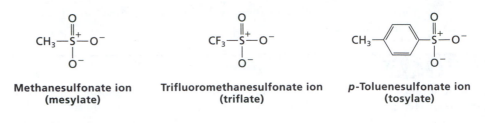

**Methanesulfonate ion
(mesylate)**   **Trifluoromethanesulfonate ion
(triflate)**   ***p*-Toluenesulfonate ion
(tosylate)**

The mesylate and tosylate ions fall somewhere between iodide and bromide ion with regard to leaving group ability; the triflate ion is better than iodide.

Leaving group abilities $CF_3SO_3^- > I^- > CH_3SO_3^- \sim CH_3C_6H_4SO_3^- > Br^- > Cl^-$

A nucleophile can react with the alkyl sulfonates by either an S_N1 or S_N2 process, depending on the structure of the substrate. *Think of an alkyl sulfonate ester exactly as you would the corresponding alkyl halide* (see Table 6.4). The following examples illustrate nucleophilic substitution reactions of alkyl sulfonate esters.

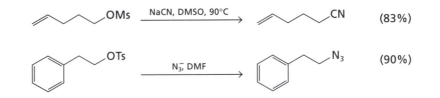

The two-step procedure, *alcohol → alkyl sulfonate → substitution product*, is usually superior to the direct reaction of an alcohol with a mineral acid, *alcohol → substitution product*, which is limited to the preparation of alkyl halides. The advantages of using an alkyl sulfonate ester and S_N2 conditions are as follows:

- The reactions are stereospecific.
- A carbocation intermediate is avoided, along with problems of rearrangement.
- A wide range of nucleophiles can be used.

The following equations illustrate the two-step procedure to carry out the substitution reaction on a secondary alcohol. Notice that formation of the sulfonate ester occurs with *retention* of configuration; the substitution step occurs with *inversion* of configuration.

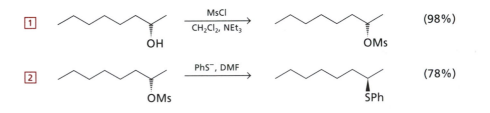

EXERCISE 7.2

Show how you would perform the following transformation:

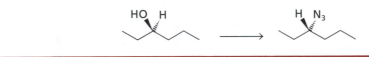

7.1c PHOSPHORUS AND SULFUR HALIDE REAGENTS ARE USED TO PREPARE ALKYL HALIDES FROM ALCOHOLS

A third way to substitute the alcohol OH group is by treating the alcohol with phosphorus tribromide (PBr_3) or thionyl chloride ($SOCl_2$), which produces alkyl bromides or chlorides, respectively, as shown in the following equations:

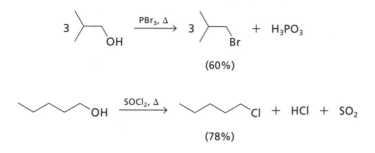

In light of the substitution processes discussed already, we can reasonably assume that these reagents are involved in two ways:

1. They convert the OH group into a good leaving group.
2. They provide a nucleophile, a halide ion, to replace the leaving group.

In the first step of the reaction with PBr_3, an alcohol oxygen atom replaces each of the bromine atoms bonded to the phosphorus. If a base like NaOH is added to neutralize the HBr that is formed, the reaction stops here.

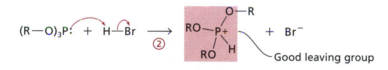

In the absence of an externally added base, however, the phosphorus atom assumes the role of the base, reacting with an equivalent of HBr. This step generates a four-coordinate, positively charged phosphorus atom, which is a good leaving group.

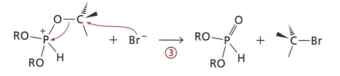

The third step in the transformation leads to substitution with formation of the very stable phosphorus–oxygen double bond.

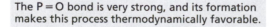

The P = O bond is very strong, and its formation makes this process thermodynamically favorable.

EXERCISE 7.3

Based on the foregoing discussion, which shows a concerted process, what is the expected product, including stereochemistry, of the transformation shown below?

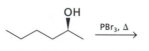

Thionyl chloride reacts with an alcohol by replacement of one of its chlorine atoms. Substitution of the leaving group generates $ClSO_2H$, which is unstable and decomposes to form SO_2 and HCl. Pyridine is sometimes added to react with the HCl that is produced. This reaction is commonly used to convert primary alcohols to primary alkyl chlorides, a transformation that is not readily accomplished using HCl.

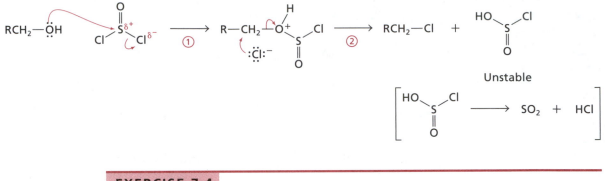

EXERCISE 7.4

What is the expected product in each of the following reactions?

a.

b.

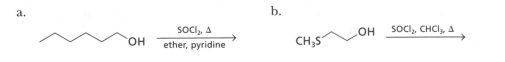

7.1d   THE MITSUNOBU REACTION IS A GENERAL METHOD FOR PERFORMING NUCLEOPHILIC SUBSTITUTION REACTIONS OF ALCOHOLS

The reagents described in the last section are limited because only halide ion nucleophiles are involved. A reagent, or combination of reagents, that could generate a phosphorus-containing leaving group in the presence of nucleophiles other than halide ion would be extremely valuable. In 1980, Oyo Mitsunobu discovered just such a reagent combination, in which triphenylphosphine, Ph_3P, diethyl azodicarboxylate (DEAD), and an acid, HX, react in the presence of the alcohol substrate. The alcohol is usually 1° or 2°, and the acids that are most effective have a $pK_a < 12$. Thus, carboxylic acids, phenols, thiols, thioacids, and inorganic compounds such as HCN and HN_3 are suitable reactants. Of particular value is carboxylate ion, normally a poor nucleophile in substitution reactions, which works well in the **Mitsunobu reaction.**

The overall reaction is represented by the following equation, but the mechanism follows the same principles we have seen earlier.

$$R-OH + HX \xrightarrow[\substack{O \quad\quad O \\ \| \quad\quad \| \\ EtO-C-N=N-C-OEt \\ (DEAD)}]{Ph_3P} R-X + O{=}PPh_3 + EtO-\overset{O}{\overset{\|}{C}}-\underset{H}{\overset{}{N}}-\underset{H}{\overset{}{N}}-\overset{O}{\overset{\|}{C}}-OEt$$

Initially, triphenylphosphine and DEAD react in the presence of the acidic component to form an adduct that reacts in Step 2 with the alcohol—as a nucleophile—to yield the corresponding (alkoxy)triphenylphosphonium ion, $[RO–PPh_3]^+$. Phosphorus, being a period 3 element, can accommodate 10 electrons, and compounds with five bonds to the phosphorus atom are common. This Mitsunobu reaction is so valuable because the counterion, X⁻, of the resulting phosphonium ion is the conjugate base of the acid that was added at the outset.

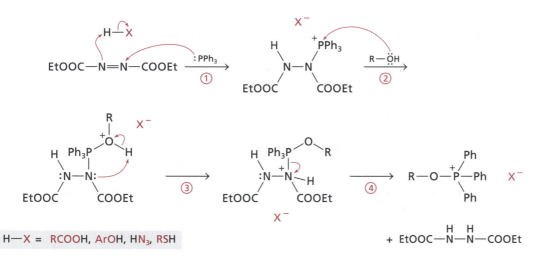

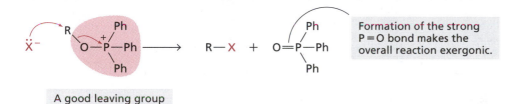

$H{-}X = RCOOH, ArOH, HN_3, RSH$

In the actual substitution step, the nucleophile X⁻ reacts with the electrophilic carbon atom that bears the leaving group, which is triphenylphosphine oxide, $Ph_3P{=}O$. Formation of the P=O bond pushes the Mitsunobu reaction to completion, just as it did in the reaction between an alcohol and PBr_3 (Section 7.1c).

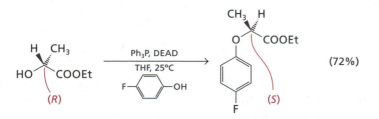

Formation of the strong P=O bond makes the overall reaction exergonic.

A good leaving group

Another valuable feature of the Mitsunobu reaction is its stereochemical course, which normally leads to inversion of configuration if the starting alcohol is chiral. Inversion occurs as the nucleophile approaches the side of the carbon atom opposite to the leaving group, as in any S_N2 process.

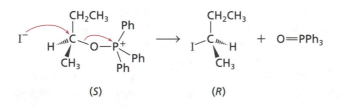

Examples of this reaction abound in the chemical literature, and new applications of the reaction appear frequently. Its great versatility is illustrated in the following equations for cases in which the conjugate base of a phenol, a thiocarboxylic acid, or a carboxylic acid functions as the nucleophile.

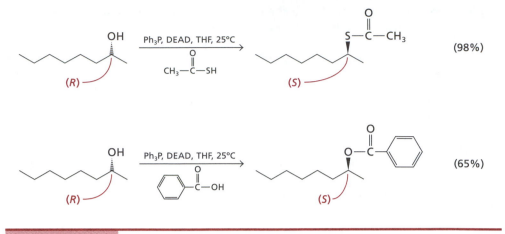

(98%)

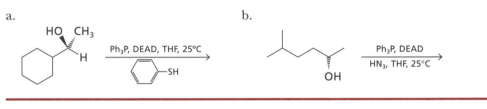

(65%)

EXERCISE 7.5

What is the expected product of each of the following Mitsunobu reactions?

a.   b.

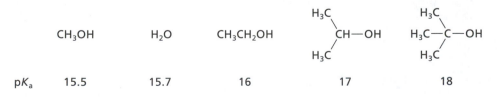

7.2 SUBSTITUTION REACTIONS OF ETHERS AND EPOXIDES

7.2a ALCOHOLS ARE READILY DEPROTONATED TO FORM ALKOXIDE IONS

Before looking at substitution reactions of ethers, it is worthwhile to consider how ethers can be made from alcohols and alkyl halides by nucleophilic substitution reactions. For the S_N1 reaction, recall that alcohols can be used just as they are, as weakly basic nucleophiles (Section 6.2a). Furthermore, alcohols are protic molecules, so they also assist with dissociation of the leaving group under S_N1 conditions.

For an S_N2 reaction, however, an alcohol molecule needs to be converted to its conjugate base to make it a better nucleophile. Using a polar, aprotic solvent would further boost an alkoxide ion's utility in any subsequent S_N2 process (Section 6.3c). Therefore, let us look at how alkoxide ions are made.

Alcohols are weakly acidic molecules with pK_a values for the OH proton between 15 and 20. Therefore, they *cannot* be readily deprotonated by hydroxide ion. The acid strength of methanol and ethanol are not much different from that of water. As more carbon and hydrogen atoms are added to the carbon skeleton, however, an alcohol becomes less acidic. Most of this effect results from poorer solvation of the resulting alkoxide ion.

	CH_3OH	H_2O	CH_3CH_2OH	$\underset{H_3C}{\overset{H_3C}{>}}CH{-}OH$	$H_3C{-}\underset{CH_3}{\overset{CH_3}{C}}{-}OH$
pK_a	15.5	15.7	16	17	18

One of the easiest ways to make an alkoxide ion is to treat the alcohol with sodium or potassium hydride in DMF, a polar, aprotic solvent. The following examples are illustrative. A particularly useful feature of this procedure is that the only byproduct is molecular hydrogen, and formation of this gas drives the reaction to completion.

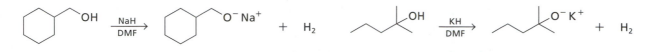

For liquid alcohols, an alternative method makes use of a redox reaction. The alcohol is treated with Li, Na, or K metal, which is oxidized during the reaction. The alcohol proton is reduced to form molecular hydrogen. Sodium metal works well for methanol, ethanol, and 2-propanol (isopropyl alcohol or isopropanol); it reacts very slowly with *tert*-butyl alcohol, so potassium metal is used in that case.

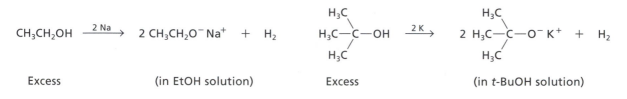

Excess   (in EtOH solution)   Excess   (in *t*-BuOH solution)

The alkoxide ion prepared by reaction with an alkali metal is usually dissolved in the starting alcohol as the solvent. If an aprotic solvent is needed for a subsequent substitution reaction, then the alcohol can be evaporated and the other solvent added.

7.2b ETHERS ARE PREPARED BY SUBSTITUTION REACTIONS

Ethers are most frequently employed in organic chemistry as inert (nonreactive) solvents, but they occasionally have other uses. The reaction between an alcohol or alkoxide ion and an alkyl halide is the best general method to make ethers. The alcohol itself is used as the nucleophile if the alkyl halide can react via the S_N1 mechanism.

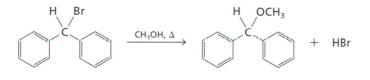

EXERCISE 7.6

As a review, write out the steps for the mechanism of the reaction shown directly above.

A more versatile method for making ethers is the **Williamson ether synthesis,** which is the S_N2 reaction between an alkyl halide and an alkoxide ion (Section 7.2a).

$$ROH \xrightarrow{\text{base}} RO^- \xrightarrow{R'X} R—O—R'$$

The Williamson reaction is general for alkyl halides and alkyl sulfonates (Section 7.1b) that readily react via the S_N2 pathway, which includes methyl, primary, and some secondary alkyl halides. [The alkoxide ion is also a strong base, so 2° substrates can give elimination products (Section 8.2).] The starting alcohol—hence, the alkoxide ion—can be of any type, and phenols can be used as well.

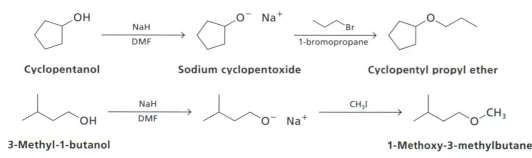

Cyclopentanol   Sodium cyclopentoxide   Cyclopentyl propyl ether

3-Methyl-1-butanol   **1-Methoxy-3-methylbutane**

7.2c INTRAMOLECULAR SUBSTITUTION REACTIONS YIELD CYCLIC PRODUCTS

Most of the substitution reactions that you have learned have been *inter*molecular, which means that two reactants are involved (Section 2.8b). If the nucleophile and leaving group are in the *same* molecule, then the reaction between them is *intra*molecular, and a ring will be formed. You saw an example of ring formation when amino alkyl halides undergo reaction (Section 6.3d). Cyclic ethers are formed when the alkoxy ion nucleophile and a halide ion leaving group are in the same molecule.

There is no major procedural or mechanistic difference between the inter- and intramolecular substitution reactions that yield ethers. The alkoxide ion is generated in the same manner, and the electrons of the nucleophilic oxygen atom react with the electrophilic carbon atom to displace the leaving group.

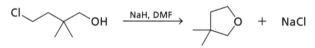

The construction of five- and six-membered rings is especially facile, as is true for most ring-forming reactions, but three-, four-, and seven-membered rings can also be made.

Generally, the alkoxide ion is prepared using NaH in DMF. Sodium hydride does not react with alkyl halides, so alkoxide ion formation is straightforward. The following example illustrates this reaction.

EXAMPLE 7.1

Propose a mechanism for the following cyclization reaction. Show electron movement with curved arrows.

Sodium hydride is a base, so it reacts in the first step with the chloro alcohol to generate the sodium salt of the alkoxide ion. Once formed, the alkoxide ion reacts at the carbon atom bearing the leaving group—chloride in this example. Displacement of chloride ion generates the ring. Make certain to count the number of carbon atoms between the oxygen and chlorine atoms. There are four, so a five-membered ring is formed, the fifth atom being oxygen.

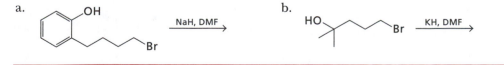

Draw the structure of the major product formed in each of the following reactions:

The intramolecular reaction that occurs when an alcohol group and a halogen atom are bonded to adjacent carbon atoms is a versatile way to make *epoxides*, which are three-membered ring ethers. Sodium hydride in DMF can be used to generate the alkoxide ion in these reactions, but a solution of sodium hydroxide in water is often sufficient to bring about ring formation by a concerted process.

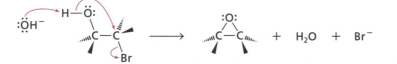

As in all reactions of the strongly nucleophilic (and basic) alkoxide ion, epoxide ring-forming reactions follow the patterns seen for typical S_N2 reactions, which means that the leaving group must be attached to a primary or secondary carbon atom. Note also that for chiral substrates, the configuration is retained at the carbon atom bearing the oxygen atom because no bond to that stereocenter is broken, but the configuration is inverted at the carbon atom with the leaving group.

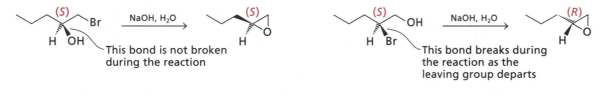

EXERCISE 7.8

Propose mechanisms for the two reactions shown directly above and confirm the stereochemical results.

7.2d   ETHERS ARE CLEAVED BY REACTION WITH HBR, HI, OR IODOTRIMETHYLSILANE

Ethers have two carbon atoms bonded to the highly electronegative oxygen atom, so there are two electrophilic centers at which reaction can occur. Nucleophilic substitution reactions of ethers can only be accomplished with a reagent that can (1) react with the oxygen atom to form a good leaving group and (2) supply a potent nucleophile. Three reagents that satisfy these criteria are HBr, HI, and $(CH_3)_3SiI$ (iodotrimethylsilane).

Ethers are cleaved using hot, aqueous hydrobromic or hydroiodic acid (sulfuric acid is sometimes added to make the solution more acidic); for example,

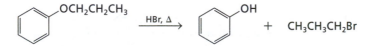

Reaction takes place after formation of a good leaving group that results from protonation of the oxygen atom, just as you saw for the reaction of an alcohol.

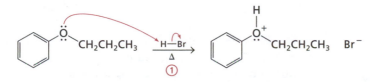

The nucleophile reacts at an adjacent carbon atom, forming 1 equiv of alkyl halide and 1 equiv of alcohol (or phenol, as in this example). Phenols do not react further because the OH group is attached to an sp^2-hybridized carbon atom. Just as aryl halides do not react in either the S_N1 and S_N2 reactions (Table 6.4), phenols are inert, too.

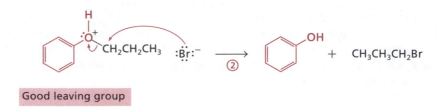

Good leaving group

When a dialkyl ether is the substrate, then two molecules of alkyl halide are formed, because the alcohol formed in the first step reacts under these same conditions (Section 7.1a). In fact, an alcohol reacts faster than an ether does.

$$CH_3CH_2-O-CH_2CH(CH_3)_2 \xrightarrow{HI, \Delta} CH_3CH_2-I + HO-CH_2CH(CH_3)_2$$

$$HO-CH_2CH(CH_3)_2 \xrightarrow{HI, \Delta} I-CH_2CH(CH_3)_2 + H_2O$$

EXERCISE 7.9

Propose a mechanism for the two reactions shown directly above. Indicate electron movement using curved arrows.

Because HBr and HI are such strong acids, their use is limited to ethers that have few other substituents. Iodotrimethylsilane, $(CH_3)_3SiI$, is a milder reagent used to cleave ethers, and its reaction with an ether also illustrates that protons are not the only electrophiles capable of converting the ether oxygen atom to a good leaving group.

In the first step, the oxygen atom reacts with the electrophilic silicon atom, displacing the iodide ion. This step generates a good leaving group. The iodide ion—an excellent nucleophile—subsequently reacts with the electrophilic carbon atom, displacing the leaving group, an alkyl trimethylsilyl ether.

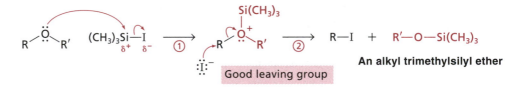

An alkyl trimethylsilyl ether

Good leaving group

If desired, the alkyl trimethylsilyl ether product can be hydrolyzed to obtain the alcohol.

$$2\ R'-O-Si(CH_3)_3 + H_2O \longrightarrow 2\ R'OH + (H_3C)_3Si-O-Si(CH_3)_3$$

A dialkyl ether with similar R and R′ groups, R–O–R′, can be cleaved by $(CH_3)_3SiI$ at either C–O bond, so a mixture of four products will be obtained, limiting the value of this transformation.

$$CH_3CH_2-O-CH_2CH(CH_3)_2 \xrightarrow[CH_3CN]{(CH_3)_3Si-I} CH_3CH_2-O-Si(CH_3)_3 + I-CH_2CH(CH_3)_2$$
$$+ CH_3CH_2-I + (H_3C)_3Si-O-CH_2CH(CH_3)_2$$

Therefore, this reagent is used to cleave ethers that react preferentially at one group. Alkyl aryl ethers constitute one of these types, because the phenyl–oxygen bond is not broken in nucleophilic substitution processes, as noted earlier in this section.

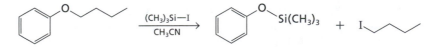

Methyl alkyl ethers constitute another type of suitable substrate. Recall that the methyl carbon atom is more reactive than any other type of carbon atom in an S_N2 reaction (Section 6.3a), so the nucleophile—iodide ion—reacts preferentially at the methyl group, which generates methyl iodide and an alkoxytrimethylsilane.

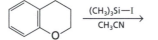

Treating the alkyl trimethylsilyl ether with water yields the alcohol.

$$2\ (H_3C)_3Si-O-CH_2CH(CH_3)_2\ +\ H_2O\ \longrightarrow\ 2\ HO-CH_2CH(CH_3)_2\ +\ (H_3C)_3Si-O-Si(CH_3)_3$$

EXERCISE 7.10

Draw the structure of the product formed in the following reaction:

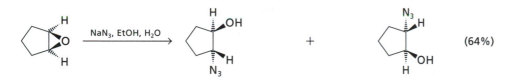

7.2e   AN EPOXIDE IS REACTIVE AND UNDERGOES RING OPENING

Ethers are notoriously poor substrates for substitution reactions, and that makes them ideal as solvents for many transformations. Epoxides, also called *oxiranes* or *oxacyclopropanes*, react readily with nucleophiles even though the leaving group is formally an alkoxide ion, which is a strong base. Epoxides are more reactive than most other ethers because of the Baeyer strain energy inherent in their three-membered rings (Section 3.2a).

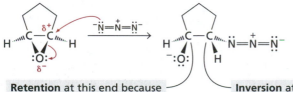

The high reactivity of an epoxide ring is illustrated by the reaction between azide ion, N_3^-, and 1,2-epoxycyclopentane, which produces a racemic mixture of azido alcohols.

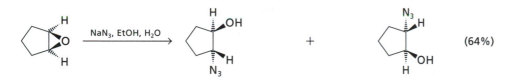

(1R,2R)-2-Azido-1-cyclopentanol   (1S,2S)-2-Azido-1-cyclopentanol

As the azide ion reacts with the electrophilic carbon atom, the carbon–oxygen bond is broken. The carbon atom at which the azide ion reacts undergoes inversion of configuration, while the stereochemistry of the other carbon atom is retained. A majority of epoxide ring-opening reactions occur with inversion of stereochemistry at one end of the ring, consistent with an S_N2 process.

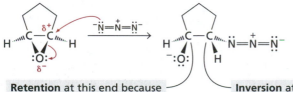

Retention at this end because the C–O bond is not broken.

Inversion at this end because the nucleophile approaches from the back.

After the ring is opened, the alkoxide ion removes a proton from water in an acid–base reaction.

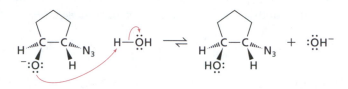

Although shown as a separate step directly above, this protonation reaction may occur simultaneously with opening of the ring. In fact, protic solvents are useful for epoxide ring-opening reactions as a way to make a better leaving group by forming hydrogen bonds to the solvent (water in the following example).

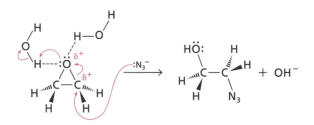

EXERCISE 7.11

Why does the reaction of 1,2-epoxycylopentane with azide ion give a racemic mixture of products?

The nucleophilic ring-opening reaction of an epoxide displays the features of many S_N2 processes. In particular, the reaction is influenced by steric hindrance at the electrophilic carbon atom, so the less highly substituted carbon atom of an unsymmetrical epoxide reacts preferentially with the nucleophile.

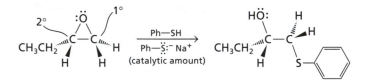

On the other hand, epoxides react under *acidic* conditions by a mechanism that is more like an S_N1 process. Protonation of the oxygen atom produces a good leaving group, so ring opening can occur to produce a carbocation intermediate.

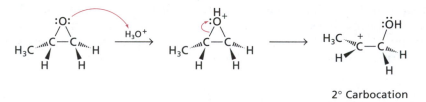

In reality, however, a pure carbocation does not form. Instead, an unshared electron pair on the oxygen atom maintains an interaction with the neighboring carbon atom, which then bears only a partial positive charge. The more highly substituted carbon atom is better stabilized by the presence of the alkyl group(s), so the nucleophile subsequently reacts preferentially (but not exclusively) at that position.

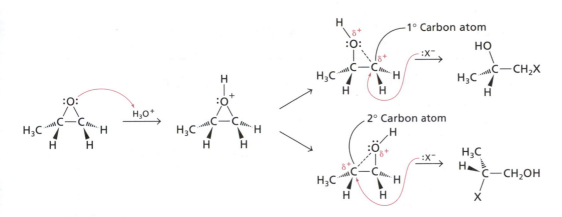

Unlike the carbocation intermediate of a true S_N1 reaction, this bridged carbocation is blocked on one face by the oxygen atom, so reaction occurs from the opposite side. Consequently, the stereochemistry is not lost completely, which would be the case if an actual carbocation were to form.

The following equations compare the reaction between an epoxide and chloride ion at different pH values. At pH 7, the reaction is nucleophilic, whereas at pH 3.8, a cationic species is generated by protonation of the epoxide oxygen atom. As a result, the proportion of reaction at the more highly substituted carbon atom is higher in the acid-catalyzed process. Notice, however, that there is still substantial reaction at the less-hindered carbon atom, leading to formation of the product with a chlorine atom attached to the primary carbon atom.

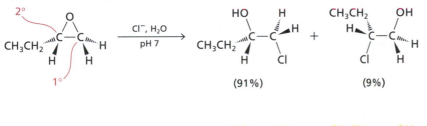

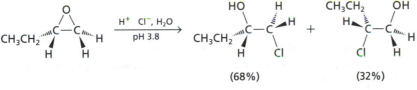

EXERCISE 7.12

Propose a reasonable mechanism for the reaction of (*S*)-1,2-epoxybutane with HCl, accounting for the formation of both products.

EXERCISE 7.13

Lithium aluminum hydride is a potent reducing reagent you will encounter several times in this text. In many instances, it acts like a source of the nucleophile H:⁻ (hydride ion). With that assumption, what is the expected product of the following reaction? Formulate a mechanism for this transformation and account for the stereochemistry of the product.

(*S*)-2-Methyl-1,2-epoxyhexane   1. LiAlH₄, Ether   2. H₂O   →   ?

The reagent reacts violently with water, so a separate step is required to quench the reaction and protonate the alkoxide ion formed.

SUBSTITUTION REACTIONS OF THIOLS
AND THIOETHERS

7.3a THIOLS ARE THE SULFUR ANALOGUES OF ALCOHOLS

Thiols are compounds that have the SH group. They are named with the functional group suffix *thiol* (Table 1.1). As a substituent, the SH group is often indicated using the prefix mercapto-, which derives from an older term for thiols—mercaptans—meaning "mercury capturing" and referring to the high affinity that many sulfur containing substances have for mercury(II) ions.

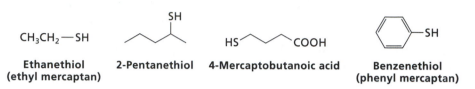

Ethanethiol	**2-Pentanethiol**	**4-Mercaptobutanoic acid**	**Benzenethiol**
(ethyl mercaptan)			**(phenyl mercaptan)**

In contrast to alcohols, thiols do not undergo substitution reactions as *substrates*, because it is difficult to make the SH group into a good leaving group. Sulfur compounds in general are less basic than their oxygen analogues because of sulfur's larger size (Section 5.2c), so protonation of a thiol SH group—unlike that of an alcohol OH group (Section 7.1a)—is more difficult. Reactions of thiols with sulfur and phosphorus reagents (Sections 7.1c and 7.1d) are also not practical, because the resulting S–S and S–P bonds are generally weaker than a C–S bond, and therefore break before substitution can occur at the carbon atom.

Because thiols are weak bases, they can be used as nucleophiles in S$_N$1 reactions. Furthermore, thiols are excellent nucleophiles, so alcohols can be used as protic solvents for these S$_N$1 reactions without any worry about competition.

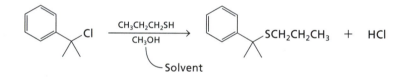

This lower basicity of sulfur versus oxygen means that molecules with SH groups are more acidic than their oxygen analogues. A comparison of four pairs of common sulfur- and oxygen-containing substances and their pK_a values is shown below.

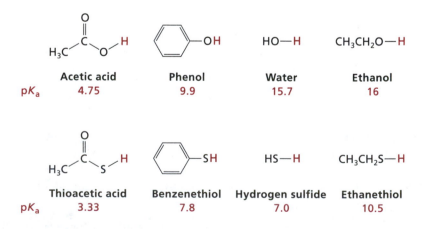

	Acetic acid	**Phenol**	**Water**	**Ethanol**
pK_a	4.75	9.9	15.7	16

	Thioacetic acid	**Benzenethiol**	**Hydrogen sulfide**	**Ethanethiol**
pK_a	3.33	7.8	7.0	10.5

A consequence of this higher acidity means that anionic sulfur-containing nucleophiles are readily prepared using hydroxide ion or a tertiary amine as the base. The conjugate base of a thiol is the **thiolate ion.**

$$CH_3CH_2CH_2CH_2SH \xrightarrow{\text{NaOH}} CH_3CH_2CH_2CH_2S^- \ Na^+ \ + \ H_2O$$

1-Butanethiol   **Sodium 1-butanethiolate**

Thiolate ions are potent nucleophiles, and they react with alkyl halides, alkyl sulfonates, and alcohols (via the Mitsunobu reaction) to form thioethers, which are compounds with two groups—alkyl or aryl—bonded to sulfur (Section 7.3b). Both intermolecular and intramolecular substitution reactions are possible and the normal S_N2 conventions apply (polar, aprotic solvents are best and inversion of configuration occurs with chiral substrates).

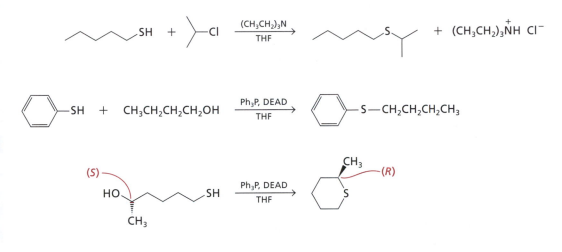

EXERCISE 7.14

Propose a reasonable mechanism for intramolecular Mitsunobu reaction shown directly above.

EXERCISE 7.15

Draw the structure of the products expected from each of the following reactions. Include stereochemistry where appropriate:

a.

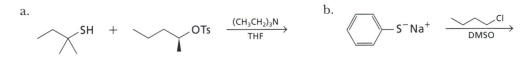

b.

7.3b   THIOETHERS ARE THE SULFUR ANALOGUES OF ETHERS

Thioethers, as the name implies, are the sulfur analogues of ethers. They are also called *sulfides*, which is reflected in their common names, formed by naming each of the groups attached to the sulfur atom, followed by the word sulfide. Cyclic sulfides are named by using the root word for the hydrocarbon with the same ring size and appending the prefix *thia—*.

$CH_3CH_2{-}S{-}CH(CH_3)_2$

Ethyl isopropyl sulfide   ***tert*-Butyl phenyl sulfide**   **Thiacyclohexane**

Thioethers do not react *with* nucleophiles, but they can react *as* nucleophiles. For example, treating dimethyl sulfide with methyl iodide in ether produces crystalline trimethylsulfonium iodide in quantitative yield. The fact that ether (the solvent) does not react with methyl iodide confirms the greater nucleophilicity of the sulfur versus the oxygen compound.

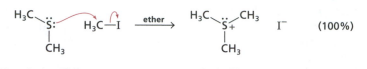

Dimethyl sulfide   **Trimethylsulfonium iodide**

Dimethyl sulfide is a weak base, which means it is a good leaving group (Section 6.1b). Thus, trimethylsulfonium iodide is a good methylating agent.

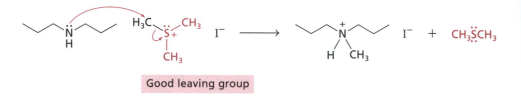

Good leaving group

This reaction has no practical value because methylation can be accomplished using methyl iodide itself instead of preparing $Me_3S^+I^-$ first. What makes this example of interest is its parallel to certain methylation reactions in Nature (Section 7.4b).

7.4   SUBSTITUTION REACTIONS IN BIOCHEMICAL SYSTEMS

7.4a   PHOSPHATE IONS FUNCTION AS GOOD LEAVING GROUPS

In the laboratory, substitution reactions provide a way to replace halogen atoms and hydroxy groups in aliphatic molecules. In biological systems, substitution processes are less important for interconverting functional groups because electrophiles might react indiscriminately with the many nucleophilic groups that are present in proteins and nucleic acids. In fact, the toxicity and carcinogenicity of many alkyl halides and epoxides stems from their propensity to undergo substitution reactions in which groups in deoxyribonucleic acid (DNA) act as nucleophiles. For a substitution reaction in the biological realm, the ground rules are the same as in laboratory reactions, however: *The substrates have good leaving groups attached to sp^3-hybridized carbon atoms and they react with nucleophiles.*

One category of leaving group in biomolecules comprises the phosphate ions, which include phosphate, PO_4^{3-}, pyrophosphate, $P_2O_7^{4-}$, and triphosphate, $P_3O_{10}^{5-}$ as well as their partially protonated analogues (HPO_4^{2-} and $H_2PO_4^-$, for example.). The formation of *S*-adenosylmethionine is an important example of a nucleophilic substitution reaction in which triphosphate ion is the leaving group, and a sulfur atom is the nucleophile. (Recall from Section 7.3b how thioethers react as nucleophiles in substitution reactions.) This biochemical transformation proceeds by an S_N2 pathway within the active site of an enzyme, where it can be shielded from random nucleophiles, including water.

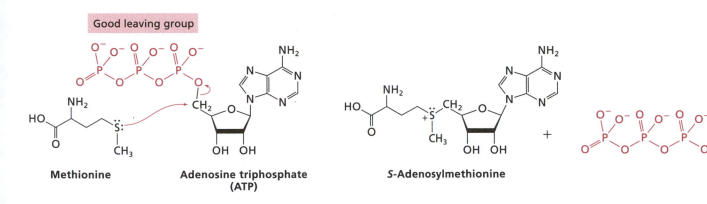

Methionine   Adenosine triphosphate (ATP)   S-Adenosylmethionine

(In Section 9.3c, we will look at some S_N1 reactions in which the pyrophosphate ion is the leaving group.)

7.4b S-ADENOSYLMETHIONINE IS A METHYLATING AGENT FOR SOME BIOCHEMICAL PROCESSES

You learned earlier that a tertiary amine reacts with methyl iodide to form the methylated, quaternary ammonium ion product (Section 6.3d). The following equation illustrates a reaction in which a tertiary amine is converted to the quaternary ammonium salt in excellent yield.

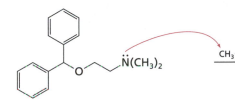 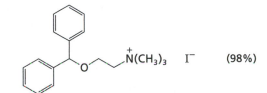

Dimethyl-1-[2-(diphenylmethoxy)ethyl]amine   Trimethyl-1-[2-(diphenylmethoxy)ethyl]ammonium iodide

Methylation of an amine in Nature does not use methyl iodide, which causes mutations in DNA that can lead to defective protein synthesis, jeopardizing the integrity of many other biological processes required for metabolism and growth. Instead, the methyl group comes from S-adenosylmethionine (Section 7.4a), which, because it is bonded to a positively charged sulfur atom, has an electrophilic carbon atom, and hence is susceptible to reactions with nucleophiles. The remainder of the molecule functions as a large, good leaving group.

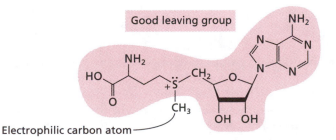

Methylation of phosphatidylethanolamine to give phosphatidylcholine in bacteria provides a specific example of a substitution reaction that occurs in Nature. This transformation is an S_N2 reaction between an amine nucleophile and S-adenosylmethionine. The leaving group is S-adenosylhomocysteine.

General:

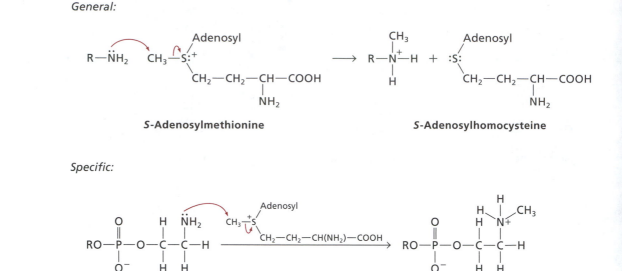

S-Adenosylmethionine   S-Adenosylhomocysteine

Specific:

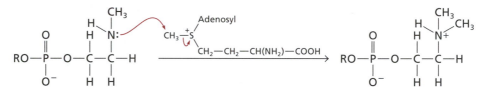

Phosphatidylethanolamine   **Phosphatidyl *N*-methylethanolammonium ion**

This same process occurs twice more to give the final product, phosphatidylcholine, a component of lipids, which are the primary constituents of cell membranes. Between each substitution step, the ammonium ion that is formed by methylation is deprotonated to liberate the nucleophilic nitrogen atom. This deprotonation is accomplished by basic sites within the enzyme that catalyzes the reaction.

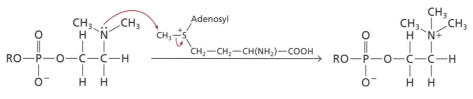

Phosphatidyl *N*-methylethanolamine   **Phosphatidyl *N,N*-dimethylethanolammonium ion**

Phosphatidyl *N,N*-dimethylethanolamine   **Phosphatidylcholine**

Besides supplying basic groups to deprotonate the ammonium ion intermediates, the active site architecture can control the access and orientation of the nucleophilic and electrophilic centers. For example, the enzyme that brings phosphatidylethanolamine together with S-adenosylmethionine might make use of methionine's amine group as a hydrogen-bond acceptor to anchor it in the active site so that the substrate's amino group can approach the electrophilic methyl group. Electrostatic interactions can also be used to anchor the reactants in the proper orientation.

Noncovalent interactions between the substrate molecules and
the amino acids that make up the enzyme active site orient the
two reactants so that only the desired transformation takes place.

Electrostatic attraction between a
protonated lysine side chain and the
negatively charged carboxylate group.

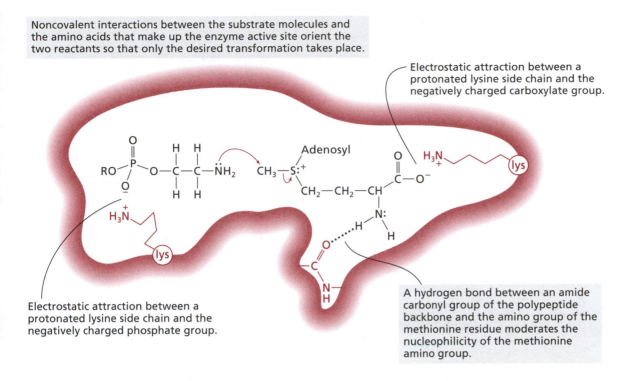

Electrostatic attraction between a
protonated lysine side chain and the
negatively charged phosphate group.

A hydrogen bond between an amide
carbonyl group of the polypeptide
backbone and the amino group of the
methionine residue moderates the
nucleophilicity of the methionine
amino group.

The reactions presented in this section are not the only ways by which biomolecules are methylated. But the reaction between S-adenosylmethionine and amines clearly shows that the S_N2 reaction occurs in Nature. Moreover, the mechanism is the same as those processes done in the laboratory with alkyl halides and related compounds that bear good leaving groups.

EXERCISE 7.16

Another transformation that makes use of S-adenosylmethionine is the biosynthesis of adrenaline, a neurotransmitter, from norepinephrine. Propose a mechanism for this transformation, showing electron movement with arrows.

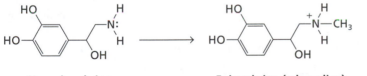

Norepinephrine   Epinephrine (adrenaline)

CHAPTER SUMMARY

Section 7.1   Substitution reactions of alcohols

- Alcohols react by an S_N1 or S_N2 pathway after the OH group is converted to a good leaving group.
- The alcohol OH group is converted to a good leaving group by reaction with a strong acid such as HCl, HBr, or HI. Substitution produces an alkyl halide.
- The most versatile method to form a good leaving group of an alcohol is to prepare its alkyl sulfonate ester derivative, which is equivalent to a halide ion in its leaving group ability. The formation of alkyl sulfonates occurs with retention of configuration if the alcohol is chiral.

- Tosylate, mesylate, and triflate groups are the most common leaving groups of the sulfonate ester type.
- Two common reagents used to convert alcohols to alkyl halides are PBr_3 and $SOCl_2$.
- The Mitsunobu reaction is a general way to substitute the OH group of an alcohol with a nucleophile.

Section 7.2   Substitution reactions of ethers and epoxides

- Alcohols are readily deprotonated by NaH in DMF or by a redox reaction with sodium or potassium metal and the alcohol itself as the solvent.
- Ethers are prepared by the Williamson ether synthesis, an S_N2 reaction between an alkyl halide and an alkoxide ion.
- Three- to seven-membered rings can be made by treating an appropriate haloalcohol with base. Cyclic ether formation is the intramolecular version of the Williamson reaction.
- Ethers are cleaved by HI, HBr, or $(CH_3)_3SiI$ to form alkyl halides. The reaction is most practical for cleaving alkyl aryl ethers to form alkyl halides and phenols.
- Iodotrimethylsilane is also used to cleave alkyl methyl ethers, which give methyl iodide and alkyl trimethylsilyl ethers. The latter can be hydrolyzed to yield the corresponding alcohols.
- Epoxides are three-membered ring ethers that react readily with a nucleophile because the strain energy in the small sized ring is relieved.
- A nucleophile normally reacts at the less highly substituted carbon atom of an epoxide ring. That carbon atom, if it is chiral, undergoes inversion of configuration. The configuration of the other carbon atom is retained during the ring-opening process.

Section 7.3   Reactions of thiols and sulfides

- Thiols are the sulfur analogues of alcohols. A thiol is less basic, more acidic, and more nucleophilic than the corresponding alcohol. The conjugate base of a thiol is a thiolate ion.
- Thioethers, also called sulfides, are the sulfur analogues of ethers. They can react as nucleophiles with methyl iodide to form methylsulfonium iodide salts.

Section 7.4   Substitution reactions in biochemical systems

- In biological systems, phosphate ions and S-adenosylhomocysteine are common leaving groups in nucleophilic substitution processes.

KEY TERMS

Section 1b
alkyl sulfonate
mesylate
triflate
tosylate

Section 1d
Mitsunobu reaction

Section 2b
Williamson ether synthesis

Section 3a
thiolate ion

Section 7.1a

Alcohols react with HX (X = Cl, Br, I) to form alkyl halides.

$$\overset{\diagup}{\underset{\diagup}{C}}-OH \quad \xrightarrow[X\,=\,Cl,\,Br,\,I]{HX} \quad \overset{\diagup}{\underset{\diagup}{C}}-X \quad + \quad H_2O$$

Section 7.1b

Alcohols react with sulfonyl chlorides to form alkyl sulfonate esters (alkyl sulfonates).

$$\overset{\diagup}{\underset{\diagup}{C}}-OH \quad \xrightarrow[R\,=\,CH_3,\,C_6H_4CH_3]{RSO_2Cl,\ base,\ CH_2Cl_2\ (solvent)} \quad \overset{\diagup}{\underset{\diagup}{C}}-O-SO_2R \quad + \quad H(base)^+\,Cl^-$$

base = pyridine or Et$_3$N

Alkyl sulfonates react with nucleophiles in the same way that alkyl halides react with nucleophiles.

$$\overset{\diagup}{\underset{\diagup}{C}}-OSO_2R \quad \xrightarrow{Nuc^-} \quad \overset{\diagup}{\underset{\diagup}{C}}-Nuc \quad + \quad RSO_3^-$$

Section 7.1c

Alcohols react with PBr$_3$ to form alkyl bromides.

$$3\ ROH \quad \xrightarrow{PBr_3} \quad 3\ RBr \quad + \quad H_3PO_3$$

Alcohols react with SOCl$_2$ to form alkyl chlorides, which is a good method by which to convert primary alcohols to primary alkyl chlorides.

$$ROH \quad \xrightarrow{SOCl_2} \quad RCl \quad + \quad HCl \quad + \quad SO_2$$

Section 7.1d

The Mitsunobu reaction. This procedure is an excellent method by which to replace the OH group of an alcohol with a nucleophile that is the conjugate base of an acid with a pK_a value < 12.

$$ROH \quad \xrightarrow[EtO_2CN=NCO_2Et]{Ph_3P,\ H-Y} \quad R-Y \quad + \quad Ph_3P{=}O \quad + \quad EtO_2CNH-NHCO_2Et$$

H—Y = H—OCOR′, H—OAr, H—N$_3$, H—SR′, H—SCOR′

Section 7.2a

Alcohols react with a metal hydride or with an alkali metal to form their conjugate bases, alkoxide ions.

$$R-OH \quad \xrightarrow{NaH,\ DMF} \quad R-O^-\,Na^+ \quad + \quad H_2 \qquad\qquad R-OH \quad \xrightarrow{KH,\ DMF} \quad R-O^-\,K^+ \quad + \quad H_2$$

$$2\ R-OH \quad \xrightarrow{2\ Na} \quad R-O^-\,Na^+ \quad + \quad H_2 \qquad\qquad 2\ R-OH \quad \xrightarrow{2\ K} \quad R-O^-\,K^+ \quad + \quad H_2$$

Section 7.2b

The Williamson ether synthesis. This reaction is the most general method for making ethers. An alkoxide ion reacts as a nucleophile with an alkyl halide or alkyl sulfonate, displacing the leaving group. The method works best when R′ is methyl, 1°, or 2°.

$$R\!-\!O^- Na^+ \ + \ R'\!-\!X \ \longrightarrow \ R\!-\!O\!-\!R'$$

Section 7.2c

Cyclic ethers are formed when the alkoxide ion and leaving group are in the same molecule. Rings with three-to-seven atoms are readily made.

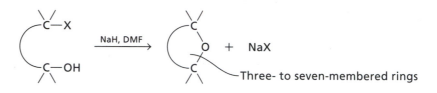

Section 7.2d

An ether is cleaved using HI, HBr, or $(CH_3)_3SiI$ to form the alkyl halides. An alkyl aryl ether forms a phenol and an alkyl halide. Alkyl methyl ethers form methyl iodide and an alkyl trimethylsilyl ether when treated with $(CH_3)_3SiI$.

$$R\!-\!O\!-\!R' \ \xrightarrow[X = Br,\ I]{HX} \ RX \ + \ R'X \ + \ H_2O \qquad\qquad Ar\!-\!O\!-\!R \ \xrightarrow[X = Br,\ I]{HX} \ ArOH \ + \ RX$$

$$R\!-\!O\!-\!CH_3 \ \xrightarrow[CH_3CN]{(CH_3)_3SiI} \ R\!-\!O\!-\!Si(CH_3)_3 \ + \ CH_3I \qquad Ar\!-\!O\!-\!R \ \xrightarrow[CH_3CN]{(CH_3)_3SiI} \ ArOSi(CH_3)_3 \ + \ RI$$

Section 7.2e

Epoxides undergo ring opening when treated with nucleophiles. The nucleophile reacts at the less highly substituted carbon atom, with inversion of configuration. The other carbon atom retains its configuration.

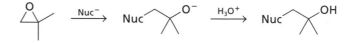

Epoxides undergo ring opening with use of acidic reagents by formation of a stabilized carbocation intermediate. A higher percentage of product with the nucleophile attached to the more highly substituted carbon atom is obtained than in the procedure that uses a nucleophile without proton assistance.

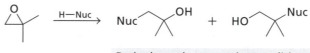

Ratio depends on reaction conditions

Section 7.3a

Thiols form thiolate ions when treated with a base. Thiols and thiolate ions are potent nucleophiles and can react in S_N1 and S_N2 reactions, respectively.

$$R\!-\!SH \ \xrightarrow{base} \ R\!-\!S^-$$

Section 7.3b

Thioethers can react as nucleophiles with methyl iodide to form methylsulfonium iodide salts.

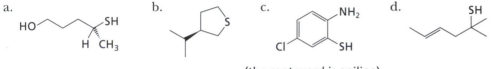

7.17.   Draw a structure for each of the following compounds:

a.   2-Methyl-2-hexanethiol

b.   Cyclobutyl methanesulfonate

c.   *trans*-2-Phenyl-1-cyclohexanethiol (draw the most stable conformation)

d.   Isobutyl phenyl sulfide

7.18.   Give a suitable name for each of the following structures. Include stereochemical descriptors where appropriate.

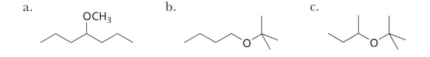

(the root word is aniline)

7.19.   Which of the following isomeric ethers, $C_8H_{18}O$, can be prepared from an alcohol and an alkyl halide by an S_N2 reaction? What starting materials would you use to prepare each?

7.20.   One way to make a sulfide was described in Section 7.3a, namely, treating an alkyl halide with a thiolate ion. To make a symmetric sulfide, you can instead add 2 equiv of an alkyl halide to a solution of Na_2S. Propose a mechanism for the following reaction that makes use of this strategy. (Even though both steps of this mechanism are S_N2 reactions, aqueous ethanol works fine as the solvent. Sulfide and thiolate ions are such potent nucleophiles that use of a protic solvent does not prevent the reaction from occurring.)

$$2 \diagup\!\!\!\diagdown\!\!\!\diagup Br \;+\; Na_2S \xrightarrow[EtOH]{H_2O} \diagup\!\!\!\diagdown\!\!\!\diagup S \diagdown\!\!\!\diagup\!\!\!\diagdown \;+\; 2\ NaBr$$

7.21.   The attempted substitution reaction of hydrosulfide ion, HS^-, with an alkyl halide is often a poor method for preparing a thiol. Instead, the symmetrical sulfide is the major product.

Attempted $\diagup\!\!\!\diagdown\!\!\!\diagup Br \;+\; LiSH \xrightarrow{THF} \diagup\!\!\!\diagdown\!\!\!\diagup SH$

Actual $\diagup\!\!\!\diagdown\!\!\!\diagup Br \;+\; LiSH \xrightarrow{THF} \tfrac{1}{2} \diagup\!\!\!\diagdown\!\!\!\diagup S \diagdown\!\!\!\diagup\!\!\!\diagdown \;+\; \tfrac{1}{2} H_2S \;+\; Br^-$

Explain how and why this result is obtained [*Hint:* It is the same situation that leads to formation of dialkyl amines from the attempted alkylation of ammonia (Section 6.3d). Which is more nucleophilic, HS⁻ or RS⁻? Why? (Consider the pK_a values of H_2S and RSH.)]

7.22. A good way to make a thiol from an alkyl halide (unlike the method shown in Exercise 7.21) is to employ potassium thioacetate, K^+ ^-S–CO–CH_3, as the nucleophile. The product is a thioester, R–S–CO–CH_3, a compound that can also be prepared by the Mitsunobu reaction. A thioester is readily hydrolyzed with dilute, aqueous KOH solution to produce the thiol, RSH, and acetic acid, after acidifying the reaction mixture with dilute aqueous sulfuric acid (H_3O^+). As described in this paragraph, write equations (see Exercise 5.17 for examples) that represent the overall processes for the transformation RBr → RSH.

7.23. Draw the structure(s) for the expected major product(s) of each of the following transformations. Indicate stereochemistry where appropriate and assign (*R*) and (*S*) to each stereogenic center. If a racemic mixture is formed, draw the structure of one enantiomer, and write the word "racemic". If no reaction occurs, write N. R.

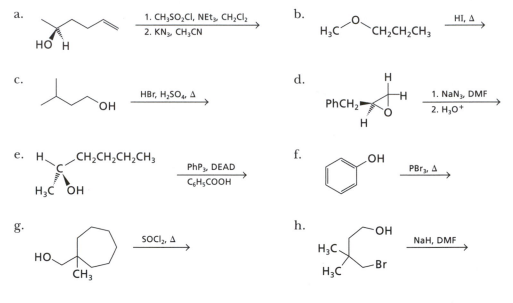

7.24. What is the expected product from each of the following Mitsunobu reactions? Indicate stereochemistry where appropriate, and assign the (*R*) or (*S*) configuration to each stereocenter.

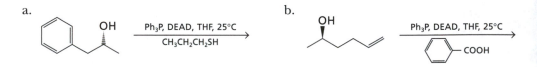

7.25. Propose a reasonable mechanism for the following S_N1 reaction:

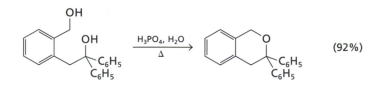

7.26. In the biosynthesis of domoic acid, which is an uncommon amino acid, the first step is the nucleophilic substitution reaction shown in the following equations [(X) in an unspecified but reactive group].

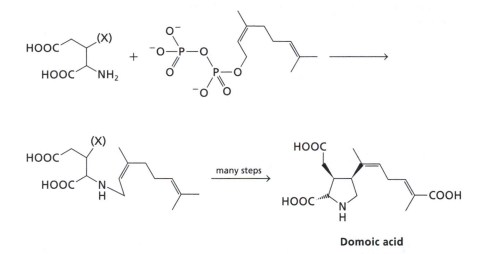

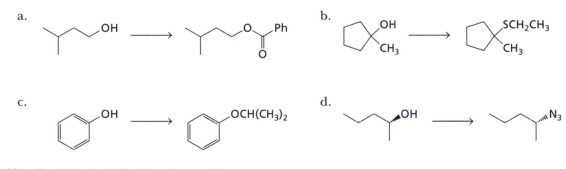

Domoic acid

a. What is the nucleophile in this first step of the overall biosynthesis pathway?

b. What is the leaving group? Propose a mechanism for the substitution process.

c. Assign (*R*) and (*S*) configurations to the stereocenters of domoic acid.

7.27. Indicate what combination of reagents and solvents can be used to carry out the following reactions:

a.

b.

c.

d.

7.28. Explain the following observations:

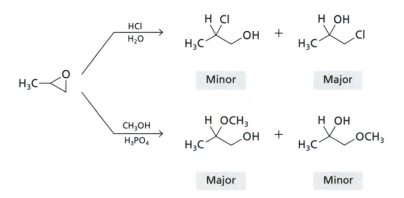

7.29. Draw the structure(s) for the expected major product(s) of each of the following transformations. Indicate stereochemistry where appropriate and assign (*R*) and (*S*) configurations to each stereogenic center. If a racemic mixture is formed, draw the structure of one enantiomer, and write the word "racemic". If no reaction occurs, write N. R.

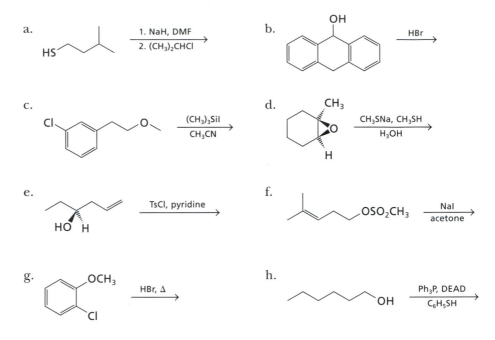

7.30. Iodotrimethylsilane reacts with alcohols and is a good reagent for making iodoalkanes. Propose a reasonable mechanism for the following reaction. Show the movement of electrons with curved arrows.

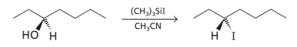

ELIMINATION REACTIONS OF ALKYL HALIDES, ALCOHOLS, AND RELATED COMPOUNDS

8.1 THE E1 REACTION

8.2 THE E2 REACTION

8.3 SUBSTITUTION VERSUS ELIMINATION

8.4 ELIMINATION REACTIONS IN BIOLOGY

CHAPTER SUMMARY

In several places, the previous chapters noted that elimination sometimes accompanies substitution, often as an unwanted side reaction (Table 6.4). Elimination reactions are important in their own right, both for the preparation of unsaturated organic molecules in the laboratory and as important steps in the metabolic pathways of biochemical systems. The formation of π bonds by elimination reactions is the topic of this chapter.

As is the case for substitution reactions, alkyl halides and alcohols are the common substrates for elimination reactions. Other types of molecules, particularly organosulfides and organoselenides, also undergo specific elimination reactions. All of these compounds are good starting materials for preparing alkenes and alkynes. Incidentally, alkenes are often called **olefins,** a name prevalent in the older literature but common enough today that the terms will be used synonymously.

We will begin by looking at the elimination reactions of alkyl halides and alcohols because you are already familiar with the substitution reactions of these substrates. The equations shown in Figure 8.1 illustrate several types of elimination reactions that are used to generate carbon–carbon double bonds. In these examples, you see two

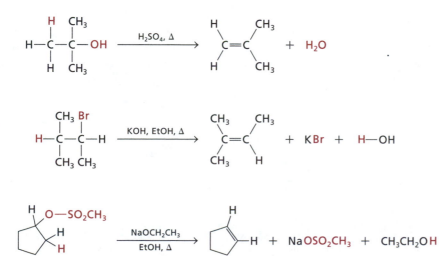

Figure 8.1
Examples of elimination reactions of alkyl halides, alcohols, and alkyl sulfonates.

types of transformations: one, an acid-catalyzed process in which an alcohol loses the elements of water; and the other, a base-promoted reaction of an alkyl halide or alkyl sulfonate ester in which the substrate loses a proton and a good leaving group.

As you learn about the features of elimination reactions that take place under both acidic and basic conditions, you will see that two limiting mechanisms exist: E1 and E2, similar to the substitution processes, S_N1 and S_N2. The stereochemical constraints are different for the two mechanisms, as are the types of molecules that react by each pathway. You will also see how these mechanisms give us insight into π-bond formation in Nature. Throughout the chapter, we will draw on information presented in Chapters 6 and 7. At the end of this chapter will be summaries of the ways in which substitution and elimination reactions are related and how they differ.

8.1   THE E1 REACTION

8.1a   DEHYDRATION OF ALCOHOLS OCCURS MAINLY BY AN E1 PROCESS

There are two common mechanisms for elimination processes that create carbon–carbon π bonds. The first is the **E1 mechanism,** which parallels the designation for the S_N1 reaction and stands for *elimination, unimolecular.*

The reaction of *tert*-butyl alcohol with sulfuric acid provides a good example of an E1 reaction, and the mechanism is similar to that for the S_N1 reaction. The oxygen atom of the alcohol functional group acts as a base, removing a proton of the mineral acid and forming a protonated alcohol.

$$\text{H}_3\text{C}-\overset{\overset{\text{H}}{|}}{\underset{\underset{\text{H}}{|}}{\text{C}}}-\overset{\overset{\text{CH}_3}{|}}{\underset{\underset{\text{CH}_3}{|}}{\text{C}}}-\ddot{\text{O}}\text{H} \underset{①}{\overset{\text{H}_2\text{SO}_4,\,\Delta}{\rightleftharpoons}} \text{H}-\text{C}-\text{C}-\overset{+}{\ddot{\text{O}}}\text{H}_2 + \text{HSO}_4^-$$

Next, a molecule of water dissociates, forming a carbocation intermediate. *Notice that this carbocation is the same intermediate that is generated in an S_N1 reaction.* Because the rate of the E1 reaction is dependent only on the concentration of substrate, dissociation of water must be the rate-determining step, just as carbocation formation in the S_N1 reaction is also the rate-limiting one.

$$\text{H}-\overset{\overset{\text{H}}{|}}{\underset{\underset{\text{H}}{|}}{\text{C}}}-\overset{\overset{\text{CH}_3}{|}}{\underset{\underset{\text{CH}_3}{|}}{\text{C}}}-\overset{+}{\ddot{\text{O}}}\text{H}_2 \underset{②}{\rightleftharpoons} \text{H}-\overset{\overset{\text{H}}{|}}{\underset{\underset{\text{H}}{|}}{\text{C}}}-\overset{+}{\underset{\text{CH}_3}{\overset{\text{CH}_3}{\text{C}}}} + \text{H}_2\text{O}$$

The appearance of a carbocation intermediate at this stage prompts a logical question: Why doesn't substitution occur? The carbocation is a Lewis acid, and therefore an electron pair acceptor (Section 5.2g). Whether it accepts electrons from the nucleophile (Lewis base) or from an adjacent sigma bond is immaterial. Steric effects make it more likely that reaction will occur at the neighboring proton, which means that the electrons will come from an adjacent C–H σ bond.

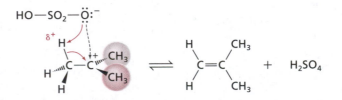

The nucleophile is attracted to the carbocation (black arrow), but this carbon atom is hindered by the three methyl groups. Reaction therefore occurs at a proton on the adjacent carbon atom, and the electron pair in the carbon–hydrogen bond moves to neutralize the positive charge (colored arrows).

Remember that the hydrogen atoms on the adjacent carbon atoms bear partial positive charges because of the bond dipoles (Section 6.2b).

Individual bond dipoles for the carbon–hydrogen bonds: The hydrogen atoms on the adjacent carbon atoms are slightly electropositive.

For secondary or tertiary alcohols, the steric bulk of the substituents especially favors elimination over substitution. The interaction between these carbocations and bulky bases such as HSO_4^- or $H_2PO_4^-$ is less favorable than reaction with the protons on the adjacent carbon atoms. Even during attempted substitution reactions with acids such as HBr, in which the conjugate base is a good nucleophile and relatively small, steric effects contribute to the formation of an alkene by elimination. We refer to the conversion of an alcohol to an alkene as **dehydration** because a molecule of water is eliminated.

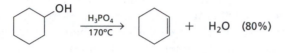

EXERCISE 8.1

Draw the structure of the major product expected from elimination of water from each alcohol in the following reactions:

a.

b.

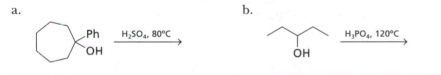

8.1b AN E1 REACTION USUALLY PRODUCES THE MORE HIGHLY SUBSTITUTED ALKENE

For the E1 reaction of an unsymmetric alcohol—2-pentanol, for example—we find that two products are formed in a ratio of 70:30. The major isomer is 2-pentene.

$$CH_3CH_2CH_2-\overset{\overset{\displaystyle OH}{|}}{CH}-CH_3 \xrightarrow{H_2SO_4, \Delta} CH_3CH_2CH=CH-CH_3 + CH_3CH_2CH_2-CH=CH_2$$

2-Pentanol   **2-Pentene**   **1-Pentene**

Ratio 70:30

More than a century ago, the Russian chemist Alexander Saytzeff recognized that these elimination reactions of alcohols under hot, acidic conditions produced the more highly substituted alkene, that is, the alkene with more groups bonded to the double-bond carbon atoms. Many subsequent studies have confirmed the generality of his initial observations.

Looking at the mechanism, we expect the first step to be protonation of the alcohol OH group, which is followed by loss of water, generating the carbocation intermediate.

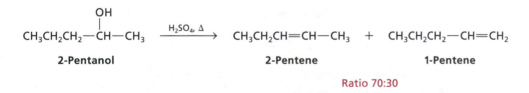

Hydrogen sulfate ion acts as a base and abstracts one of the neighboring protons to generate the products. The reaction yields 2-pentene via pathway *a*, and 1-pentene via pathway *b*.

Pathway a

$$CH_3CH_2-\underset{\underset{H}{|}}{\overset{\overset{H}{|}}{C}}-\overset{+}{C}H-\underset{\underset{H}{|}}{\overset{\overset{H}{|}}{C}}-H \quad \xrightarrow{HSO_4^-} \quad CH_3CH_2CH{=}CH-CH_3 \;+\; H_2SO_4$$

Pathway b

$$CH_3CH_2-\underset{\underset{H}{|}}{\overset{\overset{H}{|}}{C}}-\overset{+}{C}H-\underset{\underset{H}{|}}{\overset{\overset{H}{|}}{C}}-H \quad \xrightarrow{HSO_4^-} \quad CH_3CH_2CH_2-CH{=}CH_2 \;+\; H_2SO_4$$

Instinctively, you might think that reaction of the hydrogen sulfate ion would occur preferentially via pathway *b*, because the terminal carbon atom is less hindered than the secondary carbon atom. Furthermore, there are more hydrogen atoms attached to the methyl than to the methylene group, so statistically there is a 3:2 chance that reaction will occur via *b* versus *a*, ignoring other effects. How do we explain the fact that pathway *a* predominates?

The simple answer is that E1 reactions occur with **thermodynamic control,** meaning that formation of the more stable product predominates. (The other possibility is *kinetic control,* which means that the product formed more rapidly predominates.) *Alkenes that are more highly substituted are normally more stable* (Section 8.1c). [Also, (*E*)-alkenes are usually more stable that the corresponding (*Z*)-isomers, and trans isomers are more stable than analogous cis isomers.] Elimination reactions that occur under thermodynamic control are said to form the **Saytzeff product.**

The reaction coordinate diagram illustrated in Figure 8.2 helps explain thermodynamic control in the E1 reaction of 2-propanol. The first step leads to formation of the carbocation intermediate, and the second step generates the isomeric products. Note that two products are formed from the same carbocation intermediate, and their energy profiles differ.

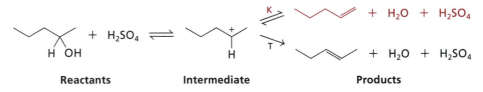

Reactants	Intermediate	Products

Thermodynamic control is important in reaction systems that satisfy three criteria:

- The process labeled K (the pathway under kinetic control) in the scheme shown directly above is an equilibrium.

- Sufficient energy is available to the system that both the forward and reverse steps of equilibrium K can and will occur.

- The product from process T (the pathway under thermodynamic control) is lower in energy than that of K. (Step T may also be an equilibrium but this feature is not required.)

You will encounter the kinetic versus thermodynamic control of a reaction again in Sections 10.3b and 22.2e.

Under thermodynamic control, any 1-pentene that *does* form will undergo the reverse reaction to regenerate the carbocation; eventually, a majority of the material will be converted to the compound with the lower energy, which in this case is 2-pentene.

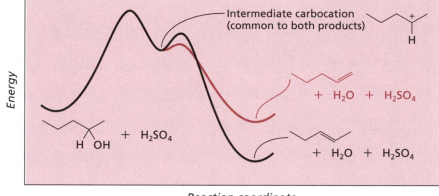

Figure 8.2
A reaction coordinate diagram for the reaction of 2-pentanol with H_2SO_4, giving 1-pentene and 2-pentene. The energy profile associated with the pathway leading to formation of 1-pentene is shown by the line in color.

EXERCISE 8.2

What are the three possible elimination products for the following reaction? (Two of the products are enantiomers.) Which one is expected to be the major product?

8.1c THE RELATIVE STABILITIES OF ALKENES CAN BE ASSESSED BY MEASURING THE CHANGE OF ENTHALPY FOR HYDROGENATION

The notion that double bonds become more stable as additional groups are attached to the sp^2-hybridized carbon atoms provides a useful guide to predict the major products of an elimination reaction. The actual data used to determine the relative stabilities of alkenes come from measuring the amount of heat evolved when molecular hydrogen adds to the olefinic π bond to create two new carbon–hydrogen σ bonds (*hydrogenation*), a transformation we will examine more closely in Chapter 11.

An instructive example compares the Pd-catalyzed hydrogenation reactions of 1-butene and *trans*-2-butene. Each alkene forms the same product—butane—therefore, each pathway finishes at the same energy. Comparing the two reactions, we find that for hydrogenation of 1-butene, 30.3 kcal mol^{-1} is evolved, whereas for 2-butene, only 27.6 kcal mol^{-1} is generated. Because reaction of the 1-isomer produces more heat, it must have had a higher enthalpy (heat content) to begin with, as shown schematically in Figure 8.3b. The 2-isomer produces less heat during hydrogenation, meaning that it was more stable to start.

If you were to examine a series of substituted ethylene derivatives, you would find that those with more alkyl groups attached at the ends of the double bonds produce less heat during the hydrogenation process. Moreover, by comparing data for many other types of alkenes, we find that tetrasubstituted alkenes are normally more stable than trisubstituted alkenes, which are more stable than disubstituted alkenes (and

a.   b.

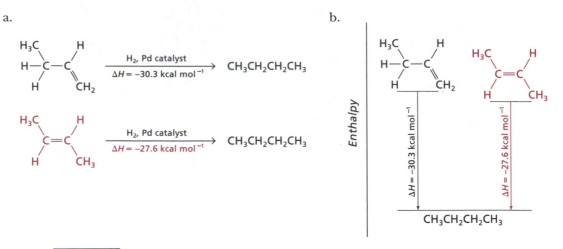

Figure 8.3
A comparison of the enthalpy changes for the hydrogenation of 1-butene and 2-butene.

trans- are more stable than cis-), which in turn are more stable than monosubstituted alkenes. These are the results we need to rationalize the formation of the Saytzeff product in the E1 reaction.

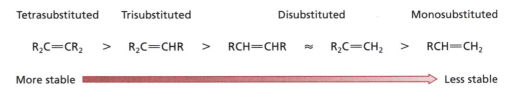

Tetrasubstituted   Trisubstituted   Disubstituted   Monosubstituted

$R_2C=CR_2$ > $R_2C=CHR$ > $RCH=CHR$ ≈ $R_2C=CH_2$ > $RCH=CH_2$

More stable ——————————————————————————————————→ Less stable

EXERCISE 8.3

Arrange the following alkenes in order of increasing amount of heat that would be evolved during hydrogenation with a Pd catalyst.

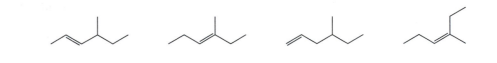

While on the topic of alkene stability, pay attention to one other common alkene type. A carbon–carbon double bond that is **endocyclic** (within a ring) is normally more stable than one that is **exocyclic** (outside the ring).

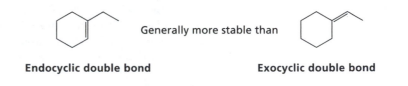

Endocyclic double bond   Generally more stable than   **Exocyclic double bond**

Exocyclic cycloalkanes are often named as derivatives of the parent cycloalkane. The substituent prefix for these substances ends in -*ylidene* except for the simplest substituent, which is called *methylene*. Illustrated below are the general structures, followed by some specific compounds and their names. Note that exocyclic double bonds can exist as (E) and (Z) isomers.

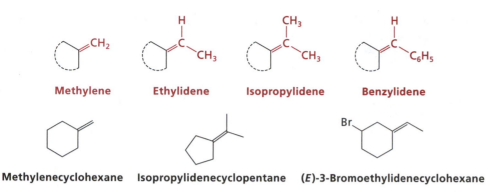

Methylene   Ethylidene   Isopropylidene   Benzylidene

Methylenecyclohexane   Isopropylidenecyclopentane   (*E*)-3-Bromoethylidenecyclohexane

---

EXERCISE 8.4

Name each of the following compounds:

a.   b.   c.

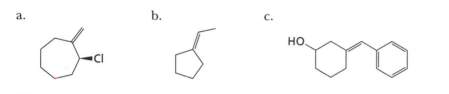

8.1d   ALKYL HALIDES ALSO UNDERGO THE E1 REACTION

So far, we have looked at the E1 reaction only as it applies to alcohols, but alkyl halides can also participate in E1 reactions. The carbon atom of an alkyl halide bears a good leaving group, so such compounds can ionize to produce a carbocation under solvolysis conditions (Section 6.2a). For example, at 65°C in aqueous ethanol, 2-bromo-2-methylpropane undergoes ionization to yield the *tert*-butyl carbocation.

$$CH_3-\underset{\underset{CH_3}{|}}{\overset{\overset{CH_3}{|}}{C}}-Br \xrightleftharpoons[①]{EtOH, H_2O} CH_3-\underset{\underset{CH_3}{|}}{\overset{\overset{CH_3}{|}}{C}}{}^+ \quad Br^-$$

Then, the competing pathways of substitution and elimination occur, producing a mixture of 2-methyl-2-propanol and 2-methylpropylene in a ratio of 64:36.

Substitution

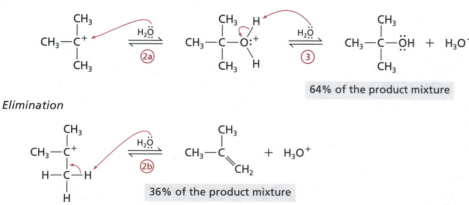

64% of the product mixture

Elimination

36% of the product mixture

   Notice that a sizable quantity of substitution product is generated, in contrast to the course of the reaction between *tert*-butyl alcohol and sulfuric acid. The reason is simple enough: Water is not as hindered as HSO_4^-, and it is also more nucleophilic. Both facets facilitate reaction at the positively charged center, which yields the substitution product.

Solvolysis conditions for the reactions of alkyl halides normally make use of solvents that are also decent nucleophiles (water and alcohols), so an E1 reaction that starts with an alkyl halide is not practical as a preparative method for alkenes. A mixture that consists of substitution and elimination products is nearly always observed. Because an alkyl halide also undergoes elimination when treated with a strong base, as described in Section 8.2, that method is preferred for making alkenes and alkynes from alkyl halides.

8.2   THE E2 REACTION

8.2a   ELIMINATION REACTIONS OF ALKYL HALIDES OR SULFONATE ESTERS ARE NORMALLY CARRIED OUT UNDER E2 CONDITIONS

When treated with a strong nucleophile, tertiary alkyl halides undergo elimination rather than substitution because steric hindrance prevents the backside approach required by the S_N2 reaction (Section 6.3a).

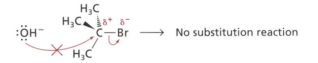

In the elimination reaction, the nucleophile acts instead as a base, approaching the protons on the carbon atom adjacent to the one bearing the leaving group. Remember that these protons are electrophilic because of the C–H bond dipoles (Section 8.1a).

As one of these hydrogen atoms is removed by the base, the electrons in the C–H bond move to generate the π bond and expel the leaving group, a concerted process called **dehydrohalogenation.** The rate of this reaction depends on the concentrations of both substrate and base, so it is bimolecular, and this pathway is the **E2 mechanism.**

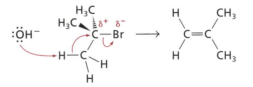

The mechanistic differences between the E1 and E2 reactions also manifest themselves in the stereochemical results. Just as you learned for the substitution reactions, stereochemistry is less important for a unimolecular process because a planar carbocation is formed, so chirality at the carbon atom with the leaving group is lost.

For an E2 reaction, the hydrogen atom and the leaving group are almost always *anti* (Section 3.1c) in the transition state, which requires this conformation to be accessible if a reaction is to occur readily. Recall that the S_N2 reaction occurs by backside reaction of the nucleophile at the carbon atom with the leaving group (Section 6.3b). In the E2 reaction, the same principle applies; what differs is the source of the electrons that actually displace the leaving group.

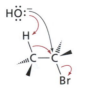

S_N2 The electrons that displace the leaving group come directly from the nuclelphile, and reacton occurs at the side of the carbon atom opposite to the C–Br bond.

E2   The electrons that displace the leaving group come from the C–H σ bond, even though they originate with the base. This displacement also occurs on the side opposite to the C–Br bond.

anti Conformation

Dehydrohalogenation is often referred to as **β-elimination.** This notation derives from the relative structural relationship (Section 5.5b) between the position of the proton and the leaving group. For a dehydrohalogenation reaction, the proton being removed is attached to the carbon atom β to the halogen atom.

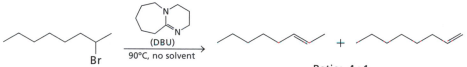

A β-elimination reaction

8.2b   DEHYDROHALOGENATION MAY PRODUCE THE LESS STABLE ALKENE

Even with the requirement that a leaving group and β-hydrogen atom be *anti,* the E2 reaction normally gives the Saytzeff product, that is, the more highly substituted and stable isomer. The following transformation illustrates this result for the E2 reaction of 2-bromooctane using the base 1,8-diazabicyclo[5.4.0]undec-7-ene (DBU), a valuable reagent in organic chemistry when a base is required. DBU is a poor nucleophile because of steric crowding of the nitrogen atoms, so substitution reactions are minimized.

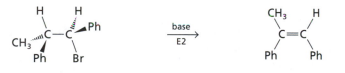

Ratio: 4 : 1

It is possible, however, for the *less* stable product to become the major one during dehydrohalogenation because the stereochemical requirements of the E2 reaction override the thermodynamic influence. Most often, this result is obtained when the only removable hydrogen atom puts two large groups cis to one another in the product. The E2 reaction of (1*S*,2*S*)-1-bromo-1,2-diphenylpropane is illustrative.

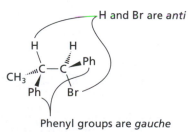

(1*S*,2*S*)-1-Bromo-1,2-diphenylpropane   **(*Z*)-1,2-Diphenylpropene**

The two phenyl substituents must be *gauche* when the bromine and hydrogen atoms are *anti.* Elimination in this conformation produces the (*Z*) isomer as the major product.

H and Br are *anti*

Phenyl groups are *gauche*

Compare that result with the course of the E1 reaction that starts with this same substrate.

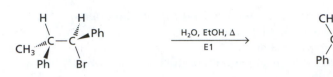

(1S,2S)-1-Bromo-1,2-diphenylpropane   **(E)-1,2-Diphenylpropene**

In the first step of the E1 pathway, a carbocation intermediate forms after dissociation of the bromide ion. (A rearrangement can occur here as well, but we will ignore that pathway.)

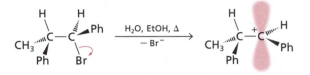

Then, rotation about the carbon–carbon σ bond occurs to move the phenyl groups apart.

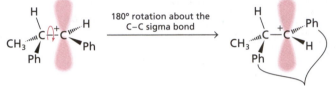

After rotation, the phenyl groups are farther apart.

Finally, deprotonation of the charged intermediate yields the (*E*) product, in which the phenyl groups are trans to each other.

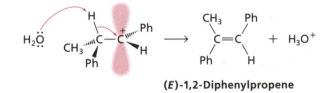

(E)-1,2-Diphenylpropene

The E1 reaction has no requirement about the alignment of the leaving group and β-hydrogen atoms. For that reason, the thermodynamically stable product predominates in nearly every E1 reaction.

EXAMPLE 8.1

Draw the structure of the product expected from the following E2 reaction. Show the mechanism for the elimination step with curved arrows.

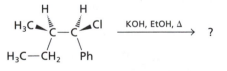

To solve this type of problem, draw the molecule in the conformation that places the leaving group (chloride ion in this case) *anti* to the proton on the adjacent carbon atom. Then show electron movement from the base to the proton to form the π bond and displace the chloride ion. Redraw the product in its standard format (i.e., flat), keeping the relative orientation of groups the same.

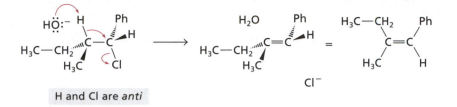

H and Cl are *anti*

EXERCISE 8.5

Draw the structure of the product of the following E2 reaction. Draw the possible staggered conformations about the central C–C bond, and then indicate which one is involved in the elimination process. Show the mechanism for the elimination step using curved arrows.

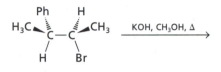

8.2c   IN THE E2 REACTION OF A CYCLOHEXYL HALIDE, THE PROTON AND LEAVING GROUP MUST BE TRANS AND DIAXIAL

The geometric constraints of the E2 reaction are even more crucial for cyclic compounds. For example, consider the reaction between bromocyclohexane and a strong base.

Bromocyclohexane   **Cyclohexene**

The hydrogen atom and leaving group can assume an *anti* orientation only if both groups are trans and diaxial. If they are not diaxial, then the cyclohexane ring must undergo a ring-flip into its other chair conformation.

    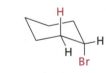

The bromine and hydrogen atoms are *gauche*.   The bromine atom is *anti* to the axial hydrogen atom.

When more than one pathway exists by which elimination can occur, the Saytzeff product (the more highly substituted alkene) will be formed preferentially, as illustrated in Example 8.2. When a substituent on the ring prevents formation of the Saytzeff product, then the less highly substituted alkene will be formed. This situation is addressed in Exercise 8.6.

EXAMPLE 8.2

What is the major product of the following reaction? Justify your answer.

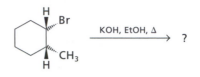

First, draw the cyclohexane in its two chair forms. For the conformation that has the leaving group in an axial position, show all of the hydrogen atoms at the positions adjacent to the carbon atom bearing the leaving group.

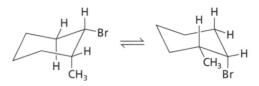

Next, consider the reaction of the base with each β-hydrogen atom trans to the leaving group, showing how the electrons move to displace the leaving group.

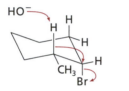   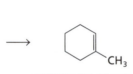

More highly substituted alkene—major product

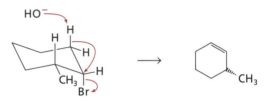

Finally, decide which will be the major product. If more than one product can be formed, then the one with the more highly substituted double bond is the major product. In this case, more 1-methylcyclohexene than 3-methylcyclohexene is formed because the 1-methyl isomer has a trisubstituted double bond, whereas the 3-methyl isomer has a disubstituted double bond.

EXERCISE 8.6

What is the major product of the following reaction? Justify your answer.

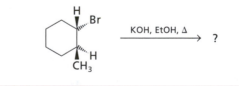

In any elimination reaction of a cycloalkyl halide, substituents on the ring can inhibit the E2 reaction if the energy barrier for a ring-flip is raised substantially. This effect is particularly clear when one substituent is the *tert*-butyl group, because of its preference to be equatorial (Section 3.3c). Conversely, a conformation that holds the halogen atom in an axial position will *increase* the ease of elimination.

Explain, using structural drawings, why *trans*-1-bromo-4-*tert*-butylcyclohexane reacts 500 times more slowly than *cis*-1-bromo-4-*tert*-butylcyclohexane when an E2 reaction is carried out using potassium *tert*-butoxide as the base.

8.2d   THE E2 REACTION IS A LIKELY PATHWAY FOR THE DEHYDRATION OF PRIMARY ALCOHOLS

Unlike an alkyl halide, an alcohol cannot undergo an E2 reaction by reaction with strong base because the OH group is a poor leaving group under such conditions. Instead, as you have seen, an alcohol undergoes elimination when treated with a strong acid, which generates a good leaving group, H_2O. For a secondary or tertiary alcohol, dehydration occurs by the E1 pathway through formation of a carbocation intermediate (Section 8.1a).

A primary alcohol cannot form a carbocation, however. After protonation of the OH group in Step 1, a primary alcohol reacts with the conjugate base of the mineral acid by β-elimination.

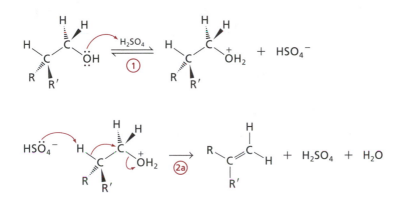

Because a primary alcohol is sterically unencumbered at the carbon atom bearing the leaving group, substitution can also take place even though the nucleophile is weak. With sulfuric acid as the reagent, an alkyl sulfate ester (a derivative of sulfuric acid in which a proton has been replaced by an alkyl group) can be formed.

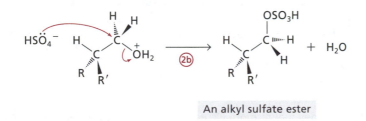

An alkyl sulfate ester

Moreover, any alcohol molecule not protonated in Step 1 can function as a nucleophile to produce the corresponding symmetrical ether.

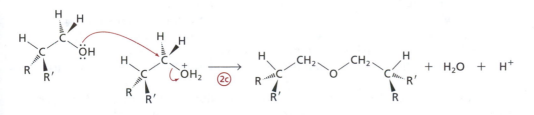

With the high temperature needed to induce elimination, these side reactions make acid-catalyzed dehydration a poor method for converting a 1° alcohol to an alkene.

Suppose you want to prepare an alkene from a primary alcohol? A preferable strategy is to convert the alcohol to its alkyl sulfonate derivative (Section 7.1b), and then carry out an E2 reaction with base, as described previously for alkyl halides. Recall that this same strategy—preparing an alkyl sulfonate ester—is the best way to perform substitution reactions of a primary alcohol (Section 7.1b).

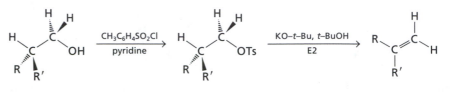

EXERCISE 8.8

What is the major product of the following reactions, including stereochemistry, under each set of conditions? Consider a rearrangement process in the acid-promoted reaction.

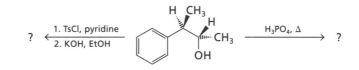

8.2e   AN ALKYNE CAN BE PREPARED BY SUCCESSIVE E2 REACTIONS OF GEMINAL AND VICINAL DIHALIDES

So far, you have seen that alkyl halides undergo elimination reactions to form alkenes. It may come as little surprise that a dihaloalkane can undergo E2 reactions as well, first to generate a vinyl halide, and then to form an alkyne.

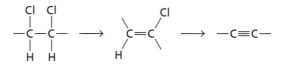

The stereochemistry of the product is not an issue because each carbon atom of the final product has only a single substituent. Therefore, the substrate can either have the halogen atoms attached to the same carbon atom (**geminal**) or to adjacent carbon atoms (**vicinal**). These types of compounds are called geminal dihaloalkanes and vicinal dihaloalkanes, respectively.

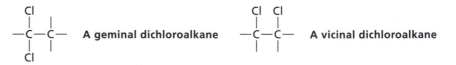

LDA

A powerful base is needed to carry out the second dehydrohalogenation reaction, and the conjugate base of an amine is often used. Sodium amide, NaNH$_2$, in liquid ammonia as the solvent (bp = −33°C), was the preferred base for many years. Now, the use of amide ions like LDA (Table 5.4 and at left), is more common. A hot, alcoholic KOH solution or an alkoxide ion, especially potassium *tert*-butoxide in *tert*-butyl alcohol, DMSO, or THF, is also used to effect elimination of HX from vinyl halides to form alkynes.

As the following examples show, these transformations can be carried out stepwise, via formation of a vinyl halide, or in one step, generating the alkyne directly.

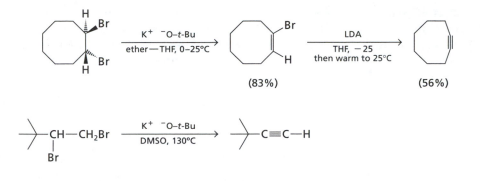

(83%)   (56%)

What is the product of the following reaction? What is the purpose of sulfuric acid in the second step?

HOOC—CH—CH—COOH $\xrightarrow[\text{2. H}_2\text{SO}_4, \text{H}_2\text{O}]{\text{1. KOH, EtOH, }\Delta}$
　　　　　|　　|
　　　　　Br　Br

8.3   SUBSTITUTION VERSUS ELIMINATION

8.3a   ALKYL HALIDES REACT PREDOMINANTLY VIA BIMOLECULAR PATHWAYS

So far Chapters 6–8 have covered several common transformations of alkyl halides, alcohols, and alkyl sulfonates. Any one of four mechanisms is possible, namely, S_N1, S_N2, E1, and E2. For alkyl halides and sulfonates, the S_N2 and E2 pathways are the most likely, depending on the specific conditions and criteria, as summarized in Table 8.1.

If the nucleophile is good, then substitution is likely when the substrate is a primary alkyl halide or sulfonate. Even strongly basic nucleophiles mainly yield substitution

Table 8.1   Elimination versus substitution reactions of alkyl halides and sulfonate esters.

Substrate	Weak base Good nucleophile[a]	Moderate base Good nucleophile[a]	Strong base[a]
1° Alkyl halide or sulfonate ester	S_N2[b]	S_N2	S_N2 With good nucleophiles; E2 with poor nucleophiles and heat
2° Alkyl halide or sulfonate ester	S_N1 or S_N2[b] Some E1 at higher temperatures	S_N2[c] E2 At higher temperatures	E2
3° Alkyl halide or sulfonate ester	S_N1 and E1 E1 predominates at higher temperatures	E2	E2

[a]See Table 6.3.
[b]These are the weak base nucleophiles shown in color in Table 6.3.
[c]The more basic the nucleophile, the more dominant the E2 pathway becomes.

products. Elimination becomes probable only when the base is hindered and the temperature is raised. In general, *higher temperatures promote elimination over substitution.*

For tertiary alkyl halides and sulfonates, the E2 mechanism predominates. Even good nucleophiles that are relatively weak bases will promote elimination over substitution for these tertiary substrates.

The ambiguous cases comprise the secondary alkyl halides and sulfonates because all four reaction pathways are possible. Good nucleophiles and low temperatures favor substitution (S_N2), but even moderately basic nucleophiles and heat favor elimination (E2). If the nucleophile is a weak base, especially under solvolysis conditions, S_N1 and E1 pathways become more important. *Remember:* The bimolecular mechanisms offer more stereochemical control.

8.3b ALCOHOLS REACT PREDOMINANTLY BY S_N1 AND E1 PATHWAYS

Unlike the reactions of alkyl halides and sulfonates, transformations of alcohols normally occur by the S_N1 and E1 pathways, as summarized in Table 8.2. In all substitution or elimination reactions of alcohols, a strong mineral acid is required first to convert the OH group into a good leaving group. By contrast, a strong base such as LDA converts an alcohol to its conjugate base, an alkoxide ion (Section 7.2a).

Primary alcohols either react slowly or require vigorous conditions such as prolonged heating. Elimination is particularly difficult to accomplish when an alcohol is primary, so conversion to a sulfonate ester derivative is a preferable strategy if you want to make an alkene from a 1° alcohol. An acid that has a strongly nucleophilic conjugate base such as HBr or HI promotes reaction of a primary alcohol via a concerted pathway rather than by formation of a 1° carbocation.

Secondary and tertiary alcohols react under acidic conditions mainly to form a carbocation intermediate. If the conjugate base of the added acid is a good nucleophile, such as Br^- or I^-, then substitution is more likely than if a weak nucleophile like HSO_4^- or $H_2PO_4^-$ is present. Remember, however, that the outcome of a reaction that occurs via a carbocation is less readily controlled than one that proceeds by a concerted pathway. For that reason, conversion of a 2° alcohol to its sulfonate ester derivative is more valuable for synthetic purposes.

Table 8.2 Elimination versus substitution reactions of alcohols.

Substrate	Weak base Poor nucleophile[a, c]	Weak base Good nucleophile[a, d]	Moderate or strong base[b]
1° Alcohol	E2 at very high temp (after reaction with HY to form a good leaving group)[c]	S_N2 (after reaction with HX to form a good leaving group)[d]	No reaction
2° Alcohol	E1 (after reaction with HY to form a good leaving group)[c]	S_N1 and E1 (after reaction with HX to form a good leaving group)[d]	No reaction
3° Alcohol	E1 (after reaction with HY to form a good leaving group)[c]	S_N1 and E1 (after reaction with HX to form a good leaving group)[d]	No reaction

[a] See Table 6.3.
[b] A strong acid is required to generate a good leaving group. Very strong bases such as LDA react with alcohols via deprotonation of their OH groups (Section 7.2a).
[c] For HY, examples include H_2SO_4 and H_3PO_4.
[d] For HX, examples include HBr and HI.

8.3c   THE TYPE OF PRODUCT BEING PREPARED HELPS TO CHOOSE WHICH MECHANISM—S_N1, E1, S_N2, OR E2—IS MOST LIKELY

When faced with the choice as to which mechanism is likely to occur for a reaction that starts with an alkyl halide or alcohol, you will likely make use of the data in Tables 8.1 and 8.2. Realize this essential point, however: *Substrates and reagents are chosen for reactions because there is a well-defined goal in mind.* Either you are trying to substitute the leaving group, or you are trying to promote elimination to form a π bond. If you want to make an alkene, for example, you will choose conditions that favor elimination—you will *not* use a reagent that is likely to give a mixture of products.

This consideration of the reaction context is most important for secondary substrates because they can react by any of four possible pathways that ultimately depend on the experimental conditions. The following example illustrates how to take the context of the reaction conditions into account.

EXAMPLE 8.3

Draw the structure(s) of the product(s) of the following reaction. Identify the major product and explain your reasoning.

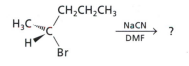

First, identify the type of substrate undergoing reaction. In this case, the compound is a 2° alkyl halide. Next, consider the reaction conditions. The nucleophile/base is cyanide ion, a strong nucleophile and a reasonably good base (its conjugate acid is HCN, which has a pK_a value of ~10). The use of a polar, aprotic solvent will promote substitution via the S_N2 pathway, which will occur with inversion of configuration. The elimination reaction will occur via the E2 pathway to produce the trans-disubstituted alkene, which is the most stable alkene product possible.

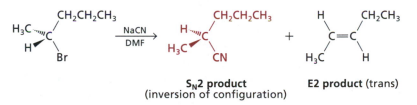

According to Table 8.1, the S_N2 and E2 reaction pathways are both possible. We designate the substitution product (shown in color above) as the major one because of context: if we had set out to make the alkene, we would have used a strong base such as hydroxide ion in a protic solvent such as ethanol. By specifying a strong nucleophile in a polar, aprotic solvent, we are evidently trying to maximize the amount of substitution product. Some elimination product will be formed, however. Cyanide ion is a strong enough base that we cannot count only on substitution.

EXAMPLE 8.4

Draw the structure(s) of the major product(s) of the following reaction. Rationalize your answer.

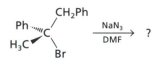

First, identify the type of substrate undergoing reaction. In this case, the compound is a benzylic, 3° alkyl halide. Next, consider the reaction conditions. The nucleophile is azide ion, a strong nucleophile as well as being a base with weak-to-moderate strength (the pK_a value of HN_3 is ~ 5). The solvent is a polar, aprotic one, which is normally used in S_N2 reactions. Because tertiary alkyl halides do not undergo substitution via the S_N2 pathway, elimination via the E2 pathway will predominate.

To predict which possible E2 product will be the major one, we expand the structure of the alkyl bromide to see which conformation will lead to formation of the most stable double bond upon dehydrohalogenation. In this example, the major product is the (E) trisubstituted alkene.

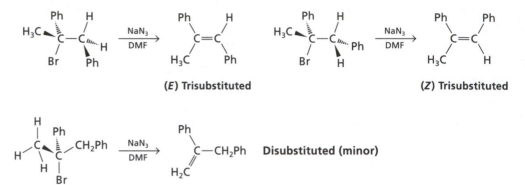

(E) Trisubstituted   (Z) Trisubstituted

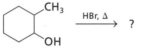

Disubstituted (minor)

EXAMPLE 8.5

Draw the structure(s) of the product(s) of the following reaction. Identify the major product and explain your reasoning.

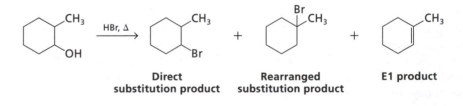

The starting compound is a 2° alcohol, and the other reactant is HBr, the conjugate base of which is a strong nucleophile. According to Table 8.2, both S_N1 and E1 reaction pathways are likely. The S_N1 reaction will occur with racemization (although no stereochemistry is shown for the starting alcohol so we can ignore stereochemistry in the product as well), and the elimination reaction will produce the trisubstituted alkene, which is the most stable alkene product possible.

Direct   **Rearranged**   **E1 product**
substitution product **substitution product**

A rearrangement is also possible in this case. Looking at the following mechanism, you can see that a secondary carbocation is produced at first. A tertiary carbocation will be formed if rearrangement occurs by a hydride ion shift.

2° Carbocation   3° Carbocation

We predict that the substitution products will predominate under these conditions. The rearranged substitution product is chosen as the major one because of the greater stability of the 3° carbocation. The justification for the choice of substitution over elimination is based again on perspective: The S_N1 and E1 reactions proceed through the same carbocation intermediate, so they are always linked. If the elimination product had been desired, then an acid such as H_2SO_4 would have been chosen instead of HBr. Notice that elimination from the 3° carbocation gives the same alkene as that formed by elimination from the 2° carbocation

EXERCISE 8.10

Complete the half-finished mechanism shown in Example 8.5, accounting for the formation of each product. Also, show that the 2° and 3° carbocations give the same elimination product, 1-methylcyclohexene.

EXERCISE 8.11

Draw the structure(s) of the product(s) for each of the following reactions. Identify the major product in each and explain your reasoning.

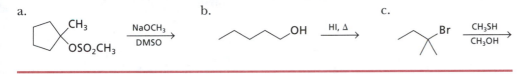

a.   b.   c.

8.4   ELIMINATION REACTIONS IN BIOLOGY

The E2 reaction is an indispensable transformation in the synthetic methodology of organic chemistry because it is a versatile way to prepare alkenes. It is worth asking whether biological systems exploit this same mechanism. If the E2 reaction is important in biological systems, three features should be evident:

1. The substrate will have a good leaving group or a group readily converted to one.

2. A base will be present to remove a proton from the carbon atom β to the leaving group.

3. The substrate molecule will be bound at an enzyme active site in the orientation that favors *anti* elimination.

Table 8.3 lists several examples of elimination reactions known to occur during metabolism. Most of these transformations occur when the substrate molecule has a proton and leaving group *anti* to each other, as determined by the stereochemistry of the product generated in each case.

Let us now consider whether an enzyme active site can satisfy the three criteria listed above. First, can a suitable base be provided? Several amino acids in the protein backbone of an enzyme have side chains that are basic, so this provision is not unreasonable. For example, the nitrogen atom in the imidazole ring of histidine (Table 5.8) often functions as a base in living systems. The imidazole ring in the side chain of histidine is strong enough to remove a proton from water at pH 7, although the equilibrium lies to the left for this simple acid–base reaction.

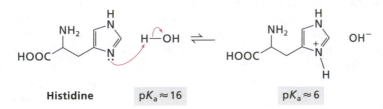

Histidine $pK_a \approx 16$ $pK_a \approx 6$

Table 8.3   Some biological elimination reactions.

Entry	Substrate	Enzyme
1.		Citraconate hydrase
2.		Fumarase
3.		2-Isopropylmalate dehydratase
4.		Enolase
5.		Phenylalanine ammonia-lyase
6.		Dehydroquinate synthase

Second, can the enzyme influence the properties of the leaving group in the substrate? Some of the substrates listed in Table 8.3 have a good leaving group already—an ammonium ion and a phosphate ion are both reasonably good. The problem lies with substrates that are alcohols. In this chapter, you have already seen that alcohols require some type of activation—usually protonation—to convert the OH group into a good leaving group. An amino acid with an acidic side chain, for example, aspartic acid (Table 5.8), might facilitate this step.

Finally, can the proper *anti* conformation of proton and leaving groups be ensured by the enzyme? Remember that a protein chain folds into a specific three-dimensional shape to create an enzyme active site (Section 5.5b). This folding places the side chains of certain amino acids at the active site where they can influence substrate binding through a combination of hydrogen bonds and electrostatic interactions. For the substrates presented in Table 8.3, an enzyme active site could, for example, employ a positively charged amino acid side chain to attract the negatively charged carboxylate ion.

EXAMPLE 8.6

Draw the structure of the product from the *citraconate hydrase* catalyzed reaction shown as entry 1 of Table 8.3. Include the rudimentary reactive centers (acids and bases) that might be provided by the enzyme.

The leaving group must be water, formed by protonation of the OH group of the substrate. An acidic site needed to supply this proton is represented by B'H$^+$ in the upper left side of the following equation. The proton to be removed from the substrate is attached to the carbon atom β to the latent leaving group. Removal of this proton requires a base, represented as :B. The products are the (*Z*)-alkene and a molecule of water. The acid is converted to its conjugate base, B': , and the base is converted to its conjugate acid, HB$^+$.

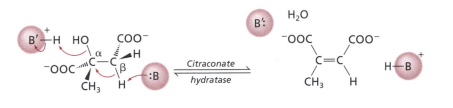

EXERCISE 8.12

Draw the structures of the products for the other entries in Table 8.3. Include the rudimentary reactive centers (acids and bases) that might be provided by the enzyme.

Looking at the structures in Table 8.3, you might assume that E2 reactions are prevalent in biochemistry, and for years, this notion was generally accepted. Dehydroamination (entry 5 in Table 8.3) may likely be an E2 reaction, but different mechanisms can operate in the other cases. (We will look at one alternative pathway—the E1cb mechanism—in Section 23.1b.)

Thinking about the parameters associated with the E2 reactions of alkyl halides and alcohols—basicity of the nucleophile, leaving group ability, and stereochemistry—can provide insights into the mechanisms by which biochemical transformations occur. If likely possibilities are considered, then understanding how enzymes catalyze chemical reactions may subsequently lead to the development of biomimetic catalysts or to the preparation of new medications to treat metabolic disorders and their associated diseases.

CHAPTER SUMMARY

Section 8.1   The E1 reaction

- The E1 (elimination, unimolecular) reaction is one pathway by which alkyl halides, sulfonate esters, and alcohols are converted to alkenes (also called "olefins").
- A secondary or tertiary alcohol readily loses water, a process termed dehydration, when it is heated with sulfuric or phosphoric acid.
- Dehydration proceeds via a carbocation, which is formed by dissociation of a molecule of water from the protonated OH group of the alcohol.
- A 2° or 3° alkyl halide undergoes the E1 reaction in competition with an S$_N$1 process.
- If more than one alkene can be formed in an E1 reaction, the more highly substituted alkene predominates because it is more stable.

- The "more highly substituted alkene" has the fewer number of hydrogen atoms attached to the double-bonded carbon atoms.
- Alkene stability is determined by the amount of heat evolved when hydrogen adds to a carbon–carbon double bond to form the corresponding C–C single bond.
- Stability decreases in the order tetrasubstituted alkene > trisubstituted alkene > disubstituted alkene > monosubstituted alkene > ethylene.

Section 8.2   The E2 reaction

- An alkyl halide (or sulfonate ester) reacts with a strong base to form an alkene by a pathway designated E2 (elimination, bimolecular). This reaction is termed dehydrohalogenation or β-elimination.
- Dehydrohalogenation is a concerted process in which a base removes a hydrogen atom that is β to the leaving group, concomitant with formation of the π bond and displacement of the leaving group.
- The proton removed during the E2 reaction is oriented *anti* to the leaving group.
- A molecule that cannot adopt a conformation in which a β-proton and leaving group are *anti* reacts more slowly than a molecule that can adopt such a conformation.
- The E2 reaction of cyclohexane derivatives requires that both the β-proton and leaving group are trans and diaxial.
- A dihaloalkane reacts with strong base via an E2 pathway to form an alkyne.
- A primary alcohol can also be dehydrated with use of a strong mineral acid, but substitution by the acid's conjugate base or by another molecule of alcohol, forming an ether, constitute competitive side reactions.
- Dehydration of a primary alcohol is better accomplished by an indirect method that involves conversion of the alcohol to its sulfonate ester derivative, followed by treatment with strong base.

Section 8.3   Substitution versus elimination

- Substitution and elimination are related because both require the presence of a good leaving group. These relationships are summarized in Tables 8.1 and 8.2.
- In general, higher temperatures favor elimination over substitution.

Section 8.4   Elimination reactions in biology

- Elimination reactions in biological systems may occur via the E2 pathway.
- The same criteria are important for an E2 reaction carried out in a laboratory and catalyzed by an enzyme: A base will be present to remove a proton from the β-carbon atom, the substrate will have a good leaving group, and the substrate will have the β-proton and leaving group *anti*.

KEY TERMS

Introduction
olefin

Section 8.1a
E1 mechanism
dehydration

Section 8.1b
thermodynamic control
Saytzeff product

Section 8.1c
endocyclic
exocyclic

Section 8.2a
dehydrohalogenation
E2 mechanism
β-elimination

Section 8.2e
geminal
vicinal

Section 8.1a

Dehydration: 2° and 3° alcohols lose water when heated with a strong mineral acid such as sulfuric or phosphoric acid. The mechanism is E1.

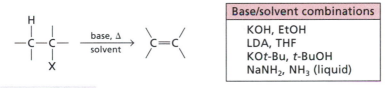

2° and 3° Alcohols

Section 8.1d

Solvolysis of 2° and 3° alkyl halides leads to E1 in competition with the S_N1 reaction.

X = Cl, Br, I

Section 8.2a

Dehydrohalogenation: alkyl halides and alkyl sulfonate esters lose HX when heated with strong base. The reaction is concerted and the mechanism is E2.

Base/solvent combinations
KOH, EtOH
LDA, THF
KOt-Bu, t-BuOH
NaNH$_2$, NH$_3$ (liquid)

X = Cl, Br, I, OSO$_2$R

Section 8.2d

High temperatures are required to promote elimination reations of primary alcohols. Inorganic esters (of sulfuric or phosphoric acid) and symmetrical ethers can be formed as byproducts.

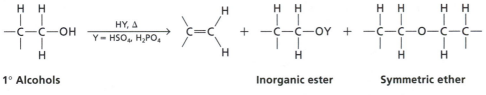

1° Alcohols   Inorganic ester   Symmetric ether

Substitution side products

Section 8.2e

When heated with strong base, geminal and vicinal dihaloalkanes form alkynes via consecutive E2 dehydrohalogenation reactions.

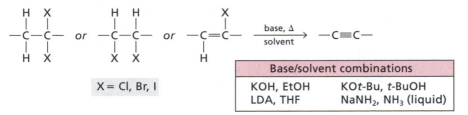

X = Cl, Br, I

Base/solvent combinations	
KOH, EtOH	KOt-Bu, t-BuOH
LDA, THF	NaNH$_2$, NH$_3$ (liquid)

ADDITIONAL EXERCISES

8.13. Draw structural formulas for the six isomers of C_5H_{10} that contain a double bond. Give the IUPAC name for each, including the stereochemical descriptor where appropriate.

8.14. Predict the product(s) (including geometric isomers) of each of the following elimination reactions. Indicate, where appropriate, which product will be the major one. Name each product.

a.   b.   c.

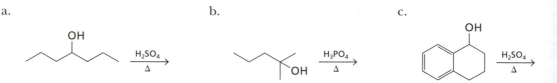

8.15. Predict the products (including geometric isomers) formed by dehydrohalogenation of the following compounds upon treatment with alcoholic KOH. For those compounds that can form more than one product, indicate which product will be the major one. Name each product.

a. 1-Bromohexane   b. 1-Bromo-1-methylcyclohexane

c. *cis*-1-Bromo-2-methylcyclohexane   d. (S)-2-Chlorobutane

e. 4-Iodoheptane

8.16. Explain why (1S,2R)-1-bromo-1,2-diphenylpropane reacts faster with a strong base than (1R,2R)-1-bromo-1,2-diphenylpropane does. What is the major elimination product formed in each reaction?

8.17. Arrange each set of compounds in decreasing order of heat evolved upon catalytic hydrogenation with a palladium catalyst.

a.

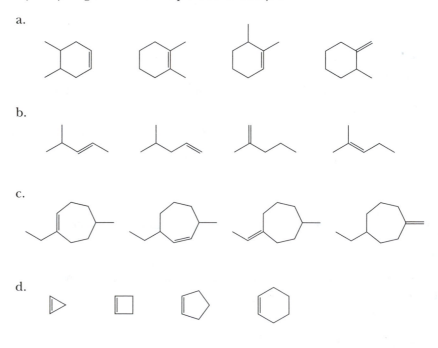

b.

c.

d.

8.18. 1-Cyclohexylethanol undergoes dehydration with sulfuric acid, yielding 1-ethylcyclohexane (among other compounds). Propose a mechanism for this dehydration reaction that explains how the endocyclic product is formed.

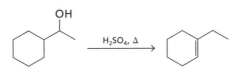

1-Cyclohexylethanol

8.19. Given the results presented in Exercise 8.18, how could you prepare ethylidenecyclohexane from 1-cyclohexylethanol by means of an elimination reaction? (*Hint:* Choose a method that avoids the formation of a carbocation intermediate; Section 8.2d.)

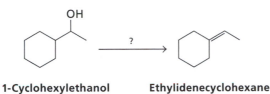

1-Cyclohexylethanol   **Ethylidenecyclohexane**

8.20. In each of the following reactions, indicate whether methoxide ion functions as a base or as a nucleophile:

a.

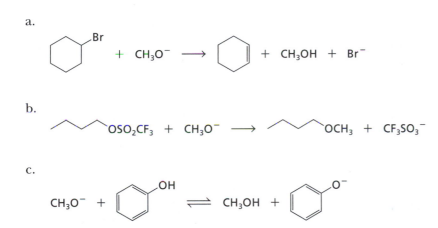

b.

c.

8.21. The reaction of an acid chloride with a non-nucleophilic base leads to formation of a ketene, $R_2C=C=O$, an unstable and reactive species. Propose a mechanism for the following reaction used to prepare dimethylketene. Show the movement of electrons with curved arrows.

8.22. An aldehyde can be converted to an oxime (Section 20.1b), which in turn can be converted to its acetate derivative. Propose a mechanism for the conversion of the oxime acetate to the nitrile under the conditions given in the following equation. Show the movement of electrons with curved arrows.

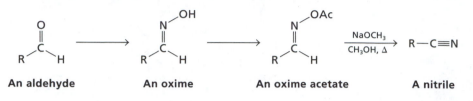

| An aldehyde | An oxime | An oxime acetate | A nitrile |

8.23. Draw the structure(s) of the major product(s) expected from each of the following reactions, which can be either substitution or elimination processes. Indicate the stereochemistry of the product as appropriate. Relative stereochemistry should be shown using wedges and dashed lines. If a racemic mixture will be formed, draw the structure of one enantiomer and write the word "racemic", or draw both enantiomeric structures. If diastereomers are formed, draw each structure; label meso compounds as such. If no reaction occurs, write N.R.

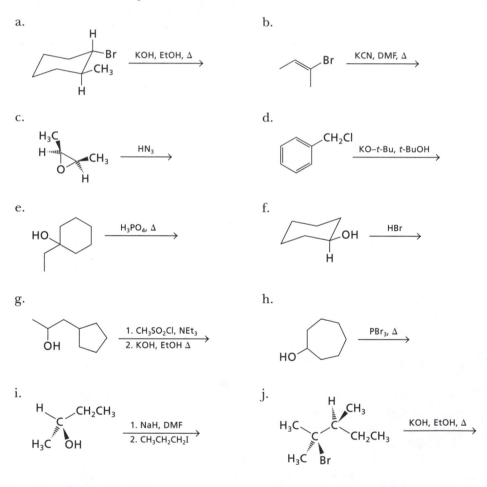

8.24. Draw a reaction coordinate diagram for the elimination of water from *tert*-butyl alcohol in sulfuric acid. Label the appropriate minima and maxima as starting materials, intermediates, and products. Mark the energy differences that correspond to ΔG° and $\Delta G^\ddagger$.

8.25. Each of the following alkenes can be made from an organohalide by an E2 reaction:

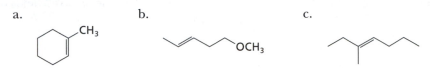

a.

b.

c.

1. Name each of these alkenes according to IUPAC rules.
2. What starting material could be used to prepare each compound as the major product of the reaction?
3. Name each starting material.

8.26. Draw the structure(s) of the major product(s) expected from each of the following reactions, which can be either substitution or elimination processes. Indicate the stereochemistry of the product as appropriate. Relative stereochemistry should be shown using wedges and dashed lines. If a racemic mixture will be formed, draw the structure of one enantiomer and write the word "racemic", or draw both enantiomeric structures. If diastereomers are formed, draw each structure; label meso compounds as such. If no reaction occurs, write N.R.

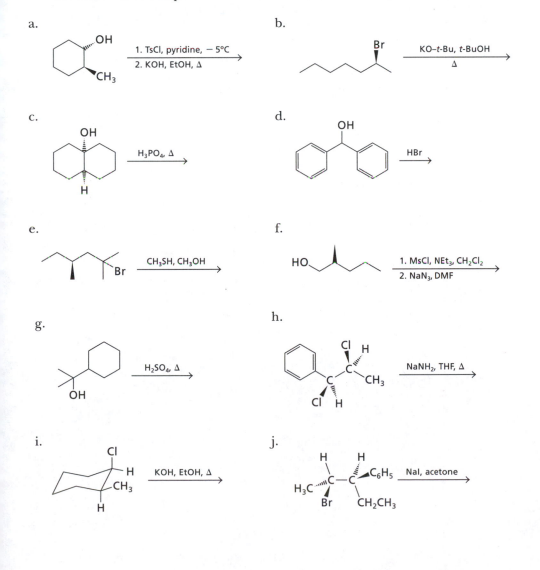

a.
1. TsCl, pyridine, − 5°C
2. KOH, EtOH, Δ

b.
KO–t-Bu, t-BuOH
Δ

c.
H₃PO₄, Δ

d.
HBr

e.
CH₃SH, CH₃OH

f.
1. MsCl, NEt₃, CH₂Cl₂
2. NaN₃, DMF

g.
H₂SO₄, Δ

h.
NaNH₂, THF, Δ

i.
KOH, EtOH, Δ

j.
NaI, acetone

8.27. Compound **A,** $C_9H_{11}Br$, is an optically active molecule that contains a benzene ring substituted at only one position. It reacts with hot, alcoholic KOH to produce **B,** which is not optically active. Stirring **A** with aqueous methanol at 35°C yields optically inactive **C.** Compound **C** is converted to **B** when it is treated first with methanesulfonyl chloride and triethylamine in dichloromethane, and then with hot, alcoholic KOH. Treating **C** with sodium hydride in DMF followed by the addition of methyl iodide gives **D,** which is also optically inactive. But treating **A** with sodium methoxide in DMF at room temperature gives optically active **D.** Draw structures for compounds **A** through **D.**

8.28. Indicate what reagents and solvents can be used to carry out the following reactions:

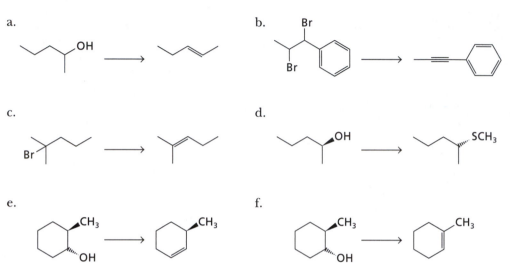

ADDITION REACTIONS OF ALKENES AND ALKYNES

In the previous three chapters (Chapters 6–8), you have been studying reactions that occur mainly by breaking σ bonds at carbon atoms with sp^3 hybridization. These processes ultimately lead to the formation of substitution or elimination products from the reactant molecule. In this chapter as well as the next (Chapter 10), you will study transformations in which the first step involves breaking a π bond at carbon atoms having sp- or sp^2-hybridization. The eventual outcome is addition to the π bond, and the common substrates are alkenes and alkynes.

Addition reactions are vital in biochemistry as one means to construct the carbon frameworks of an assortment of molecules, and some of these same transformations serve as an underpinning for the chemistry of synthetic materials such as polymers, plastics, and composites.

9.1   ELECTROPHILIC ADDITION REACTIONS OF ALKENES

9.1a   ELECTROPHILLIC ADDITION IS A TWO-STEP PROCESS THAT OCCURS VIA A CARBOCATION INTERMEDIATE

Many reagents react with an alkene by addition to its π bond; and so too, a variety of mechanisms are possible. When the first step of an addition reaction takes place between a carbon–carbon π bond and an electrophile, then a charged intermediate is often formed, and the mechanism is termed **electrophilic addition.** In a second step, this intermediate reacts with the nucleophilic portion of the original reagent.

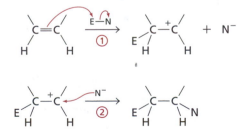

Key:

E—N is a polar reagent:
   E^+ = electrophile
   N^- = nucleophile

Proton (H⁺) as an electrophile

Halogen (X⁺) as an electrophile

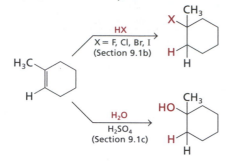

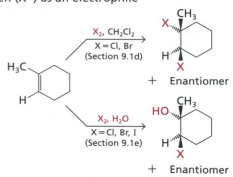

Metal ion (Hg²⁺) as an electrophile

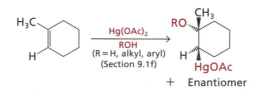

Figure 9.1
Reactions of
methylcyclohexene with
polar reagents that react
via an electrophilic
addition mechanism.

Figure 9.1 illustrates typical addition reactions that can be observed when 1-methylcyclohexene reacts with a variety of reagents. In the next sections, we will look at the details of each of these transformations to learn both the *stereochemistry* and **regiochemistry** associated with these addition processes. *Regiochemistry* refers to the orientation by which a reagent adds to a π bond. For electrophilic addition reactions, the regiochemistry is related to the carbon atom that forms a bond to the incoming electrophile. Both the stereochemical and regiochemical outcomes of electrophilic addition reactions are linked with the specific mechanism by which each transformation occurs.

9.1b   ADDITION OF HX OCCURS TO GIVE THE PRODUCT WITH THE PROTON ATTACHED TO THE LESS HIGHLY SUBSTITUTED CARBON ATOM—MARKOVNIKOV'S RULE

In 1869, the Russian chemist Vladimir Markovnikov noticed that when HX adds to a double bond, the carbon atom with fewer hydrogen atoms is more often the one to which the halogen atom of the addend becomes attached. He formulated this observation as a postulate that has come to be known as **Markovnikov's rule.**

The mechanism of the reaction reveals the basis of this rule. When the proton (electrophile) is intercepted by the π bond (nucleophile), two possible cations can form, illustrated below for 1-methylcyclohexene.

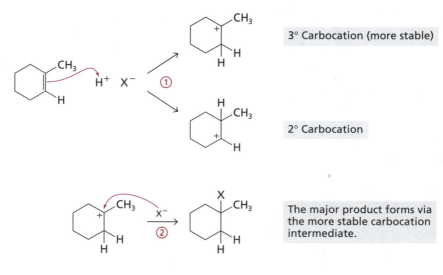

3° Carbocation (more stable)

2° Carbocation

The major product forms via the more stable carbocation intermediate.

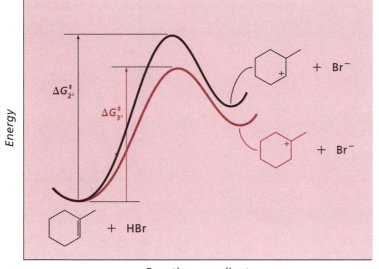

Figure 9.2
An energy diagram for the first step of the competing pathways in the electrophilic addition reaction of HBr to 1-methylcyclohexene.

According to the Hammond postulate (Section 6.2c), the free energy of activation for the pathway that produces the 3° carbocation has a lower energy than that leading to formation of the 2° carbocation, as shown in Figure 9.2 for this protonation step. The reaction via the 3° carbocation is therefore faster and will yield the 3° cycloalkyl bromide as the major product.

In **Markovnikov addition,** the proton from HX becomes attached to the alkene carbon atom that has *more* hydrogen atoms. The nucleophile, therefore, is bonded to the alkene carbon atom with *fewer* hydrogen atoms. This same regiochemistry is seen for other unsymmetric alkenes, as illustrated in the following equation:

$$CH_3CH_2CH_2-\underset{\underset{H}{|}}{C}=\underset{\underset{H}{|}}{C}-H \xrightarrow[\text{CH}_3\text{COOH}]{\text{HBr, 0°C}} CH_3CH_2CH_2-\underset{\underset{Br}{|}}{C}\underset{\underset{H}{|}}{-}CH_3 \qquad (84\%)$$

Electrophilic addition of HX is **regioselective,** which means that some of the product with the anti-Markovnikov orientation is also obtained. A **regiospecific** reaction is one in which addition occurs with only a single orientation, and such processes are rare.

Although Markovnikov formulated his rule based on the positions of hydrogen and halogen atoms in the product, the modern statement of this principle is derived from our knowledge of the reaction mechanism:

Markovnikov's rule. The electrophilic portion of a reagent adds to a π bond so that the more stable carbocation intermediate predominates.

Additional examples of this principle are illustrated in the following sections.

9.1c   WATER ADDS TO AN ALKENE IN THE MARKOVNIKOV FASHION

Water itself is not acidic enough to protonate a π bond, so a mineral acid is added to increase the acidity—hence electrophilicity—of the reaction medium. The best acids are those with a weakly nucleophilic conjugate base such as hydrogen sulfate or dihydrogen phosphate ions. When one of these ions is present, water is the principal nucleophile in the reaction mixture. The process whereby the elements of water add to an alkene is called **hydration.**

When 1-methylcyclohexene is protonated in an aqueous solution of sulfuric acid, the 3° carbocation is formed in preference to the 2° carbocation, as described in Section 9.1b.

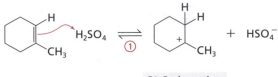

3° Carbocation

Water is a stronger base, hence a better nucleophile, than HSO_4^-, so a molecule of water intercepts the carbocation. The protonated alcohol is subsequently deprotonated to form the product, 1-methylcyclohexanol.

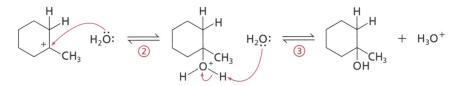

Note that the steps of this mechanism follow the opposite order of dehydration, described in Section 8.1a. Each step is in equilibrium, so the direction of the reaction can be altered by changing the reaction conditions. Using a large amount of water leads to hydration of an alkene and formation of an alcohol; removing water from the reaction mixture (usually by heating) produces an alkene *from* an alcohol.

EXERCISE 9.1

Compare the mechanism of hydration of 1-methylcyclohexene with that for the dehydration (E1 reaction) of 1-methylcyclohexanol. What are the similarities and differences?

EXERCISE 9.2

What is the major product expected from the reaction of 1,1-diphenylethene and water in the presence of sulfuric acid?

9.1d   ADDITION OF CHLORINE OR BROMINE TO AN ALKENE PROCEEDS VIA A HALONIUM ION INTERMEDIATE

An alkene reacts readily with chlorine or bromine in an inert solvent such as dichloromethane to form a vicinal dihaloalkane. Because Cl_2 and Br_2 are symmetrical, regiochemistry is not a concern in their addition reactions, so we will look in this section at the reaction of bromine with a symmetrical alkene to simplify the stereochemical issues. As for the other halogens, fluorine is extremely reactive and cleaves most C–H, C–C, and C=C bonds; CF_4 is a common product of its reactions. Iodine initially reacts with alkenes to form the cationic intermediate, but the overall reaction is endergonic, so the equilibrium is reactant-favored.

As in all electrophilic addition processes, the first step with bromine is the reaction of the π bond with an electrophile. For a symmetrical molecule like Br_2, which is not polar, the π bond polarizes the reagent to create an electrophilic center.

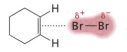

Interaction of the π bond with the bromine molecule creates a dipole because electrons surrounding the bromine nuclei are repelled by the π electrons.

Subsequently, a reaction between the π bond and Br(δ+) takes place. In the 1930s, several studies supported the formulation of the resulting cationic species as a **halonium ion** (bromonium, chloronium, or iodinium ion, specifically)—a three-membered ring that places a positive charge on the halogen atom. Spectroscopic data has subsequently been used to characterize these species and to confirm their existence.

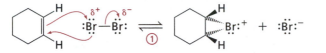

The formation of this bridged intermediate leads to the observed stereochemistry of the product, in which the halogen atoms are trans. This reaction, which is called a **trans-addition** (or **anti-addition**) to indicate that the groups have added to opposite faces of the double bond, is **stereospecific.** A *stereospecific reaction* is one that leads to generation (or destruction) of a single stereoisomer. (A **stereoselective** reaction is one that generates more of one stereoisomer than another.)

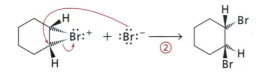

trans-**1,2-Dirbromocyclohexane**

Even though this reaction is stereospecific, it is not **enantiospecific,** nor is it even **enantioselective.** An *enantiospecific* reaction is one that produces a single enantiomer; an *enantioselective* reaction produces more of one enantiomer than the other. Here, the halonium ion can form either above or below the ring, so the bromonium ion intermediates are mirror images.

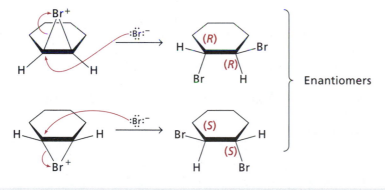

The bromonium ion can form either above or below the ring, and the bromide ion reacts at the opposite face. Therefore, the reaction is stereospecific—only the trans product is formed—but not enantiospecific—a racemic mixture is generated.

EXERCISE 9.3

What are the products from the reaction between *cis*-2-butene and chlorine? Between *trans*-2-butene and chlorine? Clearly show the stereochemistry of the products.

The electrophilic addition reaction of bromine is the basis of a simple test for the presence of a π bond. In this test, a drop or two of a Br_2/CH_2Cl_2 solution, which is red-orange, is added to the unknown compound in CH_2Cl_2. If the color is immediately discharged, then an alkene (or alkyne) group is likely present.

9.1e   HALOGENS ADD TO AN ALKENE IN THE PRESENCE OF WATER TO FORM A VICINAL HALOHYDRIN

You just learned that bromine and chlorine readily add to an alkene π bond in an inert solvent to form a dihaloalkane. In a nucleophilic solvent such as water, an alkene reacts initially with a halogen molecule in the same way and forms a bridged halonium ion. Chlorine, bromine, and iodine all react.

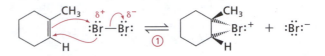

In the second step, however, the nucleophile that reacts with this intermediate is water, not the halide ion. When water is used as the solvent, its concentration is ~ 55 M, whereas the halide ion formed in Step 1 has a concentration lower than 0.01 M under typical conditions. It is more likely, therefore, that the intermediate will be intercepted in Step 2 by a molecule of water rather than by a halide ion. Because a bridged intermediate is formed and water must react at the opposite face, the stereochemistry of addition is again trans. Loss of a proton in Step 3 yields a *vicinal halohydrin*, a compound with a halogen and an OH group attached to adjacent carbon atoms.

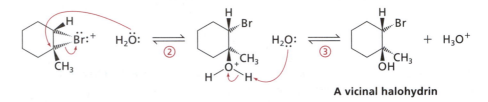

A vicinal halohydrin

The two groups that add to the π bond are not identical (Br and OH), so the regiochemistry of addition has to be rationalized. The major product from this reaction has the halogen atom (bromine in this example) attached to the carbon atom with more hydrogen atoms. This result suggests that the halonium ion is not symmetrical in its reaction with water. In fact, the more highly substituted carbon atom carries a greater share of the charge so that it can be stabilized by the electron-donating methyl group. (You saw this effect previously in Section 7.2e for epoxide ring opening reactions.) The present result is simply a broader corollary of Markovnikov's rule: *The nucleophile reacts preferentially at the carbon atom that better supports a positive charge.*

Symmetrical bonding of the bromine ion   **Unsymmetrical bonding of the bromine ion**

This form places a positive charge on the more highly-substituted carbon atom.

In summary, for bromohydrin formation from the reaction between 1-methylcyclohexene with bromine and water:

- The 3° alcohol is the major product, so the reaction is *regioselective* (Section 9.1b).
- The orientation of OH and Br are trans—the reaction if *stereospecific* (Section 9.1c).
- A racemic mixture is obtained—the reaction is *not enantioselective* (Section 9.1c).

EXAMPLE 9.1

What are the major products from the reaction between 1-hexene and chlorine in water? Show the stereochemistry of the products and assign the configurations (*R* or *S*) of any stereocenters. Comment on the regio- and stereoselectivity of the reaction.

First, draw the structure of the alkene (below, left). Chlorine reacts with this alkene to form a bridged chloronium ion, and the chlorine atom can become attached on either side of the plane of the original carbon–carbon double bond. When the front face reacts with chlorine (path a), then intermediate **A** is formed. Reaction at the back face of the double bond via pathway b yields intermediate **B**. The secondary carbon atom of the three-membered ring carries a greater share of the charge than the primary carbon atom.

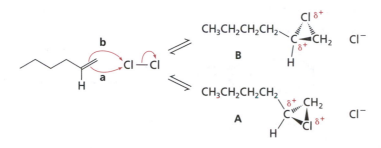

Water reacts with these intermediates in the second step, and this reaction occurs on the side of the three-membered ring away from the chlorine atom. Finally, deprotonation yields the vicinal chlorohydrin products.

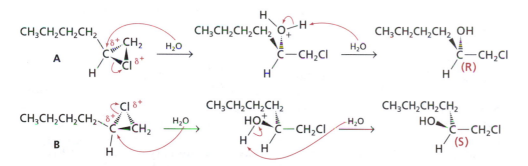

We next assign the absolute configurations of the new stereocenters. Because the reaction starts with achiral reactants and creates one new chiral center, the products must constitute a racemic mixture.

As with most electrophilic addition reactions, the overall process is *regioselective* and occurs with the Markovnikov orientation (the nucleophile reacts preferentially at the more highly substituted carbon atom). The addition step is *stereospecific* because the water reacts with the cationic intermediate anti to the position of the chlorine atom. The product does not reflect this relative stereochemistry because the product has only one chiral center and at least two are required to reveal the *relative* stereochemistry.

EXERCISE 9.4

What are the major products from the reaction between 2-phenylpropene and bromine in water? Show the stereochemistry of the products and assign the configurations (*R* or *S*) of any stereocenters. Comment on the regio- and stereoselectivity of this reaction.

9.1f   OXYMERCURATION PROVIDES AN EXAMPLE IN WHICH A METAL ION IS THE ELECTROPHILE

Metal ions constitute a third category of electrophiles with which alkenes react, and a good example of this reaction type is **oxymercuration,** the addition of OH (from water) and HgOAc to a double bond. (A more general term for this reaction is **solvomercuration,** which makes use of solvents other than water as the nucleophile.) Mercury(II) acetate, a reagent commonly used in this transformation, ionizes in water to form the (acetato)mercury(II) cation and the acetate ion, which is commonly abbreviated as OAc^- or AcO^-.

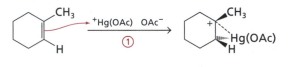

In the first step of oxymercuration, the alkene π bond reacts with the $Hg(OAc)^+$ ion. The intermediate that forms is analogous to a halonium ion, and the metal ion bridges between the two carbon atoms. For 1-methylcyclohexene, this bridged intermediate is unsymmetrical, as it is for the halonium ions (Section 9.1e), with the more highly substituted carbon atom bearing a greater positive charge.

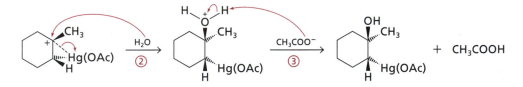

Greater stabilization of the 3°
carbocation makes the bridged
mercurinium ion unsymmetrical.

Water subsequently reacts as a nucleophile at the more highly substituted carbon atom, leading to formation of the trans product. Because the Hg(OAc) group (the electrophile) in the product is attached to the carbon atom with more hydrogen atoms, we can say that Markovnikov addition has occurred. Oxymercuration is stereospecific (anti) and regioselective (Markovnikov).

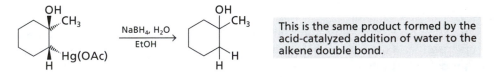

Oxymercuration would have limited utility if the mercury ion were not easily removed. In a separate reaction, the mercury–carbon bond can be cleaved using sodium borohydride, a common reducing agent in organic chemistry.

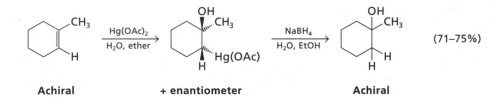

This is the same product formed by the
acid-catalyzed addition of water to the
alkene double bond.

This step, called *demercuration,* replaces the mercury atom with hydrogen and, in this instance, destroys the chirality of the molecule. The overall transformation can be written as follows:

Notice that the overall procedure of oxymercuration–demercuration yields the same product as hydration does. This two-step procedure avoids formation of a carbocation, so rearrangements are less common. For example,

Oxymercuration–demercuration

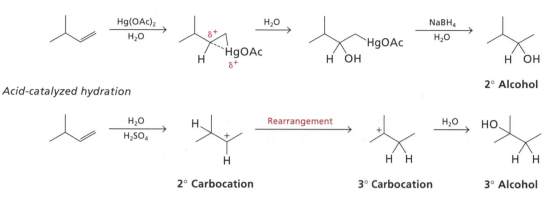

Acid-catalyzed hydration

EXERCISE 9.5

Provide the details for the mechanisms of the reactions shown in the preceding scheme by illustrating the movement of electrons with curved arrows. Ignore stereochemistry.

The more general form of this procedure that was mentioned at the beginning of this section—*solvomercuration*—can also be done in combination with the demercuration step. The following example illustrates such a transformation and also shows a case in which the product after removal of mercury retains a stereocenter.

EXAMPLE 9.2

Draw the structures, including stereochemistry, of the major products (**A**) from the reaction between 1-pentene and mercury(II) acetate in methanol. Assign the configurations (*R* or *S*) of any stereocenters. Draw the structures, including stereochemistry, of the products (**B**) obtained from demercuration of **A** with sodium borohydride ($NaBH_4$); assign the configurations of the stereocenters.

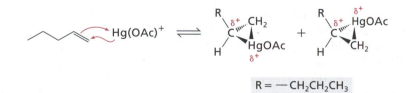

The alkene reacts with the $Hg(OAc)^+$ ion to form a bridged intermediate. Reaction can occur at each face of the double bond.

Methanol reacts with these cationic species and gives enantiomeric ethers as the products after deprotonation.

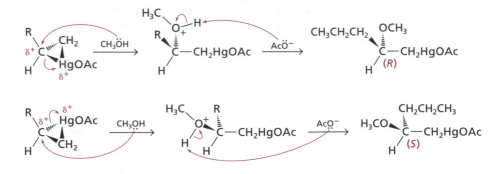

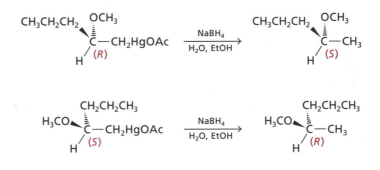

Demercuration replaces the HgOAc groups with H atoms. The stereocenter at C2 in each molecule remains, but notice that its configuration changes—not from breaking any bonds at the stereocenter, but because the group priorities are altered by removal of the Hg atom.

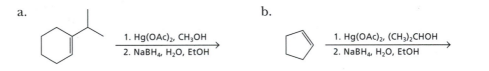

The use of an alcohol in the solvomercuration–demercuration procedure provides an alternative and complement to the Williamson ether synthesis (Section 7.2b) as a way to prepare ethers, especially those with a 3° carbon atom attached to the ether oxygen atom.

EXERCISE 9.6

What are the major products of the following reactions? Include stereochemistry.

a.

$$\xrightarrow[\text{2. NaBH}_4,\ \text{H}_2\text{O},\ \text{EtOH}]{\text{1. Hg(OAc)}_2,\ \text{CH}_3\text{OH}}$$

b.

$$\xrightarrow[\text{2. NaBH}_4,\ \text{H}_2\text{O},\ \text{EtOH}]{\text{1. Hg(OAc)}_2,\ (\text{CH}_3)_2\text{CHOH}}$$

9.2   ELECTROPHILIC ADDITION REACTIONS OF ALKYNES

9.2a   TERMINAL ALKYNES UNDERGO MARKOVNIKOV ADDITION REACTIONS

Electrophilic addition reactions of alkynes—also called *acetylenes*—occur by the same mechanisms that you have just learned for the alkene addition reactions. Because an alkyne has two π bonds, however, 2 equiv of many reagents can react with the alkyne by these addition processes.

In the first step of the reaction between a terminal alkyne and hydrogen bromide, a π bond reacts with the proton and generates what is formally a vinyl carbocation. Markovnikov regiochemistry is observed for this reaction, so the hydrogen atom ends up on the terminal carbon atom. The carbocation intermediate subsequently reacts with the bromide ion to form the vinyl bromide product.

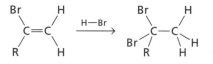

Vinyl carbocation (linear)

The alkenyl bromide can react with a second equivalent of HBr, and the product of this second step, which also proceeds with Markovnikov regiochemistry, is the geminal dibromide, which has both bromine atoms attached to the same carbon atom.

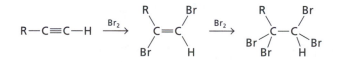

EXERCISE 9.7

Propose a reasonable mechanism for the foregoing reaction that yields the geminal dibromoalkane. Rationalize the regiochemistry of this transformation.

The vinyl carbocation formed as the first intermediate is not as stable as a secondary or tertiary carbocation (Section 6.2b). Therefore, the free energy of activation is higher for addition to a triple bond than that for a double bond. As a result, the reaction of a second equivalent of the reagent with an alkyne is often more facile than reaction of the first equivalent. It is possible to add only 1 equiv of a reagent to a triple bond if the experimental conditions are carefully controlled.

EXERCISE 9.8

What is the major product expected from this reaction?

⬡—C≡C—H   <u>HCl (xs)</u>→

The regiochemistry of addition is irrelevant when an addend is symmetrical, so bromine and chlorine add to either a terminal or internal carbon–carbon triple bond by anti addition, which initially yields the (*E*)-dihaloalkene (the higher priority Br atoms are trans). Just as in HX addition, the halogens react more slowly with the starting alkyne than with the dihaloalkene product, so the tetrahalide is the major product when an excess of the halogen is provided.

$$R-C\equiv C-H \xrightarrow{Br_2} \underset{Br}{\overset{R}{C}}=\underset{H}{\overset{Br}{C}} \xrightarrow{Br_2} \underset{Br}{\overset{R}{\underset{Br}{C}}}-\underset{H}{\overset{Br}{C}}$$

9.2b   MARKOVNIKOV ADDITION OF WATER TO AN ALKYNE PRODUCES A KETONE

An alkyne π bond is less nucleophilic than an alkene π bond, so alkynes sometimes require additional electrophilic activation in order to undergo addition reactions. This tendency is more pronounced for terminal alkynes than for internal triple bonds. To catalyze the addition of water to a triple bond, metal ions are often added. Historically, the Hg(II) ion has been used, but Pt(II) and Pd(II) ions and their complexes work as well. The important point is this: *Many metal ions are good electrophiles.*

The following equations illustrate the hydration of an alkyne in the presence of PtCl₂, which is similar to the reaction that occurs in the solvomercuration of alkenes:

1. The π bond reacts with the metal ion, forming an electrophilic intermediate.
2. A molecule of water (the nucleophile) reacts with this intermediate.

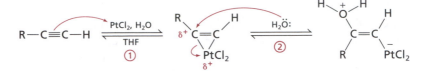

Two acid–base reactions subsequently take place to yield the initial product, a *vinyl alcohol.*

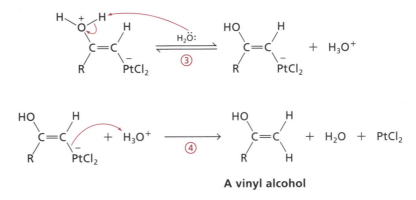

A vinyl alcohol

Vinyl alcohols are normally unstable and undergo a process called **tautomerism,** which involves a pair of acid–base reactions that interchange the positions of a proton and an adjacent π bond.

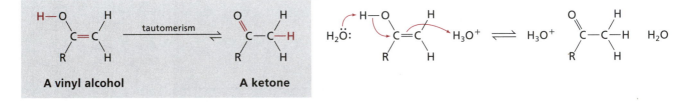

Tautomerism will be discussed in more detail in Chapter 22; but until then, we will assume that *whenever a vinyl alcohol is formed, it rearranges to the corresponding carbonyl compound.* Isomers that are interconverted by concomitant movements of a proton and a π bond are called **tautomers.**

Notice that the vinyl alcohol formed by hydration of a terminal alkene is the result of *Markovnikov addition.* The vinyl alcohol tautomerizes to form the methyl ketone, which is the product that is actually isolated.

Markovnikov regiochemistry

EXERCISE 9.9

Draw the structure of the major product(s) expected from the following reaction.

$$CH_3CH_2-C{\equiv}C-CH_2CH_3 \xrightarrow[\text{THF}]{PtCl_2,\ H_2O}$$

9.3   THE FORMATION OF CARBON–CARBON BONDS

9.3a   CARBOCATIONS ADD TO ALKENES TO FORM DIMERS

Having looked at addition reactions initiated by H^+, Br^+, and M^+ (metal ions) as electrophiles, we now consider addition reactions of carbocations, R_3C^+, which lead to the formation of carbon–carbon bonds. One reason this topic is so important derives from its relevance to pathways that create the molecular frameworks of many biomolecules. Furthermore, many valuable polymers and plastics are made by routes employing these same mechanisms.

The coupling of two molecules of isobutylene, a process called *dimerization,* is a simple reaction that illustrates the mechanism that underlies carbon–carbon bond formation. In the presence of 50% aqueous sulfuric acid at 100°C, isobutylene (2-methylpropene) is protonated to generate the *tert*-butyl carbocation.

Isobutylene
(2-methylpropene)

Markovnikov addition of the proton
gives the stable *tert*-butyl cation.

The nucleophile under these conditions is another molecule of isobutylene, and Markovnikov addition occurs: *The carbocation* (electrophile) *becomes attached to the alkene carbon atom with more hydrogen atoms.*

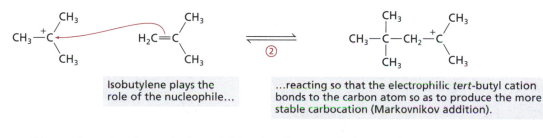

Isobutylene plays the
role of the nucleophile...

...reacting so that the electrophilic *tert*-butyl cation
bonds to the carbon atom so as to produce the more
stable carbocation (Markovnikov addition).

The carbocation formed after addition is subsequently deprotonated to form a 4:1 mixture of the two alkenes shown below.

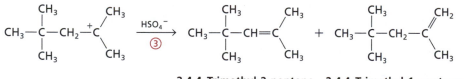

2,4,4-Trimethyl-2-pentene   2,4,4-Trimethyl-1-pentene

Ratio: 4 : 1

EXERCISE 9.10

Propose a mechanism for the elimination reaction shown directly above. What factor accounts for the product ratio (see Section 8.1b)?

9.3b   POLYMERIZATION OCCURS VIA CARBOCATION ADDITION
TO AN ALKENE IN A MARKOVNIKOV FASHION

Isobutylene reacts with acids other than sulfuric to form *polymers,* which are long-chain molecules consisting of repeating small-molecule units. (Chapter 26 presents the chemistry of polymers in more detail.) For example, when treated with $H^+[BF_3OH^-]$, isobutylene reacts to form poly(isobutylene).

To start, boron trifluoride, BF_3, reacts with water to form $H[BF_3OH]$. At low temperature, isobutylene reacts with the proton of this acid in the usual way, via Markovnikov addition.

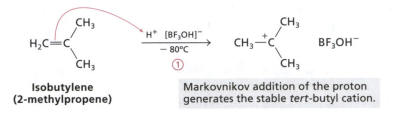

**Isobutylene
(2-methylpropene)**

Markovnikov addition of the proton generates the stable *tert*-butyl cation.

The carbocation that is generated is trapped by another molecule of isobutylene, producing a new carbocation.

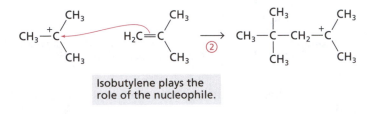

Isobutylene plays the
role of the nucleophile.

Unlike the reaction of isobutylene in the presence of sulfuric acid that was described in Section 9.3a, this process utilizing $H[BF_3OH]$ does not stop at the dimer stage, because the carbocation is stable at low temperature and elimination is slow. Instead, the addition step occurs repeatedly thousands or even millions of times to produce a cationic polyisobutylene derivative.

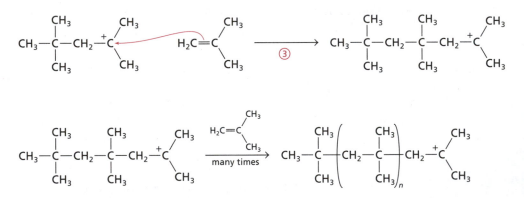

At some point during the polymerization process, elimination *does* occur, and the final polymer is formed. This termination step may occur at any time, so the product comprises a mixture of polyisobutylene molecules with a range of molecular weight values (Section 26.1c).

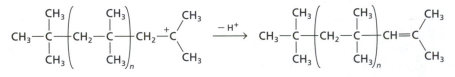

Polyisobutylene

9.3c ELECTROPHILIC ADDITION PATHWAYS ARE USED TO FORM CARBON–CARBON BONDS IN BIOCHEMICAL SYSTEMS

It may be somewhat surprising to learn that living organisms can readily generate and stabilize carbocations. The methods used to make carbocations do not employ strong acids, but once generated, these electrophiles can react with alkenes, just as isobutylene does.

Squalene, a molecule with 30 carbon atoms and six double bonds, is the biosynthetic precursor of the steroids. It is made from five-carbon (C_5) compounds (two molecules of dimethylallyl pyrophosphate and four molecules of isopentenyl pyrophosphate) by three successive addition reactions that take place between carbocations and π bonds, as shown in Figure 9.3.

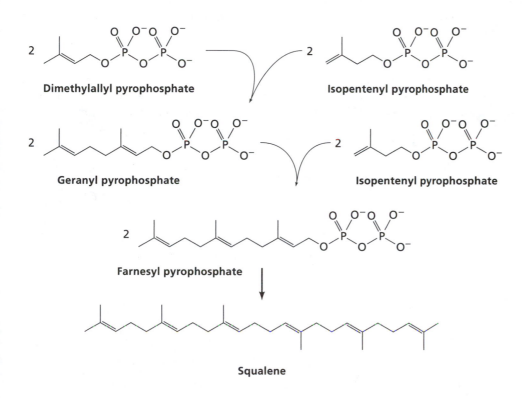

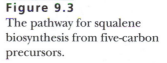

Figure 9.3
The pathway for squalene biosynthesis from five-carbon precursors.

The leaving group in all of the processes shown in Figure 9.3 (and many other biochemical reactions) is the pyrophosphate ion. (An example in which the related triphosphate ion acted as a leaving group was given in Section 7.4a.) Even though we will draw the pyrophosphate ion in its completely deprotonated form, realize that the pH of the solution and the environment of an enzyme active site can influence its degree of protonation, hence its acid strength. The monoprotonated form, $HP_2O_7^{3-}$, has a pK_a value similar to that of the ammonium ion, and the forms with additional protons are even stronger acids, which makes these ions relatively weak bases, hence reasonable-to-good leaving groups.

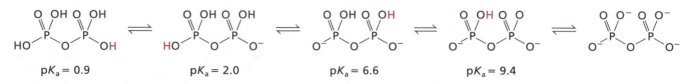

In the first step of carbon-carbon bond formation, dissociation of the leaving group from dimethylallyl pyrophosphate produces a resonance stabilized allylic carbocation.

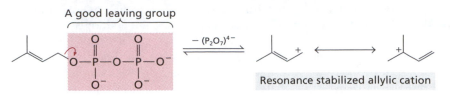

This reaction is the same as the first step in both the S_N1 (Section 6.2) and E1 (Section 8.1) mechanisms. In this biochemical transformation, which occurs within the active site of an enzyme, this carbocation may also be stabilized by interactions with negatively charged groups in addition to the usual resonance and inductive effects inherent in the allylic framework.

In the next step of squalene biosynthesis, the π bond of isopentenyl pyrophosphate reacts with the dimethylallyl carbocation to form a new carbon–carbon bond. Just as for the dimerization of isobutylene described in Section 9.3b, this process occurs in the Markovnikov fashion, with the new bond forming at the terminal carbon atom of the isopentenyl group. A tertiary carbocation is formed in this step as well.

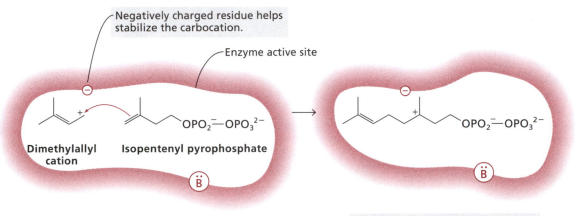

This intermediate carbocation is then deprotonated by a basic residue within the enzyme active site, producing geranyl pyrophosphate.

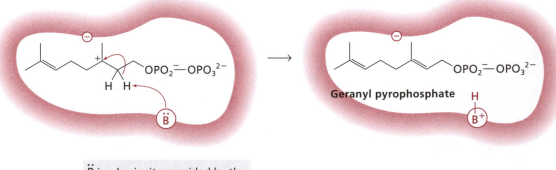

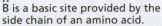
B̈ is a basic site provided by the side chain of an amino acid.

The mechanism of this biochemical dimerization process is the same as that for the sulfuric acid catalyzed dimerization of isobutylene (Section 9.3a). The carbocation is formed in different ways, but the addition step in each process occurs by the same movement of electrons and with the same regiochemistry.

The next stage in the biosynthesis of squalene is the conversion of geranyl pyrophosphate to farnesyl pyrophosphate (Fig. 9.3). Propose a mechanism for this transformation, showing the movement of electrons with curved arrows.

9.3d   ELECTROPHILIC ADDITION PATHWAYS CAN BE USED TO FORM RINGS

The electrophilic addition of carbocations to alkenes is not just used to make aliphatic compounds. If a carbocation can be generated in a molecule that also has a double bond, then a cyclic product may be formed. Recall that having a nucleophile and electrophile in the same molecule leads to intramolecular reactions (Section 7.2c). A terminal alkene forms the six-membered ring preferentially and a 2° carbocation is generated. Formation of a five-membered ring would have to form via a primary carbocation, which does not exist (Section 6.2b)

Markovnikov addition

Allylic alcohols are prone to form carbocations, so this structural unit is commonly incorporated within the substrates that are used for cyclization reactions. The OH group is protonated by formic acid, and water dissociates to generate the carbocation.

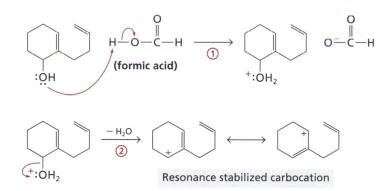

Resonance stabilized carbocation

Once formed, the carbocation is intercepted by the other double bond in the molecule. This step creates the ring and produces a second carbocation. In this example, formate ion traps this carbocation, which is more reactive than the resonance-stabilized allylic carbocation.

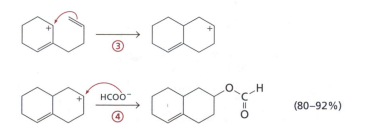

(80–92%)

EXERCISE 9.12

Propose a mechanism for the following cyclization reaction, the first step of which is protonation of the tertiary alcohol OH group and formation of a 3° carbocation. Show the movement of electrons with curved arrows.

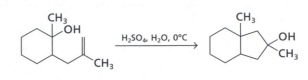

9.3e   RING FORMATION ALSO OCCURS IN NATURE

It follows logically from the material presented in Sections 9.3c and 9.3d that if carbocations can add to alkenes in living systems to form acyclic molecules such as squalene, then ring formation must be possible, too. Indeed, under appropriate conditions, geranyl pyrophosphate (Fig. 9.3) cyclizes to form products with six-membered rings. The (E) isomer of geranyl pyrophosphate can be isomerized to its (Z) isomer via the allylic carbocation formed by dissociation of pyrophosphate ion.

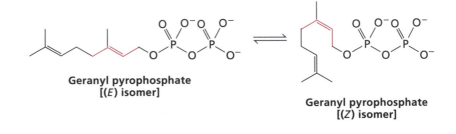

Geranyl pyrophosphate
[(E) isomer]

Geranyl pyrophosphate
[(Z) isomer]

EXERCISE 9.13

Propose a reasonable mechanism for the foregoing isomerization process of geranyl pyrophosphate. Rotation about a sigma bond in the carbocation is a required step in this process. Show the movement of electrons with curved arrows.

The allylic carbocation formed by dissociation of the leaving group can be intercepted by the other double bond in the molecule, which generates the six-membered ring. This addition to the double bond occurs with Markovnikov regiochemistry and yields a tertiary carbocation.

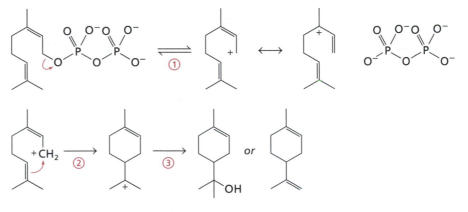

α-Terpinol   **Limonene**

Finally, this carbocation can react (Step 3) with water to form α-terpinol, a molecule found in juniper oil, or it can undergo elimination to form limonene, which is a constituent of orange and lemon oils.

EXERCISE 9.14

Propose a reasonable mechanism for both the substitution and elimination reactions that correspond to Step 3 in the preceding scheme.

9.4   HYDROBORATION REACTIONS OF π BONDS

9.4a   ALKENES REACT WITH BORANE TO FORM ORGANOBORANES

While electrophilic addition is a common mechanism by which alkenes can react, other pathways are available for the reactions between alkenes and electrophiles that are not cations. For example, borane (BH_3), which has an electron-deficient boron atom and is an electrophile, reacts with cyclohexene to form cyclohexylborane, a process called **hydroboration.**

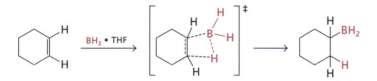

The mechanism for hydroboration does not involve formation of a cationic intermediate. Instead, this and related processes are *concerted reactions* that occur via cyclic transition states. (Recall from Section 6.3a that a species enclosed by square brackets with a superscript ‡ is a transition state.)

The product of the reaction between an alkene and borane is called an **organoborane,** first reported in 1955 by Herbert C. Brown; his leadership in developing these compounds as useful reagents in organic synthesis during the ensuing decades earned him the Nobel Prize for chemistry in 1979.

Several experimental problems had to be overcome before organoboranes were routinely used as reagents, however, the most troublesome being that diborane (B_2H_6), a dimer and the precursor of borane, is pyrophoric (spontaneously flammable in air).

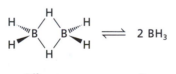

Diborane   **Borane**

Diborane is the prototype for molecules with three-center, two-electron bonds. Many boron hydrides display this type of bond in which a hydrogen atom forms a bridge between two boron atoms. Boron compounds are often good Lewis acids because the boron atom has only six electrons. The small diameter of a hydrogen atom makes the electrons in its bonds accessible, and they can be donated to a neighboring boron atom to form these B–H–B links.

> The two electrons in the B–H bond are donated into the vacant *p* orbital of the neighboring boron atom. This creates a three-center, two-electron bond.

Borane readily forms adducts with Lewis bases, accepting a pair of electrons from a heteroatom such as oxygen or sulfur. Tetrahydrofuran and methyl sulfide (Me_2S) solutions of borane are commercially available, and they have overcome the difficulty of handling the pyrophoric diborane.

Borane–THF   **Borane–Me₂S**

When an alkene is treated with $BH_3 \cdot THF$, its π electrons react with the boron atom and displace THF to form an initial adduct. This species then undergoes rupture of a boron–hydrogen bond with concomitant formation of carbon–boron and carbon–hydrogen bonds. As already mentioned, this process is called *hydroboration*.

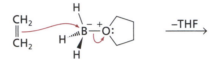

| The π bond of the alkene reacts with the boron atom... | ...leading to the formation of an intermediate... |

| ...that collapses via a four-centered transition state... | ...to the alkyborane product. The boron atom has only six electrons again, so reaction with another alkene molecule can occur. |

The reaction between an alkene and borane or organoboranes can continue as long as B–H bonds are present. Three equivalents of the alkene react with the original borane molecule to produce a trialkylborane.

$$BH_3 \;+\; H_2C{=}CH_2 \longrightarrow CH_3{-}CH_2{-}BH_2$$

Ethylborane

$$CH_3{-}CH_2{-}BH_2 \;+\; H_2C{=}CH_2 \longrightarrow (CH_3{-}CH_2)_2{-}BH$$

Diethylborane

$$(CH_2{-}CH_2)_2{-}BH_2 \;+\; H_2C{=}CH_2 \longrightarrow (CH_3{-}CH_2)_3{-}B$$

Triethylborane

Steric and electronic factors can stop the reaction of borane with an alkene at the monoalkyl- or dialkylborane stage, and many of these compounds are useful hydroborating agents themselves. For example, borane reacts with 1,5-cyclooctadiene to form 9-BBN–H (9-borabicyclo[3.3.1]nonane) and with catechol to form catecholborane. These organoboranes are especially useful because they have only one B–H bond, so the stoichiometry of any subsequent reaction with an alkene or alkyne can be readily controlled.

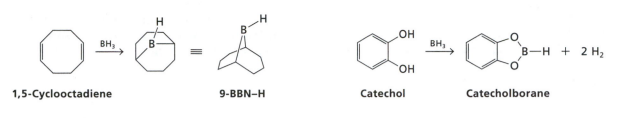

1,5-Cyclooctadiene	**9-BBN–H**	**Catechol**	**Catecholborane**

EXAMPLE 9.3

Borane reacts in THF solution with 1 equiv of 2,3-dimethyl-2-butene and forms thexylborane (thexyl is an abbreviation for *tert*-hexyl), which has the formula $C_6H_{15}B$. Draw its structure. Why is thexylborane the major product of this reaction?

First, draw the structure of the alkene (below, 2,3-dimethyl-2-butene). Borane reacts with this alkene via a four-membered ring transition state such that a hydrogen atom attaches to one carbon atom, and the boron, with its other two hydrogen atoms, attaches to the other. This alkene reacts only once with borane because thexylborane, the product, is too hindered at boron to react with another equivalent of 2,3-dimethyl-2-butene.

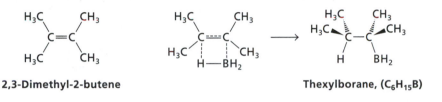

2,3-Dimethyl-2-butene		**Thexylborane, ($C_6H_{15}B$)**

EXERCISE 9.15

Borane reacts in THF solution with 2-methyl-2-butene to form the dialkylborane $(sia)_2BH$, called disiamylborane [short for di(*sec*-isoamyl)borane] in which the boron becomes attached to C3 of the starting alkene. Draw its structure.

9.4b HYDROBORATION IS A STEREOSPECIFIC AND REGIOSELECTIVE REACTION

Because the reaction between borane and an alkene occurs via a four-membered ring transition state, the addition process is a stereospecific **syn-addition** (or **cis-addition**), which means that the H and B attach to the same side of the π bond.

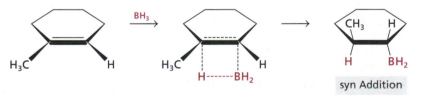

syn Addition

Hydroboration is also regioselective: The hydrogen atom from borane attaches to the carbon atom of the alkene with *fewer* hydrogen atoms, so it appears to be "anti-Markovnikov". To label hydroboration as anti-Markovnikov is misleading, however, because the mechanism is not the same as that for electrophilic addition. A major difference between borane and HCl, a typical addend that reacts via a polar mechanism, is the reversed polarity of the bond between H and the heteroatom.

$$\overset{\delta^-}{H}\!\!-\!\!\overset{\delta^+}{B}\diagdown \qquad \overset{\delta^+}{H}\!\!-\!\!\overset{\delta^-}{Cl}$$

With boron being the more electrophilic atom in the borane molecule, it makes sense that the hydrogen atom should attach to the more highly substituted carbon atom, which can support a partial-positive charge more readily.

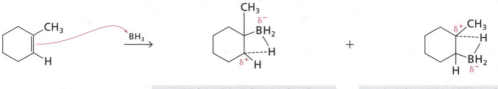

A partial positive charge develops at the 2° carbon atom (less stable).

A partial positive charge develops at the 3° carbon atom (more stable).

Furthermore, the boron atom is larger than the hydrogen atom, so the bulkier carbon atom of the alkene is the one that forms a bond to hydrogen. Both effects—steric and electronic—contribute to the regiochemistry of the hydroboration reaction.

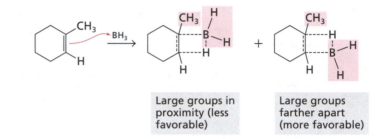

Large groups in proximity (less favorable)

Large groups farther apart (more favorable)

EXERCISE 9.16

Draw the structure of the major product expected from treating each of the following compounds with $BH_3 \cdot THF$. Based on the examples in the text, consider how many alkene groups will react with the boron (1, 2, or 3), as well as the stereochemistry and regiochemistry that are likely.

a.   b.   c.

EXERCISE 9.17

Both 9-BBN–H and catecholborane are more regioselective in their reactions with an alkene than is borane itself. Draw the structure of the product formed in the hydroboration reaction of 1-hexene with each of these reagents.

An alkyne undergoes hydroboration with the same regio- and stereochemistry as an alkene. An alkyne can react with 2 equiv of borane, however, because it has two π bonds. The reactions of alkynes with 9-BBN–H, catecholborane, or $(sia)_2BH$ (Exercise 9.15)—reagents that have only one hydrogen atom attached to boron—can be stopped after only 1 equiv of the reagent has added. The following equation illustrates the reaction between 1-hexyne and catecholborane:

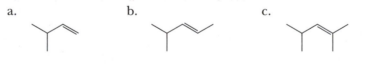

$CH_3CH_2CH_2CH_2-C{\equiv}C-H$

1-Hexyne

syn Addition occurs, and the boron is attached to the terminal carbon atom.

Draw the structure of the product formed from the reaction of 9-BBN–H with 3-hexyne.

9.4c   ORGANOBORANES CAN BE CONVERTED TO ALCOHOLS

Organoboranes are known to undergo dozens of different reactions. The most common of these transformations is one of the best methods to convert an alkene to an alcohol and a terminal alkyne to an aldehyde.

Generally, reactions of organoboranes make use of boron's Lewis acid character. An organoborane reacts with a Lewis base to form a complex in which boron has a formal negative charge. In such species, the attached carbon atoms (R groups) have nucleophilic properties.

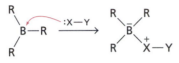

If the Lewis base that reacts with the organoborane has an appropriate leaving group, then one of the R groups can migrate, with displacement of that leaving group, in much the same way that an S$_N$2 reaction occurs.

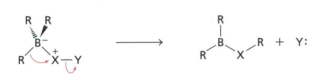

> Migration of an alkyl group from B to X occurs with tetrahedral boronate ions of this general type. Here Y is a leaving group.

To convert an organoborane to an alcohol, hydroperoxide ion (HOO$^-$) is added. This species is made from the acid–base reaction between hydrogen peroxide and base.

$$HO\overset{..}{\underset{..}{O}}-H \; + \; {}^-\!:\overset{..}{\underset{..}{O}}H \longrightarrow HO\overset{..}{\underset{..}{O}}:^- \; + \; H_2\overset{..}{\underset{..}{O}}$$

The negatively charged oxygen atom of the hydroperoxide ion reacts with boron to generate the tetrahedral adduct.

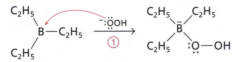

Then, one of the R groups migrates to the oxygen atom attached to boron and expels OH$^-$ as a leaving group. Hydroxide ion can function as a leaving group here because an oxygen–oxygen bond is breaking, not a carbon–oxygen bond.

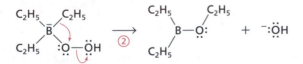

The resulting alkoxy(dialkyl)borane can subsequently react with additional hydroperoxide ion, and a second R group migrates. This process repeats with migration of the third R group.

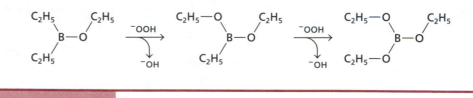

EXERCISE 9.19

Supply the details in the mechanism for each step of the foregoing scheme.

Once formed, the tri(alkoxy)borane undergoes hydrolysis by a series of acid–base reactions that involve hydroxide ion as a nucleophile. After the initial adduct formation, alkoxide ion dissociates and reacts with water to form a molecule of alcohol, which is the organic product of this transformation.

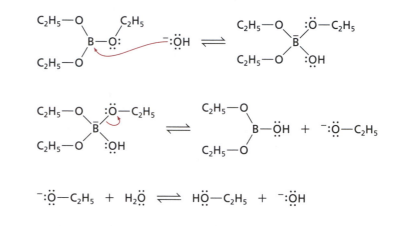

$$\text{}^{-}\!:\ddot{O}\!-\!C_2H_5 \;+\; H_2\ddot{O} \;\rightleftharpoons\; H\ddot{O}\!-\!C_2H_5 \;+\; \text{}^{-}\!:\ddot{O}H$$

Hydrolysis continues until all of the alkoxide groups have been liberated. The final boron-containing products are salts of boric acid, H_3BO_3.

$$(RO)_3B \;+\; (xs)\, NaOH_{(aq)} \;\longrightarrow\; 3\; ROH \;+\; Na_3BO_3$$

The overall process of hydroboration followed by treatment with hydrogen peroxide and hydroxide ion (hydroboration–oxidative hydrolysis) converts an alkene to an alcohol in which it appears that H and OH have added to the double bond. Note that the regiochemistry of this procedure is opposite that observed for the acid-catalyzed Markovnikov addition of water to an alkene (Section 9.1c).

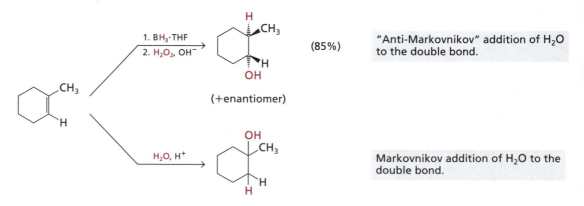

Hydroboration–oxidative hydrolysis is said to cause "anti-Markovnikov" addition of water to the double bond. The term "anti-Markovnikov" in this context provides a way to keep track of the regiochemistry of addition, but it has nothing to do with the actual mechanism of the reaction.

Migration of the alkyl group from the boron atom to the oxygen atom of the hydroperoxide group occurs with *retention of configuration*. This fact is aptly illustrated by the reaction of 1-methylcyclohexene with borane in which the methyl group and the boron atom are trans. After treating the organoborane with hydrogen peroxide and base, we obtain *trans*-2-methylcyclohexanol as the product. *The stereochemical relationship between substituents does not change during the oxidative hydrolysis step.*

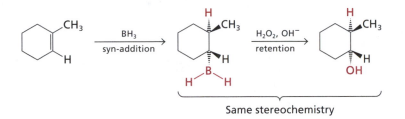

Same stereochemistry

Hydroperoxide ion also reacts with vinyl boranes, which are formed by hydroboration of alkynes, and the products are vinyl alcohols. As you learned in Section 9.2b, a vinyl alcohol is normally unstable: it tautomerizes in this case to form an aldehyde.

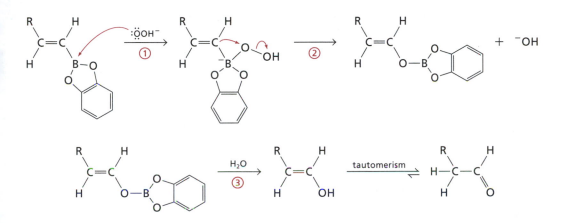

Compare this sequence with that for hydration of an alkyne, which produces a ketone.

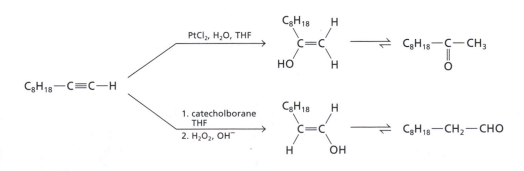

EXERCISE 9.20

Draw the structure of the major product expected from each of the following reactions:

a.   b.

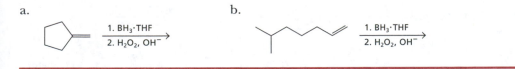

9.4d ORGANOBORANES UNDERGO PROTONOLYSIS

The migration of R from boron to an oxygen atom of the hydroperoxide ion provides a good way to make alcohols from alkenes via organoboranes. The R group can also be induced to migrate from boron to a proton. The proton cannot come from any source, however. In fact, water and even dilute mineral acids, such as hydrochloric acid, have little effect on many organoboranes except to hydrolyze the bonds between boron and heteroatoms to which it is attached. For example,

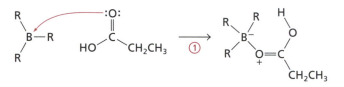

Carboxylic acids, on the other hand, react with organoboranes to cleave the C–B bond. The process that replaces a heteroatom with a proton is called **protonolysis.**

When an organoborane is heated with a carboxylic acid, a Lewis acid–Lewis base reaction occurs between the boron atom (Lewis acid) and the carbonyl oxygen atom of the carboxylic acid (a Lewis base in this transformation).

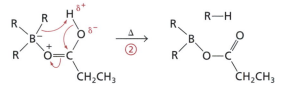

The electrons from the B–R bond subsequently move toward the proton, and then through the carboxylate group toward the oxygen atom with the positive charge. This second step generates a carbon–hydrogen bond in the molecule R–H.

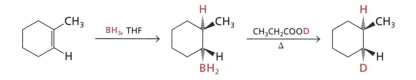

The protonolysis of a carbon–boron bond is useful in two instances. First, it provides a good way to introduce deuterium (^2H, also denoted by D) into a molecule. Notice in the following reaction that the protonolysis step occurs with retention of configuration.

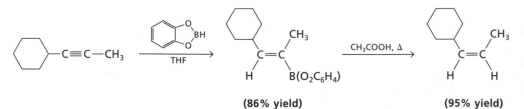

Second, hydroboration–protonolysis can be used to convert an internal alkyne to a cis-alkene.

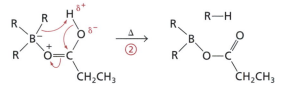

EXERCISE 9.21

Draw the structure of the major product expected from each of the following reactions:

a.

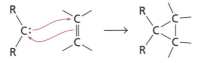

b.

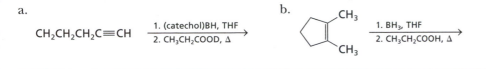

9.5   THE ADDITION OF CARBENES TO π BONDS

9.5a   A CARBENE ADDS TO AN ALKENE TO PRODUCE A CYCLOPROPANE

The last type of electrophilic reagent that we will look at in the addition reactions of alkenes comprises the *carbenes,* which are electron-deficient species with the general formula $R_2C:$ (section 2.5e). The carbene carbon atom has only six electrons, so it is a potent electrophile and reacts with the nucleophilic π bond of an alkene to form a three-membered ring. **Cyclopropanation,** as this reaction is called, is a concerted process.

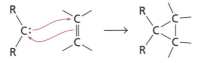

One way to make methylene, the simplest carbene, is from diazomethane either by heating or irradiating with light, a process called *photolysis*. Diazomethane is explosive, so it is difficult to work with, and methylene produced in this reaction reacts with many other functional groups besides π bonds, so it finds little use in making cyclopropane derivatives.

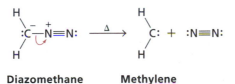

Diazomethane   Methylene

Instead of generating carbenes themselves, we often make use of *carbenoid* reagents, which react as if they were carbenes. One of the simplest procedures is the **Simmons–Smith reaction,** which involves treating CH_2I_2 with Zn that has been coated with copper.

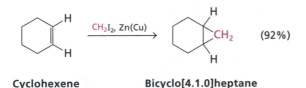

Cyclohexene   Bicyclo[4.1.0]heptane

The Simmons–Smith reaction works poorly in many cases, however, so other zinc compounds find use instead. Diethylzinc, a commercially available organometallic compound (more about organometallic compounds will be presented in Chapter 15), reacts with trifluoroacetic acid and then with diiodomethane in dichloromethane solution at 0°C to form iodomethylzinc trifluoroacetate.

$$(CH_3CH_2)_2Zn \ + \ CF_3COOH \ + \ CH_2I_2 \ \xrightarrow[0°C]{CH_2Cl_2} \ (CF_3COO)Zn-CH_2-I \ + \ CH_3CH_2I \ + \ CH_3CH_3$$

Alkenes subsequently react with this reagent to form the corresponding cyclopropane derivative. Notice that the stereochemistry of the starting alkene is retained in the cyclopropane product, a clear indication that the addition step is concerted.

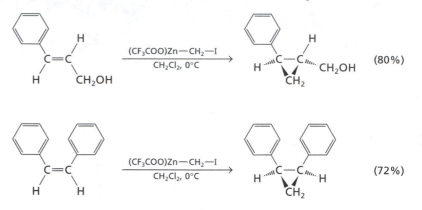

For both the Simmons–Smith reagent and the trifluoroacetate complex just mentioned, the zinc ion acts to capture the iodide ion as the carbenoid center reacts with the alkene π bond.

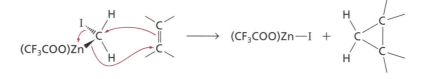

9.5b DICHLOROCARBENE IS MADE BY α-ELIMINATION

Another method for making carbenes uses the reaction between chloroform and a base like hydroxide ion in water, or *tert*-butoxide ion in *tert*-butyl alcohol. This acid–base reaction generates the trichloromethyl anion.

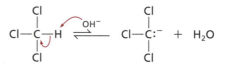

The hydrogen atom of chloroform is slightly acidic because of the highly electronegative chlorine atoms.

This anion is somewhat unstable and loses chloride ion by a process called **α-elimination.**

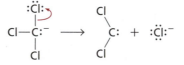

Dichlorocarbene

α-Elimination occurs when a leaving group and a proton are removed from the same carbon atom. By contrast, the E2 reaction (Section 8.2) is an example of *β-elimination* because the leaving group and proton are removed from adjacent carbon atoms.

α-Elimination   *β-Elimination*

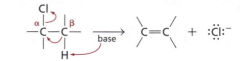

If dichlorocarbene is generated in the presence of an alkene, cycloaddition occurs to form the dichlorocyclopropane. This reaction is a stereospecific syn addition, so the stereochemistry of the alkene is retained in the cyclopropane product.

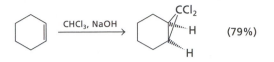

EXERCISE 9.22

Draw the structure, including stereochemistry, of the major product expected from each of the following reactions:

a.   b.

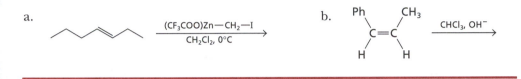

CHAPTER SUMMARY

Section 9.1   Electrophilic addition reactions of alkenes

- Electrophilic addition is a two-step process. The first step takes place between a π bond and the electrophilic portion of a reagent. This first step produces an electrophilic intermediate that, in the second step, reacts with the nucleophilic portion of the original reagent.

- Regiochemistry refers to the orientation that a reagent adopts as it adds to the π bond of an alkene.

- HX (X = halogen or OH) adds to an alkene in such a way that the H becomes attached to the carbon atom of the alkene that initially has more hydrogen atoms. This observed orientation is described in the statement of Markovnikov's rule.

- The π bond of an alkene (nucleophile) reacts with a proton (electrophile) such that the more stable carbocation is formed. This statement is another way to express Markovnikov's rule.

- Addition of X_2 (X = Cl or Br) to an alkene proceeds via a cationic intermediate called a bridged halonium ion, which has a three-membered ring.

- The stereochemistry for addition of the halogens to an alkene is anti, which means for a cyclic alkene that the halogen atoms in the product are trans.

- Addition of X_2 (X = Cl or Br) to an alkene is stereospecific, which means that only a single geometric isomer is formed.

- Addition of X_2 (X = Cl, Br, or I) to an alkene in aqueous solution occurs via a bridged halonium ion. The nucleophile that subsequently reacts with the cationic intermediate is water.

- The product of halogen addition in an aqueous solution is called a vicinal halohydrin. Addition is stereospecific (anti) and regioselective: The halogen atom is the electrophile, and Markovnikov addition is observed.

- Oxymercuration is the addition of $Hg(OAc)^+$ and OH^- to an alkene π bond. The mercury group can be replaced with H by treating the product with $NaBH_4$.

- Oxymercuration proceeds via an intermediate that has a bridged mercury atom. Addition is stereospecific (anti) and regioselective (Markovnikov orientation).

- Solvomercuration is a more general form of the oxymercuration reaction, and it involves use of a solvent other than water. It can be used to make ethers that are difficult to prepare by the Williamson method.

Section 9.2   Electrophilic addition reactions of alkynes

- Alkynes react with HX or X_2 in the same way that alkenes react. Addition is regioselective (Markovnikov). An alkyne can react with 2 equiv of the reagent.

- Water adds to an alkyne to form a vinyl alcohol, which undergoes rearrangement (tautomerism) to form a ketone. This hydration reaction is catalyzed by metal ions.

Section 9.3   Formation of carbon–carbon bonds

- Carbocation addition to an alkene π bond creates new carbon–carbon bonds during formation of dimers, polymers, or rings. Addition is regioselective (Markovnikov).

- Carbocation addition to an alkene π bond in biochemical systems occurs via an S_N1-like process and is used to prepare polyenes and ring compounds.

Section 9.4   Hydroboration reactions of π bonds

- Hydroboration is the addition of a boron-containing group and a hydrogen atom across an alkene π bond.

- Hydroboration is said to be anti-Markovnikov, which means that the H attaches to the carbon atom that initially has fewer hydrogen atoms.

- The hydroboration process is a syn-addition, also called a cis-addition, which means that for a cyclic alkene, the B group and H in the product are cis.

- An organoborane, the product formed from the reaction between an alkene and borane, reacts with hydrogen peroxide and base (oxidative hydrolysis) to form an alcohol via replacement of the boron-containing group with an OH group.

- Oxidative hydrolysis of an organoborane occurs with retention of configuration at the carbon atom bonded to boron.

- Hydroboration of an alkyne produces a vinylborane. Oxidative hydrolysis of the vinylborane yields a vinyl alcohol, which rearranges to the corresponding carbonyl compound.

- Organoboranes react with carboxylic acids to form hydrocarbons, a process called protonolysis.

Section 9.5   The addition of carbenes to π bonds

- A carbene has a carbon atom with only two bonds and six electrons.

- A carbene reacts with an alkene to form a cyclopropane derivative via a stereospecific syn addition.

- The simplest carbene, methylene, $:CH_2$, can be made from diazomethane by heating or by photolysis. Carbenoid compounds made from diiodomethane and zinc reagents furnish methylene in a more convenient form for reactions.

- Dichlorocarbene is made by treating chloroform with base. The carbanion produced initially then undergoes α–elimination of chloride ion to form $:CCl_2$.

KEY TERMS

Section 9.1a
electrophilic addition
regiochemistry

Section 9.1b
Markovnikov's rule
Markovnikov addition
regiospecific
regioselective

Section 9.1c
hydration

Section 9.1d
halonium ion
trans-addition
anti-addition
stereospecific
stereoselective
enantiospecific
enantioselective

Section 9.1f
oxymercuration
solvomercuration

Section 9.2b
tautomerism
tautomers

Section 9.4a
hydroboration
organoborane

Section 9.4b
syn-addition
cis-addition

Section 9.4d
protonolysis

Section 9.5a
cyclopropanation
Simmons–Smith reaction

Section 9.5b
α-elimination

REACTION SUMMARY

Section 9.1b

Addition of HX(X = F, Cl, Br, I) to an alkene. The regiochemistry derives from formation of the more stable carbocation (Markovnikov addition).

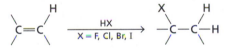

Section 9.1c

Addition of H_2O to an alkene (hydration). The regiochemistry derives from formation
of the more stable carbocation (Markovnikov addition).

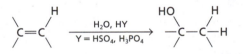

Section 9.1d

Addition of X_2 to an alkene in an inert solvent. Anti addition is observed.

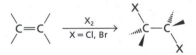

Section 9.1e

Addition of X_2 to an alkene in the presence of water. The stereochemistry is anti, and
the regiochemistry is such that the halogen atom in the product is attached to the less
highly substituted carbon atom.

Section 9.1f

The reaction of $Hg(OAc)_2$ (OAc = acetate ion) with an alkene in the presence of H_2O (oxymercuration) or ROH or RCOOH (solvomercuration). The stereochemistry is anti, and the regiochemistry is such that the Hg(OAc) group in the product is attached to the less highly substituted carbon atom.

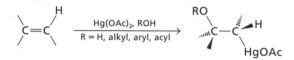

The HgOAc group can be replaced by a proton upon reaction with $NaBH_4$.

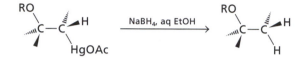

Section 9.2a

Addition of HX or X_2 to an alkyne. For HX, the regiochemistry is Markovnikov addition. Addition can occur once or twice.

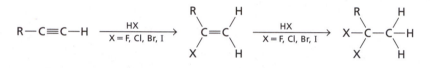

Section 9.2b

Addition of H_2O to an alkyne. Markovnikov addition is observed. The initial product is a vinyl alcohol, which tautomerizes to the corresponding carbonyl compound. Metal ions are used to activate the alkene π bond for addition of water.

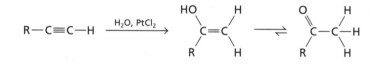

Sections 9.3a and 9.3b

Addition of a carbocation to an alkene. Depending on the acid used to generate a carbocation from an alkene, dimers or polymers can be formed. Addition of a carbocation to a π bond proceeds with Markovnikov regiochemistry.

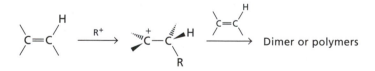

Sections 9.3d

Intramolecular addition of a carbocation to an alkene. If a carbocation intermediate is formed in a molecule with unsaturation, a ring can be formed. Substitution and elimination can take place after ring formation.

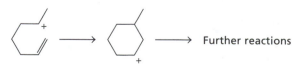

Section 9.4a and 9.4b

Hydroboration of alkenes. "Anti-Markovnikov" regiochemistry is observed in that the hydrogen atom of borane becomes attached to the more highly substituted carbon atom of the original double bond. The stereochemistry of addition is syn.

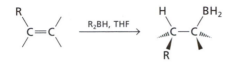

Section 9.4b

Hydroboration of alkynes. "Anti-Markovnikov" regiochemistry is observed in that the hydrogen atom of borane becomes attached C2 of the original triple bond. The stereochemistry of addition is syn.

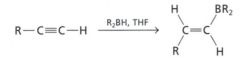

Section 9.4c

Oxidative hydrolysis of organoboranes with hydrogen peroxide and base. A vinylborane reacts to form a vinyl alcohol, which tautomerizes to the corresponding carbonyl compound. The substitution of B by OH occurs with retention of configuration.

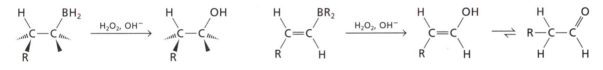

Section 9.4d

Protonolysis of organoboranes with carboxylic acids. The replacement of boron by hydrogen occurs with retention of configuration. Use of a deuterated carboxylic acid (CH_3COOD) replaces the boron atom with deuterium.

Section 9.5a

The reaction of zinc carbenoid reagents with an alkene: cyclopropanation. The stereochemistry of addition is syn.

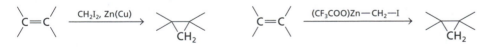

The Simmons–Smith reaction

Section 9.5b

Addition of dichlorocarbene to an alkene. The stereochemistry of addition is syn.

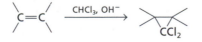

ADDITIONAL EXERCISES

9.23.   *trans*-Stilbene, C₆H₅–CH=CH–C₆H₅, reacts with bromine in methanol to yield two stereoisomers. Draw their structures.

9.24.   a. The following cyclization reaction has been observed in the mercuration–demercuration of the illustrated unsaturated alcohol. Propose a reasonable mechanism for this reaction.

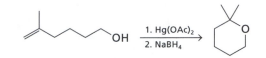

b. Reaction of the same unsaturated alcohol with bromine in dichloromethane yields a product with only one bromine atom in it. Draw its structure and propose a mechanism that supports your prediction.

9.25.   Draw the structure(s) of the major product(s) expected from each of the following addition reactions. Indicate the stereochemistry of the product as appropriate. Relative stereochemistry should be shown using wedges and dashed lines. If a racemic mixture will be formed, draw the structure of one enantiomer and write the word "racemic", or draw both enantiomeric structures. If diastereomers are formed, draw each structure; label meso compounds as such. If no reaction occurs, write N.R.

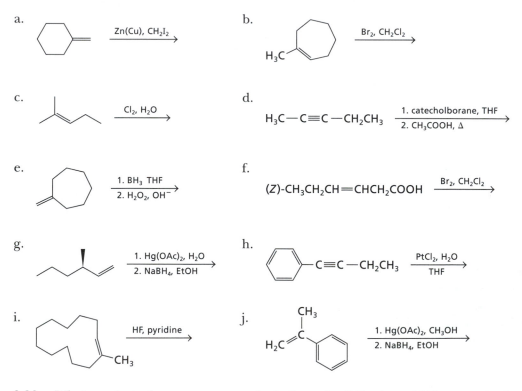

9.26.   What products do you expect to obtain from the following addition reaction? Propose a mechanism to rationalize and support your prediction.

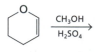

9.27. Draw and label a reaction coordinate diagram for the two possible addition products of methylenecyclohexane with aqueous sulfuric acid:

Which process has a higher transition state energy? Which process forms a more stable carbocation intermediate? Propose a complete mechanism for the favored process.

9.28. Because iodine does not add to an alkene to produce a stable product, it can be used to carry out an addition reaction in which a nucleophile other than iodide ion is present. This characteristic is particularly applicable to *iodolactonization,* illustrated by the following example:

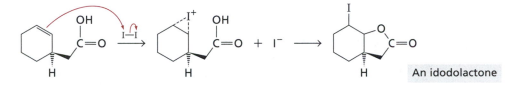

Propose a mechanism for this transformation (the first step is already shown). What is the stereochemistry of the product? Is the product chiral? (Making a model may be helpful.)

9.29. Draw the structure(s) of the major product(s) expected from each of the following addition reactions. Indicate the stereochemistry of the product as appropriate. Relative stereochemistry should be shown using wedges and dashed lines. If a racemic mixture will be formed, draw the structure of one enantiomer and write the word "racemic", or draw both enantiomeric structures. If diastereomers are formed, draw each structure; label meso compounds as such. If no reaction occurs, write N.R.

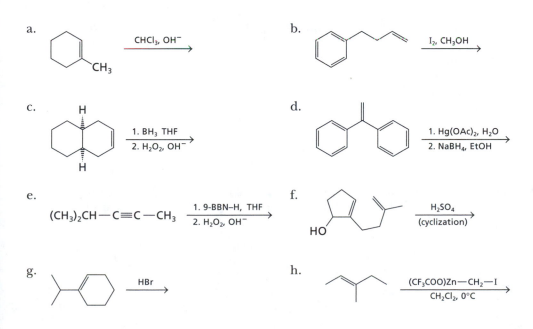

9.30. What is the major product expected from the reaction of 2-methylpropene with each of the following reagents that undergo electrophilic addition reactions with alkenes:

a. Br–CN

b. C_6H_5S–Cl

9.31. Propose a mechanism to explain the formation of compound **1** from **A** and H_3PO_4.

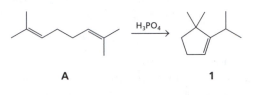

A   **1**

9.32. Indicate what reagents and solvents can be used to carry out the following reactions:

a.

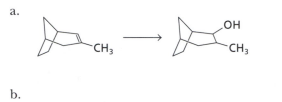

b.

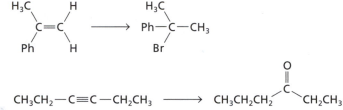

c.

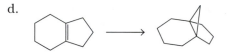

d.

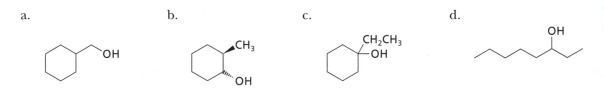

9.33. Which of the following alcohols *cannot* be made as the major product of a reaction between an alkene and borane, followed by treatment with hydrogen peroxide and hydroxide ion?

a.   b.   c.   d.

9.34. For the compounds in Exercise 9.33 that *can* be made via hydroboration–oxidative hydrolysis, draw the structure of the alkene needed as the starting material.

ADDITION REACTIONS OF CONJUGATED DIENES

As you learned in the previous chapter (Chapter 9), the π bond of an alkene is nucleophilic; so it reacts readily with electrophiles such as H^+, Br^+, and BH_3. Molecules with two (*dienes*), three (*trienes*), or even several carbon–carbon double bonds (*polyenes*) undergo the same reactions with electrophiles as alkenes, but the reactivity of each double bond can be influenced by the proximity of the other π bonds.

This chapter describes the structures and reactions of molecules having two double bonds (dienes), with particular attention paid to addition reactions of *conjugated dienes*. Besides undergoing reactions with electrophiles, conjugated dienes can participate in pericyclic reactions (Section 5.1c) because of the interactions between adjacent *p* orbitals. To appreciate how simple pericyclic reactions occur, you will also learn some essential facts about molecular orbital (MO) theory.

10.1   THE STRUCTURES OF DIENES

10.1a IUPAC NAMES OF POLYENES INCLUDE THE NUMBER, TYPES, AND POSITIONS OF MULTIPLE BONDS IN A MOLECULE

In the earlier presentation about IUPAC names (Section 1.3b), you learned that the multiple-bond index conveys how many double and triple bonds are present in a molecule. The terms *diene* or *triene*, for example, indicate that a molecule has two or three double bonds, respectively. Likewise, the suffix *–enyne* is used to denote the presence of a double and a triple bond.

If more than one double bond is present in a molecule, the compound root word in its IUPAC names refers to the longest carbon chain *that also includes all of the multiple bonds*, even if that chain is not the longest in the molecule. The locations of the π bonds are normally each indicated with a numeral; and if there is a choice, the chain is numbered in the direction that gives a lower number at the first point of difference.

Example:

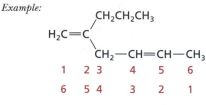

1 < 2, so: **2-propyl-1,4-hexadiene**

2-Propyl-1,4-hexadiene

5-Propyl-2,5-hexadiene

If both double and triple bonds are present, then the lower set of possible numbers is also chosen, regardless of whether the lower number refers to the double- or to the triple-bond position. In case of a tie, however, the double bond is given the lower number because it has a slightly higher priority. Regardless of numbering, "ene" precedes "yne" in the name because of alphabetical order.

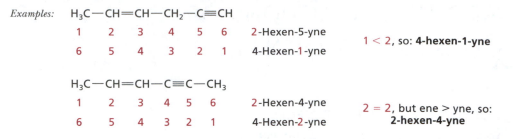

Examples:	H₃C—CH=CH—CH₂—C≡CH		
	1 2 3 4 5 6	2-Hexen-5-yne	1 < 2, so: **4-hexen-1-yne**
	6 5 4 3 2 1	4-Hexen-1-yne	

H₃C—CH=CH—C≡C—CH₃

1   2   3   4   5   6   2-Hexen-4-yne   2 = 2, but ene > yne, so:
6   5   4   3   2   1   4-Hexen-2-yne   **2-hexen-4-yne**

The names of dienes, where applicable (Section 4.1b), must also include designations for the stereochemistry of each double bond. These descriptors appear in the first field of the name, and numerals are required if more than one double bond has a descriptor.

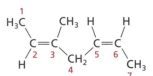

(2E,5Z)-3-Methyl-2,5-heptadiene   **(E)-2-Methyl-1,3-pentadiene**

EXERCISE 10.1

Draw a structural formula for each of the following compounds:

a. (2Z, 4Z)-4-Methyl-2,4-heptadiene   c. 2-Chloro-1,3-cyclohexadiene
b. 2-Fluoro-1-penten-3-yne   d. (2E, 4Z)-2,4-Hexadienal

EXERCISE 10.2

Give a systematic name for each of the following molecules:

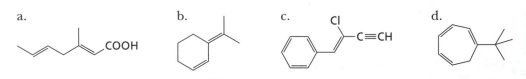

a.   b.   c.   d.

10.1b DOUBLE BONDS CAN BE ISOLATED, CONJUGATED, OR CUMULATED

For a molecule with two or more double bonds, we can classify the relationship of the double bonds as isolated, conjugated, or cumulated. **Isolated double bonds** occur when the π bonds are separated by at least one sp^3-hybridized atom, as in (E)-1,4-hexadiene.

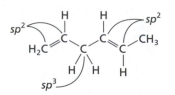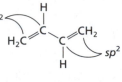

(E)-1,4-Hexadiene   **1,3-Butadiene**
(isolated double bonds)   (conjugated double bonds)

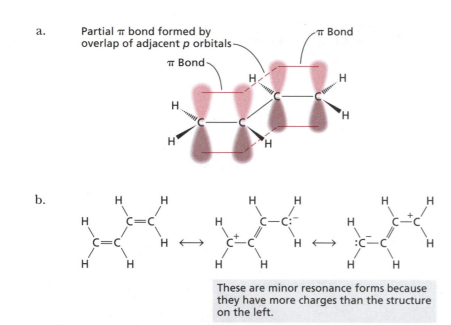

Figure 10.1
Valence bond representation (*a*) and resonance structures (*b*) for 1,3-butadiene.

These are minor resonance forms because they have more charges than the structure on the left.

Conjugated double bonds have an even number of adjacent sp^2-hybridized atoms; the prototypic conjugated diene is 1,3-butadiene. Each carbon atom has sp^2 hybridization, and its valence bond picture is shown in Figure 10.1a. Notice that four *p* orbitals are perpendicular to the plane in which the atoms lie. According to the Lewis structure shown at the left of Figure 10.1b, the bonds between C1 and C2 and between C3 and C4 are double bonds.

It seems, however, that the *p* orbitals on C2 and C3 are close enough to overlap, just as neighboring *p* orbitals overlap in benzene (Section 2.7a). We can draw a dashed line in the valence bond representation to indicate this potential overlap, but is it logical to do so? Two of the resonance forms for 1,3-butadiene are shown in Figure 10.1b, and these have a double bond between C2 and C3. Even though we might disregard their contribution to the resonance hybrid because not every atom has an octet, taken together with the valence bond picture, they suggest that partial overlap between the *p* orbitals on C2 and C3 is, in fact, reasonable. In molecules with conjugated π bonds, the electrons can be delocalized.

Cumulated double bonds have adjacent π bonds with an *sp*-hybridized carbon atom in common. The prototypic *cumulene,* as such compounds are called, is 1,2-propadiene, which is also called *allene.* The valence bond picture of allene is illustrated in Figure 10.2. Because the central carbon has *sp* hybridization, the C–C–C bond angle is 180°; but the terminal carbon atoms have sp^2 hybridization, so the H–C–H angles are 120°. The two π bonds have their overlap in perpendicular planes, which means that the hydrogen atoms at each end of the diene system are also in different planes.

One consequence of this geometry—that the groups at the ends of the cumulene π system are in different planes—is that some cumulenes are chiral. For example, 2,3-pentadiene exists as mirror-image stereoisomers that are not superimposable. Even though this molecule has no chiral carbon atoms, it exists as enantiomers.

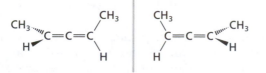

Enantiomers
Nonsuperimposable mirror images of 2,3-pentadiene

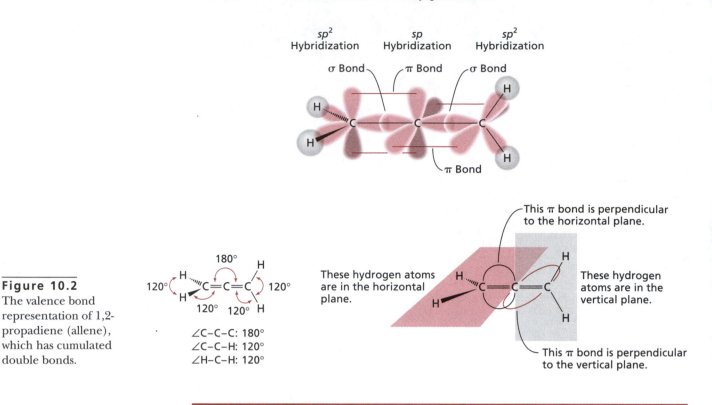

Figure 10.2

The valence bond representation of 1,2-propadiene (allene), which has cumulated double bonds.

∠C–C–C: 180°
∠C–C–H: 120°
∠H–C–H: 120°

EXERCISE 10.3

For each of the following molecules, indicate whether the double bonds are conjugated, isolated, or cumulated:

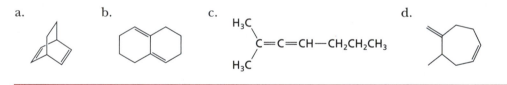

a.

b.

c.   H₃C\
       $\text{C}=\text{C}=\text{CH}-\text{CH}_2\text{CH}_2\text{CH}_3$\
  H₃C

d.

10.1C   CONJUGATED DIENES ARE MORE STABLE THAN ISOLATED AND CUMULATED DIENES

Recall from Section 8.1c that the heats of hydrogenation can be used to determine the relative stabilities of alkenes. The same procedure can be used to assess the relative stabilities of the types of dienes. If 1,2-, 1,3-, and 1,4-pentadiene were each hydrogenated to form pentane, you would find that the conjugated diene is more stable than the isolated and cumulated analogues (*Remember:* The more stable the compound, the less heat is evolved).

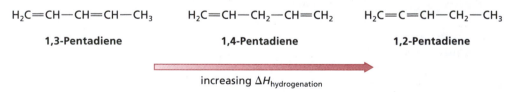

$\text{H}_2\text{C}=\text{CH}-\text{CH}=\text{CH}-\text{CH}_3$ $\text{H}_2\text{C}=\text{CH}-\text{CH}_2-\text{CH}=\text{CH}_2$ $\text{H}_2\text{C}=\text{C}=\text{CH}-\text{CH}_2-\text{CH}_3$

1,3-Pentadiene   **1,4-Pentadiene**   **1,2-Pentadiene**

increasing ΔH$_{hydrogenation}$

At least two factors contribute to the greater stability of the conjugated system. The first relates to the amount of *s* character in the carbon–carbon single bonds of each molecule. Because an electron in an *s* orbital has a higher probability of being closer to the nucleus than an electron in a *p* orbital, the overlap between hybrid orbitals with more *s* character leads to the formation of bonds that are shorter, hence stronger.

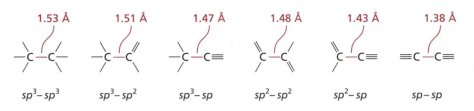

Figure 10.3

Carbon–carbon single bond lengths as a function of hybridization. A conjugated diene has at least one C–C single bond that is 1.48 Å, whereas the shortest C–C single bonds in isolated and cumulated dienes are 1.51 Å.

Bond length data for carbon–carbon single bonds are summarized in Figure 10.3. A molecule is more stable when its bonds are stronger, so upon hydrogenation, a smaller enthalpy change occurs as the product is formed.

The influence from delocalization of the π electrons is the second factor that makes a conjugated diene more stable than the other two types of dienes. This effect is especially important for aromatic compounds, where resonance effects are more significant, and this type of stabilization will be discussed in more detail in Section 17.1a. To see how bonding can be more effectively portrayed for systems with delocalized electrons, we will now look briefly at MO theory.

10.2   BONDING IN CONJUGATED DIENES

10.2a   MOLECULAR ORBITALS RESULT IN THE FORMATION OF BONDS AND ANTIBONDS

The discussion in Chapter 2 furnished details about how valence bond theory provides a suitable simplification of quantum theory to depict bonds resulting from the overlap of atomic (and hybrid) orbitals. Another way to simplify quantum theory is **molecular orbital theory,** which describes bonds by considering the electronic structure of a group of atoms as one entity. It is especially useful for molecules such as conjugated dienes, which have electrons delocalized over four atoms (Section 10.1b).

The number of MOs that are formed for a given system equals the number of atomic orbitals (AO) used to create them. The available electrons are placed into the MOs just as electrons are put into the AOs of atoms, by starting with the MO at the lowest energy. And as with AOs, only two electrons can be put into one MO. Both Hund's rule (electrons are placed one at a time into orbitals with the same energy before pairing occurs) and the Pauli exclusion principle (no two electrons can have the same four quantum numbers) apply to MOs as well as to AOs.

Molecular orbitals are mathematical constructs made by combining AOs. *When two atoms come together, each contributing an atomic orbital toward bond formation, then two molecular orbitals have to be formed.* The wave functions of each AO can either reinforce or cancel each other in the simplest case involving two atoms. Reinforcement creates a **bonding molecular orbital** and cancellation generates an **antibonding molecular orbital.**

Consider as an example the case of the hydrogen molecule, a valence bond representation of which was illustrated in Figure 2.6. If wave functions for the s orbitals of two hydrogen atoms combine, then a bonding MO (Ψ_b) is produced, which is the σ bond (Section 2.4b). When the AO wave functions cancel, an antibonding MO (Ψ_a) is formed, in which a node *between* the nuclei is introduced. Each lobe of the antibonding MO has the opposite mathematical sign, as illustrated in Figure 10.4. The antibonding combination oriented along the bond axis is called the σ **antibond** and is designated σ^*.

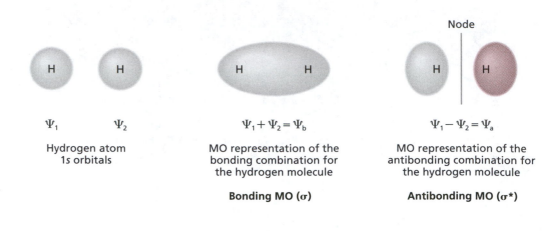

Figure 10.4
Bonding and antibonding combinations for the hydrogen molecule.

When only two atomic orbitals (including hybrid orbitals) combine to form MOs, the bonding MO results from addition and the antibonding MO results from subtraction of the wave functions. Bonding combinations will have their greatest electron density in the region between the nuclei, and the antibonding combination will have a node in this region. For every MO that is formed from atomic p, sp, sp^2, or sp^3 orbitals, there is also a node at the nucleus from the p-orbital contribution.

EXAMPLE 10.1

For the CH_4 molecule, draw representations for the bonding and antibonding MOs for one of the carbon–hydrogen single bonds formed by overlap of the sp^3 hybrid orbital of the carbon atom with the s orbital of a hydrogen atom.

A bonding MO is formed by the additive combination of the two orbitals being considered. Therefore, the larger portion of the hybrid orbital and the hydrogen 1s orbital combine as shown.

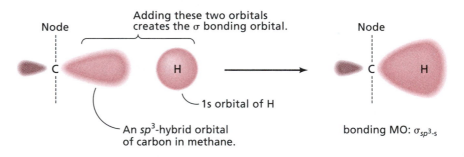

The antibonding orbital is generated by subtracting the hybrid orbital from the hydrogen atom 1s orbital. The sizes of the lobes of the hybrid orbital are inverted by this operation, and the portion of σ^* surrounding the hydrogen nucleus changes sign because it has been subtracted (the light parts in these figures are portions of the orbital having a positive sign; the dark parts are portions of the orbital having a negative sign).

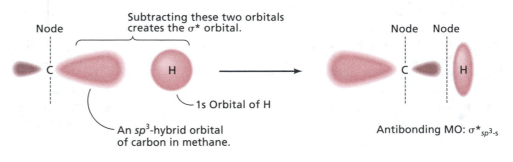

EXERCISE 10.4

Draw representations of the bonding and antibonding MOs for the carbon–carbon single bond in ethane, CH_3CH_3. These orbitals are designated $\sigma_{sp^3-sp^3}$ and $\sigma^*_{sp^3-sp^3}$.

10.2b MOLECULAR ORBITAL DIAGRAMS PRESENT BONDS AND ANTIBONDS ON THE BASIS OF THEIR DIFFERENT ENERGIES

Besides making drawings to illustrate MOs, we can keep track of MOs with diagrams that correlate their relative energies. The AOs are also included in such diagrams to indicate which orbitals have interacted to create the MOs.

Let us look again at the simple example of the hydrogen molecule, for which we can represent the energies of the orbitals by the diagram in Figure 10.5. Any orbital—whether atomic or molecular—can hold only two electrons. So the electron from each $1s$ orbital of the hydrogen atoms enters the bonding MO, which is lower in energy. The antibonding orbital (σ^*) remains empty because there are only two electrons to include, and they have already been used to fill the bonding MO (σ). In other words, the σ bond receives two electrons and σ^* is unoccupied.

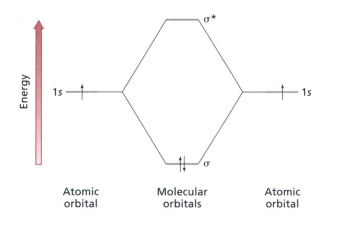

Figure 10.5
An energy diagram for the bonding and antibonding combinations that are created upon formation of the hydrogen molecule from two hydrogen atoms.

Even for complex molecules, an MO diagram will have the same form as that shown in Figure 10.5: bonding MOs are filled and antibonding MOs are empty. Because the energy difference between bonding and antibonding orbitals is often substantial, the occupancy of the orbitals does not usually change. If it does change—for example, when two electrons are added to σ^*—then the antibonding contribution negates the influence of the bonding interaction, and the σ bond is broken.

This phenomenon is what underlies the S_N2 reaction (Section 6.3a). Consider, for example, the reaction of an alkyl bromide with iodide ion, the mechanism of which is shown below.

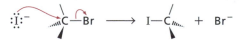

Both C—Br bonding and antibonding MOs lie along the bond axis; σ is filled; σ^* is unoccupied.

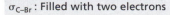
σ_{C-Br} : Filled with two electrons σ^*_{C-Br} : Unoccupied

When a nucleophile—iodide ion in this example—approaches the carbon atom, its electron pair enters the σ* MO because the σ MO is already filled and cannot accept additional electrons. In MO theory, the bond order equals the number of electrons in bonding MOs minus the number of electrons in antibonding MOs, divided by 2. No bond exists (i.e., the bond order = 0) when both bonding and antibonding combinations are filled [2 electrons(σ) − 2 electrons(σ*)/2 = 0]. Thus, if the iodide ion puts two electrons into the antibonding orbital, as illustrated below, the C–Br bond breaks because its bond order becomes zero, and the C–I bond forms.

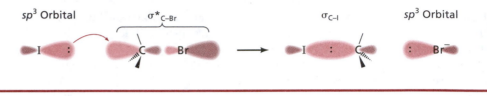

*sp*³ Orbital   σ*_{C–Br}   σ_{C–I}   *sp*³ Orbital

EXAMPLE 10.2

Draw an energy diagram for the σ and σ* molecular orbitals of a C–H bond in methane (see Example 10.1); fill in the appropriate orbitals with electrons. The *sp*³ hybrid orbital has a higher energy than the 1*s* orbital of the hydrogen atom to begin.

First, indicate the relative energy levels of the atomic orbitals that will be combining to generate the MOs. As stated, the hybrid orbital on carbon is higher in energy than the 1*s* orbital of hydrogen.

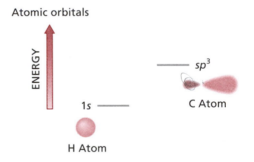

Atomic orbitals

ENERGY

*sp*³

C Atom

1*s*

H Atom

Next, show how the AOs combine to form the bonding and antibonding MOs (below, a.). The bonding combination will be lower in energy than either of the AOs. Finally, add the electrons, filling the lower energy MO first (below, b.). Each AO contributes one electron so only the bonding MO is filled.

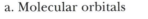

a. Molecular orbitals   b. Molecular orbitals with electrons

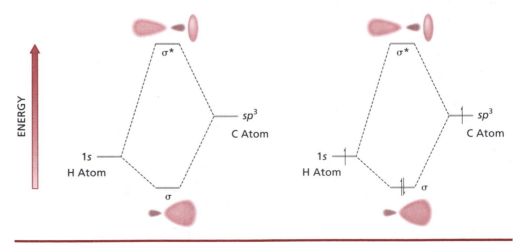

ENERGY

σ*   σ*

*sp*³   *sp*³
C Atom   C Atom

1*s*   1*s*
H Atom   H Atom

σ   σ

EXERCISE 10.5

Draw an energy diagram for bonding and antibonding MOs of the C–C bond in ethane formed by overlap of two sp^3 hybrid orbitals (Exercise 10.4); include the electrons.

10.2c BONDING AND ANTIBONDING COMBINATIONS ALSO RESULT FROM INTERACTIONS BETWEEN *p* ORBITALS

The simple MOs presented so far are sigma interactions; that is, they lie along the bond axis. For a molecule such as ethylene, $H_2C=CH_2$, the π bond interaction can also be described using MOs, which requires formation of a **π antibond (π*)**. Ethylene has four C–H σ bonds, a C–C σ bond, and a C–C π bond. In MO terms, there also have to be four C–H σ* MOs, a C–C σ* MO, and a C–C π* MO. The bonding and antibonding interactions for the two carbon atoms of ethylene are shown in Figure 10.6 (the C–H σ and C–H σ* MOs have been omitted from this figure).

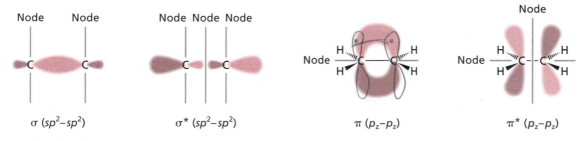

$$\sigma\ (sp^2\!-\!sp^2) \qquad \sigma^*\ (sp^2\!-\!sp^2) \qquad \pi\ (p_z\!-\!p_z) \qquad \pi^*\ (p_z\!-\!p_z)$$

Figure 10.6
Depictions of the σ, σ*, π, and π* MOs for the carbon–carbon bonds in ethylene. The light and dark portions represent lobes with different mathematical signs. The σ and σ* orbitals have a node at each nucleus as a result of a node in the sp^2 hybrid orbitals that constitute them. The σ* and π* MOs have a node between the carbon atoms as well as at the nucleus.

An energy diagram of the MOs for the carbon–carbon bond of ethylene is shown in Figure 10.7. An sp^2-hybrid orbital on each carbon atom combines to generate σ and σ* MOs, and a *p* orbital on each carbon atom combines to create π and π* MOs. Each AO contributes one electron, and as a result, the bonding orbitals are filled, creating the σ and π bonds, and the antibonding orbitals are vacant.

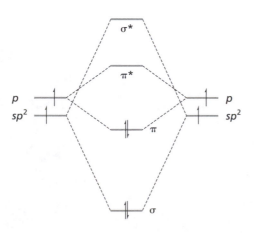

Figure 10.7
An energy diagram of the MOs for the carbon–carbon bonds in ethylene. Interactions of each carbon atom with the hydrogen atoms are not included.

10.2d CONJUGATED DIENES HAVE FOUR MOLECULAR ORBITALS FOR THE PI ELECTRONS

As molecules become more complex, so do the MO pictures and diagrams. To simplify MO representations for molecules with conjugated π systems, we only have to consider the MOs generated from the p orbitals, because the sigma bonds are not usually involved in chemical reactions. For a conjugated diene such as 1,3-butadiene, each carbon atom contributes one p orbital toward forming the MOs. Because the number of MOs equals the number of AOs, 1,3-butadiene should have four MOs to describe its π system. But unlike the interactions between only two AOs, in which addition and subtraction suffice to create the MOs, more complicated combinations are needed when there are more than two atoms.

Without worrying about the mathematics used to derive these representations, we illustrate the four MOs of 1,3-butadiene in Figure 10.8. (For comparison, the π and π* MOs of ethylene are included.) We can say this about MOs in summary:

- For any conjugated π system, the *lowest energy* MO (Ψ_1) results from the sum of n atomic orbital wave functions ($\phi_{p1} + \phi_{p2} + \phi_{p3} + \phi_{p4}$ for butadiene). It has $n-1$ bonding interactions, and no node exists between any pair of carbon atoms.

- As the MOs increase in energy, an additional node appears at each level, which corresponds to a change in sign of the orbitals on either side of the node. The number of bonding interactions decreases by one at each level.

- The highest energy wave function has the maximum number of nodes, namely, one node between each pair of atoms, and it has no bonding interactions.

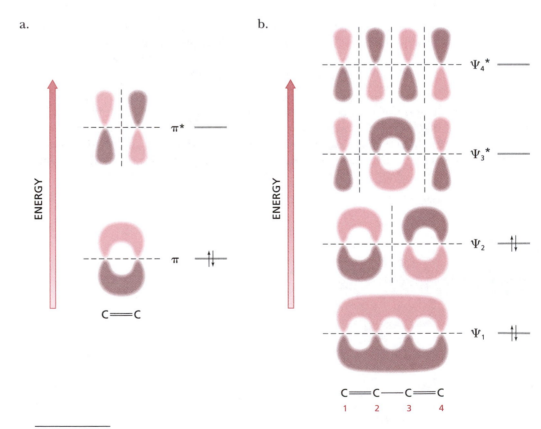

Figure 10.8
Pictorial representations of the π MOs in ethylene (a.) and in 1,3-butadiene (b.). Nodes—where an orbital changes its mathematical sign—are indicated by the dashed lines. Notice that all of the MOs have a node at each carbon atom nucleus.

In 1,3-butadiene, Ψ_1 has no node, and there are three bonding interactions (between C1 and C2, C2 and C3, and C3 and C4). For Ψ_2, two bonding interactions exist (between C1 and C2 and between C3 and C4), and there is one node (between C2 and C3). For Ψ_3, there is one bonding interaction (between C2 and C3), and there are two nodes (between C1 and C2 and between C3 and C4). The highest energy orbital has three nodes and no bonding interactions.

The atomic p orbitals of butadiene contribute a total of four electrons (one from each p orbital), and these four electrons are added to the four MOs starting with the one at lowest energy, pairing them if possible. As you can see, only Ψ_1 and Ψ_2, which are bonding orbitals, have electrons. Both Ψ_3^* and Ψ_4^* are antibonding MOs and are unoccupied.

For stable alkenes and polyenes with n MOs, $n/2$ are bonding orbitals and filled; and $n/2$ are antibonding orbitals and vacant. Such an arrangement of electrons within a set of MOs is called the *ground state*. Notice that bonding MOs have more overlapping pairs of orbitals than nodes, and antibonding MOs have more nodes than overlapping pairs.

10.3   ELECTROPHILIC ADDITION TO CONJUGATED DIENES

10.3a   CONJUGATED DIENES PARTICIPATE IN 1,2- AND 1,4-ADDITION PROCESSES

We now turn our attention to reactions of conjugated dienes. You learned in Chapter 8 that alkenes and alkynes react with electrophiles to form cationic species in the first step of many addition processes. Conjugated dienes undergo electrophilic addition in the same general way, but more than one product can form. For example, when 1,3-butadiene is treated with HBr, one of the π bonds of the diene reacts with the electrophilic proton to produce a resonance-stabilized allylic carbocation.

$$H_2C{=}CH{-}CH{=}CH_2 \xrightarrow{\ H{-}Br\ } H_2C{=}CH{-}\overset{+}{C}H{-}CH_2 \longleftrightarrow H_2\overset{+}{C}{-}CH{=}CH{-}CH_2$$

In the next step, the nucleophilic bromide ion intercepts this carbocation at either end of the allylic π system to form two products (ignoring stereoisomers). One product results from **1,2-addition,** and one is formed by **1,4-addition.**

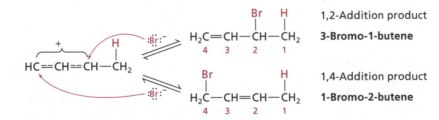

It is important to realize that the terms "1,2-" and "1,4-" refer only to the four atoms of the conjugated diene system, not to the molecule as a whole. Thus, the reaction between 2,4-hexadiene and HBr occurs as follows to give two products:

$$H_3C{-}CH{=}CH{-}CH{=}CH{-}CH_3 \xrightarrow{\ H{-}Br\ } H_3C{-}CH{=}CH{-}\underset{\underset{}{}}{CH}{-}\underset{\underset{}{}}{CH}{-}CH_3 \ + \ H_3C{-}\underset{\underset{}{}}{CH}{-}CH{=}CH{-}\underset{\underset{}{}}{CH}{-}CH_3$$

4-Bromo-2-hexene
(1,2-addition product)   **5-Bromo-3-hexene**
(1,4-addition product)

Other reagents, such as chlorine and bromine, also undergo 1,2- and 1,4-addition.

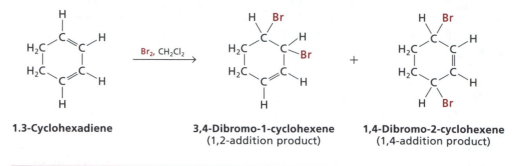

1.3-Cyclohexadiene   **3,4-Dibromo-1-cyclohexene**   **1,4-Dibromo-2-cyclohexene**
                                      (1,2-addition product)   (1,4-addition product)

EXERCISE 10.6

Draw the structures of the major products of the following reaction, which involves both 1,2- and 1,4-addition:

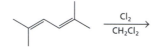

EXERCISE 10.7

Predict the major products of the following reaction, which involves both 1,2- and 1,4-addition of water:

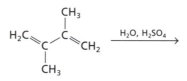

10.3b ADDITION TO A CONJUGATED DIENE EXEMPLIFIES KINETIC VERSUS THERMODYNAMIC CONTROL OF A REACTION

If we carry out the reaction between 1,3-butadiene and HBr at 0°C, we find that the ratio of 3-bromo-1-butene to 1-bromo-2-butene is ~ 70:30. If we repeat this reaction at 40°C, we obtain these same products in a ratio of 15:85.

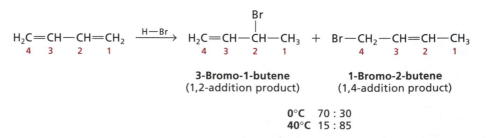

                               3-Bromo-1-butene   **1-Bromo-2-butene**
                            (1,2-addition product)   (1,4-addition product)

                                        0°C   70 : 30
                                        40°C   15 : 85

The product ratios reflect the kinetic versus thermodynamic influences on the reaction mechanism, which appear in the second step when the bromide ion reacts with the carbocation. As the bond between Br⁻ and a carbon atom begins to form in the transition state, the charge becomes more localized on the specific carbon atom at which the new bond is forming.

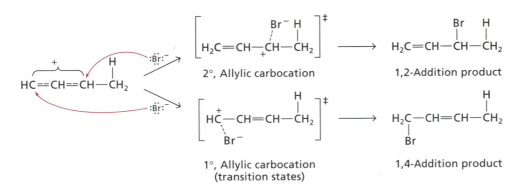

It should be apparent that these two carbocations are different: one is primary and the other is secondary (both are allylic). You know that a secondary carbocation is more stable than a primary one (Section 6.2b). Furthermore, you learned that the transition state for a reaction proceeding via a carbocation intermediate has a lower energy when the carbocation is more stable (Section 6.2c). Because the magnitude of the transition state energy correlates with the rate of reaction (lower energy = faster reaction), formation of the 1,2-addition product is said to be under "kinetic control", and 3-bromo-1-butene is the *kinetic product*. (As described shortly, the other product is the *thermodynamic product*.) Figure 10.9 shows the reaction coordinate diagram for this process and reveals the different transition state energy barriers for the second step of the mechanism.

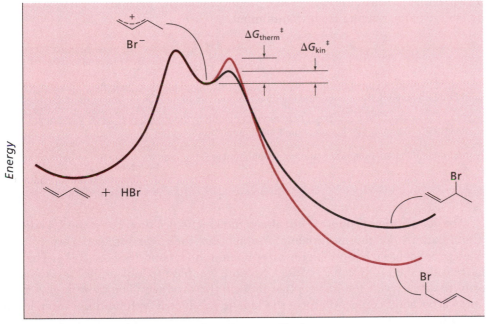

Figure 10.9
A reaction coordinate diagram for the reaction between 1,3-butadiene and HBr.

An important point about a kinetically controlled reaction is that the available energy is *not* sufficient to permit *regeneration* of the carbocation intermediate from the products; that is, no equilibrium between the intermediate and products exists. We can represent this situation as follows:

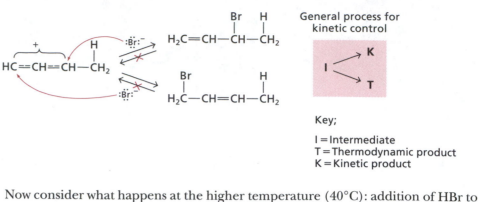

General process for
kinetic control

Key;

I = Intermediate
T = Thermodynamic product
K = Kinetic product

Now consider what happens at the higher temperature (40°C): addition of HBr to 1,3-butadiene gives predominantly 1,4-addition. The reaction is said to be under "thermodynamic control", and 1-bromo-2-butene is the *thermodynamic product*. Thermodynamic control is a term you encountered in the discussion of Saytseff product formation in the E1 reaction (Section 8.1b). Here, the meaning is the same: *The more stable product predominates* under thermodynamic control. 1-Bromo-2-butene is more stable because it has a disubstituted double bond, whereas 3-bromo-1-butene has a monosubstituted one (Section 8.1b).

At higher temperatures, sufficient energy is available for the reaction to cross the activation barrier *in either direction;* that is, an equilibrium step exists, at least for the pathway leading to formation of the 1,2-product (in this case, the kinetic product). Therefore, as 3-bromo-1-butene forms, its bromide ion can dissociate to regenerate the carbocation intermediate. Because the allylic carbocation can react with bromide ion to form either product, the more stable product eventually becomes the major one and the energy of the system reaches a minimum.

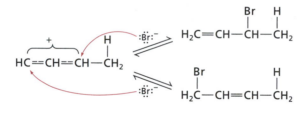

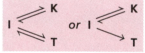

General process for
thermodynamic control

Key;

I = Intermediate
T = Thermodynamic product
K = Kinetic product

The 1,2-addition product is not always the kinetic product, nor is the 1,4-addition product always the thermodynamic product. For each reaction, you have to assess which product has the more highly substituted double bond (Section 8.1b): That substance is the thermodynamic product. Second, you have to look at the two resonance forms that react with the nucleophile to form the products: The more highly substituted carbocation center will form the kinetic product. The following example illustrates this procedure.

EXAMPLE 10.3

Draw structures for the kinetic and thermodynamic products formed in the reaction between 1-methyl-1,3-cyclohexadiene and HCl. Ignore stereochemistry.

First, consider the reaction of the diene and the electrophile (a proton here). Reaction can occur at either double bond initially, and the predominant reaction pathway is the one that generates the more stable allylic carbocation. In this reaction, the intermediate with a contribution from a 3° carbocation will predominate.

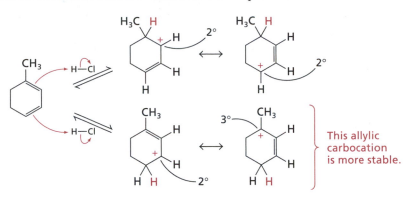

Next, draw the structures of the possible products by bringing together the nucleophile and the more stable carbocation.

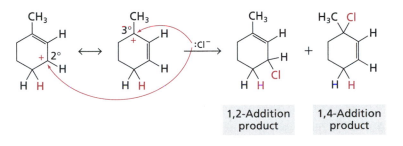

Finally, determine the thermodynamic and kinetic products by evaluating which has the more highly substituted double bond (thermodynamic product) and which is formed from the more stable carbocation (kinetic product).

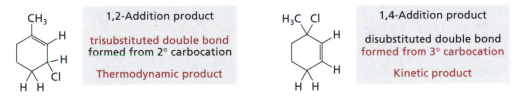

EXERCISE 10.8

Draw structures of the kinetic and thermodynamic products formed in the reaction between the following dienes and HBr. Ignore stereochemistry.

a.   b.   c.

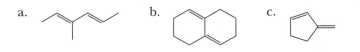

10.4   THE DIELS–ALDER REACTION

10.4a   DIENES READILY UNDERGO PERICYCLIC REACTIONS

Besides reacting with electrophilic reagents, dienes (and polyenes) can react via cyclic transitions states in transformations called *pericyclic reactions* (Section 5.1c). Three

Electrocyclic reaction
the formation of a single bond between the termini of a conjugated polyene system

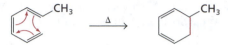

Sigmatropic reaction
the migration of a single bond adjacent to one or more π systems

Cycloaddition reaction
the formation of a ring from the reaction of two π systems

Figure 10.10
Examples of common pericyclic reactions. All of the reactions are reversible.

common types of pericyclic transformations are *electrocyclic, sigmatropic,* and *cycloaddition reactions.* An example of each is shown in Figure 10.10. We will look in detail at only one type of pericyclic reaction, the cycloaddition reaction.

10.4b THE DIELS–ALDER REACTION IS A CYCLOADDITION PROCESS THAT IS USED TO PREPARE SIX-MEMBERED RINGS

The **Diels–Alder reaction** is the prototype for literally hundreds of cycloaddition reactions. The reaction, which was discovered by Otto Diels and Kurt Alder in the 1920s—work for which they received the 1950 Noble Prize for Chemistry—is one of the most versatile methods for preparing six-membered rings.

The reactive components in the Diels–Alder reaction are a conjugated diene and a **dienophile** (diene lover), an alkene that normally has an electron-withdrawing group as a substituent. The electrons in the three π bonds move in a concerted fashion within a six-membered ring transition state, and the product is a cyclohexene derivative.

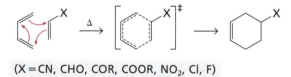

$(X = CN, CHO, COR, COOR, NO_2, Cl, F)$

The Diels–Alder reaction—also called a **[4 + 2] cycloaddition** to designate the number of π electrons provided by each reactant—is a stereospecific process, and the stereochemistry of the dienophile is retained in the product. Thus, diethyl *trans*-butenedioate reacts with 1,3-butadiene to form the trans disubstituted cyclohexene compound.

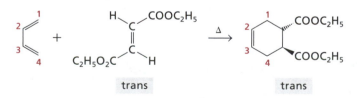

Maleic anhydride, propenenitrile (acrylonitrile), methyl propenoate (also called methyl acrylate), and dimethyl butynedioate (dimethyl acetylenedicarboxylate) are excellent dienophiles.

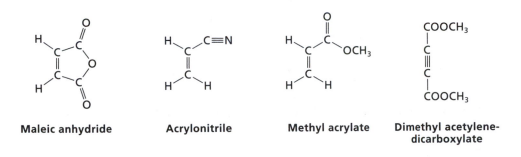

| **Maleic anhydride** | **Acrylonitrile** | **Methyl acrylate** | **Dimethyl acetylene-dicarboxylate** |

All of them have at least one electron-withdrawing group attached to an alkene carbon atom. An electron-withdrawing group is either an electronegative atom (–F, –Cl, –Br), which operates through inductive effects, or a group with unsaturation (C=O or C≡N), which influences the electron density of the π bond by resonance.

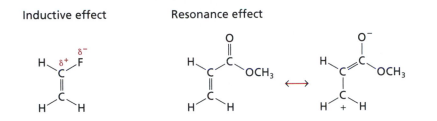

The structure of the diene can vary considerably, but it must be able to adopt the s-cis conformation; s-cis is an abbreviation for "sigma-cisoid".

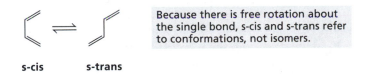

Because there is free rotation about the single bond, s-cis and s-trans refer to conformations, not isomers.

Some of the most reactive dienes have their π bonds in a ring that fixes the unsaturation in the s-cis conformation. Cyclopentadiene is one of the best dienes for the Diels–Alder reaction. It is so reactive that it reacts with itself to form a dimer called dicyclopentadiene. Upon heating, this dimer undergoes a retro-Diels–Alder reaction (**retro-** means reverse) to produce cyclopentadiene, which can be stored at low temperatures for several days. At room temperature, dimerization takes place within several hours.

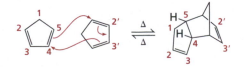

Cyclopentadiene   **Dicyclopentadiene**

EXERCISE 10.9

Draw the structure of the major product expected from each of the following reactions:

a.   b.

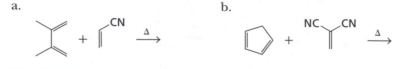

10.4c MOLECULAR ORBITALS OF A DIENE AND DIENOPHILE MUST HAVE THE PROPER SYMMETRY TO REACT

The guiding principles for understanding the Diels–Alder and other pericyclic reactions are commonly called the "Woodward–Hoffmann rules", which say that symmetry among the MOs of reactants and products must be conserved as a pericyclic reaction proceeds. "Conservation of orbital symmetry" means simply that the mathematical signs of the reacting orbitals must match each other during the course of the transformation. A detailed exposition of this topic is beyond the scope of this text, but the example provided by the Diels–Alder reaction will illustrate the basic concepts.

The analysis of orbital symmetry starts with the MOs of the reactants. For the Diels–Alder reaction, the reactants are an alkene and a conjugated diene, the MOs of which were shown in Figure 10.8. As these two reactants come together, their orbitals begin to interact. A set of line structures can be drawn to orient the σ and π bonds so that we have a framework on which to place the orbitals in the following presentation.

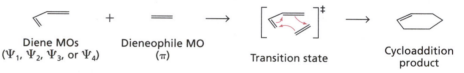

Diene MOs (Ψ_1, Ψ_2, Ψ_3, or Ψ_4)   Dienophile MO (π)   Transition state   Cycloaddition product

(Ψ_n and π are the MOs shown in Fig. 10.8)

If the π bond of ethylene (our alkene prototype) interacts with the Ψ_1 MO of 1,3-butadiene, the orbitals that come together have the same signs (circled, in the equation below). This means that they are "in phase" and have the appropriate symmetry to react. However, Ψ_1 of the diene and the π bond of ethylene are both filled. Remember that whenever two orbitals combine, two *new* orbitals have to be created, one bonding and the other antibonding. If the combining orbitals are filled already, then their electrons will populate both the new bonding *and* antibonding orbitals, which means that no net bond will result (the bond order will be 0; see Section 10.2b).

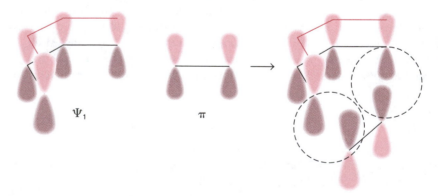

Ψ_1 π

If the π bond of ethylene interacts with Ψ_2 of 1,3-butadiene, the orbitals that need to overlap do *not* have the same sign at one junction (indicated by the arrow in the following schemes). Thus they do not have the proper symmetry to interact. If a reaction were to occur, *orbital symmetry would not be conserved*. The thesis of the Woodward–Hoffmann rules is that such a process is forbidden, and these orbitals cannot combine.

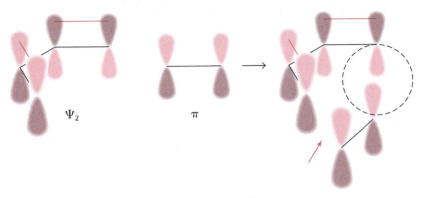

The same nonproductive situation holds for the interaction of the π bond of ethylene with Ψ_{4^*} of butadiene. These orbitals cannot combine either.

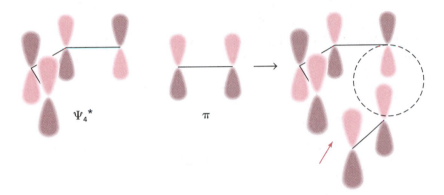

The π bond of ethylene, however, *does* have the correct symmetry to interact with Ψ_{3^*} of 1,3-butadiene, and Ψ_{3^*} is vacant. Therefore, when these orbitals combine to create new bonding and antibonding orbitals, only the bonding combinations receive electrons as transformation into the new σ bonds of the six-membered ring product takes place.

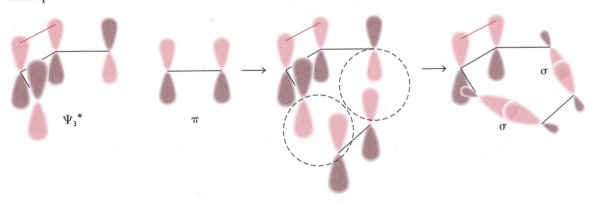

You can actually simplify an analysis of orbital interactions because new bonds will form only when a *filled* orbital of one reactant interacts with an *empty* orbital of the other. (These orbitals should also be of comparable energy to maximize their interactions.) Thus, you need only consider the symmetries of the *highest occupied molecular orbital* (HOMO) of one reactant, and the *lowest unoccupied molecular orbital* (LUMO) of the other. If these orbitals have the correct symmetry at the positions that come together to form bonds, then the reaction will proceed. Otherwise, it will not.

10.4d THE REGIO- AND STEREOSELECTIVITY OF THE DIELS–ALDER REACTION DEPENDS ON SUBSTITUENTS IN EACH REACTANT

Although the transition state of a [4 + 2] cycloaddition is dictated by the symmetries of the orbitals involved, isomeric products can still be obtained. For example, when a cyclic diene is used in the Diels–Alder reaction, two products can be obtained: endo and exo.

endo

exo

In the endo isomer, the substituent of the dienophile π bond appears on the same side of the bicyclic ring system as the bridge containing the double bond.

In the exo isomer, the substituent is on the side of the bicyclic ring system that is opposite that of the bridge containing the double bond.

In general, the endo isomer predominates. This preference results because the carbonyl group of the dienophile tends to lie under (or over) the diene π bonds, as shown below:

Enantiomers

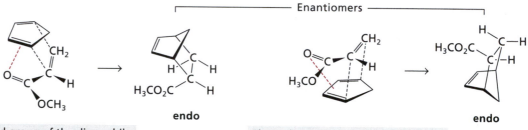

endo   **endo**

The carbonyl group of the dienophile lies under the π bonds of the diene.

The carbonyl group of the dienophile lies above the π bonds of the diene.

Isomers are also formed when both diene and dienophile are unsymmetrical. If the diene is substituted in the 1-position, then the 1,2-regioisomer of the product is formed in preference to the 1,3-disubstituted ring. For example,

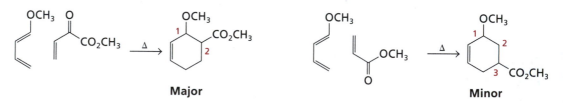

Major   **Minor**

To rationalize this regiochemistry, we look at the principal resonance forms of the two reactants.

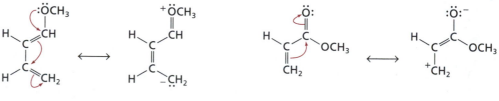

Diene   **Dienophile**

Because orbital symmetry has only to do with the mathematical signs of the MOs, either orientation is permitted; however, electronically, the better match brings the positive and negative charges of the resonance forms together, which generates the 1,2-isomer.

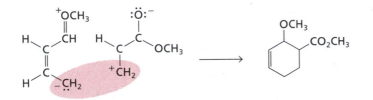

To predict the stereochemistry of this reaction, we need to orient the reactants so that the electron-withdrawing group of the dienophile lies under the π bonds of the diene, as was illustrated above for the Diels–Alder reactions that generate bicyclic products. As a result, the cis-isomer is the major product (if this stereochemical result is not obvious to you, make a model to see why the two substituents in the product are cis).

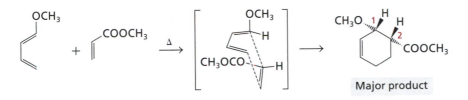

When the diene has a substituent at the 2-position, then the 1,4-regioisomer is obtained preferentially over the 1,3-disubstituted product.

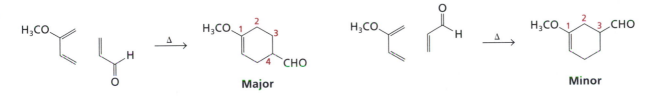

Again, we look at the resonance forms of the reactants, and then match the ends that have the positive and negative charges. In this case, we do not have to be concerned with the formation of geometric isomers because the methoxy group is attached to an sp^2-hybridized carbon atom.

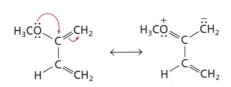

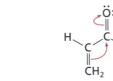

Diene   **Dienophile**

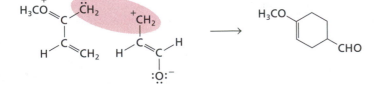

In summary, *a 1,3-disubstituted cyclohexene ring is difficult to prepare via the Diels–Alder reaction.*

An alkyne reacts with a diene to form the derivative of a 1,4-cyclohexadiene. Oxidation reactions can be used to remove two hydrogen atoms and convert these molecules into benzene derivatives, if desired.

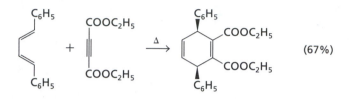

(67%)

EXERCISE 10.10

Draw the structure of the major product expected from each of the following Diels–Alder reactions:

a.   b.

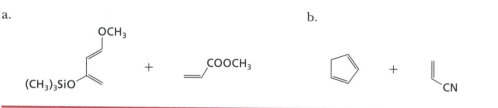

CHAPTER SUMMARY

Section 10.1   Structures of conjugated dienes

- The compound root in the IUPAC name of a polyene is the longest carbon chain that also includes all of the multiple bonds. The multiple-bond index indicates how many and what types of double or triple bonds are present.

- Dienes can be classified as isolated, conjugated, and cumulated. A conjugated diene is the most stable of the three because of a shorter sigma bond between the two π bonds and because the π electrons can be delocalized over four atoms.

Section 10.2   Bonding in conjugated dienes

- Molecular orbital theory is especially useful to depict bonding in conjugated dienes because it considers the electronic structure of a group of atoms as one unit.

- Molecular orbitals are generated from atomic orbitals (AOs), and the number of molecular orbitals (MOs) is equal to the number of AOs used to make them.

- When MOs are generated for a bond between two atoms, both bonding and antibonding combinations are generated, the former by addition and the latter by subtraction. A bonding MO has a lower energy than its corresponding type of antibonding MO.

- When MOs are generated for bonds among several atoms, the mathematical combinations are complicated. However, the lowest energy MO is the sum of the AOs.

Section 10.3   Electrophilic addition to conjugated dienes

- A conjugated diene can add 1 equivalent of HX or X_2 in two ways, called 1,2- and 1,4-addition.
- 1,2-Addition occurs by reaction at one of the two double bonds of the diene.
- 1,4-Addition occurs when the electrophile and nucleophile add to the ends of the conjugated π system; a double bond appears in the product between the second and third carbon atoms of the original four atoms of the π system.
- The kinetic product from electrophilic addition to a conjugated diene is the one that is formed faster. Predicting the structure of the kinetic product requires identifying the more stable contributor to the allylic carbocation intermediate.
- The thermodynamic product from electrophilic addition to a conjugated diene is the alkene that has the more highly substituted double bond.

Section 10.4   The Diels–Alder reaction

- Three common types of pericyclic transformations are electrocyclic, sigmatropic, and cycloaddition reactions, in which electrons move in a cyclic framework to break or form new bonds.
- The Diels–Alder reaction is an example of a cycloaddition reaction. A diene reacts with an alkene, called the dienophile, to form a six-membered ring.
- The Diels–Alder reaction is a concerted process, and the stereochemistry of the dienophile is retained in the product.
- The best dienophiles have one or more electron-withdrawing groups attached to the alkene carbon atoms.
- A cycloaddition process is allowed when the symmetry of the highest occupied molecular orbital (HOMO) of one reactant matches that of the lowest unoccupied molecular orbital (LUMO) of the other reactant.
- When a bicyclic compound is formed as the product of the Diels–Alder reaction, the predominant isomer is the endo product.
- When unsymmetric reactants participate in a Diels–Alder reaction, the major products are usually 1,2- or 1,4-disubstituted six-membered rings.

KEY TERMS

Section 10.1b
isolated double bond
conjugated double bond
cumulated double bond

Section 10.2a
molecular orbital theory
bonding molecular orbital
antibonding molecular orbital
σ antibond

Section 10.2c
π antibond

Section 10.3a
1,2-addition
1,4-addition

Section 10.4b
Diels–Alder reaction
dienophile
[4 + 2] cycloaddition
retro

REACTION SUMMARY

Section 10.3a

Addition of HX or X_2 to a conjugated diene. Depending on conditions, 1,2- and 1,4-addition products are possible.

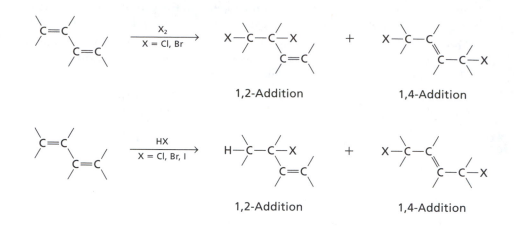

Section 10.4b

The Diels–Alder reaction. A conjugated diene reacts with a dienophile (an alkene with an electron-withdrawing substituent) to form a cyclohexene derivative. The stereochemistry of the dienophile is retained in the product.

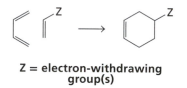

Z = electron-withdrawing group(s)

Section 10.4b

In considering the stereochemistry and regiochemistry of addition, endo is favored over exo when a bicyclic product is formed; and formation of the 1,2- and 1,4- disubstituted rings are favored over the 1,3-regioisomers.

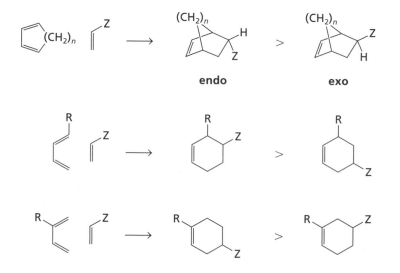

10.11. After treating each of the following cyclic dienes with 1 equiv of HCl, you isolate and purify the products that have only one chlorine atom. Draw their structures and rationalize their formation by proposing a reasonable mechanism for each reaction.

a. b. c.
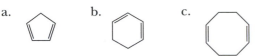

10.12. What is the expected major product, including stereochemistry, for each of the following Diels–Alder reactions?

a.   b.

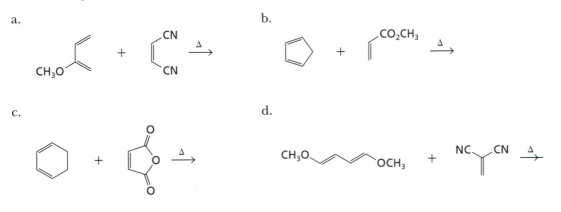

c.   d.

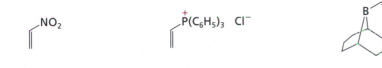

10.13. The following alkenes are all good dienophiles. Explain what effects contribute to their dienophilicity.

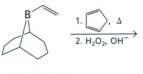

1-Nitroethylene   Vinyltriphenylphosphonium chloride   B-Vinyl-9-BBN

10.14. What is the expected product from the reaction between 2,3-dimethyl-1,3-butadiene and each of the dienophiles shown in Exercise 10.13?

10.15. What is the expected product, including stereochemistry, for the reaction sequence shown here?

B-vinyl-9-BBN

10.16. Draw the structures of the dienes and dienophiles that can be used to prepare the following compounds with a Diels–Alder reaction:

a.   b.   c.   d.

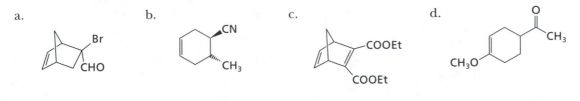

10.17. If 1-chlorocyclodecene is treated with strong base, an E2 reaction takes place, forming two products. One is a cumulene and the other is an alkyne.

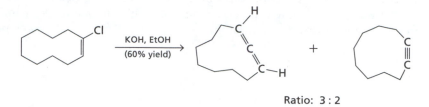

Ratio: 3 : 2

a. Assign the hybridization of each carbon atom in these molecules.

b. Rationalize the product ratio. (*Hint:* Make a model of each and assess the relative strain in each ring by considering the bond angles around the π bonds.)

10.18. Compound **A,** when treated with Br$_2$ in CH$_2$Cl$_2$, yields compound **B** in excellent yield. Heating compound **B** with sodium isopropoxide, a strong base, in a high-boiling ether solvent, causes compound **C** to distill from the reaction mixture in good yield. Compound **C** is subsequently treated with maleic anhydride, and compound **D** is isolated. Draw the structures for compounds, **A, B,** and **C,** and write equations showing the reactions that take place. Name compounds, **A, B,** and **C.**

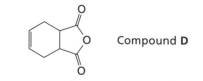

Compound **D**

10.19. Predict the major products of the following reactions, which yield both 1,2- and 1,4-addition products:

a.

b.

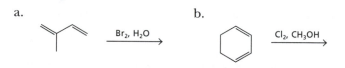

10.20. Using the information presented in Section 10.2d, draw the MOs for 1,3,5-hexatriene in the manner shown in Figure 10.8. Include the electrons and label the orbitals as bonding or antibonding.

10.21. Diethyl diazodicarboxylate, the reagent that you learned for the Mitsunobu reaction (Section 7.1d), also undergoes cycloaddition reactions with dienes. Draw the structure of product that will be formed from its reaction with cyclopentadiene.

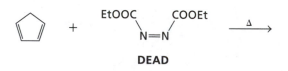

DEAD

10.22. Using the information presented in Section 10.2d, draw the MOs for the allyl system, the valence bond picture of which is shown in Figure 2.26. When a species has an odd number of carbon atoms (and *p* orbitals), the MO diagram has bonding, *nonbonding,* and antibonding orbitals. For the allyl system, there is one of each. Draw pictures of these three orbitals, then consider how the electrons are included for the allyl carbocation (2 electrons), radical (3 electrons), and carbanion (4 electrons). How do the MO diagrams correlate with the resonance forms for these three species?

10.23. Draw the structure(s) of the major product(s) expected from each of the following reactions. Indicate the stereochemistry of the product as appropriate. Relative stereochemistry should be shown using wedges and dashed lines. If a racemic mixture will be formed, draw the structure of one enantiomer and write the word "racemic", or draw both enantiomeric structures. If diastereomers are formed, draw each structure; label meso compounds as such. If no reaction occurs, write N.R.

a.

$\ce{->[HCl, CH2Cl2][-10 °C]}$

b.

$\ce{->[\Delta]}$

c.

$\ce{->[\Delta]}$

d.

Product of (c) $\ce{->[KOH, EtOH][\Delta]}$

e.

$\ce{->[Br2, CHCl3][50 °C]}$

f.

$\ce{->[1. Cl2, CH2Cl2][2. KOH, EtOH, \Delta]}$

10.24. An alternative mechanism for 1,4-addition of bromine to a conjugated diene can be written as shown below. The cyclic bromonium ion is a five-membered ring, which is expected to be quite stable. How could you tell if this pathway is likely, or even possible? (*Hint:* Consider the stereochemistry of the product formed in this scheme versus the stereochemistry from the mechanism shown in Section 10.3a).

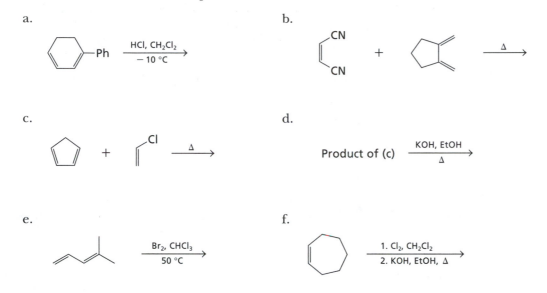

OXIDATION AND REDUCTION REACTIONS

11.1 OXIDATION STATES IN ORGANIC MOLECULES

11.2 HYDROGENATION REACTIONS

11.3 OXIDATION REACTIONS OF ALKENES

11.4 OXIDATION REACTIONS OF ALCOHOLS

11.5 OXIDATION REACTIONS OF AMINES

CHAPTER SUMMARY

So far, you have been exposed to examples that illustrate the mechanisms of five of the basic reaction types mentioned in Chapter 5: proton transfer, substitution, addition, elimination, and rearrangement. The other categories of reactions mentioned there—*oxidation* and *reduction*—are common pathways by which alcohols and alkenes react, so it makes sense to see how such processes relate to the transformations that you have already learned.

In this chapter, we will look at reduction reactions of alkenes, alkynes, and nitrogen-containing functional groups, in which molecular hydrogen adds to π bonds. We will then look at oxidation reactions of alkenes, which involve addition of one or more oxygen atom to a carbon–carbon π bond. In contrast with alkene oxidation, the oxidation reactions of alcohols occur by elimination, and a carbonyl group is produced in those transformations. The oxidation of alcohols in biochemical systems exploits the same mechanistic pathways that are important in the laboratory. This chapter concludes with the oxidation reactions of amines, which follow pathways that differ from those observed for alkenes and alcohols.

The complete mechanisms are not always discussed for the reactions presented in this chapter. Some of these mechanisms are ambiguous or not well understood; others are beyond the scope of this book. An attempt is made, however, to show the similarities among the diverse transformations that are presented. Where possible, comparisons are made between the mechanisms of oxidation and reduction reactions and the mechanisms for reactions you have already learned.

11.1 OXIDATION STATES IN ORGANIC MOLECULES

11.1a THE OXIDATION LEVEL OF CARBON ATOMS DEPENDS ON THE PRESENCE OF HETEROATOMS OR UNSATURATION

In chemistry, oxidation is normally defined as the loss of electrons from an element, and reduction is defined as the gain of electrons. If the oxidation state of an element is known both before and after a reaction takes place, then it is a straightforward decision whether oxidation or reduction (or neither) has taken place. For an organic compound, we define these processes as follows (Section 5.1b):

Oxidation is the removal of hydrogen atoms from, or the incorporation of heteroatoms into, an organic molecule.

Reduction is the addition of hydrogen atoms or the removal of heteroatoms.

The number of heteroatoms attached to a carbon atom is therefore equivalent to the **oxidation level** of carbon in organic molecules. A double or triple bond between a carbon atom and a heteroatom counts for two or three of that heteroatom, respectively, in determining oxidation level. If the oxidation level changes during a chemical reaction, then oxidation or reduction has occurred.

The range of oxidation levels for organic compounds that have one carbon atom is illustrated in Figure 11.1 using methane, methanol, formaldehyde, formic acid, and carbon dioxide as examples. These compounds range from the least oxidized to the most oxidized.

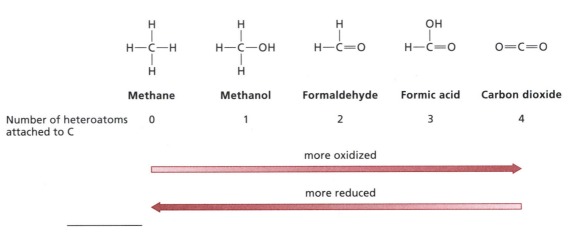

| Number of heteroatoms attached to C | 0 | 1 | 2 | 3 | 4 |

Figure 11.1
Comparison of oxidation levels for organic compounds that have one carbon atom.

The carbon atoms in bromomethane, dimethyl sulfide, and azidomethane have the same oxidation level as that in methanol, because the carbon atom forms only one bond to a heteroatom. Commonly, functional groups in which the carbon atoms have the same oxidation level are interconverted by substitution processes, as illustrated below.

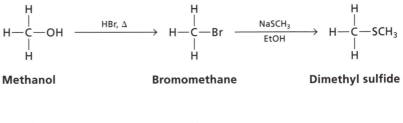

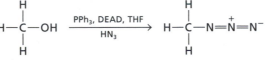

EXERCISE 11.1

Provide a reasonable mechanism for each of the foregoing transformations. Use curved arrows to show the movement of electrons.

When more than one carbon atom is present in a molecule, we can still evaluate the oxidation level of *single* carbon atoms. Thus, an alkane is oxidized at its terminal carbon atom to a primary alcohol, then to an aldehyde, then to a carboxylic acid.

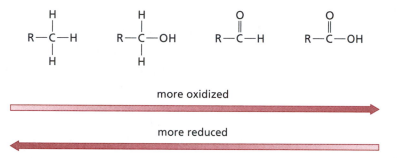

Sometimes, however, a process that changes the number of attached hydrogen or heteroatoms can involve more than one carbon atom. For those molecules, the presence of carbon–carbon π bonds also defines a more oxidized state of the molecule.

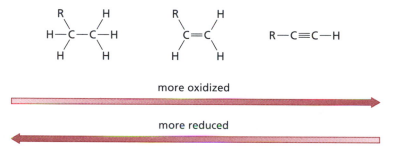

In reactions that involve carbon–carbon π bonds, the oxidation level of one carbon atom may increase, while the other decreases—a net change of zero means that reduction or oxidation has not taken place. For example, addition of water in the presence of an acid to an alkene generates an alcohol.

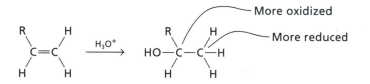

The carbon atom with the new hydrogen atom has been reduced, and the one with the OH group has been oxidized. This result is intuitively satisfying because we normally do not consider water to be either an oxidizing or reducing agent.

11.1b NITROGEN ATOMS ALSO HAVE DIFFERENT OXIDATION LEVELS

Just as we assign oxidation levels to carbon atoms according to the number of hydrogen and heteroatoms attached, we can designate oxidation levels of nitrogen atoms. Removal of hydrogen atoms or introduction of oxygen atoms leads to increased oxidation of a nitrogen atom, so an amine represents the least oxidized species, and a nitro compound contains the most oxidized type of nitrogen atom.

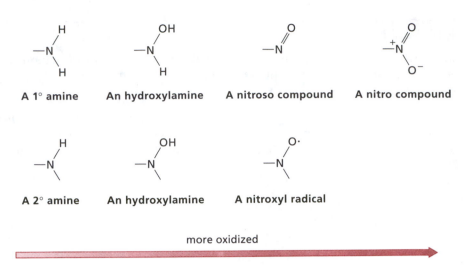

A 1° amine · An hydroxylamine · A nitroso compound · A nitro compound

A 2° amine · An hydroxylamine · A nitroxyl radical

more oxidized

An important aspect of oxidation levels for nitrogen atoms relates to their nucleophilic or electrophilic properties. Whereas an amine has a nucleophilic nitrogen atom, a nitro compound has an electrophilic one. *Oxidation* (or reduction) *therefore provides a way to change the reactivity profile of a nitrogen-containing group.*

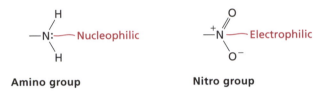

Amino group · Nitro group

For example, nitration of benzene (Section 17.2c) is feasible because NO_2^+ is a good electrophile. On the other hand, ammonia reacts with alkyl halides because it is a good nucleophile (Table 6.3).

11.2 HYDROGENATION REACTIONS

11.2a MOLECULAR HYDROGEN ADDS STEREOSPECIFICALLY TO ALKENES UNDER THE INFLUENCE OF A METAL CATALYST

The first reaction we will look at is catalytic hydrogenation, a transformation to which you were briefly introduced in Section 8.1c in the context of double-bond stabilities. Kinetically, the reaction of hydrogen gas with alkenes is slow. Being highly exothermic and exergonic, however, this reaction is thermodynamically favorable.

Like many kinetically slow processes, the reaction of an alkene with hydrogen gas, H_2, *does* occur at a useful rate when a catalyst is added. Normally, hydrogenation catalysts comprise a finely divided metal such as platinum, palladium, or nickel supported on an inert substance such as carbon or calcium carbonate; for example, we write Pd/C or Pd/CaCO₃ to represent these materials. The actual metal content is generally between 5 and 10%. *Adam's catalyst*, which is $PtO_2 \cdot H_2O$, is a useful catalyst precursor (also called a *precatalyst*) that is reduced to finely divided Pt metal—the actual catalyst—under the conditions of the reaction. For most hydrogenation reactions, any of a number of solvents can be employed; the most common ones are ethanol, acetic acid, methylene chloride, and ethers. The reaction is usually done under 1 atm of hydrogen pressure, but pressures up to 5 atm are not uncommon. At even higher pressures, other functional groups can be reduced, too.

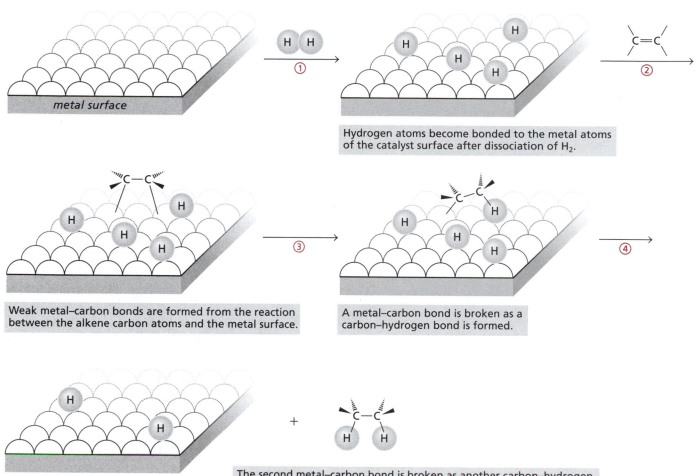

Hydrogen atoms become bonded to the metal atoms of the catalyst surface after dissociation of H_2.

Weak metal–carbon bonds are formed from the reaction between the alkene carbon atoms and the metal surface.

A metal–carbon bond is broken as a carbon–hydrogen bond is formed.

The second metal–carbon bond is broken as another carbon–hydrogen bond is formed; the reduced product dissociates from the surface.

Figure 11.2
Hydrogenation of an alkene by hydrogen atoms adsorbed on a metal surface.

The mechanism of this reaction involves a stepwise addition process. The interaction between molecular hydrogen and the surface of a metal can be pictured as illustrated in Figure 11.2, which starts with metal–hydrogen bond formation as the H–H bond of H_2 is broken (Step 1). Next, the alkene approaches the metal surface, and weak metal–carbon bonds are formed as well (Step 2). A hydrogen atom subsequently migrates to carbon, forming a C–H bond with the rupture of a C–M bond (Step 3). A second reaction of this same type yields the alkane and exposes the metal surface to react with more H_2 and π bonds (Step 4). As you might expect from the geometry with which the alkene binds to the surface, hydrogenation is a syn addition process (addition on the same side of the alkene plane), and it is stereospecific in most cases.

The following equations show examples of hydrogenation for three compounds and illustrate the stereoselectivity and chemoselectivity of the process. All three reactions proceed via syn addition. Depending on the groups attached to the alkene double bond, a meso compound, an achiral compound, and a racemic mixture can be formed as products. With regard to **chemoselectivity,** which is the differentiation between the reactivity patterns of functional groups, notice that except for the carbon–carbon double bond, many functional groups, including the benzene ring, are inert to hydrogenation under the given conditions.

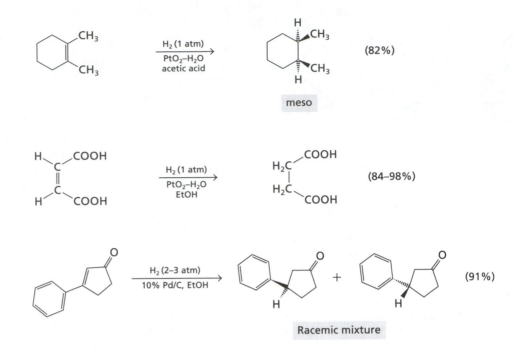

11.2b ALKYNE HYDROGENATION CAN BE STOPPED AT THE ALKENE STAGE

Alkynes are also reduced under typical hydrogenation conditions. With excess hydrogen (which is normally how these reactions are done), the corresponding alkane is formed. To stop the reaction after only 1 equiv of hydrogen has been added is difficult.

$$CH_3CH_2CH_2CH_2—C \equiv C—CH_2CH_2CH_2CH_3 \xrightarrow[\text{5\% Pd/C, EtOH}]{H_2 \text{ (1 atm)}} CH_3(CH_2)_8CH_3 \quad (96\%)$$

5-Decyne   **Decane**

If the Pd catalyst is first treated with certain substances, however, then a **poisoned catalyst** can be formed, which makes it less effective. The **Lindlar catalyst,** which is palladium on calcium carbonate to which quinoline and lead acetate have been added, is a commonly used poisoned catalyst. Its application permits the reduction of a triple-bond to the double-bond stage, but not any further. Thermodynamically, a single equivalent of dihydrogen can be added because reaction with the first π bond is more exothermic than the reaction with the resulting alkene π bond. Again, the stereochemistry of addition is syn, so a cis alkene is the major product.

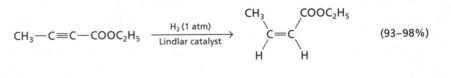

EXERCISE 11.2

What is the expected product from reaction of each of the following compounds with excess molecular hydrogen and 5% Pd/C?

a.   b.   c.

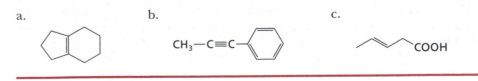

11.2c Nitrogen-Containing Groups Are Reduced by Hydrogenation

The π bonds of alkenes and alkynes are not the only functional groups reduced using hydrogen and a metal catalyst. Several nitrogen-containing groups can also be hydrogenated under similar conditions (1–5 atm of H_2). Cyano, azido, and nitro groups are readily converted to primary amines by reduction.

Cyano and azido groups are incorporated into organic compounds by substitution reactions (see Table 6.2) using primary and secondary alkyl halides or sulfonates (Section 7.1b) as substrates. The Mitsunobu reaction of an alcohol also works well as a way to introduce these groups (Section 7.1d).

Once obtained, a nitrile or an azidoalkane can be converted to a primary amine by hydrogenation.

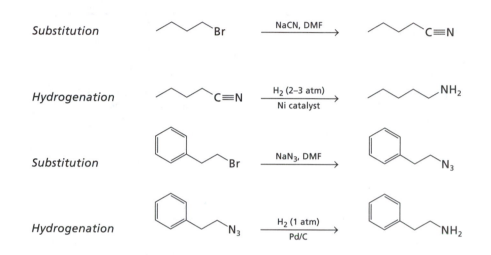

Note that the two-step procedure $RBr \rightarrow RCN \rightarrow RCH_2NH_2$ increases the length of the original chain by one carbon atom. The overall scheme $RBr \rightarrow RN_3 \rightarrow RNH_2$ leads to substitution of an amino group for the leaving group. The value of this procedure, which combines the reactions of substitution and hydrogenation, lies in the capacity to form a primary amine without contamination by secondary or tertiary amines. Formation of a mixture is the usual outcome when one tries to use ammonia or an amine as a nucleophile in the direct substitution reaction of an alkyl halide (Section 6.3d). The following example illustrates the conversion of an alcohol to an amine via the Mitsunobu reaction in the substitution step:

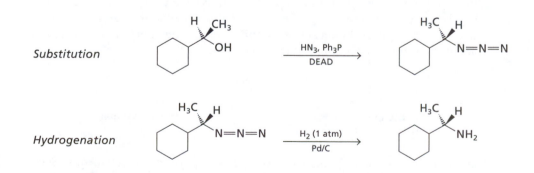

The nitro group is also hydrogenated to form an amino group using catalysts such as Pd, Pt, and Ni.

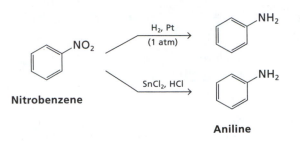

This transformation $RNO_2 \rightarrow RNH_2$ finds much use with aromatic nitro compounds, the preparation of which will be presented in Section 17.2c. Aromatic nitro compounds are readily reduced with metal ions in acid, so the combinations of $SnCl_2$ in hydrochloric acid or Zn in acetic acid are also used to form aniline derivatives.

EXERCISE 11.3

Draw the structure of the major product expected from each of the following reactions:

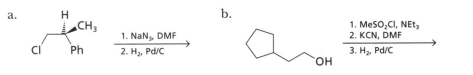

a.

b.

11.3   OXIDATION REACTIONS OF ALKENES

As described in Section 11.2c, reduction of an alkene takes place by addition of hydrogen atoms to the π bond. By analogy, we might expect that oxidation will occur by addition of oxygen or other heteroatoms to a π bond. Oxidizing agents are recognizable because they normally have several oxygen atoms or a metal ion in a high oxidation state. Examples of typical oxidants in organic chemistry include O_2, O_3, RCO_3H, $KMnO_4$, CrO_3, $K_2Cr_2O_7$, OsO_4, and $NaIO_4$. In the next sections, we will examine the reactions of several of these species.

11.3a   MOLECULAR OXYGEN IS NORMALLY INERT TOWARD REACTIONS WITH ORGANIC MOLECULES

Molecular oxygen, O_2, exists as a *triplet* state species (two unpaired electrons) and not as the species with paired electrons, which is what it would be if the octet rule were strictly obeyed. The derivative with paired electrons is called *singlet oxygen*. It is formed by photochemical excitation of O_2, often in the presence of a photosensitizer, a molecule that converts light to chemical energy and transfers the energy to other materials or molecules without undergoing a chemical change itself.

$$:\overset{..}{O}-\overset{..}{O}: \quad \xrightarrow{h\nu \,+\, \text{photosensitizer}} \quad :\overset{..}{O}=\overset{..}{O}:$$

Molecular oxygen   **Singlet oxygen**
(triplet oxygen)

Singlet oxygen reacts with unsaturated species by pericyclic mechanisms (Section 10.4a). For example, a 1,3-diene reacts with singlet dioxygen (a dienophile) to form a peroxide via cycloaddition. (This is not a commonly used transformation; it is included here only to relate the properties of singlet oxygen to the Diels–Alder reaction, with which you are already familiar.)

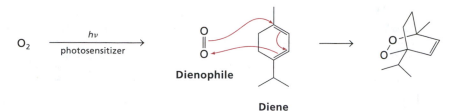

$O_2 \xrightarrow[\text{photosensitizer}]{h\nu}$

Dienophile

Diene

In contrast, the reactions between *triplet* oxygen and organic molecules, which normally have fully paired electrons, are said to be "spin-forbidden", a consequence of the rules of quantum mechanics related to the energies and interactions of electron spin states. This fortunate peculiarity of molecular structure permits life as we know it to exist—if the O_2 in the atmosphere existed solely in the singlet state, organic matter on earth would have been oxidized long ago. Reactions do occur between organic compounds and triplet oxygen, and we will consider some in Chapter 12.

11.3b OZONOLYSIS OF AN ALKENE CLEAVES THE CARBON–CARBON π BOND TO FORM CARBONYL-CONTAINING COMPOUNDS

Singlet oxygen is not the only oxygen-containing species that can react in cycloaddition reactions. Ozone reacts with alkenes through a [4 + 2]-cycloaddition process (Section 10.4b). Ozone is an example of a dipolar molecule—that is, its *principal* resonance forms display two charges, plus and minus.

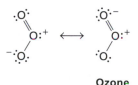

Ozone

The cycloaddition reaction between an alkene and ozone occurs within a six-electron transition state between five *atoms*. Notice that the four-electron reactant (ozone) uses two electrons from a π bond and two electrons from an unshared pair, rather than having four π electrons as in the Diels–Alder reaction (Section 10.4b).

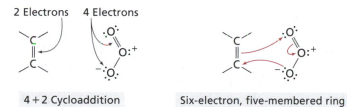

4 + 2 Cycloaddition　　　**Six-electron, five-membered ring transition state**

The first step in the reaction between an alkene and ozone produces a molozonide, which rearranges to an **ozonide.** *Ozonides* are more stable than molozonides, but they do decompose upon standing.

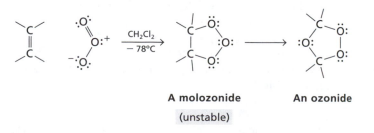

A molozonide　　　**An ozonide**

(unstable)

Ozonides can be treated with additional reagents (worked-up) to form carbonyl compounds. The overall reaction of an alkene with ozone results in the rupture of the

Table 11.1 Products formed by ozonolysis of alkenes.

Alkene	Products from oxidative workup[a]		Products from reductive workup[b]	
$\underset{H}{\overset{R}{>}}C=C\underset{H}{\overset{H}{<}}$	$\underset{HO}{\overset{R}{>}}C=O$	$O=C\underset{OH}{\overset{H}{<}}$	$\underset{H}{\overset{R}{>}}C=O$	$O=C\underset{H}{\overset{H}{<}}$
$\underset{H}{\overset{R}{>}}C=C\underset{H}{\overset{R'}{<}}$	$\underset{HO}{\overset{R}{>}}C=O$	$O=C\underset{OH}{\overset{R'}{<}}$	$\underset{H}{\overset{R}{>}}C=O$	$O=C\underset{H}{\overset{R'}{<}}$
$\underset{R}{\overset{R}{>}}C=C\underset{H}{\overset{H}{<}}$	$\underset{R}{\overset{R}{>}}C=O$	$O=C\underset{OH}{\overset{H}{<}}$	$\underset{R}{\overset{R}{>}}C=O$	$O=C\underset{H}{\overset{H}{<}}$
$\underset{R}{\overset{R}{>}}C=C\underset{R}{\overset{R'}{<}}$	$\underset{R}{\overset{R}{>}}C=O$	$O=C\underset{OH}{\overset{R'}{<}}$	$\underset{R}{\overset{R}{>}}C=O$	$O=C\underset{H}{\overset{R'}{<}}$
$\underset{R}{\overset{R}{>}}C=C\underset{R'}{\overset{R'}{<}}$	$\underset{R}{\overset{R}{>}}C=O$	$O=C\underset{R'}{\overset{R'}{<}}$	$\underset{R}{\overset{R}{>}}C=O$	$O=C\underset{R'}{\overset{R'}{<}}$

[a]Hydrogen peroxide = H_2O_2.
[b]Zn/H_2O or $(CH_3)_2S$.

carbon–carbon π bond with formation of new π bonds between carbon and oxygen atoms. The reagents used in the workup determine which products are formed. Treating the ozonide with an oxidizing agent such as hydrogen peroxide, H_2O_2, gives ketones or carboxylic acids—the former if the double-bond carbon atom is disubstituted, and the latter if this carbon atom is monosubstituted. Reductive workup, which uses zinc and water or dimethyl sulfide, produces aldehydes or ketones, again depending on whether the alkene has one or two substituents on the alkene carbon atom, respectively.

Terminal alkenes yield formic acid, HCOOH, as one of the products upon oxidative workup; and formaldehyde, HCHO, upon reductive workup. Table 11.1 summarizes the reactions of ozone with alkenes that have different substitution arrangements. Before the development of spectroscopic methods, ozone was commonly used as an analytical reagent to determine the location of double bonds within molecules. Ozonolysis is now used primarily as a means for cleaving carbon–carbon bonds on a preparative scale.

The example in the following equation illustrates reductive workup using methyl sulfide, $(CH_3)_2S$, which is converted to dimethyl sulfoxide, $(CH_3)_2S=O$, as a byproduct. Notice that the trisubstituted alkene is cleaved to produce an aldehyde and a ketone, as predicted from the data in Table 11.1.

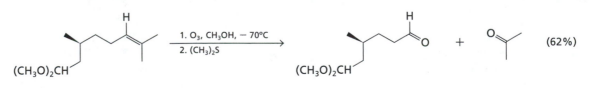

EXERCISE 11.4

What are the expected products from each of the following reactions with: (1) oxidative workup, and (2) reductive workup?

a.

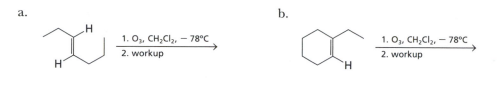

b.

11.3c OSMIUM TETROXIDE REACTS WITH ALKENES
       TO FORM METALATE ESTERS

The concerted [4 + 2] cycloaddition mechanism used to rationalize the reactions between alkenes and ozone also applies to the oxidation reactions of alkenes with certain metal oxides such as osmium tetroxide, OsO_4.

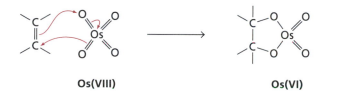

The five-membered ring osmate ester intermediate is a stable species, so a separate step is required to cleave the osmium–oxygen bonds to liberate the organic product, which is a vicinal cis-diol. The expense of osmium tetroxide (osmium is the rarest nonradioactive transition metal) led researchers to find processes that use only catalytic rather than stoichiometric amounts of Os. *tert*-Butyl hydroperoxide and *N*-methylmorpholine-*N*-oxide are reagents that reoxidize Os(VI) to Os(VIII) and liberate the diol. These reagents transfer an oxygen atom to the metal ion, yielding *tert*-butyl alcohol and *N*-methylmorpholine, respectively.

In the following examples, only a small amount of osmium tetroxide, perhaps 0.05 equiv, is added to 1 equiv of alkene and excess oxidant. This dihydroxylation reaction is *stereospecific* under these catalytic conditions; the stereochemistry of addition is syn. The second and third equations in the following scheme highlight the stereospecificity of this overall process by showing the different stereochemical results obtained when starting with the (*E*) and (*Z*) isomers of 4-octene.

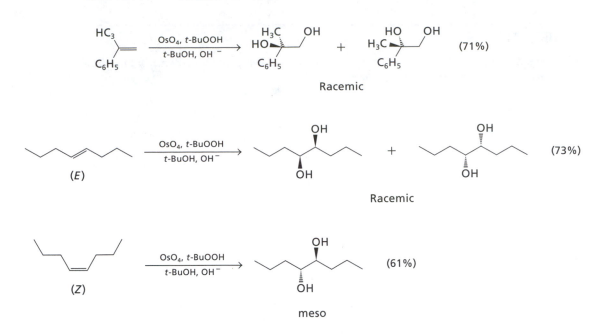

[To complement the syn-dihydroxylation of alkenes using osmium tetroxide, an anti-dihydroxylation process can be carried out by epoxidation followed by ring opening with water; this transformation will be discussed in Section 11.3e].

EXERCISE 11.5

Draw the structures of the major products, including stereochemistry, that are expected from each of the following reactions:

a.   b.

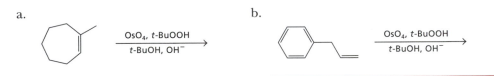

11.3d PERIODATE ION CLEAVES VICINAL DIOLS TO FORM ALDEHYDES AND KETONES

Another oxidant used to regenerate osmium tetroxide after its reaction with an alkene is periodate ion, IO_4^-. With this oxidant, however, the diol is not isolated. Instead, the carbon–carbon bond of the alkene is cleaved to generate the same types of products that are produced by ozonolysis followed by reductive workup. Thus, the products of this transformation, called the **Lemeiux–Johnson cleavage,** are aldehydes or ketones. The periodate ion oxidizes the osmium byproduct to regenerate OsO_4, which is therefore a catalyst in this procedure.

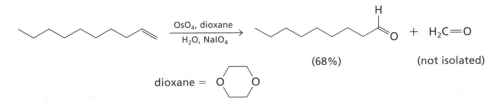

By itself, periodate ion, as well as its conjugate acid HIO_4, cleaves vicinal diols to produce carbonyl compounds. In conjunction with osmium tetroxide, periodate ion cleaves an alkene to the same aldehydes (or ketones, depending on substitution of the alkene), as shown in the following pair of equations.

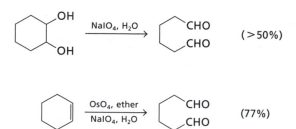

In the reaction between HIO_4 and a diol, a cyclic intermediate is formed first. Electron transfer leads to carbon–carbon bond cleavage with concomitant reduction of the iodine atom.

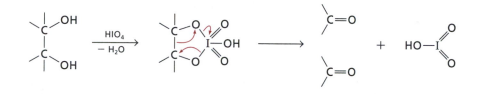

EXAMPLE 11.1

What are the products obtained from the reaction of (*E*)-3-methyl-2-hexene with: (1) ozone and oxidative workup; and (2) OsO_4 and $NaIO_4$.

An alkene is cleaved by either of the given reagents to form carbonyl-containing products, so we can simply replace the carbon–carbon double bond with two carbon–oxygen double bonds. An alkene carbon atom *without* an attached hydrogen atom yields a ketone from that half of the alkene, no matter what conditions are used. An alkene carbon atom with an attached hydrogen atom can produce either an aldehyde *or* carboxylic acid, depending on the reaction conditions.

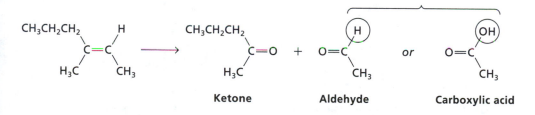

Oxidative workup after ozonolysis of an alkene yields carboxylic acid and ketone products (Table 11.1). The combination of osmium tetroxide and sodium periodate yields aldehyde and ketone products. We write the equations as follows. The left side of the alkene (as drawn) gives a ketone in each case. The right side yields either an aldehyde or a carboxylic acid.

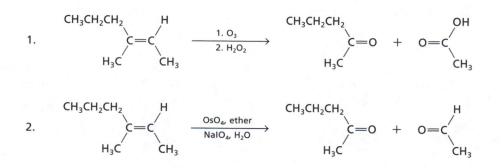

EXERCISE 11.6

What are the products obtained from each of the following substrates with: (1) ozone and oxidative workup; (2) ozone and reductive workup; (3) osmium tetroxide and periodate ion. Which reagent combinations generate the same product mixture?

a.   b.   c.

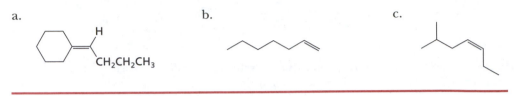

11.3e EPOXIDES ARE FORMED BY REACTIONS BETWEEN PERACIDS AND ALKENES

The last type of alkene oxidation process we will look at is epoxidation, which is conveniently done using a peroxycarboxylic acid, also called a peracid, RCO_3H. The reactive peroxy group transfers its "extra" oxygen atom directly to the π bond of alkenes under certain conditions. Peracetic acid, CH_3CO_3H, is the simplest stable peracid. The one used most commonly for epoxidation reactions for many years was 3-chloroperoxybenzoic acid, or MCPBA, but it appears to be shock sensitive and may explode. More recently, magnesium monoperoxyphthalate (MMPP) was introduced as a safer substitute less prone to thermal decomposition. Many other peracids are less stable than those mentioned here, but most can be used to make epoxides from alkenes.

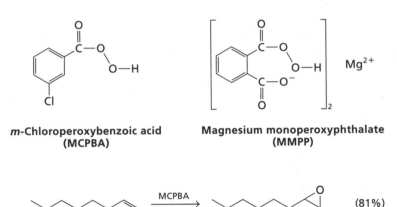

m-Chloroperoxybenzoic acid (MCPBA)   **Magnesium monoperoxyphthalate (MMPP)**

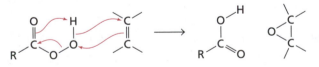

The transfer of the oxygen atom from a peracid to an alkene π bond is considered to be a concerted process, as represented by the following equation. The actual mechanistic details are not agreed upon; nevertheless, using the formalism illustrated here allows us to predict the correct product for this type of reaction, including the stereochemistry of the epoxide ring.

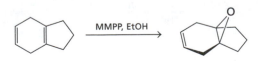

Because the alkene must present a single face of the π bond to the peracid, epoxidation is stereospecific, resulting in syn addition of the oxygen atom. Hence, the epoxide is formed with retention of the alkene stereochemistry.

Note in this example that only the tetrasubstituted double bond is epoxidized when 1 equiv of the peracid is used. In general, alkyl substitution on the carbon atoms at the ends of a double bond raises the electron density in the π system because of electron donation by the alkyl substituents (Section 6.2b), and this effect makes the π bond a better nucleophile. The oxygen atom that is transferred from a peracid is considered to be electrophilic; therefore, *more highly substituted alkenes react faster with peracids.*

Direct epoxidation works well in many cases, but an epoxide can also be formed by nucleophilic substitution. Recall that an alkene adds bromine in the presence of water to yield a bromohydrin (Section 9.1e). Treatment of a bromohydrin with base leads to deprotonation of the alcohol OH group, followed by an intramolecular substitution reaction that generates the epoxide (Section 7.2c). Because trans addition occurs in the first step to yield the bromohydrin, and because substitution proceeds by the backside replacement of the leaving group from the carbon atom, the overall process generates an epoxide with the same stereochemistry as one obtained by direct epoxidation.

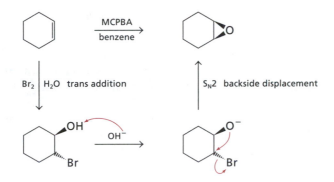

Recall from Section 7.2e that epoxides undergo ring opening when they react with nucleophiles. When the nucleophile is water, then the product is a vicinal diol. The product stereochemistry from this transformation is opposite that obtained from the osmium tetroxide dihydroxylation reaction, so these two processes provide complementary stereochemical control.

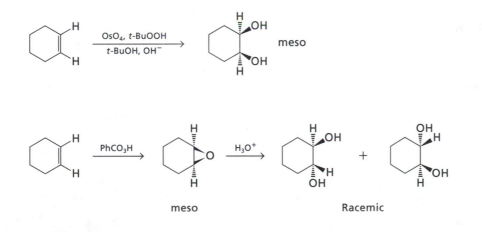

EXERCISE 11.7

What is the expected monoepoxidation product, including stereochemistry, of the following reaction.

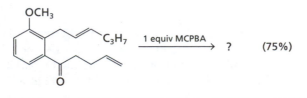

11.4   OXIDATION REACTIONS OF ALCOHOLS

11.4a OXIDATION OF A 1° OR 2° ALCOHOL OCCURS BY ELIMINATION OF TWO HYDROGEN ATOMS TO GENERATE A CARBONYL GROUP

An alcohol has an oxidation level that lies between that of an alkane and that of an aldehyde or ketone. Removing two hydrogen atoms from a primary alcohol generates an aldehyde; removing two hydrogen atoms from a secondary alcohol produces a ketone.

A 1° alcohol   **An aldehyde**   **A 2° alcohol**   **A ketone**

The fact that atoms are removed and a π bond is formed during oxidation of an alcohol suggests that elimination pathways may be suitable to carry out this type of transformation. If you consider how elimination reactions are accomplished—say, an E2 reaction—you will recall from Section 8.2 that a base and a good leaving group are required. If it were possible to replace the proton of the alcohol OH group with a good leaving group, X, then elimination of H^+ and X^- would generate a C=O double bond. In fact, several oxidizing agents work according to this principle, as you will see in the next sections.

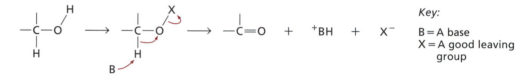

Key:
B = A base
X = A good leaving group

11.4b THE SWERN OXIDATION IS SIMILAR TO THE E2 REACTION

The **Swern oxidation** procedure is a common method for the oxidation of an alcohol. Dimethyl sulfoxide is a polar, aprotic solvent normally used to carry out substitution and elimination reactions (Section 6.3b) because of its ability to dissolve both inorganic salts and organic molecules. When combined with an "activating reagent", DMSO reacts instead as an oxidant, converting alcohols to aldehydes or ketones. The activating reagent used for the Swern oxidation is oxalyl chloride.

A reaction initially occurs between the two reagents to generate an activated DMSO complex according to the following equation. In the first step, which is conducted below –60°C, the oxygen atom of DMSO replaces one of the chlorine atoms in the oxalyl chloride molecule. The mechanism for this transformation will be discussed fully in Section 21.3a.

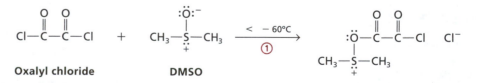

Oxalyl chloride   **DMSO**

In Step 2, the chloride ion reacts with the DMSO derived intermediate. This step may appear complicated at first, but chloride ion simply acts as a nucleophile toward the positively charged (electrophilic) sulfur atom, replacing the O-CO-CO-Cl group, which decomposes to form carbon dioxide, carbon monoxide, and chloride ion.

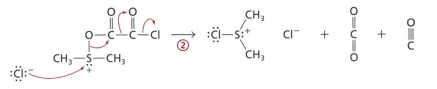

In the next step, the alcohol oxygen atom reacts as a nucleophile toward the electrophilic sulfur atom, displacing chloride ion by an S_N2 type reaction; the chloride ion removes the proton from oxygen in an acid–base reaction. The organic product is an alkoxysulfonium salt.

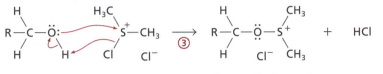

An alkoxysulfonium salt

In the second stage of the Swern oxidation, triethylamine is added to the reaction mixture. As a base, this amine removes a proton from one of the methyl groups bonded to the sulfur atom. The positive charge on sulfur facilitates this acid–base reaction, forming a species called an **ylide,** a species with a carbanion center bonded directly to a positively charged heteroatom. You will encounter ylides again in Section 20.4.

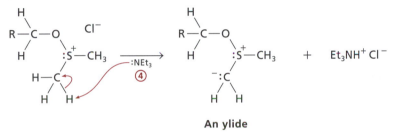

An ylide

This ylide carbanion is the base that removes the proton from the carbon atom attached to oxygen. As the positively charged sulfur leaving group departs, the C=O double bond is formed.

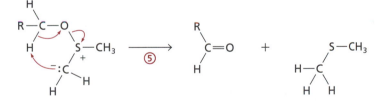

A primary difference between the Swern oxidation and an E2 reaction is the reaction stereochemistry: The Swern oxidation is a syn elimination process, whereas the E2 reaction is an anti-elimination process (Section 8.2a). Because the π bond of aldehydes and ketones cannot be classified as (E) or (Z), the stereochemical course of this oxidation reaction is not important.

EXERCISE 11.8

What is the product of the following reaction? Draw the structure of the alkoxysulfonium intermediate and show the movement of electrons that occur in the elimination step.

$$\text{cyclohexanol} \xrightarrow[\text{2. NEt}_3]{\text{1. DMSO, ClCOCOCl, CH}_2\text{Cl}_2, -78°C}$$

11.4c HIGH-VALENT METAL OXIDES ARE COMMONLY USED TO OXIDIZE ALCOHOLS TO CARBONYL-CONTAINING MOLECULES

High-valent metal oxides are routinely employed for the conversion of an alcohol to an aldehyde or ketone, a process that proceeds by a mechanism similar to that of the Swern reaction. A metal oxide can produce either an aldehyde or a carboxylic acid from a primary alcohol, depending on the specific reagent and the conditions employed. If water is present, oxidation to the carboxylic acid is facile. In the following equations, the letter "O" in brackets is a shorthand notation for oxidation; it is used when the reagent is not specified.

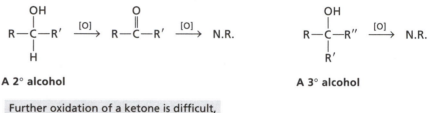

A 1° alcohol

> Depending on the reagent and conditions used, the oxidation of a primary alcohol is sometimes difficult to stop at the aldehyde stage.

A secondary alcohol is oxidized only to a ketone, which is difficult to oxidize further, as that would require breaking a carbon–carbon bond. A tertiary alcohol has no hydrogen atom attached to the carbon atom bearing the OH group, so it does not undergo these normal oxidation processes.

A 2° alcohol   **A 3° alcohol**

> Further oxidation of a ketone is difficult, so chemoselectivity is not an issue in the oxidation of 2° alcohols.

Chromium oxide reagents are commonly used to oxidize alcohols. Chromium (VI) oxide is the starting material for many reagents, and two of its derivatives find wide application. One, called **Collins's reagent,** is prepared by the addition of chromium trioxide to pyridine. The other is **Corey's reagent,** also called pyridinium chlorochromate (PCC), and is made from CrO_3, hydrochloric acid, and pyridine.

$$CrO_3 \left(N \bigcirc \right)_2 \qquad\qquad \overset{H}{\underset{}{N^+}} \bigcirc \ CrO_3Cl^-$$

Chromium oxide-pyridine   **PCC**

(Collins's reagent)   (Corey's reagent)

Both of these reagents react with primary alcohols in dichloromethane, the solvent, to produce the corresponding aldehydes in high yield. A secondary alcohol is converted to a ketone with either reagent.

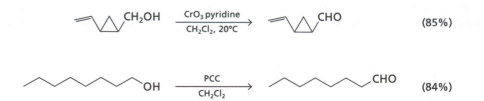

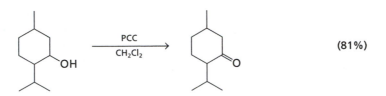

The mechanisms of these reactions are complex and comprise several steps that involve one- and two-electron changes. The first steps generate an alkoxychromium(VI) complex, as illustrated for chromium(VI) oxide in the following equations:

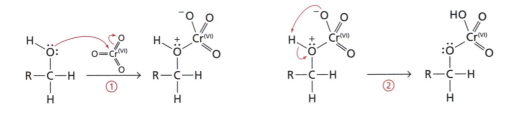

Oxidation proceeds with the removal of a proton by one of the chromium-oxo groups, which reduces the chromium(VI) ion to chromium(IV). Like the Swern oxidation, this process involves syn-elimination.

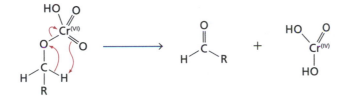

Chromium oxide can be used to make other oxidizing reagents, too. Chromic acid is made by dissolving CrO_3 in water and acid: The combination of chromic acid and sulfuric acid is called the **Jones reagent.** Acetic acid can be used in place of H_2SO_4. Unlike Collins and Corey's reagents, chromic acid is able to oxidize aldehydes as well as primary alcohols unless the reaction conditions are controlled very carefully. Therefore, chromic acid is used mainly to oxidize primary alcohols to carboxylic acids and secondary alcohols to ketones. A reagent that works better in basic solution is potassium permanganate, $KMnO_4$.

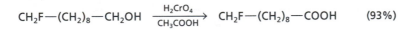

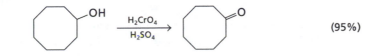

A more selective reagent than those mentioned already is manganese(IV) oxide, MnO_2, which only reacts with benzylic and allylic alcohols to yield aldehydes and ketones. Saturated 1° and 2° alcohols are inert toward this reagent, as shown in the second equation that follows.

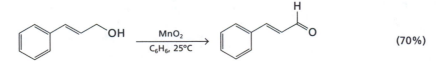

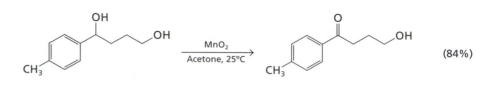

(84%)

EXAMPLE 11.2

Which metal oxide reagents can be used to carry out the following transformation?

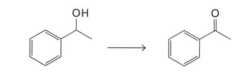

First, identify the type of alcohol represented by the starting material. This compound is a secondary, benzylic alcohol. Benzylic alcohols can be oxidized by MnO_2, so that is one reagent that can be used. Every chromium oxide reagent mentioned above will oxidize a 2° alcohol to a ketone. Overoxidation is not the problem it would be for a 1° alcohol, which can be converted to an aldehyde or carboxylic acid, depending on the reaction conditions. Therefore, any of the following can be used to perform this oxidation reaction: CrO_3–pyridine, PCC, H_2CrO_4, $KMnO_4$, or MnO_2.

EXERCISE 11.9

Which metal oxide reagents can be used to carry out each of the following transformations?

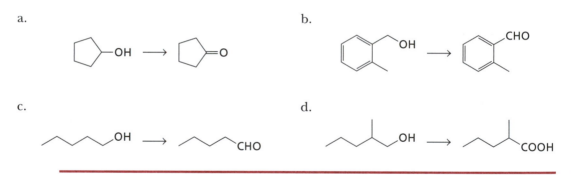

a.

b.

c.

d.

11.4d HIGH-VALENT IODINE COMPOUNDS OXIDIZE ALCOHOLS TO CARBONYL COMPOUNDS

Literally hundreds of reagents are available to oxidize alcohols to aldehydes, ketones, or carboxylic acids. The most commonly used ones are those mentioned in the previous sections. A class of reagents gaining popularity comprises high-valent organoiodine compounds. For complex substrates, these species are sometimes the only ones able to oxidize an alcohol functional group with chemoselectivity, that is, without affecting other functional groups that are present. The structures of two common oxidants are shown below: Iodoxybenzoic acid (IBX) and the Dess–Martin Periodinane (DMP), named for the two chemists who first reported its practical use as a selective oxidant. Like the related periodate ion (Section 11.3d), these reagents have a high-valent iodine atom bonded to several oxygen atoms.

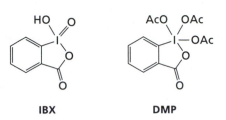

IBX   **DMP**

Both of these reagents convert primary alcohols to aldehydes and secondary alcohols to ketones. The primary difference is the solvent employed for each procedure; IBX is insoluble in halogenated solvents, hydrocarbons, and ethers, but it dissolves readily in DMSO. The DMP oxidations are conducted in methylene chloride.

The reaction mechanism follows the same course that you have seen in previous sections: The alcohol binds via its oxygen atom to the high-valent atom—iodine in these reagents—and an oxo group removes a proton with the concomitant movement of electrons to form the π bond of the carbonyl group.

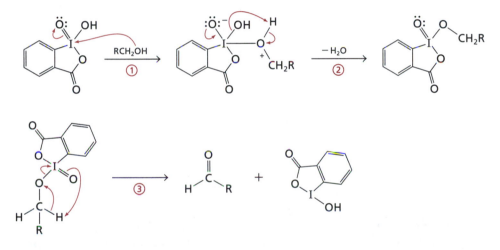

11.4e   IN BIOLOGICAL SYSTEMS, OXIDATION OF AN ALCOHOL OCCURS BY THE FORMAL DISPLACEMENT OF A HYDRIDE ION

Because alcohols play a significant role as metabolic and biosynthetic intermediates, it is not surprising to find that alcohol oxidation processes are common in biochemistry.

Before examining that transformation in detail, let us consider an alternate mechanism for alcohol oxidation, one in which a leaving group is attached to the carbon atom. Removal of the proton from the OH group along with displacement of the leaving group creates a carbon–oxygen double bond. In biochemical systems, alcohol oxidation occurs in just this manner. Astonishingly, *the leaving group is a hydride ion.* How does such a strong base function as a leaving group?

The key to this transformation is a molecule called nicotinamide adenine dinucleotide, NAD⁺, which functions as a **coenzyme** in the biochemical oxidation reactions of alcohols. A coenzyme is an organic molecule required by an enzyme to catalyze its chemical reaction. The enzyme active site serves to orient both the substrate molecule and the coenzyme in proximity so that the needed transformation can occur. *In this instance, the enzyme provides the needed basic site. No reaction occurs in the absence of either the enzyme or the coenzyme.*

As shown below, NAD⁺ contains an alkylated pyridine derivative, a compound that looks like benzene except that one CH fragment has been replaced by a nitrogen atom.

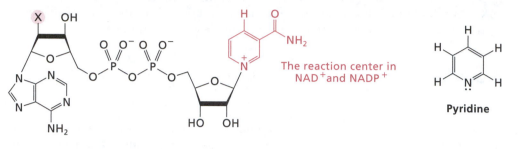

| X = OH | **Nicotinamide adenine dinucleotide (NAD⁺)** |
| X = OPO₃²⁻ | **Nicotinamide adenine dinucleotide phosphate (NADP⁺)** |

In the alkylated pyridine ring that constitutes the functional portion of NAD⁺, the nitrogen atom carries a positive charge. A related species is NADP⁺, which has an additional phosphate group. The NADP⁺ normally functions as a coenzyme in biosynthetic pathways, and NAD⁺ is utilized in degradative processes.

The alkylated pyridine ring in both of these coenzymes can act as a hydride ion acceptor, with two electrons flowing to "neutralize" the charge on the nitrogen atom. Notice that the transfer of a hydride ion is equivalent to the transfer of a proton, H⁺, and two electrons.

$$H^- = H^+ + 2\ e^-$$

The alkylated pyridinium ion in NAD⁺ is a two-electron acceptor, and it is reduced to an *N*-alkyl-4,4-dihydropyridine derivative as the hydrogen atom attached to the α-carbon atom of the alcohol is transferred to the 4-position of the ring, as illustrated in Figure 11.3. The proton attached to the alcohol OH group is removed by a basic site in the enzyme, most probably the imidazole group of a histidine residue.

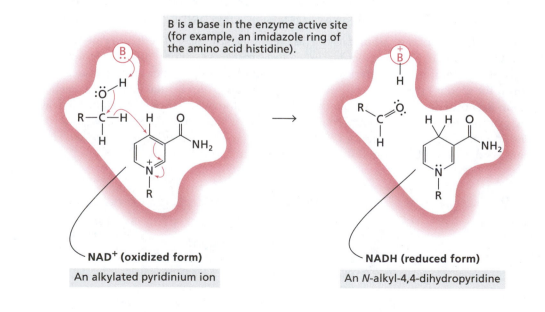

Figure 11.3

The oxidation of alcohols in biological systems by NAD⁺ within the active site of an alcohol dehydrogenase enzyme.

The process illustrated in Figure 11.3 occurs as one step in several metabolic processes, and it is reversible, as you will see later. Specific examples, which are shown in Fig. 11.4, include the oxidation of isocitrate to α-ketoglutarate via its carboxy derivative, and the oxidation of malate to oxaloacetate. These reactions constitute two steps in the **citric acid cycle** (also called the *Krebs cycle*), a pathway by which an acetyl group is converted to carbon dioxide and energy. Another important alcohol oxidation process in biochemistry yields β-ketothioesters from β-hydroxythioesters, which is a key reaction in fatty acid metabolism.

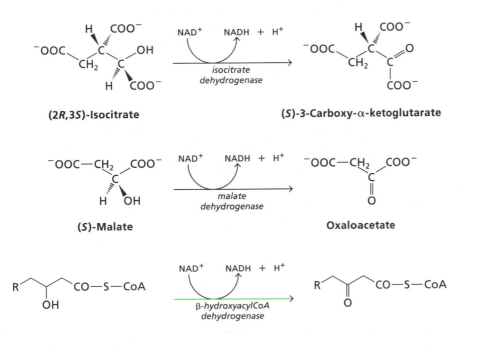

(2R,3S)-Isocitrate   **(S)-3-Carboxy-α-ketoglutarate**

(S)-Malate   **Oxaloacetate**

Figure 11.4
Some metabolic reactions in which NAD$^+$ is a coenzyme for oxidizing alcohols to ketones.

11.5   OXIDATION REACTIONS OF AMINES

Nitrogen, like carbon, has different oxidation levels (Section 11.1b). Their oxidation reactions differ, however, because a nitrogen atom often has an unshared pair of electrons, and it can readily lose an electron to form a reactive radical cation. This property of nitrogen makes oxidation at adjacent carbon atoms more difficult.

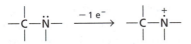   Removal of an electron from the nitrogen atom creates a radical cation.

A particularly useful reagent for the oxidation of amines is dimethyldioxirane, which is prepared by the reaction between acetone and the hydrogen persulfate ion, HSO$_5^-$.

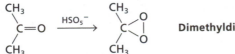   **Dimethyldioxirane**

This reagent converts *all* types of primary amines, including aryl amines, to nitro compounds.

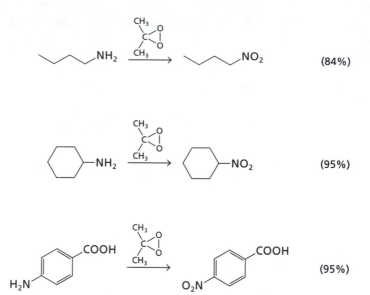

(84%)

(95%)

(95%)

By contrast, a secondary amine is oxidized cleanly by this reagent to the corresponding hydroxylamine.

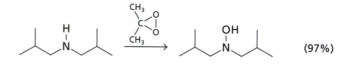

(97%)

Aniline derivatives are not always oxidized to nitro compounds, and this variability demonstrates that multiple and sometimes complex pathways exist for the oxidation reactions of nitrogen atoms. Manganese dioxide converts anilines to azo compounds, an oxidation process that provides an alternate synthetic route to dyestuffs (see Section 17.4d).

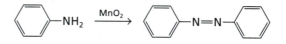

A tertiary amine undergoes a fundamentally different reaction because there are no N–H bonds. With a strong oxidant such as a peracid (Section 11.3c) or hydrogen peroxide, a tertiary amine is converted to an *N*-oxide. This process appears to involve direct transfer of an electron-deficient oxygen atom to the nucleophilic nitrogen atom, analogous with the oxygenation of a double bond to form an epoxide.

$$R'-\underset{\underset{R''}{|}}{\overset{\overset{R}{|}}{N}}: \xrightarrow[\text{H}_2\text{O}_2]{\text{RCO}_3\text{H or}} R'-\underset{\underset{R''}{|}}{\overset{\overset{R}{|}}{\overset{+}{N}}}-\ddot{\underset{..}{O}}:^{-}$$

Section 11.1   Oxidation states in organic molecules

- Oxidation is the removal of hydrogen atoms from or the incorporation of heteroatoms into a molecule.
- Reduction is the removal of heteroatoms from or the incorporation of hydrogen atoms into a molecule.
- The oxidation level of a carbon atom is assigned on the basis of the number of attached heteroatoms or π bonds that are present.
- Nitrogen atoms also have different oxidation levels, which are based on the number of attached oxygen atoms or multiple bonds.

Section 11.2   Hydrogenation reactions

- An alkene undergoes hydrogenation—addition of molecular hydrogen to the π bond—in the presence of a metal catalyst such as Pd, Pt, or Ni. The stereochemistry of addition is syn.
- An alkyne undergoes hydrogenation to form an alkane with the catalysts used for alkene hydrogenation. If a poisoned catalyst is used, the process stops at the cis-alkene.
- Azido (N_3), cyano (CN), and nitro (NO_2) groups are reduced to amino groups (NH_2) using hydrogen and a catalyst such as Pt, Pd, or Ni. Nitro groups are also reduced with $SnCl_2$ in hydrochloric acid or Zn in acetic acid.

Section 11.3   Oxidation reactions of alkenes

- Ozonolysis is a cycloaddition reaction that occurs between ozone, O_3, and an alkene. Depending on the conditions of the workup step, aldehydes, ketones, or carboxylic acids can be formed. The types of product that are formed are summarized in Table 11.1.
- Osmium tetroxide (OsO_4) reacts with an alkene to form a cyclic intermediate in which two of the oxygen atoms have added to the π bond. The stereochemistry of addition is syn.
- A vicinal diol is the product of the reaction of an alkene with OsO_4 in the presence of a cooxidant such as *tert*-butyl hydroperoxide or *N*-methylmorpholine-*N*-oxide. The overall transformation is called dihydroxylation.
- An alkene is cleaved by the reagent combination of OsO_4 and $NaIO_4$; the products are ketones and/or aldehydes.
- Epoxidation of a carbon–carbon double bond takes place when an alkene is treated with a peracid. The more highly substituted the double bond, the faster it reacts to form the epoxide derivative.

Section 11.4   Oxidation reactions of alcohols

- The Swern oxidation is a two-step procedure that exploits dimethyl sulfide as a leaving group in converting an alcohol to the corresponding carbonyl compound. An ylide intermediate acts as a base to remove a proton from the carbon atom that bears the leaving group.
- High-valent chromium oxides are the most commonly used reagents for the conversion of alcohol to carbonyl compounds.
- Oxidation of a primary alcohol with chromium oxide reagents under nonaqueous conditions produces an aldehyde. When water is present, oxidation continues to generate a carboxylic acid as the product.
- Chromium oxide reagents convert secondary alcohols to ketones.

- A tertiary alcohol is not oxidized by metal oxides under normal conditions.
- Manganese(IV) oxide converts only allylic and benzylic alcohols to their corresponding carbonyl derivatives.
- High-valent iodine compounds oxidize alcohols to carbonyl compounds.
- For oxidation of an alcohol to a carbonyl compound in biochemical systems, an enzyme uses NAD$^+$, a coenzyme, to accept a hydride ion from the carbon atom bearing the OH group.

Section 11.5   Oxidation reactions of amines

- Dimethyldioxirane is prepared from acetone and the HSO$_5^-$ ion, and it converts a primary amine to a nitro compound.
- Dimethyldioxirane converts a secondary amine to a hydroxylamine.
- Tertiary amines are oxidized to form amine *N*-oxides.

KEY TERMS

Section 11.1a
oxidation level

Section 11.2a
chemoselectivity

Section 11.2b
poisoned catalyst
Lindlar catalyst

Section 11.3b
ozonide

Section 11.3d
Lemeiux–Johnson
   cleavage

Section 11.4b
Swern oxidation
ylide

Section 11.4c
Collin's reagent
Corey's reagent
Jones reagent

Section 11.4e
coenzyme
citric acid cycle

REACTION SUMMARY

Section 11.2a

Hydrogenation of alkenes using a metal catalyst (Pd, Pt, or Ni). The reaction takes place at atmospheric pressure of H$_2$; the stereochemistry of addition is syn.

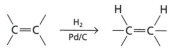

Section 11.2b

Hydrogenation of alkynes. Use of Pd/C leads to complete reduction and formation of an alkane; use of a poisoned catalyst stops the reaction at the cis-alkene stage.

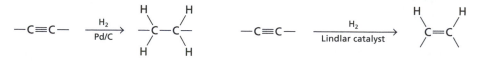

Section 11.2c

Hydrogenation of azido, cyano, and nitro groups. Each of these groups undergoes reduction to form a primary amine. Nitrobenzene derivatives react with either Zn in acetic acid or SnCl$_2$ in aqueous HCl to form aniline derivatives.

$$R-C\equiv N \xrightarrow{\text{H}_2,\text{ Ni}} R-CH_2NH_2$$

$$R-N=N=N \xrightarrow{\text{H}_2,\text{ Pd/C}} R-NH_2$$

$$R-NO_2 \xrightarrow{\text{H}_2,\text{ Pd/C}} R-NH_2 \qquad Ar-NO_2 \xrightarrow[\substack{\text{or}\\\text{Zn, HOAc}}]{\text{SnCl}_2,\text{ HCl}} Ar-NH_2$$

Section 11.3b

Ozonolysis of alkenes. The substitution pattern of the alkene, as well as workup conditions, dictate the structures of the products that are formed, which are summarized in Table 11.1.

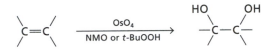

Section 11.3c

Dihydroxylation of alkenes with OsO_4. The stereochemistry of the addition is syn. A cooxidant such as *N*-methylmorpholine-*N*-oxide (NMO) or *tert*-butyl hydroperoxide is employed to reoxidize the osmium product.

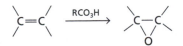

Section 11.3d

The combination of IO_4^- and OsO_4 cleaves alkenes in the same manner as ozonolysis followed by reductive workup. Vicinal diols can be cleaved using IO_4^- or HIO_4 to form the same products.

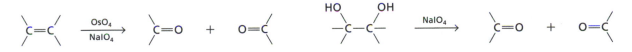

Section 11.3e

A peracid converts an alkene to the corresponding epoxide. The stereochemistry of addition is syn.

Section 11.4b

The Swern oxidation. This two-step procedure converts a primary alcohol to an aldehyde and a secondary alcohol to a ketone.

$$R-\overset{\overset{\displaystyle R'}{|}}{\underset{\underset{\displaystyle H}{|}}{C}}-OH \xrightarrow[\text{2. Et}_3\text{N}]{\text{1. ClCOCOCl, DMSO, CH}_2\text{Cl}_2,\, -60°\text{C}} \overset{\displaystyle R'}{\underset{\displaystyle R}{\diagup}}C=O$$

Section 11.4c

The oxidation of alcohols using metal oxides. Under anhydrous conditions, a primary alcohol is converted to an aldehyde. If water is present, a primary alcohol is oxidized to form a carboxylic acid. A secondary alcohol reacts to form a ketone under all conditions. A tertiary alcohol is inert toward these reagents. Manganese (IV) oxide oxidizes only benzylic and allylic alcohols to the corresponding carbonyl derivatives.

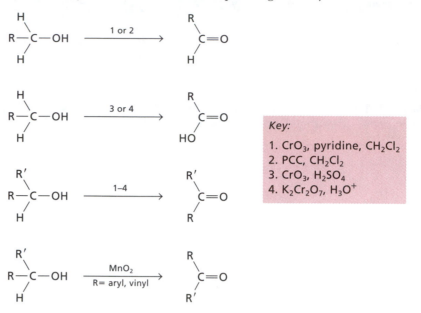

Key:
1. CrO_3, pyridine, CH_2Cl_2
2. PCC, CH_2Cl_2
3. CrO_3, H_2SO_4
4. $K_2Cr_2O_7$, H_3O^+

Section 11.4d

The Dess-Martin reaction. High-valent iodine compounds oxidize primary alcohols to aldehydes and secondary alcohols to ketones.

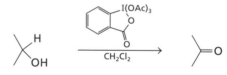

Section 11.4e

In biochemical systems, alcohols are oxidized by an enzyme plus the coenzyme NAD^+.

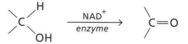

Section 11.5

Dimethyldioxirane oxidizes primary amines to nitro compounds, secondary amines to hydroxylamine derivatives, and tertiary amines to *N*-oxide derivatives. Peracids and hydrogen peroxide can also be used to oxidize tertiary amines.

$$R-NH_2 \xrightarrow{(CH_3)_2C(O_2)} R-NO_2$$

$$R-N\begin{smallmatrix}R'\\H\end{smallmatrix} \xrightarrow{(CH_3)_2C(O_2)} R-N\begin{smallmatrix}R'\\OH\end{smallmatrix}$$

$$R-N\begin{smallmatrix}R'\\R''\end{smallmatrix} \xrightarrow{PhCO_3H} R'-\overset{R}{\underset{R''}{N^+}}-O^-$$

11.10. What is the expected product from the reaction of each of the following compounds with hydrogen and 5% Pd/C? What would you expect to obtain from the reaction of the compound in part (c.) with hydrogen and a poisoned Pd catalyst.

a.   b.   c.

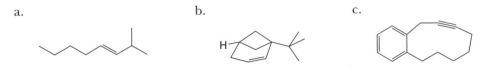

11.11. For each pair of compounds, indicate which is the more oxidized at the carbon atom indicated by the arrow. If the oxidation level is the same, indicate that fact.

a.   b.

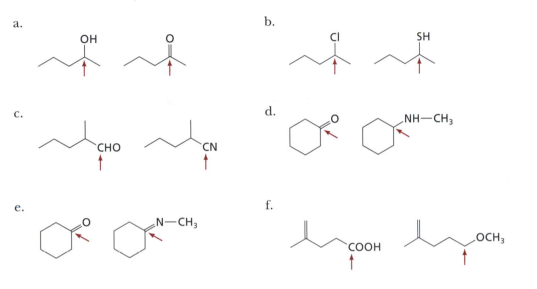

c.   d.

e.   f.

11.12. Which of the oxidation reagents described in this chapter could be used to carry out the following transformations in good yield, without significant generation of side products:

a.   d.

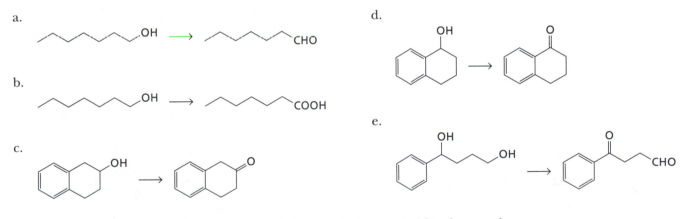

b.

c.   e.

11.13. Draw the structure of the alkene needed to react with a peracid to form each of the following epoxides:

a.   b.   c.   d.

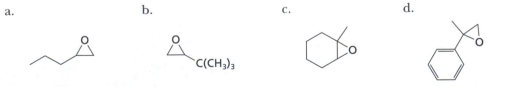

11.14.   Show how you would prepare the compounds in Exercise 11.13c and 11.13d by performing the electrophilic addition of bromine in water followed by ring closure.

11.15.   The previous two chapters (Chapters 9 and 10) have presented several reactions of alkenes. Predict the major product that would be obtained from the reaction of *cis*-2-butene and *trans*-2-butene with each of the following reagents. Specify the stereochemistry of the product by assigning (*R*) and (*S*) to any new stereogenic center that is formed and indicate whether a racemic mixture is formed.

a. $Hg(OAc)_2$ in H_2O

b. H_2 and Pd/C

c. OsO_4 and $NaIO_4$

d. O_3 followed by $(CH_3)_2S$

e. $BH_3 \cdot THF$ followed by H_2O_2, OH^-

f.   MCPBA

g. Br_2 in H_2O

h. $CHCl_3$, OH^-

11.16.   Repeat Exercise 11.15 for methylenecyclohexane.

11.17.   Rationalize the following result (*Hint:* Consider the answer to Exercise 6.25).

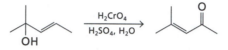

11.18.   Draw the structure of the alkene that would yield each of the following product mixtures upon ozonolysis followed by reductive workup:

a.

b.

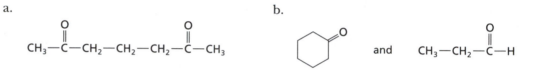

11.19.   Draw the products for each of the following reactions. Compare their stereochemical outcomes with those from the analogous reactions of cyclohexene that were presented in Section 11.3e.

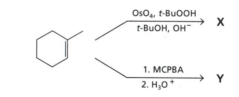

11.20.   A primary alkyl halide can be oxidized to the corresponding aldehyde by treatment with DMSO and sodium bicarbonate. The reaction is complete within 5 min at 100°C. Propose a mechanism for the following reaction:

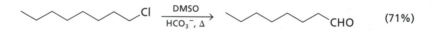

11.21.   The MO representations of ozone are illustrated below; they are the same as those in the allyl π system (Exercise 10.22). Arrange these orbitals by increasing energy. There are four electrons occupying these orbitals; indicate which orbital is the HOMO and which is the LUMO.

11.22.   Based on your answers to Exercise 11.21 as well as the illustrations in Section 10.4c showing the [4 + 2] cycloaddition process, describe which orbitals are involved in the reaction of an alkene with ozone. Based on the symmetry of the orbitals in the reacting π systems, do you expect this reaction to be symmetry allowed?

11.23.   Draw the structure(s) of the major product(s) expected from each of the following reactions. Indicate the stereochemistry of the product as appropriate. Relative stereochemistry should be shown using wedges and dashed lines. If a racemic mixture will be formed, draw the structure of one enantiomer and write the word "racemic", or draw both enantiomeric structures. If diastereomers are formed, draw each structure; label meso compounds as such. If no reaction occurs, write N.R.

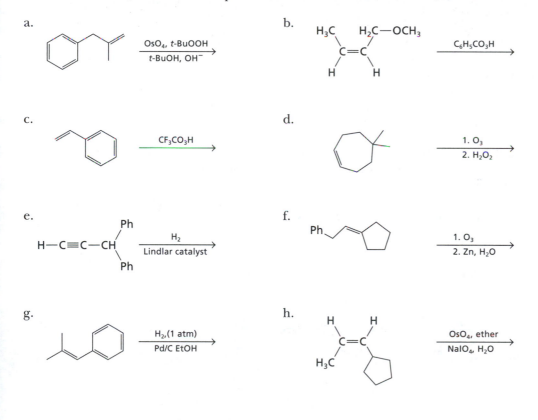

11.24. An *N*-oxide undergoes a reaction called the *Cope elimination* if the compound has a hydrogen atom attached to the β-carbon atom. Propose a mechanism for this process, which is a syn-elimination process similar to that of the Swern oxidation.

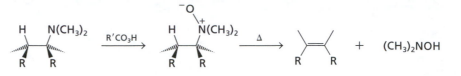

11.25. In the enzyme alcohol dehydrogenase, which is responsible for the oxidation of ethanol formed during the metabolism of glucose, NAD⁺ reacts stereospecifically in accepting a hydride ion from the alcohol. Thus, the following process takes place when 1,1-dideuterioethanol is the substrate:

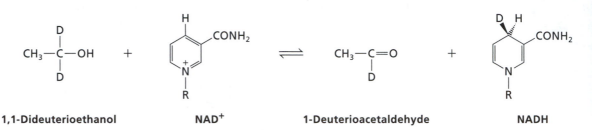

Based on this observation and on Figure 8.9, draw a picture of the alcohol dehydrogenase active site, showing the spatial relationship between the substrate and the coenzyme. What is the configuration of the new stereogenic center in the dihydropyridine product, NADH?

11.26. Methanol is extremely poisonous because the product formed during its metabolism by the alcohol dehydrogenase enzyme (Exercise 11.25) is highly reactive toward many functional groups in other proteins. Draw the structure of the product and show the mechanistic steps that lead to its formation.

11.27. Indicate all of the combinations of reagents and solvents described in this chapter that can be used to carry out the following oxidation reactions:

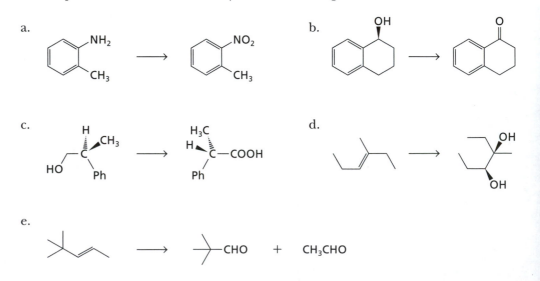

Free Radical Reactions

Organic reactions that proceed via free radical intermediates are less common than the transformations that involve polar species. Nevertheless, many important industrial processes make use of radical chain reactions, particularly those used for the halogenation, oxidation, and isomerization reactions of hydrocarbons. Furthermore, radical processes underlie numerous atmospheric chemistry reactions, especially ones implicated in the depletion of the protective ozone layer. Combustion reactions and key biochemical processes also proceed via radical intermediates.

A carbon radical carries no charge, but it is highly reactive because it has only seven electrons, which is less than a stable octet (Section 2.5e). Radicals react predominantly by three pathways:

- They abstract atoms, leading to substitution.
- They undergo oxidation or reduction reactions.
- They add to π bonds.

In this chapter, you will first learn the formalisms required to represent the mechanisms of chain reactions, which are common for radical processes. Then you will learn specific transformations that take place via radical intermediates, which include substitution reactions of hydrocarbons and their halogenated derivatives, and addition reactions of alkenes and alkynes.

The study of radical reactions constitutes a rapidly growing area of research as chemists search for more selective and general ways of making molecules, especially those with rings. The growing interest in radical species in biochemistry, especially those stabilized by metal ions, has also heightened chemists' interest in radical reactions. The details about several biochemically important radical reactions are presented in this chapter as well.

12.1 FREE RADICAL HALOGENATION REACTIONS

12.1a CHAIN REACTIONS COMPRISE THREE STEPS: INITIATION, PROPAGATION, AND TERMINATION

The first step of any radical reaction is **initiation,** which is commonly carried out using heat, UV light, or chemical reagents. The goal of initiation is to form at least one species that has unpaired electrons. Radical formation occurs by breaking a bond so that each atom of a reagent X–Y retains one electron from that bond, a process known as **homolysis,** which is usually carried out by application of heat or UV light. For radical reactions, we depict the movement of single electrons by using "fish hook" arrows.

Homolysis

Initiation can also be accomplished by treating a molecule with a reagent that has an unpaired electron, often a metal-containing species (M·).

$$M· \quad X—Y \longrightarrow M—X \ + \ Y·$$

The second stage of a chain reaction is **propagation,** which is a series of two or more steps that generate the product and regenerate a radical to continue the *chain reaction.* The actual steps depend on the reactants involved, and shortly these will be illustrated with specific examples.

The final stage of a radical chain reaction, called termination, occurs whenever two radicals combine with each other and stop the propagation process.

$$W· \ + \ Z· \longrightarrow W—Z$$

Because radicals in a reaction mixture are normally present only in low concentrations, the likelihood is much greater that they will collide with molecules of starting material than with other radicals. *Termination steps therefore occur much less frequently than propagation steps.* This difference in rate allows the reaction to continue over and over, generating the product in good yield.

12.1b THE CHLORINATION OF METHANE OCCURS VIA A RADICAL CHAIN PROCESS

Alkanes are notoriously unreactive compounds because they are nonpolar and lack functional groups at which reactions can take place. Free radical halogenation therefore provides a method by which alkanes can be functionalized. A severe limitation of radical halogenation, however, is the number of similar C–H bonds that are present in all but the simplest alkanes, so selective reactions are difficult to achieve, a fact to keep in mind as you study the following sections.

The conversion of alcohols to alkyl halides (Section 7.1) is usually the best way to make alkyl halides. The electrophilic addition of HX (X = F, Cl, Br, and I) to alkenes is another viable route by which to prepare alkyl halides. Furthermore, the fluoride and iodide derivatives can be made by addition reactions, but iodination and fluorination cannot be carried out by radical pathways. Iodine is too unreactive (see Exercise 12.2), and elemental fluorine reacts almost explosively with hydrocarbons. Fluorination of alkanes can be carried out by dilution of F_2 with an inert gas such as helium, but special equipment is still required, so this transformation is not a routine laboratory procedure.

The Cl–Cl bond of elemental chlorine undergoes homolysis when irradiated with UV light, and this process yields two chlorine atoms, also called chlorine radicals. This reaction constitutes the initiation step.

Initiation step

In the propagation stage of methane chlorination, a chlorine radical abstracts a hydrogen atom from methane to produce the methyl radical. The methyl radical in turn abstracts a chlorine atom from a chlorine molecule, and chloromethane is formed. The second step of propagation also regenerates a chlorine atom. These two steps repeat many times until termination occurs.

Propagation steps:

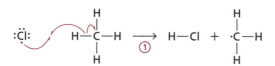

Methyl radical

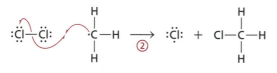

Chloromethane

Termination of this process takes place when a chlorine atom reacts with another chlorine atom to regenerate Cl_2. Or, a chlorine atom can react with a methyl radical to form chloromethane, which constitutes a minor pathway by which the product is made. Two methyl radicals can also combine to produce ethane, a very minor byproduct of this reaction.

Termination steps:

Chlorination of methane is plagued by the lack of selectivity, which becomes apparent as the reaction proceeds. As chloromethane is formed, its concentration increases. Because CH_3Cl still has C–H bonds, it can also undergo chlorination, forming dichloromethane. Eventually, the reaction mixture contains chloromethane, dichloromethane (methylene chloride), trichloromethane (chloroform), and tetrachloromethane (carbon tetrachloride).

$$CH_4 \xrightarrow[hv]{Cl_2} CH_3Cl + HCl \xrightarrow[hv]{Cl_2} CH_2Cl_2 + HCl \xrightarrow[hv]{Cl_2} CHCl_3 + HCl \xrightarrow[hv]{Cl_2} CCl_4 + HCl$$

As an industrial process, this multiple reaction sequence presents little difficulty because fractional distillation can be used to separate the mixture into its components, each of which is marketable. In the laboratory, formation of a complex product mixture can be frustrating, both because byproducts lower the yield of the desired product and because it is difficult to quantitatively separate molecules with similar molecular weights or polarities.

EXERCISE 12.1

Propose a mechanism for the free radical chlorination of chloromethane to produce dichloromethane. The initiation step comprises the homolysis of Cl_2 to form chlorine atoms.

12.1c BOND DISSOCIATION ENERGIES CAN BE USED TO CALCULATE OR ESTIMATE FREE ENERGY CHANGES OF RADICAL REACTIONS

The polarity of a solvent does not have a large influence on the propagation steps of most radical reactions because the radical intermediates carry no charge. Furthermore, the numbers of radicals and molecules are often the same on each side of an equation, so entropy changes are minimal. As a result of these two effects, the free energy of reaction is approximately equal to the enthalpy of reaction: $\Delta G° \approx \Delta H°$. To determine whether a radical reaction is favorable becomes a matter of adding or subtracting homolytic **bond dissociation energy** (**BDE**) values, which are given in Tables 12.1 and 12.2.

Table 12.1 Homolytic BDE values (kcal mol^{-1}) for bonds to heteroatoms.

X =	OH	Cl	Br	I		Bond	BDE
X—X	51	58	46	36		F—F	37
H—X	119	103	88	71		CH_3—F	108
CH_3—X	91	84	70	56		H—NH_2	103
(1°)C—X	91	81	68	53		CH_3—NH_2	51
(2°)C—X		80	68	51		H_2C=$CHCH_2$—Cl	69
(3°)C—X		79	65	50		$C_6H_5CH_2$—Cl	70
⬡—X	112	97	82			H_2C=CH—Cl	88

Table 12.2 Homolytic BDE (kcal mol^{-1}) for bonds to carbon and hydrogen.

Bonds to hydrogen		Bonds to carbon	
CH_3—H	104	CH_3—CH_3	88
(1°)C—H	98	(1°)C—CH_3	85
(2°)C—H	95	(2°)C—CH_3	84
(3°)C—H	91	(3°)C—CH_3	81
CH_3O—H	102	CH_3O—CH_3	81
C_6H_5—H	112	C_6H_5—CH_3	102
$C_6H_5CH_2$—H	70	$C_6H_5CH_2$—CH_3	72
H_2C=$CHCH_2$—H	87	H_2C=$CHCH_2$—CH_3	74
H_2C=CH—H	108	H_2C=CH—CH_3	97
$\overset{\overset{\displaystyle O}{\|\|}}{CH_3C}$—H	86	$\overset{\overset{\displaystyle O}{\|\|}}{CH_3C}$—$CH_3$	77
CH_3S—H	88		

The enthalpy change for a reaction can be calculated by taking the sum of the BDE values for the bonds being broken and subtracting the BDE values of the bonds that are being made.

$$\Delta H°_{rxn} = \Sigma(BDE_{broken}) - \Sigma(BDE_{made})$$

For chlorination of methane, the following equations show just such a calculation. The two propagation steps for this transformation, along with their BDE values, are given below:

$Cl\cdot + CH_4 \rightarrow HCl + CH_3\cdot$	$\Delta H°_1 = BDE(CH_3–H) – BDE(H–Cl)$	$= (104 – 103)$ kcal mol^{-1}
$CH_3\cdot + Cl_2 \rightarrow CH_3Cl + Cl\cdot$	$\Delta H°_2 = BDE(Cl–Cl) – BDE(CH_3–Cl)$	$= (58 – 84)$ kcal mol^{-1}
$Cl_2 + CH_4 \rightarrow HCl + CH_3Cl$	$\Delta H°_{rxn} = \Delta H°_1 + \Delta H_2$	$= -25$ kcal mol^{-1}

Adding the energy values calculated for each step gives $\Delta H°_{rxn}$, which equals –25 kcal mol^{-1}. The values for the initiation and termination steps are not included in the calculation because there are normally hundreds or even thousands of rounds of propagation steps for each initiation or termination event.

If $\Delta H° < 0$, then $\Delta G° < 0$; the reaction is product favored and proceeds as written (Section 5.4a). If $\Delta H° > 0$, then $\Delta G° > 0$; the reaction is reactant favored, and proceeds in the opposite direction. Chlorination of methane is exothermic, hence exergonic, so it proceeds readily.

EXAMPLE 12.1

Calculate the value of $\Delta H°$ for each of the propagation steps, as well as the overall reaction, involved in free radical bromination of methane, $CH_4 + Br_2 \rightarrow CH_3Br + HBr$. The propagation steps are the same as those for the chlorination of methane.

First, write equations for the two propagation steps and make sure they add to give the overall process (the radical species should cancel):

$$Br\cdot + CH_4 \rightarrow HBr + CH_3\cdot$$
$$CH_3\cdot + Br_2 \rightarrow CH_3Br + Br\cdot$$
$$\overline{CH_4 + Br_2 \rightarrow CH_3Br + HBr}$$

Next, tabulate the BDE values for each bond broken and made:

Step 1: broken $CH_3–H$ (BDE = 104 kcal mol^{-1});
    made $H–Br$ (BDE = 88 kcal mol^{-1})

Step 2: broken $Br–Br$ (BDE = 46 kcal mol^{-1});
    made $CH_3–Br$ (BDE = 70 kcal mol^{-1})

Third, calculate the $\Delta H°$ value of each step by subtracting the BDE values of the bonds made from those broken:

Step 1: $\Delta H°_1 = BDE (CH_3–H) – BDE (H–Br) = 104 – 88 = 16$ kcal mol^{-1}

Step 2: $\Delta H°_2 = BDE (Br–Br) – BDE (CH_3–Br) = 46 – 70 = -24$ kcal mol^{-1}

Finally, add the value of each step to obtain $\Delta H°_{rxn}$.

$$\Delta H°_{rxn} = \Delta H°_1 + \Delta H°_2 = 16 \text{ kcal mol}^{-1} + (-24 \text{ kcal mol}^{-1}) = -8 \text{ kcal mol}^{-1}$$

EXERCISE 12.2

Calculate the value of $\Delta H°$ for each of the propagation steps, as well as the overall reaction, involved in free radical iodination of methane, $CH_4 + I_2 \rightarrow CH_3I + HI$. What can you say about the feasibility of chlorination, bromination, and iodination of methane?

12.1d CHLORINATION OF ALIPHATIC HYDROCARBONS OFTEN GENERATES ISOMERIC PRODUCTS

The chlorination of ethane follows the same mechanism as that of methane chlorination except that six hydrogen atoms are available to be substituted. Not surprisingly, an even more complex mixture can be formed when ethane reacts than when methane does.

EXERCISE 12.3

Draw structures for the possible mono-, di-, tri-, and tetrachlorinated ethane derivatives. Name each compound.

When propane undergoes chlorination, reactivity differences between secondary and primary hydrogen atoms come into play. Looking only at monochlorinated products, we see that there are two: 1-chloropropane and 2-chloropropane (products with more than one chlorine atom are also formed).

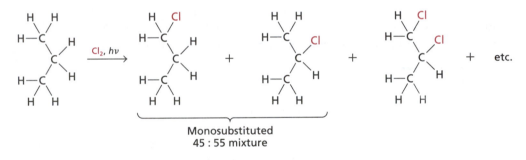

Monosubstituted
45 : 55 mixture

EXERCISE 12.4

From the BDE values given in Tables 12.1 and 12.2, calculate $\Delta H°$ for the following processes.

a. Propane + Cl· → 1-propyl radical + HCl
b. Propane + Cl· → 2-propyl radical + HCl
c. Propane + Cl_2 → 1-chloropropane + HCl
d. Propane + Cl_2 → 2-chloropropane + HCl

If the reactivity of 1° and 2° C–H bonds were identical, then the statistical ratio of 1-chloropropane to 2-chloropropane should be 3:1 because there are six primary hydrogen atoms, but only two secondary hydrogen atoms. The actual ratio of 1-isomer to 2-isomer is 45:55, which tells us that the secondary hydrogen atoms must be more reactive. The ratio between the actual percentage of product obtained and the statistical amount of product expected can be used to calculate the *reactivity ratio* for the different types of hydrogen atoms in an alkane. Thus, for chlorination of an alkane,

$$\frac{2° \text{ C–H bond reactivity}}{1° \text{ C–H bond reactivity}} = \frac{55\%/2 \text{ H atoms}}{45\%/6 \text{ H atoms}} = \frac{3.7}{1}$$

If you looked at the chlorination of a molecule with a tertiary C–H bond, you would find,

$$\frac{3° \text{ C–H bond reactivity}}{1° \text{ C–H bond reactivity}} = \frac{5}{1}$$

The reactivity of a 3° hydrogen atom is therefore even higher than that of a 2° hydrogen atom. These results parallel the carbon–hydrogen bond strengths of 98, 95, and 91 kcal mol^{-1} for 1°, 2°, and 3° C–H bonds, respectively. The 3° hydrogen atom is the easiest to remove because it has the weakest bond, and a 2° hydrogen atom is easier to remove than a 1° one.

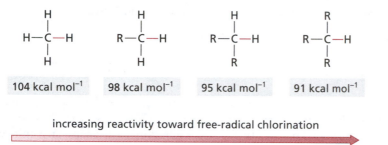

104 kcal mol^{-1} 98 kcal mol^{-1} 95 kcal mol^{-1} 91 kcal mol^{-1}

increasing reactivity toward free-radical chlorination

Just as you saw for carbocations, the stability of a radical increases in the order $CH_3 <$ 1° < 2° < 3° as a result of stabilization by electron donation and hyperconjugation (Section 6.2b).

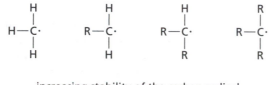

increasing stability of the carbon radical

For propane, therefore, the 2° radical forms more readily than the 1° radical, and this fact accounts for the more than statistical amount of 2-chloropropane obtained.

To predict the major product of a chlorination reaction, you have to take into account both the number of hydrogen atoms and the reactivity ratios of the different C–H bond types. Example 12.2 shows how a calculation of this type is done.

EXAMPLE 12.2

Draw the structures of the monochloro products formed by the reaction between 2-methylpropane and chlorine upon irradiation with UV light. Assuming a reactivity ratio of 5:1 for tertiary versus primary carbon–hydrogen bonds, calculate what percentage of each product is expected.

First, write the equation showing the structures of the products. 2-Methylpropane has only 1° and 3° C–H bonds, so there are only two monochloro products possible.

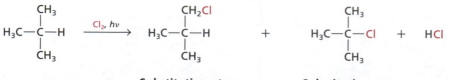

Substitution at a   **Substitution at a**
primary carbon atom   **tertiary carbon atom**

There are nine primary hydrogen atoms and only one tertiary hydrogen. The amount of each product is the number of those hydrogen atoms times the relative reactivity of each type.

$$1° \text{ C–H: } 9H \times 1.0 = 9.0 \qquad 3° \text{ C–H: } 1H \times 5.0 = 5.0$$

The percentage of each isomer is its amount divided by the *total amount* of product, multiplied by 100. In this example, the relative amounts, as calculated directly above, are 9.0 and 5.0; therefore the total amount is 14.0. Dividing each amount by the total and multiplying by 100 gives the percentage of each product.

% 1-Chloro-2-methylpropane: $[9.0/ (14.0)] \times 100 = 64\%$

% 2-Chloro-2-methylpropane: $[5.0/ (14.0)] \times 100 = 36\%$

The major product is 1-chloro-2-methylpropane.

EXERCISE 12.5

Assuming reactivity ratios of 5:3.7:1 for tertiary, secondary, and primary carbon–hydrogen bonds, respectively, what are the structures and expected percentages of monochloro products that will be formed when 2-methylhexane and chlorine are irradiated with UV light? Which is the major product?

12.1e BROMINATION OF AN ALKANE IS MORE SELECTIVE THAN IS CHLORINATION

The reaction of an alkane with bromine radicals occurs in the same way as with chlorine radicals, but bonds with bromine are generally weaker than those with chlorine (Table 12.1). As a result, the enthalpy change for the first step in the bromination reaction of methane (Example 12.1) is significantly more endothermic than the corresponding step in chlorination (section 12.1c). These data are presented graphically in Figure 12.1.

$Br\cdot + CH_4 \rightarrow HBr + CH_3\cdot$ $\quad$ $+16$ kcal mol^{-1}	$Cl\cdot + CH_4 \rightarrow HCl + CH_3\cdot$ $\quad$ $+ 1$ kcal mol^{-1}
$CH_3\cdot + Br_2 \rightarrow CH_3Br + Br\cdot$ $\quad$ -24 kcal mol^{-1}	$CH_3\cdot + Cl_2 \rightarrow CH_3Cl + Cl\cdot$ $\quad$ -26 kcal mol^{-1}
$CH_4 + Br_2 \rightarrow CH_3Br + HBr$ $\quad$ -8 kcal mol^{-1}	$CH_4 + Cl_2 \rightarrow CH_3Cl + HCl$ $\quad$ -25 kcal mol^{-1}

a.   b.

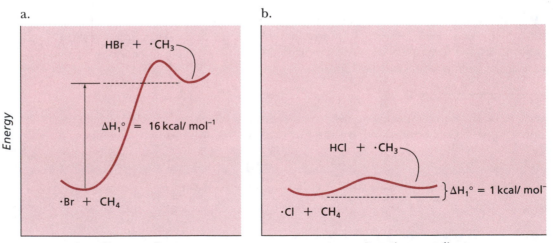

Figure 12.1
Reaction coordinate diagrams for the first steps in the radical bromination (a.) and chlorination (b.) of methane..

Reaction coordinate    *Reaction coordinate*

For these first propagation steps in each process, the Hammond postulate (Section 6.2c) can tell us something about their transition states. For bromination, the transition state will have a structure that is similar to that of the methyl radical; that is, the C–H bond will be long and weak (below, a.), and the unpaired electron will be associated more with the carbon atom than with the hydrogen atom. For chlorination, in which the energy of the radical intermediate differs little from that of the starting reactants, the transition state will have a structure in which the hydrogen atom is shared almost equally (below, b.), and the unpaired electron is more closely associated with the hydrogen atom.

a.   b.

$$\left[Br-H\text{-----}\overset{\cdot}{C}H_3 \right]^{\ddagger} \qquad \left[Cl\text{----}\overset{\cdot}{H}\text{----}CH_3 \right]^{\ddagger}$$

The significance of these structures for the halogenation of alkanes other than methane is this: For bromination, the transition state resembles the structure of the alkyl radical, so if more than one radical intermediate is possible, the pathway leading to the more stable radical will have a lower free energy of activation, and the product formed from that radical will predominate. For chlorination, however, the structure of the intermediate radical has less influence on the structure of the transition state, so the free energies of activation will differ by smaller amounts, and differentiation of the rates leading to multiple products will be less. The overall consequence is this:

> Free radical bromination occurs more selectively than does chlorination.

Experimental evidence validates this statement; for example,

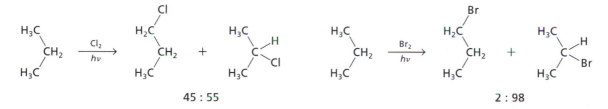

45 : 55   2 : 98

The reaction of a tertiary C–H bond is even more selective. The reactivity ratios (Section 12.1d) of chlorination versus bromination are summarized in the following table.

Chlorination of alkanes			Bromination of alkanes		
1° C—H	2° C—H	3° C—H	1° C—H	2° C—H	3° C—H
1 :	3.7 :	5	1 :	100 :	2000

Besides displaying selectivity toward different types of C–H bonds, continued reaction with bromine to produce di- and tribrominated species occurs more slowly than the monobromination step, so the degree of substitution in bromination reactions is more easily controlled than in chlorination.

To predict the major product of a free radical bromination reaction, you can assume reasonably that substitution of a tertiary hydrogen atom will predominate. If no tertiary C–H bond is present, then substitution at 2° carbon atoms will occur preferentially.

EXERCISE 12.6

What is the major monobromo product expected from the following reaction?

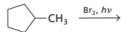

12.1f *N*-BROMOSUCCINIMIDE PROVIDES A LOW CONCENTRATION OF BROMINE FOR THE FREE RADICAL REACTIONS OF ALLYLIC AND BENZYLIC SUBSTRATES

The free radical bromination of hydrocarbons is not done routinely in the laboratory unless the molecule is highly symmetrical.

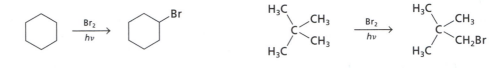

2° C–H Bonds only   **1° C–H Bonds only**

Benzylic bromides, on the other hand, are often prepared by a free radical pathway, because the benzylic radical is highly stabilized and readily formed. Unless reactive functional groups are present, elemental bromine is used.

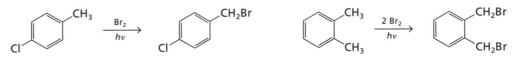

Another reagent that can be used to brominate benzylic substrates is *N*-bromosuccinimide, abbreviated NBS. The reaction is normally initiated using benzoyl peroxide $(C_6H_5COO)_2$ and heat.

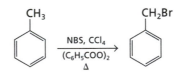

Benzoyl peroxide has a weak O–O bond with a BDE of ~ 50 kcal mol^{-1}. Upon heating, benzoyl peroxide homolyzes to generate 2 equiv of the benzoyloxy radical, which can subsequently lose carbon dioxide to produce the phenyl radical.

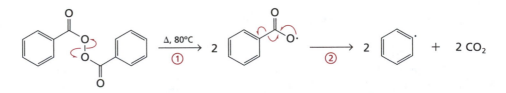

Either radical can react with bromine, which is formed from NBS and acid. Formation of the phenyl radical, as shown above, plus the two reactions that follow constitute the initiation phase and create bromine radicals.

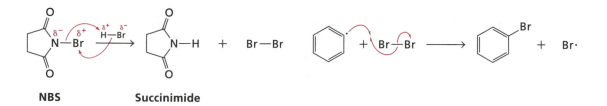

NBS   **Succinimide**

The bromine radical then participates as described in Section 12.1e, abstracting a hydrogen atom from the hydrocarbon and forming HBr. This liberated HBr reacts with NBS to generate more bromine, a process that amounts to slow release of bromine. As a result, only a low concentration of Br_2 is present at any time.

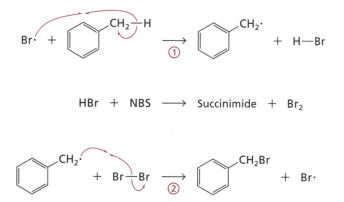

$$HBr \ + \ NBS \ \longrightarrow \ Succinimide \ + \ Br_2$$

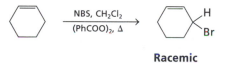

N-Bromosuccinimide bromination works well for allylic systems, too. Bromination occurs at a C–H bond alpha to the double bond.

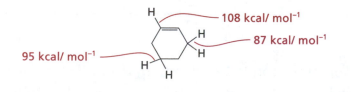

Racemic

You might also expect the double bond to undergo electrophilic addition with bromine, but the radical bromination reaction occurs faster than addition when NBS is used as the reagent.

The propensity of benzylic and allylic substrates to undergo free radical substitution is related to the stability of the benzyl and allyl radicals, which can delocalize electrons from the adjacent π bonds so as to stabilize the seven-electron carbon atom. Free radical bromination of an alkene occurs alpha to the double bond because that position has the weakest C–H bond. Vinyl carbon–hydrogen bonds are the strongest.

H —— 108 kcal/ mol^{-1}

H —— 87 kcal/ mol^{-1}

95 kcal/ mol^{-1}

EXERCISE 12.7

What is the major monobromo product(s) expected from each of the following reactions? Show its stereochemistry, too.

a.   b.

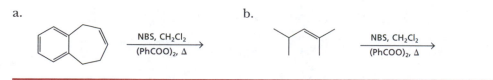

12.2   REDUCTION VIA RADICAL INTERMEDIATES

12.2a   BENZYLIC SUBSTRATES READILY UNDERGO HYDROGENOLYSIS

The addition of hydrogen to the π bond of an alkene or alkyne is called *hydrogenation* (Section 11.2a). Addition of hydrogen to a sigma bond is called **hydrogenolysis,** and it occurs via radical intermediates when benzylic halides, alcohols, and ethers are the reactants because the C–X bond (X = Cl, Br, I, OH, or OR) is relatively weak. Formation of the resonance-stabilized benzyl radical permits the bond to break readily. The carbon–oxygen bond of a nonbenzylic 1°, 2°, or 3° alcohol or ether is not cleaved under hydrogenolysis conditions.

The catalysts used for hydrogenolysis are the same as those used for hydrogenation, and hydrogen gas is a reactant, although it is adsorbed on the metal surface as atoms (see Fig. 11.2.) To depict the mechanism, we use the symbol Pd////{H} to indicate hydrogen atoms adsorbed on the surface of palladium.

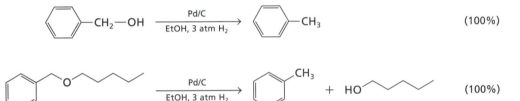

Examples of this reaction are numerous, and yields are generally high.

EXERCISE 12.8

Draw the structure of the major product expected from the following reaction:

$$\text{(structure)} \quad \xrightarrow[\text{CH}_3\text{COOH, 3 atm H}_2]{\text{Pd/C}} \quad ? \quad (72\%)$$

12.2b   DESULFURIZATION IS CARRIED OUT USING RANEY NICKEL

Carbon–sulfur bonds are also readily cleaved via formation of radical intermediates. A common way to replace sulfur with hydrogen—*desulfurization*—is by treating the substrate with **Raney nickel,** prepared by allowing an alloy of aluminum and nickel to react with sodium hydroxide solution. The aluminum dissolves, forming hydrogen gas, which is adsorbed as H atoms on the surface of the nickel (see Fig. 11.2).

$$2\text{ Al} + 6\text{ NaOH} \longrightarrow 2\text{ Na}_3\text{AlO}_3 + 3\text{ H}_2$$

Alloyed with nickel   Adsorbed on nickel

Organic sulfides, thiols, and their selenium analogues react with Raney nickel to form the corresponding alkanes and hydrogen sulfide (or hydrogen selenide).

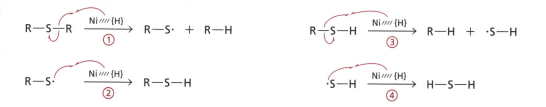

$$\text{CH}_3\text{CH}_2\text{CH}_2\text{—S—CH}_2\text{CH}_2\text{CH}_3 \xrightarrow{\text{Raney nickel}} 2\ \text{CH}_3\text{CH}_2\text{CH}_3\ +\ \text{H}_2\text{S}$$

Desulfurization likely involves radical intermediates, although the exact mechanism is not certain. The symbol Ni////{H} is used to indicate hydrogen adsorbed on Ni.

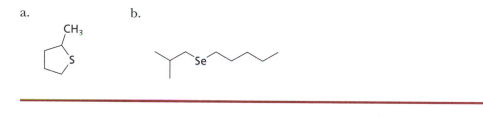

Incidentally, Raney nickel, under an atmosphere of hydrogen gas, can be used instead of Pd/C to hydrogenate alkenes and alkynes.

EXERCISE 12.9

What are the products expected from the reaction between Raney nickel and each of the following compounds?

a.

b.

12.2c AN ALKYNE IS REDUCED BY AN ALKAI METAL IN LIQUID AMMONIA

Sections 12.2a and 12.2b have illustrated that radicals are involved in reduction processes that proceed on solid catalyst surfaces via one-electron changes. A reduction reaction that occurs by transfer of electrons in a homogeneous solution is illustrated by the conversion of an alkyne to an alkene.

You have already learned that an alkyne can be reduced to a cis-alkene using a poisoned catalyst (Section 11.2b) or via hydroboration and protonolysis (Section 9.4d). When an alkyne is added to a solution of sodium or lithium metal dissolved in liquid ammonia, reduction also takes place, but the product is a trans-alkene. The reaction proceeds by a series of electron- and proton-transfer steps.

An alkali metal dissolves in liquid ammonia, forming a blue solution that contains solvated electrons.

$$\text{Na}\ +\ x\ \text{NH}_3\ \longrightarrow\ \text{Na}^+\ +\ [\text{e}^-](\text{NH}_3)_x$$

Electron solvated by ammonia molecules

These electrons can add to the antibonding π* MO of an alkyne to generate a radical-anion intermediate, which has one unpaired electron and carries a negative charge.

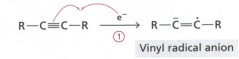

Vinyl radical anion

A radical anion is a base because it is also a carbanion, so it reacts with a proton source, which is often added to the ammonia in the guise of ethanol or *tert*-butyl alcohol. Ammonia itself will function as a proton donor if no other is added. This step yields a vinyl radical.

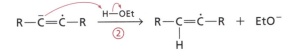

In the third step, the radical accepts another electron from solution, forming a carbanion.

Finally, the vinyl carbanion is protonated to generate the product, an alkene.

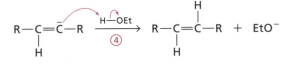

The product of this transformation, which is called a *dissolving metal reduction*, is normally the trans-alkene because protonation occurs from the side of the double bond that minimizes nonbonded steric interactions between the R groups.

$$CH_3CH_2CH_2-C{\equiv}C-CH_2CH_2CH_3 \xrightarrow[NH_3,\ -33°C]{Na}$$

(80–90%)

EXERCISE 12.10

What product is expected from reduction of 2-hexyne with sodium in liquid ammonia containing a small amount of *tert*-butyl alcohol.

12.2d TRIALKYLTIN HYDRIDE REAGENTS CONVERT ORGANOHALIDES TO THE CORRESPONDING HYDROCARBONS

A significant development in the organic chemistry of radicals during the past 20 years has been the use of tributyltin hydride as a reagent to generate specific carbon-centered radicals within a molecule. As with all reactions that take place by formation of radicals, the first stage is initiation, which is a two-step process. Heating azobis(isobutyronitrile), AIBN, causes it to homolyze and form molecular nitrogen and the 2-cyano-2-propyl radical. This radical in turn reacts with the tin hydride reagent to produce the tributyltin radical.

Initiation

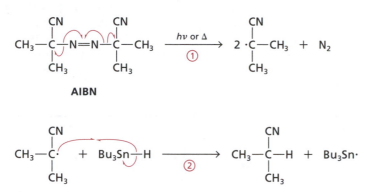

AIBN

In the propagation stage, the tributyltin radical removes a halogen atom from an organohalide, which generates a carbon-centered radical. In the second propagation step, this carbon radical abstracts a hydrogen atom from tributyltin hydride, which yields a C–H bond and regenerates the tributyltin radical.

Propagation

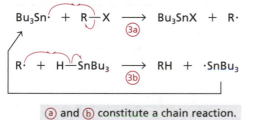

Ⓐ and Ⓑ constitute a chain reaction.

This chain reaction provides a convenient way to convert an organohalide to the corresponding hydrocarbon (R–X → R–H).

Termination occurs when any two radicals recombine; three possibilities for this stage follow:

Termination steps

$$R\cdot \; + \; \cdot R \longrightarrow R{-}R$$

$$R\cdot \; + \; \cdot SnBu_3 \longrightarrow R{-}SnBu_3$$

$$Bu_3Sn\cdot \; + \; \cdot SnBu_3 \longrightarrow Bu_3Sn{-}SnBu_3$$

In the reaction between an organohalide molecule and the tributyltin radical, an iodine atom is more easily transferred than a bromine atom, a phenylselenide group, or a chlorine atom. If a choice exists, therefore, an organobromide or iodide is selected as the substrate. Organofluorine compounds do not react with tributyltin hydride under these conditions.

$$X = \quad I \qquad Br \qquad PhSe \qquad Cl$$

decreasing reactivity of R—X toward $Bu_3Sn\cdot$

Substrate reactivity follows the expected order, namely, that the more stable radicals are formed faster in the first propagation step (allyl, benzyl > 3° > 2° > 1° > vinyl, aryl). In the second propagation step, a more stable radical reacts more slowly; an allylic radical, for example, reacts slower than a vinyl radical with tributyltin hydride. For the *overall* reaction between an organohalide and tributyltin hydride, substitution is facile no matter what the structure of the substrate is, so this procedure is an excellent way to replace a halogen atom with hydrogen.

EXERCISE 12.11

What is the expected product from each of the following reactions?

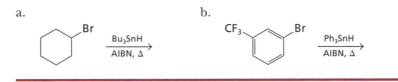

a.

b.

12.3   FREE RADICAL ADDITION REACTIONS

12.3a   HYDROGEN BROMIDE ADDS TO ALKENES VIA A RADICAL PATHWAY AND WITH APPARENT ANTI-MARKOVNIKOV REGIOCHEMISTRY

In Section 9.1b, you learned that HBr adds to an alkene by a polar mechanism that leads to formation of a product with the hydrogen atom attached to the end of the π bond with more hydrogen atoms. During the 1920s, many studies reported that the degree of Markovnikov regiochemistry varied substantially during addition reactions between alkenes and HBr. Professors Kharasch and Mayo, at the University of Chicago, ultimately discovered that *formation of the anti-Markovnikov product resulted from a completely different mechanism—radical addition of HBr to the π bond.*

Certain alkenes react readily with oxygen in the air to form peroxide impurities. Peroxides, especially when heated, homolyze to form alkoxy radicals that can initiate radical reactions. When HBr is present along with a radical initiator, hydrogen atom abstraction generates a bromine radical.

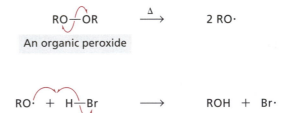

The bromine radical could conceivably remove an allylic hydrogen atom, but this step simply regenerates HBr.

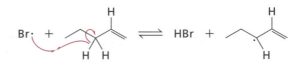

Instead, the bromine atom *adds* to the double bond, creating a carbon-centered radical. This radical subsequently abstracts a hydrogen atom from HBr and regenerates the bromine radical. These two steps constitute the propagation stage.

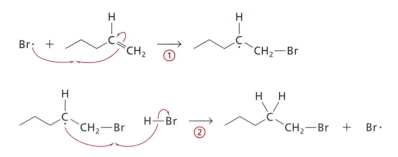

Termination occurs, as always, when any two radicals combine and stops the chain reaction.

EXERCISE 12.12

Write out the various termination steps that are possible in the foregoing discussion of the HBr addition reaction.

If you look at the intermediates formed by addition of HBr to 1-pentene, you can see why the "anti-Markovnikov" regiochemistry is observed.

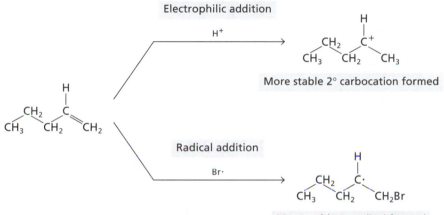

It is *not* because a less stable carbocation has somehow been generated. In the polar mechanism, the π bond reacts with an electrophile, which is H⁺. In the radical mechanism, the π bond reacts with a radical, Br·. The fact that a different atom is the first one to react with the π bond under each set of conditions means that the regiochemistry of the product appears reversed. *For each transformation, the intermediate that forms is the more stable of the two possibilities.*

Hydrogen chloride and hydrogen iodide do not undergo radical addition to alkenes. The H–Cl bond is the strongest among these three hydrogen–halogen bonds, so homolysis is slow. Hydrogen iodide forms radicals readily, but addition of I· is endothermic, and the chain reaction needed for propagation cannot be sustained.

EXERCISE 12.13

Given that the bond energy for a typical π bond is 65 kcal mol⁻¹, calculate the value of $\Delta H°$ for the following three processes, and represent them on a reaction coordinate diagram.

$$RCH{=}CH_2 \ + \ Cl\cdot \ \rightleftharpoons \ R\dot{C}H{-}CH_2Cl$$
$$RCH{=}CH_2 \ + \ Br\cdot \ \rightleftharpoons \ R\dot{C}H{-}CH_2Br$$
$$RCH{=}CH_2 \ + \ I\cdot \ \rightleftharpoons \ R\dot{C}H{-}CH_2I$$

12.3b ADDITION OF AN ALKYL RADICAL TO AN ALKENE PROVIDES A GENERAL METHOD FOR THE PREPARATION OF POLYMERS

Hydrogen bromide is not the only species that adds to alkenes by a radical mechanism. Carbon-centered radicals, like their carbocation analogues (Section 9.3) also add to π bonds to form polymers and rings.

The initiation step for radical polymerization processes is normally carried out using benzoyl peroxide or AIBN.

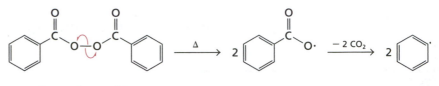

Benzoyl peroxide

Any of these radicals can add to the π bond of an alkene—ethylene in the following example—to form an alkyl radical.

The alkyl radical subsequently adds to a second molecule of the alkene, and the process repeats many times. Polymerization is terminated when any two radicals combine.

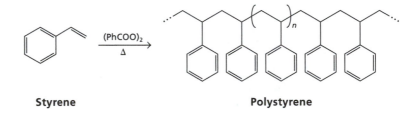

Styrene is another monomer that can be polymerized by a radical mechanism, and it yields polystyrene, a polymer used to make Styrofoam drink cups and related materials.

Styrene   **Polystyrene**

Chloroethylene (vinyl chloride) and acrylonitrile are other molecules commonly polymerized by radical methods. They form materials that you encounter routinely: poly(vinyl chloride), or PVC, is used in food wraps at the grocery store and in plumbing in most new homes; and poly(acrylonitrile) is used to make carpet fibers such as Orlon.

Chloroethylene   **PVC**   **Acrylonitrile**   **Poly(acrylonitrile)**

Radical polymerization is more commonly used than the processes that occur via carbocations (Section 9.3b) because stabilization of the reactive intermediates is more facile when radicals are involved. For example, primary radicals can be formed readily, whereas primary carbocations do not exist under normal conditions. Additional aspects of radical polymerization processes will be discussed in Section 26.2c.

EXERCISE 12.14

Show the first three steps of the polymerization of styrene that makes use of AIBN to initiate the reaction (Section 12.2d).

12.3c THE 5-HEXENYL RADICAL ILLUSTRATES A TYPICAL CYCLIZATION PROCESS THAT OCCURS WITH UNSATURATED ALKYL RADICALS

When a radical is generated within a molecule that also has unsaturation, a ring, rather than a polymer, is formed. For example, 6-bromo-1-hexene forms methylcyclopentane when treated with tributyltin hydride in the presence of AIBN as an initiator.

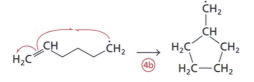

The initiation step generates the 2-cyanopropyl radical (Section 12.2d), which is relatively stable and does not itself undergo addition to π bonds. This radical does react with tributyltin hydride to produce the tributyltin radical.

Initiation

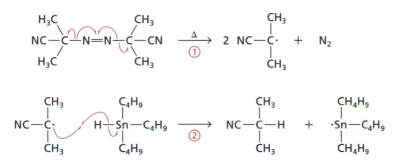

Next, the tin radical removes the halogen atom from 6-bromo-1-hexene, generating a primary radical.

There are two pathways possible for further reaction. The first is the one described in Section 12.2d: the 5-hexenyl radical reacts with tributyltin hydride to form the hydrocarbon.

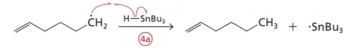

The other available pathway is cyclization, which involves the electrons in the alkene π bond. This addition process generates a different primary radical.

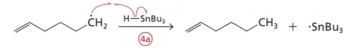

This cyclized radical *then* reacts with tributyltin hydride, yielding the saturated cyclic hydrocarbon and regenerating the tributyltin radical to propagate the chain reaction.

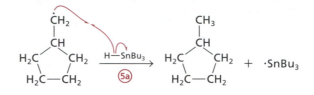

The regiochemistry of the ring-forming step is contrary to what you might expect: *The less stable radical is formed.* The other mode of addition, which produces a six-membered ring, generates the more stable secondary radical.

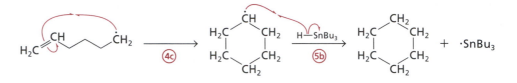

Yet the methylcyclopentane product is formed in a ratio of ~ 50:1 relative to the amount of cyclohexane product. There are several theories to explain why the five-membered ring product predominates, but none is satisfactory, so we will simply make use of the empirical results. In the competitive reactions to make other ring sizes, formation of a six-membered ring is more facile than formation of a seven-membered ring, and a five-membered ring is formed in preference to a four-membered ring.

When a radical is formed in a molecule with a π bond, a competition exists between reduction (reaction of the radical intermediate with additional tin hydride reagent) and cyclization (reaction of the intermediate radical with the π bond). Increasing the concentration of tributyltin hydride favors reduction, so it is often possible to obtain more cyclic product simply by using less of the tin hydride reagent.

EXERCISE 12.15

What is the expected major product of the following reaction under conditions that favor cyclization? Show the mechanism by which it proceeds.

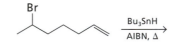

12.4   OXIDATION VIA RADICAL INTERMEDIATES

12.4a   MOLECULAR OXYGEN IS THE PRECURSOR OF OTHER RADICAL SPECIES

As noted in Section 11.3a, molecular oxygen exists as a diradical molecule instead of having all of its nonbonded electrons paired. It undergoes reduction to form the superoxide ion, O_2^-, which is a radical anion. Two-electron reduction of oxygen, or one-electron reduction of superoxide ion, leads to formation of the peroxide ion, O_2^{2-}.

$$:\ddot{O}-\ddot{O}: \quad \xrightarrow[\text{①}]{e^-} \quad :\ddot{O}-\ddot{O}:^- \quad \xrightarrow[\text{②}]{e^-} \quad :\ddot{O}-\ddot{O}:^-$$

Molecular oxygen   **Superoxide ion**   **Peroxide ion**

Peroxide ion is a good base, and it reacts with acids to form hydrogen peroxide. Like the related benzoyl peroxide (Section 12.1f), hydrogen peroxide undergoes homolysis under certain conditions to generate the hydroxyl radical, a potent and indiscriminant reactant.

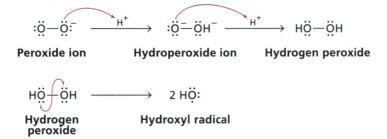

Both superoxide ion and peroxide ion are harmful to many living organisms. Superoxide ion, in particular, is believed to be a significant cause of aging. Dietary supplements such as vitamins C and E are alleged to prevent at least some detrimental effects of these oxygen-containing radicals, and food additives such as BHT (butylated hydroxy toluene) are used to retard spoilage by reacting with oxygen.

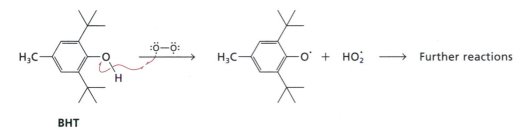

More naturally, enzymes have evolved to eliminate these species from our systems. In Nature, *superoxide dismutase* converts superoxide ion into hydrogen peroxide and oxygen. *Catalase* converts hydrogen peroxide into oxygen and water.

$$2\ O_2^{\cdot-} + 2\ H^+ \xrightarrow{\text{superoxide dismutase}} O_2 + H_2O_2$$

$$2\ H_2O_2 \xrightarrow{\text{catalase}} O_2 + 2\ H_2O$$

Not all reactions of oxygen and its reduced derivatives are harmful, however. Nature exploits these radicals to insert oxygen atoms into the carbon skeletons of organic molecules. Section 12.4c describes an important biosynthetic process of this type.

12.4b METAL IONS ARE USED TO GENERATE RADICAL INTERMEDIATES IN BIOLOGICAL SYSTEMS

When a free radical is required in biochemical systems to induce a chemical reaction, it has to be stabilized or it will react indiscriminately. Stabilization commonly occurs by placing the radical proximal to a metal ion with unpaired electrons. For example, vitamin B_{12} is a coenzyme that serves as a free radical source to initiate rearrangement reactions. Vitamin B_{12} is an organometallic compound, meaning that it has a metal–carbon bond. The cobalt ion is bonded to five nitrogen atoms, four of which are provided by a large cyclic compound called a corrin, the detailed structure of which is not important. The Co–C bond in B_{12} undergoes homolysis when the B_{12} coenzyme is bound by an enzyme in the presence of its substrate molecule. A Co(II) ion is formed along with an alkyl radical.

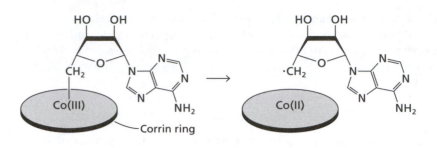

Iron and copper ions are also used to generate and to stabilize radicals. Oxo groups attached to metal ions in high oxidation states can abstract a hydrogen atom from carbon in a substrate molecule. The subsequently formed organic radical reacts with the attached hydroxyl group to produce an alcohol.

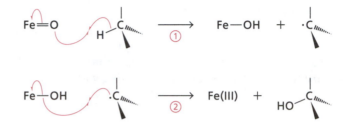

12.4c ALLYLIC OXIDATION GENERATES POTENT PHYSIOLOGICALLY ACTIVE COMPOUNDS

The ease with which allylic carbon–hydrogen bonds undergo substitution (Section 12.1f) by radical pathways is reflected in the facility of allylic oxidation processes in biochemical systems. Unsaturated fatty acids, which have long hydrocarbon chains, constitute a ubiquitous class of naturally occurring alkenes. The allylic oxidation reactions of arachidonic acid illustrate some important biochemical transformations that take place by radical intermediates.

Arachidonic acid

Oxidation of arachidonic acid produces several eicosanoids that are prostaglandins (PG), thromboxanes (Tx), or leukotrienes (LT). ("Eicosanoid" is from the Greek eikosi meaning 20; each contains 20 carbon atoms.) For example,

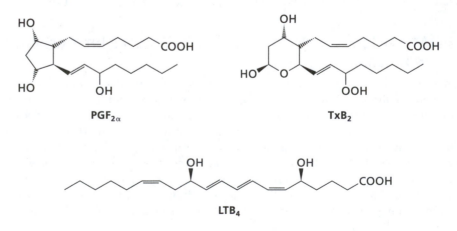

These compounds have significant physiologic activity, mediating inflammatory responses that are involved in ailments such as rheumatoid arthritis and psoriasis; production of pain and fever; regulation of blood pressure and the clotting process; control of some reproductive functions like the induction of labor; and regulation of sleep cycles. Eicosanoids act as "local hormones" that induce a response for a short time and near where they are generated. (True hormones travel in the bloodstream to the sites of their activity.)

The enzymatic oxidation steps involved in the synthesis of the prostaglandin framework constitute the *cyclic pathway* of arachidonic acid metabolism. The reaction sequence begins with the formation of a radical at an allylic position of arachidonic acid. The enzyme that catalyzes this process is *prostaglandin endoperoxide synthase,* and the radical source, Y· may be derived from the reaction of an iron(II) ion with O_2, which generates Fe–O–O or Fe–O–OH. If the latter is formed, it may undergo O–O bond homolysis to produce Fe–O· or HO·. In any event, a radical initiator is created in the enzyme active site, and abstraction of an allylic hydrogen atom starts the process.

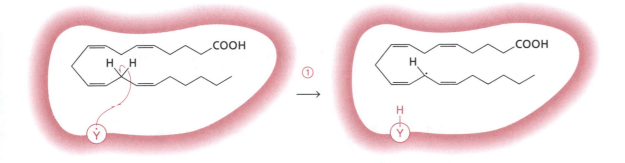

In the next step, molecular oxygen reacts with the allylic radical to generate an alkylpreroxy radical.

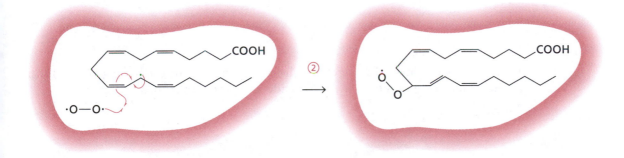

The oxygen-centered radical adds to the adjacent double bond, creating an alkyl radical.

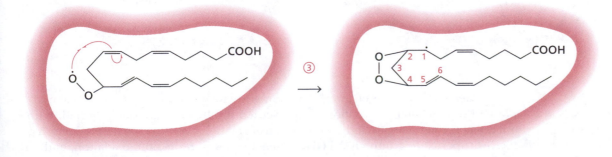

Notice that the substrate at this point has a structure *that is analogous to the 5-hexenyl radical*. As you learned in Section 12.3c, *such systems cyclize to produce a five-membered ring.*

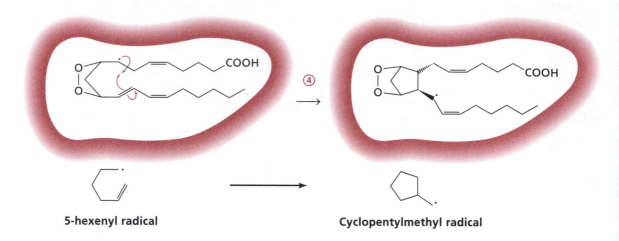

5-hexenyl radical   **Cyclopentylmethyl radical**

After cyclization, the newly formed secondary radical reacts with another equivalent of oxygen, producing a peroxy radical.

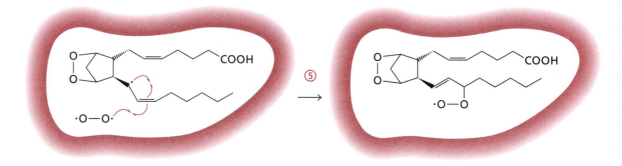

Finally, the peroxy radical abstracts a hydrogen atom from the initiating group to form a neutral molecule with both hydroperoxide and peroxide groups.

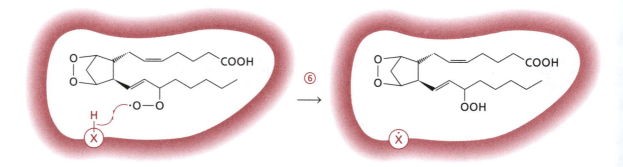

As seen earlier in this chapter, O–O bonds are weak, on the order of 50 kcal mol^{-1} (Table 12.1). Further elaboration of the cyclized product is therefore expected. Without going through details of the various transformations, you can see that reduction of the hydroperoxide group gives PGH$_2$. Then a series of reduction, oxidation, and/or elimination steps can yield other prostaglandin molecules.

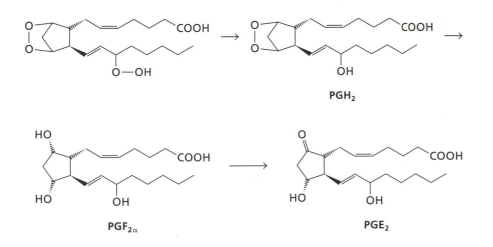

PGH₂

PGF₂ₐ

PGE₂

12.4d ENEDIYNES ARE POTENT ANTICANCER DRUGS THAT FORM DIRADICAL SPECIES THAT CAN CLEAVE DNA

In 1987, a compound called calicheamicin λ_1^I was isolated and characterized. Researchers were interested in its potent antitumor activity, which was considered promising for developing a new drug to treat cancer. Subsequent studies also focused on the unusual enediyne functionality in its 10-membered ring. Soon, several other reports of enediyne natural products appeared in the literature.

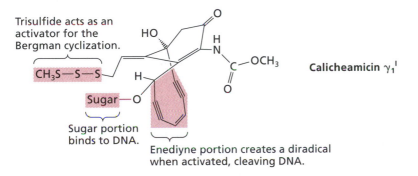

Trisulfide acts as an activator for the Bergman cyclization.

Calicheamicin γ_1^I

Sugar portion binds to DNA.

Enediyne portion creates a diradical when activated, cleaving DNA.

Enediynes were not especially novel substances in 1987. They had been synthesized many years previously, and in 1972, Professor Robert Bergman, then at the California Institute of Technology, had shown that 3-hexene-1,5-diyne undergoes cyclization to form 1,4-dehydrobenzene, a diradical. Performing the reaction in the presence of a hydrogen atom donor led to the isolation of benzene in quantitative yield.

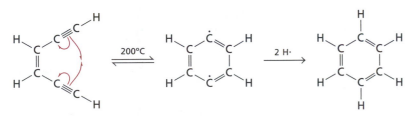

The idea that calicheamicin and related compounds might function to cleave DNA by formation of a benzene diradical directed further research into its mechanism of action. A less complex molecule that reacts much the same as calicheamicin was prepared by Professor K. C. Nicolaou and is illustrated below. Although stable at room temperature, 6,7-bis(hydroxymethyl)cyclodeca-1-ene-3,9-diyne undergoes the Bergman cyclization at 37°C and cleaves DNA by abstracting two hydrogen atoms from the nucleic acid backbone, which then reacts with molecular oxygen and causes bond cleavage.

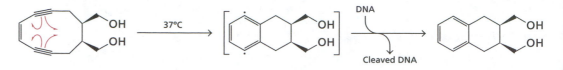

From the results of many studies, an understanding of the mechanism of action for calicheamicin soon evolved. The unusual trisulfide group is susceptible to reaction with nucleophiles that are present in cells, and this reaction leads to formation of a thiolate ion, itself a good nucleophile (Table 6.3).

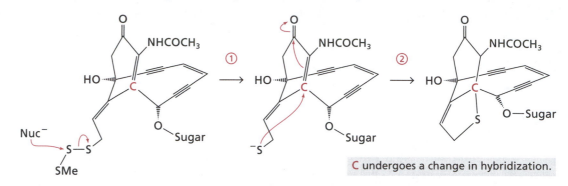

C undergoes a change in hybridization.

The thiolate ion then undergoes reaction with the α,β-unsaturated ketone portion, by what is called conjugate addition, a reaction discussed in Chapter 24. The important aspect of this reaction is the change of hybridization at the carbon atom β to the carbonyl group (shown in color in the preceding equation). The net effect is to bring the two ends of the enediyne π system together, from ~ 3.6 to ~ 3.2 Å. When the ends are closer, the Bergman cyclization can occur to generate the diradical, which in turn abstracts hydrogen atoms from DNA, leading to strand cleavage.

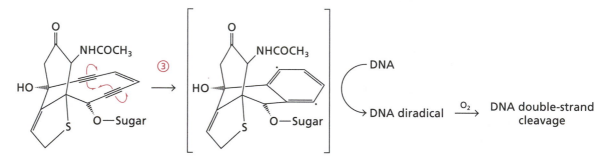

The trisulfide group acts as an activator for this reaction, allowing the enediyne unit to exist as a stable entity until the substrate interacts with DNA. Subsequent work aimed at developing enediynes as efficacious antitumor drugs has focused on synthesizing compounds that also require an activation step so that the Bergman cyclization does not occur until the drug has interacted with DNA.

EXERCISE 12.16

The compound shown below is stable toward Bergman cyclization up to 100°C, even though the distance between the ends of the enediyne unit is quite short. Explain. (*Hint:* Consider what the transition state would look like based on the mechanism of ring closure.)

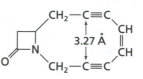

Section 12.1   Free radical halogenation reactions

- A carbon radical is a neutral, reactive species with seven electrons around carbon.

- A radical chain reaction has three stages: initiation, propagation, and termination.

- Free radical chlorination of methane and other alkanes results in substitution of a hydrogen atom by a chlorine atom. Ultraviolet light initiates the reaction.

- Values of the bond dissociation energy (BDE) can be used to estimate $\Delta H°$ (enthalpy change) of a reaction, which for radical processes is approximately equal to $\Delta G°$ (the standard free energy change) for the reaction.

- For the radical halogenation process, the relative reactivity of C–H bonds increases in the order $1° < 2° < 3°$.

- Radical chlorination is relatively nonselective and produces mixtures of products.

- Free radical bromination of alkanes results in selective substitution of hydrogen atoms by bromine atoms. The reactivity ratios for 1° versus 2° versus 3° C–H bond are 1 : 100 : 2000.

- *N*-Bromosuccinimide produces a low concentration of bromine; as a reagent it is used to carry out free radical bromination of benzylic and allylic substrates.

Section 12.2   Reduction via radical intermediates

- Benzylic alcohols and ethers are cleaved between the benzylic carbon and oxygen atoms by molecular hydrogen in the presence of a Pd catalyst. The oxygen atom is replaced by hydrogen.

- The C–S or C–Se bond of thiols, sulfides, and selenides is cleaved by Raney nickel, forming new S–H or Se–H and C–H bonds.

- An alkyne is reduced to a trans-alkene using sodium or lithium metal dissolved in liquid ammonia, a combination that generates solvated electrons.

- Trialkyltin hydride reacts with molecules that have a C–X (X = Cl, Br, or I) bond and replaces the halogen atom with hydrogen.

Section 12.3   Free radical addition reactions

- Hydrogen bromide adds to the double bond of an alkene in the presence of a radical initiator such that the hydrogen atom becomes attached to the alkene carbon atom that initially has fewer hydrogen atoms.

- An alkyl radical adds to the double bond of alkenes to produce polymers.

- A five-membered ring can be formed when a radical is generated four atoms away from an alkene double bond.

Section 12.4   Oxidation via radical intermediates

- Molecular oxygen is a diradical; it can promote or interfere with radical processes.

- Reduction products of oxygen, which include superoxide ion, peroxide ion, and hydrogen peroxide, are often harmful to living organisms.

- Biological systems employ several strategies to produce radicals to initiate reactions. A metal ion such as Fe, Co, or Cu is often used to create or to stabilize a radical center.

- Allylic oxidation that occurs by hydrogen atom abstraction is a common first step for metabolic reactions of arachidonic acid.

- Enediynes undergo cyclization to form a diradical derivative of benzene that is capable of cleaving DNA.

KEY TERMS

Section 12.1a
initiation
homolysis
propagation
termination

Section 12.1c
bond dissociation energy

Section 12.2a
hydrogenolysis

Section 12.2b
Raney nickel

REACTION SUMMARY

Sections 12.1b, 12.1d, 12.1e

Free radical halogenation of C–H bonds. Ultraviolet light is used as the initiator, and the reactivity order is 3° > 2° > 1° > methyl > aryl, vinyl. Bromination is more selective than chlorination.

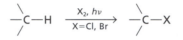

Section 12.1f

Free radical bromination of allylic and benzylic C–H bonds with NBS. Heating with peroxides normally initiates the reaction.

R = Ar, C=C

Section 12.2a

Hydrogenolysis of benzylic ethers and alcohols. The benzylic C–X bond is the weakest in the molecule, so it is readily cleaved with hydrogen and a metal catalyst.

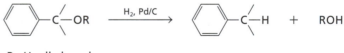

R = H, alkyl, aryl

Section 12.2b

Desulfurization and deselenization of thiols, sulfides, and selenides. Raney nickel removes S and Se from molecules, replacing C–S (or C–Se) with C–H bonds.

$$-\overset{|}{\underset{|}{C}}-S-R \quad \xrightarrow{\text{Raney nickel}} \quad -\overset{|}{\underset{|}{C}}-H \; + \; RH \; + \; H_2S$$

R = H, alkyl, aryl

Section 12.2c

Reduction of alkynes with Na or Li dissolved in liquid NH_3. An alcohol (*tert*-butyl alcohol) is sometimes added to the mixture to provide a source of protons.

$$-C\equiv C- \quad \xrightarrow{\text{Na, NH}_3 \text{(liq)}, \, t\text{-BuOH}} \quad \overset{H}{\underset{}{}}C=C\overset{}{\underset{H}{}}$$

Section 12.2d

The halogen atom is replaced by hydrogen when an organohalide is heated with tributyltin hydride and AIBN.

X=Cl, Br, I

Section 12.3a

Hydrogen bromide adds to an alkene in the presence of peroxides with apparent "anti-Markovnikov" regiochemistry, which derives from formation of the more stable radical in the first step.

Section 12.3b

Radical addition to alkenes leads to formation of polymers.

$$X \cdot \ + \ n \ \diagdown C = C \diagup \ \longrightarrow \ X \left(C - C \right)_{n-1} C = C \diagup$$

Section 12.3c

A five-membered ring is formed by the reaction of an organohalide and tributyltin hydride when the halogen atom is attached to a carbon atom four bonds away from a π bond. This cyclization process depends on the concentration of the tin hydride reagent. Replacement of the halogen atom by hydrogen competes with cyclization.

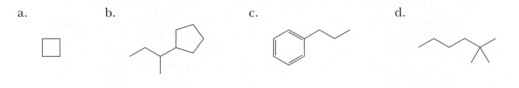

X=Cl, Br, I

ADDITIONAL EXERCISES

12.17. What would be the product ratio expected from chlorination of 2-methylpropane if all of the hydrogen atoms were to react at equal rates?

12.18. Suppose each of the following compounds were treated with 1 equiv of Br_2 under the influence of UV light. Which would produce a reasonably pure monobromo product? If more than one positional isomer will be formed, draw the structure of each expected monobromo compound.

a.   b.   c.   d.

12.19. Draw the structure of the major product expected from each of the following transformations. Draw the structure of each stereoisomer if the product contains a new chiral carbon atom.

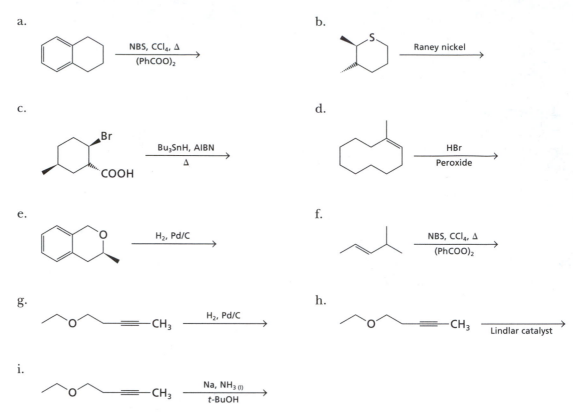

i.

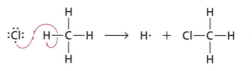

12.20. a. Draw and label a reaction coordinate diagram for the two propagation steps in the chlorination of methane. (Figure 12.1 shows only the first propagation step.) On the diagram, indicate the $\Delta H°$ values for each step.

b. Suppose the mechanism for chlorination of methane were to occur in the following way:

$$:\ddot{\text{C}}\text{l}: \quad \text{H}-\overset{\overset{\displaystyle H}{|}}{\underset{\underset{\displaystyle H}{|}}{\text{C}}}-\text{H} \longrightarrow \text{H}\cdot \;+\; \text{Cl}-\overset{\overset{\displaystyle H}{|}}{\underset{\underset{\displaystyle H}{|}}{\text{C}}}-\text{H}$$

Chloromethane

$$:\ddot{\text{C}}\text{l}-\ddot{\text{C}}\text{l}: \quad \cdot\text{H} \longrightarrow :\ddot{\text{C}}\text{l}: \;+\; \text{Cl}-\text{H}$$

Calculate the value of $\Delta H°$ for each step (use the data in Tables 12.1 and 12.2). Draw a reaction coordinate diagram for this alternative mechanism and include the $\Delta H°$ values. Which mechanism is more likely, based on the energies of each process? Making the assumption that $\Delta G° = \Delta H°$, what is the value of $\Delta G°$ for the overall reaction by each mechanism? What does this result tell you about the standard free energy change for a reaction and its mechanism?

12.21.   From BDE values given in Tables 12.1 and 12.2, calculate the values of $\Delta H°$ for the following processes.

a.   propane + Br· → 1-propyl radical + HBr

b.   Propane + Br$_2$ → 1-bromopropane + HBr

c.   Propane + Br· → 2-propyl radical + HBr

d.   Propane + Br$_2$ → 2-bromopropane + HBr

12.22.   Draw a reaction coordinate diagram like the one in Exercise 12.20 for the following two bromination processes; make use of the $\Delta H°$ values that you calculated in Exercise 12.21:

a.   Step 1   Propane + Br· → 1-propyl radical + HBr

    Step 2   1-Propyl radical + Br$_2$ → 1-bromopropane + Br·

b.   Step 1   Propane + Br· → 2-propyl radical + HBr

    Step 2   2-Propyl radical + Br$_2$ → 2-bromopropane + Br·

12.23.   From the BDE values given in Tables 12.1 and 12.2, calculate the values of $\Delta H°$ for the following processes. The C–Br bond in α-bromotoluene has a BDE of 58 kcal mol^{-1}.

a.   Toluene + Br· → benzyl radical + HBr

b.   Benzene + Br· → phenyl radical + HBr

c.   Toluene + Br$_2$ → α-bromotoluene + HBr

d.   Benzene + Br$_2$ → bromobenzene + HBr

12.24.   Draw a reaction coordinate diagram like the one in Exercise 12.20 for the following two bromination processes; make use of the $\Delta H°$ values that you calculated in Exercise 12.23:

a.   Step 1   Toluene + Br· → benzyl radical + HBr

    Step 2   Benzyl radical + Br$_2$ → α-bromotoluene + Br·

b.   Step 1   Benzene + Br· → phenyl radical + HBr

    Step 2   Phenyl radical + Br$_2$ → bromobenzene + Br·

12.25.   Anti-Markovnikov addition of HBr occurs with alkynes, too. What are the two major products of the following reaction assuming addition of 1 equiv of HBr? (They are geometric isomers.) Name the products.

$$CH_3CH_2CH_2CH_2-C\equiv C-H \xrightarrow{\text{HBr, ROOR, }\Delta} \ ? \qquad (74\%)$$

12.26.   An old bottle of tetralin (1,2,3,4-tetrahydronaphthalene) sitting in a chemical storage room was found to contain 1-hydroperoxytetralin. Propose a mechanism to account for the genesis of this impurity.

Tetralin

1-Hydroperoxytetralin

12.27. Draw the structure(s) of the major product(s) expected from each of the following reactions. Indicate the stereochemistry of the product as appropriate. Relative stereochemistry should be shown using wedges and dashed lines. If a racemic mixture will be formed, draw the structure of one enantiomer and write the word "racemic", or draw both enantiomeric structures. If diastereomers are formed, draw each structure; label meso compounds as such. If no reaction occurs, write N.R.

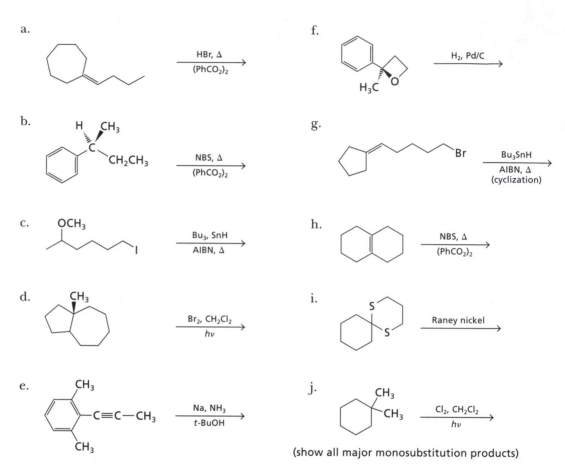

a.

$$\xrightarrow[\text{(PhCO}_2)_2]{\text{HBr, } \Delta}$$

b.

$$\xrightarrow[\text{(PhCO}_2)_2]{\text{NBS, } \Delta}$$

c.   OCH₃

$$\xrightarrow[\text{AIBN, } \Delta]{\text{Bu}_3\text{, SnH}}$$

d.   CH₃

$$\xrightarrow[\text{h}\nu]{\text{Br}_2\text{, CH}_2\text{Cl}_2}$$

e.   CH₃
      —C≡C—CH₃

$$\xrightarrow[\text{t-BuOH}]{\text{Na, NH}_3}$$
      CH₃

f.

$$\xrightarrow{\text{H}_2\text{, Pd/C}}$$
      H₃C

g.

$$\xrightarrow[\substack{\text{AIBN, } \Delta \\ \text{(cyclization)}}]{\text{Bu}_3\text{SnH}}$$

h.

$$\xrightarrow[\text{(PhCO}_2)_2]{\text{NBS, } \Delta}$$

i.

$$\xrightarrow{\text{Raney nickel}}$$

j.   CH₃
      CH₃

$$\xrightarrow[\text{h}\nu]{\text{Cl}_2\text{, CH}_2\text{Cl}_2}$$

(show all major monosubstitution products)

12.28. During petroleum refining, crude oil is subjected to a process called *pyrolysis* in which a molecule is heated so hot that some of its carbon–carbon and carbon–hydrogen bonds undergo homolysis. The radicals produced by this process recombine, creating hydrocarbons that were not necessarily present at the beginning. In some cases, carefully controlled conditions can lead to the production of specific compounds. At which bonds (either C–C or C–H) would you expect homolysis to occur most readily in each of the following molecules? Why?

a.   b.   c.   d.

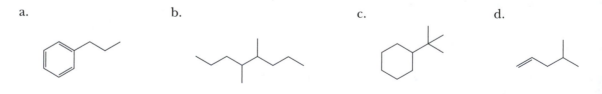

12.29. Carbonyl compounds, especially aldehydes, can react via radical addition processes. Propose a mechanism to account for the formation of the product in the following reaction:

12.30. Describe what is impractical about each of the following transformations. Give the actual major product. If no reaction occurs, write N.R.

a.

b.

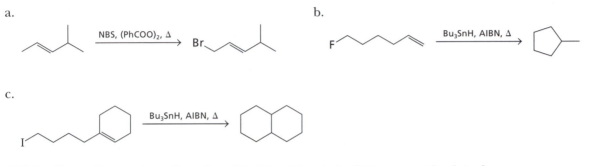

c.

12.31. Normally, reaction of an alkenyl iodide with a tin hydride reagent leads to formation of a ring:

Normal:

Rationalize the following observations by proposing a reasonable mechanism, noting the different conditions used for each transformation.

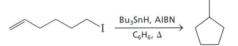

12.32. Propose a mechanism to account for formation of the aldehyde in the following reaction, which involves a radical addition step:

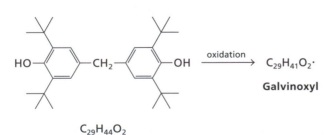

12.33. The following molecule can be oxidized to form a stable free radical called galvinoxyl, which has a deep purple color. Draw the structure of galvinoxyl and discuss what two factors make it stable (*Hint:* Many radicals are unstable because they dimerize: 2 X· → X₂.)

12.34.   *tert*-Butyl hypochlorite can be used to carry out allylic chlorination reactions. Propose a mechanism for the following reaction:

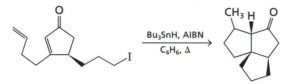

12.35.   The iodoketone shown below undergoes two radical cyclization steps to yield a product that has three five-membered rings. Propose a reasonable mechanism for this transformation, which begins with iodine atom abstraction by Bu₃Sn· .

PROTON AND CARBON NMR SPECTROSCOPY

Whether you are performing a single chemical reaction or carrying out a multistep synthetic procedure, you usually need to verify or deduce the structures of the products. Spectroscopic methods provide a straightforward means to accomplish either of these objectives. Because most organic compounds share the same structural features (Section 1.1b), few techniques are needed to differentiate their structures, and this book will describe the features of the following four types of spectra.

Proton nuclear magnetic resonance (^{1}H NMR) spectra (Sections 13.1–13.4) result from the absorption of energy that affects the spins of a molecule's hydrogen nuclei. These spectra can be used to determine a proton's local environment, which includes the attached and neighboring carbon atoms, heteroatoms, and pi bonds.

Carbon nuclear magnetic resonance (^{13}C NMR) spectra (Sections 13.5 and 13.6) are comparable to ^{1}H NMR spectra, except that the environments of the carbon atoms are detected.

Mass spectra (Section 14.2) are used to obtain the molecular weight and formula of a compound and to identify the presence of certain heteroatoms. The characteristic patterns in a mass spectrum can be used to identify substructures of the carbon skeleton.

Infrared (IR) spectra (Section 14.3) result from the absorption of energy that affects bond vibrations. Infrared spectroscopy is used to identify functional groups that are present in a molecule.

Proton nuclear magnetic resonance spectroscopy is the technique most commonly used to characterize organic compounds because it is relatively simple and does not alter or destroy the sample. Interpreting a ^{1}H NMR spectrum means that you identify the environment of every proton-containing group within a molecule and recognize how those fragments are connected. Along with knowledge about which functional groups are present, this information often allows you to deduce the complete structure of a molecule.

The use of ^{1}H NMR spectroscopy is not limited to structure elucidation. It is also a valuable tool for studying reactive intermediates and mechanisms of chemical reactions. A related technique is *magnetic resonance imaging (MRI),* which exploits ^{1}H NMR to study the composition of tissues and living organisms. In medicine, MRI provides an indispensable means for probing internal organs and structures without resorting to potentially damaging high-energy X-rays or to invasive surgery.

This chapter begins with a brief summary of the theoretical basis for the NMR phenomenon and then illustrates how ^{1}H NMR spectra can be used to identify the environments of the hydrogen atoms in an organic molecule. Procedures for interpreting the information obtained from ^{13}C NMR spectra are also presented, showing how this technique complements data obtained from the ^{1}H NMR spectra. Chapter 14 describes how data from elemental analyses and NMR, IR, and mass spectra can be used to determine molecular structures.

13.1   CHEMICAL SHIFTS AND PROTON EQUIVALENCE

13.1a   NMR SPECTRA RESULT FROM ABSORPTION OF ENERGY THAT CHANGES THE SPIN STATE OF A PROTON OR NEUTRON

Atoms that contain an odd number of protons or neutrons are like small magnets because of the spins of the nuclear particles. The simplest nucleus of this type is that of the hydrogen atom, which consists of a single proton. Its spin, which is either clockwise or counterclockwise, is associated with a spin quantum number, which is either $-\frac{1}{2}$ or $+\frac{1}{2}$. In the absence of a magnetic field, the energies of the two spin states are the same, or degenerate.

When placed in an external magnetic field, however, a proton's spin, and the associated magnetic field created by its spinning, can align either with or against the external field (Figure 13.1). If the proton's local field has the same direction as the external field, its spin quantum number is $+\frac{1}{2}$, and it is aligned with the field. If the proton spin generates a field with the opposite direction, the proton's spin quantum number is $-\frac{1}{2}$. Protons with nuclear spins aligned with the field have lower energy values than those that are opposed, and the two spin states are no longer degenerate.

When a molecule in a magnetic field is irradiated with radio waves, the proton can absorb energy (Eq. 13.1), changing its $+\frac{1}{2}$ spin to the higher, less-favorable state (spin $-\frac{1}{2}$). This absorption process creates the peaks observed in an NMR spectrum.

The frequency (ν) of radiation (units: megahertz, MHz) that is necessary to cause the spin change is related to the energy difference (ΔE) between the spin states, as given by Eq. 13.1, and to the field strength (H_0), as given by Eq. 13.2. The constant γ is called the *gyromagnetic ratio,* and it is characteristic for each kind of magnetically active nucleus.

$$\Delta E = h\nu \tag{13.1}$$

$$\nu = \frac{\gamma H_0}{2\pi} \tag{13.2}$$

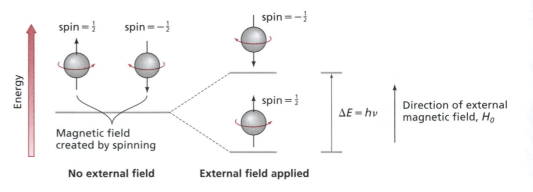

Figure 13.1

Alignment of the magnetic dipoles of a proton in a magnetic field.

13.1b PROTONS WITHIN A MOLECULE CAN BE MAGNETICALLY EQUIVALENT

If every proton in a molecule absorbed radio waves at the same frequency, NMR spectroscopy would not be possible. Fortunately, protons absorb at slightly different frequencies because of the influence of the surrounding electrons. Nevertheless, groups of protons can be **magnetically equivalent,** which means they experience identical influences when placed in an external magnetic field. The absorption frequencies for magnetically equivalent protons *are* the same.

To decide whether protons are magnetically equivalent or not, consider replacing each proton separately by another atom. Protons that are replaced to form the same compound or its enantiomer are magnetically equivalent. For example, replacement of any of the three protons in a methyl group yields the same product, as illustrated below for butane, so all three protons of this methyl group are magnetically equivalent.

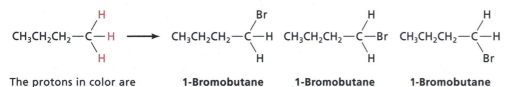

The protons in color are magnetically equivalent.   **1-Bromobutane**   **1-Bromobutane**   **1-Bromobutane**

(If the replacement process generates diastereomers, then the protons are *not* magnetically equivalent; molecules of that type will be discussed in Section 16.3b.)

Protons attached to symmetry-related atoms are also magnetically equivalent. To identify symmetry-equivalent atoms or groups, you have to look for the presence of a *center,* an *axis,* or a *plane of symmetry.* To see if one of these symmetry elements is present, expand the structure, as shown with the examples illustrated in Example 13.1, and/or make a model.

EXAMPLE 13.1

Indicate which protons in each of the following structures are magnetically equivalent to the one shown in color:

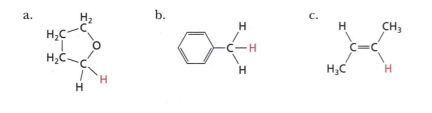

a.   b.   c.

---

a. First, consider replacing each of the protons of the methylene group with "X". These compounds (**B**) are enantiomers, so the methylene protons shown in color are magnetically equivalent.

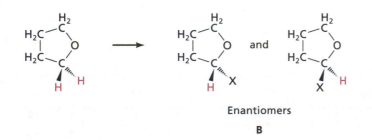

Enantiomers

B

Next, determine if the molecule has a center, axis, or plane of symmetry. This molecule has a plane of symmetry, so the protons with a mirror-image relationship are magnetically equivalent.

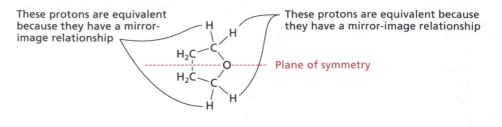

Therefore, all four protons (shown in color below) attached to the carbon atoms adjacent to the oxygen atom are magnetically equivalent.

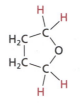

b. For this molecule, we might consider replacing each of the protons of the methyl group with "X" to see if they are equivalent, but we can also make use of the fact that a methyl group possesses a threefold axis that interrelates the positions of its protons.

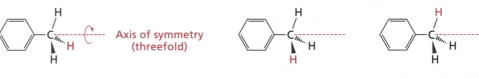

Thus, this axis of symmetry means that all three protons (shown in color below) are magnetically equivalent.

c. As in part b, we might consider replacing each alkene proton to see which are equivalent, but we can also recognize that this molecule possesses a center of symmetry (lines connecting the same atoms or groups pass through the center of the molecule).

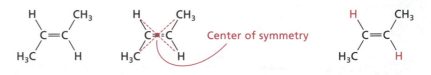

Thus, this *center of symmetry* means that the alkene protons are magnetically equivalent. (All six methyl group protons are also magnetically equivalent.)

EXERCISE 13.1

Indicate which protons in each structure are chemically equivalent to those shown in color.

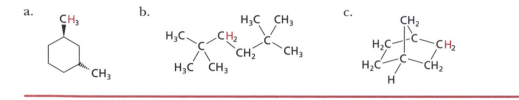

a.   b.   c.

13.1c   THE POSITION OF AN ABSORPTION PEAK IN AN NMR SPECTRUM IS CALLED ITS CHEMICAL SHIFT

Consider now the effect of a molecule's electrons on the absorption frequencies of its protons. If you have studied physics, you learned that a magnetic field causes electrons to move in a circular path, which creates a secondary magnetic field. In organic molecules, the secondary field created by a molecule's electrons opposes, and thus lowers, the externally applied magnetic field, a process known as **shielding.** Each proton therefore experiences a slightly lower magnetic influence than it would in the absence of the electrons. Because field strength and frequency are related (Eq. 13.2), the degree of shielding by the electrons leads to varied magnetic influences, and therefore to the different frequencies needed to cause resonance.

The compound CH_3OCH_2Br has the ^1H NMR spectrum shown in Figure 13.2, which was recorded with a spectrometer that operates in the 200-MHz (10^6 Hz) range.

The frequencies of the absorption peaks differ by only several hundred hertz (Hz), that is, as parts per million (ppm) of the spectrometer's frequency (MHz). To measure precise values of both field strength and frequency for each signal in an NMR spectrum is a daunting task, but gauging differences between the positions of signals is not as difficult. To make such measurements reproducible, the peak positions are specified in relation to the absorption of a reference compound, TMS, $(CH_3)_4Si$, which is defined as 0.0 ppm. The numeric value of a signal relative to the position of the TMS peak is called the **chemical shift,** and its designation is δ. Thus, the absorption peaks for the methyl and methylene protons of CH_3OCH_2Br appear at δ 3.5 and δ 5.7, respectively. Notice that "ppm" is not generally written in conjunction with the symbol δ.

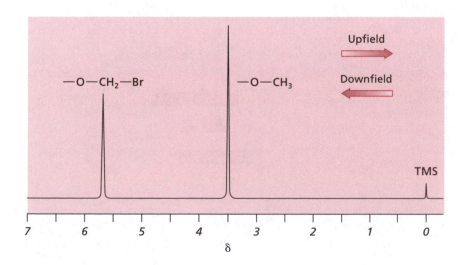

Figure 13.2
The 200-MHz ^1H NMR spectrum of CH_3OCH_2Br with an internal standard of tetramethylsilane (TMS), showing the chemical shift values.

The CH$_3$OCH$_2$Br molecule has five protons, but the spectrum displays only two signals. The two protons of the methylene group and the three protons of the methyl group are magnetically equivalent (Section 13.1b). The following point is important:

> The chemical shifts of magnetically equivalent protons are identical.

The signals observed for the methyl and methylene protons in CH$_3$OCH$_2$Br are **downfield** from the TMS signal, which means they appear to the left of the TMS signal (higher ppm value). The signal for the CH$_3$ group is **upfield** (to the right) of the signal for the CH$_2$ group but *downfield* (to the left) of the TMS signal. Notice that the terms *downfield* and *upfield* are used to denote relative positions.

Chemical shift values are related to the screening effects caused by the electrons. Protons that give signals that are further downfield are **deshielded**, which means they experience a slightly stronger magnetic field than those absorbing upfield. Conversely, proton signals upfield from others are more **shielded** and experience a slightly weaker magnetic field. Figure 13.3 shows approximate chemical shift ranges for various struc-

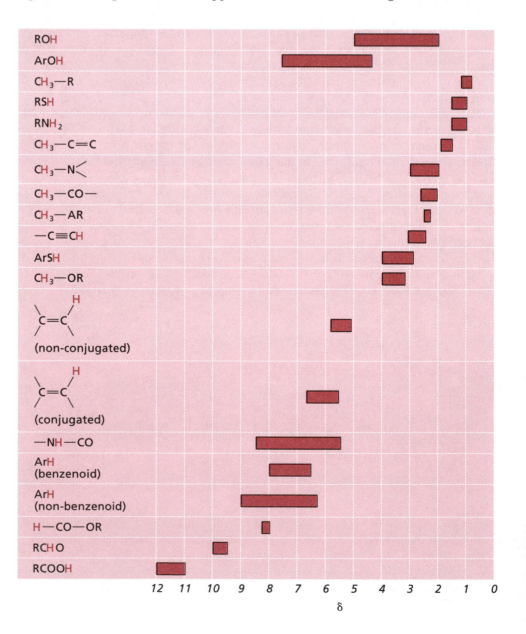

Figure 13.3
Approximate chemical shift ranges for protons.

tural units; values for aliphatic protons are given in Table 13.1. Chemical shifts are influenced mainly by inductive and magnetic anisotropy effects.

Inductive effects are related to the presence of electron-withdrawing groups on the adjacent atoms. If electron density is withdrawn from around the hydrogen atom nucleus toward a more electronegative atom, as illustrated below, the proton will be deshielded, and its signal will appear at a position farther downfield.

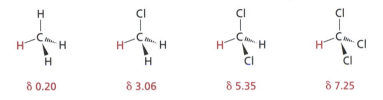

For example,

CH₃—CH₃   CH₃—Cl   CH₃—OCH₃

δ 0.26   δ 3.06   δ 3.24

Inductive effects are additive, so additional electron-withdrawing groups attached to a carbon atom will shift the proton signal farther downfield than if only one such group is present. For example,

δ 0.20   δ 3.06   δ 5.35   δ 7.25

Table 13.1 Approximate chemical shift ranges for protons bonded to aliphatic carbon atoms.

Methyl protons	δ	Methylene protons	δ	Methine protons	δ
CH₃–C	0.9	–CH₂–C	1.3	–CH–C	1.5
CH₃–C–C=C	1.1	–CH₂–C–C=C	1.7	–CH–C–C=C	1.9
CH₃–C–O	1.4	–CH₂–C–O	1.9	–CH–C–O	2.0
CH₃–C=C	1.6	–CH₂–C=C	2.3	–CH–C=C	2.2
CH₃–Ar	2.3	–CH₂–Ar	2.7	–CH–Ar	3.0
CH₃–CO–R	2.2	–CH₂–CO–R	2.4	–CH–CO–R	2.7
CH₃–CO–Ar	2.6	–CH₂–CO–Ar	2.9	–CH–CO–Ar	3.5
CH₃–CO–O–R	2.0	–CH₂–CO–O–R	2.2	–CH–CO–O–R	2.5
CH₃–CO–O–Ar	2.4	–CH₂–CO–O–Ar	2.7	–CH–CO–O–Ar	2.9
CH₃–O–R	3.3	–CH₂–O–R	3.4	–CH–O–R	3.7
CH₃–O–H	3.5	–CH₂–O–H	3.6	–CH–O–H	3.9
CH₃–OAr	3.8	–CH₂–OAr	4.3	–CH–OAr	4.5
CH₃–O–CO–R	3.7	–CH₂–O–CO–R	4.1	–CH–O–CO–R	4.8
CH₃–N	2.3	–CH₂–N	2.5	–CH–N	2.8
CH₃–NO₂	4.0	–CH₂–NO₂	4.4	–CH–NO₂	4.7
CH₃–C–NO₂	1.6	–CH₂–C–NO₂	2.2	–CH–C–NO₂	2.6
CH₃–C=C–CO	2.0	–CH₂–C=C–CO	2.1	–CH–C=C–CO	2.4
CH₃–C–Cl	1.4	–CH₂–C–Cl	1.8	–CH–C–Cl	2.0
CH₃–C–Br	1.8	–CH₂–C–Br	1.8	–CH–C–Br	1.9
CH₃–Cl	3.0	–CH₂–Cl	3.4	–CH–Cl	4.0
CH₃–Br	2.7	–CH₂–Br	3.3	–CH–Br	3.9

Notice that the additive effects of electron-withdrawing groups are not linear, so the second electron-withdrawing group produces a smaller effect than the first one.

Inductive effects apply to protons bonded to heteroatoms as well. In addition, hydrogen bonds (Section 2.8b) often influence the chemical shifts of protons attached to oxygen and nitrogen atoms by deshielding them. A combination of these effects is used to rationalize the large downfield shifts of carboxylic acid protons versus those of alcohols, for example.

Magnetic anisotropy is a term used to describe the effects caused by localized magnetic fields that are created by the bonding electrons in a molecule, especially those in π bonds. Magnetic anisotropy effects have specific directional orientations that influence the degree of shielding.

To see the effects of magnetic anisotropy, compare the resonance values for the protons in the following compounds.

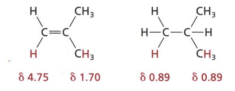

Each substituent (H and CH₃) connected to the alkene carbon atoms displays a resonance that is downfield from the corresponding group in the alkane. The π bond of the alkene creates a localized magnetic field that is deshielding toward the protons of these attached groups.

Magnetic anisotropy is particularly important for aromatic compounds. The protons attached to benzene give rise to absorptions farther downfield than those in cyclohexane.

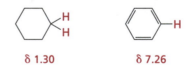

The following set of oxygen-containing molecules illustrates the effects of magnetic anisotropy caused by the presence of a carbonyl π bond. Along with the high electronegativity of the oxygen atom, the effect of magnetic anisotropy within the aldehyde group causes a significant downfield shift of its proton resonance compared with that of either the alkene or the alcohol (the protons being compared are shown in color).

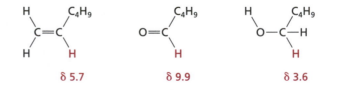

A general trend to remember is that the chemical shift value of a methyl group is usually upfield from that of a methylene group, which in turn is upfield from that of a methine group, *if the local environment is the same.* For example, in the following series of ketones, the chemical shift for the protons shown in color increases with the *decreasing* number of protons on the carbon atom α to the carbonyl group.

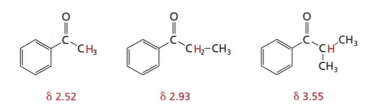

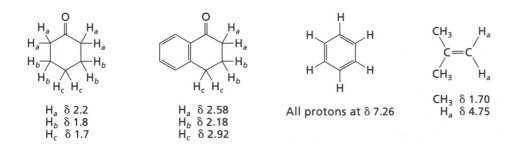

As noted already, the chemical shifts of magnetically equivalent protons are identical. Some examples of equivalent and nonequivalent protons are shown in Figure 13.4 along with their corresponding chemical shift values.

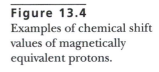

Figure 13.4
Examples of chemical shift values of magnetically equivalent protons.

EXAMPLE 13.2

For the compound shown here, specify the approximate chemical shift range for each proton or set of protons.

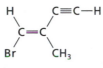

First, identify the types of protons present. This molecule contains protons attached to an alkyne carbon atom, an alkene carbon atom, and an alkane carbon atom adjacent to a *p* bond (C=C–C–H). Next, use the data in Figure 13.3 to assign approximate chemical shift values to the three types of protons.

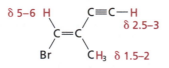

EXAMPLE 13.3

Estimate the chemical shift values for the protons in 1,1-dichloroethane.

First, we recognize that there are two types of protons—methyl and methine—so we expect to see two NMR signals. For the methine proton signal, we make use of the chemical shift values given earlier in this section for the chlorinated methane derivatives. The chemical shift for the methine proton in dichloroethane should be upfield from that of the methylene protons in dichloromethane because a methyl group is electron donating compared with a proton (Section 6.2b). We estimate that the methine proton signal will appear at ~ δ 5.2.

For the methyl proton signal, use the data in Table 13.1. Remember that a second electron-withdrawing group has less effect on the chemical shift than the first such group, so we add 0.4 instead of 0.5 (b.). The methyl proton resonances will appear at ~ δ 1.8.

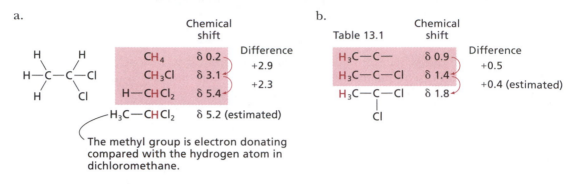

a.

	Chemical shift	
CH_4	δ 0.2	Difference
CH_3Cl	δ 3.1	+2.9
$H—CHCl_2$	δ 5.4	+2.3
$H_3C—CHCl_2$	δ 5.2 (estimated)	

The methyl group is electron donating compared with the hydrogen atom in dichloromethane.

b.

Table 13.1

	Chemical shift	
$H_3C—C—$	δ 0.9	Difference
$H_3C—C—Cl$	δ 1.4	+0.5
$H_3C—C—Cl$ (with Cl)	δ 1.8	+0.4 (estimated)

EXERCISE 13.2

For each of the following compounds, give the approximate chemical shift ranges for each proton or set of protons making use of the data in Figure 13.3 and Table 13.1:

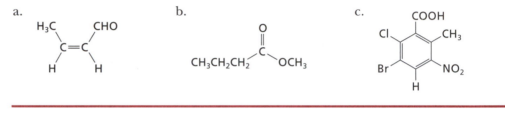

13.1d INTEGRATED INTENSITY VALUES ARE PROPORTIONAL TO THE NUMBERS OF MAGNETICALLY EQUIVALENT PROTONS

Because magnetically equivalent protons have identical chemical shift values, an absorption peak in an NMR spectrum may correspond to one, two, three, or more protons. When you record a ^1H NMR spectrum, the spectrometer calculates **integrated intensity values,** which are proportional to the areas under the peaks, and therefore to the number of protons that create those signals. The *integrated* ^1H NMR spectrum for CH_3OCH_2Br (see Fig. 13.2) is shown in Figure 13.5.

Keep in mind that integrated intensity values for a spectrum are not necessarily *absolute* values, as they are for CH_3OCH_2Br. Instead, the number of protons associated

Figure 13.5
The ^1H NMR spectrum of CH_3OCH_2Br with its integrated intensity values shown in parentheses above each peak.

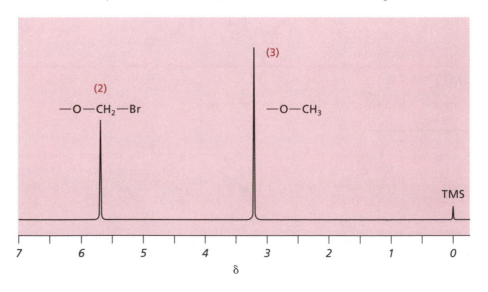

with each signal may actually be a multiple of the calculated values. For the spectra presented in this text, the integrated intensity value associated with each signal will be given as a number in parentheses above each set of peaks.

EXAMPLE 13.4

2,5-Hexanedione has the ^{1}H NMR spectrum shown below. Assign each absorption peak to the appropriate set of protons in this molecule.

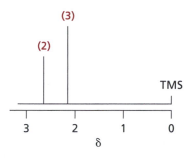

First, draw the structural formula for 2,5-hexanedione:

a.   b.

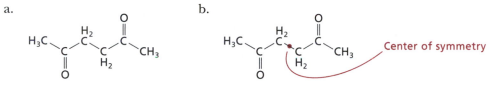

2,5-Hexanedione

The molecule has four hydrocarbon groups (a.), but the spectrum shows only two signals; therefore some of the protons must be magnetically equivalent. The next step is to determine which protons are equivalent, which is done by the procedures illustrated in Example 13.1. 2,5-Hexanedione has a center of symmetry (b.), so its two methylene groups and two methyl groups are equivalent.

The signals in the spectrum appear at δ 2.6 and 2.1, and the integrated intensity values for these signals are 2 and 3, respectively. Because integrated intensity values are ratios, these two signals correspond to 2 and 3, 4 and 6, or 6 and 9, and so on protons. Looking at the magnetically equivalent protons, we see that there are 4 methylene protons and 6 methyl protons. Therefore, the peak at δ 2.6 is the resonance for the 4 methylene protons, and the peak at δ 2.1 signal is the resonance for the 6 methyl protons.

EXERCISE 13.3

1,4-Dimethoxybenzene has the following ^{1}H NMR spectrum. Assign each absorption peak to the appropriate set of protons in this molecule.

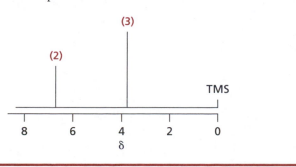

13.2   SPIN COUPLING IN SIMPLE SYSTEMS

13.2a   SPIN COUPLING OCCURS BECAUSE OF THE MAGNETIC FIELD PRODUCED BY NEIGHBORING PROTONS

The chemical shift and integrated intensity values by themselves would make NMR spectra useful for deducing the structures of molecules, yet even more structural information can be obtained from many ^{1}H NMR spectra. If you were to predict the appearance of the ^{1}H NMR spectrum for chloroacetone (CH_3COCH_2Cl), you would expect to see two signals, one for the protons of the methylene group and one for those of the methyl group. The integrated intensity values would have a ratio of 2:3. The actual spectrum, shown in Figure 13.6, confirms these expectations, displaying peaks at δ 4.2 (relative intensity 2) and δ 2.3 (relative intensity 3) for the CH_2 and CH_3 groups, respectively. The chemical shift for the $COCH_2Cl$ protons is not included in the data given in Figure 13.3 or Table 13.1, so its value has to be estimated. A CH_2 group next to a carbonyl group has a resonance at ~ δ 2.4. Adding a second electron-withdrawing group—the chlorine atom—will shift that peak another 2 ppm or so, to ~ δ 4.5, which is close to the position of the actual signal.

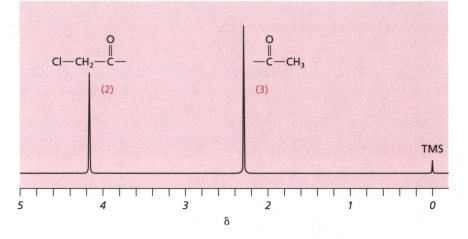

Figure 13.6
The 300-MHz ^{1}H NMR spectrum of chloroacetone, CH_3COCH_2Cl.

If you were to predict the ^{1}H NMR spectrum of propanoyl chloride, CH_3CH_2COCl, you might again expect to see two signals, one for the methylene group and another for the methyl group, with relative intensities of 2:3. The actual spectrum, shown in Figure 13.7, clearly contains many more than two peaks.

The phenomenon that gives rise to the appearance of these additional peaks results from the influence of protons attached to the *adjacent* carbon atoms. The three

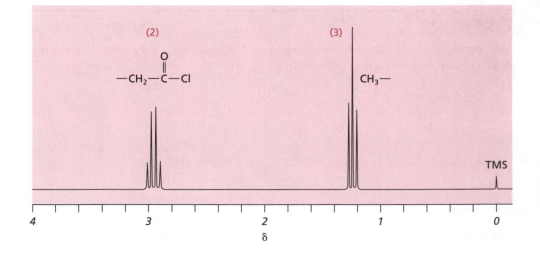

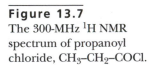

Figure 13.7
The 300-MHz ^{1}H NMR spectrum of propanoyl chloride, CH_3–CH_2–$COCl$.

equivalent protons in the methyl group of CH_3CH_2COCl should generate a single resonance when this compound is placed in a magnetic field. But the protons on the adjacent methylene group will act as small magnets that either enhance or reduce the external field. The nuclear spins of the CH_2 protons, which are aligned either with or against the external magnetic field, exist randomly within a given molecule in one of four possible spin combinations ($\uparrow$ = spin $\frac{1}{2}$ and $\downarrow$ = spin $-\frac{1}{2}$):

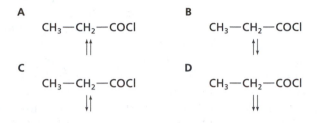

In molecule (**A**), the methyl group experiences a stronger field than it would if the methylene group were absent because the two neighboring spins create small fields that *add* to the external field. The methyl group in these molecules will display a signal slightly downfield from that for an unperturbed methyl group in the same environment.

In molecules (**B**) and (**C**), the influence of the tiny magnetic fields from the adjacent methylene group is zero because the fields cancel. [The combinations in (**B**) and (**C**) are the same energetically because the protons are magnetically equivalent.] The methyl resonances for molecules (**B**) and (**C**) will therefore occur in the same place as the resonances for an unperturbed methyl group. Finally, the methyl group in (**D**) will experience a decreased field, because the small fields generated by the methylene protons will *counteract* the external field. Its resonance signal will appear at a slightly lower frequency than that for an unperturbed methyl group, and so it will be slightly upfield.

Addition of these four combinations shown above (**A** to **D**) creates the pattern (Figure 13.8) that is observed for the methyl signal in the actual spectrum.

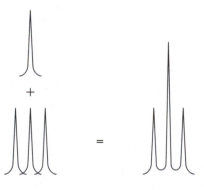

Figure 13.8
Additive effects of the magnetic fields from neighboring protons that generate the observed triplet pattern for the methyl proton signal of propanoyl chloride.

This same procedure can be used to evaluate how the small magnetic fields created by the protons of the methyl group influence the methylene proton signal. Thus, for a CH_2 group adjacent to a methyl group, four peaks will be generated for the methylene proton signal because of the four spin orientations possible from the three methyl protons (Figure 13.9).

Figure 13.9
Additive effects of the magnetic fields from the neighboring protons that generate the observed quartet pattern for the methylene proton signal of propanoyl chloride.

Thus, the resonance signal for a methylene group next to a methyl group should appear as a quartet with peak ratios of 1:3:3:1, which is the actual pattern observed for the methylene group in the spectrum of propanoyl chloride (Fig. 13.7).

Proton resonances are influenced through intervening bonds by the local fields generated by the protons attached to *adjacent* atoms. This effect is termed *spin coupling*, meaning that the spins of nearby protons couple magnetically with each other. More commonly, this phenomenon is called **spin–spin splitting** because of the appearance of the spectrum itself, in which some signals are split into several closely spaced peaks. Peak splitting is *not* observed among magnetically equivalent protons, and the influence of neighboring protons diminishes with distance so that only hydrogen atoms within three bonds of each other (i.e., H–C–C–H) have much effect. Coupling between nuclei separated by three bonds—the most common situation—is called **vicinal coupling.** Protons that are separated by only two bonds can also couple, a phenomenon known as **geminal coupling.** This type of coupling is sometimes observed for the protons of a terminal alkene with unsymmetric substitution, $R(R')C=CH_2$.

A proton resonance involved in spin coupling will be split into $n + 1$ peaks, where n is the number of equivalent protons on the *adjacent* carbon atoms. This phenomenon is called **the $n + 1$ rule,** and the observed pattern of peaks is called its **multiplicity.** In this text, any resonance, including those that have been split, will be referred to as a *feature.*

Whenever spin–spin splitting occurs, the chemical shift value for each feature is determined by finding the center of the feature. If the splitting pattern has an odd number of peaks, the chemical shift corresponds to the central and tallest peak of the splitting pattern; if spin–spin splitting generates an even number of peaks, then the chemical shift corresponds to a point halfway between the innermost peaks.

The magnitude of the splitting between peaks is designated J, the **coupling constant,** which has units of hertz (Hz), not ppm. For the vicinal coupling between similar types of protons, $J \approx 7$ Hz. (The magnitude of J values among protons of different types will be described in Section 13.4a.) Note this key fact:

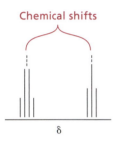

Chemical shifts

δ

> When spin coupling occurs, at least two features must show splitting, and the magnitude of the splitting in those features must be the same.

The ratios of the peak heights in a split feature will have one of the patterns summarized in Table 13.2. For example, the peaks in a doublet will have equal intensities; the peaks in a triplet will have a 1:2:1 intensity ratio; a quartet will have a 1:3:3:1 intensity pattern, and so on.

Spin–spin splitting occurs even when the protons are distributed over more than one carbon atom, as illustrated by the ^1H NMR spectrum of 2-chloropropane presented in Figure 13.10. 2-Chloropropane has two equivalent methyl groups, which means that six protons are equivalent. Therefore, the methine proton signal is split into $6 + 1 = 7$ peaks, which is better seen in the expanded spectrum (inset).

Table 13.2 Intensities of peaks in the patterns caused by spin–spin splitting.

n, No. of adjacent protons	n + 1, Feature	Intensity ratios
0	1, singlet	1
1	2, doublet	1:1
2	3, triplet	1:2:1
3	4, quartet	1:3:3:1
4	5, quintet	1:4:6:4:1
5	6, sextet	1:5:10:10:5:1
6	7, septet	1:6:15:20:15:6:1

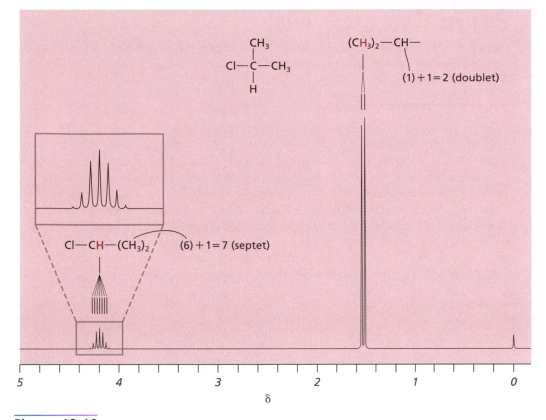

Figure 13.10
The 300-MHz ^{1}H NMR spectrum of 2-chloropropane, showing the spin–spin splitting patterns. The hydrogen atoms that are shown in color are the ones giving rise to the signal *before* splitting by the adjacent protons. *Inset*: The expanded spectrum of the methine proton region.

As you become familiar with looking at ^{1}H NMR spectra, you will begin to recognize common splitting patterns that the aliphatic protons create. The patterns observed for the ethyl, ethylene, and isopropyl groups are illustrated in Figure 13.11. As noted above, the magnitude of the splitting (the *J*-value) in the observed features is the same.

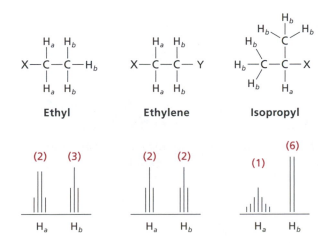

Figure 13.11
Standard splitting patterns observed for protons in common alkyl groups, along with their integrated intensity values. Substituents X and Y represent electron-withdrawing groups that do not contain protons, and therefore do not contribute to spin–spin splitting. For the ethylene group, X is more electron withdrawing than Y.

EXAMPLE 13.5

Sketch the ^{1}H NMR spectrum for 1,1-dichloroethane, showing the chemical shifts, integrated intensity values, and expected splitting pattern (singlet, doublet, triplet, etc.) for each feature.

First, we recognize that there are two types of protons—methyl and methine—so we expect to see two NMR features. We estimate their chemical shift values to be δ 1.8 and δ 5.2 (see Example 13.3) with integrated intensity values of 3 and 1, respectively.

The multiplicity of the signal for each set of protons depends on the number of protons three bonds away. When n protons are attached to the neighboring atom(s), then the splitting pattern will have $n+1$ peaks. The methine proton is three bonds away from the three protons of the neighboring methyl group, so the methine proton signal will appear as a quartet [3 + 1 = 4 (quartet)]. The methyl protons are three bonds away from one methine proton, so the methyl proton signal will appear as a doublet [1 + 1 = 2 (doublet)]. The spectrum should therefore appear as follows:

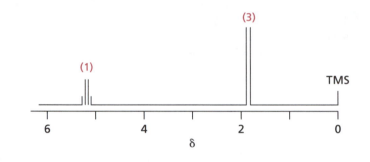

EXERCISE 13.4

Sketch the ^{1}H NMR spectrum for each of the following compounds, showing the chemical shifts, integrated intensity values, and expected splitting pattern (singlet, doublet, triplet, etc.) for each feature:

a.

b.

13.2b COUPLING IN THE AROMATIC REGION OCCURS ONLY AMONG PROTONS ATTACHED TO THE RING

The protons attached to a benzene ring are insulated from protons on groups attached to the ring because more than three bonds separate them. As a result of this separation, spin coupling occurs only between protons attached to the ring.

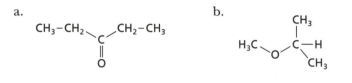
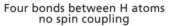
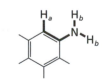

Four bonds between H atoms no spin coupling

Four bonds between H atoms no spin coupling

Three bonds between H atoms; protons engage in spin coupling

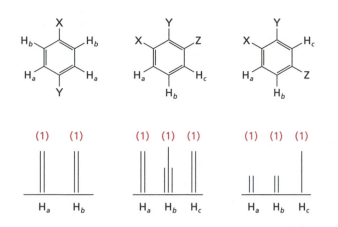

Figure 13.12
Splitting patterns observed for some benzene derivatives (X, Y, and Z are substituents other than H). The order in which the signals appear, left to right, is arbitrary.

Three common splitting patterns observed for benzene derivatives (one disubstituted and two trisubstituted) are illustrated in Figure 13.12. (Additional examples will be presented in Chapter 17.) The actual chemical shifts for each feature may differ from those shown; that is, the relative order of peaks from left to right is arbitrary. The important characteristics to be observed are the peak multiplicities, which result from coupling between protons on adjacent atoms of the ring.

The most easily recognized splitting pattern results from 1,4-disubstitution of the benzene ring in which the two substituents, X and Y, are different. The spectrum of 4-nitrotoluene, presented in Figure 13.13, shows the two-doublet pattern that is distinctive for a 1,4-disubstituted ring. Notice that the sum of integrated intensity values for the entire aromatic region reveals how many substituents are attached to the ring. When this number equals 5, the benzene ring has a single substituent; when the number is 4, the ring has two substituents; when it is 3, the ring has three substituents. In this case, the intensity value of 4 tells us that the ring has two substituents.

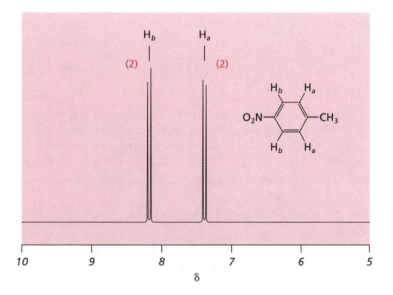

Figure 13.13
The 300-MHz ^{1}H NMR spectrum of 4-nitrotoluene.

13.3   INTERPRETING AND PREDICTING ^{1}H NMR SPECTRA

13.3a INTERPRETING ^{1}H NMR SPECTRA

Modern NMR spectrometers are computer-interfaced instruments capable of measuring spectra on very small quantities. To obtain an NMR spectrum, you dissolve the compound in a solvent that does not contain protons so that extra resonances are not present to interfere with the spectrum. Deuterochloroform, $CDCl_3$, is the most common solvent, but perdeuteroacetone, CD_3COCD_3, perdeuterobenzene, C_6D_6, and deuterium oxide, D_2O, find use as well (the prefix perdeutero- means that all of the hydrogen atoms have been replaced by deuterium). Deuterium has a nuclear spin, but its resonance frequency is different from that used to obtain proton spectra, so it does not interfere.

A ^1H NMR spectrum has four primary characteristics that can be used in the interpretation process:

1. The chemical shift of each feature.
2. The integrated intensity value of each feature.
3. The multiplicity (singlet, doublet, etc.) of each feature.
4. The J value(s) for each feature that shows splitting.

Collecting and tabulating these data is the first step in the interpretation process.

To begin, determine the chemical shift for each feature. These assignments are usually made by comparison with detailed compilations like the ones shown in Figure 13.3 and Table 13.1. At the same time, make note of the integrated intensity value of each feature.

Some types of protons are not easily identified from their chemical shift values because of the large range over which their resonance can appear. Protons attached to heteroatoms—especially oxygen—fall into this category, and these are shown at the top of Figure 13.3. Fortunately, protons of these types usually undergo **deuterium exchange** when treated with D_2O (Eq. 13.3). To detect the presence of such groups, a drop of D_2O is added to the NMR sample. The disappearance of a signal means that the group responsible for that signal usually has its proton attached to a heteroatom. Alcohols (ROH), phenols (ArOH), and carboxylic acids (RCOOH) undergo exchange rapidly. Amides ($RCONH_2$) and thiols (RSH) often need to be treated with a base like NaOD in D_2O to promote exchange. Because DOH is slightly soluble in most organic solvents, its NMR signal will also be observed and must be taken into consideration.

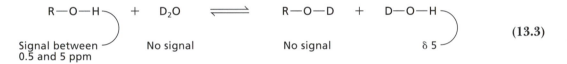

$$R\text{—}O\text{—}H \quad + \quad D_2O \quad \rightleftharpoons \quad R\text{—}O\text{—}D \quad + \quad D\text{—}O\text{—}H \tag{13.3}$$

Signal between 0.5 and 5 ppm   No signal   No signal δ 5

After you have assigned the resonances to appropriate fragments based on the chemical shifts and integrated intensity values, you construct the carbon skeleton by using the spin–spin splitting data to deduce which groups are adjacent to each other.

- Assign the multiplicity, m, of each feature (for a doublet, $m = 2$; for a triplet, $m = 3$; etc.).
- Calculate the number of neighboring protons, n, according to the following equation: $n = m - 1$
- Look for a feature with the same J value that has n protons (according to its integrated intensity value). *These two groups must be adjacent.* Make use of the patterns illustrated in Figures 13.11 and 13.12 to identify common structural fragments.
- Note that protons bonded to an oxygen atom in an alcohol or to a nitrogen atom in an amine do not always engage in spin coupling because they undergo intermolecular exchange faster than the NMR spectrometer can record the proton's $+\frac{1}{2}$ to $-\frac{1}{2}$ transition. In effect, they seem not to be bonded to a particular atom. This rate of exchange can be influenced by impurities in the sample—especially water or acids—so the manner in which the sample is prepared can affect the observations of splitting for the signals attributed to the protons bonded to an oxygen or nitrogen atom.

For simple spectra that have only alkyl and/or benzene protons, all of the J values should be about the same, namely, 7 Hz. Thus, if all of the splitting of signals looks to be equal, then $J \approx 7$ Hz. (If you want confirmation, you can readily estimate the J values: for a spectrum recorded on a 200-MHz NMR instrument, 1.00 ppm = 200 Hz; on a 300-MHz instrument, 1.00 ppm = 300 Hz; on a 400-MHz instrument, 1.00 ppm = 400 Hz; etc.)

The following examples illustrate the interpretation procedure.

EXAMPLE 13.6

Compound **A**, $C_4H_8O_2$, produces the ¹H NMR spectrum shown in Figure 13.14. Draw its structure.

The integrated intensity values are 2:3:3. Because the sum of these numbers (2 + 3 + 3 = 8) is the same as the number of hydrogen atoms in the given molecular formula, these numbers correspond to the absolute numbers of protons for each feature. This molecule has only aliphatic protons (all of the signals are between 0 and 5 ppm), so the observation of a triplet and a quartet in a ratio of 3:2 means that an ethyl group is present (Fig. 13.11). The signal farthest downfield has a chemical shift value consistent with the presence of a methylene group bonded to an oxygen atom. Thus, this molecule has the OCH_2CH_3 group.

The feature that produces the signal at δ 2.07 has an integrated intensity of 3, which means that it is a CH_3 group. Its chemical shift value suggests that the methyl group is adjacent to a carbonyl group, CH_3CO. Combining the two fragments, we conclude that this molecule is ethyl acetate.

$$CH_3-CH_2-O-\overset{\displaystyle O}{\overset{\displaystyle \|}{C}}-CH_3 \quad \textbf{Ethyl acetate}$$

Suppose, however, that you are unsure about the environment of the methyl group that produces the peak at δ 2.07. Because this feature is a singlet, you know that the associated methyl group is bonded to an atom with no proton. If you were to add together the atoms that you are certain about, namely, the OCH_2CH_3 and CH_3 groups, you would have C_3H_8O. This formula differs from the given formula by $C_4H_8O_2-C_3H_8O = CO$, which is the carbonyl group. By using this approach, you would

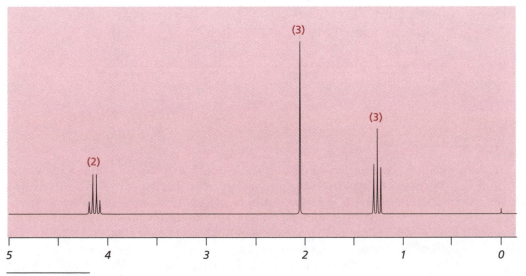

Figure 13.14
The 200-MHz ¹H NMR for compound **A**. For the spin–spin splitting patterns, $J = 7$ Hz.

conclude that the molecule has the CH_3, OCH_2CH_3, and $C=O$ groups, which are assembled as $CH_3–CO–OCH_2CH_3$, or ethyl acetate.

EXERCISE 13.5

Compound **B**, $C_5H_{10}O$, produces the ^1H NMR spectrum shown below. Draw its structure. For the spin–spin splitting patterns, $J = 7$ Hz.

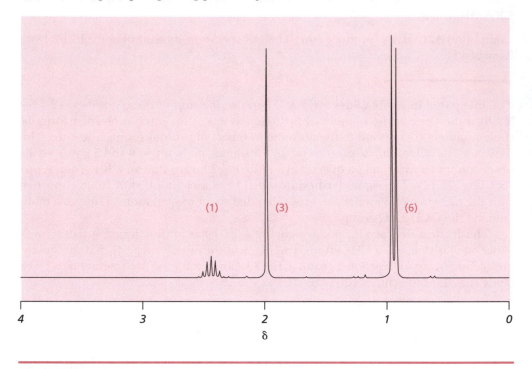

EXAMPLE 13.7

Compound **C**, C_8H_9Br, produces the ^1H NMR spectrum shown in Figure 13.15. Draw its structure.

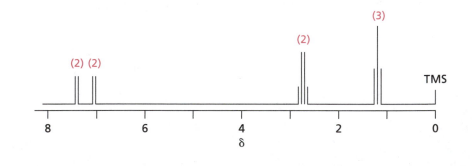

Figure 13.15
The 300-MHz ^1H NMR for compound **C**. For the spin–spin splitting patterns, $J = 7$ Hz.

This spectrum has absorption peaks in two regions: δ 7–7.5 and δ 1–3; therefore, the molecule has both aromatic and aliphatic protons. Reading left to right, the integrated intensity values are 2:2:2:3, the sum of which equals the number of protons in the given molecular formula. The *sum* of the integrated intensity values in the aromatic region is used to determine the degree of substitution on the ring (Section 13.2b). This number is 4, which indicates that the ring has two substituents; the appearance of two doublets in the aromatic region further tells us that these two substituents are attached at positions 1 and 4 (Figs. 13.12 and 13.13).

For the aliphatic protons, the observation of a triplet and a quartet in a ratio of 3:2 means that an ethyl group is present (Fig. 13.11). The signal farthest downfield has a chemical shift value that is consistent with a methylene group bonded to the benzene ring (Table 13.1). Thus, this molecule has an $ArCH_2CH_3$ group.

Adding the atoms assigned so far ($C_6H_4 + CH_2CH_3$) accounts for all of the carbon and hydrogen atoms, leaving only a bromine atom to include. This Br must be attached to the benzene ring across from the ethyl group. The molecule is $Br–C_6H_4–CH_2CH_3$, or 1-bromo-4-ethylbenzene.

EXERCISE 13.6

Compound **D**, $C_9H_{10}O$, produces the following ¹H NMR spectrum. Draw its structure. For the spin–spin splitting patterns, $J = 7$ Hz.

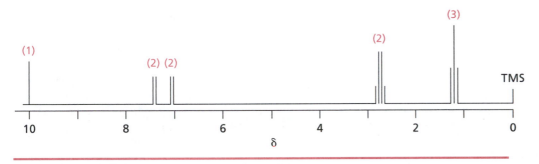

EXAMPLE 13.8

Compound **E**, $C_8H_{10}O$, produces the ¹H NMR spectrum shown in Figure 13.16. The signal at δ 2.5 disappears when the sample is treated with D_2O. Draw the structure of compound **E**.

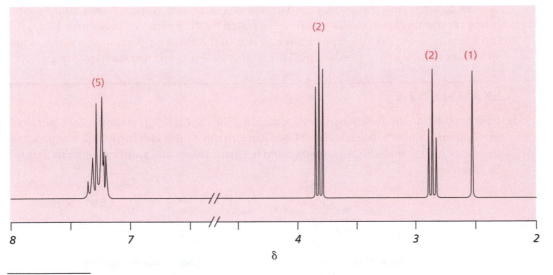

Figure 13.16
The 200-MHz ¹H NMR for compound **E**. For the spin–spin splitting patterns, $J = 7$ Hz.

This spectrum has absorption peaks in two regions: δ 7–7.5 and δ 2–4; therefore, the molecule has both aromatic and aliphatic protons. Reading left to right, the integrated intensity values are 5:2:2:1, the sum of which equals the number of protons in the given molecular formula. The integrated intensity value in the aromatic region is 5, so the benzene ring has a single substituent and we can ignore the splitting patterns.

In the aliphatic region, two triplets are observed; their integrated intensity values suggest that the molecule contains two methylene groups adjacent to each other (Fig. 13.11). Adding the atoms already accounted for ($C_6H_5 + CH_2CH_2$) and subtracting from the molecular formula ($C_8H_{10}O – C_8H_9$) reveals that a hydrogen and an oxygen atom are present, which means that the molecule has an OH group. The presence of an OH group is consistence with the disappearance of the signal at δ 2.5 when D_2O is added (Eq. 13.3). Assembling these fragments yields C_6H_5–CH_2–CH_2–OH or 2-phenylethanol.

EXERCISE 13.7

Compound **F**, $C_5H_{10}O_2$, produces the following ^1H NMR spectrum. Draw its structure. For the spin–spin splitting patterns, $J = 7$ Hz.

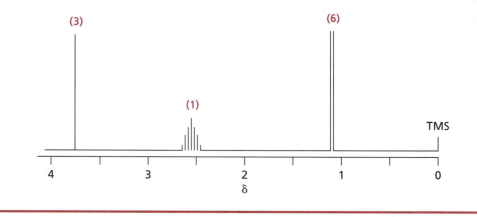

13.3b PREDICTING THE APPEARANCE OF A ^{1}H NMR SPECTRUM

If you perform a chemical reaction in the laboratory, you start with knowledge about what the structure of the product will be. Proton NMR spectra are used in those instances to verify that this expected structure was obtained. Although you could simply interpret the spectrum in the manner illustrated by the examples in Section 13.3a, it is often easier to predict the appearance of the spectrum, and then to see if the actual spectrum matches your prediction. This section will present two such examples.

EXAMPLE 13.9

Upon oxidizing the following benzylic alcohol (Section 11.4c), you expect to obtain 4-methylbenzaldehyde. Sketch the ^1H NMR spectrum of the aldehyde and comment on the expected chemical shift and integrated intensity values and splitting patterns that will be observed.

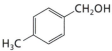
(4-Methylphenyl)methanol

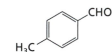
4-Methylbenzaldehyde

First, evaluate the substructures that are present and estimate the general chemical shift ranges (Fig. 13.3). The product has three parts: the aldehyde group (δ 10), the aromatic ring (δ 7–8), and an aliphatic group (δ 0–5).

Next, assign more specific chemical shift values (Table 13.1) and integrated intensity values: CHO (δ 10, 1H), ArH (δ 7–8, 4H), and Ar–CH_3 (δ 2.3, 3H).

Finally, include the effects of spin–spin splitting. The aldehyde proton has no other protons within three bonds, so it will appear as a singlet. The aromatic protons of the 1,4-disubstituted ring will appear as two doublets, each with an integrated intensity value of 2H (Fig. 13.12), and the methyl group will appear as a singlet because it has no other protons within three bonds. The expected spectrum is,

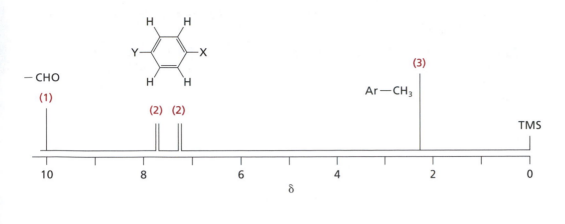

EXAMPLE 13.10

In performing the following Williamson ether synthesis (Section 7.2b), you expect to obtain ethyl isopropyl ether as the product. Sketch the ^{1}H NMR spectrum of this ether and comment on the expected chemical shift and integrated intensity values and splitting patterns that will be observed.

First, evaluate the substructures that are present. All of the protons are aliphatic, so they will all appear in the range δ 0–5.

Next, use Table 13.1 to assign more precise chemical shift values; at the same time, include integrated intensity values: –CH–O–R (δ 3.7, 1H); –CH$_2$–O–R (δ 3.4, 2H); (CH$_3$)$_2$–C–O (δ 1.5, 6H); CH$_3$–C–O (δ 1.4, 3H). Note that the substructure in Table 13.1 used to predict the chemical shifts for the methyl groups is CH$_3$–C–O, which is the same for the methyl portion of both the ethyl and isopropyl groups. Slightly different values are chosen here so that they can both be seen in the predicted spectrum. When the actual spectrum is recorded, you have to be aware that these resonances may be superimposed or perhaps reversed as to their relative positions.

Finally, include the effects of spin–spin splitting (Fig. 13.11). The protons of the isopropyl group [CH(CH$_3$)$_3$] will appear as a septet and doublet; and the protons of the ethyl group [CH$_2$CH$_3$] will appear as a quartet and triplet. The expected spectrum is,

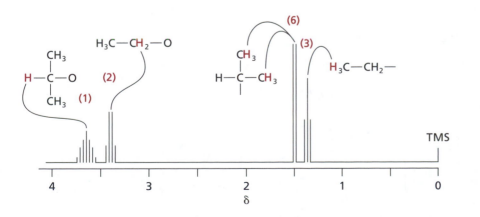

EXERCISE 13.8

In performing the following addition reaction (Section 9.1b), you expect to obtain 1-bromo-1-(4-chlorophenyl)ethane as the product. Sketch the ^{1}H NMR spectrum of this bromo compound and comment on the expected chemical shift and integrated intensity values and splitting patterns that will be observed.

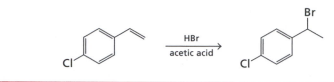

13.4 SPIN COUPLING IN MORE COMPLEX SYSTEMS

13.4a COUPLING BETWEEN DIFFERENT TYPES OF PROTONS CAN YIELD DIFFERENT J VALUES

In the previous examples, all of the J values are ~ 7 Hz, which is typical for aromatic or aliphatic protons engaged in vicinal coupling with each other. When a molecule contains an alkene, alkyne, or aldehyde group, then the J values may vary. Typical J values for several types of coupled systems, which fall roughly into one of three ranges, are summarized in Table 13.3. These values reveal the types of protons that couple with

Table 13.3   A compilation of J values for common systems that engage in spin–spin splitting

Coupled systems with J = 0–4 Hz		Coupled systems with J = 6–10 Hz		Coupled systems with J = >12 Hz	
System	Typical J value (Hz)	System	Typical J value (Hz)	System	Typical J value (Hz)
(geminal coupling)	2		7		15
(geminal coupling)	2		6		
(four bond coupling)	2		8		
(four bond coupling)	2		8		

each other (i.e., alkane/alkane; alkane/alkene; alkane/aldehyde) as well as their stereo-chemical relationships, notably those of cis- versus trans-alkenes. Notice also that coupling sometimes occurs between protons that are separated by *more* than three bonds. This type is called *long-range coupling*, and it is sometimes (but not always) observed when a molecule has a benzene ring or π bonds.

Estimating the J values in a spectrum makes use of the relationship that 1.0 ppm is equal in hertz to the magnitude of the spectrometer frequency. Thus, 1.0 ppm = 300 Hz when the spectrum is recorded on a 300-MHz instrument. In this text, either a scale or a comment in the figure caption will be provided to make it easier to estimate J values.

Many NMR instruments record and print the position (in ppm) of every peak in the spectrum. To determine the J values from such spectra is straightforward because you can calculate the numeric difference between adjacent peak positions. Multiplying that numeral by the number that corresponds to the spectrometer frequency gives the J value. Figure 13.17 shows such a calculation for a triplet feature recorded on a 300-MHz instrument.

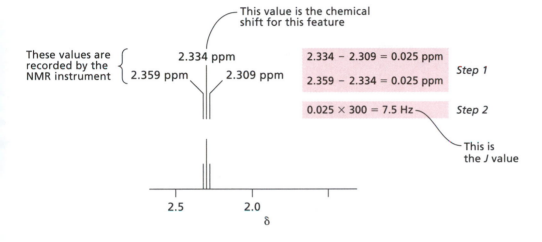

Figure 13.17
An example showing how to obtain the chemical shift and J value from a ^{1}H NMR spectrum recorded on a 300-MHz instrument that gives the numeric positions of the peaks.

Spin–spin splitting patterns created by the coupling of a set of protons with two *non*equivalent types of protons gives rise to features such as "a doublet of doublets", "a doublet of triplets", and so on (the smaller multiplicity is named first). One way to conceptualize multiple splitting patterns is to create a *splitting tree* that starts with a single peak at the chemical shift value for a given type of proton. The initial resonance is split into $n + 1$ peaks by protons on one adjacent atom, and each of *those* peaks is split into $n + 1$ peaks by the *other* set of neighboring protons. An example is presented in Figure 13.18 for the CH_2 fragment of propanal, which is split into a doublet of quartets.

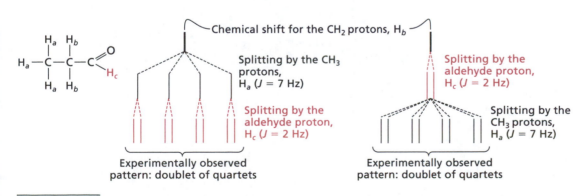

Figure 13.18
A tree structure for mapping the coupling in propanal between the aldehyde and the methyl group protons with the protons of the methylene group.

The CH_2 resonance is coupled on one side with a methyl group (3 protons, $J = 7$ Hz) and on the other side with the aldehyde proton (1 proton, $J = 2$ Hz). As illustrated in this figure, the order in which you conceptualize the splitting is irrelevant: *The resulting pattern is the same.*

Multiple splitting occurs even when the same types of protons are coupled. In these cases, however, the splitting pattern appears to follow the $n + 1$ rule because the J values are the same. For example, the central methylene group in 1-bromopropane is coupled both with a methyl and a methylene group, as illustrated in Figure 13.19. The J value for each coupling is 7 Hz, so the final pattern appears as a sextet, which is what you would predict by assuming the central methylene group is adjacent to five protons $(5 + 1 = 6)$. Thus, if the same types of protons on different atoms are involved in coupling, it is sufficient to apply the $n + 1$ rule to the *total number* of protons on the adjacent atoms. The most common examples you will encounter are those of the propyl and isobutyl groups.

Figure 13.19

A tree structure for mapping the coupling in 1-bromopropane between the methylene and methyl group protons with the protons of the central methylene group. The second splitting is offset slightly to show that the central peaks of the final pattern are actually a sum of two or three resonances with the same δ values.

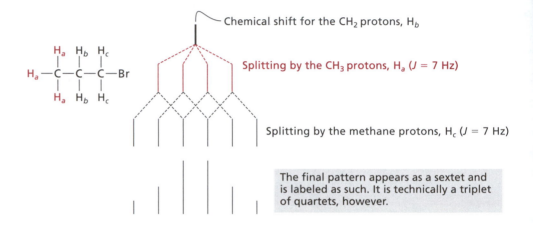

When interpreting a complicated spectrum, you may not immediately see the multiplicity of a feature. If not, you will have to "deconstruct" the pattern to determine how many different J values there are. Each J value is associated with a specific coupling system as well as the number of neighboring protons (Figure 13.20).

Figure 13.20

Diagrams illustrating how to deconstruct splitting patterns to determine the number of neighboring protons.

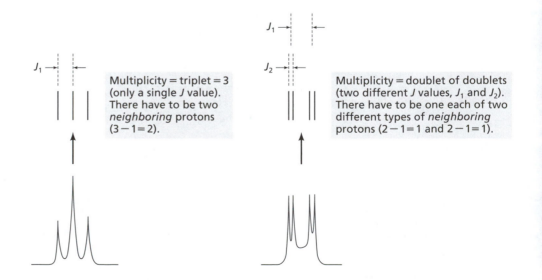

EXERCISE 13.9

Draw a splitting tree to show the patterns for each of the three features—methyl, methylene, and methine—in a generic isobutyl group, $(CH_3)_2CHCH_2X$. Assume that all of the J values equal 7 Hz.

EXAMPLE 13.11

Compound **G**, $C_9H_8O_2$, has the ^1H NMR spectrum illustrated in Figure 13.21. The signal at δ 12.4 disappears when the sample is treated with D_2O. Draw the structure of **G**.

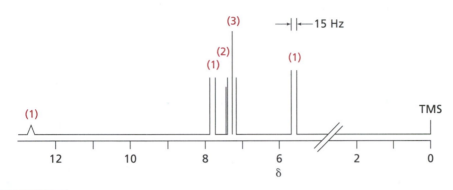

Figure 13.21
The 300-MHz ^1H NMR spectrum for compound **G**.

The ratios of integrated intensity values, as given by the NMR instrument, are 1:1:2:3:1 from left to right. This sum is equal to the number of protons in the molecular formula.

The signal at δ 12.4 is the resonance for a carboxylic acid proton. The broadness of this signal and its disappearance when the sample is treated with D_2O are consistent with the presence of a COOH group.

The sum of the integrated intensity values for the region δ 6–8 is > 5, which indicates that this molecule has both aromatic and alkene protons. The features at δ 7.77 and δ 6.45 have the same J values (~ 15 Hz), and each corresponds to one proton, so these signals are assigned to protons in a trans-alkene. The five protons associated with the remaining peaks in this region are likely those of a monosubstituted benzene ring. Assembling these fragments generates the structure of *trans*-3-phenylpropenoic acid.

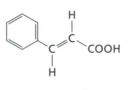

EXAMPLE 13.12

The spectrum of compound **H**, $C_6H_{10}O_2$, is shown in Figure 13.22. Draw the structure of compound **H**.

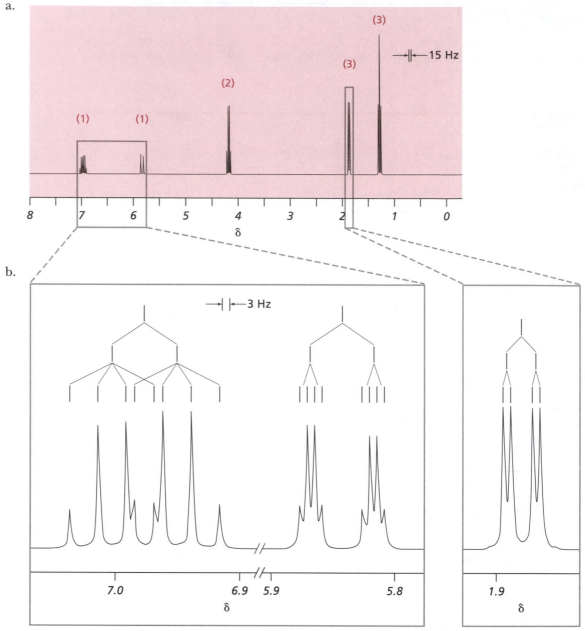

Figure 13.22
The 300-MHz ^1H NMR spectrum for compound **H.** (a.) Full range spectrum and (b.) expanded regions.

This spectrum has absorption peaks in two regions: δ 5.5–6 and δ 1–4.5; therefore, the molecule has both alkene and alkyl protons. Reading left to right, the integrated intensity values are 1:1:2:3:3, the sum of which equals the number of protons in the given molecular formula.

We can assign the features at δ 6.95 and 5.85 to protons attached to a carbon–carbon double bond (C_2H_2).

The features at δ 4.2 (two protons) and at δ 1.3 (three protons) have the same J values (7 Hz), so these two groups constitute an ethyl group (Fig. 13.11). The chemical shift of the methylene protons (δ 4.2) indicates that this ethyl group is attached to an oxygen atom, so the molecule contains the OCH_2CH_3 group.

The feature at δ 1.9 is associated with a methyl group (3 protons), and its chemical shift is consistent with bonding to an alkene carbon atom (CH_3–C=C).

Adding the atoms assigned so far (C_2H_2 + OCH_2CH_3 + CH_3) and subtracting from the molecular formula ($C_6H_{10}O_2$–$C_5H_{10}O$) leaves a carbon and an oxygen atom, which equals the carbonyl group. There are three ways to assemble all of these fragments, and these are shown in the following scheme along with estimated coupling constants—from Table 13.3—for the indicated protons (H_a, H_b, and CH_3–C=C).

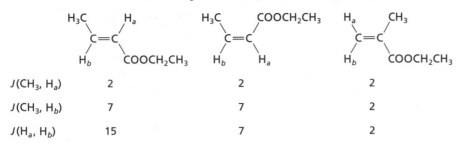

$J(CH_3, H_a)$	2	2	2
$J(CH_3, H_b)$	7	7	2
$J(H_a, H_b)$	15	7	2

The values tabulated for the trans isomer match most closely the data obtained from the actual spectrum [Fig. 13.22(b.)] which shows three different J values (3, 7, and 15 H_z), so the structure for compound **H** is ethyl *trans*-2-butenoate.

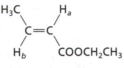

EXERCISE 13.10

Compound **J**, C_5H_9N, produces the following ^1H NMR spectrum. Draw its structure.

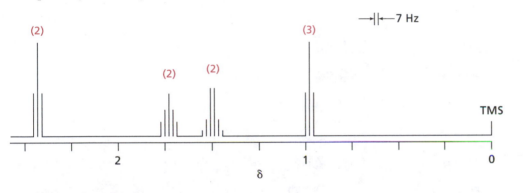

EXERCISE 13.11

Compound **K**, C_4H_8O, produces the following ^1H NMR spectrum. Draw its structure. Expanded regions of the spectrum are shown in the box.

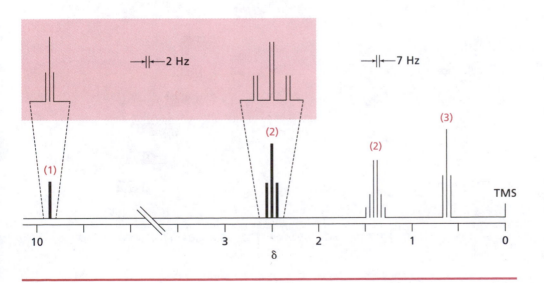

13.5 CARBON NMR SPECTRA

13.5a CARBON NMR SPECTRA RESULT FROM THE SAME ABSORPTION PROCESS THAT PRODUCES ^{1}H NMR SPECTRA

The phenomenon that gives rise to the NMR signals for ^{1}H nuclei is identical to that for ^{13}C nuclei. A ^{12}C atom has no unpaired nuclear spin, but its isotope ^{13}C has an additional neutron, for which there is an associated spin-change transition when the nucleus is placed in a magnetic field.

Another way that ^{1}H and ^{13}C NMR spectra are alike is that individual carbon atom resonances, like proton resonances, differ from one another and lead to a range of chemical shifts. These chemical shifts arise because of the electron distribution within the carbon framework, so inductive effects and magnetic anisotropy influence the frequencies at which resonances appear. The ^{13}C chemical shifts in most organic molecules cover a range of ~ 220 ppm and are measured from the resonance for the methyl carbon atoms of TMS. A summary of ^{13}C chemical shifts is presented in Figure 13.23.

Despite their similarities, differences between ^{1}H and ^{13}C NMR spectra also exist. The first difference is that the gyromagnetic ratio, γ, for carbon is about one-fourth that for a proton. The frequency needed to cause resonance will therefore be different, given the same field strength used to record ^{1}H NMR spectra. Consequently, proton and carbon resonances do not appear together in the same region of the spectrum.

Another contrast between ^{1}H and ^{13}C NMR spectra is in the strength of the signals observed. In ^{1}H NMR spectroscopy, nearly every hydrogen atom contributes to the res-

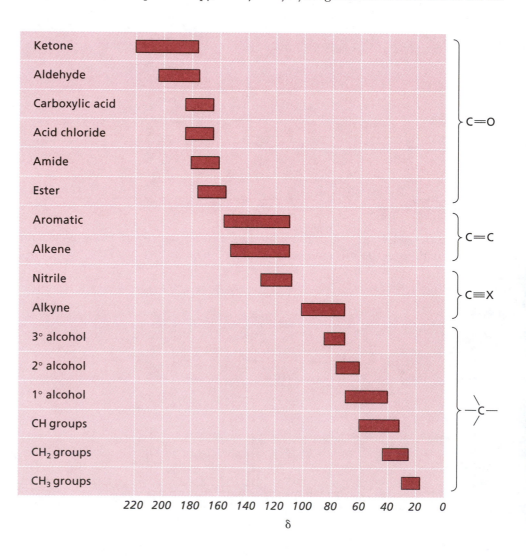

Figure 13.23
Chemical shifts for ^{13}C NMR spectra.

onance signal because the isotopic abundance of ^{1}H is almost 100%. For carbon, the natural abundance of ^{13}C is only 1.1%, which means that most of the carbon nuclei are not involved in the resonance process. Furthermore, the sensitivity of a nucleus to the spin-change transition is proportional to γ^3, so even if the isotopic abundance of ^{13}C were the same as that for ^{1}H, the sensitivity for carbon would still be only 1/64 that for hydrogen. The lower magnetic sensitivity, coupled with the low abundance of ^{13}C, means that the overall sensitivity for carbon is 6000 times less than that for hydrogen.

The low isotopic abundance of ^{13}C manifests itself in another way. Recall that the presence of neighboring protons leads to spin coupling of the proton nuclei (Sections 13.2 and 13.4). The chance is small that two neighboring carbon atoms are both present as the ^{13}C isotope. Thus ^{13}C–^{13}C splitting is not observed unless a compound is prepared using starting materials that are isotopically enriched with ^{13}C.

A final difference is that ^{13}C NMR spectra are not normally integrated.

13.5b   THE APPEARANCE OF A ^{13}C NMR SPECTRUM VARIES ACCORDING TO THE AMOUNT OF PROTON COUPLING

Proton and carbon NMR spectra look quite different. The ^{13}C NMR spectra are usually presented as a series of singlets, which is unusual for proton spectra, except of the simplest compounds. To understand why, you have to consider the effects of ^{13}C–^{1}H coupling.

The low abundance of ^{13}C accounts for the absence of ^{13}C–^{13}C spin–spin splitting. The same is true for lack of ^{13}C–^{1}H splitting in the *proton* spectrum. The likelihood that a proton is bound (or near) a ^{13}C atom is quite low. The converse is not true, however. A ^{13}C atom in a molecule will almost certainly be within three or four bonds of a proton. These nuclei can undergo spin–spin coupling, and several different J values are possible, as summarized in Figure 13.24.

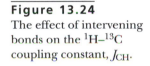

$$^1J_{CH} \qquad\qquad ^2J_{CH} \qquad\qquad ^3J_{CH}$$

$$^{13}C-H \qquad ^{13}C-C-H \qquad ^{13}C-C-C-H$$

$$J \approx 125\text{–}150\,\text{Hz} \qquad J \approx 5\text{–}10\,\text{Hz} \qquad J \approx 0\text{–}1\,\text{Hz}$$

Figure 13.24
The effect of intervening bonds on the ^{1}H–^{13}C coupling constant, J_{CH}.

The ^{13}C NMR spectrum of allyl chloride is shown in Figure 13.25. Here, all of the possible spin–spin splitting patterns between the protons and the ^{13}C nuclei are observed. Allyl chloride has three different carbon atoms, each of which can couple to the five protons. There should be at least three C–H coupling constants for each carbon

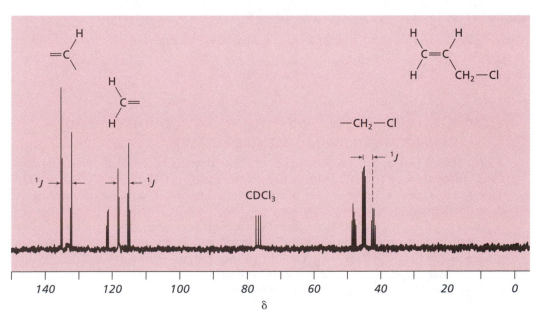

Figure 13.25
The 75-MHz ^{13}C NMR spectrum of allyl chloride without decoupling.

atom, leading to complex multiplets at every chemical shift value. The splitting of each resonance into several peaks, in addition to the low sensitivity of the carbon nuclei, explains why the spectrum has small peaks relative to the baseline "noise" (although most ^{13}C NMR spectra are not as noisy as this one). Fortunately, most ^{13}C spectra are *not* recorded with coupling to protons; otherwise spectra of compounds with even as few as 10 carbon atoms would be too complex to interpret.

13.5c   THE BROAD–BAND DECOUPLED SPECTRUM IS THE SIMPLEST FORMAT OF A ^{13}C NMR SPECTRUM

Suppose we record the ^{13}C NMR spectrum of allyl chloride as we irradiate the sample with a second radio frequency source covering the *proton* NMR range. This process, called **broad-band decoupling,** equalizes the populations of the proton's two spin states ($+\frac{1}{2}$ and $-\frac{1}{2}$), and the average proton spins influencing the carbon resonances are therefore constant. The influence of complete proton decoupling on the ^{13}C spectrum of allyl chloride is shown in Figure 13.26.

The broad-band decoupled spectrum is the most common type of ^{13}C NMR spectrum. Its simplicity is apparent, and the chemical shift for each carbon atom is readily obtained. In fact, a resonance for every carbon atom can often be observed in a broad-band decoupled spectrum, even if a molecule has 20 or 30 carbon atoms. The only disadvantage of the broad-band decoupled spectrum is that all coupling information is lost. However, coupling in the ^{1}H NMR spectrum is more useful for analyzing the interconnectivity between hydrocarbon fragments.

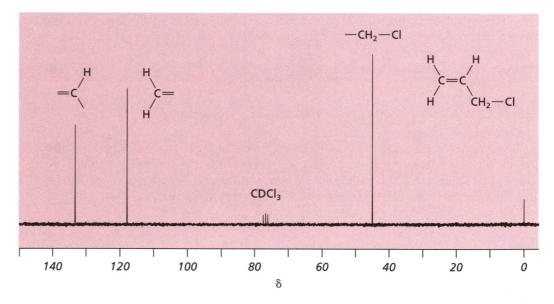

Figure 13.26
The broad-band decoupled ^{13}C NMR spectrum of allyl chloride, showing the effect of complete proton decoupling.

13.5d   OTHER TECHNIQUES ARE USED TO DISCOVER THE TYPES OF HYDROCARBON UNITS THAT ARE PRESENT

A broad-band decoupled ^{13}C NMR spectrum, like that in Figure 13.26, does not allow you to discern which types of hydrocarbon fragments (CH, CH_2, and CH_3) are present. Therefore, several techniques have been developed to obtain data about coupling between carbon and hydrogen atoms.

One method, called **off-resonance decoupling,** is similar to the broad-band decoupled method except that the sample is irradiated with a second radio frequency source that is 1000–2000 Hz away from the proton range (for broad-band decoupling, the principal frequency is in the middle of the proton range). Figure 13.27 shows the off-resonance decoupled spectrum of allyl chloride.

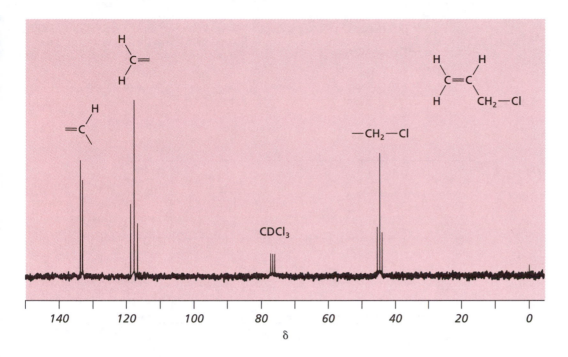

Figure 13.27
The off-resonance decoupled ^{13}C NMR spectrum of allyl chloride.

The difference between a broad band and an off-resonance decoupled spectrum is that spin–spin splitting is observed in the latter. The observed J value does not correspond to a "real" C–H coupling constant, however, because it depends on how far the decoupling frequency is from the proton range. The overall effect is to decouple all but *the attached protons* from each carbon atom, so only the one-bond coupling ($^1J_{CH}$, Fig. 13.24) is observed. Thus, the signal for a carbon atom in an off-resonance decoupled spectrum

- Appears as a singlet if it has no proton attached.
- Appears as a doublet if it has one proton attached.
- Appears as a triplet if it has two protons attached.
- Appears as a quartet if it has three protons attached.

Except for the one in methane, no carbon atom can have more than three protons attached.

Another technique used to obtain information about which hydrocarbon fragments are present is called **DEPT ^{13}C NMR spectra** (distortionless enhancement by polarization transfer = DEPT). This technique is now used more often than the off-resonance method to obtain data about the presence of CH_3, CH_2, and CH groups in a molecule. The DEPT technique is made possible by programming the NMR spectrometer's computer to pulse the sample with specific patterns of radio frequencies in both the ^1H and ^{13}C resonance ranges. These pulse sequences result in spectra in which only certain types of hydrocarbon groups can be observed. The DEPT spectra of 2-methyl-3-pentanone are shown in Figure 13.28. Notice that the carbonyl carbon atom does not have a signal in any of the DEPT spectra, but it does display a resonance at δ 216 in the broad-band decoupled spectrum.

Apart from some exercises at the end of this chapter, you will be told throughout the remainder of this text how many protons are attached to each carbon atom that produces a signal in the broad-band decoupled ^{13}C NMR spectra. Whether this number of hydrogen atoms is obtained from either DEPT or off-resonance decoupled spectra is not crucial to the interpretation process.

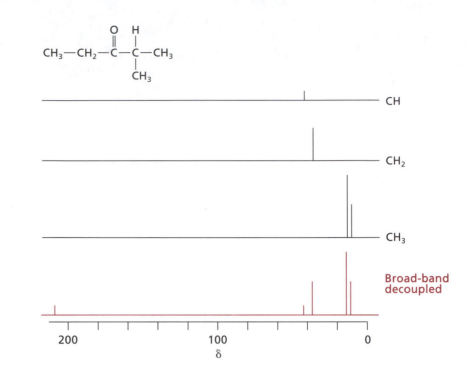

Figure 13.28
The broad-band decoupled spectrum (*bottom*) and DEPT spectra for the hydrocarbon groups in 2-methyl-3-pentanone.

13.6 INTERPRETING ^{13}C NMR SPECTRA

13.6a EXPERIMENTAL CONDITIONS FOR ^{13}C AND ^{1}H NMR SPECTRA ARE SIMILAR

The NMR spectrometers used to obtain carbon spectra are normally the same as those used to record proton spectra. The spectrum is plotted over the range from δ 250 to 0 in the same manner as for a proton spectrum, with TMS on the right and the upfield and downfield relationships defined accordingly. Integrated intensities of the peaks are not calculated.

The most important data obtained from a ^{13}C NMR spectrum are the positions of the absorptions in ppm (the chemical shifts). In addition, an off-resonance decoupled or DEPT spectrum is recorded to find how many H atoms are attached to each carbon atom.

Although solvents containing protons can be used to record ^{13}C NMR spectra, deuterochloroform ($CDCl_3$) is the most widely used. Its single carbon atom resonance is split into three peaks by coupling to the deuterium atom (deuterium has a nuclear spin $I = 1$). Because the relative abundance of ^{13}C in the solvent is the same as in the sample, symmetric compounds are used as solvents to minimize the number of extraneous peaks.

13.6b ^{13}C NMR SPECTRA CAN BE USED TO IDENTIFY FUNCTIONAL GROUPS CONTAINING CARBON ATOMS

The ^{13}C NMR spectrum provides a straightforward way to determine which carbon-containing functional groups are present in a molecule. Because only the carbon atoms can be observed with this technique, the presence of functional groups such as O–H, S–H, N–H, NH_2, N=O, NO_2, S=O, and SO_2 cannot be detected directly. To interpret the spectrum, the chemical shift data summarized in Figure 13.23 are used to identify the types of carbon atoms present. The off-resonance decoupled or DEPT spectrum is used to find the number of protons attached to each carbon atom.

EXERCISE 13.12

For each of the following compounds, indicate which of the carbon atoms marked with an arrow will have an absorption farther downfield:

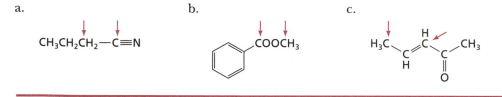

a.   b.   c.

Absorptions in the region farthest downfield from TMS indicate the presence of only a few compound types. In ^{13}C spectra, the region from δ 160 to 250 is where carbonyl carbon resonances appear. For example, Figure 13.29 shows the ^{13}C NMR spectrum of an aliphatic ketone, 2-butanone.

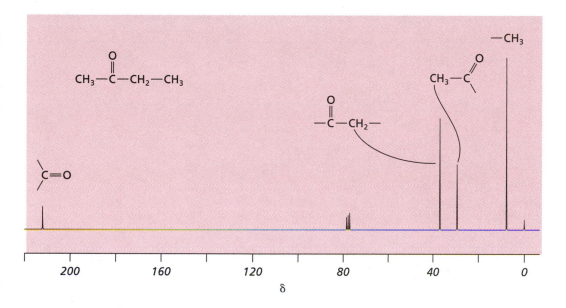

Figure 13.29
The broad-band decoupled ^{13}C NMR spectrum of 2-butanone.

The region between δ 100 and 160 is that in which compounds with sp^2 and nitrile carbon atoms display resonances. An alkene usually exhibits an even number of peaks for the double-bonded carbon atoms, unless the molecule is symmetrical or two of the carbon resonances fortuitously have the same chemical shift value. The ^{13}C NMR spectrum for a terminal alkene is unique among compounds that have a carbon–carbon double bond, because a terminal alkene has a CH$_2$ group. The off-resonance spectrum of a terminal alkene therefore has a triplet (Figure 13.30). An arene can have at most one proton attached to each carbon atom, so its peaks in an off-resonance decoupled spectrum appear only as singlets or doublets.

Another way to differentiate arenes and alkenes is by the number of resonances. A benzene ring with unsymmetrical substitution may have as many as six carbon resonances, as illustrated in Figure 13.31. You have to be careful, however, because if the substitution pattern is unsymmetrical, then only two or three peaks may be observed.

The region between δ 70 and 100 is where alkyne carbon-atom resonances appear, and the only overlap is with resonances for some types of 4° carbon atoms. For each triple bond, there will be two carbon resonances, unless the compound is symmetrical.

Finally, the region between δ 0 and 70 is that in which sp^3 carbon-atom resonances appear. Assignment of the alkyl carbon resonances to specific fragments is sometimes difficult, because there may be several possibilities for each chemical shift value. The off-resonance decoupled or DEPT spectrum may help narrow the choices.

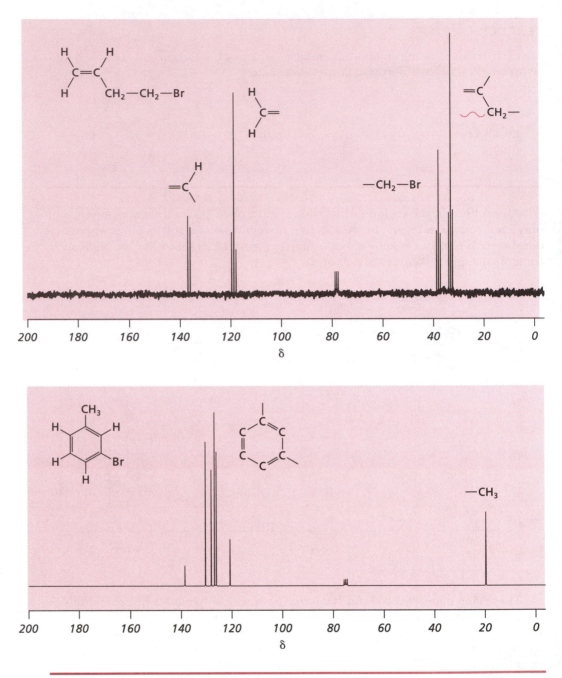

Figure 13.30
The off-resonance decoupled ^{13}C NMR spectrum of 4-bromo-1-butene, showing the expected triplet at δ 118 for the terminal –CH$_2$ group.

Figure 13.31
The broad-band decoupled ^{13}C NMR spectrum of 3-bromotoluene, showing the six peaks corresponding to the nonequivalent benzene carbon atoms.

EXAMPLE 13.13

Assign the type of carbon atom to each peak in the broad-band decoupled spectrum of compound **L** illustrated in Figure 13.32. Be as specific as possible.

From the data given in Figure 13.23, use the chemical shift values to make assignments. The resonance farthest downfield (δ 164; no attached proton according to the DEPT results) is assigned to a carbonyl carbon atom with no hydrogen atom attached. Specifically, it could be from a carboxylic acid, an acid chloride, an ester, an amide, or an acid anhydride.

The peaks in the region δ 146–130 correspond to benzene carbon-atom resonances. The substitution pattern of the ring must be unsymmetrical because all six of the carbon-atom resonances are observed. Three of the carbon atoms have an attached hydrogen atom (DEPT results), and three of the carbon atoms have a substituent besides hydrogen.

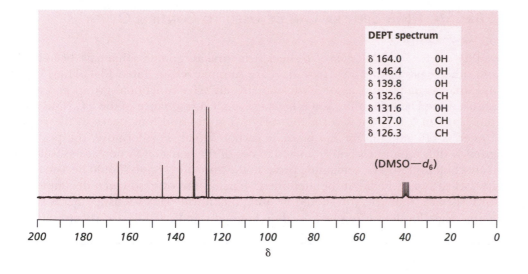

Figure 13.32
The broad-band
decoupled ^{13}C NMR
spectrum of compound **L**.
The DEPT results are
summarized in the box.

EXERCISE 13.13

Assign the type of carbon atom to each peak in the following broad-band decoupled
spectra; be as specific as possible. The DEPT results are summarized in the boxes.

a.

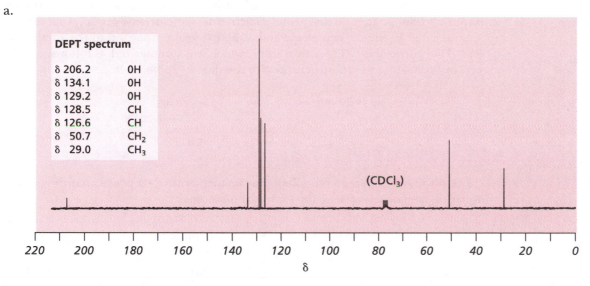

b.

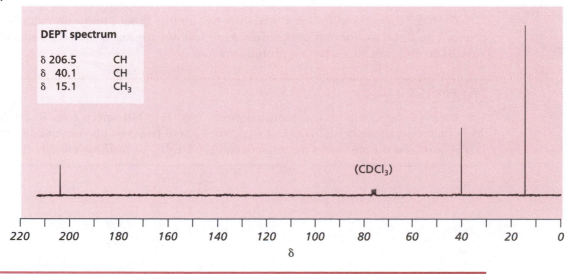

13.6a ^{13}C NMR Spectra Can Be Used to Confirm Other Assignments

Another use for ^{13}C NMR data is to check structural assignments that have been made from other types of spectra. These data are summarized in Table 13.4. Thus, if you think that a compound under study contains an ester functional group and a 1,4-didisubstituted benzene ring, you will expect to see a resonance in the ^{13}C NMR spectrum between δ 155 and 175 for the carbonyl group and four peaks in the region between δ 110 and 135 for the benzene carbon atoms. Furthermore, the peaks assigned to the carbonyl carbon atom and to two of the benzene carbon atoms will have no attached proton. The remaining peaks in the aromatic region should be associated with CH groups, which can be identified by using the off-resonance decoupled or DEPT spectra. Similarly, you should be able to confirm ^{1}H NMR assignments by observing peaks for the various hydrocarbon fragments in the DEPT ^{13}C spectrum.

EXAMPLE 13.14

For the compound identified from the ^{1}H NMR spectra in Example 13.6 (Section 13.3a), use the data in Table 13.4 to predict where peaks will be observed in the broad-band, the off-resonance decoupled, and the DEPT ^{13}C NMR spectra.

The compound identified in Example 13.6 is ethyl acetate. This molecule has an ester carbonyl carbon atom, a methyl group attached to the carbonyl group, a methylene group attached to the oxygen atom, and a methyl group attached to the methylene group. Their expected ^{13}C chemical shifts, which are summarized as follows, are deduced by correlating the substructures with the data in Table 13.4:

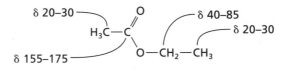

The off-resonance spectrum will display features with $n + 1$ peaks according to the number of protons (n) attached to each carbon atom.

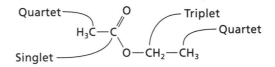

The DEPT spectra will display peaks for two methyl groups and one methylene group. The signal for the carbonyl carbon atom will not appear in the DEPT spectra, so you can conclude that it has no proton attached.

EXERCISE 13.14

For each of the compounds identified from the ^{1}H NMR spectra in Examples 13.7–13.10, use the data in Table 13.4 to predict where peaks will be observed in the broad-band, the off-resonance decoupled, and the DEPT ^{13}C NMR spectra.

Table 13.4 Expected ^{13}C resonances for confirmation of structural assignments.

Functional or structural group		Expected peak(s) in the broad-band decoupled ^{13}CNMR spectrum (δ)
Acid chloride (–CO–Cl)		1 peak, 185–165
Alcohol (–C–OH)	3°	1 peak, 85–70
	2°	1 peak, 75–60
	1°	1 peak, 70–40
Aldehyde (–CHO)		1 peak, 205–175
Alkene		2 peaks,[a] 150–110
Alkyne		2 peaks,[a] 100–70
Amide (–CO–NR$_2$)		1 peak, 180–160
Amine (–C–NR$_2$)		1–3 peaks, 75–20
Anhydride (–C–O–CO–)		2 peaks,[a] 175–150
Benzenoid compounds		2–6 peaks,[a] 155–110
Carboxylic acid (–COOH)		1 peak, 185–160
Ester (–COOR)		1 peak (C=O), 175–155 1 peak (R), 85–40
Ether (–C–O–C–)		2 peaks,[a] 85–40
Ketone (–CO–)		1 peak, 225–175
Nitrile (–C≡N)		1 peak, 130–110
Methine group (CH)		1 peak, 60–30
Methyl group (CH$_3$)		1 peak, 30–20
Methylene group (CH$_2$)		1 peak, 45–25

[a]May observe only one peak because of symmetry or pseudosymmetry in the molecule.

CHAPTER SUMMARY

Section 13.1 Chemical shifts and proton equivalence

- Nuclear magnetic resonance (NMR) spectra are produced when energy absorption by a proton or neutron leads to a change in its spin state under the influence of a magnetic field.

- Secondary magnetic fields created by the electrons within a molecule either oppose or enhance an external magnetic field, shielding the nucleus to varying degrees. These shielding effects change the energy needed to cause the spin change that underlies the resonance phenomenon.

- Protons in a molecule that experience the same influences from a magnetic field are magnetically equivalent.

- The absorption frequency for a given proton relative to that for the protons in tetramethylsilane (TMS), is the chemical shift. The chemical shift values of magnetically equivalent protons are identical.

- The relationship between chemical shift values is indicated by the terms upfield (more shielded nucleus) and downfield (less shielded nucleus).

- The two main factors that influence chemical shift values are inductive effects and magnetic anisotropy.

- Magnetic anisotropy effects result from the inhomogeneous local magnetic fields created by the distribution of bonding electrons, particularly those in π bonds.

- The areas under the peaks of equivalent protons are proportional to their total number and are called the integrated intensities of those peaks.

Section 13.2   Spin coupling in simple systems

- The spins of nonequivalent neighboring protons create yet another set of localized magnetic fields that can influence chemical shift values. These effects are smaller than those of the electrons and the external field and manifest themselves by "splitting" the primary absorption peak.

- Spin coupling is most influential when the nuclei affecting each other are separated by three or fewer bonds.

- Three-bond coupling is called vicinal coupling; two-bond coupling is called geminal coupling.

- The magnitude of the coupling is given by the J value, the coupling constant.

- Protons attached to an aromatic ring couple only with other ring protons.

Section 13.3   Interpreting ^{1}H NMR spectra

- Interpretation of a ^{1}H NMR spectrum makes use of the integrated intensity values, chemical shifts, and splitting patterns with their associated J values.

Section 13.4   Spin coupling in more complex systems

- The J value is characteristic of the environment of the protons that undergo coupling, and it can provide information about stereochemistry.

- Coupling that occurs between more than two sets of protons having different J values gives rise to complex patterns such as "doublet of doublets", "doublet of quartets", and so on.

- Coupling between more than two sets of protons with similar J values can be treated using the $n + 1$ rule for the total number of protons on the adjacent atoms.

Section 13.5   Carbon NMR spectra

- The ^{13}C NMR spectra are useful to probe the environments of carbon atoms in a molecule. The resonance effects are the same as those that generate ^{1}H NMR spectra.

- The primary difference between ^{13}C and ^{1}H NMR spectra is appearance. Carbon spectra are usually composed of single lines, one for each carbon atom.

- Coupling occurs between carbon and hydrogen nuclei, but the primary technique for obtaining ^{13}C NMR spectra, called broad-band decoupling, eliminates the effects of the proton spins toward the carbon nuclei.

- Coupling between a carbon nucleus and its attached protons is available by two techniques: off-resonance decoupling and distortionless enhancement by polarization transfer (DEPT). These techniques are used to establish the presence of CH, CH_2, and CH_3 groups.

Section 13.6   Interpreting ^{13}C NMR spectra

- Interpretation of a ^{13}C NMR spectrum makes use of both chemical shifts values and coupled spectra (off-resonance or DEPT) that reveal which types of hydrocarbon fragments are present.

Section 13.1b
magnetically equivalent

Section 13.1c
shielding
chemical shift
downfield
upfield
deshielded
shielded
magnetic anisotropy

Section 13.1d
integrated intensity values

Section 13.2a
spin–spin splitting
vicinal coupling
geminal coupling
the $n + 1$ rule
multiplicity
coupling constant, J

Section 13.3
deuterium exchange

Section 13.5c
broad-band decoupling

Section 13.5d
off-resonance decoupling
DEPT ^{13}C NMR spectra

ADDITIONAL EXERCISES

13.15. For each of the following structures, identify each group of chemically equivalent protons:

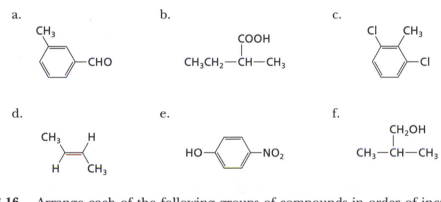

a.

b.

c.

d.

e.

f.

13.16. Arrange each of the following groups of compounds in order of increasing chemical shift of the proton shown in color:

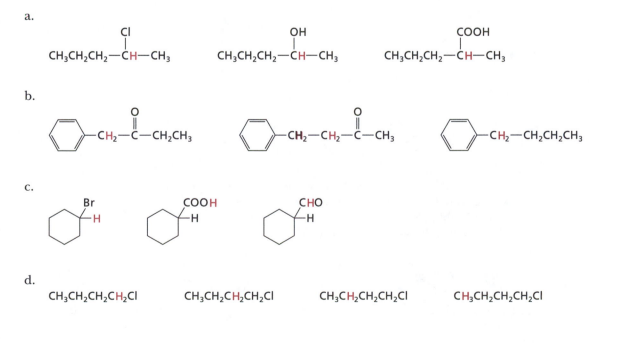

a.

b.

c.

d.

13.17. For each of the following compounds, indicate which proton will have its absorption peak farther downfield:

a.

b.

c.

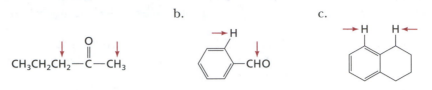

13.18. For each of the following compounds, indicate which carbon atom will have its absorption peak farther downfield:

a.

b.

c.

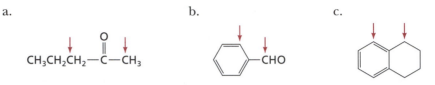

13.19. Predict the splitting pattern (singlet, doublet, triplet, doublet of doublets, etc.) that would be observed for the signal of the indicated proton.

a.

b.

c.

d.

e.

f.

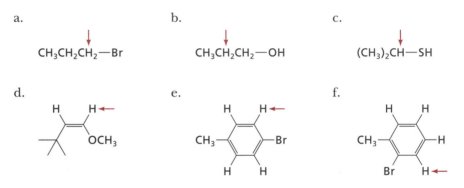

13.20. Predict the splitting pattern (singlet, doublet, triplet, etc.) that would be observed in the off-resonance decoupled ^{13}C NMR spectrum for the signal of the indicated carbon atom.

a.

b.

c.

d.

e.

f.

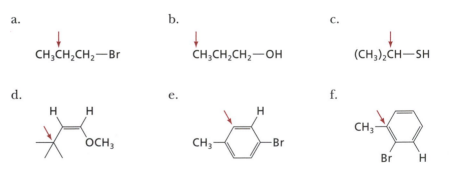

13.21. Sketch the expected ^1H NMR spectrum for each of the following compounds. Estimate the approximate values for the chemical shifts, and show splitting where applicable (assume that only vicinal coupling is observed). Indicate the integrated intensity values of each feature.

a.

b.

c.

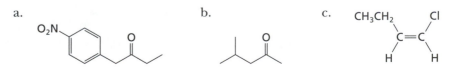

13.22. For the compounds shown in Exercise 13.21, sketch the broad-band decoupled ^{13}C NMR spectrum, estimating the approximate values of the chemical shifts.

13.23. Using ^1H NMR spectra to differentiate between each of the following pairs of isomers, explain what absorptions you would look for, commenting on the expected chemical shifts, integrated intensity values, and splitting pattern for each.

a.

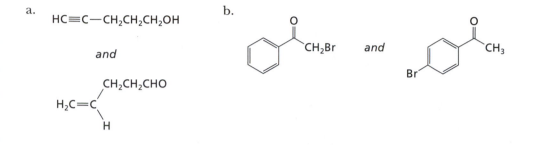

13.24. What would you look for in the broad-band decoupled ^{13}C NMR spectrum to differentiate between the isomers shown in Exercise 13.23? Explain what absorptions you would look for, commenting on their expected chemical shift values and the types of hydrocarbon groups that would be observed in the DEPT spectra.

13.25. In performing the following reactions, what differences in the ^1H NMR spectra would you expect to see? Explain what absorptions you would look for, commenting on the expected chemical shifts, integrated intensity values, and splitting patterns.

a.   b.

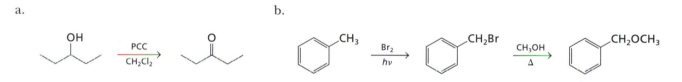

13.26. For the reactions shown in Exercise 13.25, explain what differences you will see in the broad-band and DEPT ^{13}C NMR spectra between the reactant and product in each.

13.27. The ^1H NMR spectrum of compound **Y**, $C_7H_{14}O$, is shown below. What is the structure of **Y**? Using ^{13}C NMR spectra to support your conclusion, what would you look for in the broad-band and off-resonance decoupled spectra?

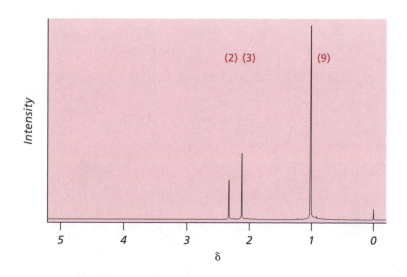

13.28. An unknown compound with the formula $C_9H_{13}N$ produces the following ^1H NMR spectrum. Draw its structure.

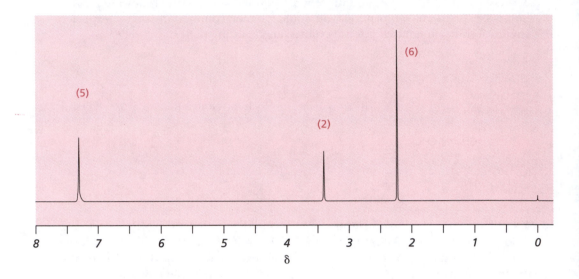

13.29. An unknown compound with the formula $C_7H_{14}O$ produces the following ^{13}C and ^1H NMR spectra. Draw its structure.

^{13}C NMR spectrum

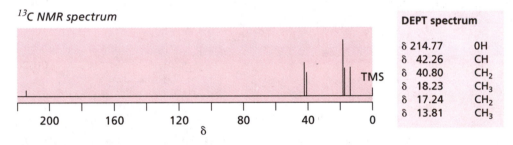

DEPT spectrum		
δ 214.77	0H	
δ 42.26	CH	
δ 40.80	CH_2	
δ 18.23	CH_3	
δ 17.24	CH_2	
δ 13.81	CH_3	

^1H NMR spectrum

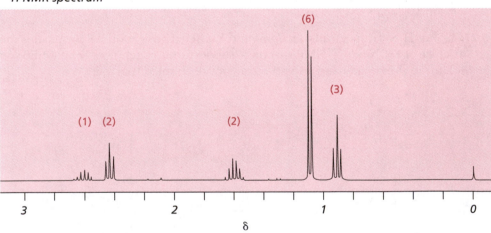

13.30. An unknown compound with the formula C_5H_9N produces the following ^{13}C and ^1H NMR spectra. Draw the compound's structure. For the spin-spin splitting patterns, $J = 7H$.

^{13}C NMR spectrum

DEPT spectrum

δ 118.89	0H
δ 26.09	CH_2
δ 25.94	0H
δ 21.77	CH_3

TMS

120   100   80   60   40   20   0
δ

^1H NMR spectrum

(6)

(2)   (1)

3   2   1   0
δ

13.31. A compound with the formula $C_6H_3Cl_2NO_2$ produces the following ^1H NMR spectrum. What is the substitution pattern of the benzene ring? For the spin-spin splitting patterns, $J = 7H$.

^1H NMR spectrum

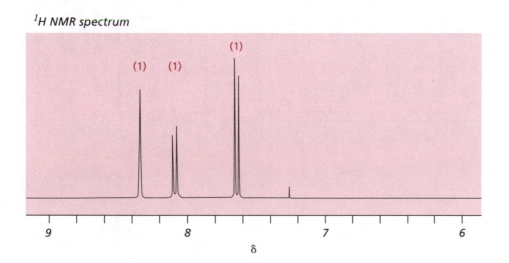

(1)   (1)   (1)

9   8   7   6
δ

13.32. The following two ^{1}H NMR spectra represent isomeric ketones with the formula $C_{10}H_{12}O$. Draw the structure of each molecule. For the spin-spin splitting patterns, $J = 7H$.

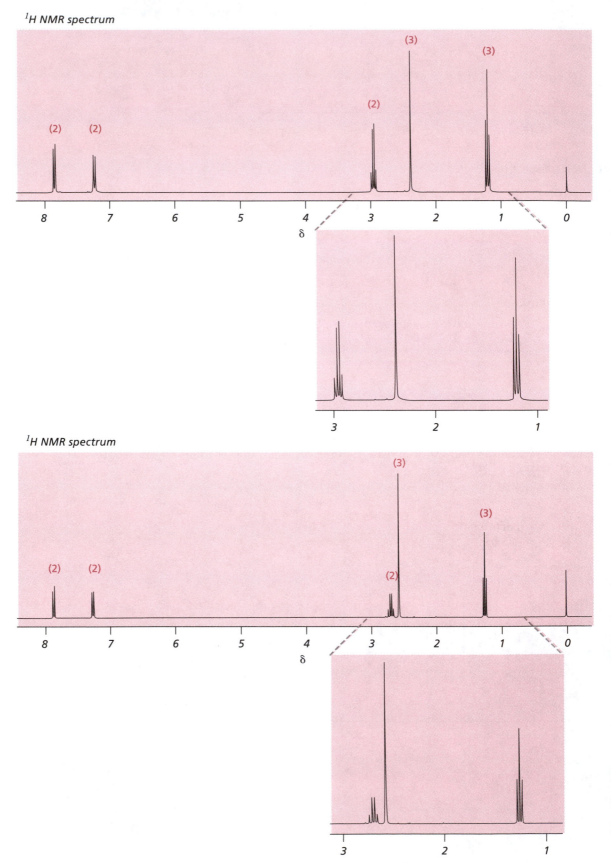

13.33. An unknown compound with the formula $C_9H_{11}Br$ produces the following ^{13}C and ^1H NMR spectra. Draw the compound's structure. For the spin-spin splitting patterns, $J = 7H$.

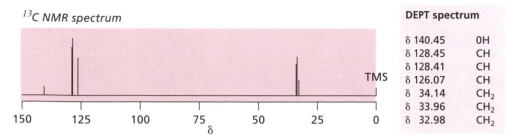

^{13}C NMR spectrum

DEPT spectrum	
δ 140.45	0H
δ 128.45	CH
δ 128.41	CH
δ 126.07	CH
δ 34.14	CH_2
δ 33.96	CH_2
δ 32.98	CH_2

^1H NMR spectrum

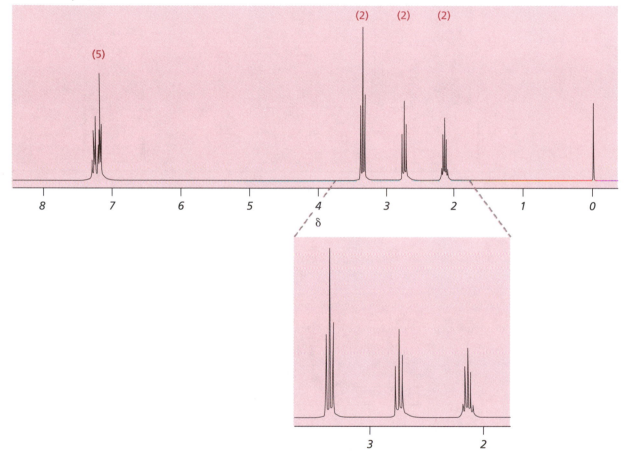

13.34.   An unknown compound with the formula $C_6H_{10}O_2$ produces the following ^{13}C and ^1H NMR spectra. Draw the compound's structure.

^{13}C NMR spectrum

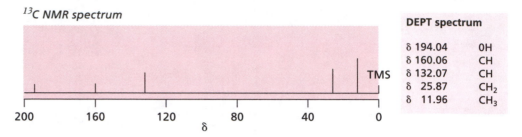

DEPT spectrum	
δ 194.04	0H
δ 160.06	CH
δ 132.07	CH
δ 25.87	CH₂
δ 11.96	CH₃

^1H NMR spectrum

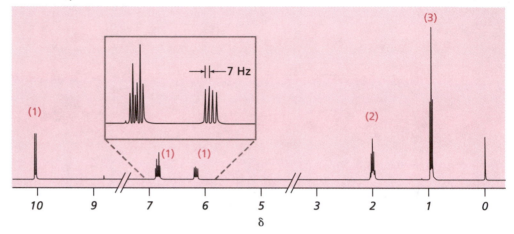

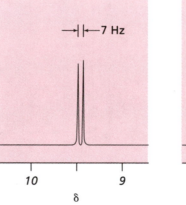 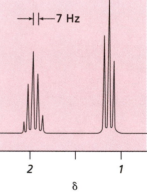

DETERMINING THE STRUCTURES OF ORGANIC MOLECULES

In the previous chapter (Chapter 13), you learned how to interpret ^{1}H and ^{13}C NMR spectra in order to establish the structures of organic molecules. Such detail about those procedures in an introductory textbook reflects a practical reality of organic chemistry in the early twenty-first century, namely, that NMR spectra are so routinely recorded and used in determining structures that even many undergraduate laboratory courses make use of NMR techniques. Infrared and mass spectra, two topics presented in this chapter, can also provide useful information about molecular structures. Their use in structure determinations is not as routine, however, so they are not always interpreted in the detailed manner applied to NMR spectra. This text will reflect that difference by discussing with less rigor the data obtained from IR and mass spectra.

This chapter begins with a discussion of analytical data other than spectra that are useful for the structure elucidation process. You will then learn what information about structures can be obtained from mass spectrometry and IR spectroscopy. The chapter concludes by presenting several examples that integrate data obtained from elemental analysis and NMR, IR, and mass spectra to determine the structures of organic molecules.

14.1 ANALYTICAL DATA IN STRUCTURE DETERMINATIONS

14.1a THE DETERMINATION OF MOLECULAR STRUCTURES REQUIRES DATA THAT IS OBTAINED ON PURE COMPOUNDS

The structure determination process can vary according to the context in which it is performed, and you are likely to encounter three common situations:

- Verification of the structure of a reaction product.
- Identification of the structure of a reaction byproduct.
- Determination of the structure of a complete unknown.

Verification is needed when you are certain about the outcome of a particular reaction. For example, the oxidation of 1-octanol to form octanal can be carried out with PCC in methylene chloride (Section 11.4c).

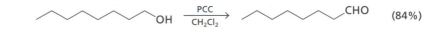

Here, you know the structures of both the starting material and product, so you can calculate the molecular formula of each substance, and you are already aware of which functional groups are present and how many hydrocarbon units each contains, for example. When you record the spectra to verify the product's structure, you will already have a good idea what to look for.

Suppose instead that you were to oxidize 1-octanol with aqueous chromium(VI) oxide, and you obtained two products:

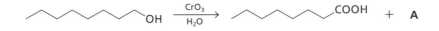

There is a reasonable chance that compound **A** has eight carbon atoms and shares many structural similarities with the starting alcohol. In this situation, you might have a good idea about the molecular weight of **A** or about the functional group that is present. You can therefore look for corroborating data in the spectra you record, keeping in mind that unexpected information may come to light as well.

The third scenario plays out when you isolate a compound from a less well defined source. For example, you might be interested in isolating compounds contained in the leaves of the Princess Tree (*Paulownia tomentose*) in your backyard, so you extract several leaves with methanol.

$$P.\ tomentosa \xrightarrow[\text{with CH}_3\text{OH}]{\text{extract leaves}} \textbf{B}$$

In this case, you are not likely to know much about compound **B**—its MW, its functional groups, or even which atoms the molecule contains. The data that you need to collect in order to determine the molecule's structure will probably be more extensive than what you need to identify the products of a chemical reaction.

In all three cases, however, the material that you isolate to analyze must be pure; otherwise, the data you record will reflect the properties of the molecule you want to identify as well as those of any impurities that are present. Crystallization, distillation, and sublimation are three physical separation methods that are routinely used in the organic laboratory to purify compounds. Chromatographic methods are sometimes more convenient to separate the desired analyte from impurities, and high-performance liquid chromatography (HPLC) and thin-layer chromatography (TLC) are two analytical techniques that are routinely used to verify the purity of compounds being investigated.

Assuming that you can perform a suitable separation, you still need an independent assessment of the purity of the isolated compound. Elemental analysis is one such method, and it can be used at the same time to determine a compound's molecular formula. Many scientific journals require elemental analysis data to be reported for the characterization of chemical compounds as a means to judge purity.

14.1b ELEMENTAL ANALYSES DATA CAN BE USED TO DETERMINE THE EMPIRICAL FORMULA

Data from elemental analyses are routinely obtained by combustion, gravimetric, and atomic absorption spectroscopy methods; the details of these techniques will not be discussed here. Suffice it to say, if analytical data are needed, we can submit a sample of the compound to a commercial analytical lab. By applying the appropriate methods, they determine the composition of the sample and report the mole-percent of each element except oxygen, the amount of which is determined by difference. By calculating the mole ratios, we can determine the compound's empirical formula by the procedure presented below. The formula of an organic compound always lists carbon first, then hydrogen, then the remaining elements in alphabetical order.

EXAMPLE 14.1

The elemental analysis of a sample containing C, H, and O gave the following results: C, 78.8%; H, 8.32%. What is the empirical formula of this compound?

First, calculate what percentage of the compound is oxygen. This computation is done by subtracting the sum of the percentage values of C and H from 100%.

$$100.00 - (78.8 + 8.32) = 12.88\% \text{ O}$$

The percentage values of the elements mean that if you have 100 g of this compound, 78.8 g would be C, 8.32 g would be H, and 12.88 g would be O. From these amounts, we calculate how many moles of each element are contained in 100 g of compound:

$$78.8 \text{ g c} \left(\frac{1 \text{ mol C}}{12.011 \text{ g c}} \right) = 6.6561 \text{ mol C} \qquad 8.32 \text{ g h} \left(\frac{1 \text{ mol H}}{1.008 \text{ g h}} \right) = 8.254 \text{ mol H}$$

$$12.88 \text{ g O} \left(\frac{1 \text{ mol O}}{16.00 \text{ g O}} \right) = 0.8063 \text{ mol O}$$

Next, we divide each number of moles by the smallest number (0.8063 mol of O in this example) to obtain the mole ratios (notice that one of these numbers will be 1.00):

$$\frac{6.6561 \text{ mol C}}{0.8063} = 8.14 \qquad \frac{8.254 \text{ mol H}}{0.8063} = 10.2 \qquad \frac{0.8063 \text{ mol O}}{0.8063} = 1.00$$

These ratios are now substituted into the generic molecular formula $C_x H_y O_z$. Rounding to integral values gives the empirical formula.

$$C_{8.14} H_{10.2} O_{1.00} = C_8 H_{10} O$$

The molecular formula will be a multiple of the empirical formula; in this example, that means

$$C_8 H_{10} O \quad \text{or} \quad C_{16} H_{20} O_2 \quad \text{or} \quad C_{24} H_{30} O_3 \quad \text{and so on}$$

EXAMPLE 14.2

The elemental analysis of a sample containing C, H, N, and O gave the following results: C, 54.8%; H, 10.1%; N, 10.5%. What is the empirical formula of this compound?

First, calculate what percentage of the compound is oxygen by subtracting the sum of the percentage values of C, H, and N from 100%.

$$100.0 - (54.8 + 10.1 + 10.5) = 24.6\% \text{ O}$$

One hundred grams of this compound would contain 54.8 g C, 10.1 g H, 10.5 g N, and 24.6 g O. From these amounts, we calculate the number of moles of each element:

$$54.8 \text{ g c} \left(\frac{1 \text{ mol C}}{12.011 \text{ g c}} \right) = 4.5624 \text{ mol C} \qquad 10.1 \text{ g h} \left(\frac{1 \text{ mol H}}{1.008 \text{ g h}} \right) = 10.020 \text{ mol H}$$

$$10.5 \text{ g n} \left(\frac{1 \text{ mol N}}{14.01 \text{ g n}} \right) = 0.7496 \text{ mol N} \qquad 24.6 \text{ g o} \left(\frac{1 \text{ mol O}}{1.008 \text{ g O}} \right) = 1.5364 \text{ mol O}$$

Next, we divide each number of moles by the smallest number (0.7496 mol of N in this example) to obtain the mole ratios:

$$\frac{4.5624 \text{ mol C}}{0.7496} = 6.09 \quad \frac{10.020 \text{ mol H}}{0.7496} = 13.4 \quad \frac{0.7496 \text{ mol N}}{0.7496} = 1.00 \quad \frac{1.5364 \text{ mol O}}{0.7496} = 2.05$$

These ratios are now substituted into the generic molecular formula $C_xH_yN_wO_z$. Rounding to integral values gives the empirical formula.

$$C_{6.09}H_{13.4}N_{1.00}O_{2.05} = C_6H_{13}NO_2$$

EXERCISE 14.1

The elemental analysis of a sample containing C, H, O, and S gave the following results: C, 35.9%; H, 5.83%; S, 23.7%. What is the empirical formula of this compound?

The use of elemental analysis to judge purity is straightforward if you know what the formula is already. In such cases, the calculated and found percentage values should be within ~ 0.3%. For example, the data for the compound in Example 14.2 would indicate that this compound is "pure".

$C_6H_{13}NO_2$ FW = 131.18   *Calculated*   C, 54.94%   H, 9.99%   N, 10.68%
                                  Found   C, 54.8%   H, 10.1%   N, 10.5%

If you are dealing with an unknown substance, then analytical chromatographic methods such as TLC or HPLC have to be used to augment the data from elemental analysis.

14.1c THE MOLECULAR FORMULA PROVIDES INFORMATION ABOUT THE PRESENCE OF DOUBLE BONDS AND RINGS IN A MOLECULE

In addition to obtaining an elemental analysis, we could also have the analytical lab determine the compound's molecular weight using freezing point depression, boiling point elevation, or vapor-phase equilibration. (Section 14.2b will describe the use of mass spectra to obtain accurate mass data.) Knowing both the empirical formula and mass of a compound tells you the compound's molecular formula, and that in turn can be used to calculate the number of **sites of unsaturation,** which are the π bonds and rings that a compound has.

For a molecule with the general formula $C_nH_m(X)_pN_qO_r$, where (X) = F, Cl, Br, or I, use the following equation to calculate the number of sites of unsaturation:

$$\text{No. of sites of unsaturation} = \frac{(2n + 2) - m - p + q}{2} \tag{14.1}$$

Thus, as shown by the following sample calculation, a molecule with the formula C_5H_9ClO has one site of unsaturation, which means that it has a single π bond or ring. Four structures (of the hundreds that are possible) are included below to illustrate that we cannot know specifically what type of π bond or what size or type of ring is actually present in the molecule.

$$C_5H_9ClO \quad \frac{2(5) + 2 - 9 - 1}{2} = 1 \text{ site of unsaturation} = 1 \, \pi \text{ bond} \quad \text{or} \quad 1 \text{ ring}$$

Examples:

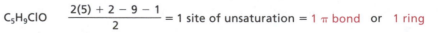

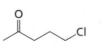

1 π Bond   1 π Bond   1 Ring   1 Ring

Still, if the ^{1}H NMR spectrum were to reveal the presence of an aldehyde group, for example, then we immediately know that the molecule cannot also contain a ring or another double bond. Therefore, this type of calculation can be used to narrow the structural possibilities that need to be considered during the elucidation process.

EXERCISE 14.2

A molecule has the formula $C_6H_{13}NO_2$. How many sites of unsaturation are present?

EXERCISE 14.3

Count the number of sites of unsaturation present in the following molecules, and then calculate the number using Eq. 14.1 to verify your answers.

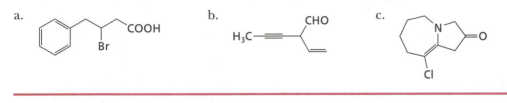

a.
b.
c.

14.2 MASS SPECTROMETRY

14.2a MASS SPECTRA ARE OBTAINED FROM GAS PHASE IONS

Mass spectrometry (MS) is a relatively old analytical method, dating from the early 1900s. (Notice that the technique is called "spectrometry" and not "spectroscopy".) As with every technique, mass spectrometry has blossomed with the development of computerized instrumentation. Mass spectrometers are not as commonly available as NMR instruments are, so they find less routine use in the laboratory. Nevertheless, for analyzing extremely small samples (as little as 10^{-12} g) or complex mixtures that can be separated by gas or liquid chromatography, mass spectrometry is sometimes the only technique that will provide any useful data to solve structural problems

A mass spectrometer accomplishes three basic tasks:

- Vaporization of molecules.
- Production of ions from gas-phase molecules.
- Separation of the gas-phase ions by their mass-to-charge ratios, m/z.

Although multiply charged organic molecules can be produced in the gas phase, the charge (z) most often equals +1, so the m/z value of an ion equals its mass.

For a typical organic molecule, vaporization is carried out by heating a sample under high vacuum (low pressure). The low pressure keeps collisions between the molecules and ions to a minimum, which simplifies the analysis.

One method used to produce ions from gas-phase molecules is by contact with a heated filament, a process called *electron impact* (EI) *ionization*. This procedure removes an electron from the molecule to produce a radical cation, a species with a positive charge because it has lost an electron and a radical because it has an unpaired electron.

$$M \rightarrow M^{+\cdot} + e^- \qquad (14.2)$$

The radical cation $M^{+\cdot}$ is called the **molecular ion** and has the same mass as the molecule.

The typical energy of the electron impact process is 70 electron volts (eV), which imparts enough vibrational energy to the molecular ion to cause **fragmentation,** as shown schematically in Eq. 14.3. One drawback of this ionization method, in fact, is that the molecular ion may fragment *too readily* and escape detection. Obviously, the inability to detect $M^{+\cdot}$ means that you will not be able to measure the compound's molecular weight.

$$M \longrightarrow M^{+\cdot} \begin{array}{l} \nearrow \quad F_1 \; + \; F_2^{+\cdot} \; \longrightarrow \; \text{and so on} \\ \searrow \quad F_3^{\cdot} \; + \; F_4^{+} \; \longrightarrow \; F_5 \; + \; F_6^{+} \; \longrightarrow \; \text{and so on} \end{array} \qquad (14.3)$$

Observing the masses of the fragments produced from a molecule provides useful structural information, however, so the failure to observe $M^{+\cdot}$ does not negate the utility of EI mass spectrometry. In the example given by Eq. 14.3, $M^{+\cdot}$ undergoes fragmentation in two ways. The first produces another radical cation ($F_2^{+\cdot}$) by loss of a neutral molecule (F_1). The second pathway occurs by expulsion of a radical ($F_3\cdot$), leaving a cation (F_4^+). Cation F_4^+ then loses a neutral molecule (F_5), producing another cation fragment (F_6^+). Both processes may continue until the species can no longer fragment, or until an especially stable cation is formed. *Species that are detected in the mass spectrum must carry a charge*, so only $M^{+\cdot}$, $F_2^{+\cdot}$, F_4^+, and F_6^+ in Eq. 14.3 will be detected.

Techniques other than electron-impact ionization can serve to produce ions in the gas phase and make the detection of $M^{+\cdot}$ more likely. *Chemical ionization* (CI) uses a stream of NH_4^+ or CH_5^+ ions to collide with the gas-phase analyte molecules. The CH_5^+ ion (protonated methane) will transfer a proton to just about any other species:

$$CH_5^+ + M \rightarrow CH_4 + MH^+ \qquad (14.4)$$

The mass of the species produced by this method is 1 amu more than the actual molecular weight, and this fact must be kept in mind. The same process can be carried out with ammonium ions, although molecules that are weakly basic—hydrocarbons, for example—will not accept a proton under these conditions.

$$NH_4^+ + M \rightarrow NH_3 + MH^+ \qquad (14.5)$$

The advantage of CI, especially with the use of ammonium ions, is that MH^+ accepts little additional energy that can lead to subsequent fragmentation. Therefore, the molecular weight of the molecule is readily determined.

For nonvolatile molecules, especially large, polar biomolecules such as proteins, nucleic acids, and polysaccharides with MW values up to 100,000 amu, vaporization can be difficult. Several techniques developed in the past two decades circumvent this volatility problem, which has led to significant advances in applying mass spectrometry to biochemical research. Some of these methods are briefly described below.

- *Field desorption* (FD). A sample of the analyte is deposited on needle tips that have been grown from a wire. An electric current and heat passing through the wire create positive ions at the tips of the needles. Because the wire also carries a positive charge, electrostatic repulsions cause the cations to leave the surface and enter the gas phase.

- *Laser desorption* (LD). The sample is bombarded with short, intense pulses of laser light with UV or IR wavelengths that correspond to the absorption maxima of the analyte. Transfer of energy to the electronic or vibrational excited states causes molecules of the analyte to enter the gas phase. *Matrix assisted laser desportion* (MALDI) is an alternate technique in which a secondary compound absorbs the laser energy and then transfers it to the analyte.

- *Fast atom bombardment* (FAB). This technique makes use of momentum transfer between accelerating xenon atoms and analyte molecules, which forces the molecule into the gas phase. The sample is first dissolved in several microliters of glycerol, and the analyte molecule enters the gas phase as MH^+, which is often the most abundant ion in the resulting spectrum. This method is especially useful for determining molecular weight values.

- *Secondary ion mass spectrometry* (SIMS). This technique is similar to the FAB method, except that momentum transfer comes from a stream of cesium ions.

Irrespective of the method used to create the gas-phase ions, a mass spectrometer generates the actual spectrum by first separating the ions—a task accomplished by a combination of electric and/or magnetic fields to control either the distance or speed that the ions move in going from the sample chamber to the detector—and then plotting the relative numbers of each ion reaching the detector (the relative intensity) versus m/z. A simple example of an EI mass spectrum is shown in Figure 14.1.

The spectrum shown in Figure 14.1 is considered "low resolution", because it separates fragments that differ by integral amu values. The region with the highest m/z values corresponds to $M^{+\cdot}$, unless the molecular ion undergoes fragmentation too readily. Because most elements have isotopic forms, **isotope peaks** that correspond to $(M + 1)^{+\cdot}$ and even $(M + 2)^{+\cdot}$ are also observed. These are described in more detail in Section 14.2b.

The largest peak in a mass spectrum—the one with a relative intensity of 100%—is the **base peak,** which corresponds to the mass of an ion that is long lived and reaches the detector in greater quantity than any other. For 2-butanone, the base peak is CH_3CO^+ (m/z 43).

Except for the molecular ion peaks, the remaining lines in the spectrum result from fragments that are formed as the molecular ion breaks down. Although the process of fragmentation may seem random, there are limited pathways available to any molecular ion. Carbocation stability (Section 6.2b) often correlates with the structures of species that might be detected as long-lived fragments in a mass spectrometer.

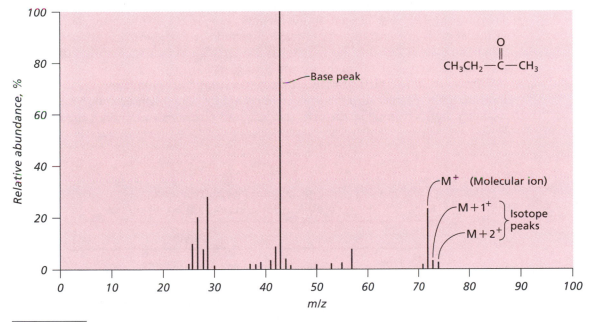

Figure 14.1
The EI mass spectrum of 2-butanone (MW 72.11) with significant peaks identified.

14.2b MASS SPECTRA ARE COMMONLY USED TO DETERMINE THE MOLECULAR WEIGHT AND FORMULA OF A COMPOUND

To determine the molecular formula of a compound from a mass spectrum, you first look at the peaks with the highest m/z values and determine the compound's MW. With a CI or FAB mass spectrum, remember to subtract 1 amu from the m/z value of the peak you consider to be $M^{+\cdot}$ because that peak actually corresponds to the species MH^+ (Section 14.2a).

When the molecular ion peak is observable, isotope peaks for $(M + 1)^{+\cdot}$, and even for $(M + 2)^{+\cdot}$ are also evident. To understand their genesis, look again at Figure 14.1. The compound that produces this spectrum is 2-butanone, C_4H_8O. For carbon, its two principal isotopes are ^{12}C and ^{13}C; for hydrogen they are ^1H and ^2H; and for oxygen, they are ^{16}O, ^{17}O, and ^{18}O (see Table 14.1). Thus, the peak at m/z 72 corresponds to the sum of the principal isotope masses, $^{12}C_4{}^1H_8{}^{16}O$ $(4 \times 12) + (8 \times 1) + (1 \times 16) = 72$. The peak at m/z 73 results from compounds that have the formulas $^{13}C^{12}C_3{}^1H_8{}^{16}O$, $^{12}C_4{}^2H^1H_7{}^{16}O$, and $^{12}C_4{}^1H8^{17}O$.

EXERCISE 14.4

Based on the isotopes given in Table 14.1 for C, H, and O, what six formulas can give rise to the $(M + 2)^{+\cdot}$ peak at m/z 74 for 2-butanone?

The intensities of the isotope peaks are related to the number of molecules containing each particular isotope and are always less intense than the molecular ion peak unless the molecule contains more than one Cl or Br atom (Section 14.2c). For example, ^{12}C and ^{13}C exist in nature in a 99:1 ratio, which means ~ 1% of the carbon atoms in any sample will be ^{13}C. Thus, only ~ 1 in 25 molecules of 2-butanone (which has four carbon atoms) will have a ^{13}C atom and a mass of 73 instead of 72. The ratio of intensities of $M^{+\cdot}$ and $(M + 1)^{+\cdot}$ should therefore be about 25:1, as observed.

Unless Cl, Br, and/or S are present (Section 14.2c), you will probably observe one of the four patterns shown in Figure 14.2 for the molecular ion region of a mass spectrum.

Pattern (a.) shows a compound with readily discernable molecular ion $(M^{+\cdot})$ and $(M + 1)^{+\cdot}$ peaks. Here, the $(M + 2)^{+\cdot}$ peak is weak and unobserved. In (b.), $M^{+\cdot}$, $(M + 1)^{+\cdot}$, and $(M + 2)^{+\cdot}$ are all seen, so the identification of $M^{+\cdot}$ is straightforward. Finally, in (c.) and (d.), *four* (or more) peaks are observed at the highest m/z values. Observation of an $(M + 3)^{+\cdot}$ peak is rare, so the peak at the highest m/z value within the group must be $(M + 2)^{+\cdot}$, and $M^{+\cdot}$ is the second peak to its left.

Once we know which peak in the mass spectrum corresponds to $M^{+\cdot}$, we can determine the molecular formula of the compound by using **high-resolution mass spectrometry.** In this technique, the mass spectrometer separates ions in the same way as in

Table 14.1 Isotopic abundance for elements commonly found in organic compounds.

Isotope	%	Isotope	%
^1H	99.98	^{31}P	100.0
^2H	0.01		
		^{32}S	95.00
^{12}C	98.89	^{33}S	0.76
^{13}C	1.11	^{34}S	4.22
		^{36}S	0.01
^{14}N	99.63		
^{15}N	0.37	^{35}Cl	75.53
		^{37}Cl	24.47
^{16}O	99.76		
^{17}O	0.04	^{79}Br	50.54
^{18}O	0.20	^{81}Br	49.46
^{19}F	100.0	^{127}I	100.0

a.

b.

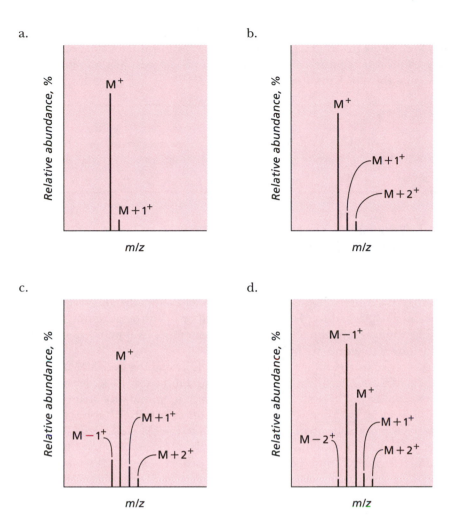

c.

d.

Figure 14.2
Typical patterns observed for
the molecular ion region in
the mass spectrum of an
organic compound that does
not contain S, Cl, or Br.

a low-resolution instrument, but the separation is carried out more precisely, so ions that differ in mass by thousandths or 10-thousandths of an amu value can be detected. Exact mass values for the principal isotope of each element are required for high-resolution analysis, and these values are given in Table 14.2.

For example, molecules with the formulas $C_3H_4N_2$, C_4H_4O, and C_5H_8 all have integral mass values of 68. Their exact mass values, calculated using the data in Table 14.2 are,

$C_3H_4N_2$	$3(12.0000) + 4(1.007825) + 2(14.0031)$	$= 68.0375$
C_4H_4O	$4(12.0000) + 4(1.007825) + 15.9949$	$= 68.0262$
C_5H_8	$5(12.0000) + 8(1.007825)$	$= 68.0626$

Table 14.2   Exact mass values in amu for the principal isotopes of common elements

Isotope	Mass (amu)	Isotope	Mass (amu)
^1H	1.007825	^{12}C	12.00000
^{14}N	14.0031	^{16}O	15.9949
^{32}S	31.9721	^{35}Cl	34.9689
^{79}Br	78.9183		

A high-resolution mass spectrometer can readily differentiate between these MW values, which allows us to determine the molecular formula of the compound being investigated. Computer software has automated the process further and can calculate the molecular formula from the mass value of $M^{+\cdot}$.

EXERCISE 14.5

The molecular ion peak in a high-resolution mass spectrum was observed at m/z 122.078. Which of the following is the likely molecular formula: $C_4H_{10}O_4$, $C_8H_{10}O$, or $C_7H_{10}N_2$?

14.2c THE PRESENCE OF NITROGEN, SULFUR, CHLORINE, AND BROMINE ATOMS ARE READILY DETERMINED FROM MASS SPECTRA

The presence of *nitrogen* in a molecule is readily established when an odd number of N atoms are present, because the molecular ion peak has an odd m/z value. This observation, called the *nitrogen rule*, is a result of the fact that a nitrogen atom bonds to one less substituent than carbon. For an even number of nitrogen atoms, of course, the mass of $M^{+\cdot}$ will be even, and a high-resolution spectrum will be needed to determine how many nitrogen atoms are present.

The presence of a *sulfur* atom manifests itself by making the $(M + 2)^{+\cdot}$ peak somewhat larger than expected. This pattern, illustrated in Figure 14.3a, occurs because the natural abundance of ^{34}S is 4.22% compared with an abundance value for ^{33}S of only 0.76% (Table 14.1). In other words, the contribution of sulfur isotopes to the intensity of the $(M + 1)^{+\cdot}$ peak is less than the contribution to the $(M + 2)^{+\cdot}$ peak. Because the $(M + 2)^{+\cdot}$ peak for most compounds is smaller than the $(M + 1)^{+\cdot}$ peak, the presence of sulfur is readily detected by observing an $(M + 2)^{+\cdot}$ peak that is ~ 1/24 the intensity of $M^{+\cdot}$.

The presence of either *bromine* or *chlorine* is also readily established from the size of the $(M + 2)^{+\cdot}$ peak (Fig. 14.3b and 14.3c). For each, the $(M + 2)^{+\cdot}$ peak is so much more intense than it is for compounds lacking a halogen atom that even a cursory glance at the mass spectrum reveals their presence. For chlorine, the relative amounts of ^{35}Cl and ^{37}Cl are 75.53 and 24.47%, respectively, and for bromine, the amounts of ^{79}Br and ^{81}Br are 50.54 and 49.46%, respectively. Thus the molecular ion region for a compound with a chlorine or bromine atom has two peaks separated by two mass units with a 3:1 or 1:1 intensity ratio, respectively.

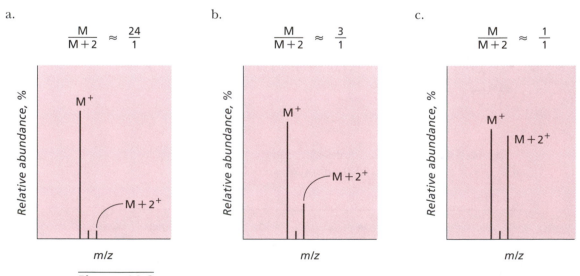

Figure 14.3

Appearance of the molecular ion region when the following heteroatom is present: (a.) sulfur, (b.) chlorine, and (c.) bromine.

EXAMPLE 14.3

Based on the following EI mass spectrum, which heteroatom –S, Cl, or Br– is present in the molecule that produces this spectrum?

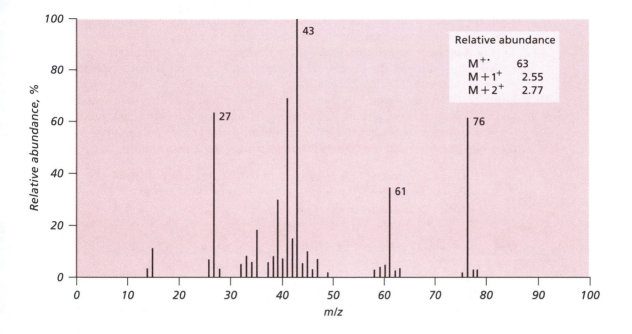

To determine whether S, Cl, or Br atoms are present, examine the region with the highest m/z values and compare the ratios of the peak heights to those in Figure 14.3 (the intensities of these peaks are also listed in the box). The intensity of $(M + 2)^{+\cdot}$ is ~ 1/24th of the intensity of $M^{+\cdot}$ (63/2.77 = 22.7), so this molecule contains S.

EXERCISE 14.6

Based on the following EI mass spectrum, which heteroatom –S, Cl, or Br– is present in the molecule that produces this spectrum?

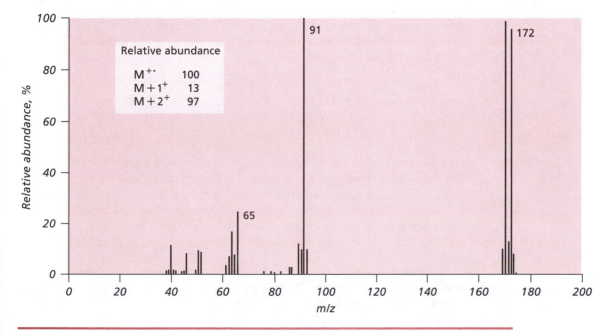

14.2d CHARACTERISTIC FRAGMENTS IN THE MASS SPECTRUM ARE ASSOCIATED WITH CERTAIN COMPOUND TYPES

In the structure identification problems in this text, mass spectra will be used sparingly, and then only to determine a compound's molecular weight, molecular formula, and whether N, S, Cl, or Br atoms are present. However, additional information *can* be obtained from a mass spectrum if the ions produced during fragmentation are identified. The loss of specific fragments from molecules in certain classes is quite predictable.

The fragments worth noting are those that produce the peaks with significant intensities, for example, the base peak, which often corresponds to an especially stable ion or cation radical. The general fragmentation patterns for many organic compounds can be summarized by the following four equations. (In these equations, the symbol ⌐⁺· is an abbreviated bracket to indicate that the molecule or fragment exists as a cation radical. Using brackets or this abbreviated bracket symbol eliminates the need to specify where the charge or unpaired electron is actually located.)

- *Cleavage at branches.* Fragmentation of the hydrocarbon portion of a molecule occurs to produce the more stable carbocation (3° > 2° > 1°). The larger group is normally lost as the radical portion.

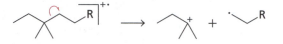

- *α,β-Cleavage.* The bonds between the atoms that are α and β to a heteroatom undergo heterolytic cleavage to form a resonance-stabilized cation. *This type of bond-breaking process for carbonyl compounds is especially important.*

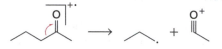

- *Loss of a neutral molecule.* A cationic fragment, formed from M⁺·, may be cleaved further by loss of a stable molecule such as CO, HCN, H_2O, HCl, NO, and so on.

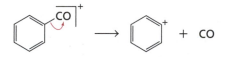

- *Rearrangement.* When π electrons are situated appropriately, substituents on the carbon framework can migrate to produce rearranged species. The following process—the **McLafferty rearrangement**—is a primary fragmentation pathway for aliphatic carbonyl compounds.

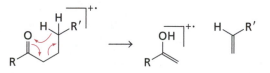

A rearrangement is detectible because in its *absence,* a molecular ion with an even mass value cleaves to give fragment ions with odd-numbered masses, and vice versa. Thus, observation of even-mass fragments from an even-mass molecular ion, or odd-mass fragments from an odd-mass molecular ion, suggests that a rearrangement has taken place. A difference of 1 amu between the expected and observed mass of a fragment usually means that migration of H has occurred.

Table 14.3 lists the general fragmentation pathways that several classes of organic compounds are known to undergo. With these data in mind, it is often possible to account for major fragment peaks in a mass spectrum.

Table 14.3 Principal fragmentation pathways for classes of organic molecules

Compound type	Principal fragmentation pathways
Alcohols	Loss of a neutral molecule (water, *m/z* 18); α,β-cleavage
Aldehydes	α,β-Cleavage; McLafferty rearrangement (aliphatic)
Alkanes	Cleavage at branches
Alkenes	α,β-Cleavage; rearrangement
Alkyl halides	Loss of a neutral molecule (HX, *m/z* 20, 36, 80, or 127)
Carboxylic acids	α,β-Cleavage; McLafferty rearrangement (aliphatic)
Esters	α,β-Cleavage; McLafferty rearrangement (aliphatic)
Ketones	α,β-Cleavage; McLafferty rearrangement (aliphatic)

EXAMPLE 14.4

3-Hepanone produces a mass spectrum with important peaks at *m/z* 114, 85, 72, and 57. Based on the general patterns listed in Table 14.3, draw the structures of the fragments that give rise to these peaks.

The overall gas-phase process can be represented by the following equation:

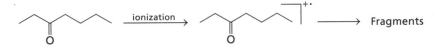

For an aliphatic ketone, α,β-cleavage and the McLafferty rearrangement are the common fragmentation pathways. These give rise to the three fragment peaks; $M^{+\cdot} = m/z$ 114.

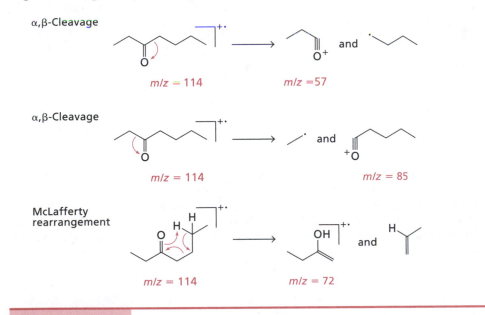

EXERCISE 14.7

2-Hexanol produces a mass spectrum with key peaks at *m/z* 102, 84, and 45. Based on the general patterns listed in Table 14.3, draw the structures of the fragments that produce these peaks.

14.3 INFRARED SPECTROSCOPY

14.3a INFRARED SPECTRA ARE GENERATED BY MEASURING THE ABSORPTION OF HEAT AS A FUNCTION OF FREQUENCY

Infrared radiation has the energy needed to affect the stretching and bending vibrations of the bonds within a molecule. Organic compounds, with many covalent bonds, often have complex IR spectra. Infrared spectroscopy is used primarily in the interpretation process as a means to identify functional groups.

The range of wavelengths in vibrational spectra from which useful information can be obtained with routine instrumentation is from 2.5 to 20 μm. It has become common practice, however, to plot IR spectra using the inverse of wavelength, $1/\lambda$, termed **wavenumbers** (cm^{-1}), which is represented by the symbol $\bar{v}$. Spectroscopists prefer this unit because it directly reflects the energies of the vibrations being measured. The range of a typical IR spectrum covers the wavenumbers between 4000 and 500 cm^{-1}.

As with most energy absorption processes in molecules, only certain frequencies of IR radiation interact with the bonded atoms. Furthermore, a vibration will only give rise to an absorption band in the IR spectrum when there is a change in the dipole moment of the bond that is vibrating. A homodinuclear molecule like H_2 does not display a hydrogen–hydrogen stretching vibration because the dipole moment does not change as the hydrogen–hydrogen bond is elongated or compressed. Likewise, symmetrically trans-disubstituted and tetrasubstituted double bonds and internal alkynes will not have a peak corresponding to the C=C or C≡C stretching mode, respectively.

If each covalent bond between two atoms is considered as a spring connecting two balls, then we can draw two general conclusions:

1. The frequency of a vibration will be inversely related to the *masses* of the atoms bonded to one another. Thus, the heavier the atoms are, the lower the frequency of the vibration will be. The following examples illustrate this point:

$$C=O \quad\quad versus \quad\quad C=S \quad\quad\quad C-H \quad\quad versus \quad\quad C-D$$
$$\bar{v} = 1700\ cm^{-1} \quad\quad 1350\ cm^{-1} \quad\quad 3000\ cm^{-1} \quad\quad 2200\ cm^{-1}$$

2. The frequency of a vibration will be directly proportional to the *strength* of the bond. For example, the stretching vibration of a triple bond appears at a higher frequency than that of either a double or single bond:

$$C≡C \quad\quad\quad\quad C=C \quad\quad\quad\quad C-C$$
$$\bar{v} = 2150\ cm^{-1} \quad\quad 1650\ cm^{-1} \quad\quad 1200\ cm^{-1}$$

These examples assume that each set of bonded atoms acts independently of others in the molecule. For an aldehyde group, the assumption is reasonable, because the strength of each bond and the masses of the atoms are quite different; so the C–H, C=O, and C–C stretching vibrations are relatively independent.

For a methyl group, however, the three hydrogen atoms have the same masses, and the C–H bonds have similar strengths, so the vibrations are *not* independent. Instead, the vibrational modes are coupled and appear as symmetric and antisymmetric CH_3 stretching vibrations.

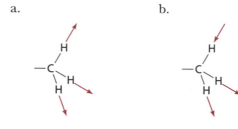

a.   b.

^{v}C–H$_{sym}$ = 2872 cm^{-1} ^{v}C–H$_{asym}$ = 2962 cm^{-1}

In the symmetric mode, elongation and compression of the bonds occur in the same manner at the same time (a.); in the antisymmetric vibration, one bond is out of phase with the other two, as illustrated in (b.).

Coupled vibrations within a group of *three* atoms (rather than four, as in a methyl group) are even more common. Some examples include the methylene, nitro, and amino groups. Each group displays two bands, corresponding to the symmetric and antisymmetric stretching modes. Coupled vibrations can affect other vibrational modes as well.

Most of the vibrations that will be discussed below are *fundamental vibrations,* which occur when a molecule absorbs IR radiation of the appropriate energy needed to reach its first vibrational excited state. Other bands also occur, however, corresponding to excitation to the second, third, or even higher excited states; these bands are called *overtone vibrations.* Although they are much weaker, overtone vibrations can be important in the characterization of certain compound types, notably benzene derivatives.

14.3b INFRARED SPECTRA PLOT TRANSMITTANCE VERSUS FREQUENCY

An IR spectrometer operates by measuring the differences in energy between two beams of IR radiation, one of which has passed through the sample, and one that has not. At frequencies where *no* absorption occurs, the beam that passes through the sample remains unaffected. The transmittance relative to the reference beam (the one that does not pass through the sample) is defined as 100%. Any absorption of radiation by the sample at a specific frequency results in a lower transmittance, leading to an absorption band, or *peak.* A strong band will extend to the bottom of the spectrum, whereas a weak band may only cause a small dip in the baseline.

Figure 14.4 shows a typical IR spectrum. The wavelength, λ, of the radiation increases from left to right as plotted, but the frequency ($\bar{v}$, cm^{-1}) decreases. The percent transmittance of IR energy through the sample is plotted so that decreased transmittance (i.e., increased absorption) of radiation is toward the bottom of the spectrum.

Three features of an IR spectrum that provide information for the identification of structural elements are, in order of decreasing importance,

1. The positions of the bands (in wavenumbers).
2. The intensities of the bands (weak, medium, strong).
3. The shapes of the bands (broad or sharp).

The total IR spectrum has two general regions. The portion of the spectrum > 1300 cm^{-1} (1300–4000 cm^{-1}) is the **functional group region.** Most bands in this region

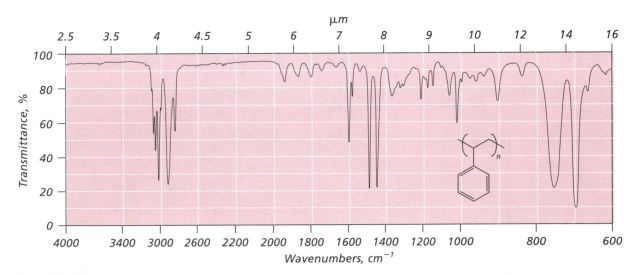

Figure 14.4
The IR spectrum of polystyrene (thin film), showing the relationship between wavelength and frequency.

correspond to bond stretching vibrations. Frequencies < 1300 cm⁻¹ define the **finger-print region,** because the pattern of bands here is *unique for each compound.* The finger-print region can be used to match the spectrum of an unknown compound with spectra of known ones. A perfect match is an unequivocal identification of a substance, just as sets of matching fingerprints identify a person. Bending vibrations produce many of the bands in the fingerprint region.

14.3c FUNCTIONAL GROUPS ARE IDENTIFIED FROM AN INFRARED SPECTRUM BY LOOKING FOR ABSORPTION BANDS IN FOUR SPECIFIC REGIONS

A typical IR spectrum can contain > 30 absorption bands, some of which are combinations of stretching and bending modes. Unlike an NMR spectrum, therefore, many of the peaks in an IR spectrum are *not* crucial for the interpretation process.

Because the majority of an organic molecule is its carbon and hydrogen framework, many bands in an IR spectrum are *not* associated with functional groups. For example, nearly every spectrum in this text has absorptions in the region ~ 3000 cm⁻¹, which are caused by C–H stretching vibrations. (When hydrogen atoms are attached to sp^2-hybridized carbon atoms, their absorptions will appear between 3000 and 3200 cm⁻¹. When the hydrogen atoms are attached to sp^3-hybridized carbon atoms, the bands appear between 2800 and 3000 cm⁻¹.) You already know that ¹H NMR spectra will readily allow you to identify the hydrocarbon groups in a molecule, so bands associated with the C–H stretching vibrations do not provide new information.

To identify functional groups, look for absorption bands in the regions *that are blank in the spectra of simple hydrocarbons.* These regions are shown in color in Figure 14.5.

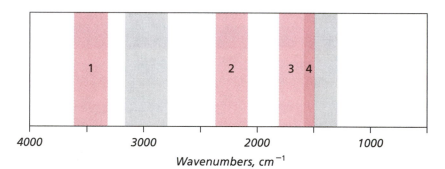

Region	Frequency range, (cm⁻¹)	Bond types	Functional groups
1	3500 – 3200	O—H N—H	Alcohol, phenol Amine, amide
2	2300 – 2100	C≡C C≡N	Alkyne Nitrile
3	1800 – 1650	C=O	Aldehyde Amide Anhydride (2 bands) Carboxylic acid Acid chloride Ester
4	1650 – 1500	C=C C=C C=N N=O	Alkene Arene Imine Nitro compound

Figure 14.5

Regions of an IR spectrum used to identify some functional groups. The most important bands appear in regions 1–4. Areas in gray are those in which peaks associated with carbon–hydrogen stretching and bending vibrations occur.

14.3d O–H AND N–H STRETCHING VIBRATIONS PRODUCE ABSORPTION BANDS IN THE REGION BETWEEN 3600 AND 3200 CM^{-1}

The highest frequency region of an IR spectrum is where bands associated with O–H and N–H stretching vibrations appear. You may see bands ~ 3500 cm^{-1} resulting from water in the sample, so first make certain that the bands in this region belong to the compound being studied. Oxygen- and nitrogen-containing functional groups tend to form hydrogen bonds, so the absorption bands associated with OH or NH groups are often broad.

An N–H stretching vibration produces a band that is usually sharper than one for an O–H stretching mode, as illustrated in Figure 14.6. If the spectrum can be recorded on a sample in dilute solution to eliminate the effects of hydrogen bonding, the free O–H stretching vibration produces an absorption between 3600 and 3800 cm^{-1}, and the free N–H stretching vibration generates a band between 3500 and 3300 cm^{-1}. Recall that OH groups can be identified in the ^{1}H NMR spectrum because they generally undergo deuterium exchange with D_2O (Section 13.3a)

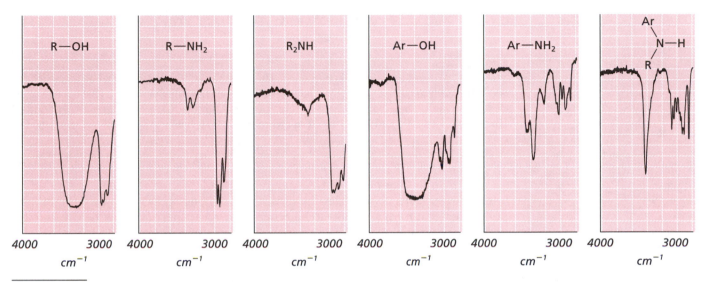

Figure 14.6
A comparison of absorption bands for the OH and NH stretching vibrations of both aliphatic and aromatic compounds: R–OH, 1-propanol; RNH$_2$, butylamine; R$_2$NH, dibutylamine; Ar–OH, o-cresol; ArNH$_2$, o-toluidine; ArNHR, N-methylaniline. The sharper bands at ~ 3000 cm^{-1} are associated with C–H stretching vibrations.

14.3e TRIPLE BONDS HAVE STRETCHING VIBRATIONS NEAR 2200 CM^{-1}

The region between 2500 and 2000 cm^{-1} is almost always blank because relatively few molecules have functional groups that absorb here. Compounds that do produce a stretching vibration in this region have triple bonds, which are of two types, C≡N and C≡C. Absorptions usually occur between 2300 and 2100 cm^{-1} (Table 14.4).

Table 14.4   Ranges of stretching frequencies for groups absorbing in the triple-bond region.

Functional group	Saturated (cm^{-1})	Aryl or α,β-unsaturated (cm^{-1})
Alkyne, terminal	2150–2120	2140–2100
Alkyne, internal	2260–2190	2240–2150
Nitrile	2260–2240	2240–2220

^{a}The intensities of bands in this region vary.

The intensity of the nitrile stretching vibration can range from medium to strong, and it is usually observed because of the dipole associated with the C≡N bond (Fig. 14.7).

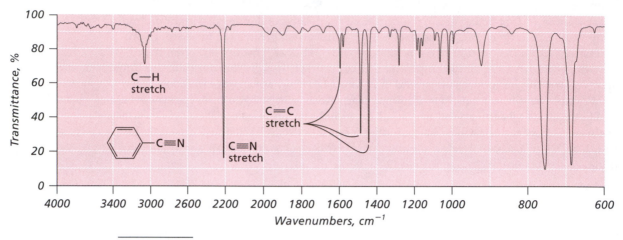

Figure 14.7

The IR spectrum of benzonitrile (thin film). The band for the C≡N stretching vibration (2200 cm^{-1}) is readily apparent. The typical bands for the C–H stretching vibrations (3100 cm^{-1}) are clearly seen also. These bands are > 3000 cm^{-1}, so the molecule has sp^2-hybridized C–H bonds. (Bands between 1600 and 1400 cm^{-1} result from the C=C stretching vibrations of the benzene ring.)

On the other hand, the C≡C stretching mode of an internal alkyne is very weak, often unobservable, because the dipole for the C≡C bond is essentially zero (Fig. 14.8). Remember that for a vibration to be observable, there must be a net dipole change when the bond interacts with IR radiation.

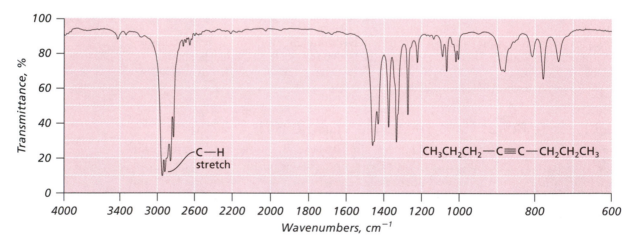

Figure 14.8

The IR spectrum of 4-octyne (thin film). No absorption for the C≡C stretching vibration appears in the region 2300–2100 cm^{-1} (cf. with the spectrum shown in Fig. 14.9.) The hydrogen atoms in 4-octyne are attached to sp^3-hybridized carbon atoms, so the typical C–H stretching vibration absorption bands appear between 2800 and 3000 cm^{-1}.

Thus, the C≡C stretching vibration of a *terminal* alkyne is usually observed. In addition, a strong absorption may be observed at 3300 cm^{-1} resulting from the ≡C–H stretching vibration (Fig. 14.9).

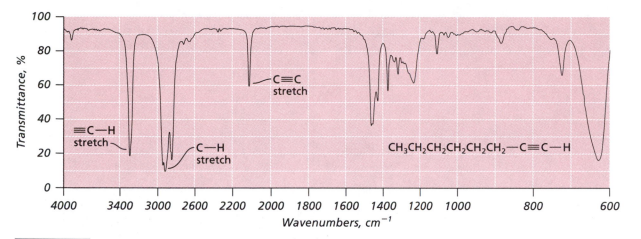

Figure 14.9

The IR spectrum of 1-octyne (thin film). The band for the C≡C stretching vibration (2100 cm^{-1}) has a medium intensity and is readily seen. The typical bands for the C–H stretching vibrations are also clearly seen: most of the hydrogen atoms are attached to sp^3-hybridized carbon atoms (2800–3000 cm^{-1}); the alkyne ≡C–H bond produces a band at 3300 cm^{-1}.

EXAMPLE 14.5

What functional group is present in the molecule that produces the following IR spectrum? What other instrumental method might be used to confirm your assignment.

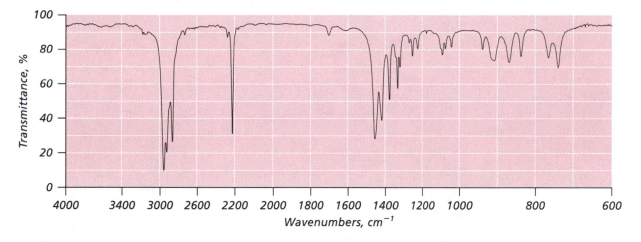

The compound that produces this spectrum has a strong absorption band at ~ 2220 cm^{-1}. The molecule is therefore an alkyne or nitrile. No absorption band is observed at 3300 cm^{-1}, so we rule out the terminal alkyne as a possibility. Because the triple-bond absorption is fairly intense, this molecule is most likely a nitrile. The mass spectrum should reveal the presence of N (odd MW). Also, the ^{13}C NMR spectrum should display a resonance for the nitrile carbon atom (see Table 13.4).

14.3f THE CARBONYL STRETCHING VIBRATION APPEARS AS AN INTENSE ABSORPTION BETWEEN 1600 AND 1800 CM^{-1}

Because the C=O stretching mode is *always* intense as a result of the large dipole moment of the carbon–oxygen double bond, you will rarely confuse it with any other absorption. Moreover, if no band is observed in the region ~ 1700 cm^{-1}, then you can immediately eliminate carbonyl-containing functional groups from further consideration

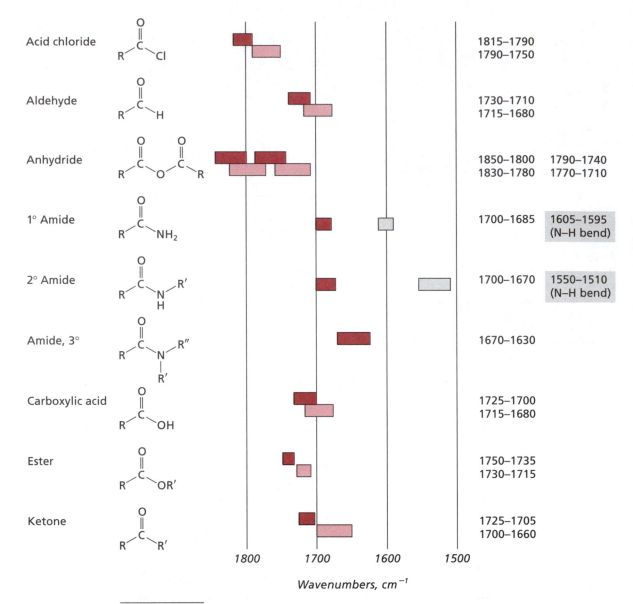

Figure 14.10
Wavenumber (cm^{-1}) ranges for the C=O stretching vibration in carbonyl compounds. The darker color blocks show base values for aliphatic compounds with the specified functional group. The lighter color blocks are the ranges for aryl and α,β-unsaturated analogues. The primary and secondary amides have an additional band in this general region that is not a C=O stretching mode. The N–H group is coupled with the carbonyl group in a bending vibration; the positions of these absorptions are indicated by the gray boxes.

(see Fig. 14.5). Figure 14.10 lists the common carbonyl-containing functional groups along with their associated stretching frequencies.

The variations in stretching frequencies listed in Figure 14.10 result from coupling between the C=O group and the other atoms attached to the carbonyl group. These combinations depend on the mass of each atom and the strength of its bonds to the carbonyl–carbon atom.

Hydrogen bonds also influence the carbonyl stretching frequency by lengthening the C=O bond and decreasing its strength, as shown in Figure 14.11. Because of hydrogen-bond formation, the carbonyl group of a carboxylic acid, which exists primarily as a dimer except in dilute solution (Section 2.8c), absorbs at a lower frequency than the carbonyl group of the structurally analogous ester. The absorption bands for a 1° (RCONH$_2$) or 2° (RCONHR′) amide are similarly affected by the formation of hydrogen bonds.

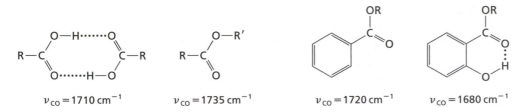

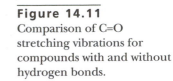

Figure 14.11
Comparison of C=O stretching vibrations for compounds with and without hydrogen bonds.

Some overlap exists between the regions associated with the different carbonyl-containing functional groups. For example, a band at 1710 cm^{-1} could be the carbonyl stretching vibration of an aliphatic ketone, a carboxylic acid, an aromatic ester, or an α,β-unsaturated aldehyde. Additional IR and ^{1}H NMR data that can be used to differentiate between these functional groups are summarized in Table 14.5, and sample spectra are shown in Figures 14.12–14.16. The position of the carbonyl resonance in the ^{13}C NMR spectrum (Fig. 13.23 and Table 13.4) is often unequivocal.

Table 14.5 Distinguishing features for carbonyl compounds.

Compound type	νCO (cm⁻¹)	Figure	Other features to look for
Acid chloride	1800		Single C=O stretching band at upper end of the carbonyl region
Aldehyde	1725	14.12	C–H stretching vibration ~ 2720 cm⁻¹; ¹H NMR: δ 10.0
Amide (1° or 2°)	1680	14.13	N–H stretching vibration ~ 3300 cm⁻¹
Amide (3°)	1650	14.14	Single C=O stretching band at lower end of the carbonyl region
Carboxylic acid	1710	14.15	O–H stretching vibration 3500–1500 cm⁻¹; ¹H NMR: δ 12.0
Ester	1735	14.16	Strong C–O stretching vibration ~ 1200 cm⁻¹; may be more intense than the C=O absorption
Ketone	1715		No outstanding features

^{a}All wavenumber values are approximate base values for aliphatic compounds (see Fig. 14.10). Representative spectra of common carbonyl compounds are presented in the indicated figure.

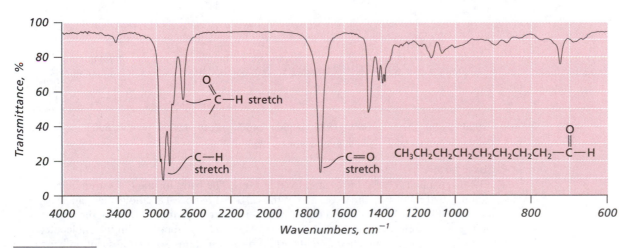

Figure 14.12
The IR spectrum of nonanal (thin film). The band for the C=O stretching vibration (1720 cm⁻¹) is in the middle of the carbonyl region, so it can be attributed to several compound types. The medium intensity band at ~ 2700 cm⁻¹ is diagnostic for the aldehyde group. Notice that this band, caused by the C–H stretching vibration of the aldehyde group (CHO), is at a lower frequency than the bands of the hydrocarbon C–H stretching vibrations (there are actually two aldehyde C–H stretching vibrations—the symmetric and antisymmetric modes—but one band is usually hidden under those of the hydrocarbon C–H stretching vibrations). In this molecule, the hydrogen atoms are attached to sp^3-hybridized carbon atoms, so their absorption bands appear between 2800 and 3000 cm⁻¹.

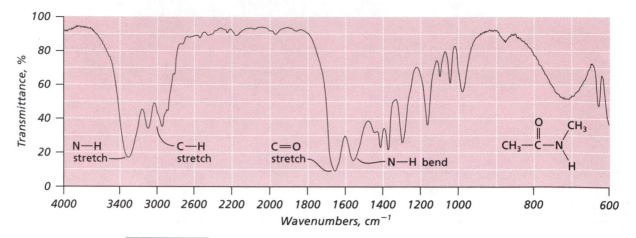

Figure 14.13

The IR spectrum of *N*-methylacetamide (melt). The band for the C=O stretching vibration (1660 cm^{-1}) is at the low end of the carbonyl region. An absorption band for the N–H stretching vibration is observed at ~ 3300 cm^{-1} and it is broad, which means that the compound is a secondary amide, RCONHR'. The N–H bending mode produces an absorption band at 1570 cm^{-1}.

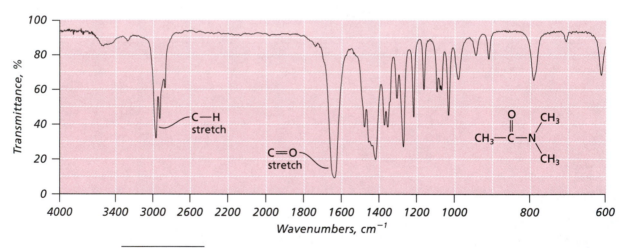

Figure 14.14

The IR spectrum of *N,N*-diethylacetamide (thin film). The band for the C=O stretching vibration (1620 cm^{-1}) is at the lower end of the carbonyl region. This molecule is a tertiary amide, which has no N–H bond, so there is no N–H stretching vibration. Thus, the region ~ 3000 cm^{-1} has only the typical bands for the C–H stretching vibrations. In this molecule, the hydrogen atoms are attached to sp^3-hybridized carbon atoms, so their absorption bands appear between 2800 and 3000 cm^{-1}.

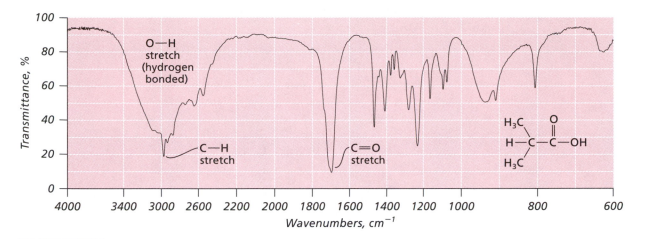

Figure 14.15

The IR spectrum of isobutyric acid (thin film). The band for the C=O stretching vibration (1710 cm^{-1}) is in the middle of the carbonyl region, so it might be attributed to several compound types. The broad and strong band that stretches from 3500 to 2500 cm^{-1} is diagnostic for the carboxylic acid group, which forms strong hydrogen bonds. Notice that the bands for the C–H stretching vibrations appear as sharp peaks between 2800 and 3000 cm^{-1} (sp^3-hybridized C–H bonds), which are superimposed on the broad band of the carboxylic acid OH stretching vibration. (The band at 1220 cm^{-1} is the C–O stretching vibration for the carboxylic acid group.)

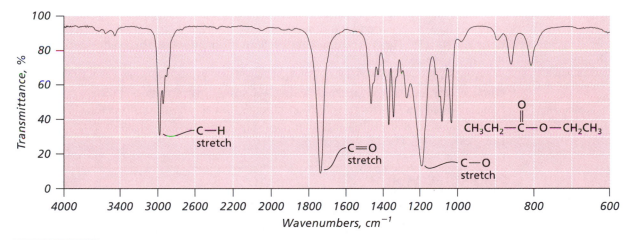

Figure 14.16

The IR spectrum of ethyl propanoate (thin film). The band for the C=O stretching vibration (1740 cm^{-1}) is in the upper region of the carbonyl region, so this compound is probably an ester. The intense band at 1200 cm^{-1} results from the two C–O stretching vibrations of the ester group. (This band may actually be more intense than the carbonyl band in some cases.) In this molecule, the hydrogen atoms are attached to sp^3-hybridized carbon atoms, so their absorption bands appear between 2800 and 3000 cm^{-1}.

EXAMPLE 14.6

Identify the type of carbonyl compound that produces the following spectrum:

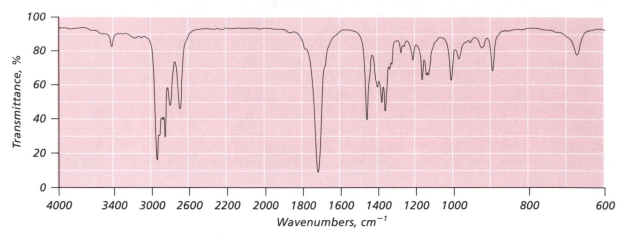

First, verify that a carbonyl group is present. The strong absorption at 1730 cm⁻¹ is unmistakable.

Next, decide which functional groups can be eliminated from further consideration on the basis of the carbonyl peak's position. The data listed in Figure 14.10 indicate that this compound is unlikely to be an acid chloride, anhydride, amide, or ester [unless it is α,β-unsaturated or aryl; the region ~ 1600 cm⁻¹ is blank, so this molecule does not have C=C bonds (Section 14.3g)]. Thus, the aldehyde, carboxylic acid, and ketone functional groups are possibilities to be evaluated.

Decide next which of the possible types of compounds is most likely. A carboxylic acid will have a very broad band between 3500 and 2500 cm⁻¹ (O–H stretching vibration; Fig. 14.15). This spectrum lacks such an absorption, so the compound is probably not a carboxylic acid.

Table 14.5 indicates than an aldehyde will have a medium intensity band at ~ 2720 cm⁻¹, which results from the aldehyde C–H stretching vibration. This spectrum does have such a band, so there is a good chance that the compound is an aldehyde. As Table 14.5 also notes, this assignment could be confirmed by observing the characteristic peak in the ¹H NMR spectrum at δ 10.0.

EXERCISE 14.8

Identify the type of carbonyl compound that produces the following spectrum:

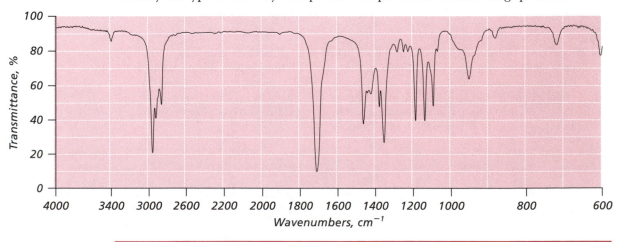

14.3g ABSORPTIONS FOR DOUBLE-BOND STRETCHING VIBRATIONS APPEAR BETWEEN 1700 AND 1500 CM^{-1}

The last region to examine in the interpretation process is that in which C=C, C=N, and N=O stretching vibrations appear. Table 14.6 summarizes the expected frequencies for functional groups that display bands in the region ~ 1600 cm^{-1}.

Many molecules contain carbon–carbon double bonds, but the best method to evaluate or elucidate the structures of aromatic compounds and alkenes will make use of the ^{1}H and ^{13}C NMR spectra, as discussed in Chapter 13.

The other types of compounds that absorb in the region between 1650 and 1500 cm^{-1} are those with C=N and N=O bonds. The nitroso (–N=O) and nitro (–NO$_2$) groups are fairly easy to identify because their bands are very intense, like absorptions for carbonyl compounds. Furthermore, the nitro group, which is the more common, has two bands (Fig. 14.17): the symmetric stretching vibration between 1600 and 1500 cm^{-1} and the antisymmetric stretching vibration between 1390 and 1300 cm^{-1}.

Table 14.6 Frequency ranges for vibrations appearing in the double-bond region.

Compound type	Bond	Frequency range^a (cm^{-1})
Alkene	C=C	1670–1640 (w–m)
Arene	C=C	1650–1400 (v) (3–4 bands)
Imine	C=N	1690–1640 (m)
Nitroso	N=O	1600–1500 (s)
Nitro	N=O	1600–1500 (s)
		1390–1300 (s)

^{a}Weak = w, medium = m, strong = s, variable = v.

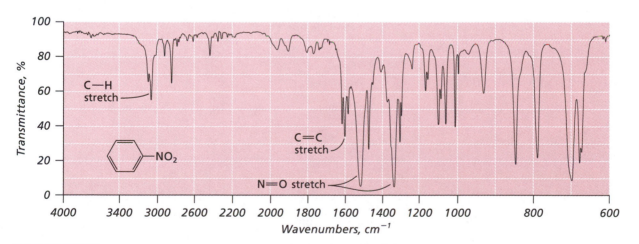

Figure 14.17

The IR spectrum of nitrobenzene (thin film). The bands for the C=C stretching vibrations appear ~ 1600 cm^{-1} and have only medium intensities. The bands for the nitro group are very strong, and appear at 1520 and 1340 cm^{-1}. The typical bands for the C–H stretching vibrations are also clearly seen: the hydrogen atoms are attached to sp^2-hybridized carbon atoms (3100–3000 cm^{-1}). The weak absorptions between 3000 and 2800 cm^{-1} result from overtone bands of the nitro group, not from C–H stretching vibrations. (It is always wise to check the ^{1}H NMR spectrum to verify the types of CH groups present.)

14.4 COMBINED STRUCTURE DETERMINATION EXERCISES

At the beginning of this chapter, you were presented with three common situations requiring the structure determination of an organic molecule (Section 14.1a). For a given compound, the information needed for this purpose—and the techniques suited to collect or make use of these data—include the following:

- The molecular formula (elemental analysis or mass spectra).
- The functional group(s) that are present (IR spectra).
- The hydrocarbon groups that are present (^{1}H and ^{13}C NMR spectra).
- The links between the substructures (^{1}H NMR spectra).

To illustrate how a combination of analytical data can be evaluated to determine the molecular structure of an unknown compound, several examples are presented below.

EXAMPLE 14.7

Draw the structure of compound **1**, which has the following IR and ^{1}H NMR spectra. Elemental analysis provides the following results: C, 88.2%; H, 11.8%. A low-resolution EI mass spectrum shows the MW = 68 amu.

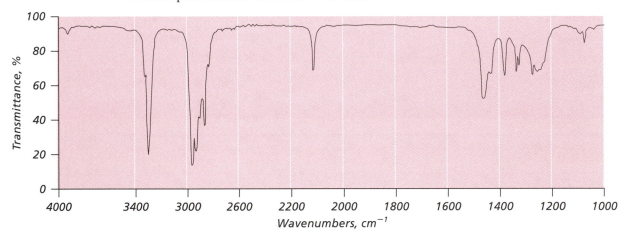

^{1}H NMR spectrum

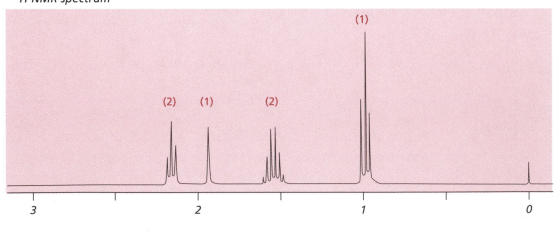

1. Use the analytical data to determine the empirical formula (Section 14.1b): **C_5H_8.**

   We know that this compound contains only C and H because their percentages total 100%. The MW value from the mass spectrum (68) tells us that the empirical formula is the same as the molecular formula.

2. Calculate the number of unsaturation sites: $[2(5) + 2 - 8]/2 = 2$ sites of unsaturation. This means that compound **1** contains two double bonds, a double bond and a ring, two rings, or a triple bond.

3. Examine the IR spectrum. The medium intensity absorption at 2220 cm^{-1} is attributed to a triple-bond stretching vibration. This molecule contains no nitrogen, so we can rule out the presence of a nitrile group. The intensity of this absorption suggests that the molecule is a terminal alkyne, and the strong absorption at 3300 cm^{-1} is the alkyne C–H stretching vibration. At this stage, we know that the following substructure is present: **–C≡C–H**

4. Examine the ^{1}H NMR spectrum. There are four features, one of which must correspond to resonance for the terminal alkyne proton (–C≡C–H). The NMR data are summarized in Table 14.7.

5. Use the multiplicity and integrated intensity of each feature to establish the connectivity between pairs of hydrocarbon groups (see Fig. 13.19). The structure of compound **1** is,

$$H_3C-CH_2-CH_2-C≡C-H$$

Table 14.7   Proton NMR data for compound **1**.

Chemical shift (ppm)	Integrated intensity	Assignment	Multiplicity	No. adjacent protons (n)	
2.15	2	≡C–CH$_2$	Triplet	2	
1.91	1	≡C–H	Singlet	0	Propyl group
1.58	2	CH$_2$	Sextet	5	
0.97	3	CH$_3$	Triplet	2	

EXAMPLE 14.8

Draw the structure of compound **2**, which has the following IR, ^{13}C NMR, and ^1H NMR spectra. Elemental analysis (C, H, O) provides the following results: C, 62.10%; H, 10.35%.

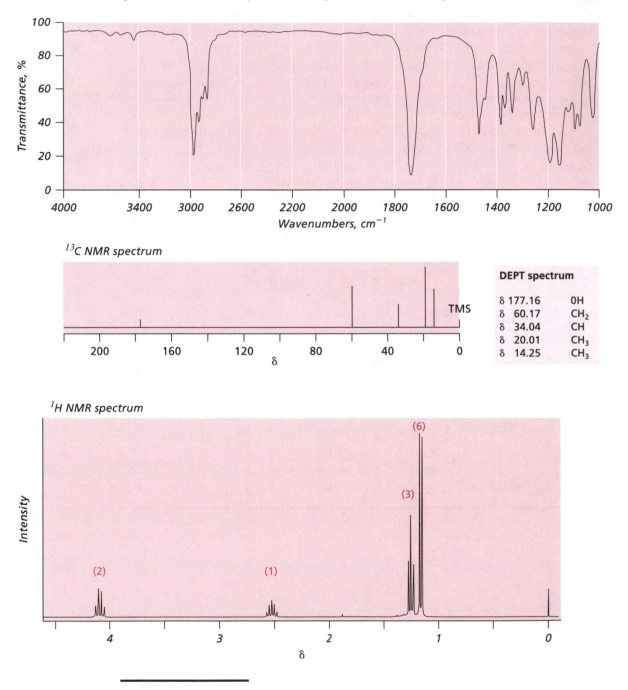

DEPT spectrum

δ 177.16	0H
δ 60.17	CH₂
δ 34.04	CH
δ 20.01	CH₃
δ 14.25	CH₃

1. Use the analytical data to determine the empirical formula (Section 14.1b): $C_6H_{12}O_2$.

2. Calculate the number of unsaturation sites: $[2(6) + 2 - 12]/2 = 1$ site of unsaturation. Compound **2** contains one double bond or one ring.

3. Examine the IR spectrum. The strong absorption at 1740 cm^{-1} is attributed to a C=O stretching vibration. The equally strong absorption at ~ 1200 cm^{-1} is consistent with the presence of an ester functional group.

4. Examine the ^{13}C NMR to determine the types of carbon atoms present. The very weak intensity resonance at δ 177 is consistent with the presence of an ester carbonyl group. All of the other carbon resonances are in the aliphatic region. All of the data are consistent with the presence of a single site of unsaturation.

5. Examine the ^{1}H NMR spectrum. There are four features with a total integration of 12 protons, consistent with the empirical formula. The ^{1}H NMR data are summarized in Table 14.8.

6. Use the multiplicity and integrated intensity of each feature to establish the connectivities between the pairs of hydrocarbon groups (see Fig. 13.11).

   The structure of compound **2** is,

$$\text{H}_3\text{C}-\overset{\overset{\displaystyle \text{CH}_3}{|}}{\text{CH}}-\overset{\overset{\displaystyle \text{O}}{||}}{\text{C}}-\text{O}-\text{CH}_2-\text{CH}_3$$

7. Verify the presence of the hydrocarbon groups using the DEPT portion of the ^{13}C NMR spectrum. Realize that the two methyl groups of the isopropyl fragment are equivalent, so the ^{13}C NMR spectrum should display only five resonances, which it does.

Table 14.8 Proton NMR data for compound **2**.

Chemical shift (ppm)	Integrated intensity	Assignment	Multiplicity	No. adjacent protons (n)	
4.12	2	CH$_2$–O	Quartet	3	
2.51	1	CH–C=O	Septet	6	Ethyl group
1.34	3	CH$_3$	Triplet	2	Isopropyl group
1.15	6	CH$_3$ (2)^a	Doublet	1	

^{a}The number in parentheses is the number of methyl groups present.

EXAMPLE 14.9

Draw the structure of compound **3,** which contains C, H, and O and has the following IR, ^{13}C NMR and ^1H NMR, spectra. From the high-resolution CI mass spectrum, MW = 178.0994.

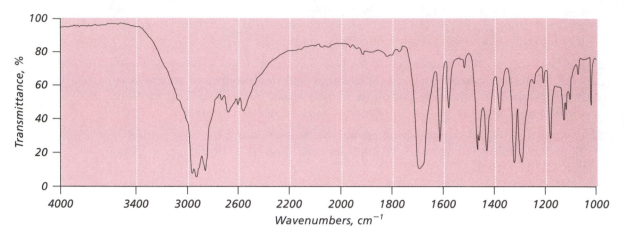

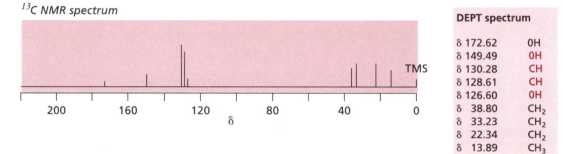

^{13}C NMR spectrum

DEPT spectrum	
δ 172.62	0H
δ 149.49	0H
δ 130.28	CH
δ 128.61	CH
δ 126.60	0H
δ 38.80	CH₂
δ 33.23	CH₂
δ 22.34	CH₂
δ 13.89	CH₃

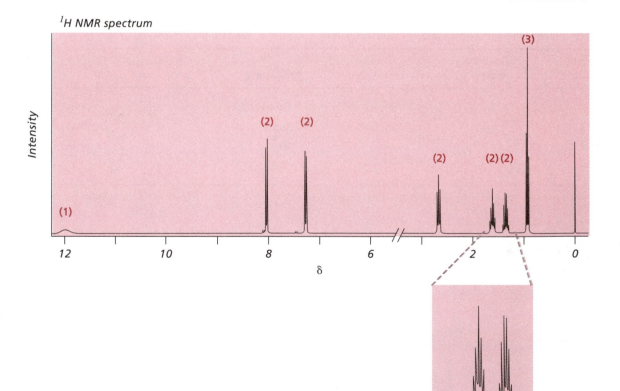

^1H NMR spectrum

1. Examine the ^{13}C NMR to determine the number and types of carbon atoms present. There are three types: C=O, aromatic (given in color in the DEPT table), and aliphatic. A benzene ring has 6 carbon atoms, and there are 5 additional resonances in the spectrum. Start with the assumption that the compound has 11 carbon atoms.

2. Examine the integrated intensity values in the ^{1}H NMR spectrum. The total is 14. Start with the assumption that the compound has 14 hydrogen atoms.

3. Use the high-resolution data to determine the molecular formula with the initial assumption that there are 11 C atoms and 14 H atoms (Section 14.2b): **C$_{11}$H$_{14}$O$_2$**. Calculated MW = 178.09935; observed MW = 178.0994. This formula is consistent with the data. (In actual practice, computer software would be used to calculate the formula from the MW value. The first three steps of this exercise are given to demonstrate that the NMR spectra of simple compounds often reflect the actual numbers of C and H atoms. Thus, with accurate mass data, you can determine the molecular formula without relying on a method of trial and error.)

4. Calculate the number of unsaturation sites: $[2(11) + 2 - 14]/2 = 5$ sites of unsaturation. The NMR data reveal that this molecule has a benzene ring, which accounts for four sites of unsaturation (three double bonds and a ring). Therefore, compound **3** contains one double bond or one ring *in addition to the benzene ring*. The ^{13}C NMR spectrum shows (Step 1) that this molecule has a carbonyl group, which is the "fifth" site of unsaturation.

5. Examine the IR spectrum. The strong absorption at 1730 cm^{-1} is attributed to a C=O stretching vibration. The broad and strong absorption between 3500 and 2500 cm^{-1} is consistent with the presence of a carboxylic acid functional group. The resonance in the ^{1}H NMR spectrum at δ 12.0 is also consistent with the presence of a carboxylic acid functional group.

6. Examine the ^{1}H NMR spectrum. There are seven features with a total integration of 14 protons. The ^{1}H NMR data are summarized in Table 14.9. The resonance for the carboxylic acid proton (δ 12.0) was already mentioned in Step 5; the aromatic region has the two-doublet pattern associated with a 1,4-disubstituted benzene ring (Fig. 13.12).

7. The multiplicities and integrated intensity values for the aliphatic protons are consistent with the presence of a butyl group. The structure of compound **3** is,

$$CH_3CH_2CH_2CH_2 - \overset{\displaystyle O}{\underset{\displaystyle C}{\|}} - OH$$

Table 14.9   Proton NMR data for compound **3**.

Chemical shift (ppm)	Integrated intensity	Assignment	Multiplicity	No. adjacent protons (n)	
12.0	1	COOH	Singlet	0	
8.05	2	ArH	Doublet	1	1,4-Disubstituted benzene ring
7.25	2	ArH	Doublet	1	
2.65	2	CH$_2$	Triplet	2	
1.62	2	CH$_2$	Quintet	4	Butyl group
1.45	2	CH$_2$	Sextet	5	
0.91	3	CH$_3$	Triplet	2	

EXAMPLE 14.10

Draw the structure of compound **4**, which contains C, H, and O and has the following ^1H and ^{13}C NMR spectra. The high-resolution CI mass spectrum reveals MW = 110.0735.

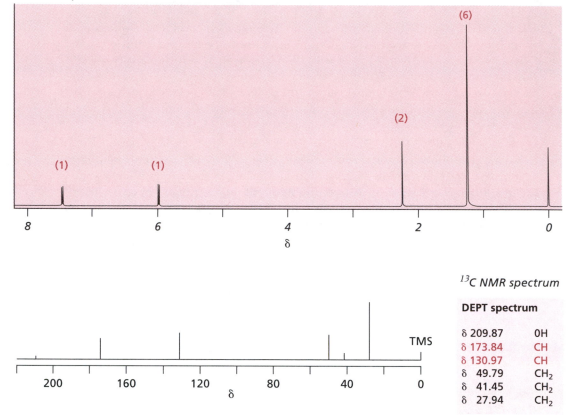

^1H NMR spectrum

^{13}C NMR spectrum

DEPT spectrum

δ 209.87	0H
δ 173.84	CH
δ 130.97	CH
δ 49.79	CH_2
δ 41.45	CH_2
δ 27.94	CH_2

1. Examine the ^{13}C NMR to determine the number and types of carbon atoms present. There are three types: C=O, alkene (given in color in the DEPT table), and aliphatic. Start with the assumption that the compound has six carbon atoms.

2. Examine the integrated intensity values in the ^1H NMR spectrum. The total is 10. Start with the assumption that the compound has 10 hydrogen atoms.

3. Use the high-resolution data to determine the molecular formula, along with the initial assumption that there are 6 C atoms and 10 H atoms (Section 14.2b): **$C_6H_{10}O$** has a MW = 98.0732, which is 12 amu less than the integral mass value given by the mass spectrum. Adding an additional carbon atom gives **$C_7H_{10}O$:** Calculated MW = 110.0732; observed MW = 110.0735.

4. Calculate the number of unsaturation sites: $[2(7) + 2 - 10]/2 = 3$ sites of unsaturation. Both ^1H and ^{13}C NMR spectra reveal that this molecule has an alkene double bond, and the ^{13}C NMR spectrum shows that this molecule contains a carbonyl group. These functional groups account for two of the sites of unsaturation. Unless the molecule is highly symmetric and has two equivalent carbonyl or alkene groups, the molecule also must contain a ring to account for the three sites.

5. Examine the ^1H NMR spectrum. There are four features with a total integration of 10 protons. The ^1H NMR data are summarized in Table 14.10.

Table 14.10 Proton NMR data for compound **4.**

Chemical shift (ppm)	Integrated intensity	Assignment	Multiplicity	No. adjacent protons (n)
7.49	1	=C–H	Doublet	1
5.95	1	=C–H	Doublet	1
2.22	2	CH₂–C=O	Singlet	0
1.23	6	CH₃ (2)	Singlet	0

6. The multiplicities show that only the alkene protons engage in coupling, and only between themselves. Using the ¹H and ¹³C NMR data, we can see that the following substructures are present. Notice that the two methyl groups shown in color are equivalent, which accounts for the appearance of only six signals in the ¹³C NMR spectrum (the numbers given below are the ¹³C NMR chemical shifts).

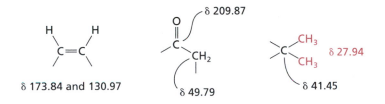

δ 173.84 and 130.97   δ 49.79   δ 41.45

7. In Step 4, we calculated that a third site of unsaturation is present as a ring. Bringing together the substructures to form a ring, we can devise two reasonable structures. However, the CH₂ group in **A** would be expected to have its resonance farther downfield than δ 2.22 because it is between two electron-withdrawing groups (carbonyl and alkene). Furthermore, the two alkene carbon atoms and protons in **A** are in similar environments, so their chemical shift values should be more similar than what is observed. Therefore, compound **4** has structure **B.**

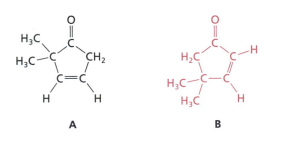

A   B

EXERCISE 14.9

Draw the structure of compound **5,** which gives a low-resolution mass spectrum having M⁺· = 152; its (M + 2)⁺· peak has about the same intensity as M⁺·. This compound also produces the IR and ¹H NMR spectra shown below.

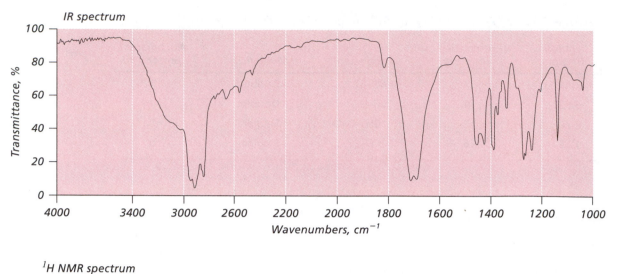

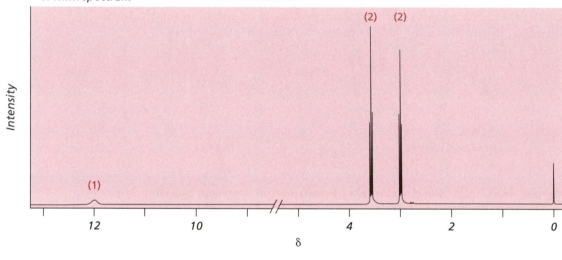

CHAPTER SUMMARY

Section 14.1   Analytical data in structure determinations

- Analytical data and spectra are often needed to identify the molecular structure of a reaction product or byproduct, or the structure of the substance isolated from a source other than a reaction mixture.

- The data used in structure determination exercises should be obtained on pure compounds.

- Elemental analysis data can be used to determine the molecular formula of an unknown compound, and the purity of a known one.

- The molecular formula can be used to calculate the number of sites of unsaturation, which is the defined as the π bonds and rings that a compound has. This information can be used to narrow the possible substructures that may be present in a molecule.

Section 14.2   Mass spectrometry

- A mass spectrometer produces ions from gas-phase molecules and then separates the gas-phase ions by their mass-to-charge ratios, m/z.

- A molecule can be ionized by electron ionization, chemical ionization, or a variety of desorption processes. The resulting charged species subsequently produce fragments to dissipate vibrational energy.

- A plot of the relative numbers of each ion reaching the detector versus m/z generates the mass spectrum.

- Removing one electron from a molecule generates the molecular ion, the mass of which equals the compound's molecular weight.

- High-resolution mass spectrometry is used to determine a compound's molecular formula.

- The m/z value for the ion produced in the highest amount is called the base peak.

- The molecular weight of a compound with an odd m/z value signifies that the molecule has an odd number of nitrogen atoms.

- Relative intensities of the isotope peaks associated with the molecular ion peak are used to determine whether S, Cl, or Br atoms are present.

- Organic compounds undergo fragmentation processes in predictable ways after formation of a radical ion species by ionization.

- Fragmentation of a molecular ion occurs predictably at branch points of a carbon chain or ring; at bonds adjacent to heteroatoms and functional groups; by loss of neutral molecules such as water, CO, HCN, and so on; and with rearrangement, especially if C=C or C=O π bonds are present.

Section 14.3 Infrared spectroscopy

- Infrared spectra result from absorption of energy that affects the vibrations of atoms bonded to each other.

- Infrared spectroscopy is used to determine the identities of the functional groups present in a molecule.

- Absorption of IR energy requires a change in the dipole moment of the bond or group that is vibrating.

- The frequency values in an IR spectrum are expressed as wavenumbers, $\bar{\nu}$, which have the units reciprocal centimeters. A typical IR spectrum spans from 4000 to 500 cm^{-1}.

- The frequency of a vibration is inversely related to the masses of the atoms bonded to one another.

- The frequency of a vibration is directly proportional to the strength of the bond.

- The spectrum has two parts: the functional group region (4000–1300 cm^{-1}) and the fingerprint region (1300–500 cm^{-1}).

- The four regions of the IR spectrum that are commonly used to identify specific functional groups are summarized in Figure 14.5.

KEY TERMS

Section 14.1c
sites of unsaturation

Section 14.2a
molecular ion
fragmentation
isotope peaks
base peak

Section 14.2b
high-resolution mass
    spectrometry

Section 14.2d
McLafferty rearrangement

Section 14.3a
wavenumbers

Section 14.3b
functional group region
fingerprint region

ADDITIONAL EXERCISES

14.10.   The elemental analysis of a sample containing C, H, O, and Cl gave the following results: C, 47.17%; H, 8.63%; Cl, 23.29%. What is the empirical formula of this compound? Calculate its exact mass.

14.11.   The elemental analysis of an unknown compound gave the following results: C, 61.3 %; H, 10.2 %; N, 28.5%. What is the empirical formula of this compound? Calculate its exact mass.

14.12.   Identify the type of carbonyl compounds represented by spectra **12A** and **12B**.

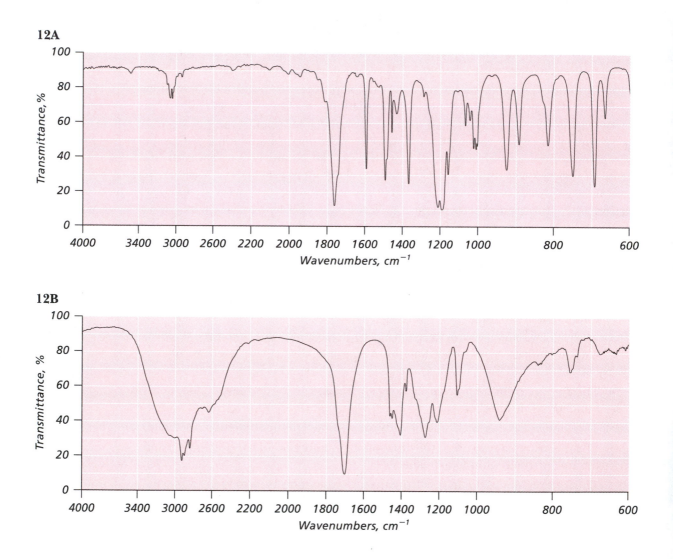

14.13. In performing each of the following reactions, indicate the principal IR absorptions that you would look for in the reactant and product to ensure that the reaction was successful. Be as specific as possible: list the position (cm^{-1}), intensity (weak, medium, or strong), and shape (sharp or broad) of the diagnostic absorption band(s). Repeat the same process for the ^1H and broad-band ^{13}C NMR spectra. List, as applicable, the chemical shifts, integrated intensity values, splitting patterns, and DEPT results.

a.

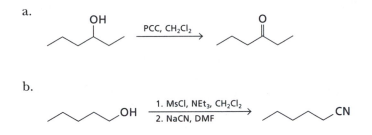

b.

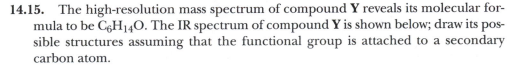

14.14. The molecular ion region of the low-resolution mass spectrum of compound **X** has two peaks with an intensity ratio of 3:1 that are two mass units apart. The larger peak appears at *m/z* 92. The IR spectrum of compound **X** is shown below; draw its possible structures.

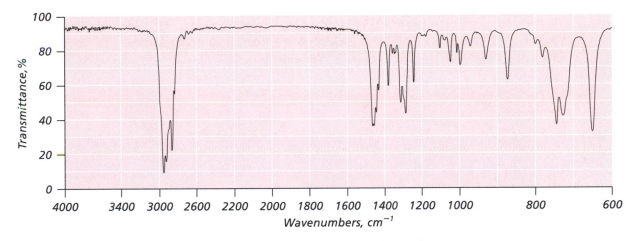

14.15. The high-resolution mass spectrum of compound **Y** reveals its molecular formula to be $C_6H_{14}O$. The IR spectrum of compound **Y** is shown below; draw its possible structures assuming that the functional group is attached to a secondary carbon atom.

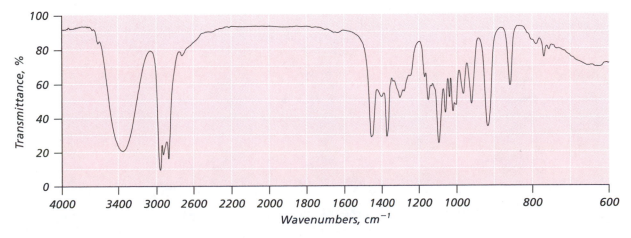

14.16.   The molecular ion peak in a high-resolution mass spectrum was observed at m/z 118.082. Which of the following is the likely molecular formula: $C_6H_{14}O_2$, $C_6H_{14}S$, or $C_5H_{12}NO_2$?

14.17.   The molecular ion peak in a high-resolution mass spectrum was observed at m/z 130.055. Which of the following is the likely molecular formula: $C_7H_{14}S$, $C_7H_{11}Cl$, or $C_6H_{12}NS$?

14.18.   The IR and ^1H NMR spectra of compound **W**, $C_8H_{10}O$, are shown below. What is its structure?

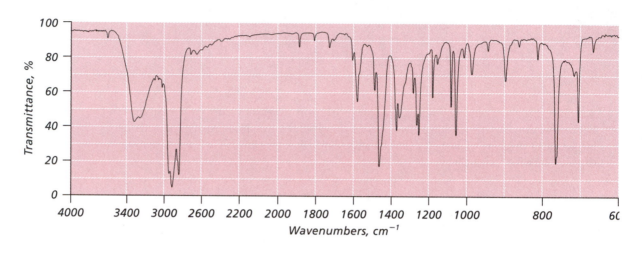

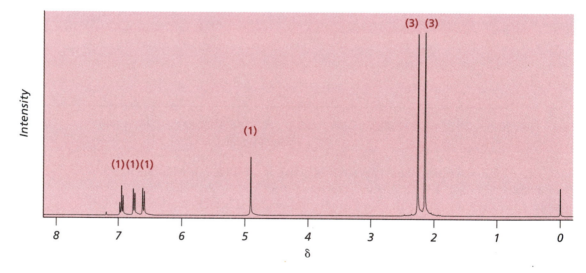

14.19.   The IR spectrum of compound **Z** is shown below. What functional group is present? If you wanted to verify your assignment, what resonance(s) would you expect to see in the broad-band decoupled ^{13}C NMR spectrum?

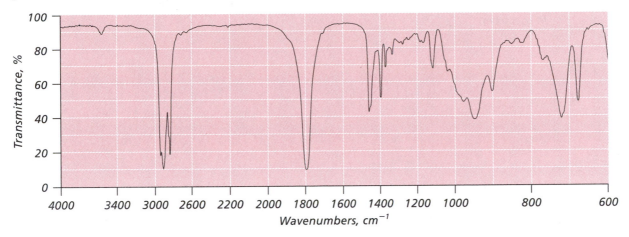

14.20.   Draw the structure of compound **R,** which has the following IR and ^1H NMR spectra. From the high-resolution CI mass spectrum, the molecular formula is determined to be $C_5H_{10}O$.

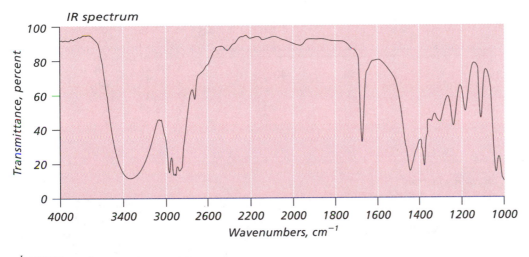

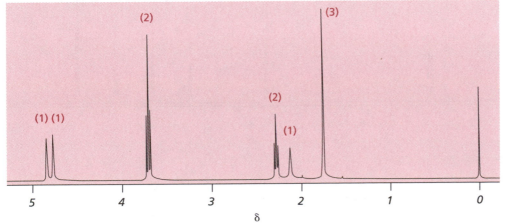

14.21. Draw the structure of compound **S,** which contains C, H, and O and has the following IR, ^{13}C NMR and ^{1}H NMR, spectra. From the low-resolution EI mass spectrum, MW = 192.

IR spectrum

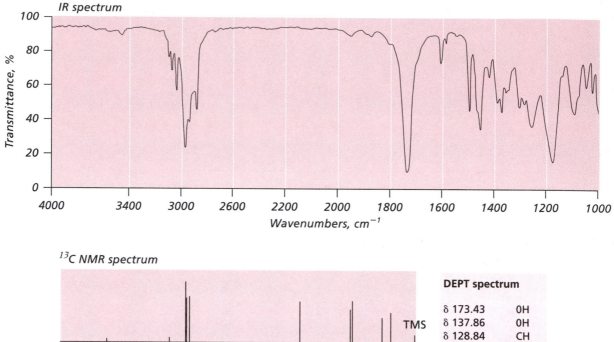

^{13}C NMR spectrum

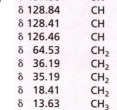

DEPT spectrum

δ 173.43	0H
δ 137.86	0H
δ 128.84	CH
δ 128.41	CH
δ 126.46	CH
δ 64.53	CH$_2$
δ 36.19	CH$_2$
δ 35.19	CH$_2$
δ 18.41	CH$_2$
δ 13.63	CH$_3$

^{1}H NMR spectrum

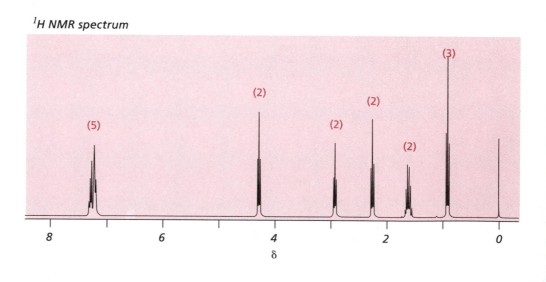

14.22.   2-Methylpentanoic acid produces a mass spectrum with key peaks at m/z 116, 99, 74, and 45. Based on the general patterns listed in Table 14.3, draw the structures of the fragments that produce these peaks.

14.23.   Draw the structure of compound **T,** which has the following IR, ^{1}H NMR, and ^{13}C NMR spectra. From the high-resolution CI mass spectrum, MW = 98.071.

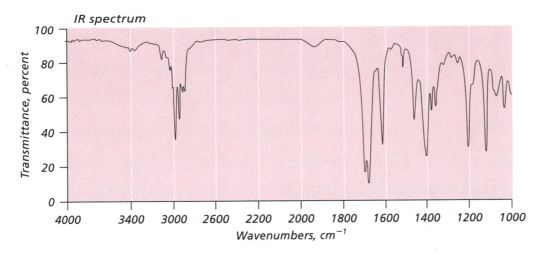

IR spectrum

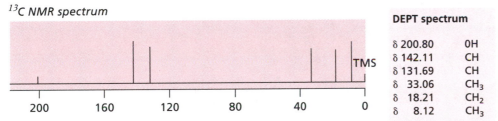

^{13}C NMR spectrum

DEPT spectrum

δ 200.80	0H
δ 142.11	CH
δ 131.69	CH
δ 33.06	CH$_3$
δ 18.21	CH$_2$
δ 8.12	CH$_3$

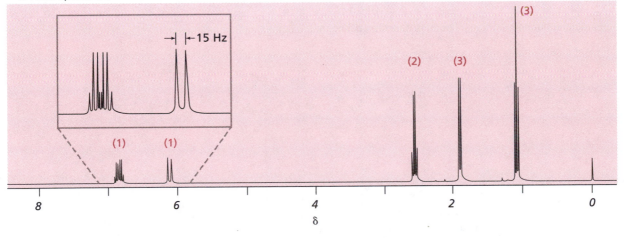

^{1}H NMR spectrum

14.24. Draw the structure of compound **U**, which has the following ^{13}C and ^1H NMR spectra. Elemental analysis provides the following results: C, 51.87%; H, 4.99%; Br, 43.22%.

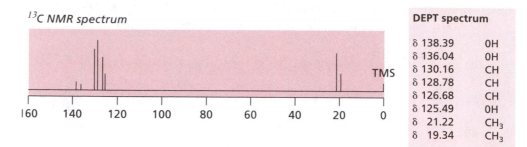

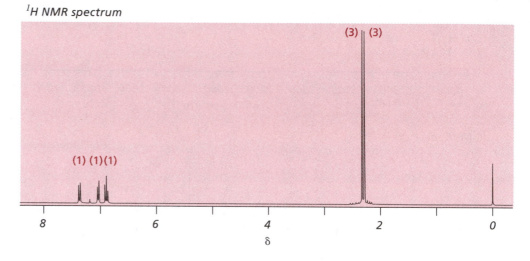

14.25. Draw the structure of compound **V**, which gives a low-resolution mass spectrum having $M^{+\cdot} = 150$ and a $(M + 2)^{+\cdot}$ peak with about equal intensity. Its ^1H NMR spectrum is shown below.

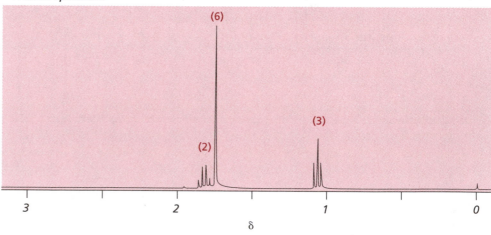

ORGANOMETALLIC REAGENTS AND CHEMICAL SYNTHESIS

Many of the previous chapters have described the chemical reactions of organic molecules, as well as their mechanisms. This chapter focuses on the application of those reactions to chemical synthesis—the construction of molecules. Chemical synthesis has been an important aspect of organic chemistry throughout its history, playing a major role in the early development during the 1800s, especially in the growth of the chemical industry based on the manufacture of dyes. Later in the nineteenth century, the research of Emil Fischer on the synthesis of carbohydrates led to a rapid evolution of stereochemical principles and an understanding of the properties of chiral substances. The early 1900s were characterized by interest in the synthesis of natural products, and the latter part of the twentieth century was distinguished by the applications of physical organic and bioorganic chemistry to synthetic strategies.

The material presented in this chapter focuses on general and versatile ways to make carbon–carbon bonds. The conceptual framework for carbon–carbon bond formation is presented first, followed by the descriptions of specific reactions that involve organometallic compounds, substances that contain a carbon atom bonded to a metal ion such as lithium, magnesium, and copper. Many versatile reagents and catalysts that are used to make carbon–carbon bonds in selective ways are organometallic compounds.

The latter part of this chapter describes some of the design strategies used in chemical synthesis. Devising synthetic protocols requires you to think about organic reactions in reverse fashion. Instead of answering the question, What compound is produced from certain reagents (i.e., $A + B \rightarrow ?$), you will have to answer the question, Which compounds are needed to prepare a given product ($? + ? \rightarrow X$)?

15.1 CARBON–CARBON BOND FORMATION

15.1a CHEMICAL SYNTHESIS RELIES ON PROCESSES FOR MAKING CARBON–CARBON BONDS AND INTERCONVERTING FUNCTIONAL GROUPS

Much of the material presented in this chapter describes ways to construct carbon–carbon bonds, which lies at the heart of organic synthesis. Once the carbon atom framework is in place, then the functional groups can be interconverted to form the desired product. The underlying principle for carbon–carbon bond formation is this:

> A reagent having a nucleophilic carbon atom ($C^{\delta-}$) reacts with a substrate that has an electrophilic carbon atom ($C^{\delta+}$). The opposite polarities of carbon in the reacting species lead to bond formation.

There are many reasons to synthesize organic substances. The product might subsequently be used for other studies, such as characterization of its physical properties or investigation of its pharmacological activity. Factors that require attention in planning a synthesis include the cost of materials, disposal of waste products, the scale on which the process can be done, and research priorities. For example, it is sometimes necessary to design a synthesis so that certain steps can be readily modified to prepare a variety of product analogues. This stipulation is particularly common in the syntheses of pharmaceutical products, where analogues of a drug may be required for testing.

15.1b CARBON–CARBON BOND FORMATION FORMALLY INVOLVES THE REACTION BETWEEN ELECTROPHILIC AND NUCLEOPHILIC CARBON CENTERS

For most organic molecules, the presence of an electronegative heteroatom renders the neighboring carbon atom electrophilic. This effect is readily apparent for alkyl halides, alcohols, and carbonyl compounds.

By exploiting the electrophilicity of carbon in an alkyl halide, you can perform a substitution reaction by supplying a nucleophilic reagent (Section 6.1a). A typical S_N2 reaction (Section 6.3a) can be represented as follows:

Nucleophilic substitution

Conceptually, the formation of a carbon–carbon bond can occur in the same manner; that is, a nucleophilic carbon atom reacts with an electrophilic carbon center, displacing a leaving group.

Conceptual formation of a C–C bond

The question is, do such processes actually occur? If so, what types of carbon nucleophiles can be used?

The simple answer is *Yes,* and an example of a substance with a nucleophilic carbon atom is cyanide ion, salts of which are readily available. Although two resonance structures can be drawn for the cyanide ion, the major contributor is the one in which the carbon atom has an octet of electrons and a formal charge of –1. The carbon atom is therefore more nucleophilic than the nitrogen atom.

$$:C{=}\ddot{N}:^- \longleftrightarrow {}^-:C{\equiv}N:$$

As a consequence of its nucleophilicity, cyanide ion reacts with alkyl halides to form nitriles, concomitant with formation of a new carbon–carbon bond. (You will learn later how the nitrile group can subsequently be converted into other functional groups.) The use of cyanide ion constitutes one method—albeit limited—for constructing the carbon framework of organic molecules.

Formation of a C–C bond

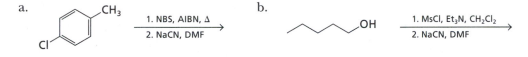

EXERCISE 15.1

Draw the structure of the major product expected from each of the following reactions:

a.

![structure]
1. NBS, AIBN, Δ
2. NaCN, DMF

b.

![structure] OH
1. MsCl, Et₃N, CH₂Cl₂
2. NaCN, DMF

15.2 ORGANOMAGNESIUM AND LITHIUM COMPOUNDS

15.2a A GRIGNARD REAGENT IS PREPARED FROM AN ORGANOHALIDE

Even though cyanide ion is an excellent nucleophile, its use is limited to those situations in which a single carbon atom is to be added to the starting compound. Nucleophilic reagents that contain more than one carbon atom should be more versatile, and there are several from which to choose.

A carbanion has a nucleophilic carbon atom by virtue of its unshared electron pair (Section 2.5e). Formally, a carbanion is the deprotonated form of a C–H fragment, but most hydrocarbons have such a high pK_a value (> 40) that an acid–base reaction of the following type has limited utility (Section 15.2f):

![carbanion mechanism, base]

A carbanion

You may recall from Section 11.4e that a hydride ion is equivalent to a proton and two electrons. In the same way, a carbanion can be considered as the equivalent of a carbocation plus two electrons.

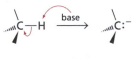

A carbocation

What makes this analogy useful is that you are already familiar with carbocations, which are intermediates in several reactions (Section 6.2). For example, you know that an alkyl halide is a good precursor of a carbocation; therefore, *adding two electrons to an alkyl halide should produce a carbanion.*

One way to reduce an organohalide to a carbanion is by its reaction with magnesium metal, a reaction discovered by Victor Grignard around the year 1900, for which he was awarded the Nobel Prize in 1912. The mechanism of this reaction is complex and involves a series of single-electron transfer reactions. Magnesium metal, Mg, is oxidized to Mg^{2+} and produces a species with the general formula RMgX, called a **Grignard reagent.**

$$H_3C-CH_2-CH_2-Br \xrightarrow[\text{diethyl ether}]{\text{Mg}} H_3C-CH_2-CH_2-Mg-Br$$

A Grignard reagent

Isolated Grignard reagents are actually polymetallic clusters, often having two, four, or six magnesium ions along with the requisite number of halide ions and organic groups. For our purposes, however, we will assume that a Grignard reagent is a discrete mononuclear compound with the formulation R–Mg–X (X = a halide ion).

The reaction that produces a Grignard reagent is general, and all sorts of organohalogen compounds can serve as starting materials. Alkyl, aryl, and vinyl chlorides, bromides, and iodides all react, although organochlorides are the least reactive.

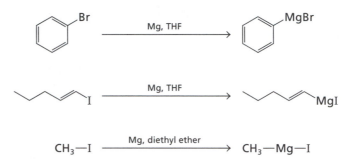

To make the methyl Grignard reagent, methyl iodide is preferred. It is a liquid, bp 41°C, so the appropriate quantity is easily measured. CH_3Cl and CH_3Br *will* form the corresponding CH_3MgX, but both are gasses at room temperature, so working with them is much less convenient.	

The solvents used for this reaction must be inert to the Grignard reagent and, at the same time, dissolve it effectively. Diethyl ether and THF meet both of these criteria, and the latter is used almost exclusively for reactions of aryl and vinyl halides with magnesium.

EXERCISE 15.2

Draw the structure of the Grignard reagent formed from each of the following compounds:

a. *p*-Bromotoluene   b. 3-Chloropentane   c. 2-Bromo-1-hexene

---

15.2b GRIGNARD REAGENTS REACT WITH ELECTROPHILES SUCH AS PROTONS AND CARBONYL GROUPS

The carbon atom of an organomagnesium compound is formally nucleophilic, but treating an alkyl halide with a Grignard reagent *rarely* leads to substitution. Grignard reagents are potent bases, and they also react as radicals. Both of these properties suppress a Grignard reagent's nucleophilic behavior.

As a base, Grignard reagents react instantaneously with protic substances to produce the corresponding hydrocarbons. For example, RMgX reacts with water, generating the hydrocarbon RH and $MgX(OH)$. Any substance having a proton with a pK_a value < 40 reacts as an acid toward Grignard reagents.

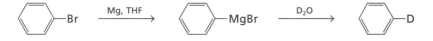

> The Grignard reagent is a very strong base. In this example, the CH_3MgI is formally CH_3^-, the conjugate base of methane. Upon reaction with a proton source, an acid–base reaction occurs to produce methane and the mixed salt, magnesium hydroxide iodide.

This reaction can be used to advantage to make compounds that contain deuterium. For example, the reaction of phenylmagnesium bromide with D_2O yields deuterobenzene.

Grignard reagents *will* react as nucleophiles with the electrophilic carbon atom of the carbonyl group. The reaction between RMgX and carbon dioxide is the prototypic transformation.

Carbon dioxide is polarized such that each of its double bonds connects an electrophilic carbon atom with a nucleophilic oxygen atom. Formally, the carbon atom of a Grignard reagent reacts as a nucleophile toward the carbon atom of CO_2, producing the halomagnesium salt of a carboxylic acid.

One reason, however, that carbonyl compounds elicit the nucleophilic properties of Grignard reagents has to do with the nucleophilicity of the carbonyl oxygen atom and its attraction toward the electrophilic metal ion.

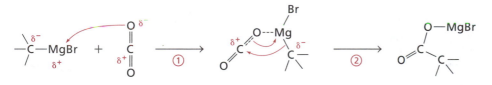

Formation of the adduct brings the carbon atoms into proximity, and the polarity difference between the two is enhanced. Reaction subsequently leads to bond formation.

A separate hydrolysis step is required to liberate the carboxylic acid, so treating the reaction mixture with dilute hydrochloric acid produces the carboxylic acid and magnesium halide. Magnesium halides are water soluble, and they can be washed away from the less polar organic molecule.

Most carbonyl compounds, as well as those with unsaturated, polar functional groups like nitriles, react with Grignard reagents. For the present, we will focus on reactions with aldehydes and ketones. (The reactions of organometallic compounds with esters are covered in Chapter 21.)

Formaldehyde, the simplest aldehyde, reacts with a Grignard reagent to yield a halomagnesium alkoxide salt. Hydrolysis with aqueous acid produces an alcohol.

$$CH_3—CH_2—CH_2—CH_2—MgBr \xrightarrow[\text{2. } H_3O^+]{\text{1. } CH_2O} CH_3—CH_2—CH_2—CH_2—CH_2—OH$$

As with the reaction between the Grignard reagent and carbon dioxide, the reaction with formaldehyde takes place between the nucleophilic RMgX reagent and the electrophilic carbon atom of the carbonyl group after adduct formation.

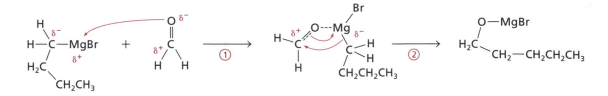

Hydrolysis with dilute mineral acid yields a primary alcohol that has one carbon atom more than the alkyl halide that was used initially.

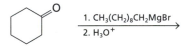

Other aldehydes react with Grignard reagents to produce secondary alcohols after hydrolysis. Ketones react with Grignard reagents to make tertiary alcohols. If needed, a weaker acid such as aqueous ammonium chloride is used to prevent E1 elimination of water from the tertiary alcohol.

EXAMPLE 15.1

What is the major product expected from the following reaction?

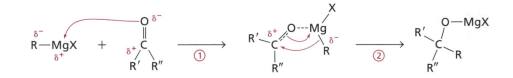

The general reaction between RMgX reagent and a ketone can be written as follows:

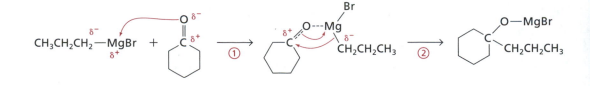

By replacing R, R′, R″, and X with the actual groups, we depict the mechanism as follows:

Acid workup generates an alcohol as the product, so the final equation is written,

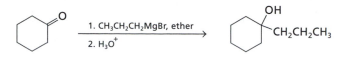

EXERCISE 15.3

What is the major product expected from the reaction shown here?

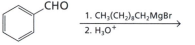

15.2c EPOXIDES UNDERGO RING OPENING WITH GRIGNARD REAGENTS

Another electrophile that is susceptible to react with Grignard reagents is the epoxide ring. The reaction of a Grignard reagent with ethylene oxide, the simplest epoxide, produces a primary alcohol with two carbon atoms more than the starting organohalide from which the organometallic species is made. This transformation is a useful supplement to the one between a Grignard reagent and formaldehyde, which yields a primary alcohol with only one carbon atom more than the original alkyl halide.

$$CH_3CH_2CH_2CH_2\!-\!Br \xrightarrow{\text{Mg, Ether}} CH_3CH_2CH_2CH_2\!-\!MgBr$$

$$CH_3CH_2CH_2CH_2\overset{\delta^-}{-}MgBr \quad \overset{\delta^+}{CH_2}\!-\!CH_2 \longrightarrow CH_3CH_2CH_2CH_2\!-\!CH_2\!-\!CH_2\!-\!O^- \ ^+MgBr$$

$$CH_3CH_2CH_2CH_2\!-\!CH_2\!-\!CH_2\!-\!O^- \ ^+MgBr \xrightarrow{H_3O^+} CH_3CH_2CH_2CH_2\!-\!CH_2\!-\!CH_2\!-\!OH$$

As in its reactions with carbonyl compounds, a Grignard reagent can form an adduct with an epoxide, which further activates the ring toward opening. This process is similar to the proton activation process that you learned in Section 7.2e.

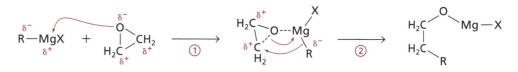

In the same manner that epoxides react with other nucleophiles (Section 7.2e), unsymmetrical epoxides react with Grignard reagents at the less hindered carbon atom of the ring. Therefore, unsymmetrical epoxides produce 2° and 3° alcohols instead of primary ones.

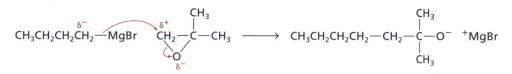

Table 15.1 A summary of reactions between a Grignard reagent (R–MgX) and electrophilic reagents.

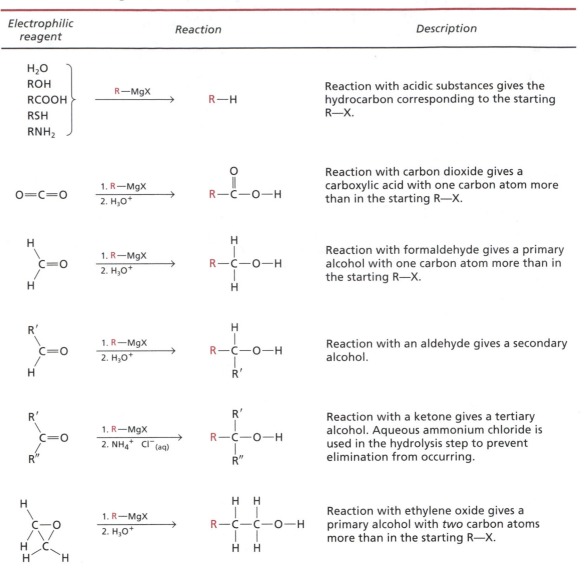

Electrophilic reagent	Reaction	Description

A summary of the reactions between Grignard reagents and compounds with electrophilic centers is given in Table 15.1.

EXERCISE 15.4

What is the expected major product of the reaction shown here?

15.2d THE FORMATION OF A GRIGNARD REAGENT IS LIMITED BY THE PRESENCE OF REACTIVE FUNCTIONAL GROUPS IN THE ORGANOHALIDE MOLECULE

Because of its highly basic nature and reactivity toward many functional groups, a Grignard reagent cannot always be prepared; that is, there are significant limitations on the structures of organic halides that can be converted to their magnesium derivatives. Specifically, any organohalide that also has *an acidic proton* (alcohol, phenol, carboxylic acid, or amine) or *a reactive functional group* (carbonyl, nitrile, epoxide, or nitro group) cannot be used to prepare a Grignard reagent. Some examples follow:

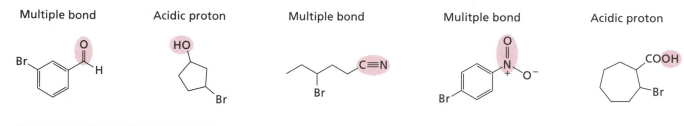

| Multiple bond | Acidic proton | Multiple bond | Mulitple bond | Acidic proton |

EXERCISE 15.5

Identify the following compound(s) for which it will be difficult to prepare a Grignard reagent. Which portion of the molecule will be problematic?

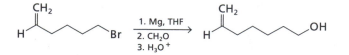

There are several ways to circumvent this limitation. One is to start with a molecule that has a protected form of the reactive center. (The use of protecting groups will be discussed in Section 15.5b.) The second choice is to prepare a different type of organometallic reagent. Some transition metal species tolerate a more diverse range of functional groups than Grignard reagents. A third alternative is to employ a functional group that can be converted subsequently to the desired one, as shown in the following example:

Suppose you wanted to prepare the Grignard reagent derived from 5-bromo-pentanal. This transformation will not be successful because the functional groups are incompatible. Even if this Grignard reagent were to form, it would react with itself.

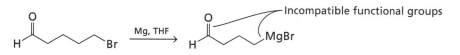

What is needed instead is a functional group that is equivalent to an aldehyde but unreactive toward Grignard reagents. One possibility is the alkene double bond, which can be converted to a carbonyl group by ozonolysis sometime later (Section 11.3b). Thus, 6-bromo-1-hexene is a practical alternative to 5-bromopentanal, and a synthetic scheme illustrating its conversion to a Grignard reagent and the subsequent reaction with formaldehyde is shown below.

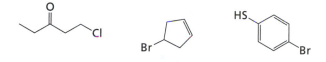

After the Grignard reaction is complete, the alkene is converted to the aldehyde group by ozonolysis.

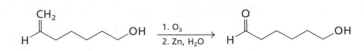

EXERCISE 15.6

From which of the following compounds can you prepare a Grignard reagent? Draw the structure of the organomagnesium compound that is formed.

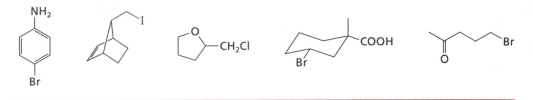

15.2e ORGANOLITHIUM COMPOUNDS ARE PREPARED FROM ORGANOHALIDES AND LITHIUM METAL

Lithium metal reacts with an organic halide in a manner similar to that observed for the reaction between magnesium and RX. The product is an **organolithium compound** and has the stoichiometry RLi; the lithium ion is closely associated with the negatively charged carbon atom. Like Grignard reagents, organolithium compounds are not monomeric but exists as tetrameric or hexameric species in solution as well as in the solid state.

$$CH_3-CH_2-CH_2-CH_2-Br \xrightarrow[\text{diethyl ether or hexane}]{\text{2 equiv of Li metal}} CH_3-CH_2-CH_2-CH_2-Li \ + \ LiBr$$

Butyllithium

Many RLi reagents with one-to-six carbon atoms are available commercially. Butyl-lithium (abbreviated BuLi) is the most commonly used of these reagents, especially when a potent base is needed.

Some commercially available alkyllithium reagents

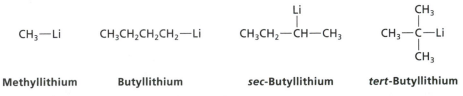

| **Methyllithium** | **Butyllithium** | ***sec*-Butyllithium** | ***tert*-Butyllithium** |

Organolithium compounds react with carbonyl compounds and epoxides to form alcohols and with carbon dioxide to form carboxylic acids, just like Grignard reagents (cf. Table 15.1). For example, the following equation depicts formation of an aryl-lithium compound and its subsequent reaction with carbon dioxide to form a carboxylic acid:

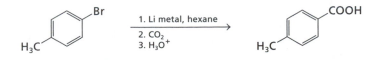

15.2f ALKYNYL ORGANOMETALLIC COMPOUNDS ARE MADE BY ACID–BASE REACTIONS

Alkynyl organometallic compounds are prepared by a route that differs from those used to make alkyl, vinyl, and aryl organometallic compounds. The carbon–hydrogen bond of a terminal acetylene is much more acidic than the carbon–hydrogen bond of other kinds of hydrocarbons because of the substantial s character in the orbital used to form the C–H bond. In fact, as you learned in Chapter 5, the pK_a value of a typical alkyne is ~ 25 (see Table 5.1). It is even possible to make the *dianion* of acetylene (ethyne).

$$Na^+ \; {}^-\!:C{\equiv}C:^- \; Na^+$$

A terminal alkyne reacts with a strong base such as butyllithium, a Grignard reagent, or sodium amide ($NaNH_2$, also called "sodamide"). The product in each case is the corresponding metal salt. Sodamide is made by dissolving sodium metal in liquid ammonia containing a small amount of iron metal. Remember that dissolving sodium metal by itself in ammonia gives solvated electrons, which can be used to *reduce* alkynes (Section 12.2c). The presence of iron metal in the reaction medium catalyzes an acid–base step that generates sodamide.

$$R{-}C{\equiv}C{-}H \xrightarrow[\text{THF}]{CH_3CH_2CH_2CH_2Li} R{-}C{\equiv}C{-}Li \; + \; CH_3CH_2CH_2CH_3$$

Alkynyl lithium compound

$$R{-}C{\equiv}C{-}H \xrightarrow[\text{ether}]{CH_3{-}Mg{-}I} R{-}C{\equiv}C{-}MgI \; + \; CH_4$$

Alkynyl Grignard reagent

$$R{-}C{\equiv}C{-}H \xrightarrow[-33°C]{NaNH_2,\; NH_{3\,(liq)}} R{-}C{\equiv}C{-}Na$$

Alkynyl sodium compound

An alkynyl carbanion reacts like any organometallic species: The carbon atom is a carbanion and a nucleophile. However, *this type of carbanion is able to react as a nucleophile via the* S_N2 *pathway, unlike the more strongly basic Grignard reagents and organolithium compounds.*

$$R'{-}C{\equiv}C:^- \; Na^+ \xrightarrow{\;R{-}Br\;} R'{-}C{\equiv}C{-}R \; + \; NaBr$$

In fact, alkynyl anions are versatile nucleophiles for making carbon–carbon bonds by reactions with alkyl halides. They are strongly basic, however, so only primary alkyl halides and tosylates react without generating significant quantities of elimination products.

The *mono*anion of acetylene is also a good nucleophile, and sodamide in liquid ammonia is the base that is often used to generate this ion. Its reactions with primary alkyl halides produce terminal alkynes.

$$H{-}C{\equiv}C{-}H \xrightarrow{NaNH_2,\; NH_{3\,(l)}} H{-}C{\equiv}C:^- \; Na^+$$

$$H{-}C{\equiv}C:^- \; Na^+ \xrightarrow[\text{2. }H_2O]{\text{1. }CH_3CH_2CH_2CH_2{-}Br,\; -70°C} CH_3CH_2CH_2CH_2{-}C{\equiv}C{-}H \qquad (86\%)$$

Substitution of a leaving group by an alkynyl carbanion is an excellent method for making carbon–carbon bonds because the initial alkyne product can be converted subsequently to a variety of compounds. In particular, partial reduction to an alkene is facile (Sections 11.2b and 12.2c), and alkenes can be converted to others by reactions such as hydroboration (Section 9.4b) or electrophilic addition (Section 9.2).

EXERCISE 15.7

What is the expected product of the following reaction sequence? Draw the structure of the compound that is formed at each step.

$$CH_3CH_2—C\equiv C—H \xrightarrow{\begin{array}{l}\text{1. NaNH}_2\text{, THF}\\\text{2. CH}_3\text{CH}_2\text{Br}\\\text{3. catecholborane (Section 9.4b)}\\\text{4. H}_2\text{O}_2\text{, OH}^-\end{array}}$$

Reactions of alkynyl carbanions are not limited to substitution. These nucleophiles also add to carbonyl compounds such as aldehydes and ketones, and they react with epoxides to form alcohols with two additional carbon atoms. Thus, they react like other types of Grignard reagents and organolithium compounds.

$$CH_3CH_2CH_2CH_2—C\equiv C—H \xrightarrow{\begin{array}{l}\text{1. C}_4\text{H}_9\text{Li, THF}\\\text{2. C}_6\text{H}_5\text{CHO}\\\text{3. H}_3\text{O}^+\end{array}} CH_3CH_2CH_2CH_2—C\equiv C—\underset{H}{\overset{OH}{C}}—C_6H_5$$

EXERCISE 15.8

What is the expected product of the following reaction sequence? Draw the structure of the intermediate that is formed at each step.

$$C_6H_5—C\equiv C—H \xrightarrow{\begin{array}{l}\text{1. CH}_3\text{MgI, THF}\\\text{2. ethylene oxide}\\\text{3. H}_3\text{O}^+\end{array}}$$

15.3 TRANSITION METAL ORGANOMETALLIC COMPOUNDS

15.3a ORGANOCUPRATES ARE PREPARED FROM ORGANOLITHIUM COMPOUNDS AND COPPER(I) SALTS

We now turn our attention to transition metal organometallic complexes. The term "complex" derives from the fact that several groups are often bonded to the metal ion. These groups are called *ligands,* and they can be alkyl groups, alkenes, alkynes, aromatic compounds, organophosphines, amines, alcohols, carbon monoxide, and small inorganic species such as water, nitric oxide, halide ions, and hydride ion.

Some transition metals and their complexes react like magnesium or lithium metal, generating an organometallic species directly from an organohalide or related substance. Other transition metal organometallic compounds have to be prepared from a Grignard or organolithium reagent. Organometallic copper compounds are among the most important compounds of this latter type.

An organolithium compound reacts with copper(I) iodide to form a lithium diorganocuprate, referred to as a **Gilman reagent.** As a specific example, butyllithium first reacts with copper iodide to produce butylcopper. Then, a second equivalent of butyllithium reacts with that species to generate lithium dibutylcuprate.

$$CH_3CH_2CH_2CH_2-Li \xrightarrow[\text{THF}]{\text{CuI}} CH_3CH_2CH_2CH_2-Cu + LiI$$

Butyllithium (BuLi)   **Butylcopper**

$$CH_3CH_2CH_2CH_2-Cu \xrightarrow{\text{BuLi}} (CH_3CH_2CH_2CH_2)_2Cu^- \ Li^+$$

Lithium dibutylcuprate
(a Gilman reagent)

Formally, organocuprates have the same charge distribution as organolithium and organomagnesium compounds—the carbon atom of each substituent carries a partial negative charge. A lithium diorganocuprate, however, is a weaker base than a Grignard reagent and a more potent nucleophile with respect to substitution reactions.

Several years ago, it was discovered that treating copper(I) cyanide with 2 equiv of an organolithium reagent produces a species that is different from a Gilman reagent.

$$2 \ BuLi + CuCN \xrightarrow{\text{ether or THF}} [(Bu)_2CuCN]Li_2$$

The best evidence reveals that the cyanide ion is not bonded to the copper ion; nevertheless, these compounds behave differently than Gilman reagents and are called **higher order cuprates.**

15.3b ORGANOCUPRATES REACT AS NUCLEOPHILES IN SUBSTITUTION REACTIONS

In *general,* organocuprates do not react with carbon dioxide or other carbonyl compounds in the same way as Grignard or organolithium reagents.

$$\left(\diagdown\diagup\diagdown \right)_2 Cu^- \ Li^+ + \text{(cyclohexanone)} \longrightarrow \text{N.R.}$$

They *do* react with many organohalides, however, and this reaction constitutes an excellent way to make a carbon–carbon bond.

$$\left(CH_3CH_2CH_2CH_2\right)_2 Cu^- \ Li^+ \xrightarrow[\text{1-iodopropane}]{CH_3CH_2CH_2-I} CH_3CH_2CH_2CH_2CH_2CH_2CH_3 + Cu(CH_2CH_2CH_2CH_3) + LiI$$

Lithium dibutylcuprate   **Heptane**

This reaction *cannot* occur by an S_N2 mechanism because aryl and vinyl iodides and bromides—which are not viable substrates in the S_N2 reaction—react with Gilman reagents to form carbon–carbon bonds. Among alkyl halides, only *primary* alkyl bromides and iodides react suitably with Gilman reagents. In all cases, organoiodides react more readily than the corresponding organobromides.

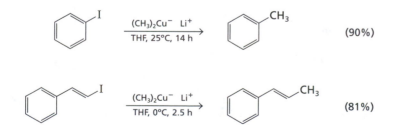

EXERCISE 15.9

Draw the structure of the major product of the following reaction:

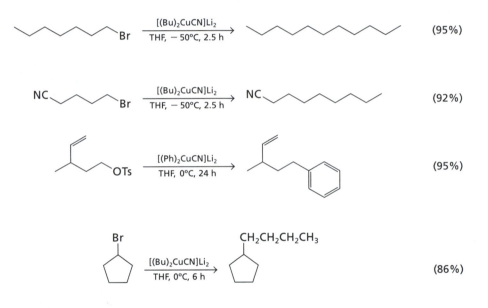

Higher order cuprates react even more readily with alkyl halides and tosylates, and they give fewer side reactions than the corresponding Gilman reagents. Furthermore, they react as quickly with organobromides as Gilman reagents do with organoiodides. A significant advantage is their use in substitution reactions with *secondary* alkyl bromides and iodides, which are poor substrates with the Gilman reagents. The following examples illustrate the utility of the higher order cuprate reagents. Notice that both primary and secondary alkyl substrates react cleanly:

Both types of organocuprates react as Grignard reagents do toward epoxides, and the following two equations show examples that make use of higher order cuprate reagents, which generally react more cleanly than Gilman reagents. Organocuprates react preferentially at the less highly substituted carbon atom of the epoxide ring, just like other nucleophiles.

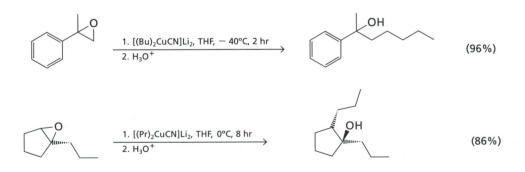

EXERCISE 15.10

What is the major product expected from each of the following reactions?

a.

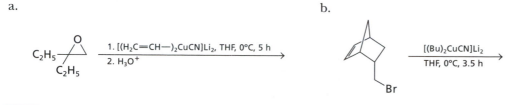

1. [(H₂C=CH—)₂CuCN]Li₂, THF, 0°C, 5 h
2. H₃O⁺

b.

[(Bu)₂CuCN]Li₂

THF, 0°C, 3.5 h

15.3c THREE FUNDAMENTAL MECHANISMS DOMINATE TRANSITION METAL ORGANOMETALLIC CHEMISTRY

The reactivity of transition metal organometallic reagents depends on three features of the metal ion:

- **Its oxidation state.**
- **Its coordination number** (C.N.), which is the number of groups bonded to the metal ion.
- **Its geometry,** which is how the ligands are arranged around the metal ion.

These three properties—oxidation state, C.N., and geometry—dictate the kinds of reactions that a metal complex can undergo, which are of three types: *oxidative addition, reductive elimination,* and *migratory insertion.*

Oxidative addition takes place when the metal ion increases its C.N. as its oxidation state also increases. Alkyl halides and H₂ are the most common reactants that participate in this process.

Oxidative addition

The metal ion has an oxidation state of *n* and a coordination number (C.N.) of 4.

The metal ion has an oxidation state of *n*+2 and a coordination number (C.N.) of 6.

Reductive elimination is the reverse of oxidative addition, so the oxidation state of the metal ion decreases, and one or two groups are eliminated from the coordination sphere of the metal ion. The elimination processes that are most important in synthesis are those that create a new C–H or C–C bond.

Reductive elimination

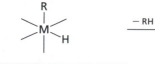
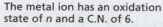

The metal ion has an oxidation state of *n* and a C.N. of 6.

The metal ion has an oxidation state of *n*−2 and a C.N. of 4.

The third mechanism type is **migratory insertion,** in which one group bonded to the metal ion inserts into a π bond of another ligand that is also bonded to the metal.

This process occurs most often with an alkene molecule or carbon monoxide as the acceptor, and there is usually no change in oxidation state. The coordination number *can* change when another ligand, often a solvent molecule, bonds to the metal ion.

Migratory insertion

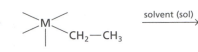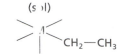

| The metal ion has an oxidation state of *n* and a C.N. of 6. | The metal ion has an oxidation state of *n* and a C.N. of 5. | The metal ion has an oxidation state of *n* and a C.N. of 6. |

15.3d ORGANOCUPRATES MAKE USE OF OXIDATIVE ADDITION AND REDUCTIVE ELIMINATION TO FORM CARBON–CARBON BONDS

The reaction between iodobenzene and lithium dipropylcuprate illustrates two of the mechanism types just presented and shows how carbon–carbon bond formation by organocuprates (Section 15.3b) can occur by a pathway that is not S_N2 in character.

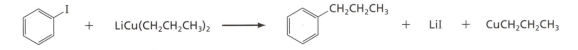

Oxidative addition occurs in the first step of this transformation, as the oxidation state of the copper ion increases from +1 to +3, and the C.N. increases from 2 to 4.

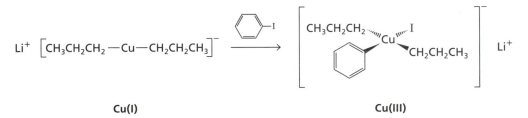

Once the phenyl ring is bonded to the copper ion, the complex is unstable, so reductive elimination occurs, forming 1-phenylpropane and a metal-containing product, $Li(I–Cu–CH_2CH_2CH_3)$, which subsequently dissociates to form LiI and $CuCH_2CH_2CH_3$.

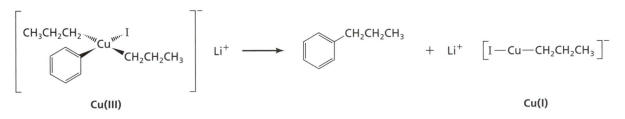

15.3e WILKINSON'S CATALYST IS USED TO HYDROGENATE ALKENES

A process that utilizes all three of the mechanism types presented in Section 15.3c is the hydrogenation of alkenes catalyzed by chloro[tris(triphenylphosphine)]rhodium(I), a complex known as **Wilkinson's catalyst.** This compound, $RhCl(Ph_3P)_3$, is soluble in many solvents, which makes it a *homogenous catalyst.* In contrast, Pd/C (Section 11.2a) is a *heterogeneous* (insoluble) *catalyst.*

Wilkinson's catalyst dissolves in a mixture of benzene and ethanol to form a reactive intermediate, which has a four-coordinate rhodium(I) ion. This species undergoes *oxidative addition* with hydrogen gas to produce a six-coordinate rhodium(III) dihydride.

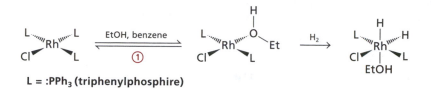

L = :PPh₃ (triphenylphosphire)

A molecule of alkene replaces the ethanol ligand in Step 2,

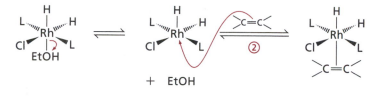

and then *migratory insertion* takes place. In this step, which is rate determining, a hydrogen atom moves from the coordination position adjacent to the alkene onto one of the alkene carbon atoms. An ethanol molecule binds to the coordination site left vacant by this migration.

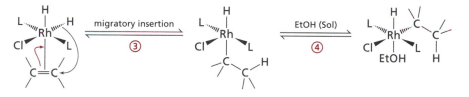

In the last step of the reaction cycle, the product alkane is formed by *reductive elimination*, which creates the second carbon–hydrogen bond. The rhodium(III) ion is reduced to Rh(I), and its coordination number decreases from 6 to 4.

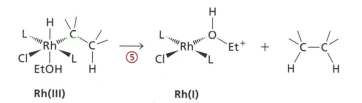

This rhodium-containing product is the same species that first reacted with H₂ in the oxidative–addition step; therefore, this mechanism can cycle through the various intermediates, leading to *catalytic* hydrogenation of the alkene double bond. Overall, this transformation differs little from heterogeneous hydrogenation, but the homogeneous reaction is more selective because the triphenylphosphine groups bonded to the rhodium ion provide steric bulk, which slows the reaction toward more highly substituted alkenes and leads to the following order of reactivity:

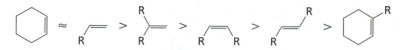

decreasing ease of reduction by H₂ and Wilkinson's catalyst

Despite differences in the alkene reactivity between the homogeneous and heterogeneous processes, the stereochemistry of the two processes is the same. If the reaction with Wilkinson's catalyst is done carefully, then cis addition of hydrogen to the double bond occurs. The primary advantage of a homogeneous catalyst over a heterogeneous one is that it can be modified to produce chiral products (Section 16.4b).

15.3f PALLADIUM COMPLEXES CATALYZE THE FORMATION OF CARBON–CARBON BONDS

Another versatile method for making carbon–carbon bonds—besides the one that uses organocuprate reagents (Section 15.3b)—is the **Suzuki reaction,** which is the palladium-catalyzed coupling of an organohalide and an organoborane. For the organohalide, R = aryl or alkenyl, but the R′ group of the organoborane can be practically anything.

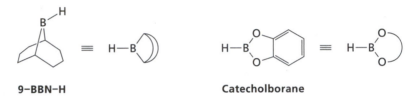

Often, the organoborane is one that has been prepared using catecholborane or 9-BBN because these reagents have a single B–H bond, and the stoichiometry of the hydroboration reaction is easy to control (Section 9.4a).

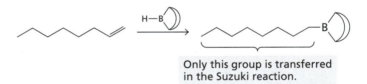

9–BBN–H   **Catecholborane**

Furthermore, with these reagents the group that derives from the alkene is the only one that is transferred in the subsequent coupling reaction.

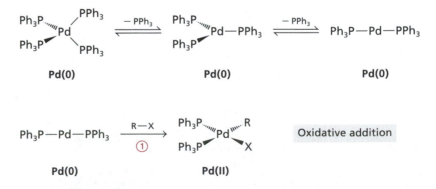

Only this group is transferred in the Suzuki reaction.

The mechanism of the Suzuki reaction is a three-step process. The first step is oxidative addition of the organohalide to the palladium(0) atom of a two-coordinate complex, formed by a series of equilibria in which triphenylphosphine ligands dissociate from Pd(Ph₃P)₄. Alkyl halides react very slowly in this oxidative addition step, which is why they are poor substrates in the Suzuki reaction.

In the next step, hydroxide ion reacts with the organoborane, and the organic group attached to boron is subsequently transferred to the palladium(II) ion, replacing the halide ion. No change in the oxidation state of Pd occurs during this substitution reaction.

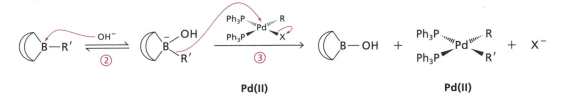

This Pd(II) complex immediately undergoes reductive elimination in Step 3, forming the coupled organic product and regenerating the two-coordinate palladium(0) complex.

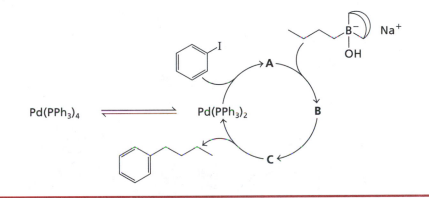

EXERCISE 15.11

The mechanism of the Suzuki reaction can be written as a cycle. Fill in structures for the letters **A, B,** and **C** in the following diagram for the coupling of iodobenzene and the NaOH adduct of butyl-9-BBN.

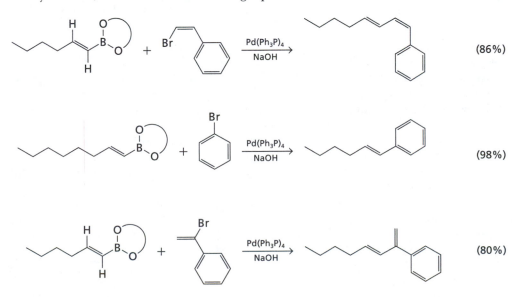

The Suzuki reaction provides a practical way to couple alkenylboranes with alkenyl or aryl halides, as shown in the following equations:

These particular transformations are valuable for two reasons. First, the stereochemistry of the resulting alkene is known with certainty because *all of the steps proceed with retention of configuration.* Second, the double bond in the product can be readily converted to other functional groups. In general, alkenes are versatile compounds in synthesis because they can be transformed into so many other types of substances.

EXERCISE 15.12

Show how you would prepare each of the organoborane reagents shown in the equations above. For starting materials, you may use $BH_3 \cdot THF$, catechol, and any alkene or alkyne.

15.4   RETROSYNTHESIS

15.4a A RETROSYNTHESIS IDENTIFIES POSSIBLE INTERMEDIATE COMPOUNDS AND STARTING MATERIALS

The most general procedure for planning a synthesis makes use of a *retrosynthetic analysis,* commonly called a **retrosynthesis.** Starting with the product molecule, a chemist considers plausible precursors that can be converted to it. These precursors themselves then become the goal of the synthesis. Working backward, the chemist eventually recognizes a compound, or perhaps several compounds, that can serve as a starting material. This whole process generates what is referred to as a **synthetic tree,** illustrated below. (Note that retrosynthetic reactions are shown with open arrows ⇒.)

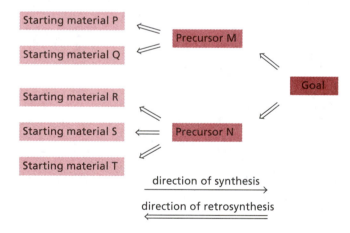

When suitable candidates for the starting materials have been identified, the planning process is reversed. Then the researcher begins to consider what reaction conditions are needed to convert the reactants to products. In so doing, alternate routes and intermediates are sometimes identified. Also during this phase, the placement of control elements such as **protecting groups** (Section 15.5b) is included. Examples that illustrate these concepts will be presented shortly.

The underlying concept behind retrosynthesis is commonly called "the disconnection approach". Manipulations of functional groups are often straightforward, *so the reactions requiring the most attention are those needed to form the carbon–carbon bonds.* In devising a retrosynthesis, the desired product must be evaluated for bonds where the disconnection between their atoms will produce fragments that can be combined by actual reactions to be used in the synthesis. Thus, if a carbon–carbon bond is to be made by bringing together a Grignard reagent and an aldehyde, for example, then the disconnection must be made at a place in the molecule that allows each portion to have its appropriate functional group. Disconnections are most suitable at or adjacent to

- A double or triple bond.
- A ring junction.
- A branch point next to a heteroatom.
- A functional group.

When a cyclic compound is to be prepared, construction of the ring system is often the overriding factor. An aromatic ring will most likely be incorporated by performing reactions with commercially available benzene derivatives.

15.4b A Preliminary Examination of the Desired Molecule Considers the Functional Groups That Are Present

Although a retrosynthesis is conceived with preparation of the carbon atom framework in mind, a primary consideration is the identity of the functional group that appears in the product molecule. *Functional group equivalents* are important to consider because they can reveal more likely disconnection points. For example, a 2° alcohol is readily converted to a 2° alkyl halide or ketone.

Similarly, a 1° alcohol can be oxidized to an aldehyde or further to a carboxylic acid (Section 11.4c).

$$R—CH_2OH \xrightarrow{[O]} R—CHO \xrightarrow{[O]} R—COOH$$

Perhaps less obvious is the equivalence of carbon–oxygen and carbon–carbon double bonds, an example of which was shown earlier (Section 15.2d). In Chapter 11, you learned reactions that can be used to cleave a double bond to produce an aldehyde, ketone and/or carboxylic acid. A carbon–carbon double (or triple) bond can therefore be a precursor for these carbonyl-containing groups.

15.4c An Alcohol Functional Group Prescribes a Logical Disconnection Site

In many of the synthesis problems in this text, you will be expected to start with any compound that has a limited number of carbon atoms (five or six). It is logical to assume that any of these compounds can be bought commercially at a reasonable price. You will also be allowed to use any "inorganic reagents" (acids, bases, metal-containing compounds, including tributyltin hydride, CuI, Mg, etc.). With these constraints in mind, the focus should be on how to construct the carbon skeleton of the desired molecule once the functional groups and their possible equivalents have been identified. *A functional group that is readily prepared is the alcohol group, because the hydrocarbon framework can be readily assembled at that carbon atom by using an organolithium or Grignard reagent.* The following examples illustrate this concept by presenting the retrosynthesis and synthesis processes involved in making several molecules.

EXAMPLE 15.2

Show how you would prepare 3-heptanol from starting materials that have five or fewer carbon atoms.

First, draw the structure of the desired product:

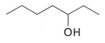

Next, ask yourself this question: What methods do I know to make a carbon–carbon bond at the carbon atom bearing an OH group? Although there are several ways to make an alcohol, a method to create a carbon–carbon bond while forming the OH group at the same time makes use of the Grignard reaction (see Table 15.1). Therefore, consider disconnection points at the positions adjacent to the OH group:

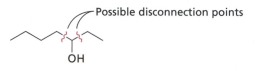

In the retrosynthetic analysis, the molecule is divided into two fragments, and the synthetic tree would have two branches, as illustrated below:

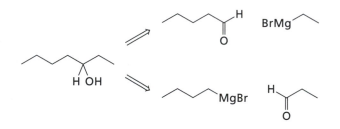

As a rule, the route to choose is the one that leads to construction of the molecule from fragments that are close to each other in size because it will be easier to purify the product from reactants that have half its mass.

This example is straightforward because, with only one disconnection, possible starting materials are identified. At this juncture, we write the synthetic scheme as follows:

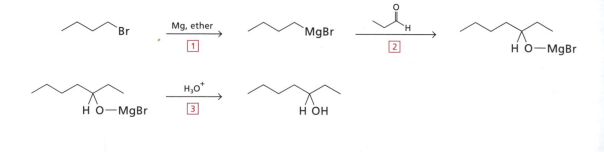

EXERCISE 15.13

Write the equations corresponding to the other conceived route to prepare 3-heptanol.

EXAMPLE 15.3

Show how you would prepare the compound shown below from starting materials that have six or fewer carbon atoms. You may also use any reagents and solvents.

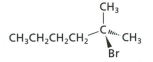

The first issue to address is how to make a 3° alkyl bromide. There are two ways that you know: substitution of an OH group via the S_N1 reaction (section 6.2a), and electrophilic addition of HBr to an alkene (Section 9.1b). Therefore, we have at least two possible routes to explore.

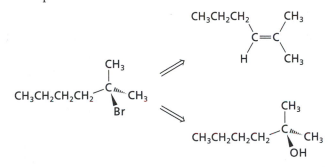

Because an alkene is often made from an alcohol or alkyl halide by elimination, the alcohol is a more logical precursor. Further retrosynthetic analysis of the alcohol leads to disconnection next to the OH group.

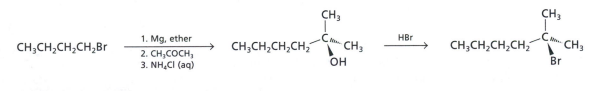

To prepare the target molecule, we start with 1-bromobutane, make its Grignard derivative, treat that compound with acetone, and use a mild acid workup to obtain the alcohol. Treating the alcohol with HBr yields the desired bromoalkane.

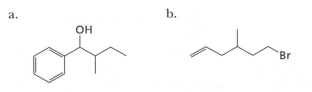

EXERCISE 15.14

Propose a synthesis for each of the following compounds starting with bromobenzene and/or any organic compounds that have five or fewer carbon atoms:

a.

b.

EXAMPLE 15.4

Show how you would prepare 1-phenylcyclohexene from cyclohexene and any other organic compounds, reagents, and solvents.

At least three retrosynthesis schemes can be envisioned. Two (**A** and **B**) make use of dehydration to form the alkene π bond, and **C** is a dehydrohalogenation reaction. The needed alcohols can be made by Grignard reactions, and the bromo compound is accessible via free radical halogenation.

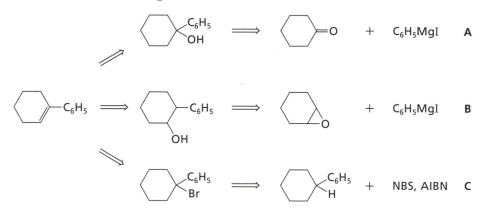

Further analysis shows that cyclohexanone in **A** can be made by oxidation of cyclohexanol, which is made by hydration of cyclohexene; the epoxide needed for **B** is prepared directly from cyclohexene, and phenylcyclohexane (route **C**) is made by the Suzuki reaction between bromobenzene and the 9-BBN derivative of cyclohexene.

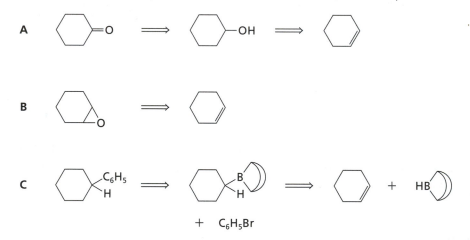

Choosing route **B** because it has the fewest number of retrosynthesis steps, we write the synthetic scheme as follows:

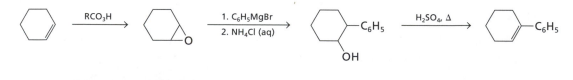

EXERCISE 15.15

Write the reactions for the syntheses of 1-phenylcyclohexane that are based on the other two retrosynthetic pathways given above.

15.5 SELECTIVITY CRITERIA IN SYNTHESIS

15.5a REACTION SELECTIVITY IS A CONSIDERATION IN THE SYNTHESIS OF A COMPOUND WITH MORE THAN ONE FUNCTIONAL GROUP

Besides making the carbon framework and manipulating the functional groups, several other issues must be addressed when planning a synthesis, and these are related to the selectivity of each reaction that is chosen. *Chemoselectivity,* which is the differentiation between the reactivity patterns of functional groups (Section 11.2a), is often a simple problem to resolve. For example, you learned that primary alcohols can be oxidized to aldehydes, but tertiary alcohols are inert to simple oxidants such as chromium trioxide (Section 11.4c). The oxidation of a primary alcohol in the presence of a 3° alcohol constitutes a *chemoselective* transformation.

Regioselectivity (Section 9.1b) refers to the orientation in addition reactions. You have seen, for example, that hydrogen bromide adds to alkenes in a Markovnikov fashion under polar conditions (Section 9.1b) and with an anti-Markovnikov orientation under radical conditions (Section 12.3a). Both are regioselective reactions.

Stereoselectivity (Section 9.1d) has to do with the stereochemistry of the reaction. You have learned several reactions that are stereospecific, like the S_N2 reaction, which goes by inversion of configuration at a stereogenic center. Hydroboration occurs by a stereospecific syn addition, and the electrophilic addition of bromine to an alkene is a stereospecific anti addition.

Enantioselectivity (Section 9.1d) has to do with the creation of new chiral centers, a topic that will be discussed further in Chapter 16.

The following example illustrates how the stereochemistry of the product impacts the planning of the synthesis scheme.

EXAMPLE 15.5

Derive a synthesis for the compound shown below, starting with organic compounds that have five or fewer carbon atoms. You may use any inorganic reagents and solvents.

First, consider what reaction will be used to create the epoxide ring. This group can be made by at least two methods that you know—direct epoxidation of a double bond, which is a syn-addition process (Section 11.3e); or an intramolecular substitution reaction of a halohydrin using base, which occurs by inversion at the carbon atom that bears the halide ion leaving group (Section 7.2c). Here is a situation in which you want to ensure that the chosen reaction yields the correct stereochemistry of the product molecule.

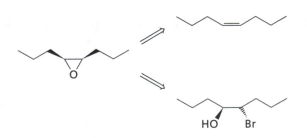

If you choose the *trans*-bromohydrin as the starting material for the epoxide, then you have to consider what is needed to make it. The starting alkene turns out to be the same cis-alkene as the one needed for the direct epoxidation process.

Next consider how to disconnect the carbon skeleton of the needed alkene in a way that leads to logical nucleophilic and electrophilic partners. There are several possibilities, two of which are shown below:

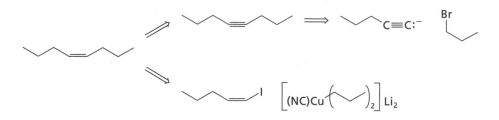

The first route has an additional step, but it begins with a nucleophilic alkynyl carbanion and an electrophilic alkyl bromide. This combination should present few difficulties. It does require the stereospecific reduction of the alkyne to form the cis-alkyne, but that transformation is straightforward with use of a poisoned catalyst (Section 11.2b). A synthesis based on the first retrosynthesis would therefore be written as follows (MMPP, Section 11.3e).

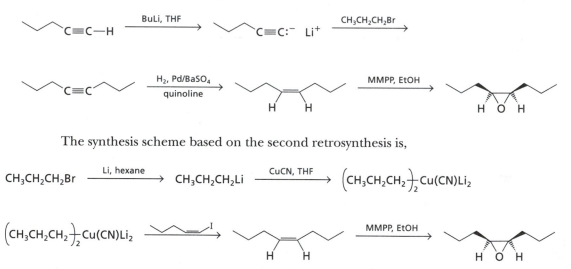

The synthesis scheme based on the second retrosynthesis is,

$$CH_3CH_2CH_2Br \xrightarrow{\text{Li, hexane}} CH_3CH_2CH_2Li \xrightarrow{\text{CuCN, THF}} \left(CH_3CH_2CH_2\right)_2 Cu(CN)Li_2$$

In the matter of selectivity, another issue has acquired importance during recent years, namely, that of waste disposal. Professor Barry Trost, of Stanford University, has challenged chemists to think about the **atom economy** of a reaction. The underlying concept is how much of the reactant(s) actually ends up in the product. In dealing with waste minimization, this concept is important. Clearly, *reactions that go by addition processes have a higher atom economy than those in which only a small portion of one reactant is transferred to the product.*

We can illustrate this objective by going back to the previous example and looking at the last step. Remember that in the first retrosynthesis, we identified two ways to make the epoxide. These are shown below as the actual reaction(s) you would perform in the laboratory. The same alkene, *cis*-4-octene, is converted to the same epoxide, *cis*-4,5-epoxyoctane.

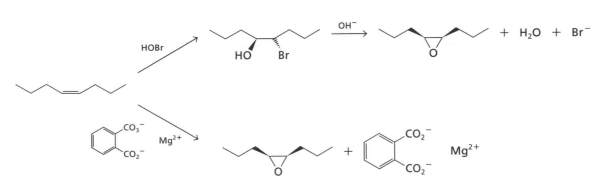

In the first pathway, the addition of HOBr has an atom economy of 100% because all of the atoms of the reactants appear in the product. During formation of the epoxide, bromine ion and water are lost, so the atom economy of the second step is less than quantitative. On the other hand, water is a benign byproduct. Bromide ion is the only waste product of the first route.

By contrast, the atom economy in the second pathway is *low*, because only one oxygen atom is transferred from magnesium monoperoxyphthalate. Most of the atoms of that reagent appear in the byproduct, magnesium phthalate. Of course, if this byproduct can be recycled to the peroxy reagent, then the overall atom economy increases dramatically. A consideration of atom economy in chemical processes will likely become more important as raw materials become scarcer and requirements for minimizing waste become more stringent.

15.5b PROTECTING GROUPS ARE USED WHEN INCOMPATIBLE FUNCTIONAL GROUPS ARE PRESENT IN THE DESIRED PRODUCT OR PRECURSORS

In the design of a synthesis scheme, you will often encounter the situation in which you want a reaction to occur at only one of two (or more) functional groups. One of these situations is the preparation of a Grignard reagent in which the starting material has a reactive second functional group (Section 15.2d). A protecting group is needed for the functional group that you want to retain.

Several commonly used protecting groups are listed in Table 15.2; the best ones have the following features:

- They can be introduced into a molecule under mild conditions.
- They are inert to the reaction conditions that will be used.
- They can be removed under conditions that do not affect other functional groups or the stereochemistry of the product.

One of the best protecting groups for an alcohol, which you have seen is a versatile functional group in synthesis, is the trimethylsilyl group (other organosilanes can also be employed). It is introduced simply by stirring an alcohol with chlorotrimethylsilane and a weak base like triethylamine.

Protection

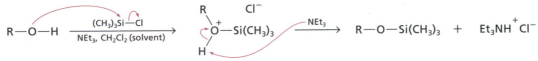

A trimethylsilyl ether, $ROSiMe_3$, is stable toward neutral and basic conditions, so it is comparable with Grignard reagents and other organometallic reagents. After reaction is complete, the trimethylsilyl group is easily removed by treatment with mild acid or,

Table 15.2 Some commonly used protecting groups in chemical synthesis.

Functional group	Reagents	Protected form	Conditions for removal
R—OH	Cl—SiMe₃, Et₃N	R—O—SiMe₃	F⁻
	Br—CH₂C₆H₅, base	R—O—CH₂—⬡	H₂, Pd catalyst (hydrogenolysis)
	⬡(O) , H⁺	R—O—⬡(O)	H₃O⁺
	Cl—CH₂—O—CH₃, base	R—O—CH₂—O—CH₃	H₃O⁺
R(R')C=O	HO—CH₂CH₂—OH , H⁺	R(R')C(O—CH₂CH₂—O)	H₃O⁺
	HS—CH₂CH₂—SH , ZnCl₂	R(R')C(S—CH₂CH₂—S)	HgCl₂, CH₃CN H₂O, CaCO₃

better, with fluoride ion, which is usually supplied as its tetrabutylammonium salt (Bu_4N^+ F^-) in aqueous solution. The reaction of this reagent with $ROSiMe_3$ produces fluorotrimethylsilane, which is an inert, gaseous substance, easily separated from the desired organic product.

Deprotection

$$R—O—Si(CH_3)_3 \xrightarrow[H_3O^+]{F^-} (CH_3)_3SiF + R—O^- \xrightarrow{H_3O^+} R—O—H + H_2O$$

Another satisfactory protecting group for alcohols is the benzyl group, which is introduced by alkylation with benzyl bromide via an S_N2 reaction (Section 6.3a). The benzyl group is readily removed by hydrogenolysis (Section 12.2a). Recognize, however, that this method of deprotection is not compatible with alkene or alkyne functional groups in a molecule because π bonds can be hydrogenated under these same reaction conditions (Section 11.2).

Protection

$$R—O—H \xrightarrow{NaH, PhCH_2Br, DMF} R—O—CH_2Ph + NaCl$$

Deprotection

$$R—O—CH_2Ph \xrightarrow{H_2, Pd/C} R—O—H + PhCH_3$$

Other protecting groups listed in Table 15.2 will be discussed later, but this list provides a preview of such compounds and how different functional groups can be protected. If a protecting group is used during a synthetic scheme, the overall procedure is lengthened by two steps, one for protection and one for deprotection. It is preferable, therefore, to design a synthesis so as to rely on as few protecting groups as possible.

The following example illustrates the use of a protecting group during a synthesis.

EXAMPLE 15.6

Show how you would prepare the following compound from substances that have six or fewer carbon atoms:

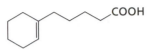

A logical disconnection site is at the ring, and at least two combinations come to mind. The curly brackets around the COOH group indicate that a protecting group or a functional group equivalent will be needed because the carboxylic acid proton will react with the organometallic center elsewhere in the scheme.

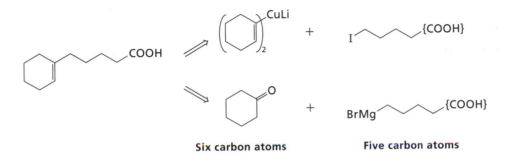

Six carbon atoms   **Five carbon atoms**

Choosing the second retrosynthesis to develop, we decide to use a protected alcohol group as a carboxylic acid equivalent (Section 15.4b). Protection is required so that we can make the needed Grignard reagent (Section 15.2d).

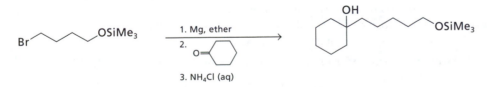

Now examine the actual synthesis. In the first step, the alcohol group is protected as its trimethylsilyl ether derivative.

Next, the Grignard reagent is generated and treated with cyclohexanone to yield the 3° alcohol.

Dehydration with phosphoric acid generates the cyclohexene ring, and the protecting group is removed from the alcohol using fluoride ion.

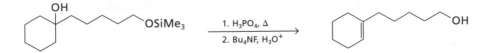

Finally, oxidation is used to obtain the final product.

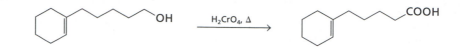

Unlike the reaction scheme described in Section 15.2d, an alkene cannot be used as an equivalent to the carboxylic acid functional group even though ozonolysis could be used to make the COOH group. In *this* scheme, the product itself has a double bond (in the ring), so ozonolysis would also take place there, as shown below.

EXERCISE 15.16

How would you synthesize the following compound from any organic starting materials with five or few carbon atoms? Any inorganic reagents and solvents may also be used.

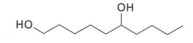

CHAPTER SUMMARY

Section 15.1   Carbon-carbon bond formation

- Chemical synthesis relies on reactions to construct carbon–carbon bonds.
- Conceptually, formation of a carbon–carbon bond can be accomplished by reaction between a nucleophilic and an electrophilic carbon atom.
- One of the simplest carbon–carbon bond-forming reactions is the S_N2 reaction of cyanide ion with an alkyl halide.
- Organometallic compounds have a nucleophilic carbon atom.

Section 15.2   Organomagnesium and lithium compounds

- The most common reactions observed for main group organometallic compounds—those of Li and Mg—are acid–base reactions, addition to a carbonyl group, and opening of an epoxide ring.
- Grignard reagents are made by the reaction between organohalides and magnesium metal.
- Organolithium compounds are made by the reaction between organohalides and lithium metal.
- Organometallic derivatives of alkynes are made by acid–base reactions.
- Organometallic derivatives of alkynes react with alkyl halides by the S_N2 pathway to form carbon–carbon bonds.

Section 15.3 Transition metal organometallic chemistry

- Organocuprate reagents are made by treating a copper(I) compound with an organolithium compound.
- Reaction between an organocuprate reagent and an organoiodide leads to carbon–carbon bond formation.
- Reactions of transition metal organometallic compounds take advantage of the multiple oxidation states available to most transition metals, as well as the ability of a transition metal ion to bind several groups at once.
- Oxidative addition, migratory insertion, and reductive elimination are the principal mechanisms by which transition metal organometallic compounds react.
- Organocuprates make use of oxidative addition and reductive elimination steps to form carbon—carbon bonds.
- Wilkinson's catalyst is a soluble Rh complex that is used to hydrogenate alkenes to alkanes. Oxidative addition, migratory insertion, and reductive elimination steps are all involved in its reaction cycle.
- The Suzuki reaction is a useful way to couple the organic group of an organoborane with an aryl halide using palladium complexes as catalysts.

Section 15.4 Retrosynthesis

- Retrosynthetic analysis is a strategy that considers the disconnection of bonds in a target molecule as a way to identify compounds likely to be used subsequently in the actual synthetic procedures.
- Disconnection of a carbon–carbon bond adjacent to an OH group is a logical and simple choice because an alcohol is the product of many reactions that make use of organometallic compounds.

Section 15.5 Selectivity criteria in synthesis

- The chemoselectivity, regioselectivity, stereoselectivity, and enantioselectivity of reactions must be considered during the design of a synthetic scheme.
- Atom economy—the amount of a reactant that appears in the product—is an important concept to bear in mind to limit costs and to minimize generation of hazardous waste.
- Protecting groups are employed when a reaction has to be conducted in the presence of additional functional groups that will either interfere with the desired reaction or react themselves with the reagents that are being used.

KEY TERMS

Section 15.2a
Grignard reagent

Section 15.2e
organolithium compound

Section 15.3a
Gillman reagent
higher order cuprate

Section 15.3c
oxidative addition
reductive elimination
migratory insertion

Section 15.3e
Wilkinson's catalyst

Section 15.3f
Suzuki reaction

Section 15.4a
retrosynthesis
synthetic tree
protecting groups

Section 15.4b
functional group
    equivalents

Section 15.5a
atom economy

REACTION SUMMARY

Section 15.2a

Grignard reagents are prepared by treating an organohalide with magnesium metal in THF or ether. The halide can be any type: methyl, 1°, 2°, or 3° alkyl, aryl, or vinyl.

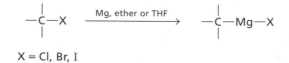

$$X = Cl, Br, I$$

Section 15.2b

A summary of the reactions of Grignard (and organolithium) reagents with electrophiles is presented in Table 15.1.

Section 15.2c

Grignard and organolithium compounds react with epoxides, and the R group of the reagent usually reacts at the less highly substituted carbon atom of the epoxide ring.

Section 15.2e

Organolithium compounds are made by treating an organohalide with lithium metal in a nonreactive solvent such as ether or a hydrocarbon.

$$-\overset{|}{\underset{|}{C}}-X \quad \xrightarrow{\text{2 Li, ether or hexane}} \quad -\overset{|}{\underset{|}{C}}-Li \quad + \quad LiX$$

$$X = Cl, Br, I$$

Section 15.2f

An alkynyl organometallic compound is made via an acid–base reaction between a terminal alkyne and an alkyl organometallic compound or sodium amide.

$$RC\equiv C-H \quad \xrightarrow{\text{R'MgX, R'Li, or NaNH}_2} \quad RC\equiv C-M$$

$$M = MgX, Li, or Na$$

Alkynyl organometallic compounds react with primary alkyl halides by an S_N2 pathway to form a carbon–carbon bond.

$$RC\equiv C-M \quad \xrightarrow{\text{R'}-X} \quad RC\equiv C-R'$$

$$M = MgX, Li, or Na$$

Section 15.3a

Organocuprate reagents, also called Gillman reagents, are made by treating a copper(I) halide with 2 equiv of an organolithium compound.

$$2 \, R'-Li \quad \xrightarrow{\text{CuX}} \quad Li[R_2Cu] \quad + \quad LiX$$

Higher order organocuprate reagents are made by treating copper(I) cyanide with 2 equiv of an organolithium compound.

$$2 \text{ R—Li} \xrightarrow{\text{CuCN}} \text{Li}_2[\text{R}_2\text{CuCN}]$$

Section 15.3b

Organocuprate reagents react with organoiodides and organobromides to form carbon–carbon bonds.

$$\text{Li}[\text{R}_2\text{Cu}] \quad + \quad \text{R}'\text{—X} \quad \longrightarrow \quad \text{R}'\text{—R} \qquad\qquad \text{Li}_2[\text{R}_2\text{CuCN}] \quad + \quad \text{R}''\text{—I} \quad \longrightarrow \quad \text{R}''\text{—R}$$

R′ = aryl, alkenyl, 1° alkyl
X = Br, I

R″ = aryl, alkenyl, 1° and 2° alkyl
X = Br, I

Organocuprate reagents react with epoxides, and the R group of the reagent reacts at the less highly substituted carbon atom of the epoxide ring.

Section 15.3e

Homogeneous hydrogenation of alkenes can be accomplished using Wilkinson's catalyst, $\text{RhCl}(\text{PPh}_3)_3$.

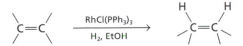

Section 15.3f

The Suzuki reaction is used to make carbon–carbon bonds by the reaction of organoboranes and either aryl or vinyl halides with a palladium catalyst.

$$\text{R}'\text{—BR}''_2 \quad + \quad \text{R}\text{—X} \quad \xrightarrow{\text{Pd}(\text{PPh}_3)_4,\ \text{OH}^-} \quad \text{R}'\text{—R}$$

R = aryl, alkenyl

Section 15.5b

The alcohol functional group is readily protected from reactions with strong bases and organometallic compounds by conversion to its trimethylsilyl ether derivative, accomplished with use of chlorotrimethylsilane and triethylamine.

$$\text{RO—H} \xrightarrow[\text{CH}_2\text{Cl}_2]{(\text{CH}_3)_3\text{SiCl, NEt}_3} \text{RO—Si(CH}_3)_3 \quad + \quad \text{Et}_3\text{NH}^+ \text{ Cl}^-$$

Alkoxytrimethylsilanes are readily converted to alcohols when treated with fluoride ion, which reacts at silicon to form fluorotrimethylsilane.

$$\text{RO—Si(CH}_3)_3 \xrightarrow[\text{CH}_3\text{CN}]{\text{Bu}_{34}\text{NF, H}_2\text{O}} \text{RO—H} \quad + \quad (\text{CH}_3)_3\text{SiF}$$

ADDITIONAL EXERCISES

15.17.   Draw the structure of the Grignard reagent derived from each of the following molecules. Show the stereochemistry of the molecules in (d) and (e).

a.   b.   c.

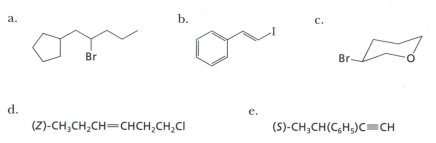

d.   e.

(Z)-CH₃CH₂CH=CHCH₂CH₂Cl   (S)-CH₃CH(C₆H₅)C≡CH

15.18.   Which of the following compounds will not readily form a Grignard reagent? Indicate which portion of the molecule causes a problem. Draw the structure of the Grignard reagent formed from compounds that do not pose problems.

a.   b.   c.   d.

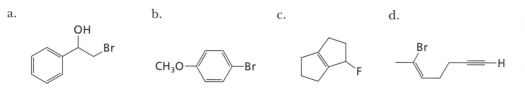

15.19.   Show how you would synthesize each of the following primary alcohols from an appropriate organohalide and formaldehyde via a Grignard reaction:

a.   b.   c. 5-Methyl-1-hexanol

15.20.   Show how you would synthesize each of the following secondary alcohols from an appropriate organohalide and an aldehyde via a Grignard reaction:

a.   b. 1-Phenyl-1-butanol   c.

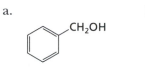   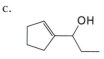

15.21.   Show how you would synthesize each of the following tertiary alcohols from an appropriate organohalide and a ketone via a Grignard reaction:

a. 3-Methyl-3-pentanol   b.   c.

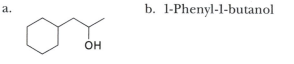

15.22.   Each of the following compounds is added to a separate ether solution of phenylmagnesium bromide. Draw the structures of the substances that will be in the solution *prior* to workup.

a.   b.   c.   d.   e.

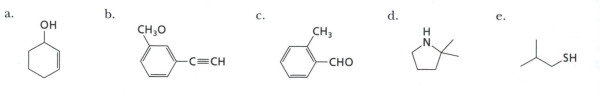

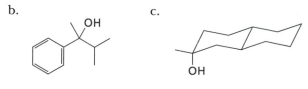

15.23. 1-Bromo-*cis*-2-butene was treated with magnesium in dry ether; that solution was then added to acetone. After careful hydrolytic workup, two isomeric products **A** and **B** with the formula $C_7H_{14}O$ could be isolated. One compound has a cis double bond, three methyl groups, and a methylene group. The other isomer has a terminal double bond, three methyl groups, and a methine group. What are the structures for compounds **A** and **B**? Explain how they are formed.

15.24. Write equations for the reactions that would be used to protect and then deprotect the following alcohols with the indicated protecting group:

a.

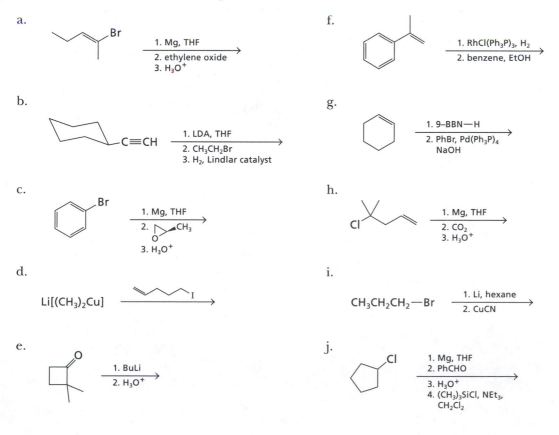

as its trimethyl silyl derivative

b. (2*R*, 3*S*)-3-Methyl-2-pentanol as its benzyl derivative

15.25. Indicate how you would prepare each of the following alcohols from the indicated starting material, an aldehyde or ketone, and any other necessary reagents and solvents.

a. 2-Cyclohexylethanol from bromocyclohexane
b. 1-Hexanol from 1-pentene
c. 1-Cyclohexylethanol from cyclohexanol
d. 3-Heptanol from 1-iodobutane

15.26. Draw the structure(s) of the major product(s) expected from each of the following reactions. Indicate the stereochemistry of the product as appropriate. Relative stereochemistry should be shown using wedges and dashed lines. If a racemic mixture will be formed, draw the structure of one enantiomer and write the word "racemic", or draw both enantiomeric structures. If diastereomers are formed, draw each structure; label meso compounds as such. If no reaction occurs, write N.R.

a.

Br

1. Mg, THF
2. ethylene oxide
3. H_3O^+

f.

1. RhCl(Ph₃P)₃, H₂
2. benzene, EtOH

b.

C≡CH

1. LDA, THF
2. CH₃CH₂Br
3. H₂, Lindlar catalyst

g.

1. 9–BBN—H
2. PhBr, Pd(Ph₃P)₄
   NaOH

c.

Br

1. Mg, THF
2. ⟨epoxide⟩CH₃
   O
3. H_3O^+

h.

Cl

1. Mg, THF
2. CO₂
3. H_3O^+

d.

Li[(CH₃)₂Cu]

⟨alkene⟩I

i.

CH₃CH₂CH₂—Br

1. Li, hexane
2. CuCN

e.

O ⟨cyclobutanone⟩

1. BuLi
2. H_3O^+

j.

Cl

1. Mg, THF
2. PhCHO
3. H_3O^+
4. (CH₃)₃SiCl, NEt₃,
   CH₂Cl₂

15.27. Show how you would prepare each of the following compounds using an organocuprate reagent and any other organic compounds with six or fewer carbon atoms. You may use any other necessary reagents and solvents.

a.   b.   c.

15.28. A Grignard reagent will react slowly with oxetane to produce a primary alcohol after acid workup. Propose a reasonable mechanism for the reaction as given below.

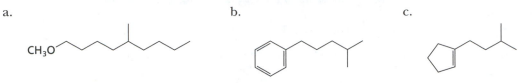

Grignard reagent   Oxetane   Salt of 1° alcohol

15.29. Each of the following pairs of transformations produces the same product. Which one has the higher atom economy, qualitatively? For each process, write an equation with the reagents on the arrow.

a. Oxymercuration of 2-methyl-1-butene followed by $NaBH_4$ reduction versus Markovnikov addition of water to 2-methyl-1-butene under acidic conditions.

b. Markovnikov addition of water to cyclohexene under acidic conditions versus hydroboration of cyclohexene followed by treatment with hydrogen peroxide and base.

c. Heating 1-hexanol with PBr_3 versus treating 1-hexanol with mesyl chloride and triethylamine followed by heating with a solution of LiBr in DMF.

d. Reduction of 1-pentyne with sodium in liquid ammonia and *tert*-butyl alcohol versus reduction of 1-pentyne with hydrogen and the Lindlar catalyst

15.30. Draw the structure of the major product expected from each of the following reactions. Ignore stereochemistry. The elemental analysis for each product is given.

In performing each of the following reactions, indicate the principal IR absorptions that you would look for in the reactant and product to ensure that the reaction was successful. Be as specific as possible: list the position (cm^{-1}), intensity (weak, medium, or strong), and shape (sharp or broad) of the diagnostic absorption band(s). Repeat the same process for the ^1H and broad-band ^{13}C NMR spectra. List, as applicable, the chemical shifts, integrated intensity values, splitting patterns, and DEPT results.

a.

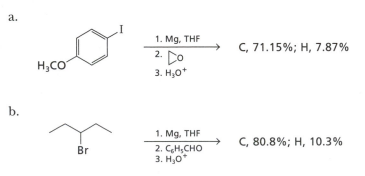

b.

15.31. In performing the following reaction to make 5-heptynoic acid, a byproduct—compound **H**—is isolated. Its ^{1}H NMR spectrum is shown. Draw the structure of compound **H**.

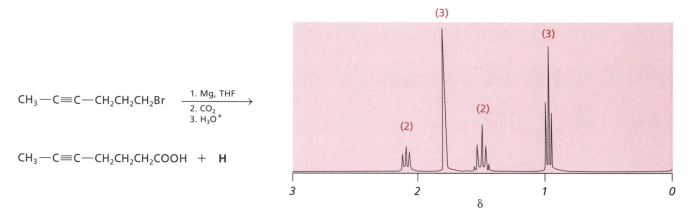

$CH_3-C\equiv C-CH_2CH_2CH_2Br$ $\xrightarrow[\text{3. }H_3O^+]{\substack{\text{1. Mg, THF} \\ \text{2. } CO_2}}$

$CH_3-C\equiv C-CH_2CH_2CH_2COOH$ + **H**

15.32. For your lab course, you decide to carry out the following reaction. Unfortunately, someone had mislabeled the bottle containing the aldehyde, so the product that you isolated was not the expected one.

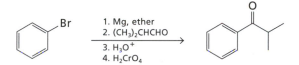

1. Mg, ether
2. $(CH_3)_2CHCHO$
3. H_3O^+
4. H_2CrO_4

Instead, the product displayed the following ^{13}C and ^{1}H NMR spectra. The IR spectrum displayed an intense absorption at 1700 cm^{-1}, and the high-resolution CI mass spectrum showed MW = 162.105. Draw the structure of the actual product, and show the correlation between the data and the structure you propose.

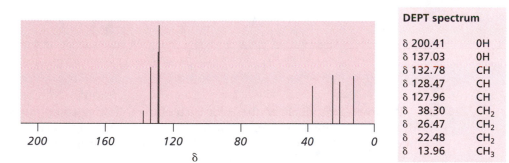

^{13}C NMR spectrum

DEPT spectrum

δ 200.41	0H
δ 137.03	0H
δ 132.78	CH
δ 128.47	CH
δ 127.96	CH
δ 38.30	CH$_2$
δ 26.47	CH$_2$
δ 22.48	CH$_2$
δ 13.96	CH$_3$

^{1}H NMR spectrum

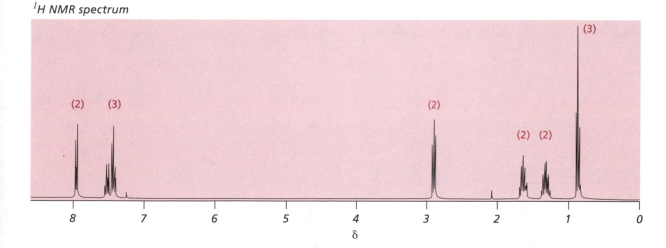

15.33. For the final exam in your lab course, you were given an unknown compound to identify. First, you obtained the ^{1}H NMR and ^{13}C NMR spectra, which are shown below. Then, you were able to have the high-resolution CI mass spectrum recorded, which showed MW = 134.110. Just when you thought you were done, you discovered that you were also required to prepare a sample of the unknown to confirm its identity. Going to the chemical storeroom for starting materials, you found that the only ones available were alkenes with 4, 5, 6, or 7 carbon atoms, as well as an old bottle containing bromobenzene. Fortunately, a number of catalysts and other reagents were available. Draw the structure of the unknown and show how you were able to prepare it from the available starting materials.

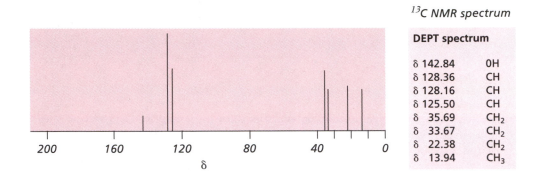

^{13}C *NMR spectrum*

DEPT spectrum

δ 142.84	0H
δ 128.36	CH
δ 128.16	CH
δ 125.50	CH
δ 35.69	CH$_2$
δ 33.67	CH$_2$
δ 22.38	CH$_2$
δ 13.94	CH$_3$

^{1}H *NMR spectrum*

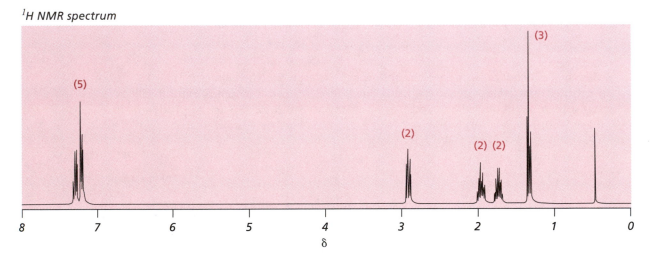

15.34. Propose a reasonable synthesis for each of the following compounds, starting with organic compounds with six or fewer carbon atoms. You may also use any other reagents and solvents. Show a brief retrosynthesis along with the synthesis.

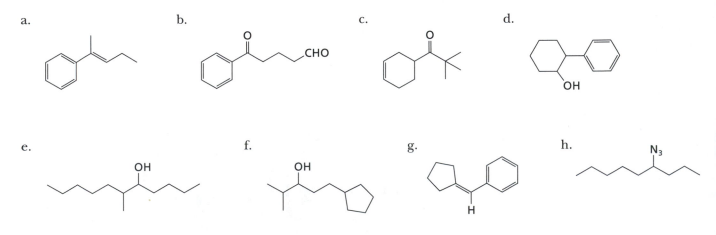

Asymmetric Reactions and Synthesis

An introduction to the concepts of chirality and how they relate to the structures and reactions of organic compounds was presented in Chapter 4. Some important concepts you learned were

- Stereoisomers that are nonsuperimposable mirror images are *enantiomers*, which have identical chemical and physical properties.
- Stereoisomers that are not enantiomers are *diastereomers*, which have different chemical and physical properties.
- A reaction that creates new chiral centers from achiral reactants yields a racemic mixture of products (or a meso compound).

This chapter develops the subject of chirality further by focusing on two topics. The first topic addresses how one quantifies and analyzes the degree of enantioselectivity of organic reactions. Nuclear magnetic resonance spectroscopy is an important analysis tool for this purpose, and it takes advantage of the dissimilar magnetic environments experienced by certain chiral groups in a magnetic field.

The second topic covers the strategies used to prepare enantiomerically pure substances. Because so many molecules in biochemical systems are chiral, an escalating emphasis on the biomedical aspects of organic chemistry has made the synthesis of chiral substances an increasingly important subject.

16.1 CHIRAL COMPOUNDS

16.1a NATURALLY OCCURRING SUBSTANCES CONSTITUTE THE SOURCE FOR MANY CHIRAL COMPOUNDS

Many naturally occurring compounds exist in chiral form, and the collection of such substances is referred to as the **chiral pool,** examples of which are shown in Table 16.1.

The *α-amino acids* are versatile, functionalized compounds that are obtained readily from the hydrolysis of proteins or by the application of microbial processes such as fermentation. Chiral *α-hydroxy carboxylic acids* are generated by many organisms during metabolism. The *alkaloids* (basic, nitrogen-containing compounds), *terpenes* (a class of

Table 16.1 Examples of molecules in the chiral pool.

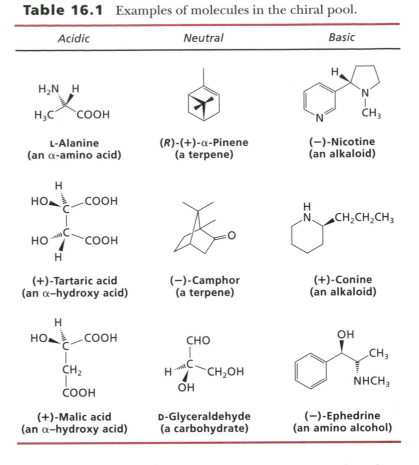

Acidic	Neutral	Basic
L-Alanine (an α-amino acid)	(R)-(+)-α-Pinene (a terpene)	(−)-Nicotine (an alkaloid)
(+)-Tartaric acid (an α–hydroxy acid)	(−)-Camphor (a terpene)	(+)-Conine (an alkaloid)
(+)-Malic acid (an α–hydroxy acid)	D-Glyceraldehyde (a carbohydrate)	(−)-Ephedrine (an amino alcohol)

alcohols, alkenes, and other hydrocarbons derived from isoprene), and many *carbohydrates* are synthesized by plants. Carbohydrates, also called sugars, are especially useful because of their highly substituted carbon skeletons. (Their chemistry will be presented in more detail in Chapter 19.)

The chiral molecules that exist in Nature are often valuable precursors for making other chiral substances. A chiral starting material that reacts by a stereospecific pathway represents a straightforward way to make an enantiomerically pure product.

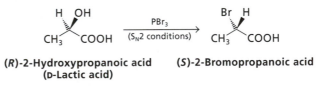

(R)-2-Hydroxypropanoic acid   **(S)-2-Bromopropanoic acid**
(D-Lactic acid)

A chemical transformation that does not affect existing stereocenters in a reactant molecule constitutes another way to make an optically active substance.

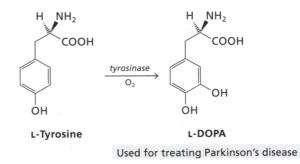

L-Tyrosine   **L-DOPA**

Used for treating Parkinson's disease

Both of these strategies are used routinely to prepare enantiomerically pure compounds. Other methods are available, however, and these will be described throughout the chapter.

16.1b Economic Factors and Physiologic Properties of Drugs Have Stimulated Interest in the Synthesis of Chiral Compounds

For much of the history of chemical synthesis, a product with stereogenic centers has *not* customarily been prepared as a single enantiomer. Instead, strategies have relied on using stereospecific reactions to construct a racemic product in which its *relative* stereochemistry was correct. If the relative stereochemistry could be established, the dogma went, one needed only to start with a chiral reactant to prepare the enantiomer with the desired *absolute* stereochemistry. For example, consider the reaction between an epoxide and a Grignard reagent (Section 15.2c). If a racemic mixture of the epoxide shown below at the left reacts with phenylmagnesium bromide (Reaction **A**), a racemic mixture of the tertiary alcohol product is formed. Notice that the relative orientation between the OH and phenyl groups is the same in each product, namely, trans. If an enantiomerically pure epoxide is used instead (Reaction **B**), then a single enantiomer of product will be formed.

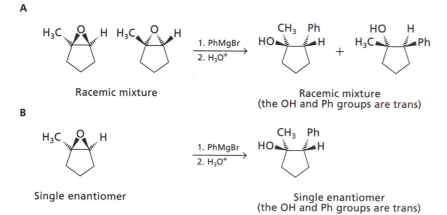

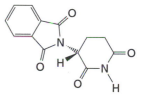

(R)-Thalidomide

For pharmaceutical products, another prevailing notion existed: If one enantiomer of a pair was efficacious in its physiologic action, a patient could simply be given twice as much of a racemic drug and the ineffective isomer would be excreted. For example, the drug (*R*)-thalidomide, shown at the right, is a sedative, which was prescribed for many women in Europe during the middle of the twentieth century. The drug was manufactured as a racemic mixture, however, and tragically, its use led to the birth of many severely deformed children because the (*S*)-enantiomer is highly mutagenic.

Apart from the need to prepare a single isomer in order to prevent unwanted side effects, the synthesis of a chiral material in pure form is desirable from an economic viewpoint, too. The following example compares the mass balance for two reactions: (1) a hydroboration process that generates a racemic mixture; and (2) a procedure using a chiral organoborane reagent that leads to formation of a single enantiomer.

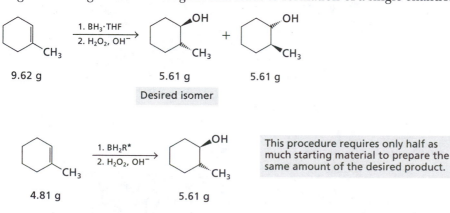

Closely associated with economic concerns are environmental issues. If an unwanted enantiomer is produced, then disposal of that substance has to be arranged.

Both of these concerns—economic and environmental—are related to the *atom economy* of a reaction (Section 15.5a).

16.1c RESOLUTION AND ASYMMETRIC SYNTHESIS ARE TWO METHODS FOR MAKING CHIRAL COMPOUNDS

Faced with the need to obtain a chiral substance from either a racemic mixture or an achiral compound, a chemist has two choices with general applicability. The first method is **enantiomeric resolution.** As noted in the introduction to this chapter, enantiomers have *identical* chemical and physical properties, but diastereomers have *different* properties. As a result, techniques such as crystallization cannot be used to separate a racemic mixture into its enantiomeric constituents. Separation by crystallization or chromatography can be accomplished on diastereomeric mixtures, however. Resolution is a process by which enantiomers are converted to diastereomeric derivatives that can be separated; then the purified compounds are converted back to the pure enantiomeric molecules.

The other route to make an enantiomerically pure compound is a *kinetically controlled asymmetric transformation* or, more simply, an **asymmetric reaction.** Such processes have attracted much interest in recent years because they allow chemists to create enantiomerically enriched products from achiral starting materials using a chiral catalyst or reagent. The 2001 Nobel Prize in chemistry was given for work on this topic.

16.2 ENANTIOMERIC RESOLUTION

16.2a RESOLUTION OF SOME RACEMIC SUBSTANCES MAKES USE OF DIASTEREOMERIC SALTS

In the process that constitutes a classical enantiomeric resolution, a racemate is treated with a chiral reagent to produce a pair of diastereomers. The diastereomers are then separated by a method that makes use of their different physical properties. Once separated, the diastereomers are treated separately with a second reagent to regenerate the resolving agent and the original substrate in its enantiomerically pure forms.

Resolution of a racemic carboxylic acid will illustrate the specific steps in this procedure. Acids and bases are often made into diastereomeric salts, which are separated by crystallization. The resolving agent in this example is 1-amino-1-phenylethane, an amine with a single chiral carbon atom.

The carboxylic acid reacts with the amine by an acid–base reaction, forming diastereomeric ammonium carboxylate salts.

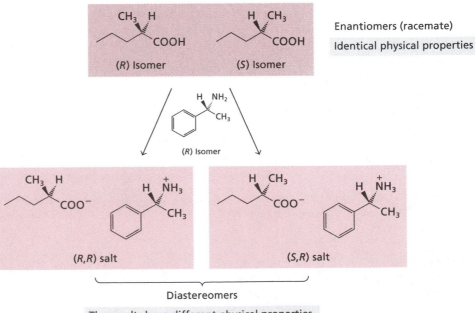

Enantiomers (racemate)
Identical physical properties

(R) Isomer   (S) Isomer

(R) Isomer

(R,R) salt   (S,R) salt

Diastereomers
These salts have different physical properties.

Fractional crystallization takes advantage of the different solubility properties of the salts. In this example, the (S,R) diastereomer remains in solution while the (R,R) isomer forms a salt that crystallizes because of its greater packing efficiency in the solid-state lattice (note that it is impossible to predict a priori which diastereomer will crystallize preferentially).

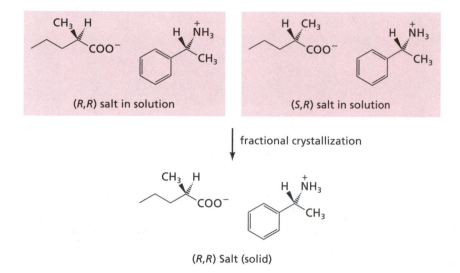

(R,R) salt in solution   (S,R) salt in solution

fractional crystallization

(R,R) Salt (solid)

After isolation by filtration, the pure salt is treated with dilute hydrochloric acid, which regenerates the enantiomerically pure carboxylic acid and liberates the chloride salt of the enantiomerically pure amine. The latter can be recovered by extraction and neutralization.

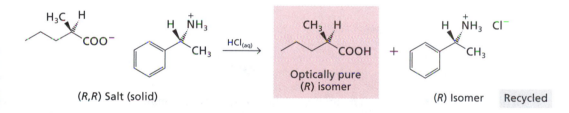

(R,R) Salt (solid)   Optically pure
                                           (R) isomer   (R) Isomer   Recycled

In practice, the resolution of a carboxylic acid is usually accomplished by using amines such as brucine or quinine, which have multiple chiral centers. The salts that form are still diastereomers and can be separated by crystallization.

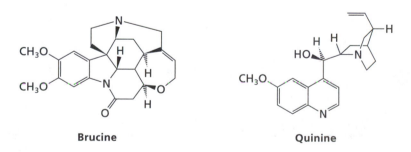

Brucine   Quinine

EXERCISE 16.1

How many stereogenic centers are there in brucine and quinine? Remember that some nitrogen atoms can be chiral (Section 4.2f).

EXERCISE 16.2

Amines and sulfonic acids react to form salts by an acid–base reaction:

$$R—NH_2 + R'SO_3H \rightarrow R—\overset{+}{N}H_3 \ R'SO_3^-$$

Show how racemic 1-amino-1-phenylethane, $PhCH(NH_2)CH_3$, can be resolved using optically active 10-camphorsulfonic acid.

What is the configuration of the carbon atom marked with the asterisk?

10-Camphorsulfonic acid

16.2b RESOLUTION CAN BE ACCOMPLISHED BY PREPARING DIASTEREOMERIC MOLECULES

Another way to carry out an enantiomeric resolution is to make a molecular derivative, rather than a salt. For example, enantiomeric alcohols can be converted to esters by treatment with an optically active acid chloride (a reaction presented in Section 21.3a).

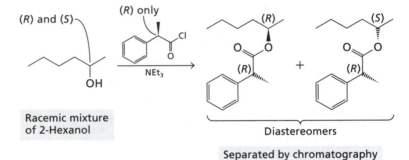

Racemic mixture of 2-Hexanol

Diastereomers

Separated by chromatography

These esters are diastereomers and can be separated by chromatography, a technique amenable for use with neutral substances. After chromatography, the ester is hydrolyzed (a reaction presented in Section 21.4d) to regenerate the enantiomerically pure carboxylic acid, as its salt. Only the recovery of the (R) enantiomer is illustrated below.

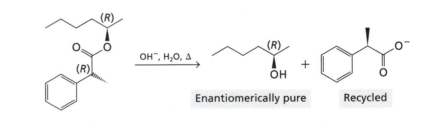

Enantiomerically pure

Recycled

16.3 ASYMMETRIC SYNTHESIS

16.3a A KINETICALLY CONTROLLED ASYMMETRIC TRANSFORMATION RELIES ON DIFFERENCES IN TRANSITION STATE ENERGIES

An *asymmetric reaction* (also called an asymmetric synthesis) is the other general method for preparing enantiomerically pure compounds from achiral precursors. Asymmetric reactions exploit chiral reagents and catalysts to create transition states with unequal energies as a result of spatial interactions between the reactants.

In the abstract, we can represent an asymmetric reaction by the following scheme, in which an achiral substance reacts with a chiral catalyst, generating diastereomeric transition states and enantiomeric products.

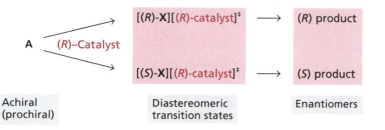

As with diastereomers, which have different physical properties, diastereomeric transition states have dissimilar energies. The enantiomeric products, however, have identical energies. The enantiomer that predominates will be generated via the transition state with the lower energy. Figure 16.1 illustrates the energy profiles for the reactions shown above, in which the (S) product is the major one.

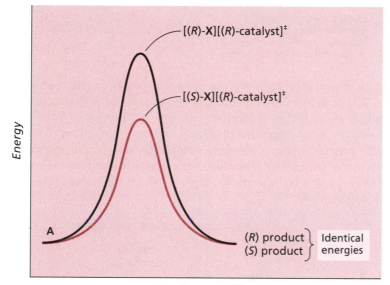

Figure 16.1

A reaction coordinate diagram for the reaction between achiral compound **A** and a chiral catalyst that preferentially yields the (S) product. The transition state structures are diastereomeric species.

16.3b A PROCHIRAL CARBON ATOM HAS TWO APPARENTLY IDENTICAL SUBSTITUENTS

Do all reactions between a chiral and achiral reactant take place via diastereomeric transition states? No! For that to happen, the achiral substance actually has to be **prochiral,** which means that it is readily converted to a chiral substance by a single substitution or addition process. A prochiral carbon atom is one in which three of its four groups are different.

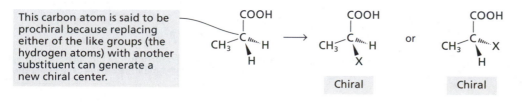

This carbon atom is said to be prochiral because replacing either of the like groups (the hydrogen atoms) with another substituent can generate a new chiral center.

A carbon atom that has three like groups is *not* prochiral; instead, it is **homotopic.**

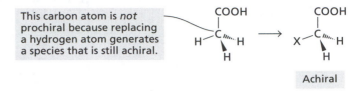

This carbon atom is *not* prochiral because replacing a hydrogen atom generates a species that is still achiral.

Achiral

In discussing prochiral substances, we differentiate the two like groups by labeling them—**pro-(R)** or **pro-(S)**—according to which configuration is produced by a particular reaction.

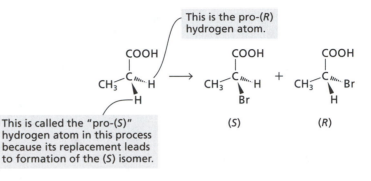

This is the pro-(R) hydrogen atom.

This is called the "pro-(S)" hydrogen atom in this process because its replacement leads to formation of the (S) isomer.

(S)   (R)

Notice that the designation of pro-(R) or pro-(S) depends on the reaction that is being carried out. The priority of the incoming group compared with that of the others present determines whether the replacement process can produce (R) or (S).

If an *achiral* molecule contains prochiral carbon atoms, then the substituents attached to the prochiral center are **enantiotopic** because replacing each of the like groups leads to the formation of enantiomers. The allylic bromination reaction (Section 12.1f) shown below is a specific example in which the organic reactant has enantiotopic hydrogen atoms.

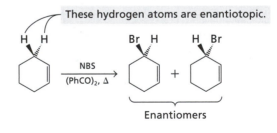

These hydrogen atoms are enantiotopic.

Enantiomers

EXERCISE 16.3

For the bromination reaction shown directly above, assign the pro-(R) and pro-(S) designations to the allylic hydrogen atoms, which are enantiotopic.

If there is already a stereogenic center in the molecule, then replacement of one of two like substituents on a prochiral carbon atom will produce diastereomers. As a result, the two allylic hydrogen atoms in the following reaction are said to be **diastereotopic.**

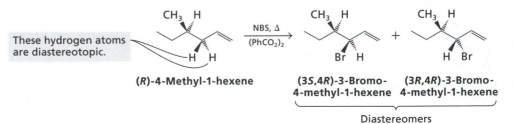

These hydrogen atoms are diastereotopic.

(R)-4-Methyl-1-hexene   **(3S,4R)-3-Bromo-4-methyl-1-hexene**   **(3R,4R)-3-Bromo-4-methyl-1-hexene**

Diastereomers

Diastereotopic groups on a prochiral carbon atom have the same identity; however, they are distinct from each other because they reside in dissimilar environments in relation to substituents attached to the chiral center(s) already present in the molecule. This nonequivalence can be seen by looking at the Newman projections for three of the limiting conformations in methyl (S)-2,3-dibromo-2-methylpropanoate. No matter which conformation is considered, the two hydrogen atoms, H_a and H_b, have different relationships (*gauche* or *anti*) relative to the groups or atoms at C2

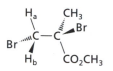

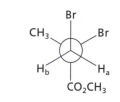

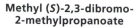

Methyl (S)-2,3-dibromo-2-methylpropanoate | H_a is *gauche* to the CH_3 group H_b is *anti* to the CH_3 group | H_a is *anti* to the CH_3 group H_b is *gauche* to the CH_3 group | H_a is *gauche* to the CO_2CH_3 group H_b is *anti* to the CO_2CH_3 group

For diastereotopic protons, this spatial nonequivalence can manifest itself by producing different chemical shifts in the ^1H NMR spectrum, which is shown for methyl (S)-2,3-dibromo-2-methylpropanoate in Figure 16.2. The chemical shift difference can range from negligible to substantial, depending on the distance between the chiral and prochiral carbon atoms.

Diastereotopic protons are not magnetically equivalent (Section 13.1b), so they can engage in spin–spin splitting *with each other*. This coupling is called *geminal coupling* because the protons are separated by only two bonds instead of the three bonds that give rise to vicinal coupling (Section 13.2a). The geminal coupling constant, ^2J, has a value that is normally between 1 and 18 Hz; for the methylene protons in methyl 2,3-dibromo-2-methylpropanoate, this value is ~ 10 Hz.

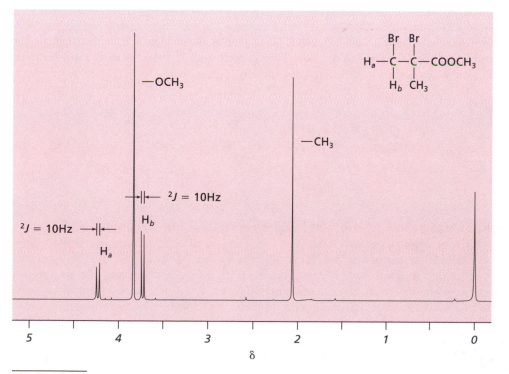

Figure 16.2
The 300-MHz ^1H NMR spectrum of methyl 2,3-dibromo-2-methylpropanoate. The assignments of the signals to H_a and H_b were made arbitrarily. The value of the geminal coupling constant, ^2J, is 10 Hz.

The terms *prochiral*, *enantiotopic*, and *diastereotopic* can also be applied to molecules with double bonds. A prochiral double bond is one that undergoes addition to yield a chiral product. Prochirality is possible with a π bond because addition can occur to either the top or bottom face. In prochiral double bonds, the face of the π bond is either enantiotopic or diastereotopic, depending on whether addition yields enantiomers or diastereomers. In the following reaction, for example, the alkene shown has an enantiotopic carbon–carbon double bond because hydroboration followed by oxidative hydrolysis leads to formation of enantiomeric alcohols.

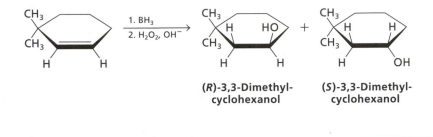

(*R*)-3,3-Dimethyl-
cyclohexanol

(*S*)-3,3-Dimethyl-
cyclohexanol

EXERCISE 16.4

For the reaction shown directly above, assign pro-(*R*) and pro-(*S*) to the two faces of the double bond.

16.3c THE ENANTIOMERIC EXCESS VALUE QUANTIFIES THE EFFICACY OF AN ASYMMETRIC REACTION

As noted in Section 9.1d, the synthesis of enantiomerically enriched compounds falls into one of two categories: *enantioselective* (more of one enantiomer than the other is formed) or *enantiospecific* (100% of one enantiomer is formed). As a way to quantify how well an asymmetric reaction proceeds, we express the degree of enantioselectivity as its **enantiomeric excess,** abbreviated **ee.** This value is defined as the proportion of the major enantiomer minus that of the minor enantiomer, expressed as a percentage. Thus, the transformation shown in the following equation occurs in 90 – 10% = 80% ee.

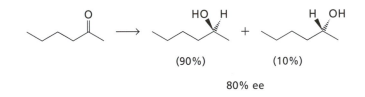

(90%)   (10%)

80% ee

A reaction that proceeds with 100% ee is an enantiospecific reaction.

If the ee of a reaction is > 90%, then it is often possible to obtain a pure enantiomer by crystallization, assuming that the product is a solid, of course. Remember, you cannot isolate a pure enantiomer from a *racemic mixture* by crystallization, but an unequal mixture of enantiomers, called a **scalemic mixture,** acts as if it were a pure compound (the major enantiomer) containing an impurity (the minor enantiomer).

EXERCISE 16.5

If a reaction produces a mixture of diastereomers, we can define *diastereomeric excess*, or *de*, in the same way that we define ee. What is the de of the following process that involves addition of HBr to the alkene double bond?

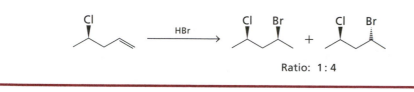

Ratio: 1 : 4

16.3d A KINETIC RESOLUTION RESULTS FROM UNEQUAL REACTION RATES BETWEEN RACEMIC MOLECULES AND A CHIRAL REAGENT

Another type of asymmetric process is **kinetic resolution,** which occurs when one enantiomer of a racemic mixture reacts faster than the other. This preferential reaction occurs when the transition state for the interaction of one enantiomer with the catalyst (or reagent) is much lower in energy than that for the other enantiomer (cf. Fig. 16.1). In a kinetic resolution, therefore, one enantiomeric starting material is converted to product, and the other isomer does not react (or reacts slowly). This process is shown schematically below:

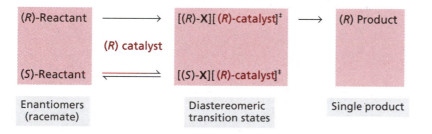

A specific example of kinetic resolution is the process by which lactate is converted to pyruvate, a reaction that occurs in the biosynthesis of sugars. Previously you have seen that enzymes oxidize alcohols to aldehydes and ketones (Section 11.4e). *Lactate dehydrogenase* (LDH) is the enzyme that catalyzes the oxidation of lactate to pyruvate, with NAD^+ as a coenzyme. If you were to mix LDH, NAD^+ and a racemic mixture of lactate together, only L-lactate would react. D-Lactate would be recovered at the end of the reaction.

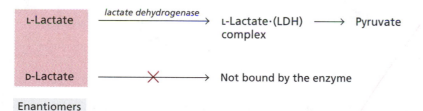

This selectivity is determined by the architecture of the LDH active site. In general, we know that enzyme active sites are chiral, because they are made from amino acids, which, except for glycine, are chiral (Section 5.5b). For *lactate dehydrogenase,* the structure of the enzyme has been determined precisely. It is known that the side chain of histidine-195 (His_{195}) forms a hydrogen bond with the OH group of L-lactate, and the side chain of arginine-171 (Arg_{171}) forms two hydrogen bonds with the carboxy group of the L-lactate ion. These three hydrogen bonds allow the enzyme to recognize and bind the substrate well. The drawing in the top part of Figure 16.3 illustrates the active site structure and shows the mechanism by which the hydrogen atom of L-lactate is transferred to the NAD^+ coenzyme.

When D-lactate enters the active site of LDH, the fit between this substrate and both His_{195} and Arg_{171} is imperfect. Two possibilities can be imagined:

- If the enzyme active site is rigid, as shown in Figure 16.3b (**A**), then the hydrogen bond between substrate and His_{195} is either weak or nonexistent, and the substrate may bind too weakly to stay long enough to react.

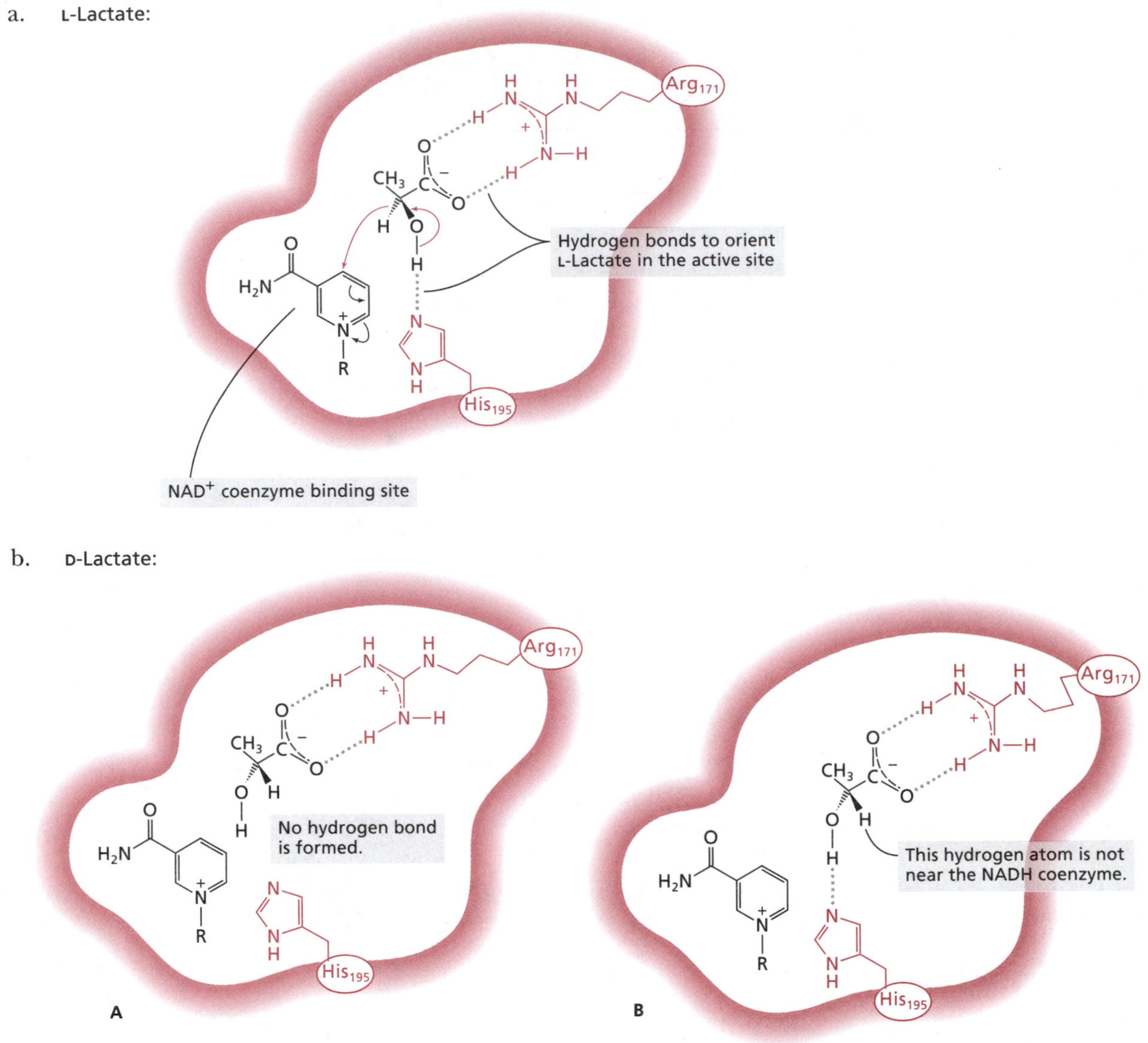

a. L-Lactate:

Hydrogen bonds to orient L-Lactate in the active site

NAD+ coenzyme binding site

b. D-Lactate:

No hydrogen bond is formed.

A

This hydrogen atom is not near the NADH coenzyme.

B

Figure 16.3
(a.) A schematic representation of the active site of *lactate dehydrogenase* and its binding to and reaction with L-lactate. (b.) Schematic representations of the possible binding modes of *lactate dehydrogenase* to D-lactate.

- If the active site can bend to accommodate the formation of both sets of hydrogen bonds, as shown in Figure 16.3b (**B**), then the hydrogen atom to be transferred to NAD+ is in the wrong position.

In either event, oxidation is either slow or does not take place, and D-lactate is recovered at the end of the reaction.

Kinetic resolution processes are widespread among reactions of chiral molecules, and it is easy to mistake kinetic resolution for a true asymmetric reaction. In each case, the ee of the reaction will be > 0%. For a kinetic resolution, however, the *chemical yield* can be only 50%, at most.

16.3e PROTON NMR SPECTROSCOPY IS A USEFUL ANALYTIC TOOL TO MEASURE THE EE OF AN ASYMMETRIC REACTION

Through the 1970s, the principal method used to judge enantioselectivity required one to measure the optical rotation (Section 4.2b) of a reaction mixture. The ratio of that number and the known value for the pure substance could be translated to a quantity called the *optical purity,* an expression that is still used. This approach to quantify the enantioselectivity of a reaction has several shortcomings, the most detrimental being that even small amounts of chiral impurities can affect the rotation of plane-polarized light, which alters the ratio being calculated. Consequently, many values reported in the older literature are in error.

The intervening years have seen the development of other ways to determine the enantiomeric purity of a mixture. For example, techniques using gas or liquid chromatography with chiral supports provide rapid methods of analysis, especially for studying a series of related compounds. Once an appropriate chromatographic support and solvent system have been selected, the product mixture can be injected onto the column and the results will be available in minutes. The analysis is based on the differential interactions of the enantiomers with the support, leading to different retention times for the two isomers. Normal detection methods are employed to quantify the amount of each isomer present. *Techniques that utilize chromatography are now the most widely used ways to measure the enantioselectivity of reactions.*

Although they are sometimes more time consuming, techniques that employ NMR spectroscopy constitute another method to determine the enantiomeric purity of a mixture. These strategies make use of the fact that diastereomeric nuclei have different chemical shift values. Analysis by NMR spectroscopy requires that the enantiomers first be transformed into diastereomeric derivatives.

One means to accomplish derivatization is to prepare covalently modified products. α-Methoxy-α-trifluoromethylphenylacetic acid, MTPA (also called "Mosher's acid"), has been the derivative of choice to analyze alcohols and amines, which react readily to form esters and amides, respectively.

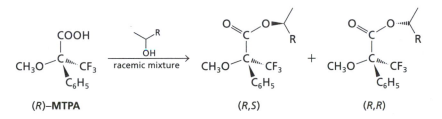

Once the MTPA derivatives have been prepared, either the ^{19}F or the ^{1}H NMR spectrum can be recorded. In the ^{19}F spectrum, the CF_3 group of each diastereomer will have a separate chemical shift, differing by as much as δ 0.3–0.7. This difference between the chemical shift values is designated $\Delta\delta$. If the separation is sufficiently distinct, then the integrated intensity values of the two signals provide the ratio between the amounts of each enantiomer. If this ratio is 1:1, then you are looking at a racemic mixture. If the ratio is 9:1, then the mixture consists of 90% of one enantiomer and 10% of the other. The ee value is the difference between these percentage values: 90 − 10% = 80% ee. Employing ^{1}H NMR spectroscopy, you can make the same type of measurement by observing the signals attributed to the CH_3 portion of the methoxy groups. In this case, the value of $\Delta\delta$ is between 0.1 and 0.2 ppm.

EXERCISE 16.6

Mosher's acid is not the only compound used to measure enantiomeric purity. A class of chiral substances having a phosphorus atom allows the use of ^{31}P NMR spectra ($\Delta\delta_P$ = 0.1 − 0.8). Assign the configurations of the three chiral centers in the chlorophosphorus reagent (below, left).

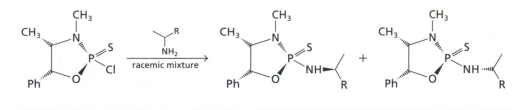

EXERCISE 16.7

A reaction that you perform yields a scalemic mixture of alcohols that is then treated with MTPA to produce diastereomeric esters. The ^{19}F NMR spectrum is recorded, and the results are shown below. What is the ee of this reaction? The integrated intensity values are given in parentheses.

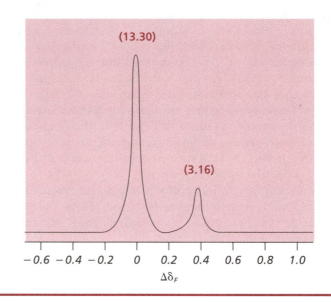

16.4   ENANTIOSELECTIVE ADDITION REACTIONS

16.4a   REAGENTS AND CATALYSTS USED FOR ENANTIOSELECTIVE REACTIONS RARELY HAVE PREDICTABLE STRUCTURES

Literally hundreds of different asymmetric reactions have been discovered and used to make chiral molecules. Our attention here will focus on the addition reactions of alkenes, which provide some of the best cases for enantioselection. A summary of the reagents and catalysts that will be described is given in Table 16.2. These reagents and catalysts take advantage of the dissimilar shapes of the two faces of a double bond with respect to the shape of the reagent. Their mutual interactions give rise to diastereomeric transition states, which provide the basis for enantioselection.

To design a reagent that will react predictably with either face of a substrate π bond is not an easy task. So far, the only way to tell whether a reagent will produce a suitable degree of ee is to prepare the reagent and test it. With results in hand, the structure of the reagent can sometimes be redesigned. There are cases in which such improvements have led to the discovery of more enantioselective reagents. Yet, even with computer programs designed to calculate the relative energies of possible interactions, the ability to design a de novo reagent is limited. Detailed explanations of how a reagent works are often little more than rationalizations, so there is much work to be done to understand and define molecular interactions that lead to enantioselectivity.

The following sections outline several useful enantioselective reagents and catalysts that have been reported in the literature. Rather than focusing on precisely which stereoisomer of a particular reagent is used to form which enantiomer of product, these presentations will deal with the general characteristics of the reagents and their reactions. The assumption is being made that the desired stereochemistry of a particular product can be obtained by using one stereoisomer or the other of the reagent.

Table 16.2 Enantioselective addition reactions and some of the reagents or catalysts that are used in asymmetric reactions.

Reaction type	Reagent or catalyst[a,b]	Alkene type 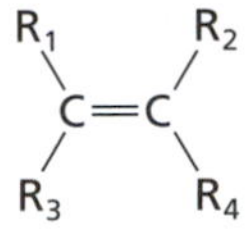
Hydrogenation (Section 16.4b)	$[Rh(*P\sim P*)_2(CH_3OH)_2]^+$ H_2, CH_3OH	R_1 or R_3 must be Ar, and R_2 must be $-NHCOR'$
	$[BINAP*]Ru(OAc)_2$ H_2, CH_3OH	One of the R groups should be $-COOH$, $-COOR'$, or $-CH_2OH$
Hydroboration (Section 16.4c)	$(Ipc*)BH_2$	cis-Disubstituted and trisubstituted
Epoxidation (Section 16.4d)	$Mn(salen*)Cl$, $NaOCl$	cis-Disubstituted and trisubstituted; best when one of the R groups is aryl
	$FR*$, HSO_5^-, H_2O, CH_3CN	trans-Disubstituted and trisubstituted
Dihydroxylation (Section 16.4e)	AD-mix*, t-BuOH, H_2O $(CH_3SO_2NH_2)$	trans-Disubstituted and trisubstituted > cis-disubstituted > terminal

Alkene type column header:

$$\underset{R_3 \quad R_4}{\overset{R_1 \quad R_2}{C=C}}$$

[a]Note: The asterisks in the reagent structure indicate where the chirality centers are located.

[b]Abbreviations used: BINAP = 2,2'-Bis(diphenylphosphino)-1,1'-binaphthyl; $(Ipc*)BH_2$ = (isopinocamphenyl)borane; salen = bis(salicylidene)ethylenediamine; FR = fructose; AD = asymmetric dihydroxylation.

16.4b ENANTIOSELECTIVE HYDROGENATION REACTIONS MAKE USE OF HOMOGENEOUS CATALYSTS

The catalytic hydrogenation of an alkene—addition of H_2 to a double bond—is a reaction type that attracted early attention for development of an enantioselective variant. William S. Knowles, who shared the 2001 Nobel Prize in chemistry for work in this area, carried out his pioneering research at the Monsanto Company during the 1960s and discovered a useful method to prepare L-DOPA, which has been used for the treatment of Parkinson's disease. He took advantage of the known homogeneous hydrogenation of alkenes (Section 15.3e) to explore how chiral phosphines (designated as *P~P* in the following equation and illustrated below at the right) attached to rhodium would influence the course of the reaction. The ligands are not the simple triphenylphosphine groups found in Wilkinson's catalyst (Section 15.3e). Instead, some are chiral at phosphorus (Section 4.2f) and others are chiral at the carbon atoms attached to phosphorus. The biggest drawback to their use is their lack of versatility: Only alkenes with an aryl ring and amido group attached to the double bond are reduced with considerable enantioselectivity.

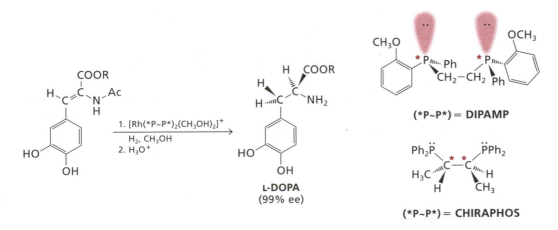

(*P~P*) = DIPAMP

(*P~P*) = CHIRAPHOS

L-DOPA (99% ee)

Rhodium is not the only metal that is useful for the homogeneous hydrogenation of alkene π bonds, however. Ryoji Noyori, Professor of Chemistry at Nagoya University in Japan and a corecipient of the 2001 Nobel Prize in chemistry, has developed other types of catalysts during the past 30 years. These catalysts contain ruthenium and have phosphine ligands different from those shown previously.

The chirality of the ligands in Noyori's complexes depends on atropisomerism—the restricted rotation between aryl rings (Section 4.3c). The naphthalene rings in BINAP cannot be coplanar because of steric interactions between the rings and the diphenylphosphino groups themselves. If you make a model, you will see that (*S*)-BINAP is not superimposable on its mirror image, (*R*)-BINAP.

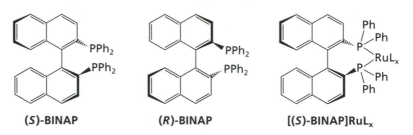

(*S*)-**BINAP**   (*R*)-**BINAP**   [(*S*)-**BINAP**]**RuL**$_x$

Either isomer of BINAP binds to a ruthenium(II) ion along with ancillary ligands such as carboxylate and halide ions and amines (all included as L$_x$ in the structure above). These complexes are useful catalysts for the catalytic hydrogenations of alkenes.

Like the rhodium-based catalysts just mentioned, these ruthenium complexes catalyze the addition of H$_2$ across π bonds by a combination of oxidative addition, migratory insertion, and reductive elimination mechanisms (Section 15.3e). The mechanisms are not as clearly delineated, however, and several possible schemes have been proposed. Besides its carbon–carbon double bond, the substrate alkene must have a group that can bind to the metal ion. Unsaturated carboxylic acids, esters, and alcohols are good substrates, and the following reactions show some specific examples:

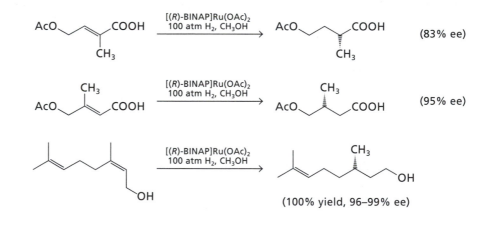

EXAMPLE 16.1

Draw the structure of the substrate and indicate which reagent you would use to prepare the following compound by means of catalytic hydrogenation. Assign the configuration of the stereogenic center in the product molecule.

To choose the substrate that forms the given product upon hydrogenation, you first identify the hydrogen atoms that will be added (assume syn addition of H$_2$ to the alkene), which in turn defines where the double bond is located. For this molecule,

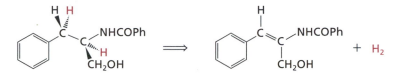

Next, consider what type of alkene the starting material is, and therefore which catalyst(s) can be used. Because this alkene is substituted with both an aryl and an amido group and because it is also an allylic alcohol, either reagent combination listed in Table 16.2 can be used. As noted in Section 16.4a, we will not specify the enantiomer of the catalyst that is needed to produce the product enantiomer shown. The assumption is made, for example, that the use of [(*S*)-BINAP]Ru(OAc)$_2$ will yield one enantiomer of the product, and [(*R*)-BINAP]Ru(OAc)$_2$ will yield the other. (In actually performing this reaction, we would use whichever catalyst yields the desired result.)

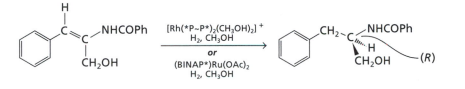

EXERCISE 16.8

Draw the structure of the substrate and indicate which catalyst you would use to prepare (*S*)-naproxen (an over-the-counter analgesic) by catalytic hydrogenation.

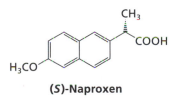

(*S*)-Naproxen

16.4c ENANTIOSELECTIVE HYDROBORATION IS A PRACTICAL WAY TO PREPARE CHIRAL ALCOHOLS FROM ALKENES

Enantioselective variations of the hydroboration reaction (Section 9.4a) have also been studied for many years. Only one hydroboration reagent is described in this section, but dozens exist, tailored to specific applications and substrates. Asymmetric hydroboration is an example of a transformation in which the chiral reagent reacts stoichiometrically, not catalytically.

When borane (BH$_3$) is treated with naturally occurring α-pinene, the molecule (Ipc*)BH$_2$ can be isolated. Because of the shape of the bicyclic ring system, the boron atom attaches to the less hindered face of this alkene, as shown in the following equation. The (+) isomer of α-pinene yields the (*R*) reagent, and (−)-α-pinene forms the (*S*) reagent.

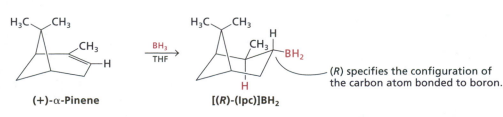

The (Ipc*)BH$_2$ reagent reacts with alkenes in the way that any borane reagent does. Because it is chiral, however, it reacts differently with the two faces of RCH=CHR, so it forms diastereomeric organoborane products of the following type: (Ipc*)BHR*. These products retain a hydrogen atom attached to boron, so they actually form dimers via three-center, two-electron bonds (Section 9.4a). An advantage to dimer formation is that the compound can be obtained as a single, pure isomer after crystallization.

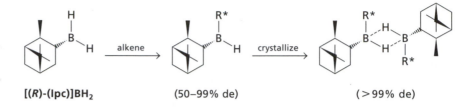

[(R)-(Ipc)]BH$_2$ (50–99% de)   (>99% de)

Once purified, the dimer is converted to monomeric chiral species by treatment with acetaldehyde. This step generates a diethyl(alkyl)boronate product and two molecules of α-pinene, which can be recycled. No loss of chirality takes place during this step, so the alkylboronate is enantiomerically pure.

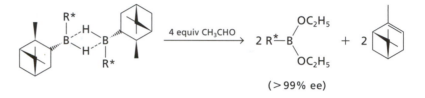

(>99% ee)

Two examples of the overall procedure are,

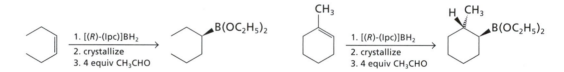

The boronates obtained by this three-step procedure are readily oxidized with hydrogen peroxide and base to yield alcohols that are also enantiomerically pure (Section 9.4c). The boronates can also be used directly in the Suzuki reaction (Section 15.3f) to form carbon–carbon bonds. Remember that replacement of the boron atom by other groups occurs with retention of configuration at the carbon atom to which the boron group is attached.

EXAMPLE 16.2

Draw the structure of the alkene needed to make the following alcohol by hydroboration followed by oxidative hydrolysis with H$_2$O$_2$/OH$^-$. Assign the configuration of the stereogenic center in the product molecule.

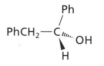

First, expand the structure of the product. Hydroboration occurs by syn-addition (Section 9.4a), and replacement of the boron group with OH occurs by retention (Section 9.4c), so the H and OH groups to remove in a retrosynthetic analysis have to be syn. Note that two combinations of H and OH have to be considered because the methylene group has two H atoms.

Next, consider the selectivity of (Ipc*)BH$_2$. This reagent gives higher enantioselection with cis-alkenes than with trans-alkenes. The reaction is therefore written as follows:

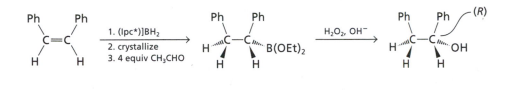

EXERCISE 16.9

Draw the structure of the alkene and write the equations showing the preparation of each of the following chiral alcohols by hydroboration followed by oxidation and hydrolysis with H$_2$O$_2$/OH$^-$:

a.

b.

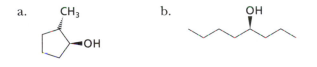

16.4d OXO-TRANSFER CATALYSTS ARE USED FOR THE ENANTIOSELECTIVE EPOXIDATION REACTIONS OF ALKENES

Epoxides are valuable reactants for use in synthesis because they undergo ring opening with nucleophiles in a regio- and stereoselective fashion (Sections 15.2c and 15.3b). Chiral epoxides are therefore important building blocks for making enantiomerically pure compounds. You learned in Section 11.3e that epoxidation of an alkene is readily accomplished by treating an alkene with a peracid. The chirality in a peracid molecule is far away from the alkene π bond during the oxygen-atom transfer step, however; so use of a chiral peracid usually results in poor enantioselection. Nevertheless, catalysts that can transfer an oxygen atom to an alkene π bond have been discovered and developed.

In 1990, Eric Jacobsen reported that a chiral manganese-containing catalyst leads to the efficient and mild epoxidation of alkenes with high ee values using the inexpensive cooxidant NaOCl. The catalyst is made using the ligand **L** (shown below and commonly referred to as "salen"), which binds manganese in the +3 oxidation state. This species, **1**, reacts with NaOCl to form the oxidizing species, formally a manganese(V)oxo compound, **2**.

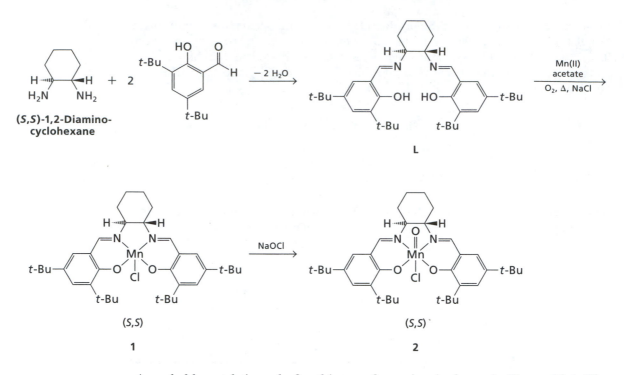

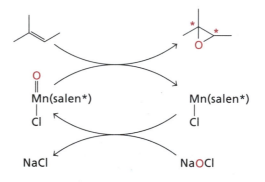

A probable catalytic cycle for this transformation is shown in Figure 16.4. The alkene reacts with the Mn(salen*)(O)Cl reagent to form the corresponding epoxide and Mn(salen*)Cl. This metal complex reacts with NaOCl to regenerate the oxo intermediate and to produce sodium chloride.

Figure 16.4
A reaction scheme for the Mn(salen*)-catalyzed epoxidation of alkenes.

Table 16.3 summarizes some specific results for epoxidation of alkenes using NaOCl and the catalyst made from the (S,S) ligand. The (R,R) catalyst reacts with these same alkenes to form their enantiomeric products.

Table 16.3   Epoxidation reactions of alkenes with NaOCl catalyzed by Mn[(S,S)-salen]Cl.

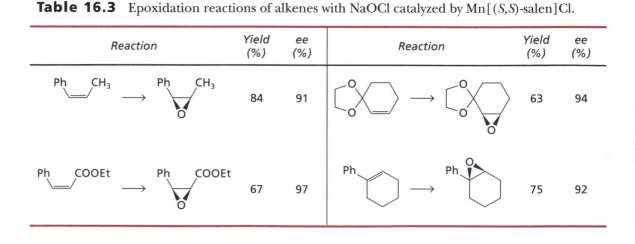

Reaction	Yield (%)	ee (%)	Reaction	Yield (%)	ee (%)
Ph, CH₃ → Ph, CH₃ (epoxide)	84	91	(dioxolane cyclohexene) → (epoxide)	63	94
Ph, COOEt → Ph, COOEt (epoxide)	67	97	Ph (cyclohexene) → Ph (epoxide)	75	92

The biggest drawback to using Mn(salen*) complexes for direct epoxidation is the lack of selectivity in the epoxidation reactions of trans-alkenes. Fortunately, a different type of reagent can be utilized to epoxidize trans- and trisubstituted alkenes.

Professor Yian Shi and co-workers at the Colorado State University have exploited chiral dioxiranes that are generated when derivatives of D- or L-fructose are treated with the persulfate ion.

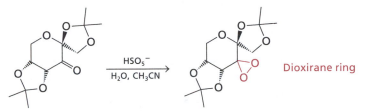

You learned in Section 11.5 that dioxiranes convert amines to nitro compounds; they are able to epoxidize alkenes, too. The great advantage of the chiral dioxiranes derived from fructose derivatives is that both enantiomers are readily prepared from the inexpensive sugars D-fructose and L-sorbose.

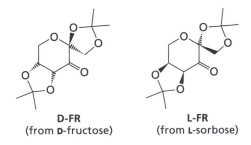

D-FR
(from **D**-fructose)

L-FR
(from **L**-sorbose)

The epoxidation reaction, which is general for many types of alkenes, is accomplished by stirring the alkene, the ketone FR*, and persulfate ion (as Oxone) in aqueous acetonitrile.

Notice that the D-isomer of the catalyst yields an epoxide product with the configuration opposite that of the epoxide formed using the L-isomer:

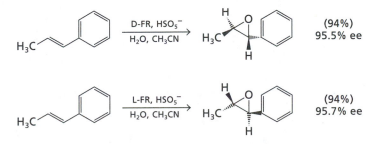

Two additional examples, shown below, demonstrate that functional groups can be tolerated and that aryl groups are not required to be present in the substrate molecule.

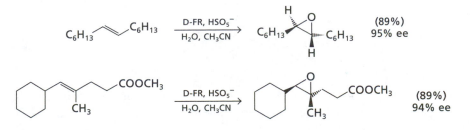

EXERCISE 16.10

Draw the structure of the alkene and specify the reagent that can be used to prepare each of the following epoxides:

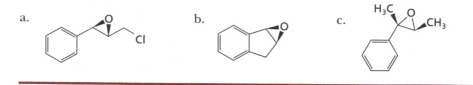

a.   b.   c.

16.4e The Asymmetric Dihydroxylation of Alkenes Makes Use of Osmium Tetroxide and Chiral Amine Ligands

The dihydroxylation of alkenes with osmium tetroxide, which produces vicinal diols (Section 11.3c), takes place with high enantioselection if chiral amine ligands are added to a reaction mixture that contains potassium osmate and potassium ferricyanide, a cooxidant that generates OsO_4 in situ. Professor K. Barry Sharpless, who shared the 2001 Nobel Prize in chemistry with Knowles and Noyori (Section 16.4b), discovered and developed this catalytic reagent. The ligands are made by coupling a linker molecule with a chiral amine made from the cinchona alkaloids quinine and quinidine. The reagents, which are commercially available, are termed AD-mix α and AD-mix β, and their actual compositions are shown below. AD means asymmetric dihydroxylation, and α and β signify which face of the double bond undergoes addition under prescribed conditions (the details of which we will ignore).

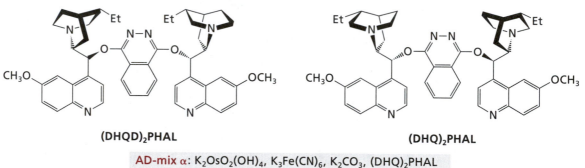

(DHQD)₂PHAL   **(DHQ)₂PHAL**

> **AD-mix α:** $K_2OsO_2(OH)_4$, $K_3Fe(CN)_6$, K_2CO_3, $(DHQ)_2PHAL$
>
> **AD-mix β:** $K_2OsO_2(OH)_4$, $K_3Fe(CN)_6$, K_2CO_3, $(DHQD)_2PHAL$

The dihydroxylation reaction is general for most alkene types, and simply using the other reagent mix switches the stereochemical outcome of the transformation. The ee values are not identical, however, because $(DHQ)_2PHAL$ and $(DHQD)_2PHAL$ are *diastereomers*, not enantiomers. The degree of enantioselectivity is sufficiently high that either isomer of the diol product can be made in enantiomerically practical yields. For example,

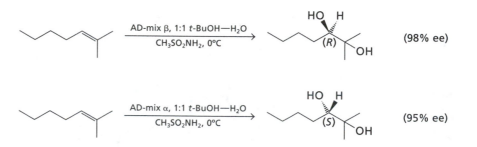

Terminal alkenes do not react as enantioselectively as di- and trisubstituted alkenes, and cis-disubstituted olefins are dihydroxylated less selectively than trans-isomers.

(More recently developed ligands give better results with terminal olefins.) The following examples will give you an idea about what results can be expected using the *commercial* AD-mix reagents. The chemical yields are very good to excellent, so they do not impose a limitation on the procedure.

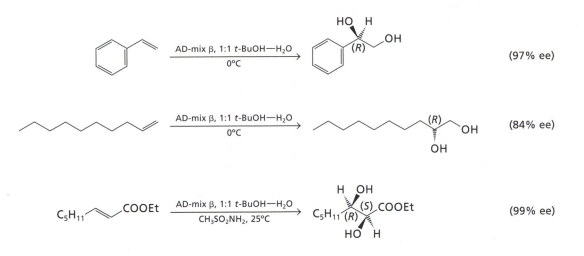

The high enantioselectivity created by these reagents can be rationalized by considering the possible shapes of the active reagent.

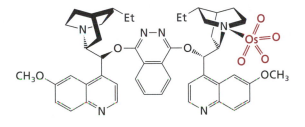

It is known that osmium binds to the *sp³*-hydbridized nitrogen atom of the ligand. This coordinated species is much more reactive than osmium tetroxide itself, adding oxygen atoms across the π bond of the alkene according to the mechanism presented in Section 11.3c. The ligand is thought to create a cavity in which the olefin can add at only one of its faces to the osmium center. Note that the ligand has approximate twofold symmetry, so it does not matter which *sp³*-hydbridized nitrogen atom binds to the osmium tetroxide molecule, as long as the stoichiometry of metal to ligand is 1:1. The reagent formed after binding the metal ion no longer has twofold symmetry, so the chiral pocket that forms will interact with the alkene in a single orientation to create the enantioselective bias.

16.4f   ENZYMES CAN BE USED IN THE LABORATORY TO CARRY OUT ENANTIOSELECTIVE TRANSFORMATIONS

The previous sections have described a handful of reagents that have applications for asymmetric synthesis. Besides processes that take advantage of these synthetic reagents, there are cases in which an enzyme can be used in vitro (i.e., outside of its natural living environment) to perform desired transformations. The principal advantages found in utilizing enzymic catalysts to make organic molecules are the high degrees of chemo-, regio-, and enantioselectivity that can be expected.

There are also disadvantages, however. One is the limited number of substrates that many enzymes will accept. In fact, certain substrates and products act as inhibitors for some reactions. Other limitations relate to the nature of the solvents that can be used. Most enzymes function best in water or aqueous buffer at a specific pH value, and these conditions may not be compatible with the solubility properties of an organic compound. Discovering ways to employ enzymes in organic solvents is an active area of research.

In the context of the reactions you have just seen, consider the dihydroxylation of benzene and its derivatives. This process is difficult to do in the laboratory, but microbial oxidation employing a particular strain of an organism called *Pseudomonas putida* is specific for the formation of the isomer shown in the following equation.

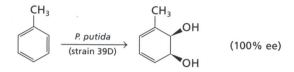

Chloro- and bromobenzene react in a similar fashion, and the diol products can be used to make a variety of substances such as carbohydrates. Notice that if you want to make the enantiomer of the product, you cannot simply alter the enzyme as you can a reagent. Instead, you would have to search for a different enzyme, perhaps even from another organism. Moreover, such an enzyme might not even exist.

16.5 SYNTHETIC DESIGN OF CHIRAL MOLECULES

16.5a RETROSYNTHETIC DESIGN OF CHIRAL MOLECULES REQUIRES ADDITIONAL PLANNING

With even the limited repertoire of asymmetric reactions and concepts of reaction chemistry that have been covered in this chapter, you can consider how to incorporate these ideas into the planning of synthetic strategies. The concepts of retrosynthesis presented in Section 15.4 need only to be modified to consider when it is best to incorporate the stereogenic carbon atom(s) that appear in the product. There are at least three possibilities:

1. Begin with a chiral starting material and employ stereospecific reactions.
2. Carry out an asymmetric reaction during the course of the synthesis scheme.
3. Prepare a racemic mixture and perform an enantiomeric resolution.

The decision to use one of these methods depends mainly on the availability and cost of starting materials and reagents. As a result, even products with similar structures might be prepared using alternate strategies. Certain functional groups lend themselves to enantiomeric resolution better than others, so option (3.) might not always be viable.

16.5b SPECIFIC EXAMPLES ILLUSTRATE THE PLANNING FOR THE SYNTHESIS OF ENANTIOMERICALLY PURE SUBSTANCES

The best way to illustrate the design concepts in the synthesis of an enantiomerically pure compound is to look at examples that utilize the three methods listed in Section 16.5a. Those approaches have to be considered in the context of the retrosynthetic analysis (Section 15.4), which leads us to adopt the following procedure:

1. Draw the structure of the desired product if it is not given.
2. Consider whether the functional group(s) that are present can be used to carry out an enantiomeric resolution. If so, propose a synthesis of the *racemic* product by normal protocols (retrosynthesis via disconnections; functional group equivalents; protecting groups: review Sections 15.4 and 15.5)
3. Consider whether the functional group(s) that are present can be incorporated into the molecule by an enantioselective reaction. The data in Table 16.2 suggest that alcohols and epoxides are the most common functional groups that can be

prepared by asymmetric reactions, but hydrogenation can be used to make carboxylic acids, alcohols, and amides in some cases.

4. If the functional group in the product is *not* easily incorporated by an enantioselective reaction, consider whether there are functional group equivalents that can
be introduced. If so, consider following an enantioselective reaction with stereospecific reactions to interchange the functional group.

5. If the carbon skeleton needs to be built before introducing a functional group by
an asymmetric reaction, find logical disconnection sites at which the carbon skeleton can be constructed based on the functional groups identified in Steps 3 or 4.

EXAMPLE 16.3

Show how you would prepare (*S*)-3-aminohexane starting with an alkene that has six or
fewer carbon atoms. You may also use any commercially available reagents and solvents.

---

1. Draw the structure of the desired product:

2. Amines are amenable to enantiomeric resolution (see Exercise 16.2), so that option can be exploited in this instance. Amines can be made by hydrogenation of
substitution products from alcohols (Section 11.2c), which in turn can be made
from alkenes by hydroboration with oxidative hydrolysis. The product has six carbon atoms, and use of a six-carbon alkene is permitted in this exercise, so you need
only think about functional group interconversions.

Retrosynthesis

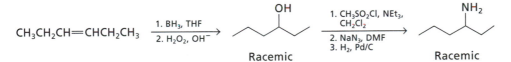

Synthesis

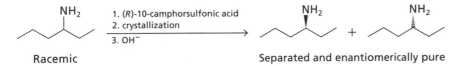

Resolution

3. Amines are *not* readily made by the asymmetric reactions discussed in this chapter,
so you need not consider this possibility further.

4. Alcohols are functional group equivalents of amines (via substitution reactions),
and alcohols are readily prepared by enantioselective reactions. By making use of
the retrosynthesis strategy outlined directly above, the required alcohol can be
made in chiral form by asymmetric hydroboration followed by oxidative hydrolysis
(Section 16.4c). A stereospecific substitution reaction (S$_N$2) yields the chiral amine.

Notice that the alcohol must be made with its configuration opposite that of the amine product because an S_N2 reaction proceeds with inversion of configuration.

Asymmetric synthesis

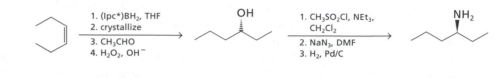

EXAMPLE 16.4

Show how you would prepare (1*S*,2*S*)-2-methyl-1-phenylcyclohexanol from organic compounds that have six or fewer carbon atoms. You may use any commercially available reagents and solvents.

---

1. Draw the structure of the desired product:

2. Enantiomeric resolution is not practical in this case because the alcohol is tertiary and benzylic, so it is likely to undergo dehydration (Section 8.1a).

3. Alcohols can be made by enantioselective hydroboration, so consider what alkene is needed. Hydroboration of the required alkene will be difficult because the alkene is tetrasubstituted, so this route is a poor choice.

    Retrosynthesis

4. Alcohols can also be made readily from epoxides, which can be prepared by enantiomeric processes (Section 16.4d). Use of an epoxide also allows you to attach one of the substituents to the ring while generating the alcohol group at the same time.

    Retrosynthesis

    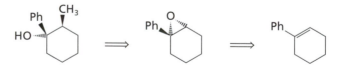

    Asymmetric synthesis

    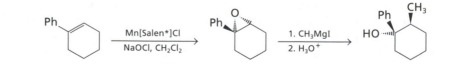

5. Working back to compounds having six or fewer carbon atoms, retrosynthetic analysis of the required alkene reveals a logical disconnection site between the six-membered ring and the phenyl group. The synthesis can be done by a Grignard reaction, followed by dehydration. (It can also be done via the Suzuki reaction.)

Retrosynthesis

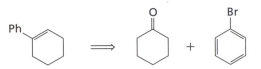

Synthesis

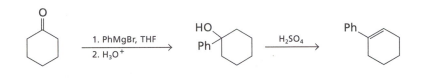

EXERCISE 16.11

Rationalize the regio- and stereochemistry of the Grignard reaction that is used in the asymmetric synthesis route shown in Example 16.4:

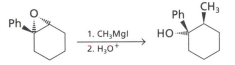

EXAMPLE 16.5

Show how you would prepare (*S*)-2-methylcyclohexanone from organic compounds that have six or fewer carbon atoms. You may use any commercially available reagents and solvents.

---

1. Draw the structure of the desired product:

2. Ketones are not amenable to enantiomeric resolution, so this option in not viable.
3. Ketones are not readily made by the asymmetric reactions described in this chapter.
4. Recognize that a ketone can be prepared by oxidation of the corresponding alcohol (Section 11.4), and alcohols can be made enantioselectively. Also note that either diastereomer of the alcohol can be used because the chiral carbon atom bearing the OH group loses its stereochemistry during oxidation. Asymmetric hydroboration is well suited for the transformation of an alkene to a chiral alcohol with two stereogenic centers. In the actual synthesis, asymmetric hydroboration followed by Swern oxidation (Section 11.4b) gives the desired product.

Retrosynthesis

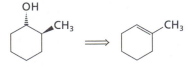

Asymmetric synthesis

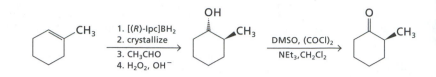

5. To start with compounds having six or fewer carbon atoms, retrosynthetic analysis of the alkene reveals a logical disconnection site between the six-membered ring and the methyl group. The synthesis can be done using an organocuprate reagent (Section 15.3b), which is prepared from commercially available methyllithium and copper(I) cyanide (Section 15.3a).

Retrosynthesis   *Synthesis*

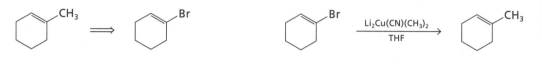

EXERCISE 16.12

Draw the structure of the product after each step in the asymmetric synthesis given in Example 16.5.

EXAMPLE 16.6

Show how you would prepare (*S*)-3-methylhexanoic acid from any achiral organo-halide. You may use any commercially available reagents and solvents.

---

1. Draw the structure of the desired product:

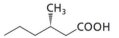

2. A carboxylic acid is readily obtained in chiral form by enantiomeric resolution of a racemic mixture (Section 16.2a). The racemic carboxylic acid is available via the Grignard reaction between an organohalide—the specified starting material—and carbon dioxide.

Synthesis

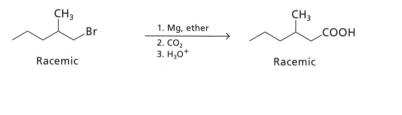

Resolution

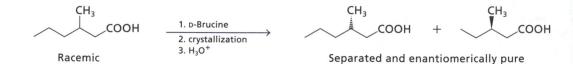

3. To carry out an asymmetric synthesis, hydrogenation of an unsaturated carboxylic acid with Noyori's BINAP catalyst is a good choice (Section 16.4b).

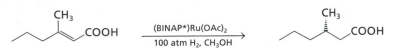

4. The unsaturated carboxylic acid is available via a Grignard reaction between a bromoalkene and carbon dioxide.

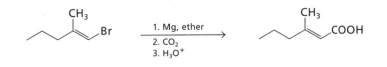

EXAMPLE 16.7

Show how you would synthesize the compound shown here from any alkene starting materials and any other needed reagents and solvents.

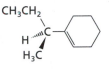

1. The structure of the desired product is given.

2. An alkene is not easily resolved, and there is no other functional group present.

3. A chiral alkene is not easily prepared by an asymmetric synthesis, so you have to consider functional group equivalents.

4. An alkene can be considered as a dehydrated derivative of an alcohol. However, you may also recognize that an alkene can be made by the Suzuki reaction (Section 15.3f), which makes use of an organoboron compound in combination with an alkenyl halide. The boron compound can be made by an asymmetric reaction (Section 16.4c). The retrosynthesis identifies two alkenes (one is a bromoalkene), so you do not have to disconnect the reactants further.

Retrosynthesis

Asymmetric synthesis

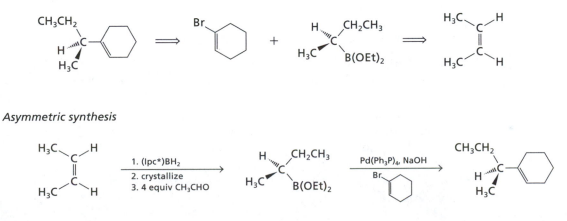

EXERCISE 16.13

Starting with any alkene, propose a synthesis for each of the following compounds. You may use any commercially available reagents and solvents.

a.   b.

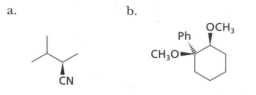

CHAPTER SUMMARY

Section 16.1 Chiral compounds

- Many compounds that occur naturally are chiral, and this property has influenced ideas about the synthesis of organic compounds, especially those that have medicinal and pharmacological uses.

- Two common strategies for obtaining chiral compounds from achiral starting materials are resolution and asymmetric synthesis.

Section 16.2 Enantiomeric resolution

- Resolution makes use of the dissimilar chemical and physical properties of diastereomers. The diastereomeric substances can either be salts or molecules.

- After the formation of diastereomeric derivatives, a physical method—crystallization or chromatography—is used to separate them. Another chemical reaction is used to regenerate the original substance, now enantiomerically pure.

Section 16.3 Asymmetric synthesis

- In an asymmetric synthesis, diastereomeric transition states lead to the preferential formation of one enantiomer over the other because the rates of reaction to form each enantiomer are different.

- Compounds that can react via diastereomeric transition states have a prochiral carbon atom or double bond (usually C=C or C=O), one that can be converted to a chiral product by a single substitution or addition reaction.

- Prochiral atoms or double bonds that lead to formation of enantiomers are said to be enantiotopic.

- The group of a prochiral center or the face of a double bond that leads to formation of a chiral center with the (*S*)-configuration is called the pro-(*S*) group or face. The one giving the (*R*)-isomer is pro-(*R*).

- Prochiral atoms or double bonds that lead to the formation of diastereomers are said to be diastereotopic.

- Diastereotopic protons are nonequivalent and can give rise to two different chemical shift values in the ^{1}H NMR spectrum. Geminal coupling can occur between diastereotopic protons.

- Enantiomeric excess refers to how much of one enantiomer is produced relative to the other, and the value is expressed as a percentage (% major isomer – % minor isomer).

- A kinetic resolution occurs when one enantiomer of a racemic mixture reacts preferentially versus the other. The maximum chemical yield of an enantiomerically pure compound via kinetic resolution is 50%
- Proton NMR spectroscopy is one way to measure the ee of an asymmetric reaction after the product is converted to a diastereomeric derivative.

Section 16.4   Enantioselective addition reactions

- The asymmetric catalytic hydrogenation of alkenes makes use of homogeneous Rh or Ru catalysts with chiral phosphine-containing ligands.
- Chiral Ru(BINAP) catalysts are able to hydrogenate a wide range of olefins that have a binding group such as COOH or OH.
- Enantioselective hydroboration is carried out with a chiral borane derived from α-pinene. Oxidative hydrolysis leads to the formation of chiral alcohol molecules, or a Suzuki reaction can be used to make carbon–carbon bonds.
- The enantioselective epoxidation of alkenes, especially cis- and aryl-trisubstituted alkenes, can be accomplished with manganese complexes and NaOCl as the oxidant.
- Chiral dioxiranes, formed in situ by the reaction between the persulfate ion and chiral ketones derived from fructose, can be used to catalyze the enantioselective epoxidation of alkenes, especially those that are trans- or trisubstituted.
- The Sharpless asymmetric dihydroxylation converts an alkene to a 1,2-diol using chiral osmium complexes. The chiral ligands are derivatives of cinchona alkaloids, naturally occurring chiral nitrogen-containing compounds.
- Enzymes can sometimes be employed in the laboratory to carry out enantioselective reactions.

Section 16.5   Synthetic design of chiral molecules

- The retrosynthetic design of chiral compounds makes use of three strategies: starting with a chiral compound and using stereoselective reactions, employing an asymmetric reaction during the synthetic scheme, or carrying out an enantiomeric resolution of a racemic product at the end of a synthesis.

KEY TERMS

Section 16.1a
chiral pool

Section 16.1c
enantiomeric resolution
asymmetric reaction

Section 16.3b
prochiral
homotropic
pro-(R)
pro-(S)
enantiotopic
diastereotopic

Section 16.3c
enantiomeric excess, ee
scalemic mixture

Section 16.3d
kinetic resolution

REACTION SUMMARY

Sections 16.2a

Resolution of carboxylic acids with chiral amines via formation of diastereomeric salts. The diastereomers are separated and converted back to the enantiomerically pure carboxylic acid by treatment with dilute hydrochloric acid.

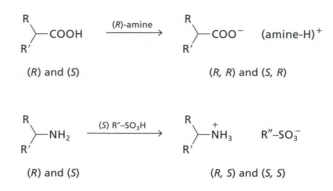

(R) and (S) → (R, R) and (S, R)

(R) and (S) → (R, S) and (S, S)

Section 16.4b

Enamides are hydrogenated by a rhodium catalyst with chiral phosphine ligands to form enantiomerically enriched amido esters.

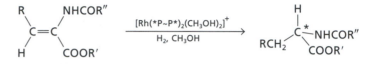

Alkenes undergo enantioselective hydrogenation reactions with Ru(BINAP) catalysts, where BINAP is a di(phosphine) ligand with a binaphthyl backbone that is chiral because it exists as atropisomers (Section 4.3c).

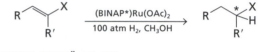

X = COOH, COOR″, CH₂OH

Section 16.4c

Alkenes undergo enantioselective hydroboration reactions with (Ipc*)BH₂ reagents, which are formed from the reaction between BH₃ and enantiomers of α-pinene. The organoborane produced by hydroboration with this reagent is treated with acetaldehyde to produce an enantiomerically pure organoboronate.

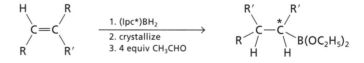

Section 16.4d

Jacobsen epoxidation. Alkenes—especially cis-alkenes—undergo enantioselective epoxidation reactions when treated with a chiral manganese catalyst and NaOCl.

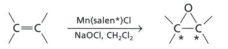

Alkenes—especially trans- and trisubstituted alkenes—undergo enantioselective epoxidation reactions when treated with the persulfate ion and a chiral ketone derived from fructose.

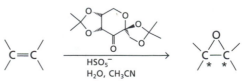

Section 16.4e

The Sharpless asymmetric dihydroxylation converts an alkene to a 1,2-diol using chiral osmium complexes. The chiral ligands are derivatives of cinchona alkaloids, naturally occurring chiral nitrogen-containing compounds.

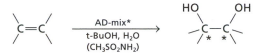

16.14. Show the scheme by which you would obtain enantiomerically pure 3-methylpentanoic acid by resolution of the racemic material.

16.15. For the following reactions that you have learned, assign pro-(R) and pro-(S) designations to the hydrogen atoms that are replaced or to the face of the π bond to which addition has occurred. Draw the three-dimensional structure of each enantiomer in the product mixture.

a.

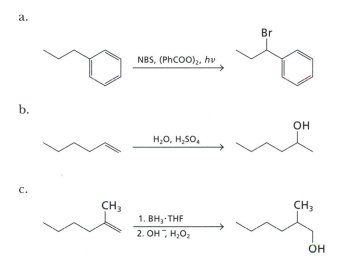

b.

c.

16.16. Many biochemically important molecules have prochiral carbon atoms. For each of the following, assign pro-(R) and pro-(S) (or neither) to each set of screened hydrogen atoms, assuming a hydroxyl group will replace one H. Label each indicated prochiral center as either enantiotopic or diastereotopic.

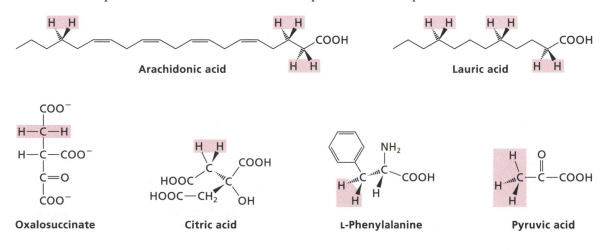

16.17. The metabolism of citrate in many organisms proceeds through a number of steps: A key step is the hydration of aconitate, which produces (2*R*,3*S*)-isocitrate. For the following equation, draw the full stereochemical structures of these two molecules and indicate which face of the double bond in (*Z*)-aconitate is pro-(*R*) and which is pro-(*S*). Or, are these designations even valid in this instance? Explain.

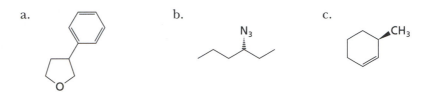

(*Z*)-Aconitate   (2*R*,3*S*)-Isocitrate

16.18. Show how you would prepare each of the following compounds from any alkene, (Ipc*)BH$_2$, and any other necessary reagents and solvents:

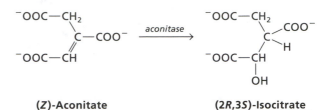

a.   b.   c.

16.19. Calculate the ee or de for each of the following processes:

a.

Ratio: 65 : 35

b.

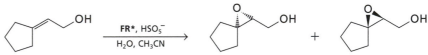

16.20. The 300-MHz ^{1}H NMR spectrum of 3-methyl-2-pentanone is shown below. Assign each of the observed signals to the protons in this compound, and account for the spin–spin splitting patterns that are observed.

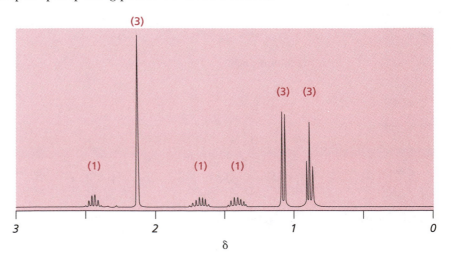

16.21. Propose a reasonable synthesis for each of the following chiral compounds from any alkene or bromoalkene that has eight or fewer carbon atoms. You may also use bromobenzene and any necessary reagents and solvents.

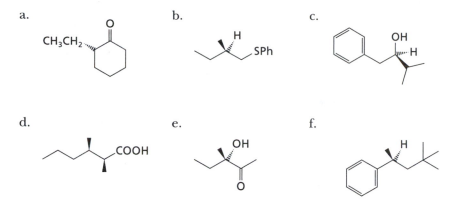

a.   b.   c.

d.   e.   f.

16.22. An unspecified asymmetric amination procedure is used to make 3-aminohexane (compound **1**). This amine is subsequently converted to a mixture of diastereomeric phosphoramides **2** according to the following equation:

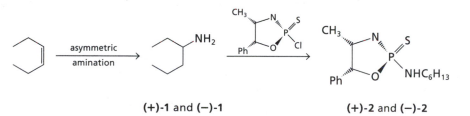

(+)-**1** and (−)-**1**   (+)-**2** and (−)-**2**

a. Draw structures for the two diastereomers of **2**.

b. A ^{31}P NMR spectrum of the mixture of (+)-**2** and (−)-**2** is illustrated below. The ^{31}P signal associated with (+)-**2** occurs upfield from the peak assigned to (−)-**2**. What is the ee of the amine product in the asymmetric amination procedure shown in the first step of the preceding equation?

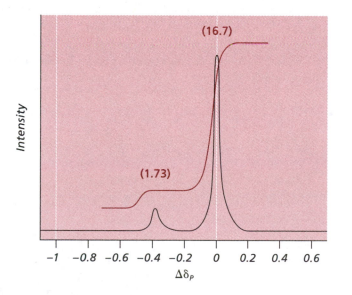

16.23. Rationalize the stereochemical result of the following reaction sequence:

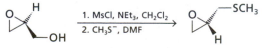

16.24. For each of the following transformations, draw the structure of the isomeric products and assign the absolute configuration for each of the chiral centers present. What is the ratio of isomers formed in each reaction scheme? The *relative* amounts of each enantiomer are indicated at the far right side of each equation.

a.

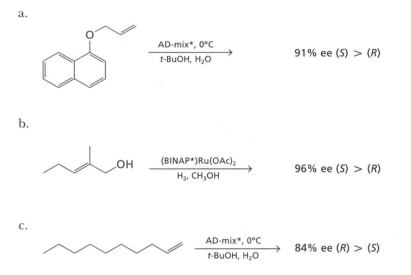

AD-mix*, 0°C
───────────→
t-BuOH, H₂O

91% ee (S) > (R)

b.

$$\text{(BINAP*)Ru(OAc)}_2$$

OH   ──────────────→   96% ee (S) > (R)
      H₂, CH₃OH

c.

AD-mix*, 0°C
──────────→   84% ee (R) > (S)
t-BuOH, H₂O

16.25. Compound **A**, C₆H₁₂, was treated with ozone followed by zinc and water, yielding only **B**, the NMR spectrum of which showed a triplet at δ 9.9 (*J* = 2 Hz, intensity = 1H), a doublet of quartets at δ 2.3 (*J* = 2 Hz and 7 Hz, 2H), and a triplet at δ 1.0 (*J* = 7 Hz, 3H). When **A** was treated with either of the epoxidizing reagents described in Section 16.4d, meso-compound **C** was formed. Treating **C** with ethylmagnesium bromide followed by hydrolysis gave racemic compound **D**, C₈H₁₈O. Draw the structures of compounds **A** through **D**, including stereochemistry.

16.26. Based on the discussion in Section 16.3a about diastereomeric transition states, explain how an enzyme takes advantage of the prochirality of its substrate to produce a chiral product.

16.27. The metabolism of citrate was mentioned in Exercise 16.17. If you were to perform this metabolic reaction in the laboratory with the following ¹⁴C-labeled citrate, you would discover, after you isolated the isocitrate molecule, that the labeled carbon atom was still present as a methylene group. Based on your answer to Exercise 16.26, rationalize this result.

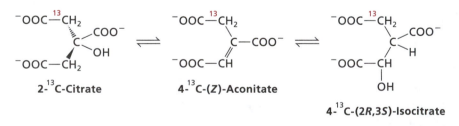

Notice that the central carbon atom in citrate is prochiral because it is bonded to two CH₂COO⁻ groups.

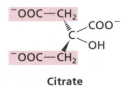

Citrate

THE CHEMISTRY OF BENZENE AND ITS DERIVATIVES

During the 1800s, many of the compounds isolated from coal tar were referred to as "aromatic compounds" because of their characteristic fragrances. Scores of these substances were subsequently shown to be derivatives of benzene, a compound identified by Michael Faraday in 1825. The structure of benzene presented a challenge to chemists in the nineteenth century because its high degree of unsaturation could not be reconciled with its low reactivity toward many reagents that readily transformed alkenes and alkynes into other types of molecules.

Several chemists—Andrew Couper in England, Joann Josef-Loschmidt in Austria–Hungary, and August Kekulé in Germany—proposed structures for benzene during the early 1860s. Kekulé, who was a renowned professor and author of many textbooks, is generally credited with proposing the correct structure in 1865 after purportedly receiving inspiration from a dream. In his proposal, benzene existed as an equilibrium mixture between two cyclic forms with three alternating single and double bonds, as shown below at the left. The modern structure (Section 2.7a) is ostensibly similar (below, right), but benzene's structure is now viewed as a resonance hybrid of two equivalent forms, the important distinction being that there are no alternating single and double bonds, hence all of the bond lengths are equal.

Kekulé proposal:
interconverting isomers

Modern understanding:
resonance hybrid

Still, we pay tribute to Kekulé's insight that added to the development of bonding theories by referring to the resonance forms of benzene as *Kekulé structures*.

Benzene is only one member of the group of compounds that are classified as *aromatic*. This chapter begins with a discussion of the characteristics that define aromatic compounds, which are also called *arenes*. Following the subject of aromaticity is a brief review of nomenclature and NMR spectroscopy as they apply to benzene and its derivatives, and then the remainder of the chapter describes in some detail the types of reactions that benzene and its derivatives undergo.

17.1   STRUCTURAL ASPECTS OF AROMATIC MOLECULES

17.1a   BENZENE IS MORE STABLE THAN RELATED COMPOUNDS WITH ISOLATED OR CONJUGATED DOUBLE BONDS

You learned in Section 8.1c how the enthalpy of hydrogenation, $\Delta H°_{hydrog}$, could be used to evaluate the stabilities of alkenes. The same type of measurement can be applied to benzene; note, however, that the hydrogenation of benzene requires both heat and high pressures of H_2, which means that its reaction with hydrogen is slower than those of alkenes and alkynes.

When H_2 adds to cyclohexene to form cyclohexane, $\Delta H°_{hydrog} = -28.4$ kcal mol^{-1} (Figure 17.1). If we could carry out the same reaction on the imaginary molecule 1,3,5-cyclohexatriene, the value of $\Delta H°_{hydrog}$ should be three times that amount (three double bonds vs. one for cyclohexene), or about -85.2 kcal mol^{-1}. For benzene, however, the value of $\Delta H°_{hydrog}$ is -49.3 kcal mol^{-1}.

The energy difference between $\Delta H°_{hydrog}$ (benzene) and $\Delta H°_{hydrog}$ (1,3,5-cyclohexatriene) of about 36 kcal mol^{-1} is called the **resonance energy,** which must represent the degree of stabilization of the benzene ring relative to the imaginary triene, because the final product is the same for each reaction. Benzene clearly has an exceptional degree of stability relative to that of the other six-membered ring polyenes.

17.1b   AN AROMATIC COMPOUND HAS (4N + 2) π ELECTRONS IN A PLANAR RING

The property that accounts for the unusual stability of benzene is called **aromaticity,** and molecules that possess significant resonance stabilization of the type described in Section 17.1a are called **aromatic compounds.** The basis for aromaticity is rationalized by a theory proposed by Erich Hückel in 1931, a formalism that has come to be known as **Hückel's rule** or the **4n + 2 rule:**

> A molecule is aromatic when it has a planar, uninterrupted, and cyclic π system that contains $4n + 2$ electrons, where $n = 0, 1, 2, 3, 4$, and so on.

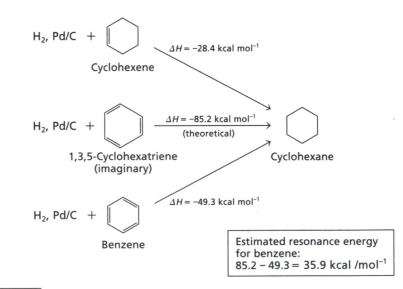

Figure 17.1
The reactions used to compare the $\Delta H°_{hydrog}$ values for cyclohexene, the imaginary molecule 1,3,5-cyclohexatriene, and benzene. The numeric difference between $\Delta H°_{hydrog}$ for cyclohexatriene and for benzene is the estimated resonance energy for benzene.

The basis of Hückel's rule has to do with the relative energies of the MOs in aromatic compounds. As you learned in Section 10.2a, the number of π MOs equals the number of atomic p orbitals required to construct the π system. To generate the MOs needed for aromaticity, overlap has to occur between p orbitals on adjacent atoms. Because aromatic compounds have to be cyclic, this overlap will occur only when the ring is planar and each p orbital is aligned with its neighbors. Furthermore, there cannot be a "gap" in the overlap among the orbitals, so every atom must have a p orbital.

Hückel recognized that when these criteria are met, the resulting MOs have relative energy values that are ordered as shown in Figure 17.2. The MOs that are lowest in energy—the bonding orbitals—are filled for molecules with an odd number of electron pairs ($4n + 2$ π electrons), which means that the π system has 2, 6, 10, 14, and so on electrons. Benzene has six carbon atoms, hence six MOs, and adding its six π electrons (three pairs) to the MOs will fill the bonding orbitals and leave the antibonding ones vacant.

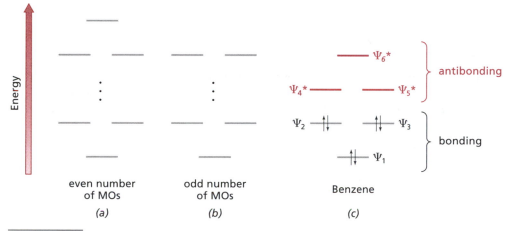

Figure 17.2
Relative energy levels for the molecular orbitals generated from uninterrupted, overlapping atomic p orbitals in planar, cyclic molecules. The vertical dotted lines signify that additional orbitals may be present at these intermediate levels. *(a)* The energy levels of MOs in molecules with an even number of atoms. *(b)* The energy levels of MOs for molecules with an odd number of atoms. *(c)* The energy levels of MOs for benzene showing that the bonding orbitals are completely filled by the six available π electrons.

17.1c Cyclobutadiene and Cyclooctatetraene Are Not Aromatic

Some cyclic compounds with alternating single and double bonds, which as a class are called **annulenes,** have reactivity properties that differ from that of benzene, even though at first glance, their structures look similar. Cyclobutadiene, with four carbon atoms, is unstable except at very low temperatures. Cyclooctatetraene, with eight carbon atoms, reacts like a typical alkene. Its ring is not planar, and the bond lengths are not all equal. Neither molecule has $4n + 2$ π electrons, and neither is aromatic.

A simple mnemonic allows us to arrange the order of the energy levels for a compound that we think might be aromatic. We start by drawing the ring with one atom pointing down. Then, for every vertex of the ring, we draw a horizontal line that corresponds to an energy level. For cyclobutadiene, this drawing is made as shown below:

Once the energy levels are drawn, the electrons are added to the MOs beginning with the ones at lowest energy. Cyclobutadiene has four MOs and four available electrons, so the MOs are filled as shown below (a.). This representation assumes that cyclobutadiene *has a delocalized π system like benzene.* The two MOs at the second levels are not filled, however, so this molecule is actually **antiaromatic,** which is *destabilized* relative to the analogous cyclic polyene. When a molecule is classified as antiaromatic, it will instead have a structure that removes the orbital degeneracy (the orbitals at the same energy levels) and allows the electrons to pair. The actual structure of cyclobutadiene has *two carbon–carbon π bonds of equal energies,* as illustrated in (b.), and the bond lengths are not all equal.

a.   b.

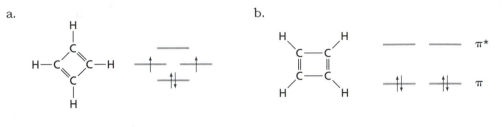

EXERCISE 17.1

Using the mnemonic mentioned above, show the relative energy levels that would be used to assess whether cyclooctatetraene is aromatic or antiaromatic. Add the eight electrons to the MOs, and explain why this compound is antiaromatic. Make a model of cyclooctatetraene and predict its actual shape.

17.1d THE CYCLOPENTADIENYL ANION IS AROMATIC

Compounds with five and seven carbon atoms cannot form an uninterrupted π system because one carbon has sp^3 hybridization. Cyclopentadiene can be deprotonated to form the cyclopentadienyl anion, which now has five carbon atoms with sp^2 hybridization. The pK_a value of the parent hydrocarbon, C_5H_6, is ~ 16, which makes it nearly as acidic as water. The factor that accounts for its acidity is the aromaticity of the anion.

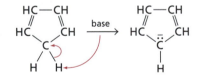

The relative energy levels of the MOs for the cyclopentadienyl system can be drawn as we did for cyclobutadiene. With one vertex pointing down, we draw a horizontal line that corresponds to the energy levels of the MOs (Fig. 17.3a). The cyclopentadienyl (abbreviated Cp) *anion* is aromatic because the three lowest MOs are filled: *It is planar and has six π electrons, just like benzene.* The cyclopentadienyl radical and the cyclopentadienyl cation are antiaromatic and unstable.

The Cp⁻ ion can be isolated in the form of stable salts. Transition metal ions form "π complexes" with this ion, and ferrocene is the prototype for literally thousands of organometallic compounds with the Cp⁻ ion (Figure 17.3b).

EXERCISE 17.2

Generate the corresponding energy level diagram expected for the MOs of the seven-membered ring systems. Classify the cycloheptatrienyl anion, radical, and cation as aromatic or antiaromatic.

---

a.

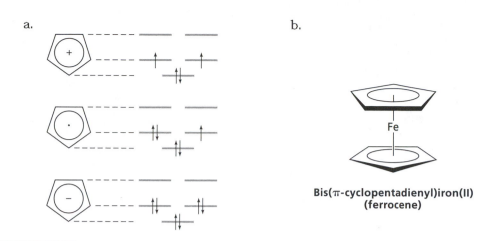

b.

**Bis(π-cyclopentadienyl)iron(II)
(ferrocene)**

Figure 17.3
(*a*) The energy levels for the MOs of the cyclopentadienyl cation (*top*), radical (*middle*), and anion (*bottom*). (*b*) The structure of ferrocene, which has two cyclopentadienyl anions bonded to an iron(II) ion.

17.1e CYCLIC COMPOUNDS WITH MULTIPLE RINGS OR HETEROATOMS CAN ALSO BE CLASSIFIED AS AROMATIC

Hückel's rule strictly applies only to compounds with a single ring, but it is often used for those with multiple rings as well. For example, naphthalene has 10 π electrons in two rings, and anthracene and phenanthrene each have 14 π electrons among three rings. All three react in ways that are similar to benzene and its derivatives, and they are considered to be aromatic. (More will be said about the properties and reactions of these molecules in Section 25.1.)

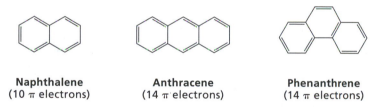

Naphthalene   **Anthracene**   **Phenanthrene**
(10 π electrons)   (14 π electrons)   (14 π electrons)

A **heterocycle,** which is a cyclic compound that has at least one heteroatom within its ring, can also be classified as aromatic if the ring is planar and has an uninterrupted π system that contains $4n + 2$ electrons. To count the electrons correctly, first realize that an atom can have only one p orbital that will be aligned with the others to create the MOs and π system. The electrons to be counted in the assessment of aromaticity are as follows:

- If an atom forms a π bond, then its two electrons are included in the total.
- If an atom forms a π bond and has an unshared electron pair, only the two electrons of the π bond are included in the total.
- If an atom has only unshared pairs, then only one pair (two electrons) is included in the total.

To illustrate the foregoing points, Figure 17.4 shows the valence bond representations of three typical aromatic heterocycles as well as the electrons that constitute the aromatic π systems. (More about the properties and reactions of these and related molecules will be presented in Sections 25.2 and 25.3.)

a.

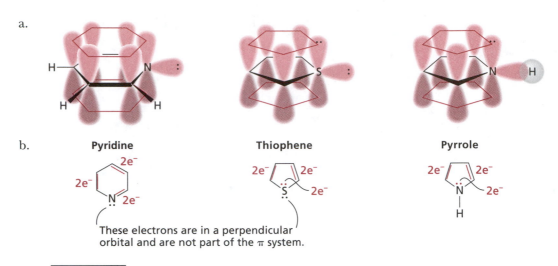

b.

Pyridine   **Thiophene**   **Pyrrole**

These electrons are in a perpendicular
orbital and are not part of the π system.

Figure 17.4
(*a.*) The valence bond representations of pyridine, thiophene, and pyrrole. (*b.*) Structures
showing the six electrons that constitute the aromatic π system.

EXERCISE 17.3

Specify how many π electrons each of the following molecules has, and circle the compounds that are expected to be aromatic:

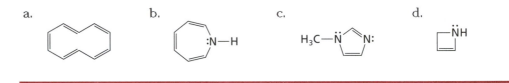

a.   b.   c.   d.

17.1f DISUBSTITUTED BENZENE ISOMERS ARE NAMED USING THE PREFIXES ORTHO, META, AND PARA

Like their parent compound, substituted derivatives of benzene are aromatic because they have six π electrons. Even though the substituents attached to the benzene ring can influence the electron density at different carbon atoms of the π system, rarely do they exert enough effect to disrupt the compound's aromatic character.

To review briefly the nomenclature of benzene derivatives (Section 1.3a), recall that many such compounds are named with the root *benz-*, followed by a suffix that designates the principal functional group. Prefixes identify the substituents attached to the ring, and numerals are used to indicate the positions of attachment.

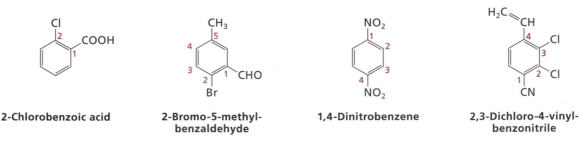

2-Chlorobenzoic acid   **2-Bromo-5-methyl-benzaldehyde**   **1,4-Dinitrobenzene**   **2,3-Dichloro-4-vinyl-benzonitrile**

When only two groups are attached to a benzene ring, the positional relationship between them can be specified by the prefixes **ortho** (*o*-) for 1,2-disubstitution, **meta** (*m*-), for 1,3-disubstitution, and **para** (*p*-), for 1,4-disubstitution. For example,

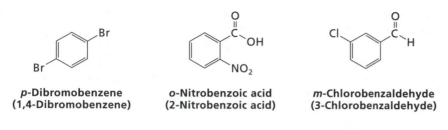

p-Dibromobenzene	*o*-Nitrobenzoic acid	*m*-Chlorobenzaldehyde
(1,4-Dibromobenzene)	(2-Nitrobenzoic acid)	(3-Chlorobenzaldehyde)

Some names use a root word different from benz-, and these include phenol, anisole, aniline (Section 1.3a), and acetophenone. These once common names were incorporated into IUPAC system. For example,

m-Chlorophenol   5-Bromo-2-ethylanisole   2,3-Dichloroaniline   *p*-Methoxyacetophenone

The names of most methylated benzene derivatives do not include the root word benz-. The common root for the isomers of dimethylbenzene is *xylene*, which *requires* use of the *o*-, *m*-, or *p*- prefix.

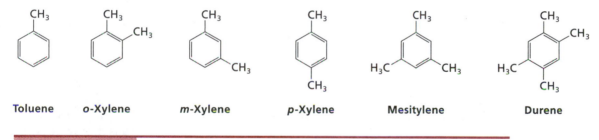

Toluene   o-Xylene   *m*-Xylene   *p*-Xylene   Mesitylene   Durene

---

EXERCISE 17.4

Draw a structure corresponding to each of the following names:

   a. 3-Chloro-2-methylphenol

   b. *m*-Hydroxybenzoic acid

   c. *o*-Dichlorobenzene

   d. 2,6-Dimethoxytoluene

   e. α,α′-Dibromo-*m*-xylene (for benzene derivatives with common root names, substitution on a carbon atom *adjacent* to the ring is indicated by "α").

EXERCISE 17.5

Name the following compounds.

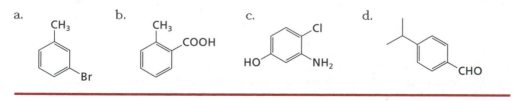

17.1g PROTON NMR SPECTRA ARE USED TO DETERMINE THE SUBSTITUTION PATTERN OF A BENZENE RING

You learned in Section 13.2b that spin coupling between protons attached to a benzene ring creates splitting patterns in the aromatic region of a ^{1}H NMR spectrum that can be used to determine the substitution pattern of the ring. The most prominent splitting patterns result from **ortho coupling,** that is, coupling between protons that are ortho to one another. The magnitude of this coupling constant, J, is ~ 7–8 Hz. For a disubstituted benzene ring, the three possible patterns that you will observe are shown in Figure 17.5. (Two of the common splitting patterns for trisubstituted benzene derivatives were presented in Figure 13.12 along with that for the para-disubstituted benzene ring.)

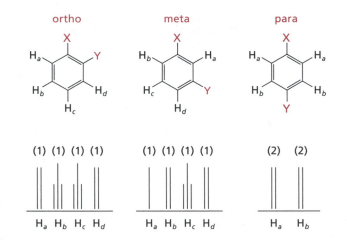

Figure 17.5
Splitting patterns observed for disubstituted benzene derivatives (X and Y are substituents other than H). The order in which the signals appear, left to right, is arbitrary.

Remember that the sum of integrated intensity values for the entire aromatic region reveals how many substituents are attached to the ring, so a total value of 4 indicates that the ring has two substituents.

Even though the *principal* pattern you observe will be one of those in Figures 13.12 or 17.5, be aware that long-range coupling can occur between protons that are meta to each other (Section 13.4b and Table 13.3). These splitting patterns have smaller J values—usually < 2 Hz—so they are generally easy to spot, as illustrated in Figure 17.6 for the ^{1}H NMR spectrum of 2-nitro-4-methylphenol.

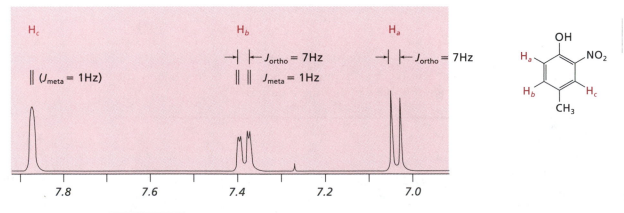

Figure 17.6
The ^{1}H NMR spectrum of 2-nitro-4-methylphenol in the aromatic region. Considering only coupling between H$_a$ and H$_b$, which are ortho, this trisubstituted benzene derivative should display two doublets and a singlet (see Fig. 13.12). The smaller splitting observed (especially on the peak at δ 7.39) is the long-range meta coupling (J = 1 Hz) between H$_b$ and H$_c$.

EXERCISE 17.6

Shown in (a.) is the aromatic region in the ^{1}H NMR spectrum of benzoic acid. Match the resonances to the aromatic protons in the structure given in (b.), and rationalize the appearance of the splitting patterns.

a.   b.

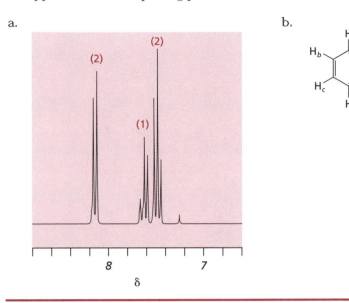

17.2   ELECTROPHILIC SUBSTITUTION REACTIONS OF BENZENE

17.2a BENZENE REACTS READILY WITH ELECTROPHILES AND UNDERGOES SUBSTITUTION

As noted in the introduction to this chapter, the reactivity of benzene was initially puzzling because the ring was known to be highly unsaturated, yet addition reactions did not occur. Instead, the hydrogen atoms could be replaced by a variety of groups.

As we now understand benzene, its π system is electron rich; therefore, at its core, benzene is a nucleophile. As expected, benzene reacts with electrophiles. The general reaction mechanism, called **electrophilic aromatic substitution,** comprises two steps.

Step 1   Generation of the carbocation intermediate; this step is rate determining.

Step 2   Regeneration of the aromatic system by deprotonation of the cation intermediate.

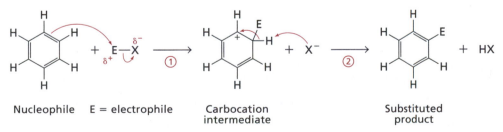

Notice that the carbocation intermediate formed in Step 1 is resonance stabilized:

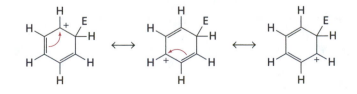

This delocalization of electrons in the intermediate has a distinct influence on the reactions of benzene *derivatives* (Section 17.3), so make certain that you are familiar with these structures.

The energy profile for an electrophilic substitution reaction of benzene is illustrated in Figure 17.7 and is typical for an exergonic reaction that proceeds via an intermediate (Section 5.4c).

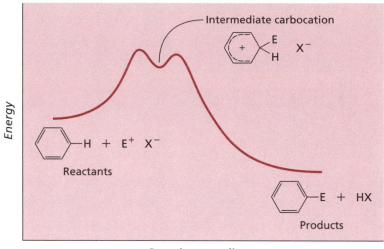

Figure 17.7
The reaction coordinate diagram for an electrophilic substitution reaction of benzene.

The types of electrophiles that react with benzene vary widely in their structures, and five common ones are listed in Table 17.1 along with the actual reagents that are used and the products that are obtained. The following sections will discuss specific aspects of each reaction.

Table 17.1   Electrophiles used to convert benzene into substituted derivatives

Electrophile	*Reagent*	*Product*
Cl⁺, Br⁺ (Section 17.2b)	Cl_2, $FeCl_3$ Br_2, $FeBr_3$	Chlorobenzene Bromobenzene
NO_2^+ (Section 17.3c)	HNO_3, H_2SO_4	Nitrobenzene
HSO_3^+ (Section 17.2d)	SO_3, H_2SO_4	Benzenesulfonic acid
R⁺ (Section 17.2e)	RCl, $AlCl_3$	Alkylbenzenes
RCO⁺ (Section 17.2e)	RCOCl, $AlCl_3$	Acylbenzenes

17.2b CHLORINE AND BROMINE REACT WITH BENZENE IN THE PRESENCE OF A LEWIS ACID

Benzene does not react with electrophiles in the same way that alkenes do because the resonance energy of the aromatic system has to be overcome in order to form the carbocation intermediate. Therefore, chlorine and bromine do not react with benzene unless a catalyst is supplied. The iron(III) halides are good Lewis acids, so they react with X_2 to form X^+ and $[FeX_4]^-$ (X = Cl or Br). Benzene then reacts with the electrophilic halogen ion by the two-step process shown in Section 17.2a.

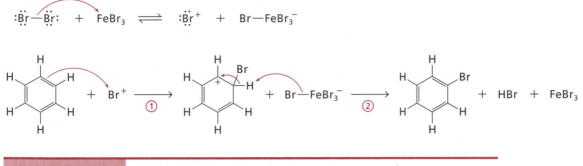

EXERCISE 17.7

Propose a reasonable mechanism for the reaction of benzene with chlorine and the Lewis acid aluminum chloride, $AlCl_3$, to form chlorobenzene.

17.2c NITRIC ACID IS USED TO ATTACH A NITRO GROUP TO THE BENZENE RING

When benzene reacts with nitric acid, the electrophile is the **nitronium ion,** NO_2^+, not H^+, as you might first think. The reaction to generate NO_2^+ is commonly promoted by sulfuric acid according to the following equation:

$$HNO_3 + H_2SO_4 \rightleftharpoons NO_2^+ + HSO_4^- + H_2O$$

The nitronium ion is a reactive electrophile, and its resonance forms show that the nitrogen atom is the electrophilic center. Benzene reacts with NO_2^+ to form nitrobenzene.

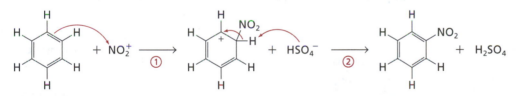

Stable salts such as nitronium tetrafluoroborate, $NO_2^+ BF_4^-$, are commercially available and can be used instead for the nitration process.

Recall that the nitro group can be converted to the amino group by reduction (section 11.2c).

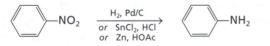

Aminobenzenes, which are named as derivatives of *aniline*, are versatile intermediates for the synthesis of functionalized arenes via diazonium salts. (These transformations will be discussed in Section 17.4c.)

17.2d BENZENE CAN BE CONVERTED TO BENZENESULFONIC ACID

The electrophile HSO_3^+ is generated either from the reaction of sulfuric acid with itself at high temperatures or by dissolving sulfur trioxide in sulfuric acid, a mixture called "oleum". Benzene reacts with HSO_3^+ to form benzenesulfonic acid.

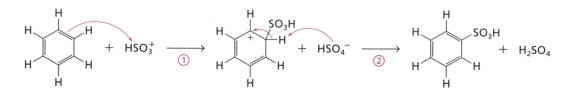

$$2 \ H_2SO_4 \ \rightleftharpoons \ HSO_3^+ \ + \ HSO_4^- \ + \ H_2O$$

$$SO_3 \ + \ H_2SO_4 \ \rightleftharpoons \ HSO_3^+ \ + \ HSO_4^-$$

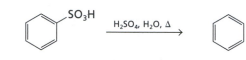

EXERCISE 17.8

Sulfonation is reversible, so when benzenesulfonic acid is heated with aqueous sulfuric acid, the sulfonic acid group is replaced by a proton. Propose a reasonable mechanism for this transformation.

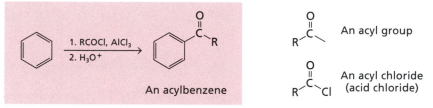

17.2e FRIEDEL–CRAFTS REACTIONS ARE USED TO ATTACH CARBON-CONTAINING SUBSTITUENTS TO THE BENZENE RING

Carbocations are potent electrophiles, so it makes sense that a carbocation might be useful for attaching an alkyl group to the benzene ring. In the presence of a Lewis acid—most often aluminum chloride—alkyl and acyl chlorides react with benzene to form alkyl- and acylbenzene derivatives, respectively. These reactions are called **Friedel–Crafts alkylation** and **Friedel-Crafts acylation** in honor of the two chemists—Charles Friedel and James Mason Crafts—who discovered and studied them.

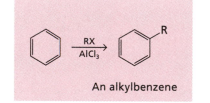

The utility of the Friedel–Crafts alkylation reaction is limited because of a number of undesirable side reactions. When RX is a primary alkyl chloride, rearrangement is common (see Section 6.2e), so it is not possible to prepare alkyl benzene derivatives having a CH_2 group adjacent to the ring by this route. Even 2° and 3° alkyl chlorides will undergo rearrangement when treated with aluminum chloride, so alkylation is limited to small and highly symmetric alkyl substituents such as isopropyl and *tert*-butyl groups.

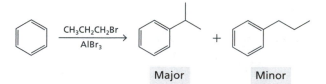

EXERCISE 17.9

Propose a mechanism to account for the products of the reaction directly above. (*Hint:* Review Section 6.2e.)

A second problem of the Friedel–Crafts alkylation procedure is polyalkylation. As you will see, an alkyl group makes the ring more reactive than benzene itself (Section 17.3b), so after an alkyl group is attached, the product becomes more reactive than the starting compound, which leads to further alkylation.

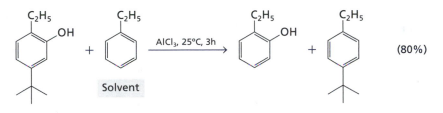

A third problem is that alkylation is reversible, and an alkyl group can migrate from one molecule to another, producing a mixture of products. Sometimes, this process can be used to advantage, for example, to transfer a *tert*-butyl group from one arene to another.

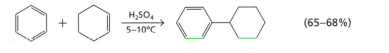

EXERCISE 17.10

An alkene forms a carbocation when it is treated with strong acid (Section 9.1b). Propose a reasonable mechanism for the following variant of Friedel–Crafts alkylation:

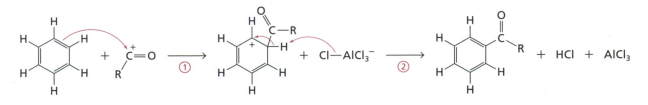

The Friedel–Crafts acylation reaction is a more useful synthetic method for attaching a carbon-containing group to the benzene ring. In the first step, chloride ion is abstracted from the acid chloride by complexation with aluminum chloride. This step generates an acyl cation, which is resonance stabilized. Such species are relatively stable and are *not* prone to rearrangement.

Benzene reacts with this electrophile in typical fashion:

A slight complication is that the ketone product is a Lewis base, and aluminum chloride is a Lewis acid, so they react with each other. A separate hydrolysis step is therefore required to react with the aluminum chloride and liberate the ketone product.

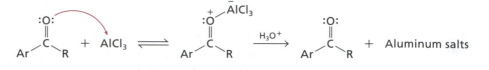

Acylation of benzene has two advantages versus alkylation. The first has been mentioned already: Rearrangement of the incoming group does not occur. The second is that an acyl group deactivates the ring, which means polysubstitution is not a problem with acylation.

An acyl group can be converted to an alkyl group by several methods, four of which are depicted in Figure 17.8.

Figure 17.8
Four methods used to reduce acylbenzene derivatives to alkylbenzenes.

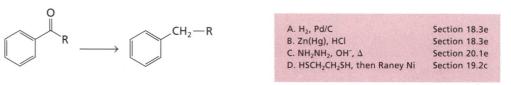

A. H$_2$, Pd/C	Section 18.3e
B. Zn(Hg), HCl	Section 18.3e
C. NH$_2$NH$_2$, OH$^-$, Δ	Section 20.1e
D. HSCH$_2$CH$_2$SH, then Raney Ni	Section 19.2c

Acylation followed by reduction therefore provides a suitable way to attach an RCH$_2$ group to a benzene ring.

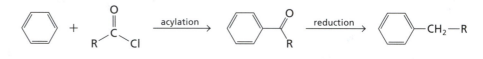

EXERCISE 17.11

By making use of a Friedel–Crafts acylation reaction, show how you would prepare each of the following compounds from benzene, an acid chloride, and any reagents and solvents needed:

a.

b.

17.3 ELECTROPHILIC SUBSTITUTION REACTIONS
OF BENZENE DERIVATIVES

17.3a SUBSTITUTED BENZENE DERIVATIVES REACT TO FORM ISOMERIC PRODUCTS

The electrophilic substitution reactions of benzene are straightforward because the six positions are equivalent. If you know what electrophile is present to react with benzene, it is a simple matter to predict what product will be formed. When a substituent is attached to the benzene ring, however, the course of a substitution reaction depends on several factors. Some groups are *activating*, which means that such compounds react faster than benzene; and some groups are *deactivating*, which means that those compounds react more slowly than benzene.

Furthermore, a substituent "directs" an incoming electrophile to react at specific positions, so regiochemistry becomes important. The following structure summarizes the designation of each position relative to the carbon atom that bears a substituent, X.

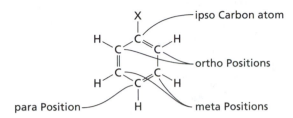

In electrophilic substitution reactions of benzene derivatives, a substituent directs the incoming group either to the ortho and para positions, or to the meta position. Reaction at the ipso carbon atom does not occur normally. By surveying the reactions of hundreds of benzene derivatives, you could sort the substituents into five categories, which are summarized in Table 17.2 and discussed in the following sections.

Substituents influence the transition state of electrophilic substitution reaction by three principal means: *steric effects* (Section 3.1c), *inductive effects* (Section 5.2d), and *resonance effects* (Section 5.2c). These effects lead to the positional and reactivity preferences summarized in Table 17.2.

Steric effects are most important for substituents that generate ortho/para substitution products. If a substituent is bulky, then more of the para-substituted product will be formed relative to amounts of the ortho-isomer. Note that there are generally no *electronic* preferences for para versus ortho substitution.

Table 17.2 Reactivity profiles of ⟨benzene⟩—X toward electrophilic aromatic substitution.

Section	X	Reactivity effect of X versus X = H[a]	Position of substitution	Electronic factors influencing reactivity
17.3b	—alkyl	Activating	ortho/para	Inductive only
17.3c	$-\overset{}{O}\underset{Y}{}$, $-\overset{Y}{\underset{Y}{N}}$ (Y = H, alkyl, aryl, –COR)	Activating	ortho/para	Resonance > inductive
17.3d	—X (X = Cl, Br, I)	Deactivating	ortho/para	Inductive > resonance
17.3e	—NO$_2$, $-\overset{O}{\underset{Y}{C}}$, $-\overset{O}{\underset{Y}{S}}$, —C≡N (Y = H, OH, OR, alkyl, aryl)	Deactivating	meta	Resonance = inductive
17.3e	—$\overset{+}{N}R_3$ (R = H, alkyl, aryl)	Deactivating	meta	Inductive only

[a]A substituted benzene derivative with an activating group will react faster than benzene. A derivative with a deactivating group will react more slowly.

Inductive effects are related primarily to differences in the electronegativity values for the carbon atom in the ring and the atom or group attached to it. Most heteroatoms are more electronegative than carbon, so they withdraw electron density from benzene, making the ring more electopositive, and therefore less nucleophilic. A less nucleophilic ring will react more slowly than benzene toward electrophiles.

Resonance effects can stabilize the π system itself, or they can influence stabilization (or destabilization) of the cation intermediate. The following sections will illustrate how inductive and resonance effects relate to the reactivity patterns summarized in Table 17.2.

17.3b THE ALKYL GROUP IS AN ORTHO/PARA DIRECTOR AND ACTIVATOR

Table 17.2 indicates that an alkyl group activates the benzene ring toward reaction with electrophiles. A simple example that illustrates this effect is the reaction of toluene with bromine. Little of the meta-substituted product is formed.

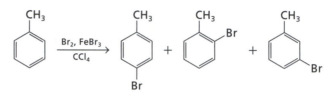

Ratio: 60 : 40 : <1

To understand the observed substitution pattern, look at the possible intermediates formed by the reaction of toluene with an electrophilic reagent. When this reaction occurs, the electrophile can become attached at a position ortho, meta, or para to the methyl group. We use an arrow that originates in the middle of the ring to indicate that a π MO reacts with the electrophile, not an isolated double bond of a Kekulé form.

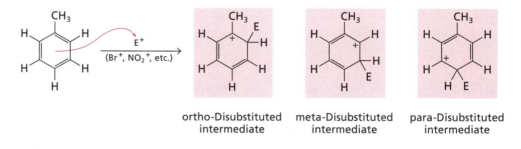

ortho-Disubstituted intermediate   meta-Disubstituted intermediate   para-Disubstituted intermediate

With these three structures before us, we now ask the question, What are the relative stabilities of the carbocation intermediates? For the ortho-disubstituted structure, three good resonance forms can be drawn. The more good resonance structures possible, the more stable the intermediate will be and the more likely it is that products will form via that intermediate.

ortho

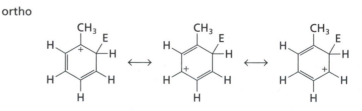

However, three good resonance structures can be written for the other two possible intermediates too.

meta

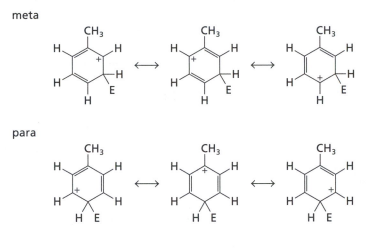

para

It makes sense that resonance effects have little influence on the reactions of toluene because the methyl group has no π bond or unshared electron pair with which to stabilize the positive charge of the intermediate through delocalization. If resonance effects are not important, then inductive effects must govern the electrophilic substitution reactions of toluene and other alkylbenzene compounds. Recall that alkyl groups are considered to be "electron donating" (Section 6.2b), so just as alkyl groups stabilize carbocation centers in aliphatic compounds, they can stabilize the positive charge of the intermediate in electrophilic aromatic substitution *when the charge appears on the* ipso *carbon atom.* This situation occurs only when the incoming electrophile reacts at the ortho and para positions.

Figure 17.9 shows the reaction coordinate diagram for the first step in the reaction of toluene with an electrophile. The Hammond postulate (Section 6.2c) predicts that the pathway leading to formation of the ortho and para isomers will have a lower energy of activation than the one leading to the meta-disubstituted product as a result of stabilization by the electron-donating methyl group. The lower energy of activation means that the ortho- and para-disubstituted products are formed faster, which is consistent with the experimental results.

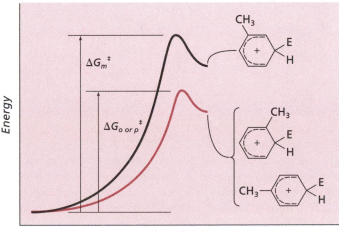

Figure 17.9

A reaction coordinate diagram for the first step of an electrophilic substitution reaction of toluene.

EXERCISE 17.12

Draw the structures of the major products expected from the reaction shown here.

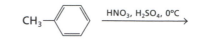

17.3c SUBSTITUENTS THAT HAVE AN ELECTRON PAIR ADJACENT TO THE RING ARE ORTHO/PARA DIRECTORS

Phenol, anisole, and acetanilide react faster than benzene and undergo substitution primarily at the ortho and para positions. The feature that is common to these substances is the presence of an unshared electron pair on the heteroatom attached directly to the ring.

Looking at the possible intermediates formed in the first step of the reaction between anisole and an electrophile, we see that resonance stabilization *by the methoxy group* is possible only when an ortho- or para-substituted intermediate is formed. Shown here are the resonance forms for the carbocation generated by reaction of the electrophile in the position ortho to the methoxy group.

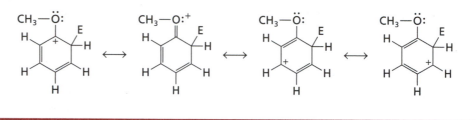

EXERCISE 17.13

Draw the corresponding resonance forms for the para-substituted intermediate.

Delocalization of the positive charge onto the methoxy oxygen atom is not possible if the incoming group attaches meta to the substituent, and only three good resonance forms can be drawn. As in the case of toluene, the free energy of activation for the pathways leading to ortho- and para-disubstituted products will be lower than that leading to the meta-disubstituted product.

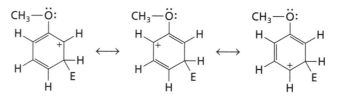

How do we rationalize the observation that anisole reacts faster than benzene? To explain this characteristic, we have to consider how the *starting* compound is influenced by electronic effects.

Consider the inductive effects first. Oxygen is more electronegative than carbon, so a methoxy group should attract electrons from the aromatic π system, making the ring less nucleophilic than benzene itself. Therefore, if inductive effects were important in the electrophilic reactions of anisole, it would react more slowly than benzene.

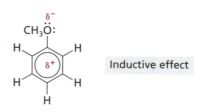

Inductive effect

Now consider the effects of resonance *prior to the reaction* between anisole and an electrophile. We can draw five structures in which each atom has an octet. In three of the structures, a carbon atom within the ring bears a formal charge of −1, and the oxygen atom has a formal charge of +1.

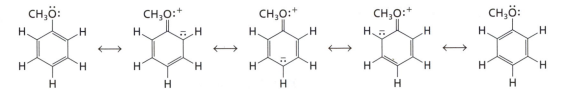

The effects of resonance are therefore consistent with the notion that the π system in anisole is more nucleophilic (i.e., has greater electron density, as denoted by the δ− symbols below) than the one in benzene, which means that anisole should react faster than benzene. The experimental findings lead us to conclude that the resonance effects outweigh the inductive effects in their influence on the reactivity of anisole and related compounds.

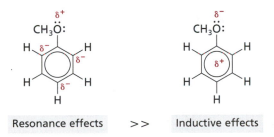

Resonance effects   >>   Inductive effects

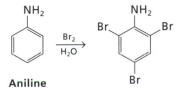

Substituted benzene derivatives that have an unshared electron pair adjacent to the ring can sometimes be so highly activated that monosubstitution is difficult to control. Aniline, which has the ortho/para directing amino group, is one such compound. Aniline reacts with bromine, even without a Lewis acid catalyst, to form 2,4,6-tribromoaniline as the major product.

To carry out the monobromination of aniline, the amine is first converted to acetanilide by the reaction with acetic anhydride (this reaction will be described in Section 21.3b). The acetamido group is an ortho/para-directing group because the nitrogen atom still has an unshared pair of electrons. However, the acetyl group is electron withdrawing, so it moderates the electron-donating influence of the nitrogen atom. After bromination, which produces *p*-bromoacetanilide and *o*-bromoacetanilide (the ortho isomer is not shown in the following equation), the acetyl group is removed by hydrolysis (Section 21.5a).

The acetamido group is an o, p director because the nitrogen atom retains an unshared electron pair.

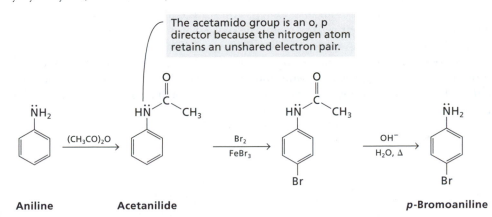

Aniline   Acetanilide   *p*-Bromoaniline

Phenol, like aniline, is also highly reactive in electrophilic substitution reactions; bromination, for example, does not require the use of a Lewis acid. The degree of substitution can be controlled in many instances by varying the temperature. Lower temperatures yield monosubstituted products, whereas higher temperatures lead to the formation of di- and trisubstituted products.

EXERCISE 17.14

Nitration of phenol with excess nitric and sulfuric acid yields an explosive material called picric acid, which has three nitro groups attached to the ring. Draw its structure.

17.3d HALOGEN ATOMS ARE ORTHO/PARA DIRECTORS BUT DEACTIVATING

A halogen atom appears to fit into the category of the ortho/para directors described in Section 17.3c because it has an unshared electron pair. As the following equation shows, substitution does occur in the ortho and para positions. However, halobenzenes react more slowly than benzene.

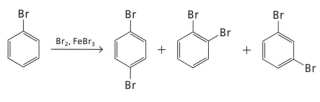

Ratio: 85 : 13 : 2

EXERCISE 17.15

Draw resonance structures for each of the three possible cation intermediates **A, B,** and **C** in the following scheme:

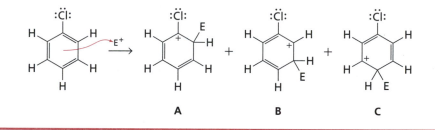

A   B   C

The slower rate of reaction for chlorobenzene versus benzene must mean that inductive effects are more significant than resonance effects; otherwise, its reactivity profile would be the same as that of anisole. One reason for the different influences of oxygen and chlorine relates to their relative sizes: The oxygen atom is smaller than the chlorine atom, so its orbitals overlap more effectively with the orbitals of benzene's π system.

The reactivity of the halobenzenes toward electrophilic substitution can be thought of in this way: A withdrawing inductive effect deactivates the ring, but resonance effects offset this deactivation somewhat at the ortho and para positions. This idea accounts for the overall slower reactions compared with those of benzene, while rationalizing the formation of ortho- and para-disubstituted products.

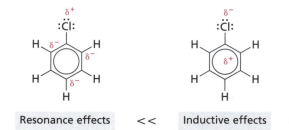

Resonance effects   <<   Inductive effects

17.3e META-DISUBSTITUTED PRODUCTS ARE FORMED WHEN A SUBSTITUENT HAS A POSITIVE OR PARTIAL POSITIVE CHARGE ADJACENT TO THE RING

Compounds such as nitrobenzene, acetophenone, and benzoic acid react with electrophiles at a much slower rate than benzene, and the major product is the meta-disubstituted isomer. For example, in nitration of benzoic acid to form *m*-nitrobenzoic acid,

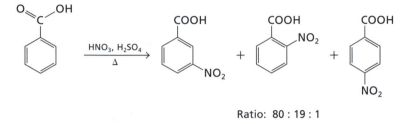

Ratio: 80 : 19 : 1

Any compound with a double (C=N, C=O, N=O, S=O) or triple (C≡N) bond or a positive charge adjacent to the benzene ring displays this reactivity profile. The common feature among these compounds is an electrophilic atom attached directly to the ring.

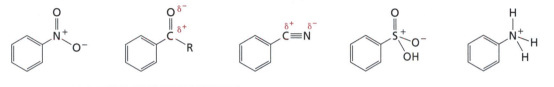

R=H, OH, OR, NR$_2$, alkyl, aryl

These substituents are often called meta directors, which is a misnomer because the incoming electrophile reacts at the meta position by default. Because of resonance effects, most of these groups deactivate the ortho and para positions, as illustrated below for nitrobenzene. Notice the positive charges at positions 2, 4, and 6, relative to the ipso carbon atom.

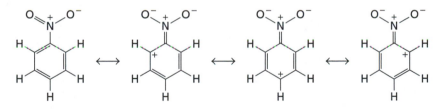

Resonance structures for nitrobenzene put a positive charge at the ortho and para positions.

For substituents in this category, *resonance and inductive effects operate in the same direction* and both decrease the electron density in the ring. These combined effects make the ring with an attached meta director so unreactive that Friedel–Crafts alkylation and acylation cannot be done on these substrates. In fact, nitrobenzene is used frequently as a solvent for Friedel–Crafts reactions.

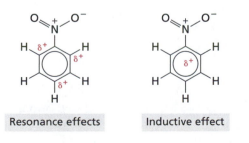

Resonance effects   Inductive effect

EXERCISE 17.16

Benzoic acid undergoes bromination with iron(III) bromide as a catalyst to produce *m*-bromobenzoic acid. Draw the structures of the possible intermediate carbocations formed in the first step of this transformation, and then rationalize why the meta isomer is the major one.

An ammonium ion places a positive charge adjacent to the ring so it is also a meta-director. Like the methyl group, the $-NH_3^+$ group influences electrophilic substitution only through inductive effects because there is no *p* orbital on nitrogen in which to delocalize the electrons in the ring. An ammonium ion substituent is often formed when aniline derivatives are treated with strong acid.

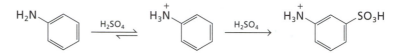

To circumvent the acid–base reaction of aniline that forms the ammonium ion, aniline can be converted to its acetamido derivative by reaction with acetic anhydride, as mentioned in Section 17.3c. The acetamido group is an ortho/para director because the nitrogen atom retains an unshared pair of electrons. But an amide is a weak base and will not be protonated under the reaction conditions. Hydrolysis converts the amido group back to the amino group.

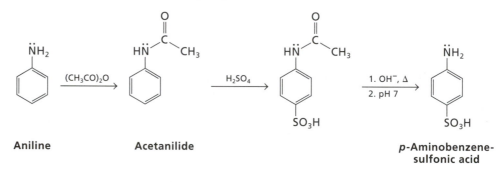

EXERCISE 17.17

α,α,α-Trifluorobenzene ($CF_3C_6H_5$) gives predominantly meta-disubstituted products in electrophilic substitution reactions, and the influence is mainly inductive. Rationalize this result.

17.3f For Polysubstituted Benzene Rings, the Effects of Each Substituent Must Be Evaluated

If two (or more) groups are attached to a ring, electrophilic substitution will be influenced by a combination of three factors:

1. The positions to which the electrophile will be directed by each substituent (ortho/para versus meta).
2. The relative strength of activation or deactivation by each substituent.
3. Steric effects.

Table 17.2 is a summary of the general directing and activating properties of various substituents. With regard to activation and deactivation, substituents can be more precisely classified according to their relative strengths, as shown in Figure 17.10. Two statements about deactivating substituents are in order:

a.

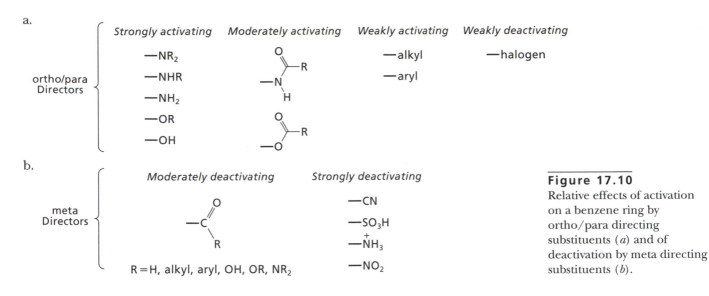

| | Strongly activating | Moderately activating | Weakly activating | Weakly deactivating |

ortho/para Directors

b.

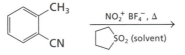

Figure 17.10
Relative effects of activation on a benzene ring by ortho/para directing substituents (*a*) and of deactivation by meta directing substituents (*b*).

- When two meta-directing substituents are attached to a ring, electrophilic substitution occurs with difficulty because the ring is too deactivated.

- Friedel–Crafts alkylation and acylation do not occur if the ring has only a meta-directing substituent. (As noted in Section 17.3e, nitrobenzene is often used as a solvent for the Friedel–Crafts reactions.)

Steric effects result from the repulsions between the incoming electrophile and the substituent and are important mainly for substituents that are ortho/para directors. Generally, larger substituents produce more para- than ortho-disubstituted product. Another manifestation of steric hindrance is that an incoming group rarely enters between two groups that are meta to each other.

To illustrate how these influences operate, several examples follow.

EXAMPLE 17.1

Draw the structures of the major monosubstitution products expected from the reaction shown here. (Note the use of nitronium tetrafluoroborate instead of nitric acid to provide NO_2^+.)

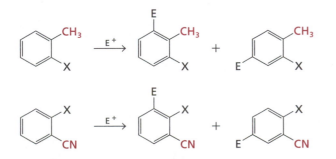

Start by evaluating the effects of each substituent separately: the CH_3 group is an o,p director and activating; the CN group is a meta director and deactivating.

Because each substituent directs the incoming electrophile to the same positions of the ring, we expect the reaction to proceed as follows:

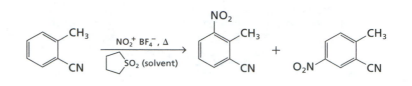

EXAMPLE 17.2

Draw the structures of the major monosubstitution products expected from the following reaction.

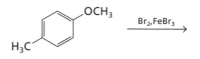

By evaluating the influence of each substituent separately, we write the following equations:

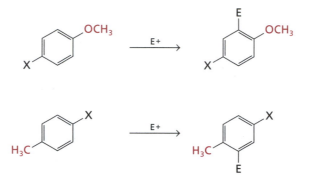

> Substitution at the para position is not considered because that position already bears a substituent in each case.

Both substituents are ortho/para directors, so the magnitude of activation by each must be considered: *The group that is more activating will direct the substitution process.* A methoxy group is more activating than a methyl group (Fig. 17.10), so its influence will be greater. The actual experimental result confirms the greater influence of methoxy versus methyl:

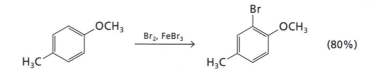

EXAMPLE 17.3

What are the major monosubstitution products of the following reaction?

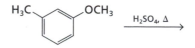

By evaluating the influence of each substituent separately, we write the following equations:

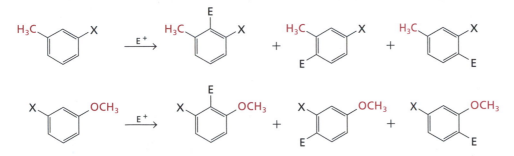

The substituents direct the incoming electrophile to the same positions. Because an incoming group rarely enters between two groups that are meta to each other, we can ignore the first product in each group. The overall reaction is,

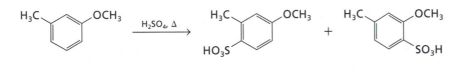

EXAMPLE 17.4

Draw the structure of the expected monosubstitution product for the following reaction.

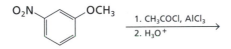

When only a meta directing substituent is attached to the ring, Friedel–Crafts reactions do not occur. However, the activating methoxy group will offset the influence of the deactivating nitro group, so we evaluate the possible products.

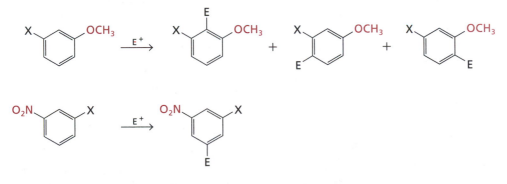

The methoxy group will direct the incoming electrophile, which is the acetyl group. Two products will be formed because the acetyl group will not enter between the two substituents already present.

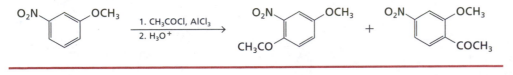

EXERCISE 17.18

Draw the structures of the monosubstitution products expected from each of the following reactions:

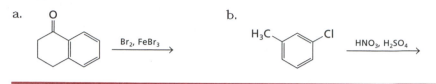

a.   b.

17.3g THE CHEMICAL SHIFTS OF THE AROMATIC PROTONS IN ^{1}H NMR SPECTRA ARE INFLUENCED BY THE SUBSTITUENTS

Even though the discussion in the previous sections has focused on how the substituents attached to a benzene ring affect the course of electrophilic substitution reactions, any properties of an aromatic compound that can be influenced by electronic effects will reflect the nature of the substituents that are present. The chemical shifts of proton resonances in the NMR spectra of aromatic compounds are particularly prone to such influences, and the effects can be summarized as follows:

- The chemical shift of the proton resonance of benzene is δ 7.25.
- The chemical shift for the resonance of a proton ortho or para to an activating group appears upfield from δ 7.25.
- The chemical shift for the resonance of a proton ortho or para to a deactivating group appears downfield from δ 7.25.
- Protons that are ortho to a given substituent produce resonances that are farther from δ 7.25 than those that are para to the same substituent, an influence called the **ortho effect.**

Figure 17.11 illustrates these effects for two benzaldehyde derivatives. The splitting patterns appear as illustrated in Figure 13.12; that is, the identities of the substituents have no effect on the J values.

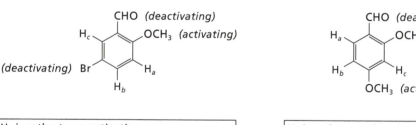

CHO *(deactivating)*
OCH$_3$ *(activating)*
H$_c$
(deactivating) Br   H$_a$
H$_b$

CHO *(deactivating)*
OCH$_3$ *(activating)*
H$_a$
H$_b$ H$_c$
OCH$_3$ *(activating)*

H$_a$ is ortho to an activating group
H$_b$ is ortho to a deactivating group and para to a deactivating group
H$_c$ is ortho to two deactivating groups

H$_a$ is ortho to a deactivating group
H$_b$ is ortho to an activating group and para to an activating group
H$_c$ is ortho to two activating groups

Figure 17.11
The ^{1}H NMR spectra of two, 1,2,4-trisubstituted benzene derivatives in the aromatic region. Each compound displays the singlet and two doublet features expected for this substitution pattern (Fig. 13.12). What differ are their chemical shifts. The J value for each doublet is 8 Hz.

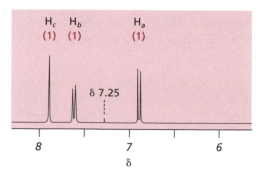

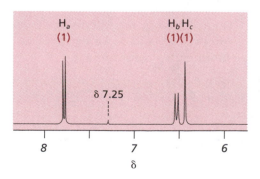

EXAMPLE 17.5

A derivative of benzene produces the NMR spectrum shown here. What is the substitution pattern on the ring, and what are the natures of the substituents?

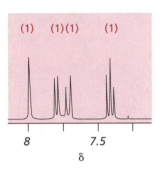

The total integrated intensity for the signals in the aromatic region equals 4, so the ring is disubstituted. From the data shown in Figure 17.5, we deduce that the two substituents are meta, because a singlet, two doublets, and a triplet are observed. Because all of the signals have chemical shifts downfield from δ 7.25, we conclude that the two substituents are deactivating.

EXERCISE 17.19

A derivative of benzene produces the NMR spectrum shown here. What is the substitution pattern on the ring, and what are the natures of the substituents?

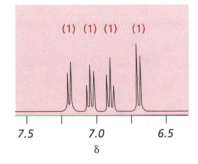

17.4   OTHER REACTIONS OF BENZENE AND ITS DERIVATIVES

17.4a   ALKYLBENZENES REACT WITH OXIDIZING AGENTS

The electrophilic reactions of benzene and its derivatives provide compelling evidence that these molecules do not react like alkenes. Therefore, it should come as no surprise that the benzene ring is inert toward many of the oxidizing agents that react with alkenes (Section 11.3).

On the other hand, methyl and 1° and 2° alkyl groups attached to the benzene ring can be oxidized to form benzoic acid derivatives. These reactions occur because radical intermediates can be formed at the benzylic carbon atom by transfer of a hydrogen atom to an oxidizing agent, and the intermediate is stabilized by resonance.

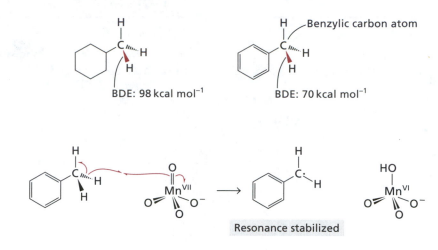

The radical that is formed initially undergoes single-electron transfer and oxygen-atom transfer reactions with the reagent, which is generally potassium permanganate ($KMnO_4$) or a salt of the dichromate ion ($Na_2Cr_2O_7$ or $K_2Cr_2O_7$).

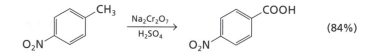

(84%)

Permanganate ion requires the use of basic conditions, so a separate acidification step is necessary to liberate the carboxylic acid during workup.

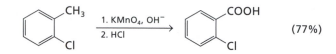

(77%)

EXERCISE 17.20

What will be the major product from each of the following reactions?

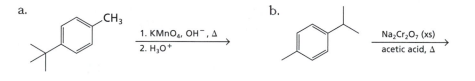

a.

1. $KMnO_4$, OH^-, Δ
2. H_3O^+

b.

$Na_2Cr_2O_7$ (xs)
acetic acid, Δ

17.4b ANILINE DERIVATIVES ARE CONVERTED TO DIAZONIUM COMPOUNDS

Even though electrophilic substitution reactions provide general ways to attach different groups to the ring of benzene and its derivatives, some electrophiles react with the substituent itself instead of the ring. In particular, the nitrosonium ion, NO^+, reacts with aniline and its derivatives to form **diazonium compounds,** which can be explosive. While aryl diazonium compounds can be handled safely enough in solution, they are rarely isolated.

Nitrous acid, HONO, is formed when aqueous sodium nitrite is treated with a mineral acid. Nitrous acid is stable only below ~ 5°C, so it is made in situ (i.e., in the reaction mixture, not beforehand). The HONO itself reacts with the mineral acid to generate NO^+. The most commonly used acid is HCl, but H_2SO_4 and HBF_4 are also employed.

$$\boxed{1}\quad HCl \;+\; NaNO_2 \;\xrightarrow[\text{0-5°C}]{H_2O}\; HONO \;+\; NaCl$$

$$\boxed{2}\quad HONO \;+\; HCl \;\rightleftharpoons\; NO^+\,Cl^- \;+\; H_2O$$

When a derivative of aniline is treated with HONO, the nucleophilic amino group reacts with the electrophilic NO^+. A series of acid–base reactions occur to produce the diazonium ion, the counterion of which derives from the mineral acid that is initially used.

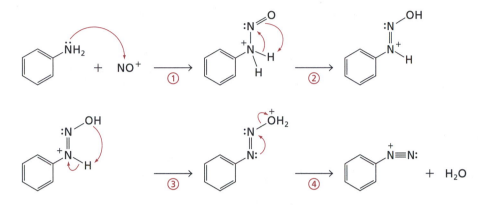

17.4c   AN ARYL DIAZONIUM COMPOUND IS CONVERTED TO AN ARENE DERIVATIVE

Molecular nitrogen is arguably the best possible leaving group that exists, so you might expect N_2 to dissociate from a diazonium ion to form a phenyl cation.

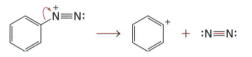

If a nucleophile were present, it could react and form benzene derivatives by substitution. Substitution *does* take place, as summarized in Figure 17.12, but evidence suggests that radical intermediates are actually involved.

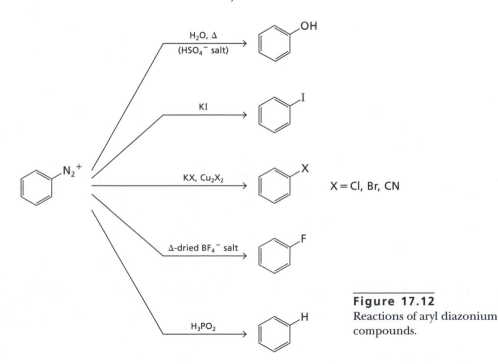

Figure 17.12
Reactions of aryl diazonium compounds.

Phenols are made from the reaction of diazonium salts with water. The diazonium salt is formed using sodium nitrite and sulfuric acid so that no chloride ion is present to react and form chlorobenzene (chloride ion is a good enough nucleophile to compete with water).

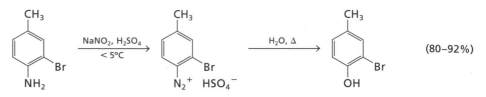

Iodide ion is a potent nucleophile (Section 6.1c), and it reacts with diazonium compounds to form iodobenzene derivatives. Direct iodination of arenes is not as facile as chlorination and bromination (Section 17.2b), so this method is a good way to prepare aryl iodides.

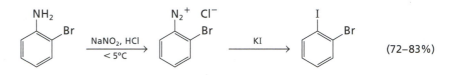

Chloride, bromide, and cyanide ion can replace the N_2 leaving group, but a copper(I) salt must be supplied to the reaction mixture. The copper(I) ion likely reduces the diazonium salt to form an aryl radical intermediate. This procedure, called the **Sandmeyer reaction,** produces the corresponding arenes in high yields.

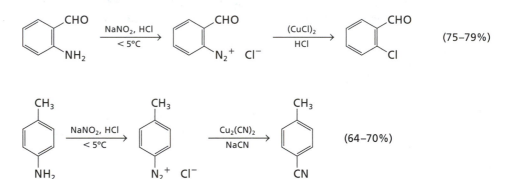

If HBF_4 is added to the diazonium reaction mixture, then the BF_4^- salt often precipitates from solution.

H_2N—⟨benzene⟩—⟨benzene⟩—NH_2   1. HCl,NaNO_2, < −5°C / 2. HBF_4 → BF_4^- $^+N_2$—⟨benzene⟩—⟨benzene⟩—N_2^+ BF_4^-

Diazonium salts having the BF_4^- anion, which are more stable than the chloride or hydrogen sulfate salts, can be dried in air without fear of explosion. When this dried salt is heated, decomposition occurs to form the aryl fluoride, N_2, and BF_3. These byproducts are gases, so the aryl fluoride often distills from the flask in pure form.

BF_4^- $^+N_2$—⟨benzene⟩—⟨benzene⟩—N_2^+ BF_4^- $\xrightarrow[-N_2, -BF_3]{\Delta}$ F—⟨benzene⟩—⟨benzene⟩—F   (54–56%)

Hypophosphorus acid reduces the *carbon–nitrogen* bond, replacing the diazonium group with a hydrogen atom.

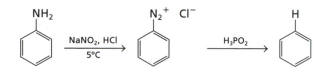

This transformation is used to exploit the activating nature of an amino group, which is subsequently removed from the ring. For example, *m*-aminobenzoic acid can be tribrominated.

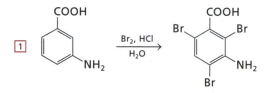

The amino group is then replaced with a proton by forming the diazonium ion and treating it with hypophosphorus acid.

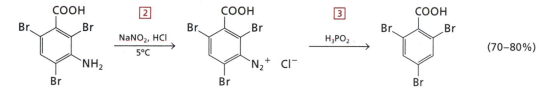

Because the COOH group is a meta director and deactivating, it would not be possible to obtain 2,4,6-tribromobenzoic acid by directly brominating benzoic acid.

EXERCISE 17.21

Show how you would prepare the following compounds from any monosubstituted benzene derivative using electrophilic substitution or a diazonium compound, as appropriate:

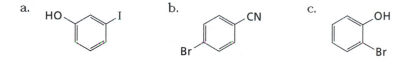

a.

b.

c.

17.4d DIAZONIUM COMPOUNDS UNDERGO COUPLING WITH SOME ARENES

The diazonium ion has two resonance forms, one of which places the positive charge on the terminal nitrogen atom.

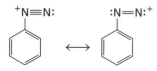

If an *activated* aromatic compound is added to the solution of a diazonium salt, then electrophilic substitution can occur. Only the most activated substrates—derivatives of aniline and phenol—can be employed in this reaction, because the diazonium ion is a weak electrophile. The reaction follows the typical electrophilic substitution mechanism:

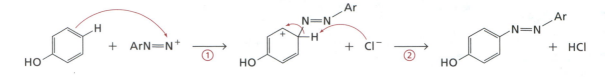

This coupling reaction produces molecules that are brightly colored as a result of extended conjugation of the π systems across the nitrogen–nitrogen double bond. Such compounds find applications as dyes. For example, Alizarin Yellow R is employed for dyeing wool, and *p*-(dimethylamino)azobenzene was once utilized as a colorant for margarine.

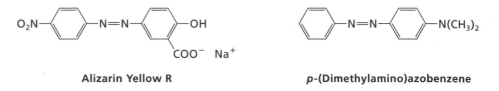

Alizarin Yellow R   ***p*-(Dimethylamino)azobenzene**

EXERCISE 17.22

Show how you would prepare Alizarin Yellow R starting with aniline and salicylic acid (*o*-hydroxybenzoic acid).

17.4e CERTAIN HALOBENZENES UNDERGO NUCLEOPHILIC SUBSTITUTION

It is vital to begin this section by emphasizing a point made in Section 6.1d: *An aryl halide does not undergo nucleophilic substitution reactions by the S_N1 or S_N2 mechanism.*

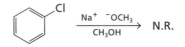

If a nitro group is in the position ortho or para to a halogen atom, however, then the carbon atom bearing the halogen atom is electrophilic and susceptible to reactions with nucleophiles. The chloride ion is a good leaving group, so it can be replaced.

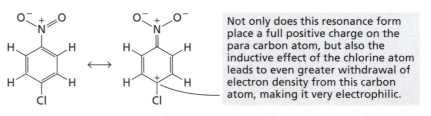

Not only does this resonance form place a full positive charge on the para carbon atom, but also the inductive effect of the chlorine atom leads to even greater withdrawal of electron density from this carbon atom, making it very electrophilic.

The reaction between *p*-chloronitrobenzene and methoxide ion typifies this transformation, which can be designated as the **S_NAr mechanism** (substitution, nucleophilic, aromatic) to differentiate if from the S_N1 and S_N2 mechanisms. In the first step, an intermediate is formed that has two heteroatoms attached to the same carbon atom. This intermediate is resonance stabilized.

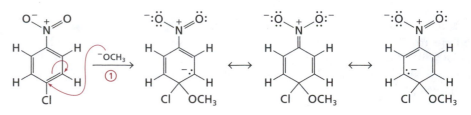

In the second step, the chloride ion is displaced to regenerate the aromatic system.

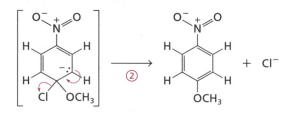

EXERCISE 17.23

Propose a mechanism for the following reaction, showing how the intermediate is stabilized by resonance:

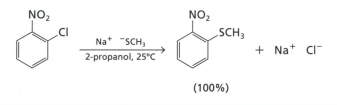

(100%)

One way we can be certain that the S_NAr mechanism differs from the S_N1 or S_N2 mechanisms is that fluoride ion is a good leaving group in the S_NAr process, but it is *not* a good leaving group in the S_N1 and S_N2 mechanisms. Aryl fluorides are actually better substrates than the corresponding aryl chlorides for the S_NAr reaction because the highly electronegative fluorine atom makes its attached carbon atom more susceptible to reaction with a nucleophile. The nitro group is still required to be present, and it must be ortho or para to the fluorine atom.

17.4f BENZYNE CAN BE MADE BY AN ELIMINATION REACTION

While on the topic of nucleophilic substitution reactions of arenes, consider the reaction shown here, which was originally thought to occur by a nucleophilic substitution pathway (S_N2 or S_NAr).

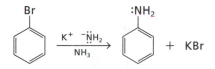

When the bromo benzene molecule was labeled with ^{14}C at the ipso carbon atom, the product was found to have the labeled atom in two places.

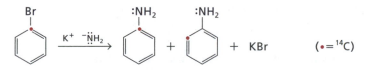

In fact, *elimination* occurred under the reaction conditions of strong base, producing an unsaturated benzene derivative called **benzyne.**

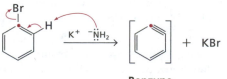

Benzyne

Addition of ammonia to benzyne accounted for the formation of aniline and explained why only 50% of the product has a nitrogen atom attached to the original ipso carbon atom.

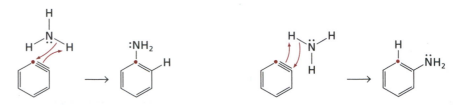

Benzyne, the parent member of a class of compounds called **arynes,** is highly reactive. For example, it reacts with 1,3-dienes by a [4 + 2] cycloaddition process (Section 10.4b) to produce a fused-ring system. Other methods now exist to make benzyne derivatives, and many of them require milder conditions than the one that utilizes potassium amide.

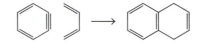

CHAPTER SUMMARY

Section 17.1   Structural aspects of aromatic molecules

- Benzene and other aromatic compounds have a special stability that manifests itself with an unexpectedly low enthalpy of hydrogenation.
- The energy difference between the experimental and theoretical values of $\Delta H°_{hydrog}$ for aromatic compounds is called the resonance energy.
- Aromatic compounds, also called arenes, have a planar ring with an uninterrupted π system that contains $4n + 2$ electrons.
- The stability of an arene is related to the number of filled molecular orbitals, which results when the π system has an odd number of *pairs* of electrons.
- Aromatic compounds can have multiple rings and/or heteroatoms in the ring. Molecules of this latter type are called heterocycles.
- Benzene compounds are named with the root *benz* or a common name that has been incorporated into the IUPAC system.
- When a benzene ring has only two substituents, the positional relationships can be specified by the prefixes ortho (*o-*) for 1,2-disubstitution, meta (*m-*), for 1,3-disubstitution, and para (*p-*), for 1,4-disubstitution.
- The positional relationships of substituents on benzene can be determined by looking at the splitting patterns of the peaks in the aromatic region of the ^{1}H NMR spectrum. The principal pattern results from coupling between protons on adjacent carbon atoms, which is called ortho coupling.

Section 17.2   Electrophilic substitution reactions of benzene

- Benzene and its derivatives undergo substitution of a hydrogen atom rather than addition to their π bonds.
- The mechanism of electrophilic aromatic substitution involves two steps. In the first step, the π system reacts with the electrophile, forming a cationic intermediate. In step two, a base removes a proton to restore aromaticity.
- Typical electrophiles are Br^+, Cl^+, NO_2^+, HSO_3^+, R^+, and RCO^+. Reactions of benzene with these electrophiles produce bromobenzene, nitrobenzene, benzenesulfonic acid, alkylbenzenes, and acylbenzenes, respectively.

- Rearrangement occurs under the conditions of Friedel–Crafts alkylation reactions, so a primary alkyl group cannot be appended to the ring by this method. Polysubstitution and reversibility are other problematic side reactions.

- Friedel–Crafts acylation followed by reduction is the method used to append an alkyl chain having a CH_2 group adjacent to the ring.

Section 17.3   Electrophilic substitution reactions of benzene derivatives

- A substituted benzene reacts to form products in which the incoming electrophilic group is attached either ortho and para or meta to the substituent.

- There are five categories of substituents. Their influences on electrophilic aromatic substitution are summarized in Table 17.2.

- When two or more substituents are attached to the benzene ring, a competition exists between them in their influences on product formation. Factors to be considered include the positions to which the electrophile will be directed by each substituent (ortho/para versus meta), the relative strength of activation or deactivation by each substituent, and steric effects.

Section 17.4   Other reactions of benzene and its derivatives

- Benzene is inert toward oxidizing agents, but alkylbenzene derivatives that have at least one benzylic hydrogen atom are oxidized to the corresponding benzoic acid derivatives.

- Aniline and its derivatives react with nitrous acid to form diazonium ions, ArN_2^+. These species are unstable but react readily with nucleophilic reagents by substitution of the N_2 group. Reactions of diazonium compounds are summarized in Figure 17.12.

- Diazonium ions can also react as electrophiles with activated arenes to form highly colored substances that are commonly used as dyes.

- A halobenzene compound having a nitro group ortho or para to the halogen atom can undergo nucleophilic substitution, designated as the S_NAr mechanism. Fluoride ion is the best leaving group in these reactions.

- Under the influence of strong base, a halobenzene compound can undergo elimination to form a reactive intermediate called benzyne that formally has a triple bond.

KEY TERMS

Section 17.1a
resonance energy

Section 17.1b
aromaticity
aromatic compounds
Hückel's rule
$4n + 2$ rule

Section 17.1c
annulenes
antiaromatic

Section 17.1e
heterocycles

Section 17.1f
ortho
meta
para

Section 17.1g
ortho coupling

Section 17.2a
electrophilic aromatic
   substitution

Section 17.2c
nitronium ion

Section 17.2e
Friedel–Crafts alkylation
Friedel–Crafts acylation

Section 17.3g
ortho effect

Section 17.4b
diazonium compound

Section 17.4c
Sandmeyer reaction

Section 17.4e
S_NAr mechanism

Section 17.4f
benzyne
aryne

REACTION SUMMARY

Section 17.1a

Hydrogenation of benzene and its derivatives requires high temperature and pressure of hydrogen; the product is a cyclohexane derivative.

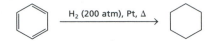

Section 17.2b

Halogenation. Bromine and chlorine can be substituted for hydrogen on the benzene ring in the presence of a Lewis acid.

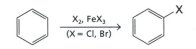

Section 17.2c

Nitration. A mixture of nitric and sulfuric acids is normally used to introduce the NO_2 group. Nitronium salts can also be used as reagents to avoid strong acid conditions.

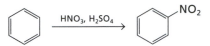

Section 17.2d

Sulfonation. Concentrated sulfuric acid or "oleum", a solution of SO_3 in sulfuric acid, is used to replace a hydrogen atom with the sulfonic acid group.

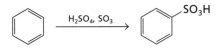

Section 17.2e

Friedel–Crafts alkylation. An alkyl chloride reacts with aluminum chloride to form a carbocation intermediate, which subsequently reacts to replace a hydrogen atom with an alkyl group. The alkyl group can undergo rearrangement, especially when a primary alkyl halide is used.

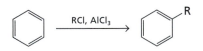

Friedel–Crafts acylation. An acyl chloride reacts with aluminum chloride to form an acyl cation intermediate, which subsequently reacts to replace a hydrogen atom with an acyl group. Methods to reduce the carbonyl group to a methylene group are summarized in Figure 17.8. This reduction provides a way to attach a primary alkyl chain to the benzene ring without the problem of rearrangement.

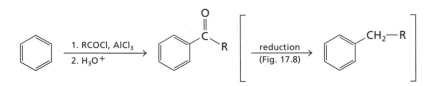

Section 17.4a

Oxidation of alkylbenzene derivatives: When the benzylic position bears a hydrogen atom, an alkyl group can be converted to the carboxylic acid group.

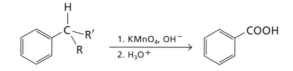

Section 17.4b

Diazonium compounds are formed by the reaction of anilines with HONO at 0–5°C. An aqueous solution of HONO is made from $NaNO_2$ and a mineral acid, usually HCl.

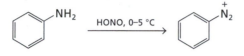

Section 17.4c

Substitution of the diazonium group by nucleophiles: The N_2 group is readily replaced using different reagents (Fig. 17.12), many of which are good nucleophiles.

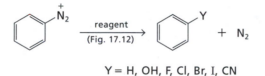

Y = H, OH, F, Cl, Br, I, CN

Section 17.4d

Electrophilic aromatic substitution occurs with diazonium ions and activated arenes (phenols, ethers, and aniline derivatives). The product is an azo dye.

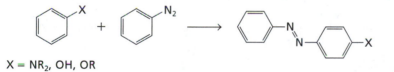

X = NR_2, OH, OR

Section 17.4e

Nucleophilic aromatic substitution of nitrohaloarenes: The nitro group must be ortho or para to the leaving group. Fluoronitrobenzene derivatives are the most reactive.

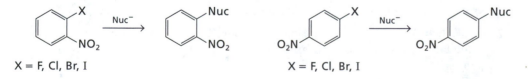

X = F, Cl, Br, I   X = F, Cl, Br, I

Section 17.4f

Benzyne formation: A halobenzene reacts with strong base by elimination to form a reactive species that formally has a triple bond.

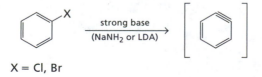

X = Cl, Br

ADDITIONAL EXERCISES

17.24.   Give a systematic name for each of the following compounds:

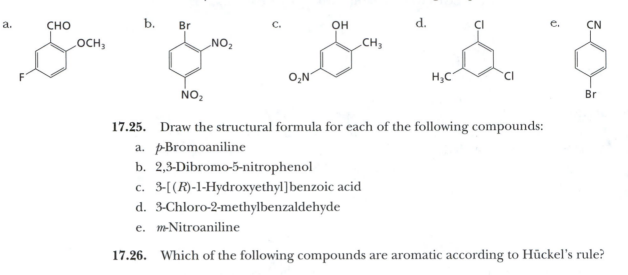

17.25.   Draw the structural formula for each of the following compounds:

a.   *p*-Bromoaniline
b.   2,3-Dibromo-5-nitrophenol
c.   3-[(*R*)-1-Hydroxyethyl]benzoic acid
d.   3-Chloro-2-methylbenzaldehyde
e.   *m*-Nitroaniline

17.26.   Which of the following compounds are aromatic according to Hückel's rule?

a.   The cyclobutadienyl dianion   b.   The cyclooctatetraenyl dianion

c.   The cycloheptatrienyl anion   d.   e.

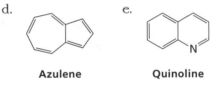

Azulene   **Quinoline**

17.27.   Which compound in each pair will be more reactive toward electrophilic bromination? If the reactivity of the two compounds is the same, indicate neither.

a.   Bromobenzene or toluene   b.   *p*-Xylene or *o*-xylene
c.   Nitrobenzene or chlorobenzene   d.   Anisole or toluene

17.28.   Which of the following groups are *o,p*-directing substituents?

a.   —CHO   b.   —Br   c.   —OH   d.   —N⟨

e.   —C(=O)—OCH₃   f.   —O—C(=O)—CH₃   g.   —N(H)—C(=O)—CH₃   h.   ⬡—

17.29.   Which isomer of xylene forms only a single monosubstituted product when treated with a mixture of nitric and sulfuric acids? Write equations showing the reactions.

17.30.   What is the expected mononitro product obtained by treating each of the following compounds with a mixture of nitric and sulfuric acids? There may be more than one major product for some parts of this exercise.

a.   b.   c.

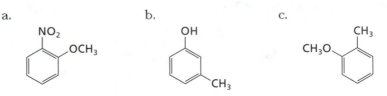

17.31. What is the expected monobromo product obtained by treating each of the following compounds with a mixture of bromine and iron(III) bromide? There may be more than one major product for some parts of this exercise.

a.   b.   c.

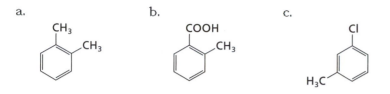

17.32. What is the expected mono(sulfonic acid) product obtained by treating each of the following compounds with hot sulfuric acid? There may be more than one major product for some parts of this exercise.

a.   b.   c.

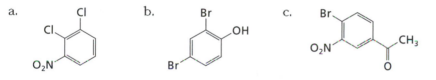

17.33. Which of the following compounds can be prepared as a major product via electrophilic substitution of a disubstituted benzene derivative? Show the reaction as an equation having the reagents with the arrow. If the compound cannot be made in a single step, explain what the problem is, and then propose a method to prepare it otherwise.

a.   b.   c.

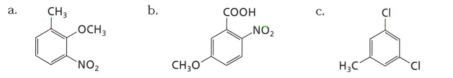

17.34. Repeat Exercise 17.33 for the following substances:

a.   b.   c.

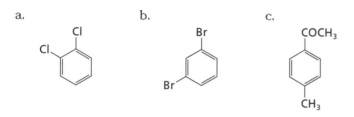

17.35. The NO group in nitrosobenzene (below) is an o,p director, in contrast with the NO$_2$ group in nitrobenzene.

Nitrosobenzene

Draw resonance structures for the cationic intermediates formed by the possible reactions of nitrosobenzene with E$^+$, and explain why PhNO and PhNO$_2$ react differently.

17.36. Styrene reacts with aqueous sulfuric acid according to the following equation. Propose a reasonable mechanism for this transformation, which involves protonation of the alkene double bond in the first step.

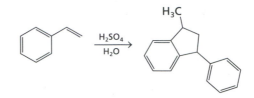

17.37. The *Gatterman–Koch aldehyde synthesis* is used to make benzaldehyde derivatives, and it involves treating an aromatic hydrocarbon with carbon monoxide and hydrogen chloride in the presence of aluminum chloride. This procedure is the equivalent of a Friedel–Crafts reaction with formyl chloride, HCOCl, which itself is unstable. Propose a reasonable mechanism for the following reaction:

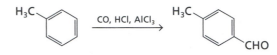

17.38. Activated benzene derivatives with a methoxy group attached to the ring will react with *N*-iodosuccinimide, NIS, to yield iodinated compounds. Propose a mechanism for the following reaction: (*Hint:* Consider how the N–I bond is polarized.)

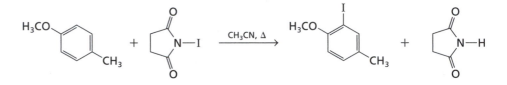

17.39. How would you prepare each of the following compounds from benzene or toluene and any other reagents? More than one step may be necessary.

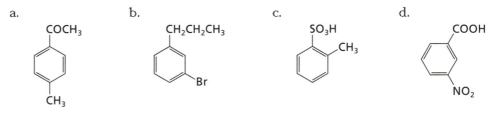

17.40. Compound **X,** an aromatic compound with the formula C_7H_7Br, can be nitrated with a mixture of nitric and sulfuric acids, producing two isomeric nitro compounds, **Y** and **Z**. These compounds are reduced with Zn in acetic acid, giving two aniline derivatives **P** and **R**. Reaction of the aniline compounds with excess bromine gives compounds **Q** and **T,** both with the formula $C_7H_7Br_2N$. What is compound **X?**

17.41. Another way to make benzyne is to generate the diazonium salt of anthranilic acid. Propose a mechanism for this process that also produces carbon dioxide and nitrogen.

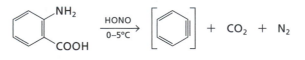

Anthranilic acid

17.42. Draw the structure(s) of the major product(s) expected from each of the following reactions. Indicate the stereochemistry of the product as appropriate. Relative stereochemistry should be shown using wedges and dashed lines. If a racemic mixture will be formed, draw the structure of one enantiomer and write the word "racemic", or draw both enantiomeric structures. If diastereomers are formed, draw each structure; label meso compounds as such. If no reaction occurs, write N. R.

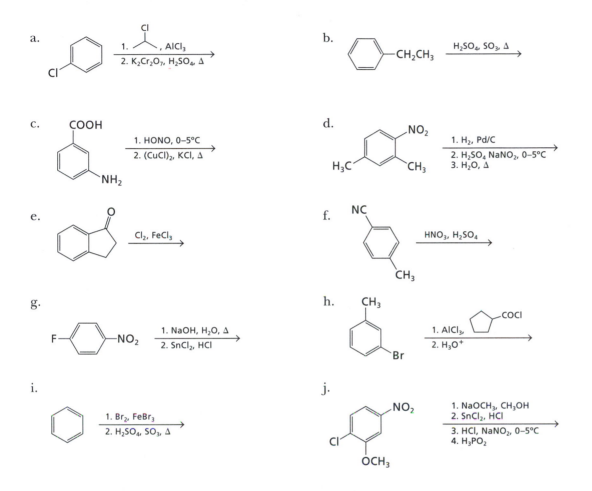

17.43. Tertiary amines generally react reversibly at *nitrogen* with NO⁺.

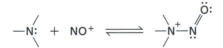

N,N-Dialkylanilines, however, undergo substitution with this electrophile at the ortho and para positions of the benzene ring. Propose a reasonable mechanism for this transformation.

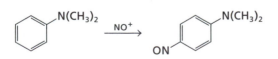

17.44.   Draw the structure of compound **A**, which has the following IR, ^{13}C NMR, and ^{1}H NMR spectra. From the high-resolution CI mass spectrum, MW = 166.068.

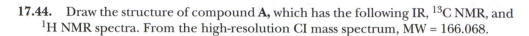

IR spectrum

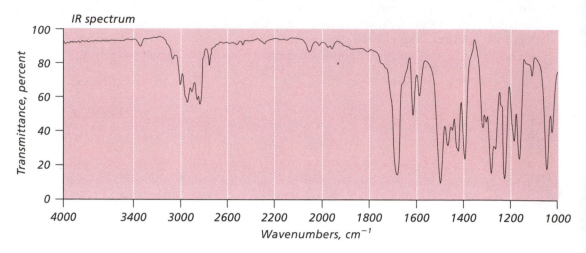

^{13}C NMR spectrum

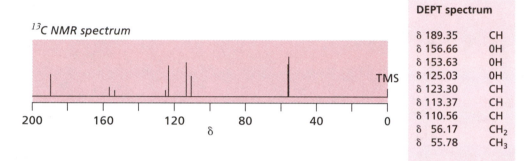

DEPT spectrum

δ 189.35	CH
δ 156.66	OH
δ 153.63	OH
δ 125.03	OH
δ 123.30	CH
δ 113.37	CH
δ 110.56	CH
δ 56.17	CH$_2$
δ 55.78	CH$_3$

^{1}H NMR spectrum

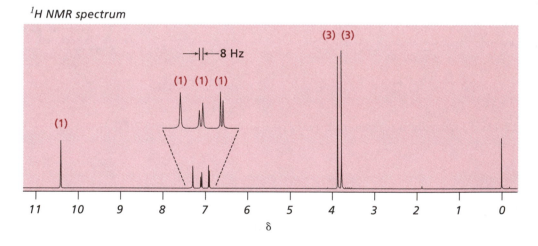

17.45.   Draw the structure of the molecule with the formula C_6H_6BrN that produces the following 300-MHz ^1H NMR spectrum; all of the J values are ~ 8 Hz.

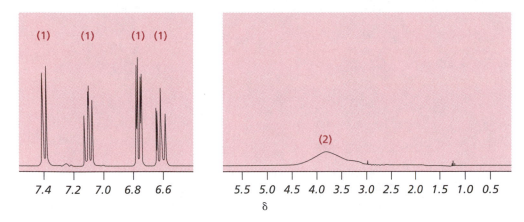

17.46.   Draw the structure of compound **B,** which gives a low-resolution mass spectrum having $M^{+\cdot} = 191$. The relative intensity values for the $M^{+\cdot}$, $(M+2)^{+\cdot}$, and $(M+4)^{+\cdot}$ peaks are 3 : 4 : 1 [*Hint:* Consider what atoms are required in order for a molecule to have a sizable $(M + 4)^{+\cdot}$ peak.] The ^1H NMR spectrum of **B** is shown below; there are no peaks except those in the aromatic region.

^1H NMR spectrum

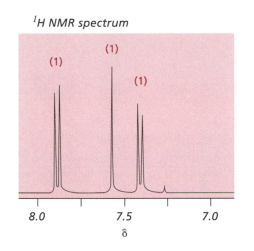

NUCLEOPHILIC ADDITION REACTIONS OF ALDEHYDES AND KETONES

The carbonyl group comprises the carbon–oxygen double bond, arguably the most important substructure found in organic molecules. The variety of transformations associated with functional groups that have a C=O bond accounts for the presence of the carbonyl group in a variety of molecules that are involved in biochemical processes. The reactions of the carbonyl group in aldehydes and ketones occur as they do because the carbonyl substituents are either carbon or hydrogen atoms. (In contrast, the carbonyl group of carboxylic acids and their derivatives is bonded to at least one heteroatom, which can function as a leaving group, as you will learn in Chapter 21.)

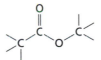

An aldehyde   **A ketone**   **A carboxylic acid**   **A carboxylic acid ester**

An acid chloride   **An amide**   **A thioester**   **A carboxylic acid anhydride**

In this chapter, we examine addition reactions that take place with the C=O double bond of aldehydes and ketones. These transformations include the addition of cyanide ion, water, the hydrogen halides, organometallic compounds, hydride ion, and some oxidizing agents. The products of these reactions usually have an OH group bonded to what was the original carbonyl carbon atom.

18.1 GENERAL ASPECTS OF NUCLEOPHILIC ADDITION REACTIONS

18.1a THE CARBONYL GROUP IS THE STRUCTURAL CENTER OF ALDEHYDES AND KETONES

Before we look at the reactions that aldehydes and ketones can undergo, let us briefly review some structural aspects of these two molecule classes. The carbonyl group in aldehydes is attached to a hydrogen atom and to a carbon-containing group. The only exception is formaldehyde, the common name for methanal, in which the carbonyl group is bonded to two hydrogen atoms. The names of aliphatic aldehydes have the suffix *al* or *carbaldehyde* (Table 1.1), depending on whether the compound is acyclic or cyclic, respectively. In acyclic compounds, the aldehyde group must be at the end of the chain, and its carbonyl carbon atom defines which end is designated C1. In cyclic compounds, the point of attachment of the CHO group defines C1 of the ring. Aldehyde derivatives of benzene are named using the root word *benzaldehyde*.

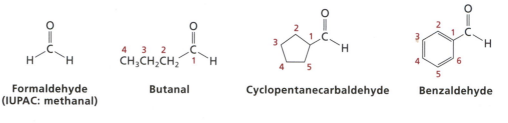

Formaldehyde
(IUPAC: methanal)   **Butanal**   **Cyclopentanecarbaldehyde**   **Benzaldehyde**

Ketones have two carbon-containing groups bonded to the carbonyl carbon atom, and their names have the suffix *one* (Table 1.1). Ketones having a benzene ring as one of the carbonyl substituents are often known by common names, for example, acetophenone, propiophenone, and benzophenone. In the IUPAC system, these compounds have the carbonyl group at the end of the chain, so its carbon atom is C1. (Notice that ketones with two aliphatic groups cannot have the carbonyl group at C1 because such compounds would be classified as aldehydes.)

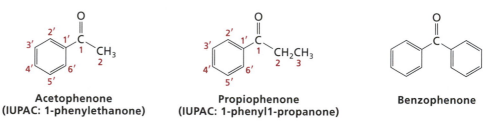

Acetophenone
(IUPAC: 1-phenylethanone)   **Propiophenone**
(IUPAC: 1-phenyl1-propanone)   **Benzophenone**

An older system for naming ketones specifies the two substituents that are attached to the carbonyl group (smaller substituent first), followed by the work *ketone*. This text will not use these names, but you may encounter them in other places.

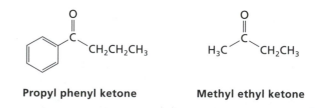

Propyl phenyl ketone   **Methyl ethyl ketone**

The primacy of the carbonyl group in the structures of aldehydes and ketones makes spectroscopic identification of these molecules straightforward. The ^{13}C NMR spectra of aldehydes and ketones have an absorption peak in the downfield region (δ 160–250); in fact, the ketone carbonyl group displays a signal farther downfield than those of other carbonyl-containing molecules (Fig. 13.23 and Table 13.4). The signals in the NMR spectra of aldehydes are also unique: In the ^{13}C NMR spectrum, the aldehyde carbonyl group is the only type that is coupled to a proton, which can be observed in the off-resonance and DEPT spectra (Section 13.5d). In the ^{1}H NMR spectrum, the aldehyde proton appears ~ δ 10, which is farther downfield than any other signal for a proton attached to a carbon atom (Fig. 13.3).

Infrared spectroscopy is another technique well suited to identify the presence of the aldehyde and ketone functional groups. Both compound classes display intense bands in their IR spectra ~ 1700 cm^{-1} that are assigned to their carbonyl stretching vibrations (Section 14.3f). Aldehydes also have absorption bands ~ 2720 cm^{-1} that are generated by the C–H stretching vibration of the CHO group.

The preparative methods for aldehydes and ketones most commonly employ starting materials with the alcohol, alkene, or alkyne functional groups. Aromatic ketones can be made by the Friedel–Crafts acylation reaction. The methods that have been presented so far are summarized in Table 18.1.

Table 18.1   A summary of preparative methods for aldehydes (**A**) and ketones (**K**).

Procedure	Section
Electrophilic addition of water to alkynes (**K**)	9.2b
Hydroboration/oxidative hydrolysis of alkynes (**A**)	9.4c
Ozonolysis of alkenes (**A, K**)	11.3b
Lemeiux–Johnson cleavage of alkenes (**A, K**)	11.3d
Oxidation of alcohols: 1° (**A**) and 2° (**K**)	11.4
Friedel–Crafts acylation of aromatic compounds (**K**)	17.2e

18.1b THE CARBONYL GROUP IS THE REACTIVITY CENTER OF ALDEHYDES AND KETONES BECAUSE ITS CARBON ATOM IS ELECTROPHILIC

The polarity of the carbonyl group makes aldehydes and ketones versatile precursors for the construction of other organic molecules. The Grignard reaction, for example, is used to make new carbon–carbon bonds by addition of a carbanion to the carbonyl group (Section 15.2b). The carbon atom of a carbonyl group bears a partial positive charge that results from the polarity of the C=O double bond; furthermore, we can draw a resonance form, shown below at the right, that places a positive charge on the carbon atom.

The electrophilic carbon atom is susceptible to reactions with nucleophiles.

When present in solution with an aldehyde or ketone, a nucleophile reacts with the electrophilic carbon atom of the carbonyl group. Electrons in the π bond move to the oxygen atom to generate a nucleophilic site.

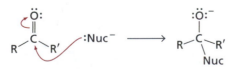

Aqueous acid workup leads to protonation of the oxygen atom, but electrophiles other than H$^+$ can be used to react with the oxygen atom if desired.

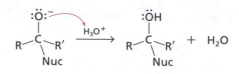

For example, organochlorosilanes, which have an oxyphilic silicon atom, react readily to form silylated derivatives.

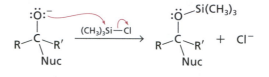

18.1c NUCLEOPHILIC ADDITION PROCESSES ARE OFTEN FACILITATED BY PROTONATION OF THE CARBONYL OXYGEN ATOM

If the nucleophile used for addition to a carbonyl group is the conjugate base of a moderately strong acid, a second general mechanism is available for the reactions of aldehydes and ketones. In this pathway, the conjugate acid of the nucleophile serves to protonate the carbonyl oxygen atom. This acid–base reaction *activates* the carbonyl group by making its carbon atom more electrophilic.

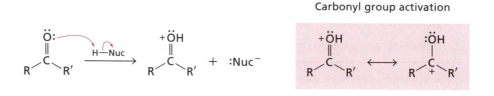

Because a protonated carbonyl group is more electrophilic than its neutral form, it is more susceptible toward reaction with a nucleophile that is present.

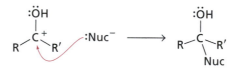

18.2 NUCLEOPHILIC ADDITION REACTIONS

18.2a HYDROGEN CYANIDE ADDS TO A CARBONYL GROUP UNDER BASIC CONDITIONS TO PRODUCE A CYANOHYDRIN

Direct addition of a nucleophile to a carbonyl group is exemplified by the reaction between an aldehyde and HCN, which proceeds readily under basic conditions, yielding a **cyanohydrin.**

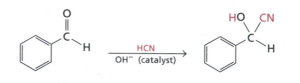

Benzaldehyde   **Benzaldehyde cyanohydrin**
                                    (2-hydroxy-2-phenylethanenitrile)

Hydrogen cyanide is a polar molecule and a weak acid. The equilibrium for the acid–base reaction between HCN and a carbonyl group lies predominantly to the left, and further reaction does not take place.

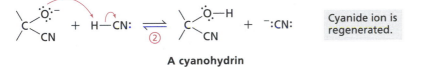

This acid-base equilibrium lies far to the left.

With base present, however, HCN forms the cyanide ion, a good nucleophile that reacts at the electrophilic carbon atom of the carbonyl group to generate a cyano (alkoxy) intermediate.

$$HCN + OH^- \rightleftharpoons CN^- + H_2O$$

This species deprotonates HCN, forming the cyanohydrin and regenerating cyanide ion, which functions as a catalyst for the reaction.

A cyanohydrin

Cyanide ion is regenerated.

Notice that all of the steps in this overall mechanism are equilibria, so the reaction is reversible. For ketones, which are more sterically hindered than aldehydes, these equilibria lie toward the side of the starting material, even when cyanide ion is present; so only ketones with small groups form appreciable amounts of the cyanohydrin product.

EXERCISE 18.1

Acetone, CH_3COCH_3, reacts with hydrogen cyanide in the presence of sodium cyanide to form the cyanohydrin product. Propose a mechanism for this transformation showing electron movement with curved arrows.

18.2b WATER AND THE HYDROGEN HALIDES ADD TO THE CARBONYL GROUP TO FORM GEMINAL DIOLS AND HALOHYDRINS, RESPECTIVELY

Hydrogen cyanide is not the only molecule that adds to carbonyl groups under basic conditions. Water undergoes addition to form an unstable **geminal diol** (*gem*-diol; see Section 8.2e). An aqueous solution at high pH contains the hydroxide ion, which is both a base and a nucleophile. Its reaction with a C=O double bond generates an alkoxide ion intermediate.

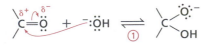

The alkoxide ion, itself a strong base, removes a proton from water, the solvent, forming the geminal diol, also called a **hydrate.** This step regenerates hydroxide ion, which functions as a catalyst.

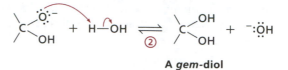

A *gem*-diol

Hydration (the addition of water) of a carbonyl group can also be conducted under acidic conditions. If mineral acid is present, the carbonyl group is protonated at its oxygen atom, which increases the electrophilicity of the carbon atom (Section 18.1c).

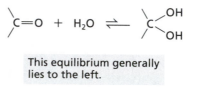

Water, a weakly basic nucleophile (Table 6.3), subsequently reacts with the electrophilic carbon atom to form the protonated geminal diol. This intermediate is deprotonated in turn by the solvent to form the *gem*-diol.

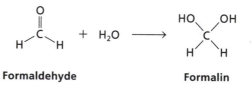

Carbonyl hydration, which is catalyzed either by acid or by base, is reversible, and the equilibrium normally lies to the side with the ketone or aldehyde.

This equilibrium generally
lies to the left.

Some hydrates are reasonably stable, however, and this condition is more likely to be observed for aldehydes than for ketones, which means that the size of the carbonyl substituents is one factor that affects the equilibrium. Formaldehyde, the smallest aldehyde, exists in aqueous solution solely as its hydrate, formalin, a substance that is commonly used to preserve biological samples.

Formaldehyde   **Formalin**

If one of the R groups attached to the carbonyl group is electron withdrawing, the *gem*-diol form will also be stabilized. Trichloroacetaldehyde, for example, exists primarily as the compound chloral hydrate. Hexafluoroacetone also forms a stable *gem*-diol when exposed to water.

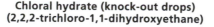   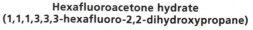

Chloral hydrate (knock-out drops)   **Hexafluoroacetone hydrate**
(2,2-trichloro-1,1-dihydroxyethane)   **(1,1,1,3,3,3-hexafluoro-2,2-dihydroxypropane)**

EXERCISE 18.2

Cyclopropanone forms a stable hydrate. Why? (*Hint:* Consider the bond angles in both starting ketone and product hydrate.)

The hydrohalic acids (HF, HCl, HBr, or HI) add to the carbonyl group of ketones and aldehydes in the same manner as water under acidic conditions. This reaction has little practical significance because it is reversible, and like hydration, addition of HX is an equilibrium process that lies in the direction of the carbonyl compound.

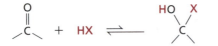

(X = F, Cl, Br, I)   **A geminal halohydrin**

EXERCISE 18.3

The double bond of a haloalkene undergoes Markovnikov addition of water in the presence of acid by the mechanism you learned in Section 9.1c. Using curved arrows to depict the movement of electrons, propose a reasonable mechanism for the following reaction.

18.2c ORGANOMETALLIC COMPOUNDS ADD TO ALDEHYDES AND KETONES TO PRODUCE ALCOHOLS

You have already learned that Grignard reagents and organolithium compounds add to the carbonyl groups of aldehydes and ketones (Section 15.2b), transformations that can be used to prepare a diverse variety of alcohols. The RMgX adds to formaldehyde to form a primary alcohol; it adds to any other aldehyde to afford a secondary alcohol; and it adds to a ketone to produce a tertiary alcohol (all after hydrolysis).

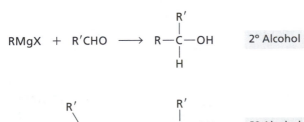

To review briefly, the nucleophilic portion of the organometallic reagent, R, reacts at the electrophilic carbon atom of the carbonyl group after association of the oxygen atom with the Mg ion. This process is an example of carbonyl activation (Section 18.1c) by a Lewis acid rather than by H⁺.

Because these reactions are carried out in aprotic solvents such as ether or THF, a second step—hydrolysis—is required to form the alcohol product. A mildly acidic solution of ammonium chloride is often employed in the hydrolysis step of ketone reactions to prevent elimination of water from occurring with the tertiary alcohol product.

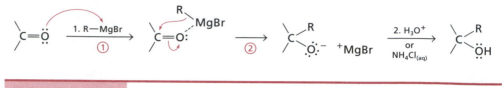

EXERCISE 18.4

What is the major product expected from each of the following reactions?

a.

b.

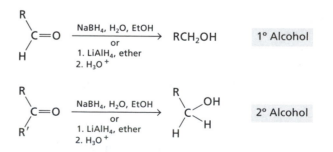

18.3 REDUCTION REACTIONS OF ALDEHYDES AND KETONES

18.3a BORON AND ALUMINUM HYDRIDES PROVIDE A SOURCE OF HYDRIDE ION, WHICH ADDS TO THE CARBONYL GROUP TO GENERATE AN ALCOHOL

Just as the nucleophilic R group of a Grignard reagent adds to a carbon–oxygen double bond, the nucleophilic hydride ion can add to an aldehyde or ketone carbonyl group, producing an alcohol, after hydrolysis.

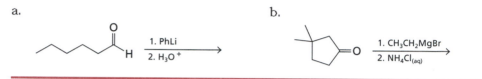

This transformation is the reverse of the oxidation process that converts an alcohol to a carbonyl compound, a reaction that was discussed in Section 11.4.

The two reagents most commonly used for carbonyl group reduction are sodium borohydride, $NaBH_4$, and lithium aluminum hydride, $LiAlH_4$, both of which convert ketones and aldehydes to the alcohols in high yields. Aldehydes are reduced to form primary alcohols, and ketones to form secondary alcohols.

Sodium borohydride is the less reactive of the two reagents, and its reactions are frequently carried out using aqueous ethanol as the solvent. Besides the reaction between BH_4^- and the carbonyl group (Step 1, following), the borohydride ion reacts slowly with the protons in water to generate H_2 as a byproduct; the basicity of the solvent gradually increases as hydroxide ion is formed. Hydrolysis (Step 2, below) occurs because water is already present in the reaction medium.

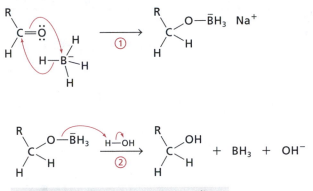

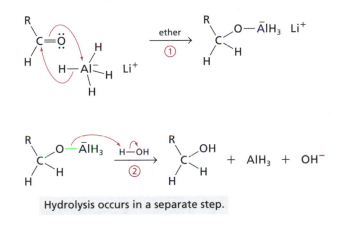

Hydrolysis occurs in the reaction medium.

Reactions of aldehydes and ketones with LiAlH$_4$, on the other hand, are carried out in THF or ether as solvents. LiAlH$_4$ is an extremely powerful reducing agent, and *reacts explosively with water and alcohols.* It must be kept away from a proton source until the reaction with the aldehyde or ketone is complete. A separate hydrolysis step is included to remove the aluminum-containing group from the oxygen atom of the alcohol. Any aluminum hydride byproducts undergo hydrolysis during this workup step.

Hydrolysis occurs in a separate step.

$$AlH_3 \; + \; 3 \; H_2O \; \longrightarrow \; Al(OH)_3 \; + \; 3 \; H_2$$

In the mechanisms illustrated above, only one hydride ion of the reagent is shown to react. And while reduction reactions can be conducted in this fashion—with equimolar amounts of carbonyl compound and reagent—such conditions are wasteful. All four hydride ions have nucleophilic properties, so up to 4 equiv of the ketone or aldehyde can be reduced using only 1 equiv of LiAlH$_4$.

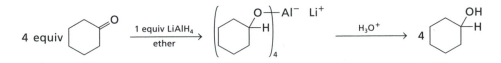

Under typical experimental conditions, between 2 and 3 equivalents of the carbonyl compound are reduced for each equivalent of reagent that is used.

Lithium aluminum hydride reduces many other functional groups, too. For example, LiAlH$_4$ reduces carboxylic acids and esters to alcohols, but sodium borohydride does not react with these functional groups. (This aspect of reagent chemoselectivity will be discussed further in Chapter 21.)

EXERCISE 18.5

What is the expected product from each of the following transformations?

a.   b.

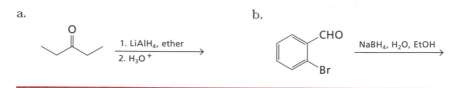

18.3b KETONES CAN BE REDUCED ENANTIOSELECTIVELY

When an aldehyde is reduced to form a primary alcohol, the carbon atom that bears the OH group has two hydrogen atoms, so it is achiral.

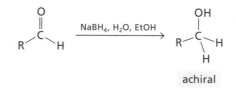

When an unsymmetrical ketone is reduced, however, a new stereogenic center is formed in the product molecule, so a racemic mixture is obtained.

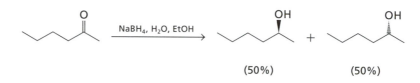

(50%)   (50%)

Much effort has been spent to develop ways to perform *enantioselective* reduction reactions of ketones. Such transformations complement the methods used to prepare chiral 2° alcohols from alkenes, for example, via enantioselective hydroboration followed by oxidative hydrolysis (Section 16.4c). The reagents that will be discussed in this section are summarized in Table 18.2. The stereochemistry of the alcohol product is readily predictable with these reagents, but in planning synthetic schemes, we will assume that the desired stereochemistry of the product can be obtained with one stereoisomer or the other of the reagent or catalyst.

Table 18.2   Reagents used for the enantioselective reduction reactions of ketones

Reaction type	*Reagent combination[a]*
Hydride addition	oxazaborolidine*–BH_3
Hydrogenation	RuX_2(phosphine*)(diamine*)
	$(CH_3)_3COK$, 2-propanol, H_2

[a]Note: The asterisks in the reagent structures indicate where the chirality centers are located.

In 1981, Itsuno and co-workers first reported that chiral amino alcohols in combination with borane, BH_3, could reduce a ketone with high enantioselectivity and in excellent chemical yield. In 1987, E. J. Corey's group at Harvard University reported that the actual reagent is a chiral **oxazaborolidine** complex of borane. These researchers subsequently developed oxazaborolidine complexes as both stoichiometric and catalytic reagents for reducing ketones enantioselectively.

Extensive work in this area has led to the development of several related reagents. For example, the amino acid proline can be converted to the oxazaborolidine **A,** shown below, which is subsequently used to prepare the complex **A·BH₃**. This complex is the actual reducing agent (D-proline can be used to prepare the enantiomer of **A**).

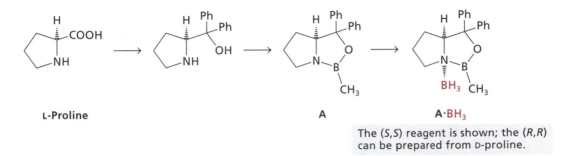

L-Proline   **A**   **A·BH₃**

The (*S*,*S*) reagent is shown; the (*R*,*R*) can be prepared from D-proline.

A ketone can be reduced enantioselectively with **A·BH₃** as a reagent or with an amount of **A** as a catalyst in the presence of excess BH_3·THF. Under stoichiometric conditions, only two of the three hydride ions attached to boron are reactive. Notice that for alkyl aryl ketones in which the Cahn–Ingold–Prelog priority of the aryl group > alkyl group, the (*S*,*S*)-reagent yields the (*R*)-alcohol, and vice versa.

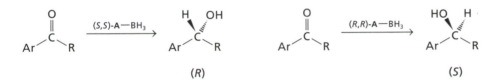

(*R*)   (*S*)

The enantioselective reduction of acetophenone is illustrated below as a specific example.

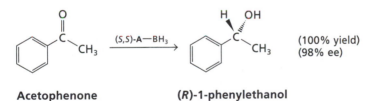

Acetophenone   (*R*)-1-phenylethanol

(100% yield)
(98% ee)

The high enantioselectivity associated with this reagent can be rationalized by considering structures of the diastereomeric transition states that are involved during the reaction (Figure 18.1). To attain selectivity, both the ketone and the reducing agent (the BH_3 unit) must interact within a rigid complex oriented so that only one face of the carbonyl group reacts. One boron atom of the reagent acts as a Lewis acid to bind to the basic oxygen atom of the carbonyl group, and the other boron atom provides the hydride ion that reduces the C=O double bond.

When the oxazaborolidine–borane complex is used as a catalyst, the same diastereomeric transition states exist. Under catalytic conditions, additional borane–THF provides the hydride ions that reduce the carbonyl group. When two of the hydride ions on the borane attached to the nitrogen atom have reacted with the ketone, as they do under stoichiometric conditions, then the excess borane in solution displaces the

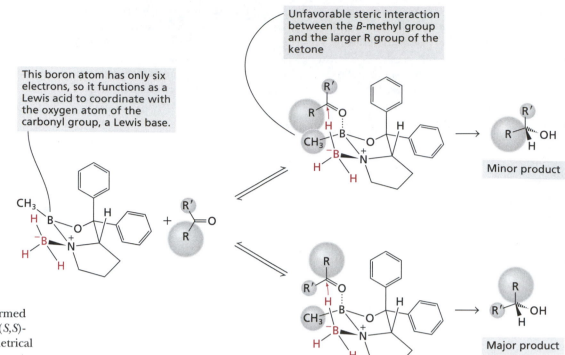

Figure 18.1
The transition states formed during the reaction of (*S,S*)-A·**BH₃** with an unsymmetrical ketone. If the (*R,R*) reagent were used, these structures would be mirror images of the ones shown here.

one attached to the nitrogen atom, and this step regenerates the chiral reactant. The reaction continues until the ketone is gone.

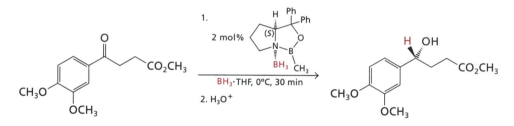

It is important to recognize one factor that can limit the use of catalysts of this type: *The ketone must not react with uncomplexed borane.* If a reaction does occur with borane–THF itself, then racemic products will be formed, which will lower the ee of the transformation. Fortunately, the oxazaborolidine reagent *activates the carbonyl group toward nucleophilic addition by prior coordination of its boron atom to the carbonyl oxygen atom.* Therefore, the ketone is more reactive when it is in proximity to the complexed BH₃ than when it is in solution with BH₃·THF.

EXAMPLE 18.1

What is the product expected from reduction of the ketone shown here using (*R,R*)-A·BH₃?

Unsymmetrical ketones in which the carbonyl group is attached to a large group (usually an aryl ring) and a smaller alkyl group can be reduced enantioselectively using the oxazaborolidine reagents described in the text. The (*R*,*R*) reagent yields the (*S*) alcohol, *when the aryl group has a higher Cahn–Ingold–Prelog priority than the alkyl group.* [The (*S*,*S*) reagent produces the (*R*) alcohol in such cases.] In this reaction, the alkyl group has a higher priority than the phenyl group of the starting ketone, so the stereochemistry of the product will be opposite that normally obtained. The product is (*R*)-1-phenyl-2-chloroethanol.

EXERCISE 18.6

What is the product expected from the reduction of each of the following ketones using (*S*,*S*)-**A**·BH$_3$?

a.

b.

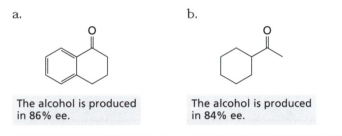

The alcohol is produced in 86% ee.

The alcohol is produced in 84% ee.

A second practical method for enantioselective reduction of ketones makes use of chiral hydrogenation catalysts. Ryoji Noyori, who developed the chemistry of ruthenium BINAP catalysts for alkene reduction (Section 16.4b), discovered that modification of the catalyst structure and reaction conditions makes it possible to hydrogenate carbonyl groups.

The catalysts for ketone reduction have a chiral diamine and two halide ions bonded to the ruthenium ion in addition to a chiral diphosphine. Many such catalysts have been studied, for example, the one shown below.

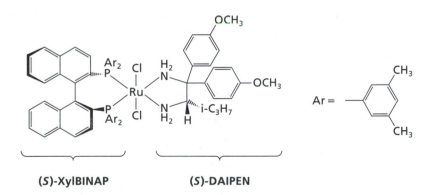

(*S*)-XylBINAP   (*S*)-DAIPEN

In the presence of a strong base like potassium *tert*-butoxide and with 2-propanol as the solvent, alkyl aryl ketones are reduced to form secondary alcohols in high chemical yield and with high ee values. When the aryl group has a higher Cahn–Ingold–Prelog priority than the alkyl group, the (*S*,*S*) catalyst yields the (*R*) alcohol, and the (*R*,*R*) catalyst yields the (*S*)-alcohol.

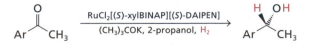

This catalyst tolerates a variety of groups in the ketone structure. For example, the following alcohols have been made from the corresponding ketones:

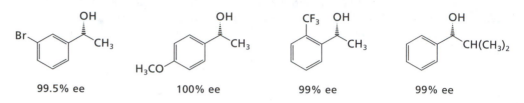

99.5% ee   100% ee   99% ee   99% ee

The following equation illustrates the preparation of an intermediate in the synthesis of the antidepressant Prozac (which is actually sold as a racemic mixture). Note that the amino group in the alkyl chain does not interfere with the enantioselectivity of the reduction process.

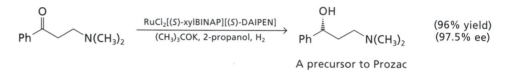

A precursor to Prozac

(96% yield)
(97.5% ee)

Unfortunately, ketones with two alkyl groups are not reduced with high selectivity by either the ruthenium or oxazaborolidine catalysts described above. This problem is one that still awaits discovery of a practical solution.

EXERCISE 18.7

What is the product expected from the following reaction? It is formed in 95% ee.

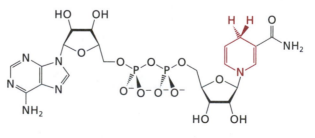

18.3c NADH AND NADPH PROVIDE A SOURCE OF HYDRIDE IONS IN BIOCHEMICAL SYSTEMS

Let us now turn our attention to the reduction processes that occur in biochemical systems. In Chapter 11, you learned about biochemical oxidation processes that are used to convert alcohols to carbonyl compounds (Section 11.4e). Those transformations utilize NAD^+ as a coenzyme to accept a hydrogen atom and its electrons from the carbon atom bearing the OH group, concomitant with deprotonation of the alcohol OH group. As you might surmise, reduction reactions in biological systems occur in the opposite way, that is, by transferring a hydride ion from NADH (or NADPH) to a carbonyl group.

Binding an aldehyde or ketone at the active site of the enzyme that also has the NADH coenzyme present brings the hydrogen atom of the dihydropyridine ring of NADH into proximity with the carbonyl carbon atom.

**Nicotinamide adenine dinucleotide
(reduced form: NADH)**

Transfer of a hydride ion occurs as the oxygen atom is protonated by an acid side chain of an appropriate amino acid, resulting in reduction of the ketone or aldehyde to an alcohol. In the illustration shown below, B is a basic site such as the imidazole group in the side chain of the amino acid histidine. Notice the similarity of this mechanism to that of laboratory methods using a hydride reagent followed by acid workup.

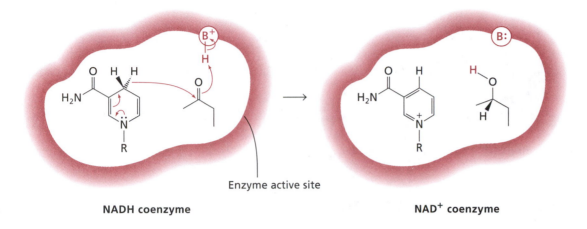

NADH coenzyme   **NAD⁺ coenzyme**

This transformation is common in biochemistry. For example, pyruvate is reduced by NADH to lactate when insufficient oxygen is present to convert pyruvate to acetyl coenzyme A (Acetyl CoA, see Section 21.3c), a situation that is common during strenuous exercise when the level of oxygen in muscle tissues is low. The pain of sore muscles results partly from lactate accumulation.

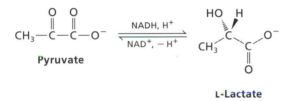

Pyruvate   **L-Lactate**

Lactate dehydrogenase, the enzyme that catalyzes this reversible reaction (cf. Section 16.3d), is well characterized: A representation of the active site structure (Figure 18.2) summarizes the important structural features that are required for these transformations to occur. These features include binding sites for the coenzyme and substrate, as well as a source of the proton that is used to form the OH group of the alcohol.

18.3d NADH AND ENZYMES CAN BE USED TO REDUCE KETONES ENANTIOSELECTIVELY IN THE LABORATORY

A prominent area of research in organic chemistry that is ripe for development is the use of enzymes to carry out reactions in the laboratory. The enantioselective reduction of ketones (Section 18.3b) is a particularly fruitful field of investigation because such processes are sometimes difficult to carry out using synthetic reagents. Moreover, many enzymes that perform reduction reactions are readily available.

Alcohol dehydrogenases, a specific example of which is the *lactate dehydrogenase* enzyme mentioned in Section 18.3c, utilize NADH to reduce ketones and NAD⁺ to oxidize alcohols. This reversibility is a distinct benefit if an enzyme is to be used as a reagent in the laboratory, because the needed coenzyme can be regenerated by performing both reactions in the same flask. The following scheme shows the enzyme catalyzing the reduction of a ketone (in the box) while at the same time oxidizing

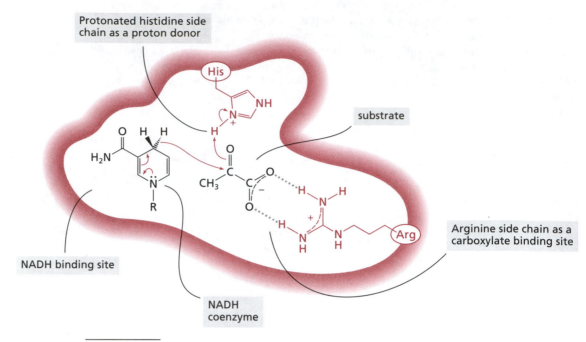

Figure 18.2

A representation of the active site structure of *lactate dehydrogenase,* the enzyme used to reduce pyruvate to lactate. Shown are the hydrogen bonds that orient the substrate molecule as well as the curved arrows that depict the mechanism of the hydride and proton-transfer steps.

2-propanol to acetone. The net result is to replenish NADH so that reduction of the ketone continues. Excess 2-propanol drives reduction of the ketone to completion.

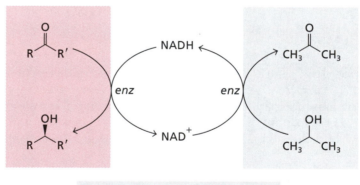

Key: *enz* = *alcohol dehydrogenase*

The following equations illustrate specific examples that make use of an alcohol dehydrogenase enzyme (PADH-1) isolated from a bacterial *Pseudomonas* species. Of note are the high chemical yields and enantioselectivity associated with reduction.

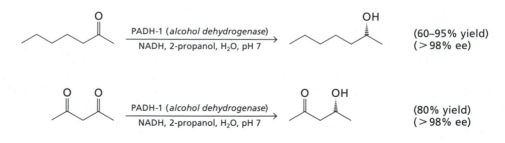

The PADH-1 enzyme appears to be especially well suited for regeneration of the NADH coenzyme by 2-propanol. Alcohol dehydrogenases from other organisms also selectively reduce ketones, and some generate the (*S*) isomer preferentially.

18.3e ALDEHYDES AND KETONES CAN BE FULLY REDUCED BY AMALGAMATED ZINC IN HYDROCHLORIC ACID

Hydride ion is not the only species that will add to a carbonyl group to produce an alcohol. The combination of hydrogen gas and a metal catalyst (Section 11.2a) also reduces the C=O double bond (Section 18.3b). If the enantioselectivity of reduction is not important, then heterogeneous catalysts suffice. For example,

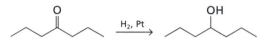

If the resulting alcohol group is adjacent to a benzene ring, then hydrogenolysis occurs (Section 12.3a), to replace the ketone carbonyl group with two hydrogen atoms.

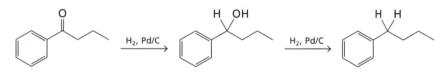

Another way in which the carbonyl group can be converted to a methylene group is to use zinc and hydrochloric acid, a reaction known as the **Clemmensen reduction.** This reaction works best when the carbonyl group is adjacent to a benzene ring, which is why it was first mentioned in Section 17.2e in the context of Friedel–Crafts acylation reactions.

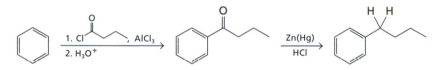

When zinc is stirred with aqueous mercury (II) chloride, the surface of the zinc metal is coated with a small amount of mercury to produce an amalgam, Zn(Hg).

$$Zn\ (xs)\ +\ HgCl_2\ \longrightarrow\ Hg\ +\ ZnCl_2$$

$$Zn\ +\ Hg\ \longrightarrow\ Zn(Hg)$$

Amalgamated zinc reacts more slowly with hydrochloric acid than elemental zinc does, thereby minimizing the formation of hydrogen gas.

$$Zn\ +\ 2\ HCl\ \longrightarrow\ H_2\ +\ ZnCl_2\ (rapid)$$

$$Zn(Hg)\ +\ 2\ HCl\ \longrightarrow\ \text{Slow formation of } H_2 \text{ and } ZnCl_2$$

The mechanism of the Clemmensen reduction is not known precisely, but it probably comprises a series of one-electron addition and proton-transfer steps. The first step is electron transfer that produces a benzyl radical. Such a radical is stabilized best when R is benzene or one of its derivatives, which is why the reaction works well for aromatic ketones. The zinc atom supplies the electrons, and it is oxidized to Zn^{2+} during the course of the reaction. A series of proton- and electron-transfer reactions lead to formation of the corresponding alcohol, which can be reduced further via a carbocation intermediate that is also stabilized by the adjacent aromatic ring.

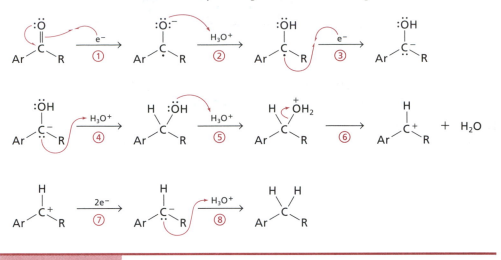

EXERCISE 18.8

What is the expected product of the Friedel–Crafts acylation of benzene with benzoyl chloride followed by Clemmensen reduction of that product? How might you prepare the same product from bromobenzene and benzaldehyde plus other reagents?

18.4 OXIDATION REACTIONS OF ALDEHYDES AND KETONES

18.4a AN ALDEHYDE IS READILY OXIDIZED TO A CARBOXYLIC ACID

An aldehyde has an oxidation level that lies between those of the corresponding alcohol and carboxylic acid molecules (Section 11.1a).

$$RCH_2OH \qquad RCHO \qquad RCOOH$$

more oxidized

To oxidize a primary alcohol to an aldehyde without further oxidation to the carboxylic acid, an anhydrous reagent combination such as CrO_3–pyridine, DMSO–oxalyl chloride, or the Dess-Martin periodinane (Section 11.4) is normally used. Aqueous oxidants lead to formation of the carboxylic acid product, and these same conditions are used to oxidize aldehydes to carboxylic acids:

$$RCHO \xrightarrow[\text{2. } H_3O^+]{\text{1. } KMnO_4,\ OH^-_{(aq)}} RCOOH \qquad\qquad RCHO \xrightarrow[H_3O^+]{K_2Cr_2O_7} RCOOH$$

When water is present, an aldehyde—whether by itself or from oxidation of a primary alcohol—can form a hydrate, as described in Section 18.2b.

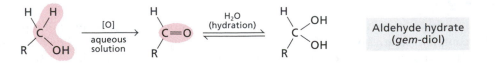

Aldehyde hydrate
(*gem*-diol)

This hydrate can subsequently undergo oxidation because it has the requisite H–C–OH unit needed to form the carbonyl group, C=O. In aqueous solution, therefore, a carboxylic acid is the product formed by oxidizing either a primary alcohol or an aldehyde.

This transformation works better under basic conditions, which are not well suited for the chromium oxides reagents. Potassium permanganate is therefore the reagent of choice for oxidizing a primary alcohol or aldehyde to the carboxylic acid, even though chromium oxides in acid can be used in certain cases (Section 11.4c).

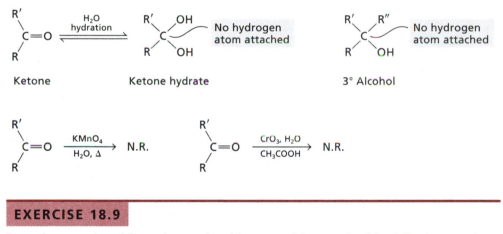

Ketones have no hydrogen atom attached to the carbonyl carbon atom, nor do their hydrate derivatives. As is the case for tertiary alcohols, further oxidation of ketones by metal oxide reagents does not normally occur. In fact, acetone is often used as a solvent for oxidation reactions because it is inert to many oxidants.

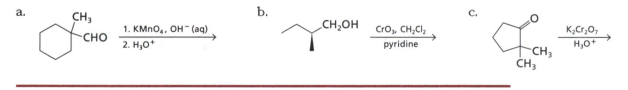

EXERCISE 18.9

Draw the structure of the major product(s) expected from each of the following reactions:

a.

CH₃

—CHO

1. KMnO₄, OH⁻ (aq)
2. H₃O⁺

b.

CH₂OH

CrO₃, CH₂Cl₂
pyridine

c.

O

CH₃
CH₃

K₂Cr₂O₇
H₃O⁺

18.4b KETONES CAN BE CONVERTED TO ESTERS USING PERACIDS

Ketones are generally inert toward metal oxide reagents such as CrO_3 and $KMnO_4$, as well as being unaffected by Swern (Section 11.4b) and Dess-Martin (Section 11.4d) conditions. The oxidation level of a ketone is intermediate between that of a secondary alcohol and an ester, however, so oxidation is theoretically feasible.

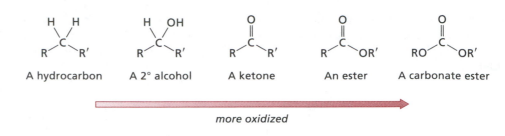

more oxidized

Ketones can be converted to esters by the **Baeyer–Villiger reaction,** which begins with addition of a peracid, RCO₃H (Section 11.3e), to the carbonyl group, the mechanism of which follows the steps outlined in Section 18.1c involving acid-catalyzed carbonyl group activation.

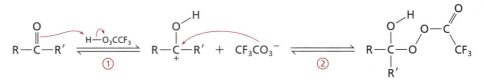

A peroxyacyl compound

After formation of the peroxyacyl compound, an electron pair on oxygen moves to regenerate the C=O double bond. This movement causes one of the R groups with its pair of electrons to migrate to an oxygen atom of the peracid group, displacing carboxylate ion as a leaving group. The breaking of the weak O–O bond with formation of the much stronger C=O double bond is the primary factor that drives this rearrangement to completion. Note, too, that this migration is nearly identical to the one that occurs after an organoborane reacts with hydroperoxide ion (Section 9.4c).

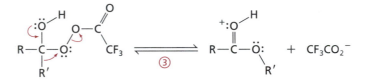

The displaced carboxylate ion subsequently deprotonates the cationic intermediate.

The Baeyer–Villiger reaction is especially useful for converting cyclic ketones to lactones (cyclic esters).

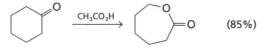

Many ketones, cyclic or otherwise, are unsymmetrical, so formation of two products is possible. The "migratory aptitude" of the R groups in the Baeyer–Villiger oxidation is H > 3° > 2° and aryl > 1° > methyl. This order means that the "new" oxygen atom inserts into the bond between the carbonyl group and the more highly substituted carbon atom of an unsymmetrical ketone. The reaction proceeds with retention of configuration if the migrating carbon atom is chiral.

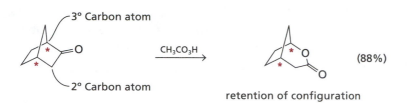

retention of configuration

The migratory aptitude of the different groups means that methyl ketones produce acetate esters. The acetate group can be readily hydrolyzed to yield the corresponding alcohol, as you will learn in Chapter 21. This overall transformation provides a way to convert a methyl ketone to an alcohol with two fewer carbon atoms.

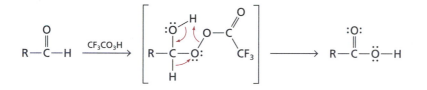

The Baeyer–Villiger reaction of an aldehyde proceeds in the same manner as it does for a ketone. The hydrogen atom migrates preferentially, so a carboxylic acid, rather than an ester, is the major product.

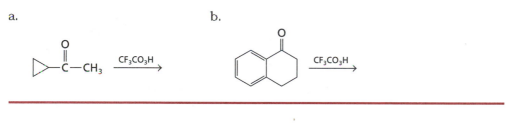

EXERCISE 18.10

What is the major product expected from each of the following reactions?

a.

b.

CHAPTER SUMMARY

Section 18.1   General aspects of nucleophilic addition reactions

- Good nucleophiles add to the carbonyl group of an aldehyde or ketone by reaction at the electrophilic carbon atom.
- The carbonyl group of aldehydes and ketones can be activated by protonation
  at oxygen, which makes the carbon atom more electrophilic and able to react
  with species that are weakly nucleophilic. Activation also occurs with Lewis acids
  besides H^+.

Section 18.2   Nucleophilic addition processes

- Hydrogen cyanide (HCN) adds to a carbonyl group of aldehydes under conditions of cyanide ion catalysis. The product of the reaction is a cyanohydrin.

- Water adds to a carbonyl group under conditions of either acid or base catalysis. The product is a *gem*-diol, also called a hydrate, which is normally unstable toward loss of water to regenerate the carbonyl group.
- The hydrate of an aldehyde or ketone is stable if the R group is strongly electron withdrawing.
- Addition of HX (X = F, Cl, Br, or I) to a carbonyl group yields an unstable *gem*-halohydrin. This equilibrium lies in the direction of the carbonyl compound.
- Organometallic compounds such as Grignard reagents add to aldehydes and ketones to produce an alcohol after workup with aqueous acid.

Section 18.3 Reduction reactions of aldehydes and ketones

- Sodium borohydride in aqueous alcohol or LiAlH$_4$ in ether or THF will reduce an aldehyde or ketone to its corresponding alcohol. The use of LiAlH$_4$ requires a separate hydrolysis step.
- Oxazaborolidines constitute a class of reagents and catalysts that permit the facile enantioselective reduction of ketones to secondary alcohols.
- Chiral ruthenium catalysts can be used to hydrogenate ketones to chiral secondary alcohols.
- Both NADH and NADPH are coenzymes that act as hydride ion donors in biochemical systems to reduce ketones or aldehydes to the corresponding alcohols.
- Alcohol dehydrogenase enzymes, along with NADH, can be used in the laboratory to perform the enantioselective reduction of ketones.
- The Clemmensen reduction, which makes use of amalgamated zinc and hydrochloric acid, converts a carbonyl to a methylene group.

Section 18.4 Oxidation reactions of aldehydes and ketones

- An aldehyde is oxidized to a carboxylic acid with metal oxides in aqueous solution. These reactions occur via the hydrate form of the aldehyde.
- A ketone can be oxidized to an ester by using a peracid. This transformation, called the Baeyer–Villiger reaction, occurs with migration of an R group to an oxygen atom of the adduct formed after nucleophilic addition of the peracid to the carbonyl group.
- An unsymmetrical ketone undergoes the Baeyer–Villiger oxidation with migration of the more highly branched R group. An aryl group migrates in preference to a primary or methyl group.
- An aldehyde is oxidized to a carboxylic acid under Baeyer–Villiger conditions. The hydrogen atom migrates preferentially.

KEY TERMS

Section 18.2a
cyanohydrin

Section 18.2b
geminal diol
hydrate

Section 18.3b
oxazaborolidine

Section 18.3e
Clemmensen reduction

Section 18.4b
Baeyer–Villager reaction

Sections 18.2a

Addition of HCN: Aldehydes and low molecular weight ketones form cyanohydrins when treated with HCN and cyanide ion as a catalyst.

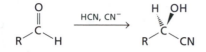

Section 18.2b

Aldehydes and ketones react with water (hydration) to form a *gem*-diol. The hydrate is normally isolable only when R is a strong electron-withdrawing group (i.e., when it contains halogen atoms).

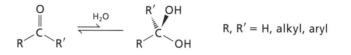

R, R' = H, alkyl, aryl

The reaction between aldehydes or ketones and HX is an equilibrium process that lies toward the carbonyl form.

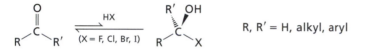

R, R' = H, alkyl, aryl

Section 18.2c

Aldehydes and ketones react with organometallic compounds (RMgX or RLi) to form alcohols after aqueous acid workup.

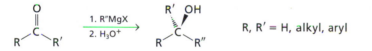

R, R' = H, alkyl, aryl

Section 18.3a

Aldehydes and ketones are reduced to alcohols with LiAlH$_4$ or NaBH$_4$.

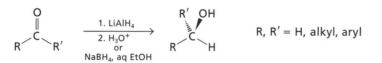

R, R' = H, alkyl, aryl

Section 18.3b

The enantioselective reduction of ketones is possible using chiral oxazaborolidines as reagents or catalysts in combination with BH$_3$·THF.

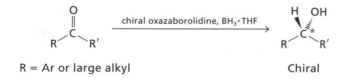

R = Ar or large alkyl   Chiral

The enantioselective reduction of ketones can be accomplished by hydrogenation reactions that use chiral ruthenium catalysts.

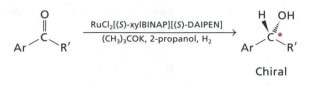

Section 18.3d and 18.3e

Aldehydes and ketones can be reduced using NADH and enzyme catalysts. Unsymmetrical ketones yield chiral products.

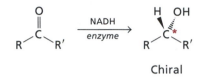

Section 18.3e

The Clemmensen reduction: This reaction, which converts a carbonyl to a methylene group, is most facile when R is aryl.

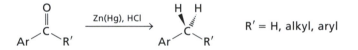

Section 18.4a

Oxidation of aldehydes to carboxylic acids: Chromium(VI) reagents and permanganate ion in water oxidize the hydrated form of the aldehyde to a carboxylic acid.

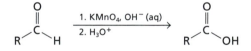

Section 18.4b

The Baeyer–Villiger reaction: With peracids, ketones are converted to esters, and aldehydes react to form carboxylic acids.

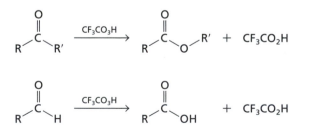

18.11. Draw structures for each of the following ketones and aldehydes:

a. (*Z*)-3-Chloro-2-pentenal

b. 2-Bromo-3-nitrobenzaldehyde

c. (*R*)-3-Phenyl-4-pentene-2-one

d. 4,4-Dimethyl-2-cyclopentenecarbaldehyde

18.12. Predict the appearance of the ^{1}H and ^{13}C NMR spectra for compounds (a.) and (b.) in Exercise 18.11.

18.13. Give an acceptable IUPAC name for each of the following aldehydes and ketones:

a.   b.   c.

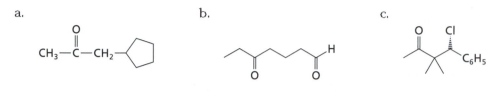

18.14. Write an equation for the reaction between 1-phenyl-1-propanone and each of the following reagents:

a. Zn(Hg), HCl

b. NaBH$_4$ in aqueous ethanol

c. HNO$_3$, H$_2$SO$_4$

d. LiAlH$_4$ in ether followed by H$_3$O$^+$

18.15. Show how you would prepare butanal from each of the following compounds: Several steps may be required.

a. 1-Butanol

b. 1-Bromobutane

c. 1-Butene

d. 1-Pentene

18.16. Show how you would prepare each of the following compounds using an organometallic reagent. You may use any hydrocarbons, organohalides, and carbonyl compounds that have six or fewer carbon atoms. You may also use any needed reagents and solvents.

a.   b.   c.

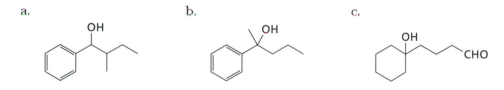

18.17. The pinacol rearrangement the occurs when a vicinal diol is treated with acid. From your knowledge about the reactions and properties of carbocations, propose a reasonable mechanism for the following reaction, which converts pinacol to pinacolone. Give IUPAC names to the compounds shown.

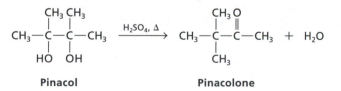

18.18. Draw the structure(s) of the major product(s) expected from each of the following reactions. Indicate the stereochemistry of the product as appropriate. Relative stereochemistry should be shown using wedges and dashed lines. If a racemic mixture will be formed, draw the structure of one enantiomer and write the word "racemic", or draw both enantiomeric structures. If diastereomers are formed, draw each structure; label meso compounds as such. If no reaction occurs, write N.R.

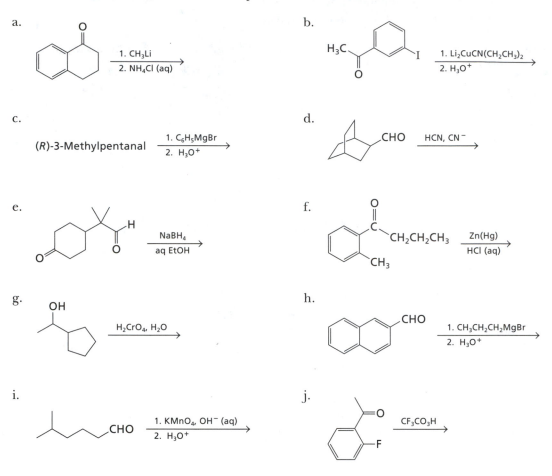

a.

   1. CH₃Li
   2. NH₄Cl (aq)

b.

   1. Li₂CuCN(CH₂CH₃)₂
   2. H₃O⁺

c.

(*R*)-3-Methylpentanal   1. C₆H₅MgBr / 2. H₃O⁺

d.

CHO   HCN, CN⁻

e.

   NaBH₄
   aq EtOH

f.

CH₂CH₂CH₃   Zn(Hg) / HCl (aq)

g.

OH   H₂CrO₄, H₂O

h.

CHO   1. CH₃CH₂CH₂MgBr / 2. H₃O⁺

i.

CHO   1. KMnO₄, OH⁻ (aq) / 2. H₃O⁺

j.

CF₃CO₃H

18.19. What is the expected product from the Baeyer–Villiger reaction of each of the following ketones?

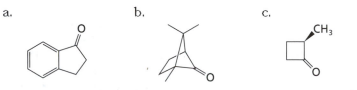

a.   b.   c.

CH₃

18.20. Cyanotrimethylsilane, (CH₃)₃SiCN, reacts with aldehydes and ketones in the presence of a catalytic amount of cyanide ion to form the trimethylsilyl derivative of the cyanohydrin. Propose a reasonable mechanism for the following reaction:

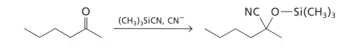

   (CH₃)₃SiCN, CN⁻   NC O—Si(CH₃)₃

Predict the stability of the trimethylsilyl cyanohydrin toward (a) HF; (b) OH⁻, and (c) F⁻. Compare these reactions with those of a cyanohydrin under each set of conditions.

18.21. The compound HN$_3$ adds to a aldehyde in the same way as HCN. Propose a reasonable mechanism for the following transformation:

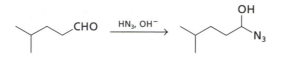

18.22. Draw the structure of the major product expected from each of the following reaction sequences. Indicate stereochemistry where appropriate. If no reaction occurs, write N.R.

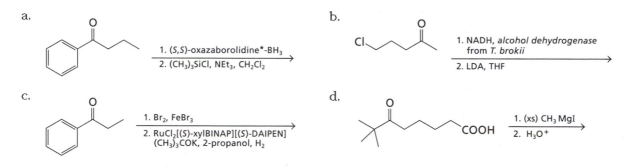

a.

1. (S,S)-oxazaborolidine*-BH$_3$
2. (CH$_3$)$_3$SiCl, NEt$_3$, CH$_2$Cl$_2$

b.

1. NADH, *alcohol dehydrogenase from T. brokii*
2. LDA, THF

c.

1. Br$_2$, FeBr$_3$
2. RuCl$_2$[(S)-xylBINAP][(S)-DAIPEN]
   (CH$_3$)$_3$COK, 2-propanol, H$_2$

d.

1. (xs) CH$_3$MgI
2. H$_3$O$^+$

18.23. Show how you would prepare each of the following compounds starting with any hydrocarbon, organohalide, and/or carbonyl compound that has six or fewer carbon atoms. You may also use any additional reagents and solvents.

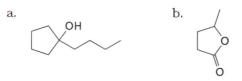

a.

b.

18.24. Show how you would prepare each of the following compounds starting with any hydrocarbons having six or fewer carbon atoms. You may also use any additional reagents and solvents.

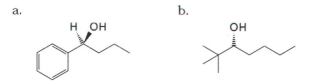

a.

b.

18.25. When B-allyl-9-BBN is treated with benzaldehyde, an addition product is obtained according to the following equation:

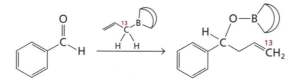

The mechanism is not "simple" addition to the carbonyl group however, because incorporating ^{13}C in the allyl group leads to the product with the label atom at the terminal position of the allyl group. Propose a mechanism that is consistent with these results. (*Hint:* The first step is an acid–base reaction, and the second step is a pericyclic reaction.)

18.26. Compound **X**, C_9H_8O, a ketone, has the following ^1H NMR spectra. When treated with the (S,S)-oxazaborolidine-borane reagent, **X** is converted to optically active **Y**. Compound **Y** is dehydrated by heating with concentrated sulfuric acid to form **Z**. Jacobsen epoxidation of **Z** with the chiral manganese epoxidation catalyst shown in Section 16.4d yields optically active **W**, and catalytic hydrogenation of **W** with 10% Pd/C produces optically inactive **V**. Draw a structure for each of the compounds **V** through **Z**.

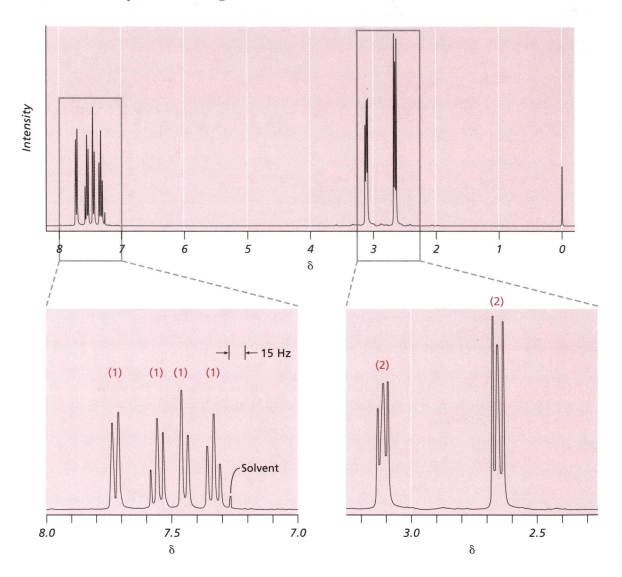

18.27. Identify the following two compounds from their IR and ^1H and ^{13}C NMR spectra. Compound **27A** has the formula C_7H_5ClO and **27B** has the formula $C_6H_{12}O$. The *J*-values for **27B** are all 7 Hz.

27A.

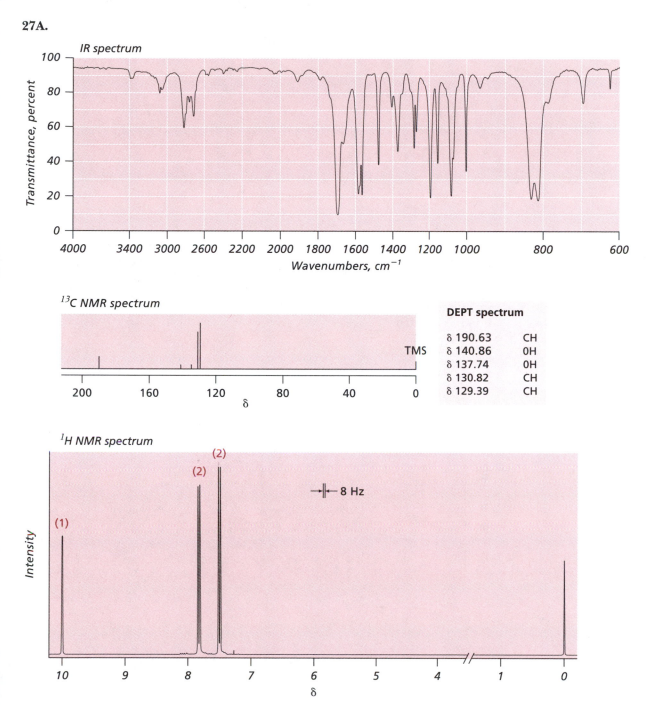

DEPT spectrum	
δ 190.63	CH
δ 140.86	0H
δ 137.74	0H
δ 130.82	CH
δ 129.39	CH

27B.

IR spectrum

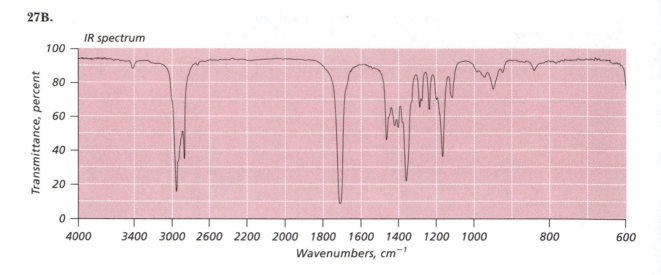

¹³C NMR spectrum

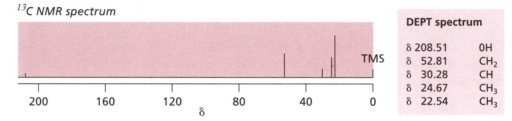

DEPT spectrum

δ 208.51	0H
δ 52.81	CH_2
δ 30.28	CH
δ 24.67	CH_3
δ 22.54	CH_3

¹H NMR spectrum

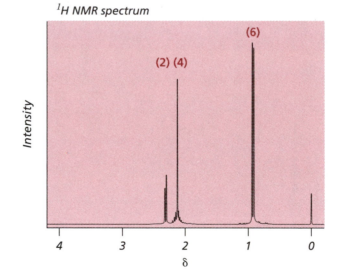

ADDITION–SUBSTITUTION REACTIONS OF ALDEHYDES AND KETONES: CARBOHYDRATE CHEMISTRY

As described in the previous chapter (Chapter 18), a nucleophile readily adds to the carbonyl group of an aldehyde or ketone to form a product with an OH group (Sections 18.1b and 18.1c). This OH group can be protonated to create a good leaving group, and substitution or elimination reactions can subsequently take place.

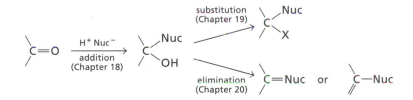

This chapter will describe several transformations that occur when the OH group of a carbonyl-addition product undergoes substitution. An especially common reaction of this type takes place when the original nucleophile is an alcohol molecule. The overall addition–substitution processes are essential reactions of carbohydrates and are used in Nature to make polysaccharides such as cellulose and starch. The addition–substitution reactions between alcohols and aldehydes or ketones are also important for synthetic chemistry in their use as protecting groups (Section 15.5b).

Besides describing addition–substitution reactions, this chapter will present the structural characteristics of carbohydrates, starting with their open-chain forms that bear either the aldehyde or ketone functional group. Fischer projections, first mentioned in Section 4.4, will be used extensively to depict carbohydrate structures, but other representation methods will also be used. Because carbohydrates play a central role in biochemistry as structural elements in the nucleic acids and as intermediates in many metabolic pathways, a knowledge of aldehyde and ketone addition–substitution reactions is required to understand the roles of these important biochemical substances.

19.1 HEMIACETALS AND ACETALS

19.1a ALCOHOLS REACT WITH ALDEHYES AND KETONES TO FORM HEMIACETALS, WHICH ARE NORMALLY UNSTABLE

Just as water adds reversibly to a carbonyl group (Section 18.2b), so an alcohol reacts with aldehydes and ketones. The mechanism of this acid-catalyzed addition reaction has three steps:

1. Protonation of the carbonyl oxygen atom to form a carbocation intermediate.

2. Reaction of the nucleophilic oxygen atom of the alcohol molecule with this carbocation.

3. Deprotonation to yield the neutral product.

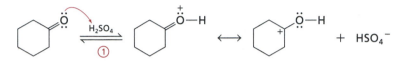

Protonated carbonyl groups are stabilized by resonance.

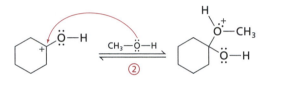

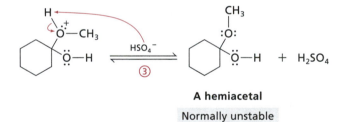

A hemiacetal

Normally unstable

The product is a **hemiacetal,** which has an OH and an OR group on what *was* the carbonyl carbon atom.

A hemiacetal (like a *gem*-diol, Section 18.2b) is normally unstable, which means the equilibria that are shown lie in the direction of the carbonyl and alcohol reactants. The mechanism for regenerating the carbonyl group follows the reverse of the steps shown directly above.

EXERCISE 19.1

Show the mechanistic steps to regenerate benzaldehyde from its methyl hemiacetal derivative.

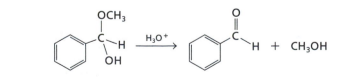

Cyclic hemiacetals, which are generated from hydroxy aldehydes or hydroxy ketones, are exceptions to the rule about instability. The equilibrium for these types of molecules lies toward the right if a five- or six-membered ring can be formed.

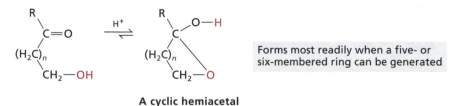

> Forms most readily when a five- or six-membered ring can be generated

A cyclic hemiacetal

Because of this reaction, carbohydrates, which have both carbonyl and hydroxy groups, exist primarily as cyclic molecules, as described in Sections 19.3–19.5.

19.1b AN ACETAL IS FORMED BY SUBSTITUTING THE OH GROUP OF A HEMIACETAL WITH AN OR GROUP

Hemiacetals have *two* sites that can react with acids. In the following structure, protonation at **A** leads to the reverse of hemiacetal formation; protonation at **B** converts the OH group into a good leaving group.

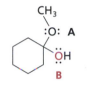

> **A** and **B** represent two basic sites in this hemiacetal molecule. Protonation at **A** reverses hemiacetal formation.

When the OH group is protonated, a pair of electrons on the alkoxy oxygen atom displaces the water molecule to generate a resonance-stabilized cationic intermediate.

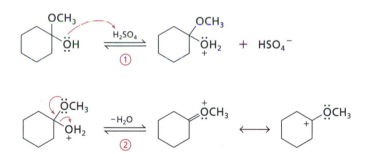

A molecule of alcohol intercepts the carbocation, which, after deprotonation, yields an **acetal,** a geminal dialkoxy compound.

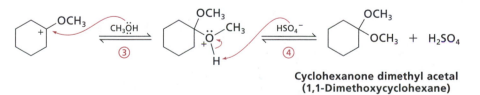

**Cyclohexanone dimethyl acetal
(1,1-Dimethoxycyclohexane)**

The overall reaction between a ketone or aldehyde and alcohol is summarized by the following equilibria. The reaction is conducted using alcohol as the solvent, which means that it is present in large excess. The transformation is acid catalyzed, and TsOH, an abbreviation for *p*-toluenesulfonic acid, $CH_3C_6H_4SO_3H$, is commonly used as the catalyst because it is soluble in organic solvents yet is as strong an acid as sulfuric acid.

By conducting the reaction under conditions that remove water (e.g., by heating the mixture and distilling water from solution), the acetal forms readily.

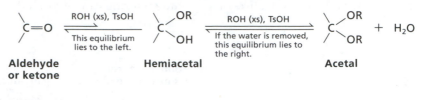

Benzaldehyde reacts with ethylene glycol (1,2-dihydroxyethane) in the presence of TsOH to yield a cyclic acetal. Show the steps for the mechanism of its formation.

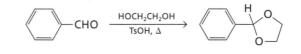

19.1c VINYL ETHERS FORM HEMIACETALS VIA HYDRATION

Ethers are generally unreactive molecules, undergoing reaction only when treated with potent acids such as HI, HBr, or $(CH_3)_3SiI$ (Section 7.2d).

$$\text{\textbackslash}O\text{/} \xrightarrow{HI, \Delta} 2\ CH_3CH_2CH_2I\ +\ H_2O$$

Vinyl ethers (also called **enol ethers**) are not like their saturated analogues, however, and they react readily even with aqueous acid.

$$\text{\textbackslash}O\text{=} \xrightarrow{H_3O^+, \Delta} CH_3CH_2CH_2OH\ +\ CH_3CHO$$

The difference between the two, of course, is the presence of a double bond in the vinyl ether structure, and hydration of this alkene produces an unstable hemiacetal.

The first step of vinyl ether hydrolysis is protonation of the π bond and formation of a resonance-stabilized carbocation. Water reacts with the positively charged carbon atom, and deprotonation yields a hemiacetal. (This stage of the transformation is simply Markovnikov addition of water to the double bond, Section 9.1c.)

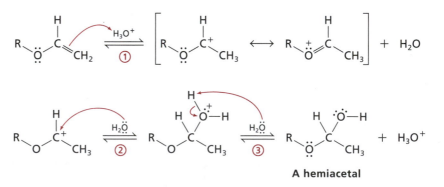

As noted in Section 19.1a, hemiacetals are normally unstable, so the products of this overall transformation comprise the aldehyde (in this example) and the alcohol.

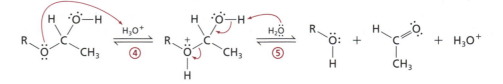

Ketones are formed if the double bond of the vinyl ether has a carbon-containing substituent attached to the carbon atom bearing the O atom. For example,

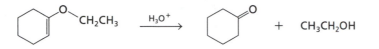

EXERCISE 19.3

Vinyl ethers react with *alcohols* to form *acetals*. Propose a reasonable mechanism for the following transformation showing the movement of electrons with curved arrows:

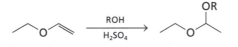

19.1d ACETALS UNDERGO HYDROLYSIS TO FORM ALDEHYDES OR KETONES

Each step of acetal formation is reversible, so an acetal can be hydrolyzed to regenerate the aldehyde or ketone from which it was made. Protonation of one oxygen atom occurs as the first step, and then a molecule of alcohol is displaced to produce the resonance-stabilized carbocation intermediate.

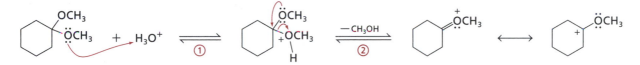

The carbocation is subsequently intercepted by a molecule of water, and deprotonation yields the hemiacetal.

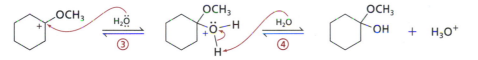

In the remaining steps of acetal hydrolysis, the other molecule of alcohol is displaced from the hemiacetal. Its mechanism is the basis of the following exercise.

EXERCISE 19.4

Using curved arrows to show the movement of electrons, propose a reasonable mechanism for hydrolysis of the hemiacetal, shown above at the right, to form cyclohexanone.

In terms of their mechanisms, acetal *hydrolysis* is just the reverse of acetal *formation*. If **molecular sieves** (mol sieves), which are inorganic aluminosilicates with small channels that only water can enter, are combined with a carbonyl compound, excess alcohol, and acid, then an acetal will be formed; that is, the following equilibrium lies toward the right and is driven by the *removal* of water:

$$\text{C=O} \ + \ 2\,\text{ROH} \ \underset{\text{acid, (xs) H}_2\text{O, }\Delta}{\overset{\substack{\text{acid, (xs) ROH,}\\ \text{mole sieves, }\Delta}}{\rightleftharpoons}} \ \text{C}\overset{\text{OR}}{\underset{\text{OR}}{}} \ + \ \text{H}_2\text{O}$$

If an acetal is heated in aqueous acid (i.e., with excess water), then this equilibrium lies toward the left. Hydrolysis of an acetal does require heating, however: An acetal is not so unstable toward aqueous acid that hydrolysis occurs rapidly at room temperature.

19.2 ACETALS AS PROTECTING GROUPS

19.2a ACETALS ARE GOOD PROTECTING GROUPS FOR ALDEHYDES AND KETONES

The central carbon atom of an acetal, like the carbonyl carbon atom, is electrophilic.

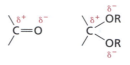

An acetal does not normally react with nucleophiles, however, because substitution of an alkoxide ion would have to occur, and RO^- is a strong base, hence, a poor leaving group. *Whereas a carbonyl group undergoes addition reactions with nucleophiles, an acetal is generally inert.* Thus, an acetal is a good protecting group for the carbonyl group of an aldehyde and ketone, a point that was first made in Section 15.5b.

A protecting group must be readily introduced and removed, and the following example shows the previously described reactions in the context of using acetals as protecting groups. Even though many alcohols can be used to make acetals, ethylene glycol (1,2-dihydroxethane) is often chosen to make an acetal for carbonyl protection because formation of a cyclic acetal is highly favored. By using a diol, both of the OR groups in the acetal product are present in the starting alcohol molecule.

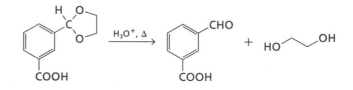

Consider now the preparation of a Grignard reagent from *m*-bromobenzaldehyde. The presence of an aldehyde group is incompatible with Grignard reagent formation (Section 15.2d), so the acetal derivative, which is stable toward reactions with most organometallic reagents, is made first. Next, the Grignard reagent is prepared and—in this example—treated with carbon dioxide followed by dilute acid to form the carboxylic acid.

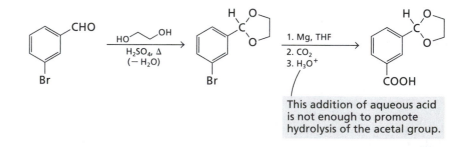

This addition of aqueous acid is not enough to promote hydrolysis of the acetal group.

The acetal is subsequently hydrolyzed to liberate the aldehyde group of the product.

19.2b ACETALS ARE ALSO GOOD PROTECTING GROUPS FOR ALCOHOLS

An acetal group is not only the masked form of a carbonyl group, it is also the masked form of an alcohol. To protect an alcohol group as an acetal, a vinyl ether is used. An especially useful vinyl ether is the cyclic molecule 3,4-dihydropyran (DHP). The product from the reaction between DHP and an alcohol is the corresponding 2-alkoxytetrahydropyran molecule, also called a THP ether or the THP derivative of the alcohol.

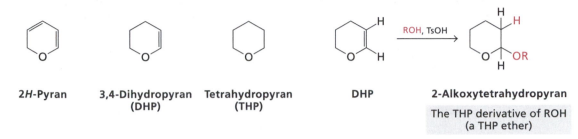

| 2*H*-Pyran | 3,4-Dihydropyran (DHP) | Tetrahydropyran (THP) | DHP | 2-Alkoxytetrahydropyran |

The THP derivative of ROH
(a THP ether)

As an example to illustrate the use of DHP to protect an alcohol group, consider the preparation of the Grignard reagent of 4-bromobutanol, which contains the acidic and incompatible OH group. If the alcohol group is first converted to its THP derivative, then the organomagnesium compound can be made. In this example, the Grignard reagent is again used to prepare a carboxylic acid.

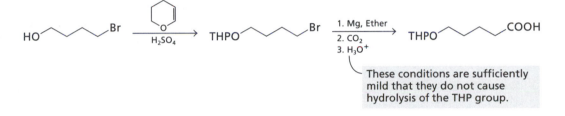

These conditions are sufficiently mild that they do not cause hydrolysis of the THP group.

A separate hydrolysis step removes the protecting group, yielding the hydroxy carboxylic acid and the stable hemiacetal molecule, 2-hydroxytetrahydropyran.

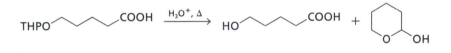

EXERCISE 19.5

Propose a reasonable mechanism for hydrolysis of a generalized THP ether.

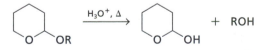

EXERCISE 19.6

Another acetal derivative used to protect alcohols is the methoxymethyl group (MOM), which is introduced by the S_N2 reaction between an alkoxide ion and chloromethyl methyl ether, $ClCH_2OCH_3$. Show the steps with their mechanisms for the synthesis of the MOM derivative of cyclopentanol (*Hint:* See Sections 7.2a and 7.2b).

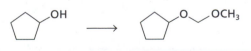

19.2c THIOACETALS ARE ALSO USED AS PROTECTING GROUPS

Thiols (Section 7.3a) react with aldehydes and ketones in the same way as alcohols. The products are **thioacetals,** which are more resistant to hydrolysis than acetals, and therefore find use as protecting groups when hydrolysis might be a problem.

A Lewis acid such as $ZnCl_2$ is often used instead of a protic acid to catalyze thioacetal formation. The carbocation formed in Step 1 is intercepted by a molecule of thiol, and a proton is transferred from sulfur to the oxygen atom along with dissociation of $ZnCl_2$.

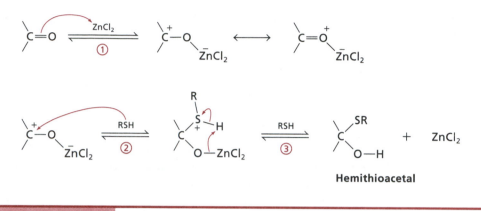

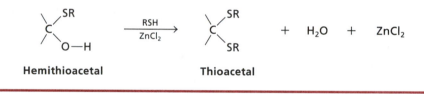

With the use of curved arrows to show the movement of electrons, propose a mechanism for the steps involved in the conversion of a hemithioacetal to a thioacetal, as shown in the following equation:

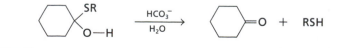

The hydrolysis of thioacetals makes use of mercury salts, which are highly thiophilic (sulfur loving) and can bind to sulfur to form a good leaving group.

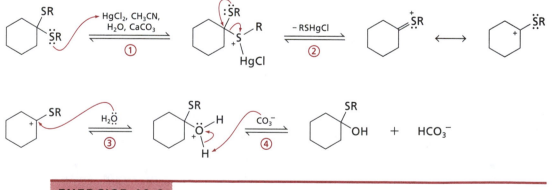

EXERCISE 19.8

Show the mechanism for the remaining step of thioacetal hydrolysis:

The thioacetal group is valuable for another transformation as well: the conversion of a carbonyl to a methylene group. This procedure makes use of the Raney nickel desulfurization reaction (Section 12.2b), and the sequence ketone → thioacetal → hydrocarbon is one of the ways to convert an acyl to an alkyl group (Section 17.2e) under essentially neutral conditions.

$$\begin{matrix} \diagdown \\ \diagup \end{matrix} C{=}O \ + \ 2\ RSH \ \xrightarrow{ZnCl_2} \ \begin{matrix} \diagdown \\ \diagup \end{matrix} C \begin{matrix} SR \\ SR \end{matrix} \ \xrightarrow{Raney\ Ni} \ \begin{matrix} \diagdown \\ \diagup \end{matrix} C \begin{matrix} H \\ H \end{matrix} \ + \ 2\ RH \ + \ 2\ H_2S$$

EXERCISE 19.9

Draw the structure of the product after each step of the reaction series shown here.

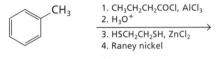

1. $CH_3CH_2CH_2COCl$, $AlCl_3$
2. H_3O^+
3. $HSCH_2CH_2SH$, $ZnCl_2$
4. Raney nickel

19.3   CARBOHYDRATES

19.3a   CARBOHYDRATES ARE POLYHYDROXY ALDEHYDE AND KETONE DERIVATIVES

Carbohydrate—literally "hydrate of carbon"—is a term that reflects the molecular formulas of these important biomolecules: $C_n(H_2O)_n$. Their structures, however, do not contain water molecules per se; instead, these derivatives of aldehydes and ketones acquire their formulas by having an abundance of alcohol OH groups. A carbohydrate is also referred to as a *sugar* or **saccharide.**

The nomenclature of carbohydrates, like that of many compounds originally isolated from natural sources, is based on common names, which means that you have to be familiar with their structures and names or look them up as you need them. The chiral carbohydrates illustrated in Figures 19.1 and 19.2 are *monosaccharides,* and their absolute configuration is designated as "D", which means that when drawn as a Fischer projection (Section 4.4), the chiral carbon atom farthest from the carbonyl group has its OH group on the right side of the vertical. The corresponding L carbohydrate has this OH group to the left, *and the configuration of every other chiral carbon atom is inverted.* If only the configuration of the last chiral carbon atom is changed, then the new carbohydrate is L, but it is a diastereomer of the original—not an enantiomer—and its root name will be different.

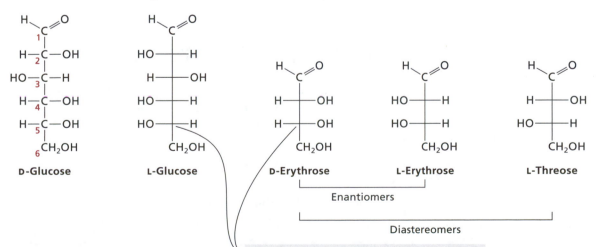

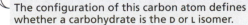

The configuration of this carbon atom defines whether a carbohydrate is the D or L isomer.

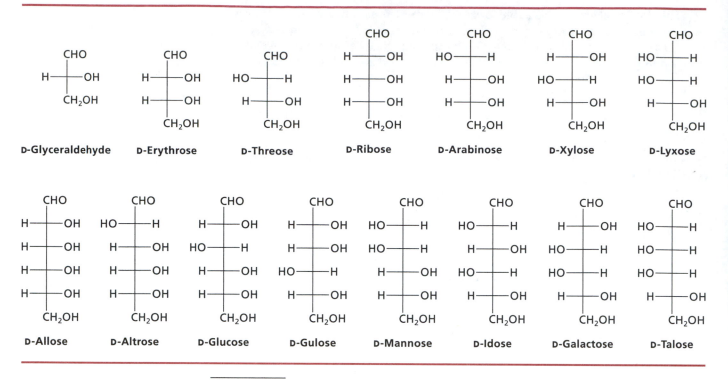

Figure 19.1
Structures and names of the common aldoses as their Fischer projections.

The common name of a sugar includes information about the length of the carbon chain, the functional group(s) that are present, *and* the substituents. Therefore, the *absence* of an OH group has to be expressed if it is missing, and *deoxy* is a common prefix in the names of sugars, meaning that an OH group has been replaced by a hydrogen atom.

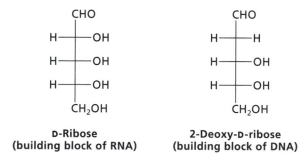

D-Ribose
(building block of RNA)

2-Deoxy-D-ribose
(building block of DNA)

Carbohydrates that have an aldehyde functional group are called **aldoses,** and those with ketone groups are **ketoses.** We further identify a general type of carbohydrate by the number of carbon atoms it has: an *aldohexose* has six carbon atoms and an aldehyde group, a *ketotriose* has three carbon atoms and a ketone group, and so on.

EXERCISE 19.10

Draw the Fischer projection for each of the following compounds:

a. L-Arabinose   b. D-2-Deoxygalactose   c. L-3,4-Dideoxytalose

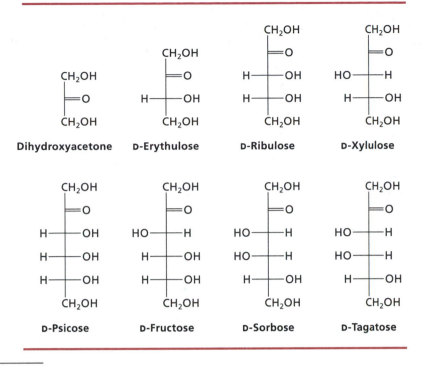

Figure 19.2
Structures and names of the common 2-ketoses as their Fischer projections.

19.3b CARBOHYDRATES USUALLY EXIST AS CYCLIC HEMIACETALS

For many of the sugars shown in Tables 19.1 and 19.2, a hydroxy group three or four carbon atoms away from the carbonyl group can react to form a cyclic hemiacetal by the mechanism shown in Section 19.1a. Even traces of acid catalyze this reaction, so many carbohydrates exist as a mixture of forms at pH 7, and equilibria exist between an open chain and both five- and six-membered ring structures.

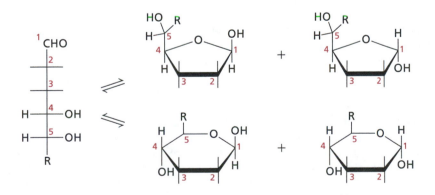

Each form has to be specified by a different name, which requires modification of the name of the parent sugar. D-Glucose will be used to correlate names with structures.

The five-membered ring hemiacetal form of D-glucose is denoted by appending the suffix **furanose** to its root name, hence D-glucofuranose. This suffix is derived from the resemblance of the carbohydrate ring to the compound *furan,* which has a five-membered ring.

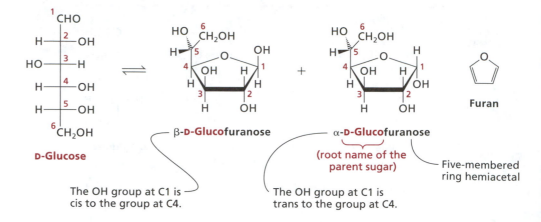

When D-glucose forms the six-membered ring hemiacetal, the compound is denoted by appending the suffix **pyranose** to its root name, hence D-glucopyranose. This suffix is derived from the resemblance of the carbohydrate ring to the compound *pyran,* which has a six-membered ring.

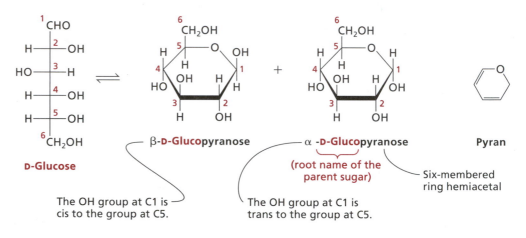

You have undoubtedly noticed that formation of these hemiacetals creates a new stereocenter at C1. The OH group at C1 can be either cis or trans in its relation to the substituent at C4 (in the furanoses) or C5 (in the pyranoses)—the OH group is designated α if it is trans and β if it is cis to this other substituent. This new stereocenter, C1, has a special name: the **anomeric carbon atom.** The two isomers—α and β—are called **anomers.** (If these cyclic forms are drawn with the anomeric carbon atom on the right side, as they are here, then α means that the OH group is below the plane of the ring, and β means the OH group is above the ring.)

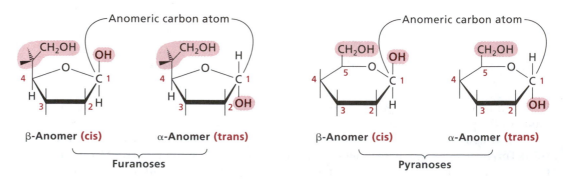

The cyclic forms that are illustrated above are the ones commonly presented in biochemistry texts and are called **Haworth projections.** These representations show a perspective view of the ring, with the anomeric carbon atom at the right and the oxygen atom within the ring at the back. Because the open chain structures (Figures 19.1 and 19.2) are shown as Fischer projections, it is important in drawing the Haworth forms to know how to attach the groups correctly. The following table summarizes these spatial relationships.

When converting from a Fischer to a Haworth projection, atoms and groups are placed as follows:

Fischer projection		Haworth projection
Left side of the vertical	→	**above** the ring
Right side of the vertical	→	**below** the ring

Even though the cyclic forms are normally shown as their Haworth forms, they can also be represented as Fischer projections by drawing a line from the oxygen atom of the C4 or C5 OH group to the carbonyl carbon atom, approaching this C1 or C2 atom from above the structure. The anomeric OH group that is created is α if it is on the right and β if it is on the left. We can also draw the chair forms (conformational structures) of these cyclic hemiacetal forms. The groups above and below the plane in the Haworth projection occupy the same orientations in the corresponding chair form. Note that the chair form of β-D-glucopyranose has all of the nonhydrogen groups in the equatorial positions.

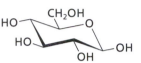

β-D-Glucopyranose

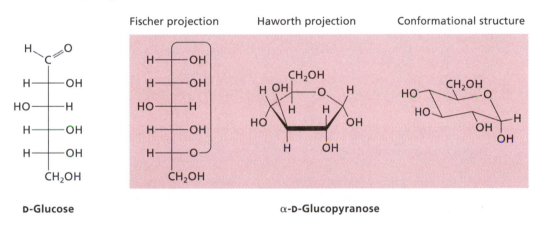

Fischer projection   Haworth projection   Conformational structure

D-Glucose   **α-D-Glucopyranose**

EXAMPLE 19.1

For α- D-fructofuranose, draw the (a.) Fischer projection and (b.) the Haworth structure.

a. Fischer projection

1. First, reproduce the Fischer projection of the parent sugar, D-fructose. Its structure, from Table 19.2, is shown at the right. Number the carbon atoms, starting at the end nearest the carbonyl group.

2. Interpret the name of the structure to be drawn. The suffix is "furanose", which signifies a five-membered ring hemiacetal.

3. Find the alcohol group attached three carbon atoms away from the carbonyl carbon atom. This OH group—at C5— is the one that adds to the carbonyl group to form the hemiacetal.

4. Draw a bond from the C5–OH group to the carbonyl carbon atom. The carbonyl oxygen atom becomes the *anomeric* OH group of the hemiacetal ring. If it is α, as in this example, orient it to the right of the vertical in the Fischer projection, as shown in the following scheme at the right.

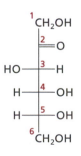

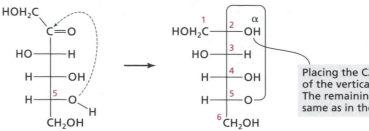

Draw a line starting from the C5–OH group and approach the carbonyl group from directly above the structure.

Placing the C2–OH group on this side of the vertical makes the α anomer. The remaining chiral centers stay the same as in the open chain structure.

b. Haworth structure

To convert the foregoing Fischer projection to the Haworth form:

1. Draw a flattened five-membered ring. The oxygen atom in the ring is placed at the back. (For a pyranose, the ring oxygen atom is back and to the right.)

2. Add the anomeric OH group: By convention, the anomeric carbon atom is placed at the right side of the ring in a D-carbohydrate.

3. Add the C6 carbon atom. For a D-carbohydrate, C6 (if present) is above the ring and at the back left.

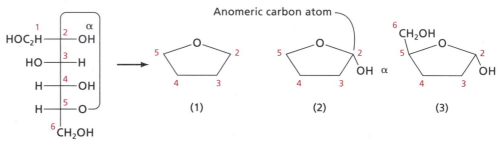

Fischer projection

4. Add the remaining non-hydrogen groups: Those on right side of the vertical in the Fischer projection are placed below the plane of the ring in the Haworth form. For a 2-ketose—which fructose is—the carbon atom at C1 is placed in the position opposite that occupied by the anomeric OH group. The anomeric OH group is down, because it is α, so C1 is attached above the plane of the ring.

5. Add the hydrogen atoms.

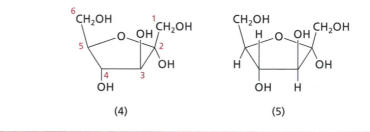

When drawing the structure of an L-carbohydrate, orient the Haworth projection with the anomeric carbon atom to the *left* and attach C6 (in a pyranose, or C5 in a furanose) at the *right* and *above* the plane of the ring. The carbon atoms in the ring are numbered counterclockwise. The anomeric OH group is still classified as α when it is trans to the C6 group, and β when it is cis. Note, however, that correlations between the substituents in the Fischer and Haworth structures are reversed: For L-carbohydrates, a group on the **right** (Fischer) is **up** (Haworth) and one on the **left** (Fischer) is **down** (Haworth).

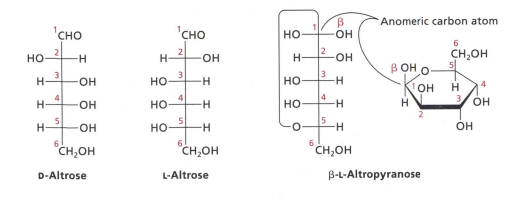

D-Altrose   **L-Altrose**   **β-L-Altropyranose**

EXERCISE 19.11

Draw the structure for each of the following compounds as both Fischer and Haworth projections:

a. β- D-Arabinofuranose   c. α- L-Ribofuranose   e. β- D-Lyxofuranose

b. α- D-Galactopyranose   d. β- D-Threofuranose   f. α- D-Manopyranose

---

19.3c HEMIACETAL ISOMERS OF CARBOHYDRATES (ANOMERS) EQUILIBRATE

Pyranoses and furanoses are hemiacetals that form by addition of the OH groups to the carbonyl groups of aldoses or ketoses. The overall mechanism occurs as follows:

1. Protonation of the carbonyl oxygen atom to form a carbocation intermediate.

2. Reaction of the nucleophilic oxygen atom of the appropriate alcohol group (at C5 in the following example) with this carbocation.

3. Deprotonation to yield the neutral product.

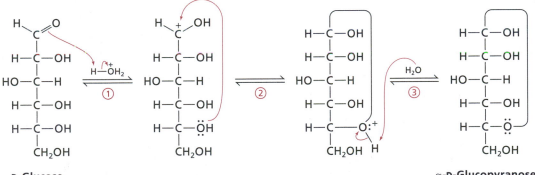

D-Glucose   **α-D-Glucopyranose**

The preceding mechanism shows only the formation of the α-anomer of the pyranose form, but as noted at the beginning of Section 19.3b, anomers of both furanoses and pyranoses may be formed, as summarized in the following scheme:

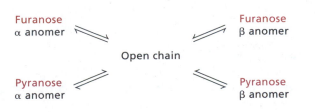

While all five isomers can exist in equilibria with each other, it is helpful to simplify this scheme by making the following assumptions based on the major isomers that are actually observed:

Aldotetrose or **2-ketopentose:**	Furanose α anomer	⇌	Open chain	⇌	Furanose β anomer
Aldopentose or **2-ketohexose:**	Furanose α anomer	⇌	Open chain	⇌	Furanose β anomer
Aldohexose:	Pyranose α anomer	⇌	Open chain	⇌	Pyranose β anomer

For example, ribose, an aldopentose, forms furanoses, which are the structures present in the backbone of RNA, in NADH (Section 18.3c), and in ATP.

Even though the preceding mechanism is shown starting with the open-chain form, equilibria are established from any of the constituents. Therefore, when a pure anomer is placed in solution, it will be converted to its isomer through participation of the open-chain form. For D-glucose—an aldohexose—the mixture at equilibrium consists of the two pyranose forms with almost twice as much of the β anomer as the α anomer. The furanose forms are essentially nonexistent, which is consistent with the simplified equilibria summarized directly above.

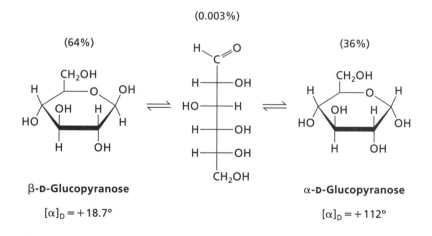

β-**D**-Glucopyranose
$[\alpha]_D = +18.7°$

(0.003%)

α-**D**-Glucopyranose
$[\alpha]_D = +112°$

This equilibration process can be followed by observing the rotation of plane-polarized light by a solution of the sugar (Section 4.2b). For D-glucose, the specific rotation eventually stabilizes at a value of $[\alpha]_D = 56.8°$, increasing from +18.7° if the solution is prepared from the pure β anomer, or decreasing from a value of +112° if starting with pure α anomer. This process of equilibration is called *mutarotation*.

Mutarotation, shown by the equilibria in the following equation, is an example of **epimerization,** which is the change in the configuration of a single chiral center.

α Anomer ⇌ Open chain ⇌ β Anomer

(If a molecule has only one chiral center, then the process is normally referred to as *racemization*.) Isomers that differ at the configuration of only one chiral center are called **epimers.**

Anomers constitute a special kind of epimer in which the configuration of the anomeric carbon atom (C1) is changed. D-Allose and D-glucose are also epimers, but they are not anomers: they are C3 epimers (notice that only the configuration of C3 is inverted). *Epimers are diastereomers.*

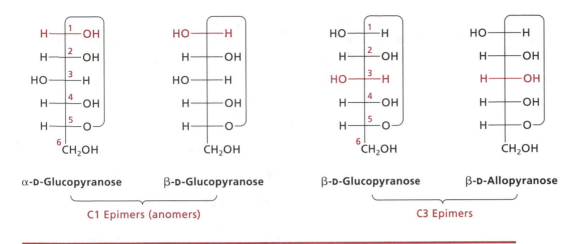

α-D-Glucopyranose   β-D-Glucopyranose   β-D-Glucopyranose   β-D-Allopyranose

C1 Epimers (anomers)   C3 Epimers

EXERCISE 19.12

Using curved arrows to depict the movement of electrons, show the steps in the mechanism for the conversion of α-D-glucopyranose to D-glucose in aqueous solution.

19.4   GLYCOSIDES

19.4a   ACETAL DERIVATIVES OF CARBOHYDRATES CAN BE PREPARED

The OH group attached to the anomeric carbon atom is *not* the alcohol functional group, but part of a hemiacetal. Therefore, it will react like any hemiacetal does, undergoing substitution when treated with a nucleophile and acid (Section 19.1b). For example, D-glucose reacts with methanol and acid to form diastereomeric acetal derivatives.

In the first step, protonation of the anomeric OH group produces a good leaving group. Expulsion of water in Step 2 generates a resonance-stabilized carbocation. If any other OH group in the carbohydrate were protonated and lost a molecule of water, only an unstabilized 1° or 2° carbocation could be produced.

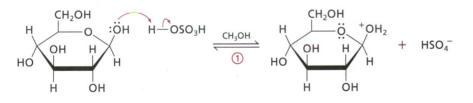

The carbocation is intercepted in the next step by methanol, and two isomers are formed because the methanol can react from either above or below the planar carbocation.

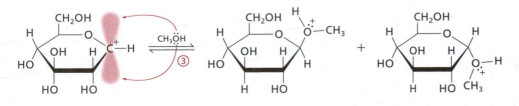

These intermediates are then deprotonated by the solvent to form acetals (only the acid–base reaction to form the α anomer is shown below). As a class, acetal derivatives of carbohydrates are called **glycosides.**

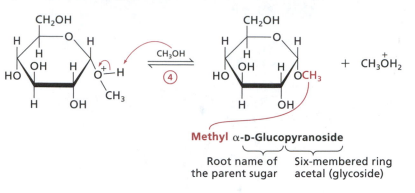

Methyl α-D-Glucopyranoside

Root name of   Six-membered ring
the parent sugar   acetal (glycoside)

Glycosides are named as **pyranosides** or **furanosides,** depending on whether a six- or five-membered ring is formed. The name of the group attached to the anomeric oxygen atom appears in the name as an unnumbered prefix (methyl, in this example). The root word of the carbohydrate is the same as that of the parent compound. The anomer designations, α and β, are the same as for the hemiacetal forms.

EXERCISE 19.13

Draw the structure for each of the following compounds as both Fischer and Haworth projections:

a. Ethyl β-D-fructofuranoside   c. Methyl α-D-ribofuranoside
b. Isopropyl α-D-galactopyranoside   d. Phenyl β-D-allopyranoside

---

19.4b ACETAL DERIVATIVES CAN BE FORMED BY COUPLING TWO OR MORE CARBOHYDRATE MOLECULES

If the alcohol that reacts to form a glycoside derivative is the OH group of another carbohydrate molecule (rather than a simple molecule such as methanol), then formation of dimeric and polymeric sugars can result. This type of transformation *cannot* be done directly by treating a sugar with acid because there are too many alcohol OH groups that can react. Specific coupling reactions are performed in Nature by enzymes and produce carbohydrates known as disaccharides, trisaccharides, oligosaccharides (having a relatively small number of coupled monosaccharides), and polysaccharides (having a large number of coupled monosaccharides). β-Maltose, shown below, is an example of a disaccharide.

Acetal (glycoside)   Hemiacetal

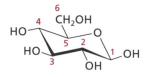

α-D-Glucopyranose
(hemiacetal component)

β-D-Glucopyranose
(alcohol component)

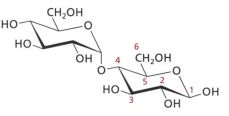

4-O-(α-D-Glucopyranosyl)-β-D-Glucopyranose
(β-maltose, a disaccharide)

When drawing the structures of disaccharides and larger oligomers, conformational forms, which tend to be more compact, are often used instead of the flat Haworth structures. The systematic name of a disaccharide lists the glycoside as a *substituent* on the oxygen atom of the sugar that provides the OH group to form the acetal. For β-maltose, the glycoside is α-D-glucopyranoside, and it is attached as a substituent to the oxygen atom of the alcohol group at C4 of β-D-glucopyranose, hence the prefix "4-*O*".

For maltose, the sugar unit on the right is a hemiacetal. As such, it undergoes mutarotation via its open-chain form. The acetal center of the disaccharide (at the left) does not isomerize because the anomeric carbon atom is not in equilibrium with the aldehyde carbonyl group (the open chain form).

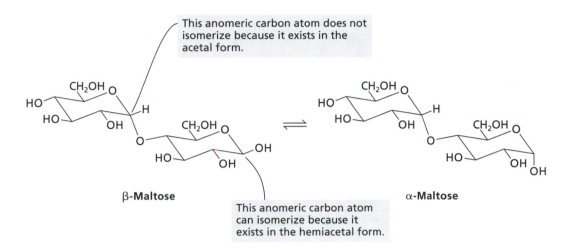

This anomeric carbon atom does not isomerize because it exists in the acetal form.

β-Maltose   α-Maltose

This anomeric carbon atom can isomerize because it exists in the hemiacetal form.

In Nature, oligomers of carbohydrates abound, and several are illustrated in Figure 19.3. The substance called "sugar" (table sugar, sucrose) is actually a disaccharide made from D-glucose and D-fructose. In this disaccharide, the linkage occurs through the anomeric carbon atom of *each* carbohydrate component. Lactose (milk sugar) is a disaccharide comprising D-glucose and D-galactose.

The substances known as cellulose, starch, and glycogen are polymers of D-glucose. The principal difference has to do with the stereochemistry of the linkage at the anomeric carbon atoms. Starch and glycogen also differ from cellulose by having branched structures.

EXERCISE 19.14

Draw the structure of β–lactose, the anomer of α–lactose shown in Figure 19.3.

19.4c GLYCOSIDES ARE HYDROLYZED USING AQUEOUS ACID OR ENZYMES

Glycosides, being acetals, are readily hydrolyzed in aqueous acid solution. The mechanism follows the reverse of the steps by which glycosides are made. Polysaccharides, which are glycosides (acetals), also undergo hydrolysis in dilute aqueous acid. For example, cellulose yields up to 95% of the theoretical amount of glucose upon hydrolysis with aqueous hydrochloric acid. A similar mechanism occurs in the enzyme-catalyzed hydrolysis of glycosides, except that the acidic proton comes from the side chain of an amino acid at

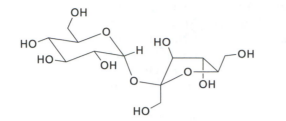
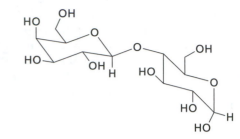

Sucrose
(α-D-Glucopyranosyl-α-D-fructofuranoside)

α-Lactose
[4-*O*-(β-D-Galactopyranosyl)-α-D-glucopyranose]

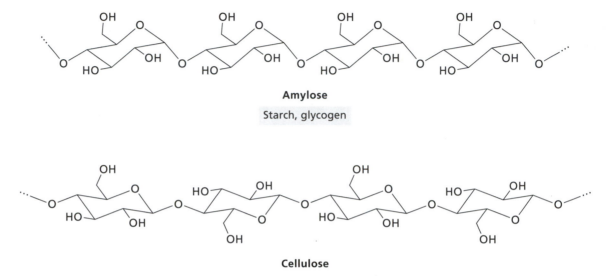

Amylose

Starch, glycogen

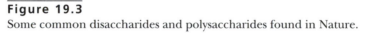

Cellulose

Figure 19.3
Some common disaccharides and polysaccharides found in Nature.

an enzyme active site. *Lysozyme* (Section 5.5c) is a prototype for such enzymes, and it catalyzes the cleavage of bacterial cell walls, which have a high carbohydrate content.

Bacterial cells walls consist of the aminocarbohydrates *N*-acetylglucosamine (NAG) and *N*-acetylmuramic acid (NAM), both of which are structural relatives of glucose.

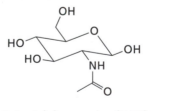
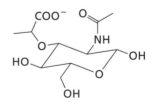

***N*-Acetylglucosamine (NAG)**   ***N*-Acetylmuramic acid (NAM)**

These two building blocks are linked together, as depicted below, in alternating fashion through acetal linkages like those in cellulose and other glucose-derived polymers.

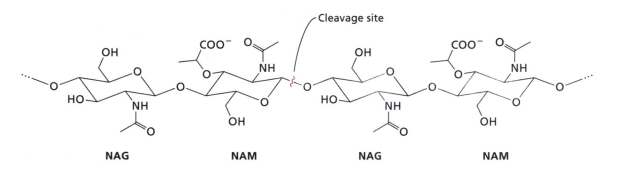

Chitin, which forms the hard exoskeleton of insects such as cockroaches, is a related substance with the formula (NAG)$_n$. Chitin is also hydrolyzed by the action of *lysozyme*.

In the active site of *lysozyme,* the carboxylic acid side chain of glutamic acid furnishes the proton that reacts with the acetal oxygen atom.

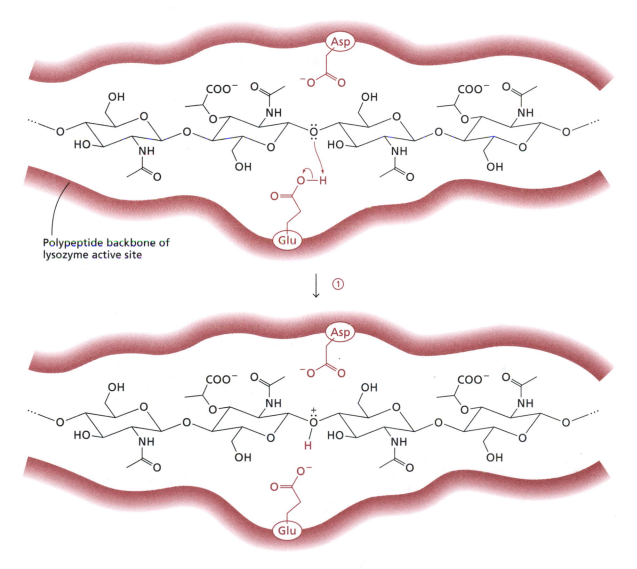

An oxycation is subsequently formed after displacement of the alcohol that corresponds to the right half of the substrate molecule.

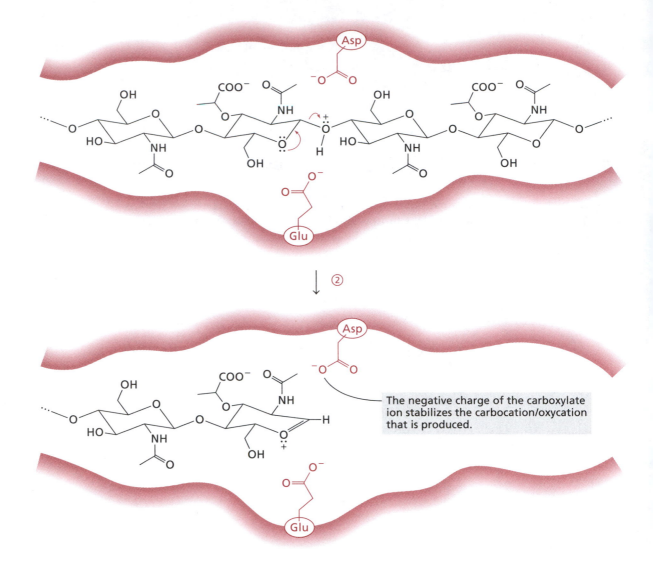

The negative charge of the carboxylate ion stabilizes the carbocation/oxycation that is produced.

This oxycation and its related carbocation provide resonance stabilization for the positive charge that remains when the polysaccharide chain is cleaved. This type of stabilization occurs any time an acetal (glycoside) is hydrolyzed. Within the enzyme active site, however, an additional influence—the negative charge of an aspartate side chain—helps to stabilize this positive charge. The additional stabilization by the enzyme accounts for much of the rate enhancement of this catalyzed reaction.

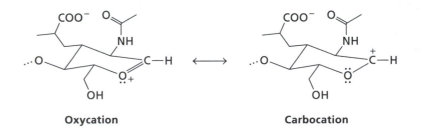

Oxycation   **Carbocation**

Once this carbocation–oxycation has been formed (Step 2), a molecule of water enters the active site. The oxygen atom of the water molecule reacts with the carbocation in Step 3, while the glutamate side chain acts as a base to remove one of the protons from H_2O. The substrate molecule dissociates from the enzyme active site in Step 4.

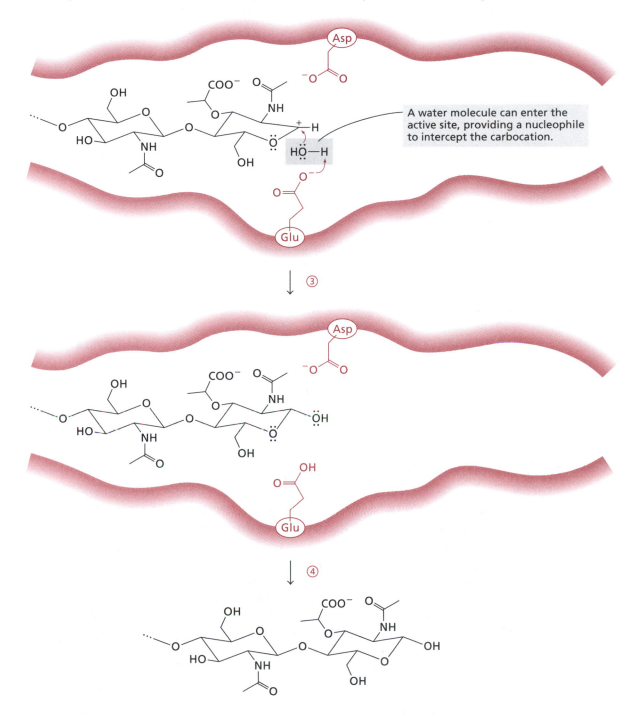

Overall, *lysozyme* catalyzes the hydrolysis of a glycoside (an acetal), which here is part of a bacterial cell wall. The process can be summarized by the following equation, where the bond that is broken is indicated by the colored cleavage symbol:

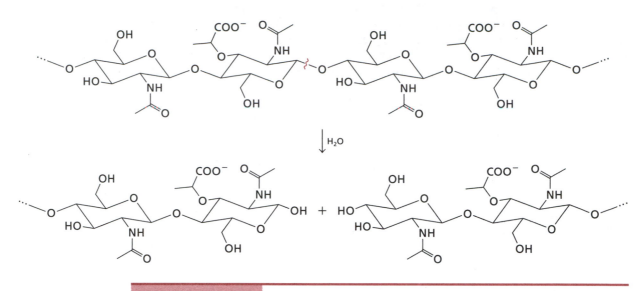

EXERCISE 19.15

Propose a mechanism for the hydrolysis reaction shown directly above, but assuming that the reaction takes place in aqueous acid rather than an enzyme active site (*Hint*: see Section 5.5c).

EXERCISE 19.16

Lysozyme works best at pH values between 3 and 8. At pH 1, both aspartate and glutamate residues at the active site will be protonated. How would this condition affect the mechanism? At pH 10, both aspartate and glutamate residues will be deprotonated. How would this condition affect the mechanism?

19.5 OXIDATION AND REDUCTION REACTIONS OF CARBOHYDRATES

19.5a THE CARBONYL GROUP OF A CARBOHYDRATE IS REDUCED TO A POLYALCOHOL CALLED AN ALDITOL

At its core, a carbohydrate is an aldehyde or ketone, and the carbonyl portion of a sugar can undergo reactions typical of those functional groups. Reduction is one such process (Section 18.3), and sodium borohydride successfully reduces the carbonyl to an alcohol group. When the carbohydrate exists in its hemiacetal form, the open-chain (carbonyl) form is still accessible, so reduction takes place. The general class of product from reduction is called an **alditol**. The actual name is generated by appending the suffix -itol to the carbohydrate root name.

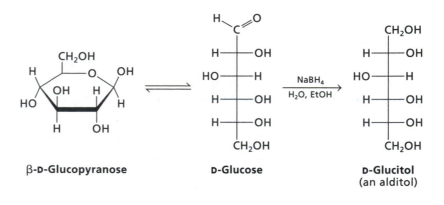

β-D-Glucopyranose   D-Glucose   D-Glucitol
(an alditol)

The stereochemical consequences of reduction vary. When a 2-ketose is reduced, a new chiral center is generated at C2 and diastereomers are formed. For example,

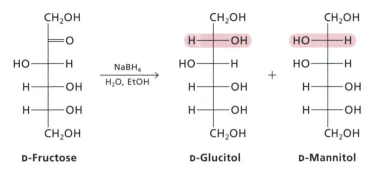

These alditol products are C2 epimers, and they have the same structures—hence names—as the alditols derived from D-glucose and D-mannose, respectively.

Because the end groups of the alditols are identical, namely CH_2OH, the product of carbohydrate reduction reactions may be a meso compound. For example, D-xylose is reduced to *meso*-xylitol, which has an internal mirror plane.

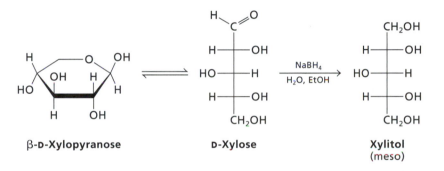

A meso-compound can also be formed as one product from the reduction of a 2-ketose. D-Sorbose is reduced to form two C2 epimers, D-talitol and *meso*-galactitol, which is also called sorbitol.

EXERCISE 19.17

An unknown D-aldohexose treated with sodium borohydride yields a meso alditol. What are the possible structures of the starting aldohexose? (Consult Fig. 19.1.)

19.5b CARBOHYDRATES ARE SUSCEPTIBLE TO OXIDATION

Because a carbohydrate has alcohol and (in some cases) aldehyde functional groups, oxidation processes should be facile but complex. Selective reagents are therefore needed if a single product is desired.

In the early days of sugar chemistry, copper(II) or silver(I) ions were used to test for the presence of carbohydrates. These reactions were easy to follow because of the appearance of a brick-red precipitate [copper(I) oxide] or a silver mirror on the bottom of the test tube (metallic silver). The reactions occur because the aldehyde group of the sugar is oxidized to a carboxylic acid group (Section 18.4a). Carbohydrates that give positive results—a precipitate or mirror—are called **reducing sugars** because they reduce the added reagent (note that the carbohydrate itself is oxidized).

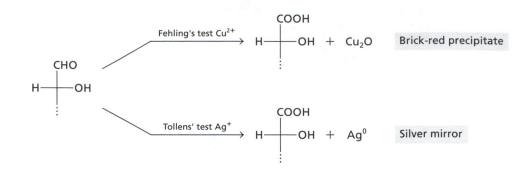

We now understand that carbohydrates in their hemiacetal forms are in equilibrium with the open-chain aldehyde compounds, which are readily oxidized. (Some ketoses also react because they can be isomerized to an aldose under the reaction conditions.) Carbohydrates that exist as glycosides (acetals) do not react under these conditions.

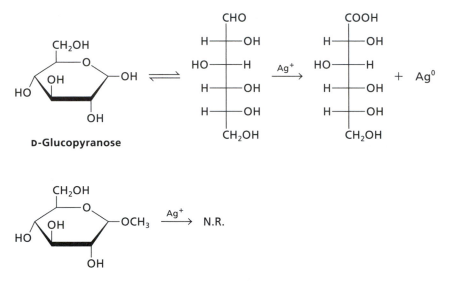

D-Glucopyranose

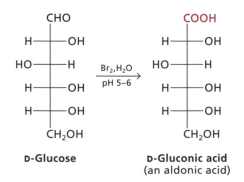

Methyl-D-glucopyranoside

For preparative purposes, reagents other than metal ions are used. Bromine in water at pH 5–6 is mild enough to oxidize only the aldehyde group without affecting the alcohol groups appreciably. The product acids, as a general class, are called **aldonic acids,** and they can exist in a cyclic form called a *lactone* (discussed in Section 21.2b). At present, it is sufficient to recognize that aldonic acids cyclize to form the corresponding lactone (both five- and six-membered rings can be formed) just as an aldose cyclizes to form a hemiacetal.

For example, D-gluconic acid forms both five- and six-membered ring lactones.

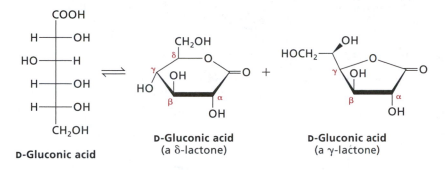

| | D-Gluconic acid (a δ-lactone) | D-Gluconic acid (a γ-lactone) |

D-Gluconic acid

If an aldose is treated with nitric acid, then the aldehyde group as well as the primary alcohol group at the end of the chain can be oxidized to carboxylic acid groups. This class of compound is an **aldaric acid** (also called a saccharic acid), and aldaric acids exist as lactones too. For example, D-glucose gives D-glucaric acid, as shown here.

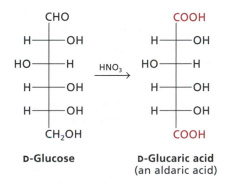

D-Glucose   **D-Glucaric acid** (an aldaric acid)

Like alditols, described in Section 19.5a, aldaric acids can be meso because the two end groups are the same, namely, COOH. For example, D-galactose is converted to galactaric acid, which is meso.

EXERCISE 19.18

D-Glucose, when oxidized with nitric acid to give D-glucaric acid, forms a dilactone that has two five-membered rings. Draw this dilactone's structure.

EXERCISE 19.19

An unknown D-aldopentose is treated with nitric acid to give a meso aldaric acid. What are possible structures for the starting aldopentose? (Consult Fig. 19.1.)

CHAPTER SUMMARY

Section 19.1   Hemiacetals and acetals

- A molecule of alcohol undergoes the acid-catalyzed addition to a carbonyl group of an aldehyde or ketone to form a hemiacetal.
- A hemiacetal is normally unstable with respect to regeneration of the carbonyl component and a molecule of alcohol.
- A hemiacetal that exists as a five- or six-membered cyclic compound is often stable.
- An acetal is formed by the acid-catalyzed substitution of a hemiacetal OH group by a molecule of alcohol.

- Addition of water to a vinyl ether yields a hemiacetal.
- An acetal is readily hydrolyzed in aqueous acid to form an aldehyde or ketone and a molecule of alcohol.

Section 19.2   Acetals as protecting groups

- Acetals are stable toward reactions with nucleophiles and bases, so they are useful protecting groups for the carbonyl function of an aldehyde or ketone.
- Acetals can be made by addition of a molecule of alcohol to a vinyl ether.
- An acetal prepared from a vinyl ether, especially DHP (3,4-dihydropyran), is a suitable protecting group for the alcohol OH group.
- Thiols react with aldehydes and ketones to form thioacetals, which are also useful as protecting groups. A thioacetal can be hydrolyzed, or it can removed with use of Raney nickel, which replaces the sulfur atoms with hydrogen atoms.

Section 19.3   Carbohydrates

- Carbohydrates are aldehyde and ketone derivatives that also have several OH groups attached to the carbon chain.
- The nomenclature of carbohydrates is based on common names known for more than a century.
- Carbohydrates normally exist in cyclic hemiacetal forms called anomers.
- The anomeric forms of carbohydrates comprise five- and six-membered rings and are called furanoses and pyranoses, respectively.
- There are several ways to depict cyclic carbohydrate structures. These include the Fischer and Haworth projections, as well as conformational drawings. The six-membered ring form of a carbohydrate exists in the chair form.
- Anomers of a carbohydrate exist in equilibrium with the open-chain form, so they readily interconvert.
- Anomers are a category of epimers, which are diastereomers that differ in the configuration of a single chiral center.

Section 19.4   Glycosides

- Acetal derivatives of carbohydrates are readily prepared by reactions between an anomer and an alcohol under the influence of an acid catalyst.
- An acetal derivative of a carbohydrate is called a glycoside.
- Glycosides can be formed by substitution of the anomeric OH group of one sugar molecule by an alcohol OH group of another carbohydrate.
- A glycoside, like any acetal, is readily hydrolyzed with aqueous acid into its carbonyl and alcohol components.
- Lysozyme is an enzyme that catalyzes the hydrolysis of specific oligosaccharides.

Section 19.5   Oxidation and reduction reactions of carbohydrates

- The carbonyl group of a carbohydrate is readily reduced with use of sodium borohydride. The product of the reaction is called an alditol.
- The carbonyl group of an aldose is readily oxidized with use of bromine in water at pH 5–6. The product is an aldonic acid, and the carbohydrate is called a reducing sugar.
- Both the carbonyl and the primary alcohol groups of an aldose are oxidized upon treatment with nitric acid. The product is an aldaric acid.

KEY TERMS

Section 19.1a
hemiacetal

Section 19.1b
acetal

Section 19.1c
vinyl ether
enol ether

Section 19.1d
molecular sieves

Section 19.2c
thioacetal

Section 19.3a
carbohydrate
saccharide
aldose
ketose

Section 19.3b
furanose
pyranose
anomeric carbon atom
anomer
Haworth projection

Section 19.3c
epimerization
epimers

Section 19.4a
glycoside
pyranoside
furanoside

Section 19.5a
alditol

Section 19.5b
reducing sugar
aldonic acid
aldaric acid

REACTION SUMMARY

Sections 19.1a

Aldehydes and ketones undergo addition of alcohols to their carbonyl groups to form
hemiacetals. Unless it exists in a five- or six-membered ring, the hemiacetal form is less
stable than its two constituents.

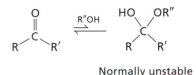   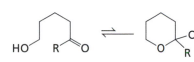

Normally unstable   Normally stable
                            (five-membered ring also)

Section 19.1b

Aldehydes and ketones react with alcohols in the presence of an acid catalyst to form
acetals. Removal of water drives the reaction to completion.

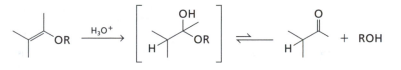

Section 19.1c

Vinyl ethers undergo addition of water in aqueous acid to form hemiacetals, which convert to molecules of the corresponding carbonyl compound and alcohol.

Section 19.1d

An acetal undergoes hydrolysis in aqueous acid to regenerate its carbonyl and alcohol
constituents.

Section 19.2b

Vinyl ethers react with alcohols in acid to form acetals. Dihydropyran reacts with alcohols to form tetrahydropyran derivatives called THP ethers, which are used as protecting groups for alcohols.

DHP

Section 19.2c

Aldehydes and ketones react with thiols and a Lewis acid catalyst to form thioacetals.

Thioacetals undergo hydrolysis, a reaction assisted by mercury(II) ions, which react with the thiol groups.

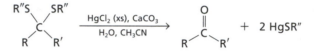

Thioacetals, when treated with Raney nickel, are desulfurized: The sulfur atoms are replaced by hydrogen atoms.

Section 19.4a

Carbohydrates react with alcohols in acid to form acetal derivatives, called glycosides. Anomeric products are formed regardless of which anomeric hemiacetal is used as the starting material.

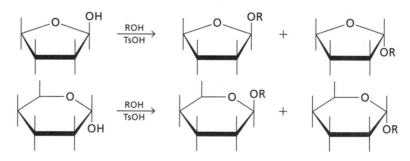

Section 19.5a

Carbohydrates undergo reduction by sodium borohydride to form alditols.

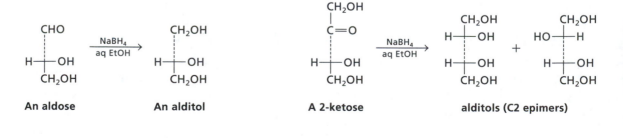

Section 19.5b

Carbohydrates undergo oxidation with aqueous bromine to form aldonic acids and with nitric acid to form aldaric acids.

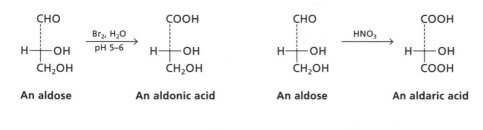

| An aldose | An aldonic acid | An aldose | An aldaric acid |

ADDITIONAL EXERCISES

19.20. Give a systematic name for each of the following carbohydrate derivatives. Consult Figures 19.1 and 19.2 as needed. Circle the carbohydrates that are reducing sugars.

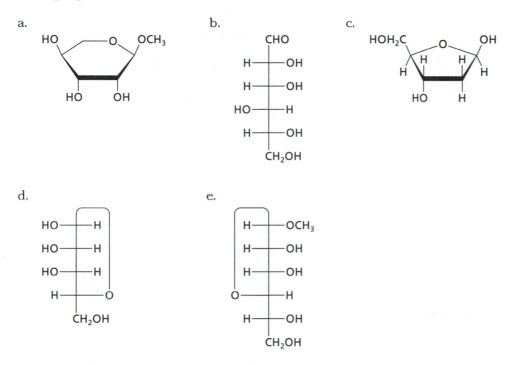

a.

b.

c.

d.

e.

19.21. Draw the Fischer projection for each of the following compounds:

  a. L-Glucose b. D-2-Deoxythreose c. L-3-Deoxymannose

19.22. Without looking up actual names, identify each of the following as D or L, and classify each as an aldo- or ketohexose, pentose, etc.

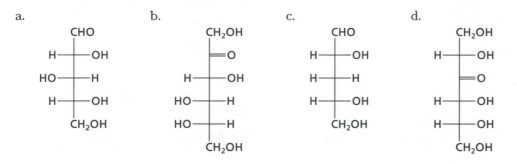

a.

b.

c.

d.

19.23.   Draw the Fischer and Haworth projections for the following compounds:

    a. β-D-Glucofuranose

    b. α-D-2-Deoxymannopyranose

    c. α–D–Talopyranose-6-phosphate

    d. Methyl β-D-2-galactofuanoside

19.24.   Draw the anomer and enantiomer for each of the following compounds:

a.

b.

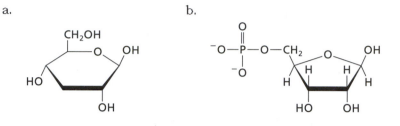

19.25.   An unknown D-aldopentose treated with sodium borohydride yields an optically active alditol. What are possible structures for the starting aldopentose?

19.26.   An unknown L-aldohexose is treated with nitric acid a give a meso aldaric acid. What are possible structures for the starting compound? Draw the structures of the meso aldaric acids that can be produced from an L-aldohexose.

19.27.   Draw the structure of the major product(s) expected from each of the following reactions. Show stereochemistry where appropriate. If no reaction occurs, write N.R.

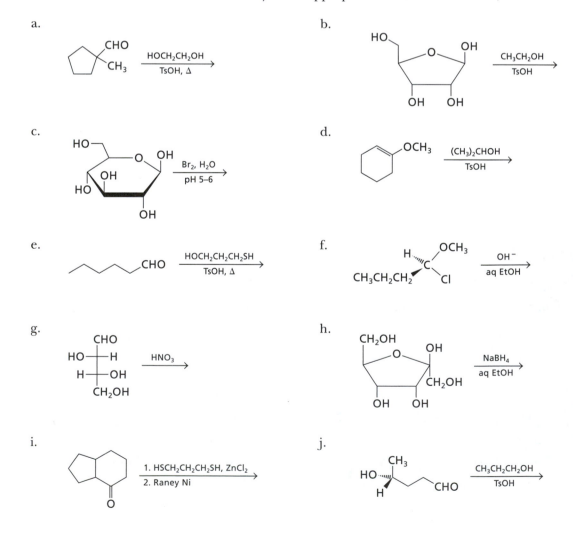

a.

b.

c.

d.

e.

f.

g.

h.

i.

j.

19.28. Borane–THF is commonly used to carry out hydroboration reactions of alkenes (Section 9.4a). It is also a good reducing agent for aldehydes and ketones (cf. Section 18.3b). As a result, one cannot conduct hydroboration reactions of an alkene if the molecule also has an aldehyde or ketone functional group. Show how you would prepare each of the following products from the given starting material. You may use any other reagents or solvents.

a.

b.

ADDITION–ELIMINATION REACTIONS OF ALDEHYDES AND KETONES

In the previous two chapters (Chapters 18 and 19), you have learned several reactions that start with addition of a nucleophile to the carbonyl group of an aldehyde or ketone. When the nucleophile is a molecule of an alcohol or a thiol, *substitution* can take place after the initial addition step (Chapter 19). In this chapter, you will learn several transformations in which an *elimination* step occurs after the initial addition reaction.

The nucleophiles that engage with aldehydes and ketones in addition–elimination reactions are amines, derivatives of ammonia such as hydrazine and hydroxylamine, and ylides. *Ylides* are molecules in which a carbanion is stabilized by bonding to a heteroatom bearing a positive charge; only the chemistry of phosphorus and sulfur ylides will be discussed here. Phosphorus ylides are particularly effective reagents for making carbon–carbon bonds, so they have important applications in synthesis.

Primary amine

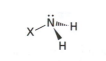

Hydrazine (X = NH₂)
Hydroxylamine (X = OH)

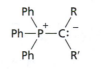

Secondary amine   **A triphenylphosphonium ylide**

A dimethylsulfonium ylide

20.1 COMPOUNDS WITH CARBON–NITROGEN DOUBLE BONDS

20.1a A PRIMARY AMINE REACTS WITH AN ALDEHYDE OR KETONE TO PRODUCE AN IMINE, ALSO CALLED A SCHIFF BASE

The reaction of a primary amine with an aldehyde or ketone provides a straightforward example of an addition–elimination process. This reaction is a crucial step during the course of several metabolic pathways, and its importance in biochemistry would be difficult to overemphasize.

The nucleophilic nitrogen atom of a primary amine reacts with the electrophilic carbon atom of the carbonyl group with the concomitant movement of the π electrons onto the oxygen atom. This pattern of electron movement in Step 1 is the common pathway that you have already learned (Section 18.1b). The nucleophilic addition step is followed by an acid–base reaction to form a **hemiaminal** (a geminal amino alcohol), which is normally unstable and either reverts to starting materials or reacts further.

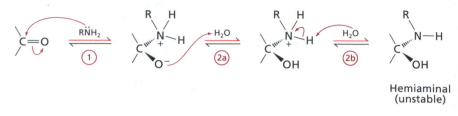

Hemiaminal
(unstable)

If the OH group of a hemiaminal is protonated, a molecule of water can dissociate to generate a resonance-stabilized cationic intermediate.

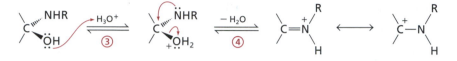

This cationic species does not react with a second molecule of the amine because deprotonation occurs more readily, forming a neutral, stable product. This product is an **imine,** also called a **Schiff base.**

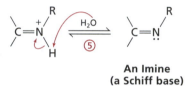

**An Imine
(a Schiff base)**

Imine formation is often carried out under acidic conditions—often with aqueous ethanol as the solvent—to increase the rate of Step 3, which generates the H_2O leaving group for Step 4. Remember, however, that amines are bases (Section 5.2d), so they can react with the acid too. Protonation of the amine lowers its concentration as a nucleophile in the solution. Thus, this acid–base side reaction will slow the formation of the imine if the pH of the solution is too low. The optimum pH for imine formation lies between 6 and 7, as illustrated in Figure 20.1.

EXERCISE 20.1

Draw the structure of the product from the reaction of cyclohexanone with 2-amino-2-methylpropane. Propose a mechanism for this reaction catalyzed by sulfuric acid.

Figure 20.1
A plot of pH versus rate for imine formation from an aldehyde and a primary amine.

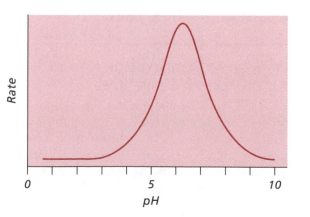

20.1b HYDRAZINE AND HYDROXYLAMINE CONVERT CARBONYL COMPOUNDS TO STABLE DERIVATIVES CALLED HYDRAZONES AND OXIMES

Hydrazine is similar in structure to primary amines, and its NH_2 group reacts analogously. Aldehydes and ketones react with hydrazine to form derivatives called **hydrazones,** which have a C=N double bond.

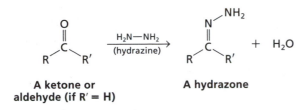

A ketone or aldehyde (if R′ = H)   **A hydrazone**

Before the advent of spectroscopic methods, substituted hydrazines such as phenylhydrazine and 2,4-dinitrophenylhydrazine were employed to make solid derivatives with well-defined melting points as a way to characterize aldehydes and ketones.

Phenylhydrazine   **2,4-Dinitrophenylhydrazine**

The mechanism for hydrazone formation is the same as that for imine formation: Nucleophilic addition of one hydrazine nitrogen atom to a carbonyl group is followed by proton transfer steps.

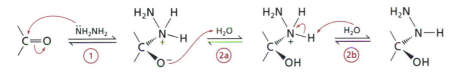

Next, the hydroxy intermediate is protonated at its OH group, and that species loses water and a proton to generate the hydrazone product.

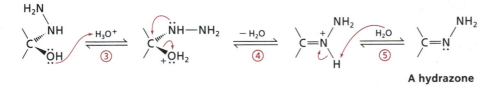

A hydrazone

Hydrazone formation is still used occasionally to facilitate isolation of an aldehyde or ketone from a reaction mixture. This application is especially important when the carbonyl compound is a volatile liquid and its substituted hydrazone derivative is a crystalline solid. A hydrolysis step, described in Section 20.1c, is subsequently carried out to regenerate the carbonyl compound.

An **oxime** is the product of the reaction between hydroxylamine, NH_2–OH, and a carbonyl compound. Like a hydrazone, an oxime is often a crystalline solid.

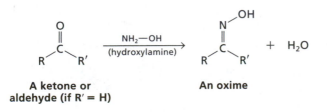

A ketone or aldehyde (if R′ = H)   **An oxime**

EXERCISE 20.2

Propose a reasonable mechanism for the reaction between 2-pentanone and hydroxyl-amine in the presence of a small amount of sulfuric acid as a catalyst.

20.1c IMINES, HYDRAZONES, AND OXIMES CAN BE HYDROLYZED TO REGENERATE THE CORRESPONDING CARBONYL COMPOUND

Compounds with a C=N bond can be hydrolyzed by aqueous acid, and the mechanism follows in the reverse order the steps required for their formation:

1. Protonation at nitrogen, forming a cationic intermediate.
2. Reaction of the cation at its carbon atom with a molecule of water.
3. Deprotonation to form the neutral hydroxy amino compound.
4. Protonation at the nitrogen atom to generate a good leaving group.
5. Regeneration of the C=O double bond with concomitant elimination of an amine (X = R), hydrazine (X = NH$_2$), or hydroxylamine (X = OH).

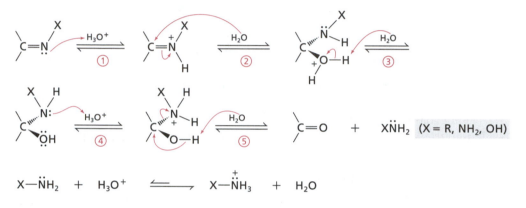

$$X-\ddot{N}H_2 + H_3O^+ \rightleftharpoons X-\overset{+}{\ddot{N}}H_3 + H_2O$$

The equilibria of these hydrolysis processes are driven toward formation of the carbonyl compound by a final reaction between the liberated amino compound and aqueous acid. This acid–base step lowers the concentration of the nucleophilic XNH$_2$ species in solution, which inhibits its readdition to the C=O double bond.

EXERCISE 20.3

Propose a reasonable mechanism for the reaction shown here. What is the identity of the amine product?

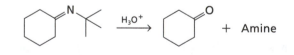

20.1d OXIMES CAN UNDERGO DEHYDRATION OR REARRANGEMENT

Besides hydrolysis, oximes can undergo reactions that take advantage of the presence of the OH group attached to the nitrogen atom. For the oxime derivative of an *aldehyde*, a proton and an OH group are bonded to adjacent atoms (C and N, respectively). Loss of water occurs when an aldehyde oxime reacts with a dehydrating agent such as P$_4$O$_{10}$. A nitrile is the product of this transformation.

$$\underset{R}{\overset{O}{\parallel}}\underset{H}{\overset{C}{}} \quad \xrightarrow[-H_3O]{NH_2OH} \quad \underset{R}{\overset{N\diagup OH}{\overset{\parallel}{C}}}\underset{H}{} \quad \xrightarrow[-H_2O]{P_4O_{10}} \quad R-C\equiv N$$

When an oxime is treated with acid at a higher concentration than that used in its formation (which requires only a catalytic amount), it *rearranges* to form an amide, a reaction that is facilitated by the presence of the weak N–O bond.

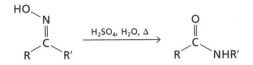

Oximes can exist as isomers if R and R′ are different in the structure shown directly above. These are labeled as syn- [or (Z) Section 4.1b] and anti- (or E). In rearrangement reactions of oximes, the group anti to the OH group is the one that usually migrates. In certain cases, the syn and anti isomers can rapidly interconvert, but for our purpose, we will assume that pure isomers can be isolated and used.

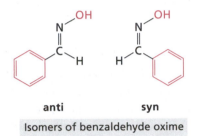

anti   **syn**

Isomers of benzaldehyde oxime

When an oxime is heated with sulfuric acid or PCl₅ (a Lewis acid), then a good leaving group is generated from the OH group. [The OH group can also be converted to its sulfonate ester derivative (Section 7.1b) to create a good leaving group.] Once the leaving group is formed, the group anti to it (R′ in the following equations) migrates from carbon to nitrogen, displacing the leaving group to generate a cationic intermediate. This species is then trapped by a molecule of water in Step 3.

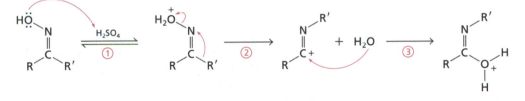

The final product is subsequently formed by two more acid–base reactions (Steps 4–6, below). This overall conversion of an oxime to an amide is the **Beckmann rearrangement.**

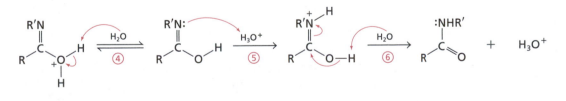

EXERCISE 20.4

Propose a reasonable mechanism for the following reaction:

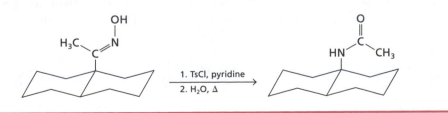

The Beckmann rearrangement also works well with cyclic ketones, producing the corresponding cyclic amides, known as *lactams*. For example, this transformation is employed commercially to convert cyclohexanone to ε-caprolactam, the precursor of Nylon 6 (Section 26.3c).

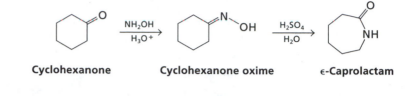

Cyclohexanone   **Cyclohexanone oxime**   **ε-Caprolactam**

EXERCISE 20.5

Propose a reasonable mechanism for the conversion of cyclohexanone oxime to ε-caprolactam.

20.1e THE WOLFF–KISHNER REACTION COMBINES HYDRAZINE AND BASE TO CONVERT A CARBONYL TO A METHYLENE GROUP

At least four general routes exist for the conversion of a carbonyl to a methylene group (Fig. 17.8). Three of those processes proceed via single electron-transfer pathways: hydrogenolysis (Section 18.3e), Raney nickel desulfurization of a thioacetal (Section 19.2c), and the Clemmensen reduction (Section 18.3e). The fourth method is the **Wolff–Kishner reaction,** and it proceeds by a series of acid–base reactions from the hydrazone derivative.

The Wolff–Kishner reaction

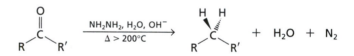

The Wolff–Kishner procedure is carried out in a single flask into which is placed the ketone (or aldehyde), hydrazine hydrate, sodium hydroxide, and a polar, high-boiling solvent like diethylene glycol, $HOCH_2CH_2OCH_2CH_2OH$, or DMSO. The hydrazone, generated by the usual pathway in Step 1, undergoes an acid–base reaction with the hydroxide ion in Step 2, yielding a resonance-stabilized anion.

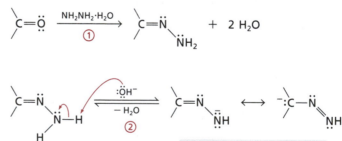

Resonance-stabilized anion

This anion is subsequently protonated to form an intermediate diazene (Step 3), which itself can undergo reaction with base to produce a second anion with loss of molecular nitrogen (Step 4). The resulting carbanion (a very strong base) is rapidly protonated in Step 5 to form the hydrocarbon product.

The Wolff–Kishner reaction is general for both aliphatic and aromatic ketones, but aldehydes are not reduced as cleanly, and side reactions sometimes predominate. The experimental conditions of the reaction make it compatible with protecting groups like acetals, which are stable toward base. Many other functional groups are likewise tolerated by the combination of reagents. For example, the carboxylic acid groups in the following molecule do not react except with hydroxide ion. The second step—dilute aqueous acid—is needed to protonate the carboxylate ions and restore the COOH groups.

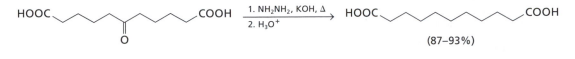

(87–93%)

EXERCISE 20.6

What is the product expected at each step of the reaction sequence shown here? The acid workup (Step 2) is meant to hydrolyze the acetal group (Section 19.1d).

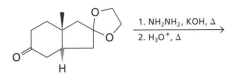

20.2   IMINE CHEMISTRY AND BIOCHEMISTRY

20.2a   AN IMINE IS REDUCED BY HYDRIDE ION TO FORM AN AMINE

The C=N double bond reacts with nucleophiles in the same way as a C=O double bond. Therefore, many of the nucleophilic addition reactions that you learned in the previous chapters apply to the reactions of imines, as well.

Sodium borohydride and $LiAlH_4$ are two reagents used to reduce an imine to an amine. The mechanism is the same as that for reduction of the carbonyl group of aldehydes and ketones, in which a hydride ion reacts at the electrophilic carbon atom of the functional group (Section 18.3a). The anion that forms is subsequently protonated in a second step to generate the secondary amine. Remember that $LiAlH_4$ requires a separate step for workup because the reagent is incompatible with protic solvents; with sodium borohydride, the solvent supplies the proton.

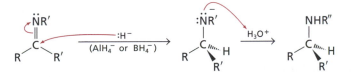

A common procedure for making an amine through reduction of an imine derivative is a transformation called **reductive amination,** which starts with an aldehyde or ketone and generates an imine in situ. The reagent used for this transformation is sodium cyanoborohydride ($NaBH_3CN$), which has a cyano group in place of one of the protons of the borohydride ion, making the reagent less nucleophilic and more stable toward acid.

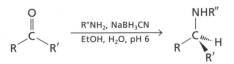

Under the slightly acidic conditions used for the reductive amination procedure, the carbonyl compound reacts with the amine to form an imine derivative (*amination*). The imine is a better base than its carbonyl precursor, so it is readily protonated near pH 7, making the carbon atom more electrophilic, hence reactive toward the cyanoborohydride ion (*reduction*). The carbonyl compound is *not* protonated appreciably under these conditions, so it is not activated (Section 18.1c) and does not react with the hydride ion. (At pH 3–4, an aldehyde or ketone *is* reduced by $NaBH_3CN$.)

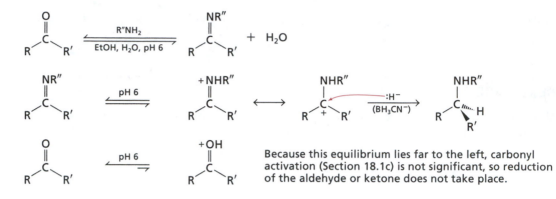

Another procedure used to convert an aldehyde or ketone to an amine makes use of the oxime derivative, which reacts with $LiAlH_4$, a reagent that is able to reduce both C=N and N–O bonds. This transformation is used to convert an aldehyde or ketone to a primary amine without contamination by secondary amine byproducts (see Section 6.3d).

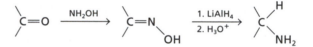

EXERCISE 20.7

The C=N double bond can also be reduced using hydrogen and a catalyst. What is the product expected from each of the following reactions?

a.   b.

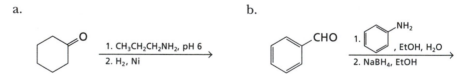

20.2b IMINES PLAY AN IMPORTANT ROLE IN STRUCTURAL BIOLOGY, FORMING CROSS-LINKS THAT CAN STABILIZE PROTEINS

The C=N double bond has several substantive roles in biochemical systems. For example, an imine bond is crucial for the success of several reactions that take place during the metabolism of carbohydrates (Section 23.2) and amino acids (Section 20.2c). Imines are also used to provide structural support within biomolecules.

Proteins, as you have seen throughout this text, are constructed from amino acids linked together by peptide bonds (Section 5.5b). In addition, hydrogen bonds and other noncovalent interactions (Section 2.8) help define a protein's shape by influencing how the polypeptide chain folds. Proteins that have structural roles within organisms—for example, the ones that make up hair, skin, bones, and the like—rely on covalent bonds to ensure that their strength and shapes are maintained, even when the proteins are subjected to a variety of stresses.

Collagen, the most abundant protein found in vertebrates, constitutes connective tissues such as bone, cartilage, tendons, skin, and blood vessels. It exists as bundles of three, helical polypeptide chains (Figure 20.2) arranged in larger arrays called fibrils.

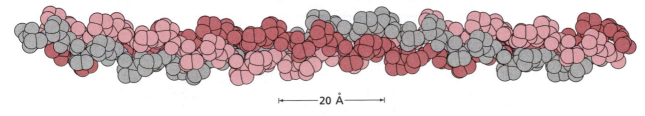

—20 Å—

Figure 20.2
The three intertwined polypeptide chains of a collagen bundle.

Collagen's highly ordered structure, which, for example, can provide sites for bone nucleation, is maintained by bonds that are **cross-links** between the polypeptide chains.

Many cross-links in collagen are imine bonds. For example, an imine can be formed by the reaction between the aldehyde group of allysine and the amino group of lysine.

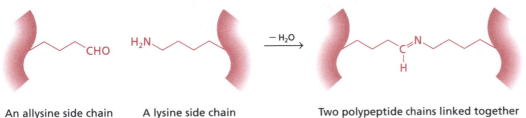

An allysine side chain   A lysine side chain   Two polypeptide chains linked together

Allysine is itself formed from the imine derivative of lysine, which is hydrolyzed to generate the aldehyde functional group, as shown in the following scheme:

Side chain
of lysine

Side chain
of allysine

Another connective tissue protein is elastin, the name of which derives from its elastic properties. It is found in the lungs, in large blood vessels like the aorta, and in some ligaments such as those in the neck. Unlike collagen, elastin forms a three-dimensional network of fibers, lacking the regular superstructure of fibrils that collagen has. Nevertheless, covalent cross-links maintain elastin's overall structure while providing elasticity.

In elastin, imine bonds between lysine and allysine residues are reduced to form amino groups. Just as you saw for the biochemical reduction reactions of carbonyl groups (Section 18.3c), NADH can reduce C=N double bonds as well. The advantage achieved by reduction of an imine group is that the cross-link is no longer subject to hydrolysis. In the aqueous environment of living systems, this reduced sensitivity to water adds yet another measure of stability to the protein structure.

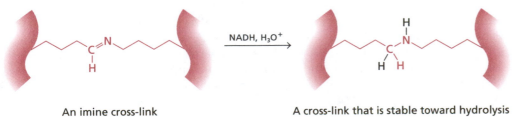

An imine cross-link   A cross-link that is stable toward hydrolysis

Write out the individual steps in the mechanism by which lysine and allysine condense to form the imine cross-link.

20.2c IMINE BOND FORMATION AND HYDROLYSIS PLAY SIGNIFICANT ROLES IN AMINO ACID METABOLISM

You have already learned about several enzymes that perform their functions in the presence of coenzymes. For example, the reduction of aldehyde and ketone carbonyl groups makes use of NADH as a coenzyme (Section 18.3c). A coenzyme bound within the active site of an enzyme is often crucial to orient electrophilic centers so that a substrate can react appropriately. As might be expected, covalent attachment of the coenzyme to an amino acid residue in the active site provides one means to create a proper bonding site. An imine bond is especially suitable because it is readily hydrolyzed when the coenzyme needs to be released or renewed.

The *coenzyme* utilized during the metabolism of amino acids is *pyridoxal-5′-phosphate* (PLP), a derivative of vitamin B_6. The PLP is converted during this process to pyridoxamine-5′-phosphate (PMP), which is recycled to PLP.

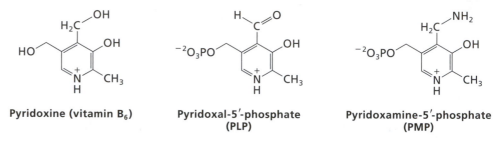

Pyridoxine (vitamin B_6) **Pyridoxal-5′-phosphate (PLP)** **Pyridoxamine-5′-phosphate (PMP)**

The aldehyde group of PLP is covalently attached as an imine bond with the side chain of a lysine residue in the enzyme active site.

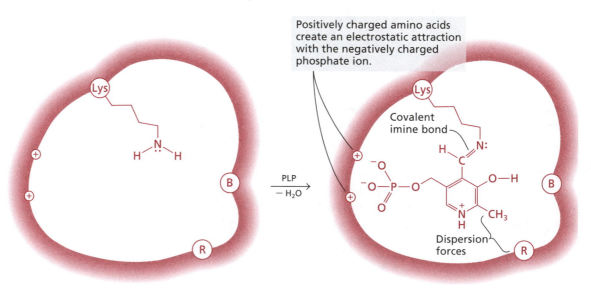

Positively charged amino acids create an electrostatic attraction with the negatively charged phosphate ion.

Covalent imine bond

Dispersion forces

Besides this imine bond, however, several noncovalent interactions hold PLP at the active site. This anchoring of the coenzyme is crucial during amino acid metabolism *because the imine bond is broken during the catalytic reaction*. Without other anchors, the coenzyme might dissociate from the enzyme active site during the process.

After PLP becomes attached via an imine bond at the enzyme active site, a proton is transferred from the OH group on the heterocycle to the nitrogen atom of the imine, as shown below. *Protonation of the nitrogen atom near pH 7 is an especially important feature of an imine and is crucial for making the carbon atom of the imine group more electrophilic.* This proton-transfer step is a reflection of the basicity of imines that makes reductive amination successful when $NaBH_3CN$ is employed as a reagent (Section 20.2a).

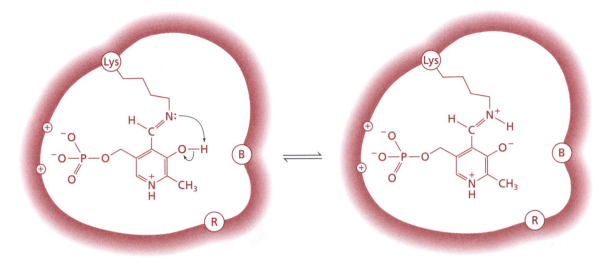

The enzyme–coenzyme complex is now ready to catalyze its reaction, and the amino acid substrate (as its conjugate base) enters the active site and undergoes transamination with the existing protonated imine group.

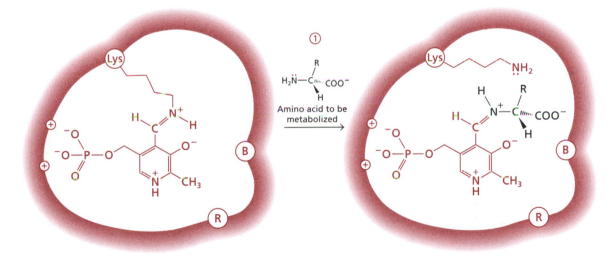

Transamination, the mechanism of which is shown by the following equation, occurs because the protonated imine, having an electrophilic carbon atom, undergoes addition to its C=N π bond by the amino group of the amino acid. After transfer of a proton, the lysine amino group is displaced, leaving the PLP bonded via an imine bond to the amino acid.

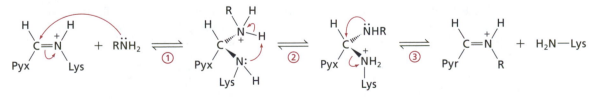

Key: RNH_2 = Amino acid
         Pyx = Pyridoxal-5-phosphate ring
         Lys = Lysine residue in the enzyme active site

Once the C=N bond between the amino acid and PLP has been established, a base site in the enzyme removes a proton from the position α to the imine nitrogen atom. *This reaction is another example of how a positive charge on the imine influences neighboring groups: The proton attached to the carbon atom is made more acidic because of the charge on the nitrogen atom.* At the same time, the heterocycle acts as an electron acceptor, drawing the flow of electrons from the C–H bond in the amino acid toward the nitrogen atom of the pyridine ring that bears a positive charge. This electron movement corresponds to oxidation of the α-carbon atom of the amino acid, and the heterocycle is acting in the same way that NAD⁺ functions as an oxidizing reagent (Fig. 11.3).

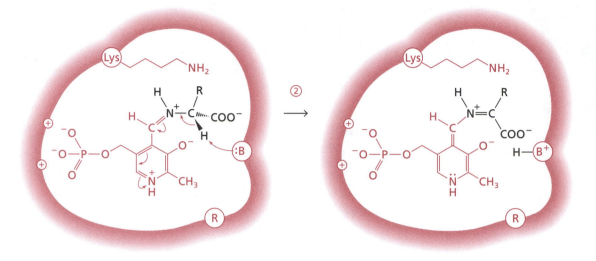

At this stage, electrons can flow in the other direction, resulting in protonation (and reduction) of the carbon atom adjacent to the pyridine ring.

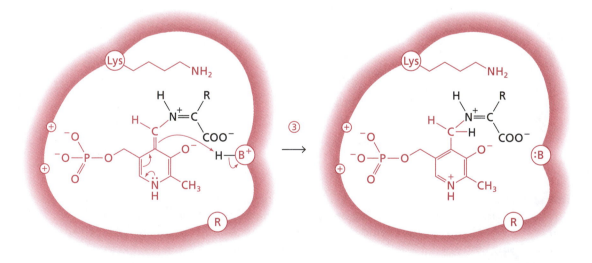

Hydrolysis of the imine bond leads finally to formation of PMP plus the α-keto acid that is the structural analogue of the original amino acid. The α-keto acid dissociates from the enzyme to participate in other metabolic reactions.

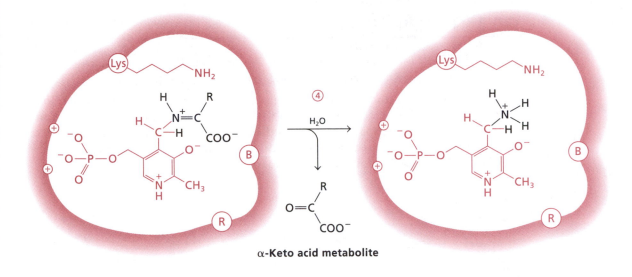

α-**Keto acid metabolite**

EXERCISE 20.9

Propose a mechanism for the hydrolysis reaction shown directly above as Step 4.

   The scheme that has been presented is the general route by which α-amino acids are converted during metabolism to α-keto acids. The metabolic process is incomplete, however, until PMP is converted back to PLP. That conversion takes place by a *reverse* of the steps that have just been shown, starting with formation of an imine bond between PMP and a molecule of α-ketoglutarate. By this pathway, α-ketoglutarate is converted to glutamate, the conjugate base of glutamic acid. Glutamate can be deaminated by an independent route not described here, so it functions as a catalyst in the deamination of the other amino acids.

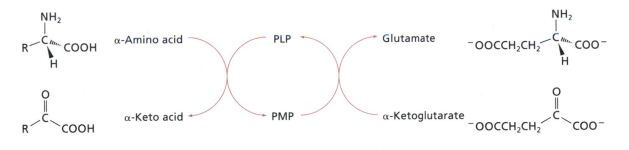

EXERCISE 20.10

Draw the structures of the α-keto acids from the following amino acids (see Figure 27.1 for the structures of the common amino acids). Name the α-keto acids using IUPAC nomenclature.

a. glutamic acid   b. alanine   c. phenylalanine

d. aspartic acid   e. valine   f. leucine

20.3   ENAMINES

20.3a   AN ENAMINE IS THE PRODUCT OF THE REACTION BETWEEN A SECONDARY AMINE AND AN ALDEHYDE OR KETONE

So far, this chapter has focused on the reactions between primary amines and carbonyl compounds. Now, consider what happens when a *secondary* amine adds to a carbonyl group. The first four steps are the same as those that occur when a primary amine reacts. An acid catalyzes this reaction, and as with acetal formation (Section 19.1b), TsOH is often used because it is soluble in organic solvents.

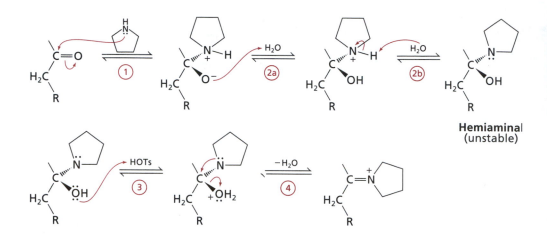

A cationic species is generated with the elimination of a molecule of water in Step 4. The nitrogen atom does not have an attached proton, however, so removal of a proton occurs instead from the adjacent carbon atom, which yields a C=C double bond. This product is an **enamine** (ene + amine).

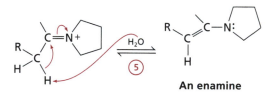

You might think that the cation intermediate could undergo substitution rather than elimination because the carbon atom can bear the positive charge by resonance delocalization, as illustrated below:

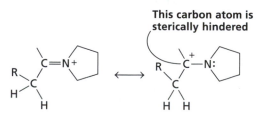

This carbocation is sterically hindered, however, and that leads instead to reaction at a proton on the *adjacent* carbon atom. Recall that a tertiary carbocation often undergoes elimination rather than substitution because steric factors make reactions at the neighboring carbon atom more favorable (Section 8.1a). Furthermore, notice that the amine used to make the enamine is cyclic. Pyrrolidine, morpholine, and piperidine are the secondary amines most often employed to make enamines because the addition process in Step 1 is slow if the nitrogen atom is hindered. Even simple dialkylamines such as $(CH_3CH_2)_2NH$ are more hindered than these cyclic amines.

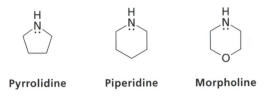

Pyrrolidine   Piperidine   Morpholine

EXERCISE 20.11

What is the product of the following reaction? Propose a reasonable mechanism for its formation.

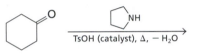

20.3b ENAMINES ARE NUCLEOPHILES AND BASES

Even though an enamine is a stable compound, it can act as a base or a nucleophile because delocalization of the unshared electron pair on the nitrogen atom imparts carbanion character to the β-carbon atom.

As a base, an enamine becomes protonated at its β-carbon atom, which generates an iminium ion. This species has an important resonance contributor that places a positive charge on the α-carbon atom. If aqueous acid is being used, then a molecule of water can intercept this carbocation (Step 2), which leads—after proton transfer from oxygen to nitrogen—to formation of a carbonyl group and displacement of the secondary amine molecule.

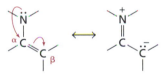

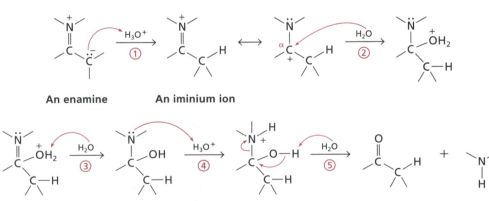

An enamine   An iminium ion

Enamines will also react as nucleophiles with the electrophilic carbon atom of reactive organohalides, which include methyl, benzylic, and allylic halides, and α-halo carbonyl compounds. The anionic carbon atom of the enamine displaces the halide ion, forming a substituted iminium ion. This cationic iminium ion product can be hydrolyzed in a separate step to form the carbonyl group.

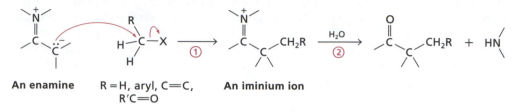

An enamine   R = H, aryl, C=C,   **An iminium ion**
                        R'C=O

The reaction of an enamine with a reactive alkyl halides is one part of a three-step procedure—enamine formation, alkylation, and iminium ion hydrolysis—that can be used to alkylate aldehydes and ketones. (Other ways to alkylate ketones will be presented in Chapter 22.)

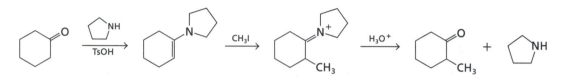

EXERCISE 20.12

Show the mechanism for each of the three stages of the preceding scheme.

20.4 YLIDES

20.4a Phosphorus Ylides are Prepared from Phosphonium Salts

The focus in this chapter on the addition–elimination reactions of amines has been deliberate and is based on the importance of the nitrogen analogues of carbonyl groups throughout biochemistry. But other substances, notably *ylides,* react with aldehydes and ketones by addition–elimination pathways, too. It is to the reactions of ylides that we turn our attention for the remainder of this chapter.

An **ylide** is a substance that has a carbanion center bonded to a heteroatom that bears a positive charge. The most common ylides are those having either phosphorus or sulfur as the heteroatom, but ylides with other heteroatoms also exist. You were introduced to ylides in the discussion of the Swern oxidation (Section 11.4b).

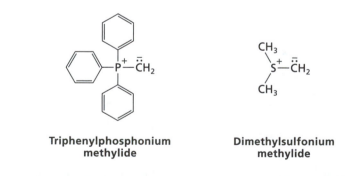

Triphenylphosphonium   **Dimethylsulfonium**
methylide   **methylide**

The most useful ylides in synthesis are those derived from triphenylphosphine, and literally hundreds are known. They are readily prepared from the corresponding alkylphosphonium salts by treatment with a base such as sodium hydride, butyllithium, or LDA (Section 8.2e). The pK_a value of the proton adjacent to the positively charged phosphorus atoms has been estimated to be ~ 35.

The solvents used for this reaction include THF, diethyl ether, and DMSO. Actually, DMSO reacts with organolithium compounds to produce a solution of CH_3SOCH_2Li. This conjugate base of DMSO, $CH_3-SO-CH_2^-$, is called the *dimsyl anion* and is a strong base.

The phosphonium salts themselves are prepared by the S_N2 reaction between an alkyl halide and triphenylphosphine.

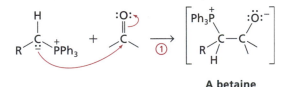

As is true for any nucleophilic displacement reaction, a primary alkyl iodide or bromide works best. Many secondary halides can also react with little difficulty. Triphenylphosphine is not an especially strong base, so elimination is rarely a troublesome side reaction.

EXERCISE 20.13

What is the product formed from each of the following reaction sequences?

a.

b.

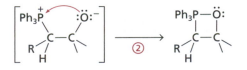

20.4b THE WITTIG REACTION COMBINES A PHOSPHORUS YLIDE WITH AN ALDEHYDE OR KETONE TO PRODUCE AN ALKENE

Phosphorus ylides react with a carbonyl group by nucleophilic addition of the carbanion to the C=O double bond.

A betaine

In the alleged intermediate, called a *betaine,* an intramolecular reaction takes place between the oxyanion and the positively charged phosphorus atom, producing an oxaphosphetane.

Oxaphosphetane

There is some controversy about the intermediacy of the betaine, and the best evidence implies that such a species does not actually exist. As a *formal* mechanism, however, nucleophilic addition of the ylide to the carbonyl group provides a convenient way to predict the course of the reaction, so we will make use of it.

The oxaphosphetane, which does exist but is stable only at low temperatures (−78°C), collapses upon warming via a four-membered cyclic transition state to yield the observed products: *an alkene and triphenylphosphine oxide.*

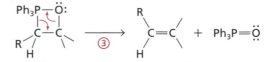

This transformation, known as the **Wittig reaction,** is general. *It is one of the most powerful methods for making alkenes because you know exactly where the double bond will be positioned.* For example, if we were to carry out an elimination reaction (E2) of 2-bromo-2-methylpentane to make an alkene, the major product will have the more highly substituted double bond (Section 8.1b). This E2 reaction is only regioselective, not regiospecific, so more than one product will be formed.

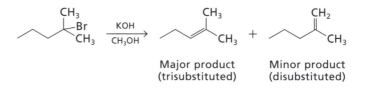

By employing a phosphorus ylide, we can generate the product with the less highly substituted double bond. Moreover, this reaction is *regiospecific:* Only one isomer is formed. *The Wittig reaction constitutes the best way to make a terminal alkene.*

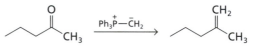

The regiospecificity of the Wittig reaction means that the double bond will be placed where the carbonyl group of the aldehyde or ketone is initially. For example,

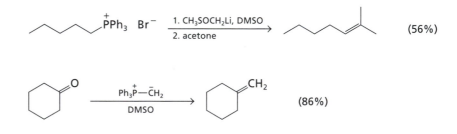

This predictability means that retrosynthetic analysis is straightforward, as illustrated in the following example.

EXAMPLE 20.1

Using the Wittig reaction, show how you would prepare the alkene illustrated here.

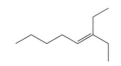

Because the double bond of an alkene made by the Wittig reaction coincides in its position with that of the carbonyl group of the aldehyde or ketone that is used in its synthesis, we consider which possible carbonyl components are needed by splitting the molecule at the double bond (a colored box is used below to show which portion comes from the aldehyde or ketone):

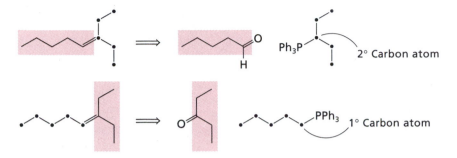

The ylide reactant comprises the remaining carbon atoms. If possible, the phosphorus atom should be bonded to a primary carbon atom. For this molecule, we therefore use the second retrosynthetic possibility. The synthesis is as follows:

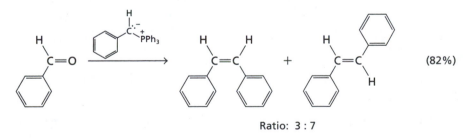

EXERCISE 20.14

Using the Wittig reaction, show how you would prepare each of the following alkenes [ignore the double-bond stereochemistry in (a)]:

a.

b.

20.4c THE WITTIG REACTION IS NOT STEREOSELECTIVE

The reactions shown so far have ignored the stereochemistry of the double bond in the product. Considering cases in which both (*E*) and (*Z*) isomers can be generated, however, we would find in most instances that both isomers are formed.

Ratio: 3 : 7

The exact proportion of each isomer depends on the structure of the ylide as well as the reaction conditions, including such variables as the solvent, the temperature, the base, and the concentration of salts dissolved in the solution. In planning a synthesis, it is best to expect that both isomers will be obtained.

20.4d PHOSPHONATE YLIDES ALSO CONVERT ALDEHYDES AND KETONES TO ALKENES

Ylides that have a carbonyl or nitrile group adjacent to the carbanion are much less reactive than those without unsaturation because the electrons are delocalized. For example,

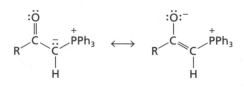

To carry out the Wittig reaction with such species, we use a reagent with a *phosphonate* group rather than a triphenylphosphonium ion attached to the carbanion. The presence of the oxo and alkoxy groups in an alkyl phosphonate compound renders the phosphorus atom neutral rather than charged, so while delocalization can still occur in the carbonyl portion, the charge on the overall species is –1, not 0, which makes it a better nucleophile than the ylide shown above. Notice that an alkyl phosphonate anion is not technically an ylide because the P atom does not carry a +1 charge.

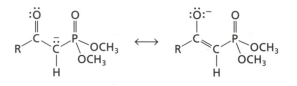

The synthesis of an alkyl phosphonate is readily accomplished by treating a reactive organohalide with trimethyl- or triethylphosphite, a process called the **Arbuzov reaction.** Benzylic and allylic halides and compounds with a halogen atom α to carbonyl, nitrile, or sulfonyl groups are the best substrates for this transformation. The first step is a typical S_N2 reaction, in which the phosphorus nucleophile displaces the halide ion.

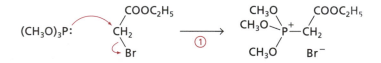

The product itself is ideally configured to undergo an S_N2 reaction as well; a methyl group bonded to an oxygen atom provides the electrophilic center, and the alkyl phosphonate product is a good leaving group.

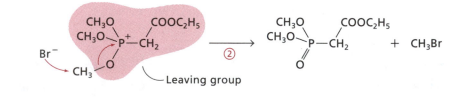

EXERCISE 20.15

Propose a reasonable mechanism for the Arbuzov reaction between benzyl chloride and trimethylphosphite.

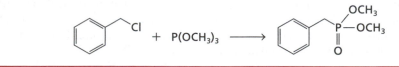

An alkyl phosphonate obtained from the Arbuzov reaction can be deprotonated in a separate step to form the corresponding carbanion. Bases like those used for the Wittig reaction can be employed, but weaker ones usually suffice because two electron-withdrawing groups are attached to the methylene group. An alkoxide ion in alcohol is a suitable combination of base and solvent.

$$CH_3O-\overset{\overset{O}{\|}}{P}-CH_2-\overset{\overset{O}{\|}}{C}-OCH_2CH_3 \quad \xrightarrow[\text{EtOH}]{\text{NaOEt}} \quad CH_3O-\overset{\overset{O}{\|}}{P}-\overset{..}{C}H-\overset{\overset{O}{\|}}{C}-OCH_2CH_3$$
$$\underset{CH_3O}{} \qquad\qquad\qquad\qquad\qquad \underset{CH_3O}{}$$

The resulting carbanion adds readily to the carbonyl groups of aldehydes and ketones, and addition is followed by elimination of $(RO)_2PO_2^-$. All of the steps are analogous to those of the Wittig reaction itself. The product of this transformation, known as the **Horner–Emmons reaction,** is an alkene.

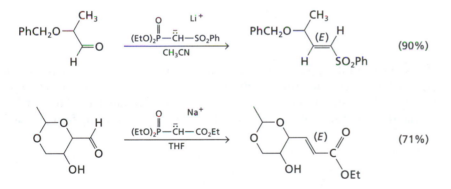

$$CH_3CH_2O-\overset{\overset{O}{\|}}{P}-\overset{..}{C}H-\overset{\overset{O}{\|}}{C}-OC_2H_5 \text{ Na}^+ \quad \xrightarrow[\text{EtOH}]{} \qquad\qquad (66\%)$$
$$\underset{CH_3CH_2O}{}$$

One obvious difference between the Horner–Emmons and Wittig reactions is the nature of the alkyl halide used. Their stereochemical outcomes are also different. The Wittig reaction is rarely stereospecific, but the Horner–Emmons reaction between an aldehyde and an alkyl phosphonate carbanion prepared from an α-halo carbonyl, nitrile, or sulfonyl compound produces the (E)-alkene. This stereochemical outcome is influenced mainly by formation of a double bond that is conjugated with the unsaturated functional group in the product. For example,

PhCH₂O—[CH₃, CHO] + $(EtO)_2P(=O)-\overset{..}{C}H-SO_2Ph$ Li⁺ $\xrightarrow{CH_3CN}$ PhCH₂O—[CH₃, (E), H, H, SO₂Ph] (90%)

[dioxane-CHO] + $(EtO)_2P(=O)-\overset{..}{C}H-CO_2Et$ Na⁺ $\xrightarrow{THF}$ [dioxane-(E)-C(=O)OEt] (71%)

EXERCISE 20.16

How would you prepare each of the alkenes shown here from compounds that have six or fewer carbon atoms?

a.

b.

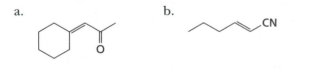

20.4e SULFUR YLIDES CONVERT ALDEHYDES AND KETONES TO EPOXIDES

No doubt, phosphorus ylides are the most useful reagents among ylide-derived substances. Sulfur ylides are next in importance as synthetic reagents. They can be made by deprotonating trialkylsulfonium compounds, which in turn are prepared by alkylating a thioether with an alkyl halide (Section 7.3b). In practice, only the methyl derivative is *routinely* employed, as shown in the following equation:

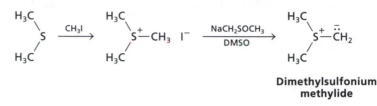

Dimethylsulfonium methylide

The positive charge on sulfur makes the protons of the attached methyl groups relatively acidic, and the conjugate base of DMSO is a strong enough base to bring about ylide formation. A sulfur ylide is important in the Swern oxidation (Section 11.4b), where it acts as a base to promote the elimination step that creates a C=O bond.

Like a phosphorus ylide, the nucleophilic carbon atom in a sulfur ylide can add to the carbonyl group of an aldehyde or ketone.

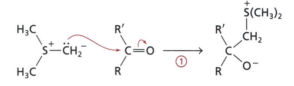

Unlike the betaine that is purportedly formed by addition of a phosphorus ylide to a carbonyl group (Section 20.4b), the sulfur analogue does *not* collapse to generate a four-membered ring intermediate.

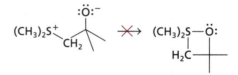

Instead, the nucleophilic oxygen atom reacts at the carbon atom bearing the sulfur substituent, which displaces dimethyl sulfide as a leaving group and generates an epoxide. The structure of the sulfonium intermediate should be familiar to you from Section 7.3b—recall that an organosulfide, R_2S, can be a good leaving group in S_N2 reactions.

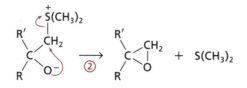

The overall reaction is a general one, and methylene epoxides are readily prepared as shown in the following examples:

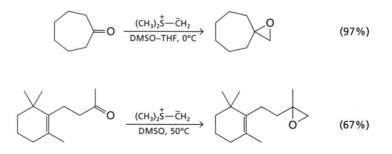

Other analogues of these ylides are not as stable, but when they can be prepared, the corresponding epoxide is also produced. Notice that the ylide in the following example is derived from diphenyl sulfide. Unless the three substituents are identical, only one of them can have a proton *alpha* to the sulfur atom. Otherwise, a mixture of products will be formed.

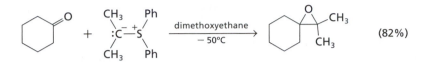

(82%)

EXERCISE 20.17

The following cyclopropyl ylide is stable and undergoes an addition–elimination reaction. What is the structure of epoxide **A** that is formed?

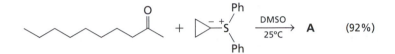

When **A** is treated with acid, a cyclobutane product is formed. Propose a reasonable mechanism for this rearrangement.

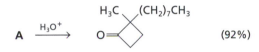

(92%)

CHAPTER SUMMARY

Section 20.1   Compounds with C=N bonds

- A primary amine undergoes addition to the carbonyl group of an aldehyde or ketone to form a hemiaminal; the acid-catalyzed loss of water generates an imine with formation of a C=N double bond.

- Hydrazine or its derivatives reacts with an aldehyde or ketone to form a hydrazone; hydroxylamine reacts with aldehydes and ketones to form oximes.

- Compounds with a C=N double bond (imines, hydrazones, or oximes) can be hydrolyzed with aqueous acid to regenerate the carbonyl compound and X–NH₂.

- The oxime derivative of an aldehyde can be dehydrated to form a nitrile.

- An oxime undergoes rearrangement when the OH group is converted to a good leaving group. The R group trans to the OH group migrates from carbon to nitrogen to form an amide.

- An unsubstituted hydrazone reacts with hydroxide ion at elevated temperatures replacing the =N–NH₂ group with two hydrogen atoms. The transformation >C=O → >C=N–NH₂ → >CH₂ is called the Wolff–Kishner reaction.

Section 20.2   Imine chemistry and biochemistry

- Like its carbonyl precursor, an imine is reduced with NaBH₄ or LiAlH₄ (followed by acid workup) to yield the corresponding amine. An oxime is reduced to a primary amine by using LiAlH₄ followed by acid workup.

- Sodium cyanoborohydride (NaBH$_3$CN) can be used to prepare a secondary amine from an aldehyde or ketone and a primary amine, a process called reductive amination.
- The carbon–nitrogen double bond is important in biological systems. An imine is a cross-linking group in proteins, especially those that have structural roles.
- Because the carbon–nitrogen double bond is readily hydrolyzed, an imine group is useful to attach a coenzyme covalently within an enzyme active site.
- Pyridoxal-5′-phosphate (PLP) is the coenzyme used for the metabolic conversion of amino acids to α-ketoacids. An imine bond is formed between the amino group of the amino acid and the aldehyde group of pyridoxal phosphate.
- Pyridoxal-5′-phosphate functions as an oxidizing agent in much the same way that NAD$^+$ operates.

Section 20.3   Enamines

- A secondary amine undergoes addition to the carbonyl group of an aldehyde or ketone; the acid-catalyzed loss of water produces an enamine.
- An enamine is readily hydrolyzed to its carbonyl and amine components.
- An enamine can be alkylated at its β-carbon atom using reactive alkyl halides.

Section 20.4   Ylides

- An ylide is a compound with a carbanion bonded to a heteroatom bearing a positive charge.
- The most common ylides are those that have phosphorus or sulfur atoms.
- A phosphorus ylide is made by the S$_N$2 reaction between triphenylphosphine and an alkyl halide, followed by deprotonation with a strong base.
- A phosphorus ylide reacts with a carbonyl group, replacing the oxygen atom with the carbanion portion of the ylide. This transformation is called the Wittig reaction, and it constitutes a general way to make alkenes.
- The Wittig reaction is not stereoselective.
- Alkyl phosphonates can be deprotonated to form carbanions, which react with aldehydes and ketones to form alkenes in the Horner–Emmons reaction.
- A phosphonate carbanion is used when the original alkyl halide is reactive (benzylic, allylic, α-keto-).
- Sulfur ylides are prepared from trialylsulfonium ions. They react with aldehydes or ketones to form epoxides. Sulfur ylides are also intermediates in the Swern oxidation as described in Section 11.4b.

KEY TERMS

Section 20.1a	Section 20.1e	Section 20.4a
hemiaminal	Wolff–Kishner reaction	ylide
imine		
Schiff base	Section 20.2a	Section 20.4b
	reductive amination	Wittig reaction
Section 20.1b		
hydrazone	Section 20.2b	Section 20.4d
oxime	cross-link	Arbuzov reaction
		Horner–Emmons reaction
Section 20.1d	Section 20.3a	
Beckmann rearrangement	enamine	

Section 20.1a

Aldehydes and ketones react with primary amines to form imines. The reaction is catalyzed by acid.

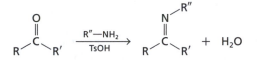

Section 20.1b

Aldehydes and ketones react with hydrazines to form hydrazones and with hydroxylamine to form oximes. The reaction is acid catalyzed.

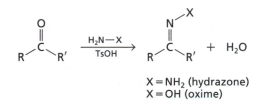

X = NH$_2$ (hydrazone)
X = OH (oxime)

Section 20.1c

Species with the >C=N–X group undergo hydrolysis in aqueous acid to regenerate the carbonyl compound and X–NH$_2$.

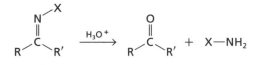

X = alkyl, aryl, NH$_2$, OH

Section 20.1d

An aldehyde oxime can be dehydrated by heating with P$_4$O$_{10}$ to form a nitrile.

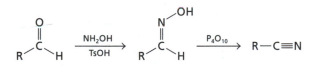

The Beckmann rearrangement. An oxime reacts with strong acid and undergoes rearrangement to form an amide. The R group anti (trans) to the OH group migrates to the nitrogen atom.

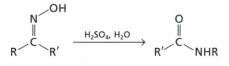

Section 20.1e

The Wolff–Kishner reaction. With a mixture of hydrazine and sodium hydroxide, ketones and some aldehydes are deoxygenated: The carbonyl group is converted to a CH$_2$ group.

Section 20.2a

Imines are reduced by hydride reagents to form amines.

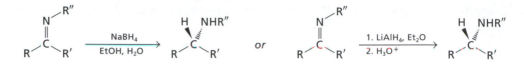

Reductive amination: Sodium cyanoborohydride can be used to form secondary amines directly from aldehydes or ketones and primary amines.

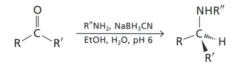

An oxime is reduced by lithium aluminum hydride to form a primary amine.

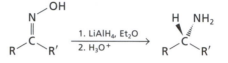

Section 20.2c

Amino acid metabolism. The coenzyme and PLP forms an imine bond with an amino acid and converts the amino acid to an α-keto acid.

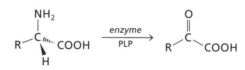

Section 20.3a

Aldehydes and ketones react with cyclic secondary amines to form enamines. The reaction is acid catalyzed.

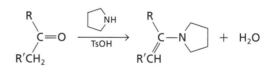

Section 20.3b

Enamines undergo alkylation with reactive alkyl halides or hydrolysis in aqueous acid to form a carbonyl compound and the secondary amine.

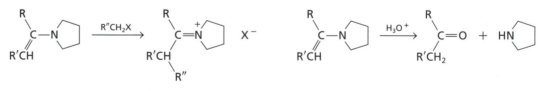

R″ = H, aryl, C=C, RC=O

Section 20.4a

Alkylphosphonium salts are prepared by the reaction between triphenylphosphine and an alkyl halide. The corresponding triphenylphosphonium ylide is made by an acid–base reaction between the alkylphosphonium salt and a strong base in an aprotic solvent.

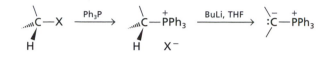

X = Cl, Br, I, OMs, OTs

Section 20.4b

The Wittig reaction. A triphenylphosphonium ylide reacts with aldehydes and ketones to form alkenes. The reaction is not stereospecific.

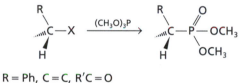

Section 20.4d

The Arbuzov reaction. Trimethylphosphite reacts with a reactive alkyl halide to form an alkylphosphonate.

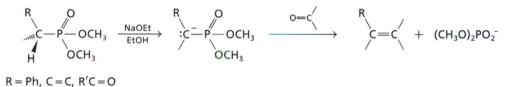

R = Ph, C=C, R'C=O

The Horner–Emmons reaction. An alkyl phosphonate reacts with base to form a phosphonate carbanion, which reacts with aldehydes and ketones to form alkenes.

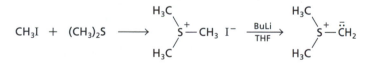

R = Ph, C=C, R'C=O

Section 20.4e

Sulfur ylides are made from the reaction between a thioether and an alkyl halide (most commonly, methyl iodide) followed by an acid–base reaction with strong base in an aprotic solvent.

$$CH_3I + (CH_3)_2S \longrightarrow$$

Aldehydes and ketones react with sulfur ylides to form epoxides.

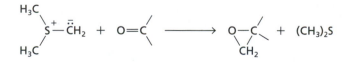

ADDITIONAL EXERCISES

20.18. Propose a reasonable mechanism for the addition–elimination reaction between pentanal and semicarbazide, $H_2NCONHNH_2$, in the presence of a small amount of sulfuric acid as a catalyst.

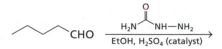

20.19. Draw the structures of the imines that will be made from the reactions of each amine in the first column and each aldehyde or ketone in the second column, for a total of nine structures.

1-Aminobutane	Pentanal
p-Toluidine (4-methylaniline)	3-Methylcyclohexanone
trans-2-Methyl-1-aminocyclopentane	*p*-Nitrobenzaldehyde

20.20. Draw the structures of the enamines that will be made from the reactions of each amine in the first column and each aldehyde or ketone in the second column, for a total of nine structures.

Morpholine	Cyclohexanone
Pyrrolidine	Acetophenone
Diethylamine	Hexanal

20.21. Write an equation for the reaction between 1-phenyl-1-propanone and each of the following reagents:

a. CH_3Li in ether—hexane, followed by aqueous acid workup

b. $NaBH_4$ in $EtOH_{aq}$

c. NH_2NH_2, KOH, diethylene glycol, Δ

d. Br_2, $FeBr_3$

e. $Ph_3P^+{-}^-CHCH_3$

f. $(CH_3)_2S^+{-}^-CH_2$ followed by addition of CH_3MgI in ether then H_3O^+

g. Morpholine and TsOH

h. NH_2OH, TsOH

20.22. The tranquilizer Valium, considered one of the most overprescribed medications of modern times, can be made by the intramolecular formation of an imine bond. Assuming that the final step in the synthesis of Valium is formation of the Schiff-base linkage, draw the structure of the immediate precursor. Propose a mechanism for ring closure.

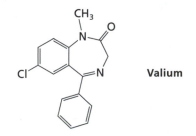

Valium

20.23. The amino acid proline is made in living systems by an intramolecular imine bond-forming reaction. Glutamate is first reduced in a two-stage process to yield glutamate-5-semialdehyde. What two steps are needed to convert the aldehyde to proline? Propose a reasonable mechanism for these last two steps.

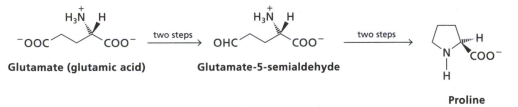

20.24. Draw the structure(s) of the major product(s) expected from each of the following reactions. Indicate the stereochemistry of the product as appropriate. Relative stereochemistry should be shown using wedges and dashed lines. If a racemic mixture will be formed, draw the structure of one enantiomer and write the word "racemic", or draw both enantiomeric structures. If diastereomers are formed, draw each structure; label meso compounds as such. If no reaction occurs, write N.R.

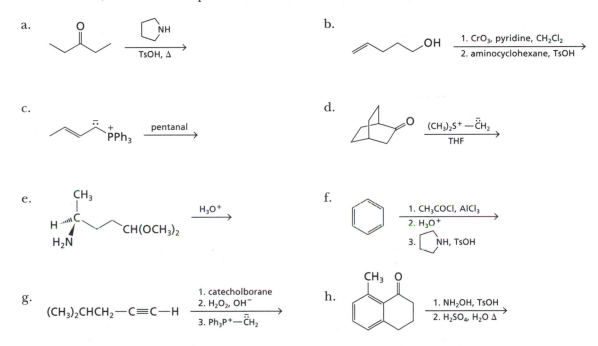

20.25. In performing each of the following reactions, indicate the principal IR absorptions that you would look for in the reactant and product to ensure that the reaction was successful. Be as specific as possible: list the position (cm^{-1}), intensity (weak, medium, or strong), and shape (sharp or broad) of the diagnostic absorption band(s).

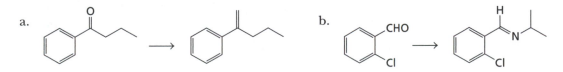

20.26.   For the reactions shown in Exercise 20.25, repeat the process for the ^{1}H and broad-band ^{13}C NMR spectra. List, as applicable, the chemical shifts, integrated intensity values, splitting patterns, and DEPT results. Make a sketch of the ^{1}H NMR spectra of the starting carbonyl compounds.

20.27.   Show the reagents needed to carry out each reaction in Exercise 20.25.

20.28.   Show how you would prepare each of the following alkenes using a Wittig reaction and starting with benzene, toluene, cyclohexene, triphenylphosphine, and any other organic compounds that have five or fewer carbon atoms. You may also use any other reagents and solvents.

a.   b.   c.

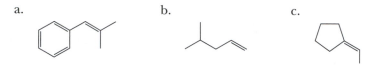

20.29.   Show how you would prepare each of the following compounds from benzene, cyclohexanol, and any other organic compounds that have four or fewer carbon atoms. You may also use any other reagents and solvents.

a.   b.   c.

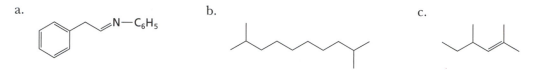

20.30.   A triphenylphosphonium ylide is nucleophilic, so it can react with a primary alkyl halide by the S$_N$2 mechanism. Draw the structure of the product formed after each step in the following sequence.

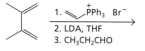

$$Ph_3\overset{+}{P}-CH_3 \ \ Br^-$$

1. NaH in DMSO
2. C$_6$H$_5$CH$_2$Br
3. NaH in DMSO
4. C$_6$H$_5$CHO

20.31.   Vinyltriphenylphosphonium bromide is a reasonable dienophile in the Diels–Alder reaction (Section 10.4b). Draw the structure of the product formed after each step in the following sequence.

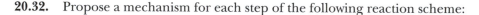

1. $\overset{+}{P}Ph_3 \ \ Br^-$
2. LDA, THF
3. CH$_3$CH$_2$CHO

20.32.   Propose a mechanism for each step of the following reaction scheme:

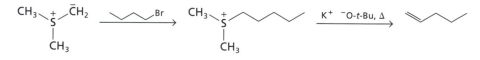

20.33. Compound **X,** $C_{12}H_{16}$, undergoes ozonolysis followed by reductive workup to produce a C_4 aldehyde plus **Y,** which has the IR and ^1H NMR spectra shown below. Stirring **X** with aqueous sulfuric acid followed by heating with potassium dichromate gives **Z,** which can also be made by a Friedel–Crafts reaction from toluene and $(CH_3)_2CH_2COCl$. Draw the structures of compounds **X, Y,** and **Z.** What reactions can be used to convert compound **Y** to **X?**

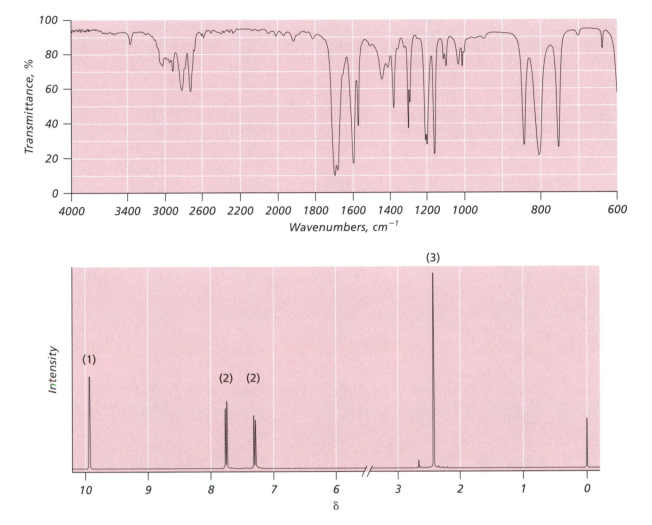

ADDITION–ELIMINATION REACTIONS OF CARBOXYLIC ACIDS AND DERIVATIVES

Like aldehydes and ketones, carboxylic acids and some of their derivatives are important as intermediates in metabolic processes. The COOH functional group is ubiquitous in Nature, occurring in amino acids, which are the building blocks of proteins; in fatty acids, which are used to store energy and provide components for membranes; and as parts of many hormones and related regulatory substances such as the prostaglandins (Section 12.4c).

As the name indicates, a carboxylic acid is acidic, and its proton can be removed by reaction with a base, which generates the corresponding carboxylate ion. This aspect of carboxylic acid chemistry was discussed in Section 5.2c. This chapter will describe how carboxylic acids and their derivatives, which include acid chlorides, esters, thioesters, amides, and nitriles, can be interconverted. Amides, which are nitrogen-containing derivatives of carboxylic acids, are also common in biochemistry. For example, the amide group is used to link amino acids to form proteins (Section 5.5b). Furthermore, amides form hydrogen bonds (Section 2.8b) that stabilize the structures of proteins and give them their three-dimensional shapes.

Carboxylic acids and their derivatives are also important molecules in the field of polymer chemistry. Polyesters such as polyethylene terephthalate (PETE or PET) and polyamides such as nylon and Kevlar are materials that have profoundly affected our daily lives. The chemistry of these polymers is presented in Chapter 26.

21.1   GENERAL ASPECTS OF STRUCTURE AND REACTIVITY

21.1a THE CARBONYL GROUP OF CARBOXYLIC ACIDS AND THEIR DERIVATIVES IS BONDED TO A HETEROATOM

The reactions of carboxylic acids and their derivatives differ from those of aldehydes and ketones because the carbonyl group bears a heteroatom and not just carbon- and hydrogen-atom substituents. The presence of a heteroatom means that these functional groups must be at the end of the carbon chain in aliphatic compounds. The names of carboxylic acids, amides, and nitriles, which were presented in Chapter 1, have the suffixes *–oic acid, –amide,* and *–nitrile,* respectively (Table 1.1). For cyclic compounds, the point of attachment of the COOH, $CONH_2$, or CN group defines C1 of the ring, and the alternate suffixes *carboxylic acid, carboxamide,* and *carbonitrile* are used. The parent names of the benzene compounds with these functional groups are benzoic acid, benzamide, and benzonitrile.

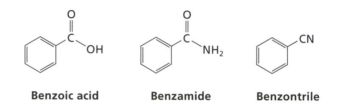

Benzoic acid   **Benzamide**   **Benzontrile**

The names of amides with substitution at their nitrogen atoms have an upper case, italicized *N-* instead of a numeral, which is used for substitution of the carbon chain. For example,

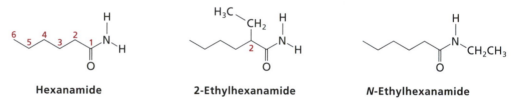

Hexanamide   **2-Ethylhexanamide**   ***N*-Ethylhexanamide**

The nomenclature of two common types of carboxylic acid derivatives—esters and acid chlorides (Section 21.1b)—was not discussed previously, although you have doubtless seen several names already. Ethyl acetate, a common solvent, is an ester (Fig. 6.7).

When an alkyl group is attached to a carbon chain as a substituent, its point of attachment is specified by a numeral, as previously noted. Occasionally, you will encounter a name that has a prefix with no associated number; furthermore, this prefix is separated from the rest of the compound's name by a space. The presence of an ester functional group is denoted by the suffixes "*oate*" or "*carboxylate*". When you see one of these suffixes, the next step is to look for an unnumbered prefix, which specifies the identity of the alkyl group (R′ in the general structure given below at the left) attached to the oxygen atom of the ester functional group.

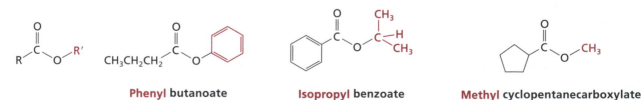

Phenyl butanoate   **Isopropyl** benzoate   **Methyl** cyclopentanecarboxylate

The following example illustrates the interpretation process.

EXAMPLE 21.1

Draw the structural formula of isobutyl 3-chloropentanoate.

The compound root is "pent" (5 carbon atoms), the multiple bond index is "an", which means no carbon–carbon double or triple bonds, and the suffix is "oate". The compound is an ester, a functional group that must be at the end of the chain. The unnumbered prefix "isobutyl" is the name of the alkyl group attached to the oxygen atom.

Looking at the rest of the name, we see that a chlorine atom is attached at C3. We finish by adding hydrogen atoms to every carbon atom to give each a total of four bonds.

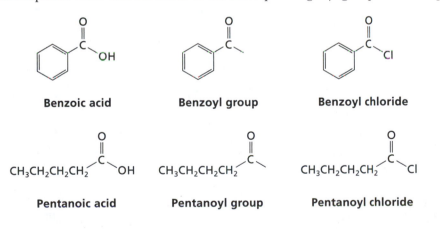

EXERCISE 21.1

Draw the structural formula of *tert*-butyl (*E*)-2-hexenoate.

21.1b ACYL GROUPS ARE DERIVED FROM CARBOXYLIC ACIDS

The **acyl group,** mentioned previously in Section 17.2e, is the general unit of a carboxylic acid formed by removing the OH fragment. The names of acyl groups derived from the IUPAC names of carboxylic acids are created by changing the suffix from "oic" to "oyl". The names of acid chlorides are then generated by including the suffix "chloride" as a separate word with the name of the corresponding acyl group. For example,

Benzoic acid　　**Benzoyl group**　　**Benzoyl chloride**

Pentanoic acid　　**Pentanoyl group**　　**Pentanoyl chloride**

The names of the two smallest acyl groups—*formyl* and *acetyl* (Table 1.5)—originate with the common names formic acid and acetic acid (IUPAC names: methanoic acid and ethanoic acid, respectively).

Formic acid　　Formyl　　**Acetic acid**　　Acetyl
(IUPAC: methanoic acid)　　group　　**(IUPAC: ethanoic acid)**　　group

These common roots, "form" and "acet", appear in the names of many other molecules with multiple bonds (especially those with carbonyl groups) having one or two carbon atoms, which are illustrated in Figure 21.1.

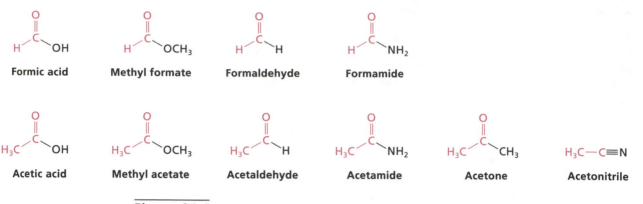

Figure 21.1
Common names derived from the roots "form" and "acet".

21.1c THE REACTIONS OF CARBOXYLIC ACID DERIVATIVES PROCEED BY WAY OF A TRANSIENT TETRAHEDRAL INTERMEDIATE

When a carboxylic acid derivative is treated with a nucleophile, addition occurs to the π bond of the carbonyl group to form a **tetrahedral intermediate,** so-called because the carbon atom changes its hybridization from sp^2 (trigonal) to sp^3 (tetrahedral) as the π bond is broken. The tetrahedral intermediate is normally unstable with respect to the form having the carbonyl group, so an electron pair on the oxygen atom displaces the group X to regenerate the carbon–oxygen π bond. *This displacement occurs most readily when X is a good leaving group.* This overall process, which results in interconversion of carboxylic acid derivatives, is classified as an **addition–elimination reaction.**

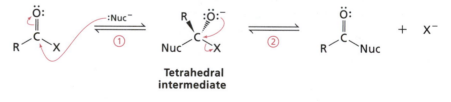

Tetrahedral intermediate

The addition–elimination reactions of carboxylic acids differ from those observed for aldehydes and ketones (Chapter 20). In the latter, *no* leaving group is attached to the carbonyl carbon atom. For aldehydes and ketones, elimination occurs instead by loss of the oxygen atom of the original carbonyl group. For example,

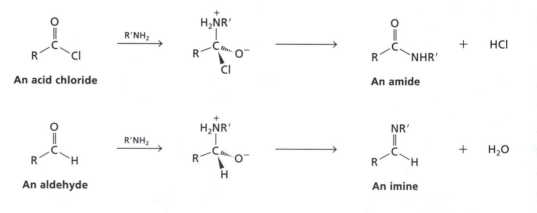

21.1d THE REACTIVITY OF THE CARBOXYLIC ACID DERIVATIVES VARIES IN RELATION TO THE BASICITY OF THE CARBONYL SUBSTITUENT

Even though each carboxylic acid derivative reacts by nucleophilic addition to its carbonyl group, the rate of addition to form the tetrahedral intermediate is tempered by the relative basicity of the heteroatom substituent attached to the carbonyl carbon atom.

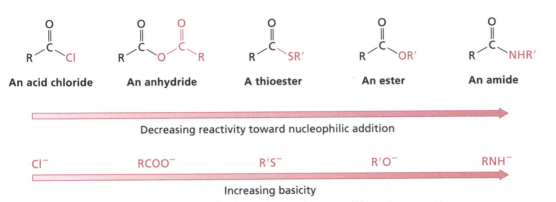

If you think of a carboxylic acid derivative as the product of a Lewis acid–Lewis base reaction between an acyl cation and the heteroatom-containing group, then this equilibrium will lie in the direction of the neutral molecule (to the right) as the basicity of the heteroatom increases. The net effect of this attraction is to lower the effective electrophilicity of the carbon atom, which makes it less susceptible toward reaction with a nucleophile, hence slower to react.

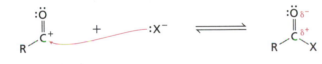

The rate of the second step of the addition–elimination mechanism—the one in which the tetrahedral intermediate collapses to regenerate the carbonyl group—will be faster as its leaving group becomes better (weak base). This trend is the same as for the addition step:

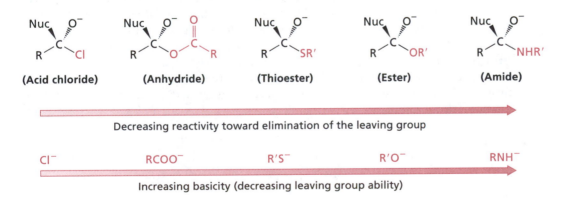

Therefore, the overall rate of addition–elimination reactions of carboxylic acid derivatives decreases with increasing basicity of the heteroatom substituent.

21.1e THE ADDITION REACTIONS OF CARBOXYLIC ACIDS ARE SUPPRESSED BY THE PRESENCE OF THEIR PROTONS

Like its derivatives, the carbonyl carbon atom of a carboxylic acid carries a partial positive charge, which makes it electrophilic. In the COOH group, however, *the proton is the most electrophilic atom*, so an acid–base reaction—rather than addition—normally occurs when a carboxylic acid is treated with a nucleophile, which is also a Lewis base (Section 5.3c).

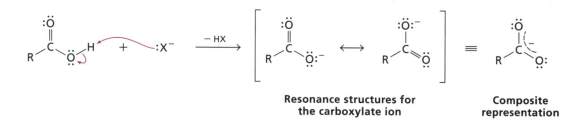

The proton is the most electrophilic atom of the COOH group.

When a nucleophile undergoes an acid–base reaction with a carboxylic acid, removal of the proton generates the carboxylate ion, which is stabilized by resonance. A composite structure of the resonance hybrid (below, right) shows the negative charge distributed over the entire group. Many nucleophiles cannot overcome the repulsive effect created by the negative charge of the carboxylate ion, so the further reaction between a nucleophile and a carboxylate ion is suppressed.

Resonance structures for the carboxylate ion   **Composite representation**

21.2   REACTIONS OF CARBOXYLIC ACIDS

21.2a ESTERS CAN BE MADE DIRECTLY FROM CARBOXYLIC ACIDS AND ALCOHOLS IN THE PRESENCE OF AN ACID CATALYST

Addition to the carbonyl group of a carboxylic acid has to be carried out under acidic conditions so that the acidity of the COOH proton is less important, and the carbonyl group can exert its electrophilic character. In acid solution, however, a nucleophile can also be protonated, compromising its nucleophilic character. Reactions of carboxylic acids in the presence of a stronger acid are therefore limited. Fortunately, alcohols are very weak bases, so an alcohol will react with a carboxylic acid in the presence of a stronger acid to form an ester, a process called **esterification.**

The carbonyl oxygen atom of a carboxylic acid reacts with a strong acid, such as sulfuric acid in the present example, to form a cation intermediate that is stabilized by resonance delocalization.

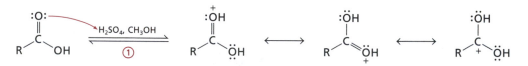

You have seen this effect previously, in the reactions of aldehydes and ketones that take place in acid (Section 18.1c). The carbonyl group is activated by protonation to make it more susceptible toward reactions with nucleophiles.

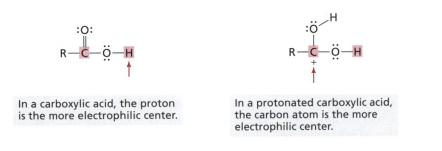

In a carboxylic acid, the proton is the more electrophilic center.

In a protonated carboxylic acid, the carbon atom is the more electrophilic center.

After the carbonyl group is activated, the alcohol, which is often used as the solvent in esterification reactions, reacts to form an intermediate having one alkoxy and two hydroxy groups.

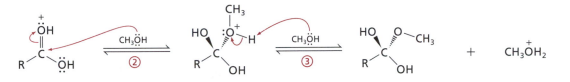

This intermediate can be protonated at one of its OH groups, which forms a good leaving group, namely, water. A proton is then removed from the *other* OH group to generate the carbonyl group of the product as a molecule of water is displaced.

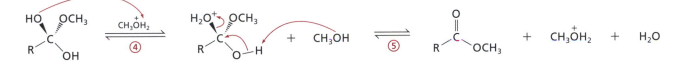

In the overall process of esterification, a molecule of water is formed:

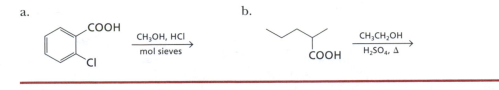

To facilitate esterification, a dehydrating agent is often added to the reaction mixture to drive the equilibrium toward ester formation, and *molecular sieves* (Section 19.1d) work particularly well. Heating the reaction mixture to remove water (along with some alcohol), using a large excess of alcohol, or using a hygroscopic acid such as sulfuric acid are other ways to shift the equilibrium in favor of the ester product.

Not all esters can be made by this method. If the alcohol is sensitive to strong acid, then the ester must be made from an acid chloride (Section 21.3a). Tertiary alcohols, for example, can be protonated at their oxygen atoms and undergo E1 reactions (Section 8.1a). The acid-catalyzed synthesis of the widely used methyl and ethyl esters from carboxylic acids works well, however.

EXERCISE 21.2

What is the major product expected from each of the following reactions?

a.

b.

21.2b HYDROXY CARBOXYLIC ACIDS FORM LACTONES

When a molecule contains both alcohol and carboxylic acid functional groups separated by three or four carbon atoms, a cyclic ester, called a **lactone,** forms readily. This intramolecular process is analogous to that for cyclic hemiacetal formation from hydroxy aldehydes and ketones (Section 19.1a).

A general designation for lactones uses a Greek letter to denote the position of the OH group relative to the COOH group in the corresponding hydroxy acid. The β-lactone (four-membered ring) is reactive but isolable. The γ-lactone (five-membered ring) and δ-lactone (six-membered ring) are stable and form readily.

A β-Lactone   **A γ-Lactone**   **A δ-Lactone**

The mechanism for lactone formation is the same as that for acid-catalyzed esterification, and even traces of acid (indicated by HX in the following mechanism) will catalyze this reaction if a five- or six-membered ring is formed.

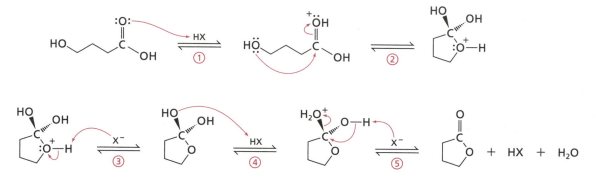

Lactones were previously mentioned when you learned about the oxidation of carbohydrates to aldaric and aldonic acids (Section 19.5b). One of the hydroxy groups of an oxidized carbohydrate can react with the carboxylic acid group to yield a lactone. Note that a furanose or pyranose can be oxidized directly, a transformation that converts the hemiacetal to the lactone.

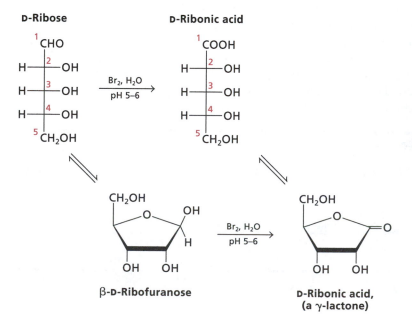

EXERCISE 21.3

A byproduct of oxidizing 1,4-butanediol in aqueous acid is 4-butanolactone. Propose a mechanism for its formation. (*Hint:* Review alcohol oxidation in Section 11.4d).

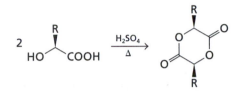

EXERCISE 21.4

α-Hydroxy acids react to produce lactides (six-membered ring diesters) rather than α-lactones because of the ring strain that would be created by formation of a three-membered ring. Propose a reasonable mechanism for the following transformation:

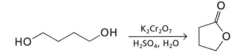

21.2c AN ACID CHLORIDE IS READILY PREPARED
FROM A CARBOXYLIC ACID

An acid chloride is the most reactive derivative of a carboxylic acid, and the best general method used to prepare one involves heating a carboxylic acid with thionyl chloride. You have already encountered this reagent for the conversion of alcohols to alkyl chlorides (Section 7.1c).

The carbonyl oxygen atom of a carboxylic acid reacts with thionyl chloride, displacing chloride ion. This first step effectively activates the carbonyl group for chloride ion addition in Step 2, which generates a neutral tetrahedral intermediate.

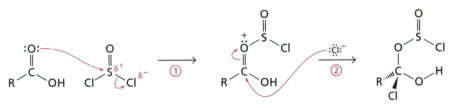

The concerted movement of electrons within this compound regenerates the carbonyl group, at the same time producing a molecule each of HCl and SO_2, both of which are gasses. Their loss from the reaction mixture drives the reaction toward formation of the acid chloride and makes the workup simple.

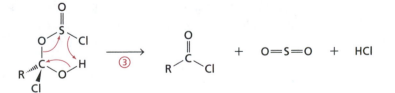

The procedure is thus: The carboxylic acid and thionyl chloride ($SOCl_2$) are heated until gas evolution ceases, and then the excess thionyl chloride is evaporated. The yields are generally very good to excellent. For example,

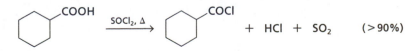

21.3 THE CHEMISTRY OF ACID CHLORIDES, THIOESTERS, AND ANHYDRIDES

21.3a ACID CHLORIDES REACT WITH NUCLEOPHILES TO FORM OTHER ACID DERIVATIVES

Having learned how carboxylic acids can react to form esters or acid chlorides, you can now consider the reactions of those derivatives. Remember that acid chlorides have the most reactive carbonyl group (Section 21.1c), and a variety of nucleophiles will react with acid chlorides to form other derivatives.

If an acid chloride is treated with an alcohol, the oxygen atom of the OH group is the nucleophile and reacts at the electrophilic carbonyl carbon atom to form a tetrahedral intermediate. This species expels chloride ion, a good leaving group, to regenerate the C=O double bond. A base such as pyridine is often added to the reaction mixture to remove the proton from the original alcohol group.

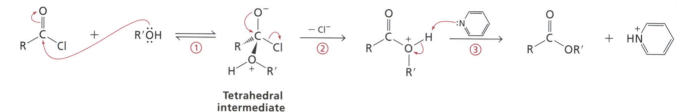

Tetrahedral intermediate

EXERCISE 21.5

Propose a synthesis of *tert*-butyl pentanoate from pentanoic acid and any other necessary reagents. Remember that a 3° alcohol cannot be used in the acid-catalyzed esterification procedure (Section 21.2a).

To prepare an amide, an acid chloride is treated with an amine, which functions as the nucleophile. Ammonia is used to make a primary amide; a primary amine is used to make a secondary amide; and a secondary amine is used to make a tertiary amide.

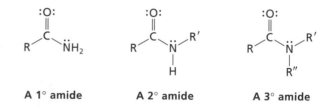

A 1° amide   **A 2° amide**   **A 3° amide**

The mechanism for the reaction between an acid chloride and an amine begins with addition of the amine to the carbonyl group to form a tetrahedral intermediate. This species expels chloride ion to regenerate the C=O double bond. Because an amine is also a base, a second molecule of amine removes a proton from the charged nitrogen atom.

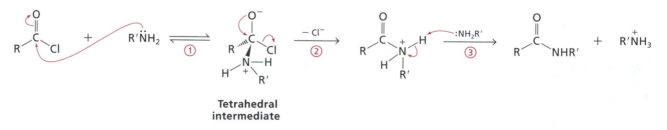

Tetrahedral intermediate

The stoichiometry requires that 2 equiv of the amine be present, one to add to the carbonyl group and one to deprotonate the penultimate product.

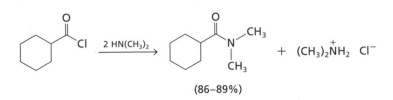

(86–89%)

The **Schotten–Baumann procedure** uses hydroxide ion as the base. The amine is generally more nucleophilic toward the electrophilic carbon atom than hydroxide ion itself, so the amine reacts preferentially with the acid chloride. (The Schotten–Baumann procedure can be used to prepare esters too.)

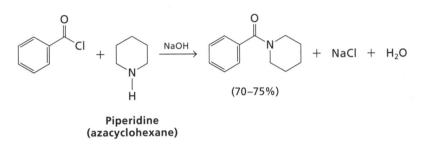

(70–75%)

**Piperidine
(azacyclohexane)**

EXAMPLE 21.2

Show how you would prepare the amide illustrated here starting with any organic compounds that have six or fewer carbon atoms and contain only C, H, N, and/or O.

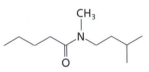

An amide is constructed from an amine and an acid chloride, which in turn is made from a carboxylic acid. The second step of the synthesis can either make use of the amine with aqueous NaOH or an excess of the amine.

Retrosynthesis

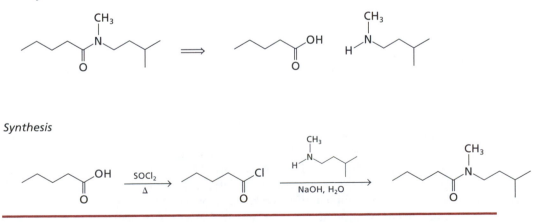

Synthesis

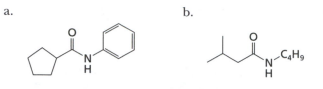

EXERCISE 21.6

Show how you would prepare the following amides. You may use any organic compounds that have six or fewer carbons.

a.   b.

EXERCISE 21.7

Hydrazine (Section 20.1b) reacts to form the *hydrazide* derivative of a carboxylic acid. How would you prepare $CH_3CH_2CH_2CONHNH_2$ from butanoic acid? Propose a reasonable mechanism for its formation.

21.3b ANHYDRIDES ARE NEARLY AS REACTIVE AS ACID CHLORIDES

Anhydrides are reactive carboxylic acid derivatives that rival the acid chlorides in their rates of reactions with nucleophiles. Formally, an anhydride can be made by removing a molecule of water from 2 equiv of a carboxylic acid (the term *anhydride* means "without the elements of water"). This procedure works well only for certain diacids like maleic and phthalic acid, which, when heated to temperatures > 100°C, release water and form five- or six-membered ring anhydrides. Chemical dehydrating agents can also be used to prepare molecules of this type.

Like any acid derivative, an anhydride itself reacts by forming a tetrahedral intermediate, shown in the following reaction with an amine as the nucleophile. The leaving group of an anhydride is the carboxylate ion. The products are the amide and 1 equiv of the carboxylic acid that constitutes the anhydride. Notice that an unsymmetric anhydride will have two different reactive sites, so most anhydrides used in reactions are symmetrical.

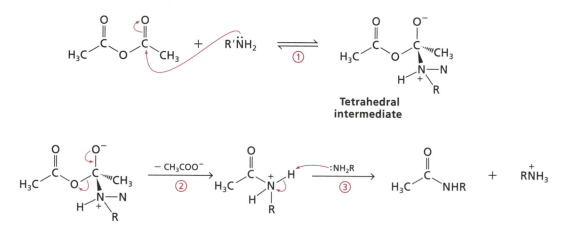

An anhydride is about as reactive as an acid chloride, but half of an anhydride's mass is lost as the leaving group, which limits its usefulness, except for low molecular weight ones such as acetic anhydride. Acetic anhydride is prepared on an industrial scale from ethylene, carbon monoxide, and methanol, and it is a common reagent for making acetate and acetamide derivatives from alcohols and amines, respectively. (This reaction has been mentioned twice before in that context; see Sections 17.3c and 17.3e.)

EXERCISE 21.8

Propose a mechanism for the following reaction, which is used to make aspirin:

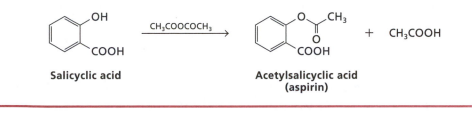

Salicyclic acid   **Acetylsalicyclic acid**
                                        (aspirin)

21.3c   A THIOESTER IS THE BIOCHEMICAL COUNTERPART
OF AN ACID CHLORIDE

In living organisms, esters are often made from the reaction between alcohols and
thioesters, which are reactive carboxylic acid derivatives (Section 21.1d). Acid chlorides
and anhydrides would react indiscriminately with the many nucleophilic sites that are
present in proteins and nucleic acids, as well as in other substances, but thioesters display a practical balance between reactivity and stability.

Acetyl coenzyme A is the most common thioester in living systems, and it functions
to transfer the acetyl group during the course of several important metabolic processes.

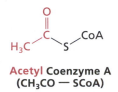

Acetyl Coenzyme A
($CH_3CO - SCoA$)

Coenzyme A (abbreviated CoA-SH) is the *thiol* that makes up thioester derivatives in
biochemical systems.

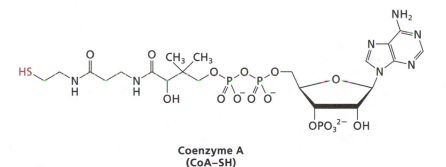

Coenzyme A
(CoA–SH)

A specific example of ester formation is the coupling between a fatty acid—an
aliphatic carboxylic acid that has between 14 and 20 carbon atoms—and dihydroxyacetone phosphate.

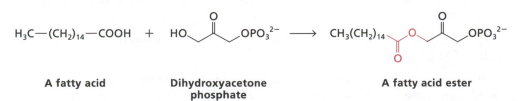

A fatty acid   **Dihydroxyacetone**   **A fatty acid ester**
                        phosphate

Like most biochemical reactions, this one requires an enzyme catalyst. More important
is the fact that the carboxylic acid component reacts as its CoA thioester derivative. The

transformation illustrated below constitutes an important step in the biosynthesis of phospholipids, one of the principal components of cell membranes in animals.

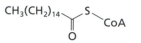

An acyl Coenzyme A derivative

The enzyme active site provides a base to remove the proton of the alcohol OH group as it adds to the carbonyl group of the thioester and forms a tetrahedral intermediate.

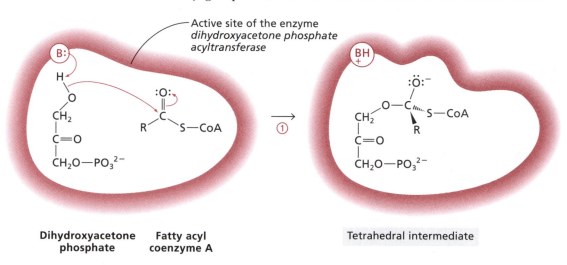

| Dihydroxyacetone phosphate | Fatty acyl coenzyme A | Tetrahedral intermediate |

Formation of the tetrahedral intermediate is followed by regeneration of the carbonyl group and expulsion of the thiolate ion derivative of CoA as the leaving group, which yields the ester.

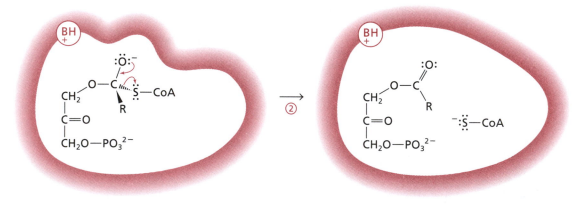

This ester product can undergo subsequent transformations either to produce diacylglycerol phosphate, a precursor of phospholipids, or to form the triester of glycerol, commonly known as fat.

EXERCISE 21.9

In the laboratory, thioesters are prepared from acid chlorides and thiols. Based on your knowledge of ester formation, propose a reasonable mechanism for the synthesis of propyl thioacetate, $CH_3COSCH_2CH_2CH_3$, from acetyl chloride and 1-propanethiol.

21.4 THE CHEMISTRY OF ESTERS

21.4a DIAZOMETHANE CONVERTS A CARBOXYLIC ACID TO THE CORRESPONDING METHYL ESTER

You have already learned two ways to make esters: direct esterification of a carboxylic acid (Section 21.2a) and the reaction between an acid chloride and an alcohol (Section 21.3a). A third method makes use of diazomethane, a reactive compound with a strongly nucleophilic carbon atom. Resonance structures show why: *The carbon atom in one form bears an electron pair and a negative charge,* so it can act as either a base or a nucleophile.

$$H_2C=\overset{+}{N}=\overset{\cdot\cdot}{\underset{\cdot\cdot}{N}} \longleftrightarrow H_2\overset{\cdot\cdot}{C}-\overset{+}{N}\equiv N:$$

Diazomethane

When treated with a carboxylic acid, diazoemethane removes the acid proton, which produces a carboxylate ion and the methyldiazonium ion. This cation is a potent alkylating agent because its methyl group is bonded to an excellent leaving group, N_2. The carboxylate ion participates as a nucleophile in an S_N2 reaction (Section 6.3) on the highly reactive methyldiazonium ion, and the product is the methyl ester derivative.

$$R-\overset{\overset{\textstyle :O:}{\|}}{C}-\overset{\cdot\cdot}{\underset{\cdot\cdot}{O}}-H \; + \; H_2\overset{\cdot\cdot}{C}-\overset{+}{N}\equiv N: \quad \underset{①}{\longrightarrow} \quad R-\overset{\overset{\textstyle :O:}{\|}}{C}-\overset{\cdot\cdot}{\underset{\cdot\cdot}{O}}:^- \; + \; H_3C-\overset{+}{N}\equiv N: \quad \underset{②}{\longrightarrow} \quad R-\overset{\overset{\textstyle :O:}{\|}}{C}-\overset{\cdot\cdot}{\underset{\cdot\cdot}{O}}-CH_3 \; + \; N_2$$

This transformation is one of the best for making methyl esters, especially with small samples. Yields are close to quantitative and the only byproduct is gaseous molecular nitrogen. On larger scales, the problem of generating sizable quantities of the potentially explosive diazomethane limits its utility.

21.4b ESTERS ARE READILY CONVERTED TO AMIDES

On the reactivity scale of carboxylic acid derivatives (Section 21.1d), esters lie between acid chlorides and amides with respect to the rate of their addition–elimination reactions. Therefore, the conversion of esters to amides is favorable.

Unlike the transformation of an acid chloride to an amide, no base is needed when an ester is used as one of the starting materials. In this situation, the tetrahedral intermediate is formed in the usual way by nucleophilic addition of the amine nitrogen atom to the carbonyl group. The departing alkoxide group subsequently removes the proton from the nitrogen atom, producing a molecule each of the alcohol and of the amide.

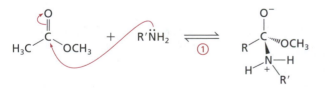

Tetrahedral intermediate

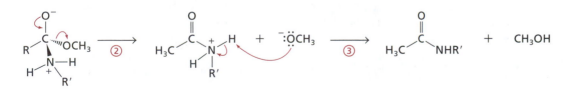

21.4c ESTERS UNDERGO HYDROLYSIS UNDER ACIDIC CONDITIONS TO FORM CARBOXYLIC ACIDS

The addition–elimination reaction that esters undergo with water is called *hydrolysis.* Addition of an aqueous solution of a mineral acid to an ester leads to protonation of the carbonyl oxygen atom and formation of a cation intermediate. When water is present, this charged intermediate is trapped to produce a *gem*-diol after deprotonation in Step 3.

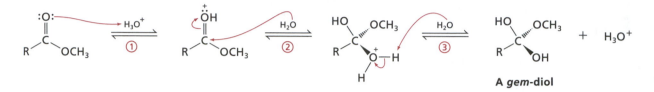

At this stage, protonation can occur on the oxygen atom of the OCH_3 group to generate a good leaving group. Displacement of ROH occurs as a proton is removed and a C=O double bond is formed. The products comprise a molecule each of a carboxylic acid and an alcohol (methanol in this example).

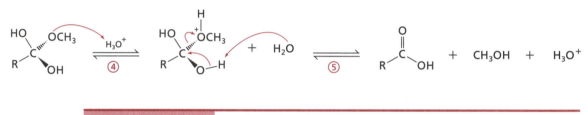

EXERCISE 21.10

Write the steps in the mechanism for the acid-catalyzed hydrolysis of phenyl propanoate.

The mechanism of ester hydrolysis comprises the same steps, in reverse order, as those in acid-catalyzed esterification (Section 21.2a). Because each step can go in either direction—that is, all of the steps are equilibria—the reaction conditions are chosen to optimize formation of the desired product.

- To prepare an ester from a carboxylic acid and alcohol, a dehydrating agent (e.g., molecular sieves) is added, or the solution is heated to remove water.

- To hydrolyze an ester to a carboxylic acid, water is present in excess. A cosolvent such as alcohol is often added to ensure that the reactants are soluble.

21.4d ESTERS UNDERGO HYDROLYSIS UNDER BASIC CONDITIONS, A PROCESS CALLED SAPONIFICATION

Even though esters are readily hydrolyzed under acidic conditions, aqueous base can be used as well. The hydrolysis of an ester in base is called **saponification.** Lipids and other fats that occur in Nature are di- and triesters of glycerol. Fats undergo hydrolysis reactions with base according to the mechanism outlined below. The term "saponification" originated for the process used in the manufacture of soap, which was once carried out by cooking animal fat with potassium or sodium hydroxide to produce the salts of the long-chain fatty acids by hydrolysis.

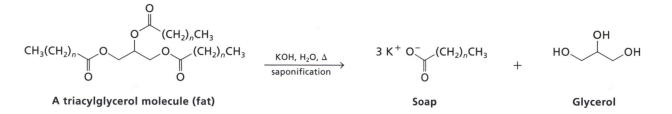

A triacylglycerol molecule (fat)   **Soap**   **Glycerol**

In saponification, the nucleophilic hydroxide ion is naturally attracted to the electrophilic carbon atom of the carbonyl group. The tetrahedral intermediate, which is produced in Step 1, subsequently collapses in the second step to regenerate the C=O double bond by displacing the leaving group, R'O⁻. Because this leaving group is a strong base, assistance is provided by the aqueous solvent, which can transfer a proton to the base as it leaves. Finally, an acid–base reaction takes place.

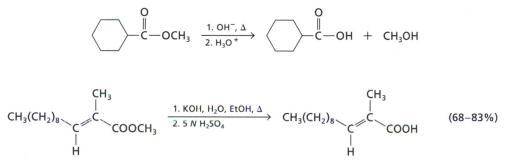

Notice that hydrolysis reactions conducted under basic conditions require use of a full equivalent of OH⁻. The mixture usually has to be heated for several hours, and an alcohol cosolvent is sometimes employed to keep all of the reactants dissolved. To obtain the carboxylic acid itself, aqueous acid is added during workup.

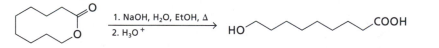

EXERCISE 21.11

Propose a reasonable mechanism for the following reaction, hydrolysis of a lactone:

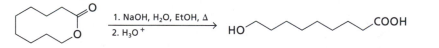

21.4e TRANSESTERIFICATION CONVERTS ONE ESTER TO ANOTHER

Esters can undergo reactions with alcohols in the same way that they do with water. The procedure does not yield a carboxylic acid; instead, it results in exchange of the alcohol groups, a process called **transesterification**. With an acid catalyst—HCl in the following example—the mechanism follows the common steps shown previously, which results in exchange of RO for CH₃O.

1. The carbonyl group is protonated to activate it.
2. The alcohol molecule intercepts the cation intermediate.
3. A hydroxyacetal intermediate is formed by deprotonation.
4. The methoxy group is protonated.

5. Elimination of methanol accompanies regeneration of the carbonyl group.

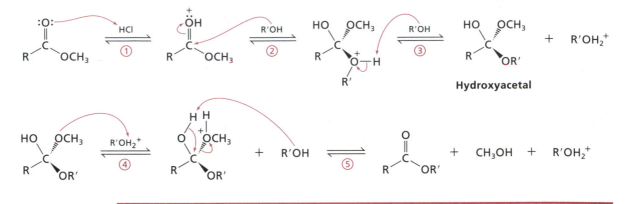

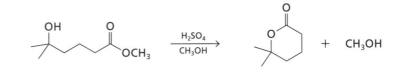

EXAMPLE 21.3

Show the steps in the mechanism for transesterification of the following hydroxy ester to form the corresponding lactone. The reaction is catalyzed by a trace of H_2SO_4.

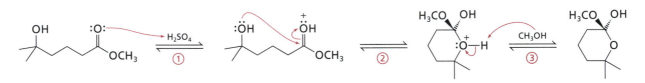

In any mechanism of a reaction between a carboxylic acid derivative and sulfuric acid, the first step is protonation of the carbonyl group, which serves to activate it. The alcohol oxygen atom on the δ-carbon atom is the nucleophile in the second step, and it reacts with this first-formed cation. Deprotonation yields the neutral hydroxy intermediate.

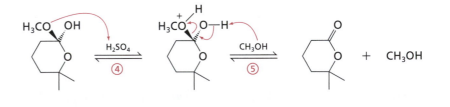

Next, the alkoxy group is protonated and solvent (methanol) removes the proton from the OH group to regenerate the carbonyl group and to displace a molecule of methanol.

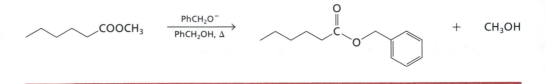

EXERCISE 21.12

Transesterification can also occur under basic conditions. Propose a reasonable mechanism for the following reaction:

21.5   THE CHEMISTRY OF AMIDES

21.5a   AMIDES ARE HYDROLYZED SLOWLY

Amides, being the least reactive of the carboxylic acid derivatives, undergo few inter-conversion reactions, but they can be hydrolyzed to form their parent carboxylic acids. Under basic conditions, hydroxide ion adds to the carbonyl group and forms a tetra-hedral intermediate in the usual way. Regeneration of the carbonyl group displaces the strongly basic amide ion, a process that is assisted by removing a proton from water as it leaves. The hydroxide ion reacts with the carboxylic acid in a final acid–base step; the carboxylic acid itself can be obtained with acid workup.

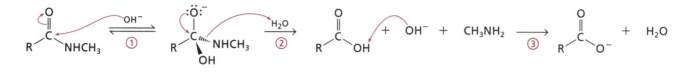

Hydrolysis of an amide is more often done under acidic conditions, which circumvents the problem with having a strongly basic leaving group. Protonation of the amide carbonyl group leads to addition of water across the C=O double bond and formation of a *gem*-diol intermediate. In the subsequent steps, the amino group is protonated to form a decent leaving group.

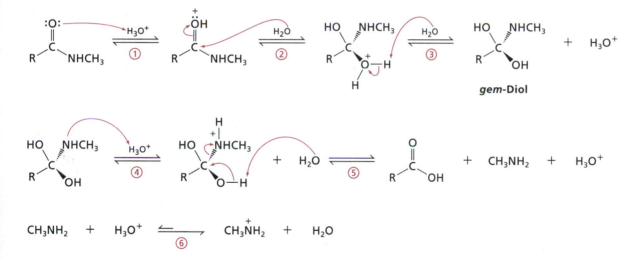

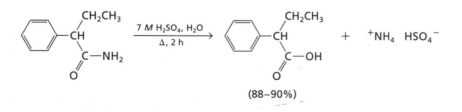

One factor that drives the reaction to completion is a final acid–base reaction between the displaced amine (methylamine in this example) and aqueous acid to form an am-monium ion that is no longer nucleophilic. The following equation shows the overall reaction that takes place between an amide and aqueous sulfuric acid to form the car-boxylic acid product and ammonium hydrogen sulfate.

EXERCISE 21.13

Propose a reasonable mechanism for the acid-catalyzed hydrolysis of the following lactam (cyclic amide):

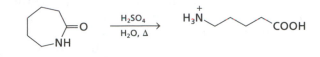

21.5b THE ENZYME-CATALYZED HYDROLYSIS OF PEPTIDE BONDS (AMIDE GROUPS) IS FACILITATED BY STABILIZATION OF THE TETRAHEDRAL INTERMEDIATE

The hydrolysis of proteins provides one of the chief sources of energy for animals. Enzymes that digest proteins have evolved an active site structure that accelerates the rate of this reaction, which is amide hydrolysis, by many orders of magnitude compared with the rate of reaction in solution, even with acid or base. Considering that amides are the least reactive of the acid derivatives, whatever process catalyzes their hydrolysis in Nature clearly has important lessons for us to design catalysts to carry out such transformations in the laboratory.

Let us look at the process catalyzed by the enzyme *chymotrypsin,* the prototype of a large number of enzymes called **serine proteases.** This name derives from the fact that the enzyme catalyzes the cleavage of protein peptide bonds, and it accomplishes this transformation by making use of the serine side chain as a nucleophile. We can portray the *chymotrypsin* active site with the drawing in Figure 21.2. The most important features are listed below.

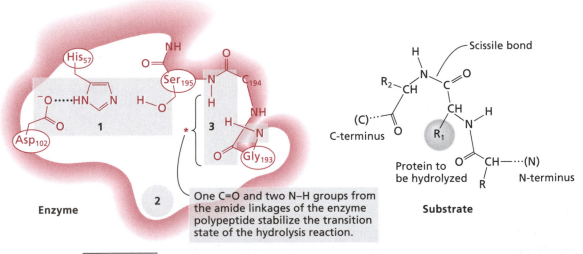

Figure 21.2

A diagram of the active site of *chymotrypsin,* showing some important structural features, and the structure of a general substrate molecule.

1. A group of three amino acids [with their positions from the N-terminus given as subscripts, Asp_{102}, His_{57}, and Ser_{195}] define what is called the *catalytic triad.* The side chains of these amino acids create a nucleophilic region that ultimately cleaves the amide C–N bond. The bond in the substrate that is broken has a special name, the *scissile bond.*

2. A pocket, created by the folding of the enzyme backbone, will bind one of the side chains of the protein substrate (shown as R_1 in Fig. 21.2). This pocket has specifically developed to bind the hydrophobic, aryl side chains of phenylalanine, tryp-

tophan, and tyrosine. The scissile bond, therefore, is the amide bond between the carboxy group of an aromatic amino acid and the amine group of the next residue.

3. Three parts of the polypeptide backbone (shown within the gray patch in Fig. 21.2)—a peptide carbonyl group of Gly$_{193}$, the amide NH group of Gly$_{193}$, and the amide NH group of Ser$_{195}$—can form strong hydrogen bonds during the course of the reaction. *These hydrogen bonds stabilize the transition state of the reaction and account for most of the acceleration of the reaction rate.*

The first step of the reaction is formation of the enzyme–substrate complex, shown below, which results mainly from binding the large side chain of the substrate (R$_1$). Most of the stabilizing influences for this binding are London forces Section 2.8a).

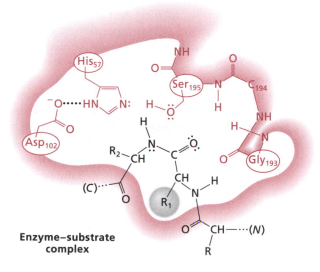

Enzyme–substrate complex

Once formed, the enzyme–substrate complex begins to undergo reaction as a result of nucleophilic addition of a serine alkoxide ion to the carbonyl group, resulting in formation of a tetrahedral intermediate. The movement of electrons, as well as the formation of the stabilizing hydrogen bonds, can be represented as follows:

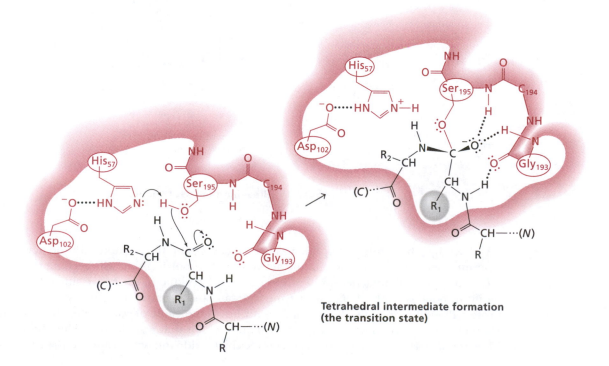

Tetrahedral intermediate formation (the transition state)

The tetrahedral anionic intermediate is stabilized by hydrogen bonds with the N–H amide groups of Ser_{195} and Gly_{193}. Histidine (57) removes the proton from the OH group of Ser_{195} to generate the nucleophile. Aspartic acid (102), in its carboxylate form, is important for keeping the histidine side chain oriented so that its nitrogen atom is able to accept the proton from the serine OH group. One other interaction stabilizes the transition state: The C=O unit of Gly_{193}, which is part of the enzyme's polypeptide backbone, forms a new hydrogen bond with the N–H group of the substrate polypeptide, one residue toward the N-terminus away from the scissile bond. It is important that the carbonyl group at the scissile bond in the substrate does not interact appreciably with the enzyme active site residues until the tetrahedral intermediate forms. *This selective interaction ensures that the transition state is stabilized, not the unreacted substrate itself.*

The tetrahedral intermediate, once formed, will subsequently regenerate the carbonyl group, expelling the leaving group (the polypeptide bearing the new amino terminus) according to the mechanism that you are already familiar with. Remember from the previous discussion that one of the problems with base hydrolysis of an amide in solution is that the amide ion is a strong base. The enzyme circumvents this problem by transferring a proton from His_{57} to the amine as it leaves. This proton originally came from the serine side chain. Notice the efficiency of this process: The histidine residue removed the proton to assist the serine alcohol group to become more nucleophilic; now it uses the proton as an acidic center to assist the amine to become a better leaving group. The fact that histidine can serve as both an acid (in its protonated form) and as a base near pH 7 accounts for its involvement at the active site in many enzymes.

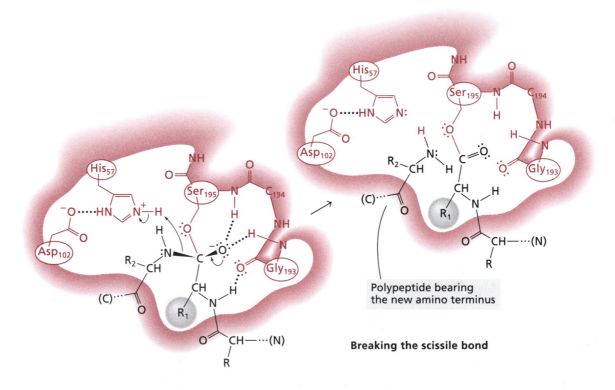

Breaking the scissile bond

Polypeptide bearing
the new amino terminus

At this stage, the substrate peptide has been cleaved, and the portion with the new N-terminus dissociates from the enzyme active site. The portion that will become the new C-terminus is still attached *covalently* to the serine residue in the active site. Obviously, if this bond is not broken, the enzyme is "dead" and cannot function any longer. In protein digestion, though, the reaction continues. A molecule of water enters the active site in place of the portion of the polypeptide chain that was cleaved. The His_{57} residue now acts as a base once again and creates hydroxide ion, which adds to the carbonyl group. A tetrahedral intermediate is formed once more, which is stabilized by the same portions of the active site.

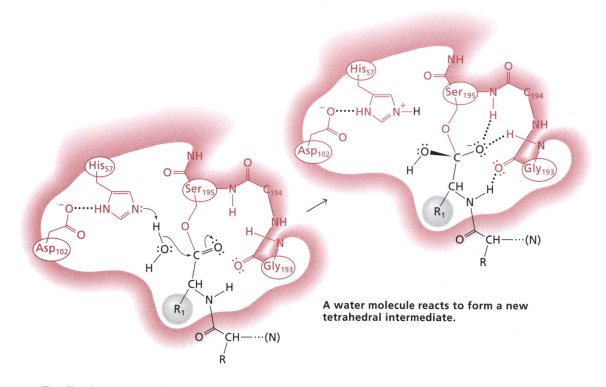

A water molecule reacts to form a new tetrahedral intermediate.

Finally, the intermediate collapses to generate the carbonyl group, expelling the oxygen atom of the serine side chain, which, in turn, takes back a proton from the histidine residue. The product of this reaction is the carboxylic acid that constitutes the new C-terminus of the cleaved substrate polypeptide. After the polypeptide dissociates from the active site (not shown in the following scheme), the enzyme is free to bind the next molecule of substrate.

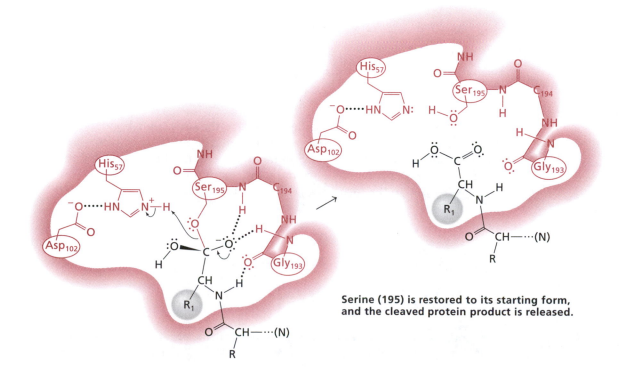

Serine (195) is restored to its starting form, and the cleaved protein product is released.

Some important details about the mechanism of action for *chymotrypsin* were elucidated by studying the hydrolysis reaction of *p*-nitrophenyl acetate. One of the products is *p*-nitrophenol, which is acidic enough that is ionizes in solution to form a highly colored anion. Trace steps for hydrolysis of this substrate by *chymotrypsin.*

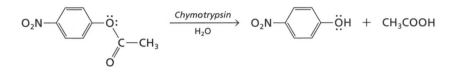

21.6   THE CHEMISTRY OF NITRILES

You can easily see that an ester is a derivative of a carboxylic acid because each has a carbonyl group attached to an oxygen-containing group (OH and OR, respectively). What is not so apparent is the relationship between a nitrile and a carboxylic acid. Like all acid derivatives, however, a nitrile is susceptible to both acid and base hydrolysis. These reactions differ from those of ester hydrolysis because a nitrile lacks a carbonyl group. The relationship of a nitrile to a carboxylic acid is like that of an imine to a ketone.

In aqueous acid, protonation of a nitrile group occurs at the nucleophilic nitrogen atom. Water reacts with this cation, forming a hydroxy imine after deprotonation. The hydroxy imine then undergoes two proton-transfer reactions that convert it to an amide. The mechanism for hydrolysis from there on follows that described for amides (Section 21.5a).

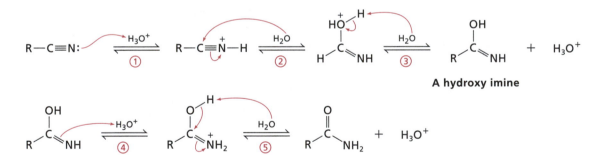

Under basic conditions, hydroxide ion adds to the C≡N triple bond of a nitrile by reaction at the electrophilic carbon atom. The negatively charged nitrogen atom is then protonated by the aqueous environment, generating the hydroxy imine. Subsequent proton-transfer reactions convert the hydroxy imine to the amide, which can be further hydrolyzed to the carboxylic acid and amine.

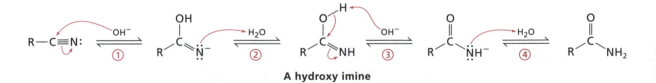

In both the acid- and base-catalyzed routes, a nitrile adds one molecule of water to form an amide. It would seem, therefore, that *dehydration* of a primary amide might yield a nitrile. Indeed, heating an amide with P_4O_{10}, a dehydrating agent, yields the nitrile, demonstrating the ready interconversion between these two carboxylic acid derivatives.

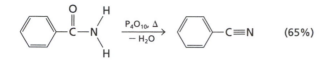

Recall that a nitrile can also be prepared by the S_N2 reaction of cyanide ion with an alkyl halide (Table 6.2) and by dehydration of an oxime, prepared from an aldehyde (Section 20.1d).

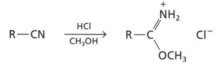

EXERCISE 21.15

A nitrile reacts with HCl in methanol to form an imidate ester as its HCl salt. Propose a reasonable mechanism for the following transformation:

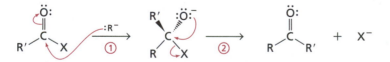

An imidate ester

21.7   REACTIONS WITH ORGANOMETALLIC COMPOUNDS

21.7a   ACID CHLORIDES CAN BE CONVERTED TO KETONES

Because they have carbonyl groups, acid derivatives can be used to form carbon–carbon bonds by addition reactions of organometallic reagents. Organomagnesium, organolithium, and organocuprate reagents all react with the carbonyl group by the same general two-step process you have become familiar with in this chapter: *Addition of the nucleophile is followed by the expulsion of the leaving group to regenerate the carbonyl group.*

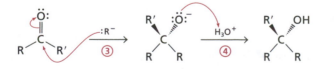

A new wrinkle appears in these reactions, however: *The organometallic compound can also react with the product, which has a carbonyl group* (Table 15.1).

Organocuprate complexes (Section 15.3a) are unreactive toward ketones, so Step 3 above does not occur. In fact, organocuprates react only with the most reactive carboxylic acid derivatives, so they are particularly useful for the conversion of acid

chlorides to ketones. It you think of an organocuprate reagent as the source of a nucleophilic alkyl group, then addition to the carbonyl group of an acid chloride followed
by elimination of chloride ion as the leaving group yields a ketone.

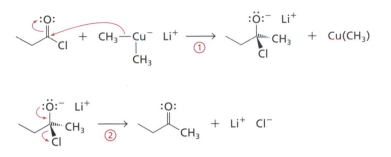

This reaction gives good yields of product, even when the organocuprate reagent has
hindered alkyl groups. The following example shows the use of a *mixed cuprate*, a
reagent with a heteroatom-containing group that is not transferred. The *tert*-butyl
group readily replaces the chlorine atom of benzoyl chloride in the following example:

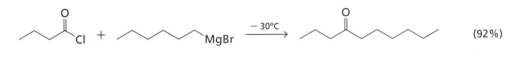

In this mixed cuprate, the alkyl
group is transferred in
preference to the PhS group.

Even though organocuprates are the preferred reagents for making ketones, acid
chlorides *can* react with Grignard reagents to form ketones under suitable conditions.
To maximize formation of the ketone product, the reaction is conducted at low temperature to magnify reaction rate differences. Ketones are less reactive than acid chlorides toward nucleophilic addition.

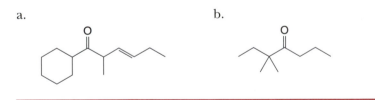

EXERCISE 21.16

Show how you would prepare the following compounds. You may use any carboxylic acid
precursor, any organometallic reagent, and any other inorganic reagents or solvents.

a.   b.

21.7b AMIDES CAN BE CONVERTED TO ALDEHYDES AND KETONES

In spite of its overall low reactivity toward addition, an amide can be used to prepare a
ketone or aldehyde because the tetrahedral intermediate also decomposes slowly. Only
a tertiary amide can be used with a strongly basic organometallic compound, however.
A primary or secondary amide has at least one proton attached to the nitrogen atom,
and each can react as an acid in acid–base reactions (Section 22.1b).

A procedure to prepare aldehydes makes use of the reaction between a Grignard reagent and DMF. Addition to the carbonyl group in the first step generates the tetrahedral intermediate. The amino group is not eliminated readily because it is such a strong base, and the solvent (THF) has no protons to assist loss of the amine.

When aqueous acid is added for workup, a hemiaminal is formed and any excess Grignard reagent that might still be present is destroyed. The unstable hemiaminal (Section 20.1a) decomposes to form an aldehyde and dimethylamine as products.

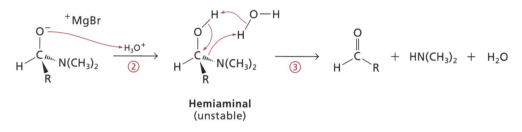

Hemiaminal
(unstable)

Amides other than formamide derivatives will yield ketones when treated with a Grignard reagent followed by aqueous acid workup. The following exercise is an example of this transformation.

EXERCISE 21.17

Propose a mechanism for the following reaction that illustrates the synthesis of a ketone from a tertiary amide and a Grignard reagent:

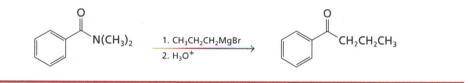

21.7c ESTERS REACT WITH ORGANOMETALLIC COMPOUNDS TO FORM ALCOHOLS

Aldehydes and ketones can be made from the most and least reactive carboxylic acid derivatives, namely, acid chlorides and amides, respectively. When an *ester* reacts with organometallic compounds, however, an alcohol is usually formed. The product is a tertiary alcohol with at least two groups alike. (Formate esters produce secondary alcohols.)

Addition of R⁻ to the carbonyl group of an ester in the first step forms the tetrahedral intermediate. The alkoxy group (methoxy in this example) is eliminated because even though it is a strong base, magnesium ions have a relatively high affinity for oxyanions, which moderates their basicity properties somewhat.

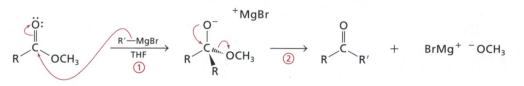

As the concentration of the ketone increases, it competes with the ester for the organometallic reagent that has not yet reacted. Because ketones undergo nucleophilic addition faster than esters, any unreacted Grignard reagent will react preferentially with the ketone. To ensure that the ester reacts completely, at least 2 equiv of the organometallic compound are normally supplied to the reaction mixture, and the product, after aqueous acid workup, is a tertiary alcohol with at least two groups alike.

$$\text{Ph—COOCH}_2\text{CH}_3 \xrightarrow[\text{2. H}_2\text{O}]{\text{1. 2 PhMgBr}} (\text{Ph})_3\text{C—OH} \quad + \quad \text{CH}_3\text{CH}_2\text{OH} \qquad (89\text{–}93\%)$$

If a formate ester is used, then a 2° alcohol is formed. For example,

$$\underset{\textbf{Ethyl formate}}{\overset{\displaystyle O}{\underset{\displaystyle H}{\overset{\|}{C}}}\!\!-\!\text{OCH}_2\text{CH}_3} \xrightarrow[\text{2. H}_3\text{O}^+]{\text{1. CH}_3\text{CH}_2\text{CH}_2\text{MgBr}} \underset{\text{CH}_3\text{CH}_2\text{CH}_2 \quad \text{CH}_2\text{CH}_2\text{CH}_3}{\overset{\text{H} \quad \text{OH}}{C}} \quad + \quad \text{CH}_3\text{CH}_2\text{OH}$$

EXERCISE 21.18

Using curved arrows to depict the movement of electrons, show the steps in the mechanism of the preceding reaction between ethyl formate and propylmagnesium bromide.

EXERCISE 21.19

What is the expected product of the following reaction?

$$\xrightarrow[\text{2. H}_3\text{O}^+]{\text{1. CH}_3\text{MgBr (XS)}} \quad ? \qquad (57\%)$$

21.8 REDUCTION REACTIONS OF CARBOXYLIC ACIDS AND DERIVATIVES

21.8a LITHIUM ALUMINUM HYDRIDE CONVERTS ESTERS OR ACIDS TO 1° ALCOHOLS

All of the carboxylic acid derivatives described in this chapter can be converted either to alcohols or amines by addition of hydride ion to the carbonyl group. Lithium aluminum hydride in ether or THF is the most commonly used reagent for these reduction reactions. The mechanism for reduction of an acid chloride, anhydride, ester, or thioester proceeds by hydride ion addition to form a tetrahedral intermediate, followed by expulsion of the leaving group with concomitant regeneration of the carbonyl group.

$$\underset{}{\overset{\displaystyle :O:}{\overset{\|}{\text{—C—X}}}} \xrightarrow[]{\overset{①}{\text{H—}\bar{\text{A}}\text{lH}_3 \;\; \text{Li}^+}} \underset{\overset{|}{\text{H}}}{\overset{\displaystyle :\overset{..}{O}:^-}{\underset{}{\text{—C—X}}}} \xrightarrow[]{\overset{②}{- \; \text{X}:^-}} \underset{}{\overset{\displaystyle :O:}{\overset{\|}{\text{—C—H}}}}$$

X = Cl, OR, OCOR, SR

Under these conditions, the resulting aldehyde is also reduced by LiAlH$_4$, yielding the alcohol after aqueous workup, as you saw in Section 18.3a.

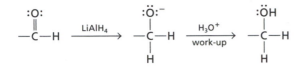

Sodium borohydride is not potent enough to reduce the carbonyl group of an ester. It is therefore used to reduce an aldehyde or ketone group in the presence of an ester or carboxylic acid group.

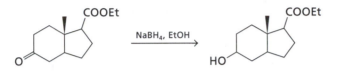

Carboxylic acids themselves are reduced with LiAlH$_4$ to the corresponding alcohol, but the mechanism is different from that for reduction of carboxylic acid derivatives because a natural leaving group is not present. The first step of the reaction between a carboxylic acid and LiAlH$_4$ generates a carboxylate ion by an acid–base reaction.

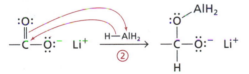

The lithium salt of a carboxylate ion retains substantial C=O double-bond character, so hydride ion is able to add in a second step. The aluminum atom, rather than a lithium ion, bonds with the second oxygen atom, making a good leaving group, H$_2$Al–O (this is the conjugate base of H$_2$Al–OH, which is expected to be a fairly strong acid).

Elimination occurs to produce an aldehyde. Reduction continues, giving the alcohol after hydrolysis.

21.8b LITHIUM ALUMINUM HYDRIDE CONVERTS AMIDES TO AMINES

Because an amide contains both nitrogen and oxygen atoms, it might eliminate either heteroatom. While this is possible, an amide is usually reduced to form an amine, with complete removal of the oxygen atom from the product. The mechanism of reduction by LiAlH$_4$ is similar to that shown for a carboxylic acid, which includes complexation of the oxygen atom by aluminum and reductive cleavage of the carbon–oxygen bond.

The first step is an acid–base reaction, which is followed by addition of hydride ion to the carbonyl group. An amide is relatively acidic, compared with most nitrogen-containing substances (see Table 5.1).

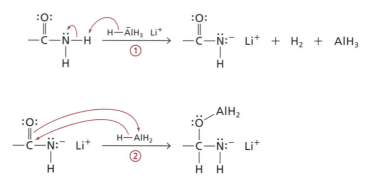

Once the aluminoxy group has been formed, elimination occurs to give the unsubstituted imine. This species is reduced by hydride ion to the conjugate base of the amine (Section 20.2a).

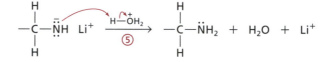

Protonation during acid workup gives the final product of reduction, the amine.

Even though a tertiary amide does not undergo an acid–base reaction in the first step, it nevertheless generates an amine as the product from reduction by LiAlH₄, presumably by forming a similar type of aluminoxy intermediate.

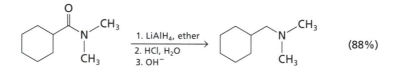

21.8c ACID CHLORIDES CAN BE CONVERTED TO ALDEHYDES

Just as the addition of an organometallic compound to a carboxylic acid derivative can be stopped at the ketone stage, an aldehyde can be isolated if reduction of an acid chloride is carried out using a reagent that does not also react rapidly with the aldehyde carbonyl group. One reagent used for this transformation is lithium tri(*tert*-butoxy) aluminum hydride, in which the hydride ion is hindered by the presence of the *tert*-butoxy groups. The reagent is prepared by carefully treating LiAlH₄ with *tert*-butyl alcohol. When an acid chloride is treated with this reagent at low temperature, the aldehyde can be isolated in reasonable yields.

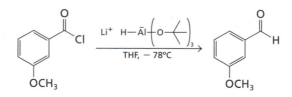

Section 21.1   General aspects of structure and reactivity

- The names of esters have an unnumbered prefix to identify the R group attached to the oxygen atom of the COOR unit. Substitution on the nitrogen atom of amides is designated by an italicized uppercase *N-* .

- The acyl group is the general unit of a carboxylic acid formed by removing the OH group. Derivatives of carboxylic acids have different heteroatom-containing groups attached to the acyl fragment.

- Anionic nucleophiles add to the carbonyl group of carboxylic acid derivatives to form intermediates in which the oxygen atom has a negative charge and the carbon atom is tetrahedral. This tetrahedral intermediate expels a leaving group to regenerate the C=O double bond. This two-step process constitutes an addition–elimination mechanism.

- Relative reaction rates for carboxylic acid derivatives are influenced by the basicity of the heteroatom-containing substituent.

- The order of reactivity toward addition–elimination is acid chloride > anhydride > thioester > ester > amide > carboxylic acid.

Section 21.2   Reactions of carboxylic acids

- A carboxylic acid does not readily undergo addition–elimination reactions because the proton is the most electrophilic atom of the COOH group. Therefore, carboxylic acids normally participate in acid–base reactions.

- An ester can be made by treating a mixture of a carboxylic acid and alcohol with strong acid. This procedure is called esterification.

- A carboxylic acid that has a hydroxy group four or five carbons atoms away from the COOH group readily forms a cyclic ester called a lactone.

- As with any hydroxy acid, the aldonic and aldaric acids (oxidized monosaccharides) form lactone derivatives.

- An acid chloride is prepared by heating a carboxylic acid with thionyl chloride.

Section 21.3   The chemistry of acid chlorides, thioesters, and anhydrides

- Acid chlorides react readily with nucleophiles to form other carboxylic acid derivatives. Addition–elimination is facile because chloride ion is a good leaving group.

- Anhydrides are made by dehydrating carboxylic acids, either by heating or with other reagents. Reactions of anhydrides are analogous to those of acid chlorides.

- In living systems, a thioester is the functional equivalent of an acid chloride. The most common thioesters are derivatives of coenzyme A (CoA).

Section 21.4   The chemistry of esters

- A methyl ester can be made by treating a carboxylic acid with diazomethane.

- Esters react with amines to form amides.

- An ester can be hydrolyzed under acidic or basic conditions, forming a carboxylic acid and an alcohol. The reaction under basic conditions is called saponification.

- Reaction of an ester with an alcohol results in apparent exchange of the R group attached to oxygen, a process called transesterification.

Section 21.5   The chemistry of amides

- An amide is hydrolyzed under acidic or basic conditions, forming a carboxylic acid and an amine. The acid-catalyzed reaction is the more facile of the two procedures, but an amide still undergoes hydrolysis more slowly than other carboxylic acid derivatives.

- Enzymes readily catalyze the hydrolysis of amides (peptides bonds) by stabilizing the tetrahedral intermediate via formation of hydrogen bonds.

Section 21.6   The chemistry of nitriles

- Nitriles are carboxylic acid derivatives that lack a carbonyl group.

- A nitrile adds water to form an amide, which can be further hydrolyzed to the corresponding carboxylic acid and amine.

Section 21.7   Reactions with organometallic compounds

- Acid chlorides react with organocuprates or with Grignard reagents at low temperature to form ketones.

- Tertiary amides react with Grignard reagents to form ketones after acid workup. Formamide derivatives react with Grignard reagents to form aldehydes after acid workup.

- Esters react with Grignard reagents to form tertiary alcohols after acid workup. Formate esters yield secondary alcohols. The products in all cases have two of the same R group attached to the alcohol carbon atom.

Section 21.8   Reduction reactions of carboxylic acids and derivatives

- Lithium aluminum hydride reduces an ester, acid chloride, or carboxylic acid to the corresponding alcohol.

- Lithium aluminum hydride reduces an amide to the corresponding amine.

- An aldehyde can be formed by partial reduction of an acid chloride using lithium tri(*tert*-butoxy)aluminum hydride.

KEY TERMS

Section 21.1b
acyl group

Section 21.1c
tetrahedral intermediate
addition–elimination
  reaction

Section 21.2a
esterification

Section 21.2b
lactone

Section 21.3a
Schotten–Baumann
  procedure

Section 21.3c
Acetyl coenzyme A
coenzyme A

Section 21.4d
saponification

Section 21.4e
transesterification

Section 21.5b
serine proteases

Section 21.2a

Carboxylic acids react with alcohols in the presence of an acid catalyst to form esters.

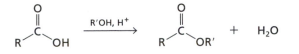

Section 21.2b

Carboxylic acids with a hydroxy group situated several carbon atoms away from the carbonyl group react to form cyclic esters, which are called lactones. The reaction is catalyzed by acid, and five- and six-membered ring lactones form readily.

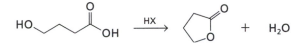

Section 21.2c

Carboxylic acids react with thionyl chloride to form acid chlorides.

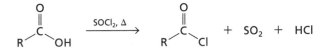

Section 21.3a

Acid chlorides react with amines and alcohols to form amides and esters, respectively.

Section 21.3b

Cyclic anhydrides can be made by dehydration of dicarboxylic acids. This reaction works best for the formation of five-membered ring anhydrides.

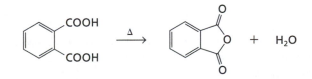

Anhydrides react with alcohols or amines to form esters and amides, respectively.

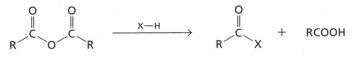

Section 21.4a

Methyl esters are prepared by reactions between carboxylic acids and diazomethane.

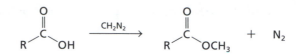

Section 21.4b

Amides can be prepared by reactions between esters and amines.

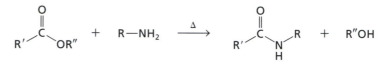

Section 21.4c

An ester undergoes hydrolysis in aqueous acid to form a carboxylic acid and an alcohol.

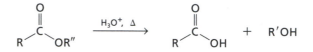

Section 21.4d

An ester undergoes hydrolysis in aqueous base to form a carboxylic acid and an alcohol, a process called saponification.

Section 21.4e

Alcohol groups can be interchanged when an ester is heated with an alcohol and an acid catalyst.

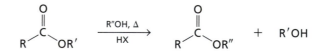

Section 21.5a

An amide undergoes hydrolysis in aqueous acid to form a carboxylic acid and the salt of an amine; an amide undergoes hydrolysis in aqueous base to form the salt of a carboxylic acid and an amine.

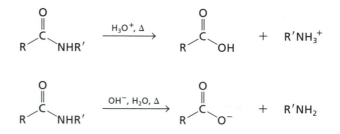

Section 21.6

Nitriles add a molecule of water under the influence of an acid catalyst to form amides.

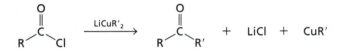

A nitrile undergoes hydrolysis in aqueous acid to form a carboxylic acid and the ammonium ion; a nitrile undergoes hydrolysis in aqueous base to form the salt of a carboxylic acid and ammonia.

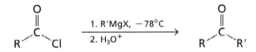

Section 21.7a

Acid chlorides react with organocuprates to form ketones.

Acid chlorides react with Grignard reagents at low temperature to form ketones.

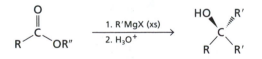

Section 21.7b

Amides react with Grignard reagents to form ketones after hydrolysis. Aldehydes are the products of this transformation when a formamide derivative is used.

Section 21.7c

Esters react with Grignard reagents to form tertiary alcohols after hydrolysis. At least two of the R groups attached to the alcohol carbon atom are the same.

Section 21.8a

Carboxylic acids and esters are reduced by LiAlH₄ to form primary alcohols after hydrolysis.

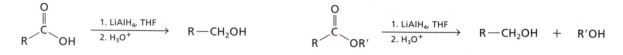

Section 21.8b

Amides are reduced by LiAlH₄ to form amines after hydrolysis.

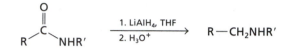

Section 21.8c

Acid chlorides react with LiAlH(O–t-Bu)₃ to form aldehydes.

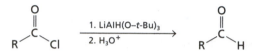

ADDITIONAL EXERCISES

21.20.   Draw the structure for each of the following compounds:

a.   *N,N*-Diethylacetamide

b.   Isobutyl (*S*)-3-methylpentanoate

c.   Ethyl 3,3-difluorocyclopentanecarboxylate

d.   (*E*)-3-Chloro-2-butenoic acid

e.   3-Bromo-2-nitrobenzoic acid

21.21.   Give an acceptable systematic name for each of the following compounds:

a.   b.   c.

   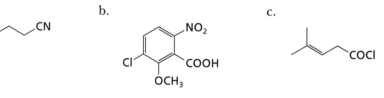

d.

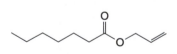

21.22.   Write an equation for the reaction between pentanoic acid and each of the following reagents:

a.   (CH₃)₂CHCH₂OH and TsOH, with heating.

b.   Lithium aluminum hydride followed by aqueous acid workup.

c.   Thionyl chloride with heating, then 1-aminobutane followed by workup with dilute aqueous NaOH.

d. Lithium aluminum hydride followed by aqueous acid workup, then PCC (Section 11.4b) in dichloromethane.

e. Thionyl chloride with heating, then LiAlH(O–*t*-Bu)₃ at low temperature.

f. Thionyl chloride with heating, then CH₃OH, then excess CH₃MgI in ether, then aqueous NH₄Cl.

21.23. Arrange the following compounds from most-to-least reactive for the hydrolysis reaction with aqueous NaOH:

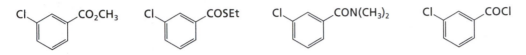

21.24. Oxalyl chloride reacts with a carboxylic acid in the presence of triethylamine, first to generate a mixed anhydride, then to undergo decomposition to produce carbon monoxide, carbon dioxide, triethylamine hydrochloride, and a new acid chloride. Propose a mechanism for this transformation.

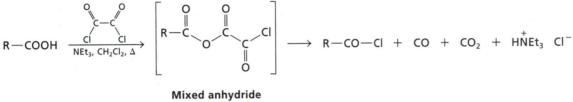

Mixed anhydride
(not isolated)

21.25. A cyclic anhydride reacts with an alcohol to produce a half-acid ester. Propose a mechanism to rationalize how this product is formed.

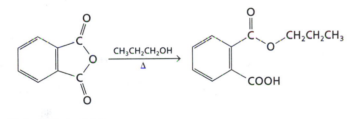

Phthalic anhydride

21.26. In performing each of the following reactions, indicate the principal IR absorptions that you would look for in the reactant and product to ensure that the reaction was successful. Be as specific as possible: list the position (cm⁻¹), intensity (weak, medium, or strong), and shape (sharp or broad) of the diagnostic absorption band(s).

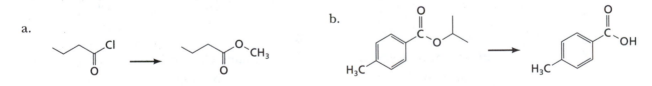

a.

b.

21.27. For the reactions shown in Exercise 21.26, repeat the process for the ¹H and broad-band ¹³C NMR spectra. List, as applicable, the chemical shifts, integrated intensity values, splitting patterns, and DEPT results.

21.28. Indicate what reagents are needed to carry out each reaction in Exercise 21.26.

21.29. Draw the structure(s) of the major product(s) expected from each of the following reactions. Indicate the stereochemistry of the product as appropriate. Relative stereochemistry should be shown using wedges and dashed lines. If a racemic mixture will be formed, draw the structure of one enantiomer and write the word "racemic", or draw both enantiomeric structures. If diastereomers are formed, draw each structure; label meso compounds as such. If no reaction occurs, write N.R.

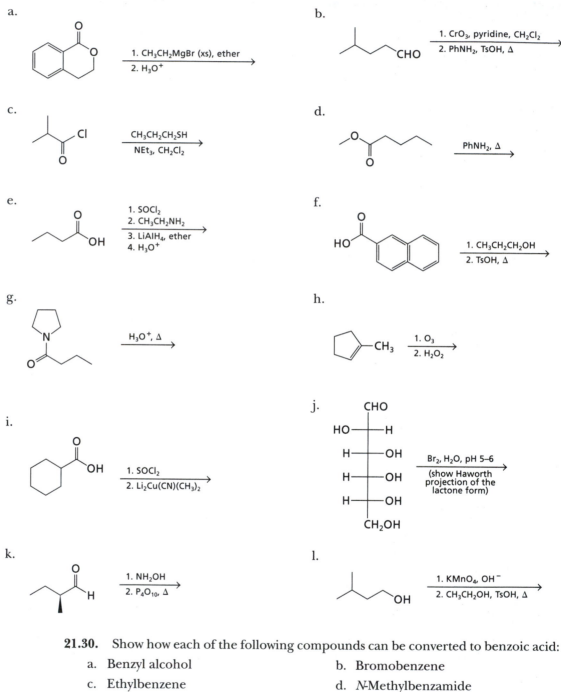

a.

1. CH₃CH₂MgBr (xs), ether
2. H₃O⁺

b.

1. CrO₃, pyridine, CH₂Cl₂
2. PhNH₂, TsOH, Δ

c.

CH₃CH₂CH₂SH
NEt₃, CH₂Cl₂

d.

PhNH₂, Δ

e.

1. SOCl₂
2. CH₃CH₂NH₂
3. LiAlH₄, ether
4. H₃O⁺

f.

1. CH₃CH₂CH₂OH
2. TsOH, Δ

g.

H₃O⁺, Δ

h.

1. O₃
2. H₂O₂

i.

1. SOCl₂
2. Li₂Cu(CN)(CH₃)₂

j.

Br₂, H₂O, pH 5–6
(show Haworth projection of the lactone form)

k.

1. NH₂OH
2. P₄O₁₀, Δ

l.

1. KMnO₄, OH⁻
2. CH₃CH₂OH, TsOH, Δ

21.30. Show how each of the following compounds can be converted to benzoic acid:

a. Benzyl alcohol

b. Bromobenzene

c. Ethylbenzene

d. *N*-Methylbenzamide

e. Benzonitrile

f. Styrene

21.31. By identifying the absorption bands that correspond to the principal functional group, match the IR spectra **31A–31C** with the appropriate type of acid derivative in the following list.

    a. RCN   b. RCOCl   c. RCOOH   d. RCOOR′   e. RCONH$_2$

31A.

31B.

31C.

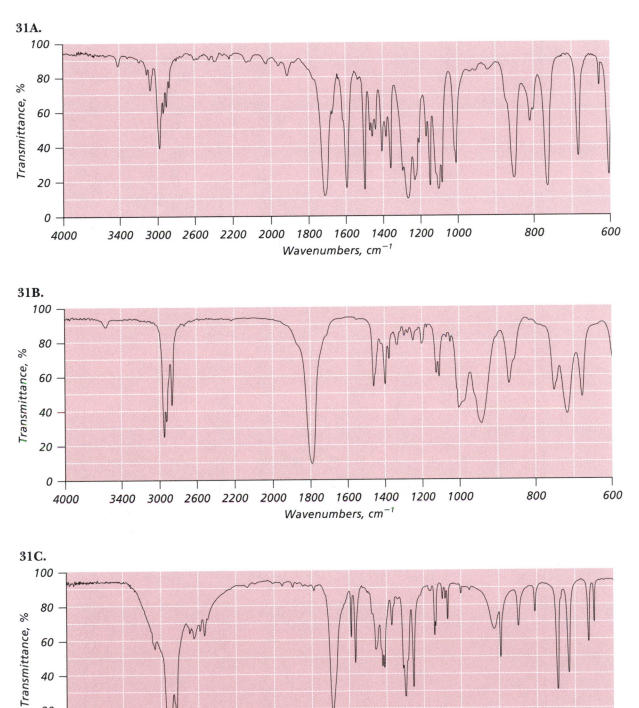

21.32. By identifying the resonances that correspond to the principal functional group, match the ^{13}C NMR spectra **32A–32C** with the appropriate type of acid derivative in the following list.

a. RCOOH   b. RCOOR′   c. RCONH$_2$ d. RCN

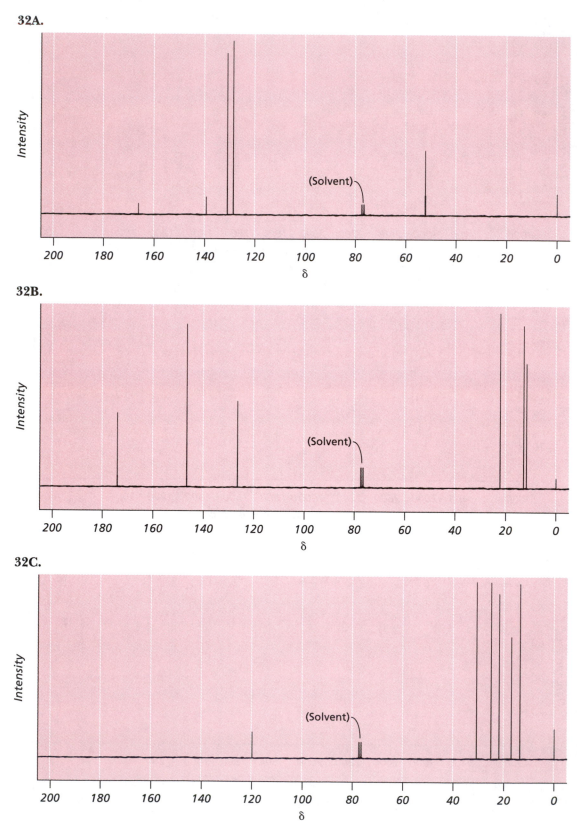

21.33. Draw the structure of the product expected from each of the following reduction reactions. If no reaction occurs, write N.R. Indicate the stereochemistry of the product where applicable.

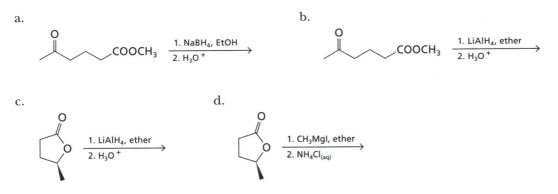

a.   b.

c.   d.

21.34. Show three different ways to make 1-phenyl-1-butanone from an acid chloride by routes that use an organometallic reagent or electrophilic aromatic substitution.

21.35. Show how you would prepare each of the following carboxylic acids from any organic compounds that have six or fewer carbon atoms, making use of an organometallic reagent. You may also use other reagents and solvents.

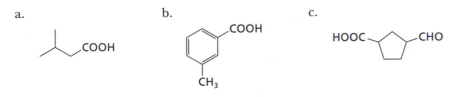

a.   b.   c.

21.36. The Wohl degradation is a procedure for converting an aldose to its homologue with one less carbon atom. You have seen all of these steps in different sections of this text—here they are collected into one reaction sequence. Propose a mechanism for each step.

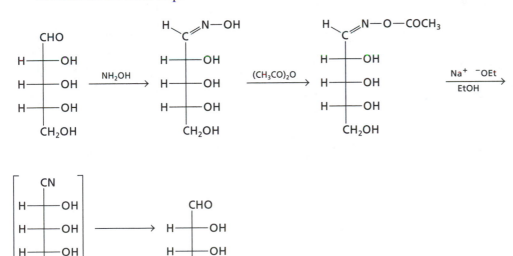

Not isolated
(why not?)

21.37. What carbohydrate is produced if the Wohl degradation (Exercise 21.36) is carried out starting with each of the following aldoses?

    a. D-Mannose   b. D-Ribose   c. L-Arabinose

21.38. What carbohydrate could be used in the Wohl degradation (Exercise 21.36) to prepare each of the following aldoses?

    a. D-Glyceraldehyde   b. L-Threose   c. D-Erythrose

21.39. Identify each of the following compounds, which are carboxylic acids or derivatives, from their IR and ^{1}H NMR spectra.

a. $C_4H_6O_2$

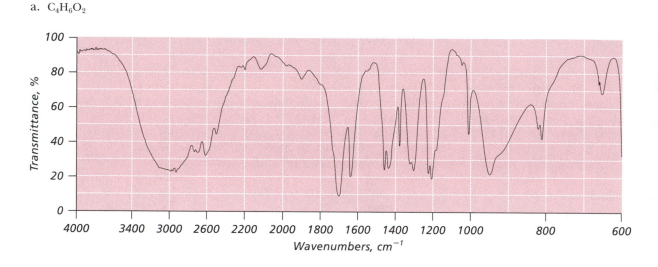

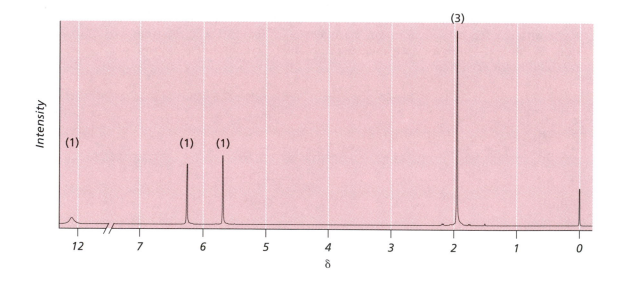

b. $C_6H_{11}ClO_2$: All of the J-values = 7 Hz.

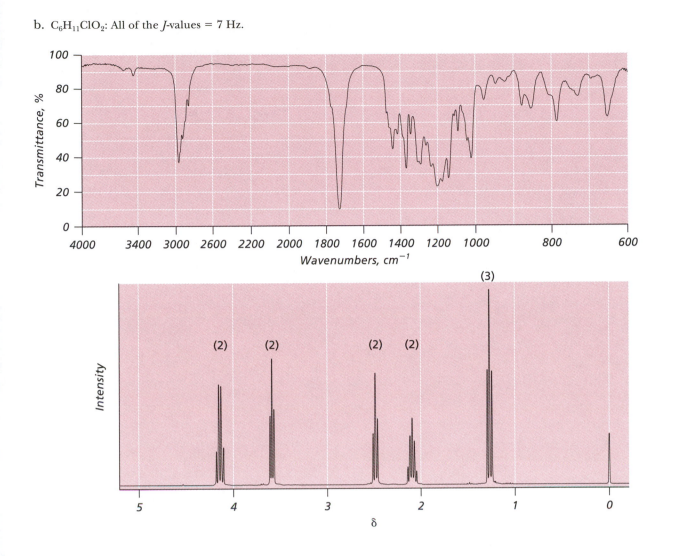

21.40. Anhydrides can be used instead of acid chlorides as reactants in the Friedel–Crafts acylation reaction (Section 17.2e). When a cyclic anhydride is used, the product is a keto acid. Propose a reasonable mechanism for the following reaction between benzene and succinic anhydride:

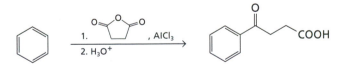

chapter 22

THE ACID–BASE CHEMISTRY OF CARBONYL COMPOUNDS

22.1 ACIDITY OF CARBONYL COMPOUNDS

22.2 ENOLS AND ENOLATE IONS

22.3 REACTIONS OF ENOLATE IONS

22.4 DICARBONYL COMPOUNDS

22.5 REACTIONS OF AMIDE AND IMIDE IONS

CHAPTER SUMMARY

Acid–base chemistry is a fundamental part of reaction mechanisms, and you have seen how factors such as electronegativity, hybridization, and resonance effects influence the acidity of functional groups (Section 5.2). This chapter describes the chemistry of anions that are stabilized primarily by delocalization of electrons with a carbonyl group, which is related to the acidity of protons α to the carbonyl group of aldehydes, ketones, and esters. Deprotonation at the position adjacent to a carbonyl group generates a resonance-stabilized *enolate ion*. For example, from a ketone,

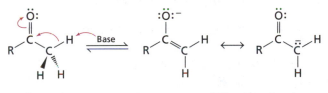

A ketone   **Resonance-stabilized enolate ion**

An enolate ion reacts like any nucleophile, although its high basicity sometimes causes troublesome side reactions.

Anions associated with a nitrogen atom adjacent to a carbonyl group are also discussed in this chapter. Some of these species undergo rearrangement reactions.

22.1 ACIDITY OF CARBONYL COMPOUNDS

22.1a THE ACIDITY OF CARBOXYLIC, SULFONIC, AND PHOSPHORIC ACIDS RESULTS FROM RESONANCE STABILIZATION OF THE CORRESPONDING ANION

For the dissociation of an acid HA into a proton and its conjugate base A⁻, the dissociation constant, K_a, and its related quantity pK_a are defined as follows:

$$HA_{(aq)} + H_2O \rightleftharpoons A^-_{(aq)} + H_3O^+_{(aq)} \qquad (22.1)$$

$$K_a = \frac{[A^-_{(aq)}][H_3O^+_{(aq)}]}{[HA_{(aq)}]} \qquad pK_a = -\log K_a \qquad (22.2)$$

The stronger the acid, the lower the pK_a value. Mineral acids such as HCl have negative pK_a values, but most organic compounds have pK_a values between 0 and 50.

A carboxylic acid has the tendency to undergo acid–base reactions with many nucleophiles (Section 21.1e). The pK_a value for a typical carboxylic acid is ~ 5, and its acidity is influenced through inductive effects by electron-withdrawing or -donating substituents (Section 5.2c).

Sulfonic ($pK_a < 0$) and phosphoric ($pK_a \approx 2$) acids are two other types of organic acids you have already encountered. Resonance stabilization of their conjugate bases accounts for their tendency to transfer a proton to a base that is present in solution, and perhaps the most important acid–base reaction is the one in which water acts as a base.

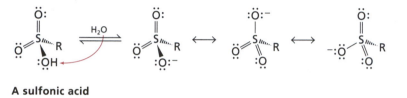

A sulfonic acid

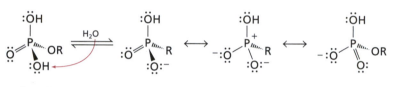

A phosphoric acid

A phosphoric acid can lose a second proton when its concentration is low and the pH of the solution is near 7, for example, under physiological conditions.

EXERCISE 22.1

Draw appropriate resonance structures for the dianion of a phosphoric acid formed by the dissociation of two protons ($RO–PO_3^{2-}$).

You may have noticed that in the biochemical processes shown in this text, carboxylic acids have been represented and referred to by the structures and names of their conjugate bases, which are the principal forms that exist under physiological conditions.

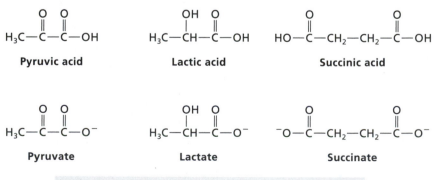

Predominant forms in water under physiological conditions

Phosphoric acid derivatives of carbohydrates and the nucleic acids (Chapter 28) exist predominantly in their phosphate ion forms under physiological conditions as well.

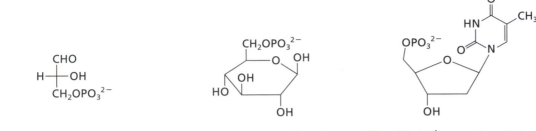

D-Glyceraldehyde-3-phosphate   **β-D-Glucopyranose-6-phosphate**   **Thymidine-5′-monophosphate**

22.1b INDUCTIVE EFFECTS INFLUENCE THE ACID STRENGTHS OF CARBOXYLIC ACIDS

As described in Section 5.2c, acetic acid, with a pK_a value of 4.75, provides the reference mark for a "typical" carboxylic acid. Substitution at the carbon atom adjacent to the carbonyl group places electron-donating or -withdrawing groups near the acidic proton, and these substituents affect the compound's acidity through inductive effects. Recall that inductive effects become less important as the distance between the substituent and the proton increases.

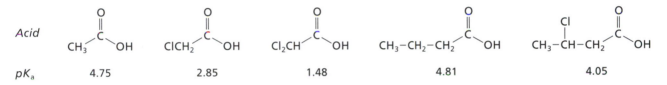

Acid					
pK_a	4.75	2.85	1.48	4.81	4.05

A carboxylic acid group attached to a benzene ring is influenced less than you might expect. The primary stabilization of the carboxylate ion results from delocalization of electrons through the CO_2 group; the aromatic π system has little additional influence. As with the aliphatic carboxylic acids, the acidity of benzoic acid derivatives can be rationalized mainly on the basis of inductive effects.

Benzene itself functions as a weak electron-withdrawing group, so benzoic acid is more acidic than acetic acid. Electron-withdrawing groups attached to the benzene ring increase the acidity even more. For example, *o*-, *m*-, and *p*-nitrobenzoic acid are all stronger acids than benzoic acid. As expected, electron-donating groups (that is, the activating groups shown in Fig. 17.10) such as methyl or OH make the acid weaker. Table 22.1 lists the pK_a values of benzoic acid and some derivatives.

Benzoic acid derivatives with a substituent at the 2-position are more acidic than benzoic acid, no matter whether the group attached to the ring is electron withdrawing or electron donating. This influence, known as the *ortho effect*, undoubtedly results from steric interactions between the substituent and the carboxylate group.

A compound with two carboxy groups is known as a *diprotic acid*. Because of the electron-withdrawing effect of the neighboring carbonyl group, dissociation of the first proton from a dicarboxylic acid is greater than for acetic acid itself, as summarized in Table 22.2. As the carboxylic acid groups are moved farther apart, the effect of the second COOH group on the dissociation of the first proton drops dramatically, just as you saw for substituents like chlorine. Dissociation of the second proton occurs at a higher pK_a value: The negative charge of the first carboxylate group renders the second COOH group less acidic as a result of inductive and solvation effects.

Table 22.1 The pK_a values of benzoic acid and its derivatives.

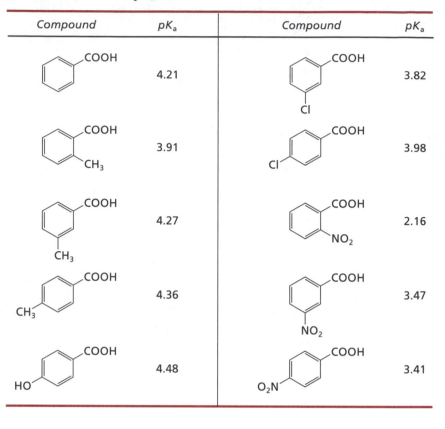

Compound	pKₐ	Compound	pKₐ
COOH	4.21	COOH (Cl meta)	3.82
COOH (CH₃ ortho)	3.91	COOH (Cl para)	3.98
COOH (CH₃ meta)	4.27	COOH (NO₂ ortho)	2.16
COOH (CH₃ para)	4.36	COOH (NO₂ meta)	3.47
COOH (HO para)	4.48	COOH (O₂N para)	3.41

EXERCISE 22.2

The dissociation of the first proton in oxalic and malonic acids might also be enhanced by formation of a strong hydrogen bond in the monoanion. Draw structures of these species and explain why they should be more stabilized than the corresponding monoanions generated from the other diacids shown in Table 22.2.

22.1c A CARBOXAMIDE IS MORE THAN AN AMINE

If you shift attention from carboxylic acids to their amide derivatives (carboxamide), you would find that the same factors affecting the acidity of carboxylic acids—resonance and inductive effects—are equally important when comparing the acid–base properties of amides with those of amines. An amine has a pK_a value between 35 and 40, so it can be deprotonated only with a very strong base like an organolithium reagent. For example, the nonnucleophilic base LDA is prepared by deprotonating diisopropylamine. Lithium diisopropylamide has many uses in preparing carbanions that you will encounter throughout this chapter, so make special note of this reaction. (Note that the word "amide" can refer both to the conjugate base of an amine as well as to the carboxylic acid derivative containing nitrogen; the context of the discussion will usually make clear which substance is involved. The word carboxamide will be used in many cases to prevent confusion.)

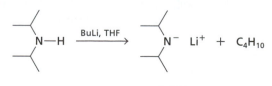

(LDA)

Table 22.2   The pK_a values of dicarboxylic acids.

Compound	pK_1	pK_2
Oxalic acid	1.23	4.19
Malonic acid	2.83	5.69
Succinic acid	4.16	5.61
Glutaric acid	4.34	5.41
Adipic acid	4.43	5.41

The pK_a value of a carboxylic acid amide is ~ 17. It can be deprotonated by bases much weaker than butyllithium.

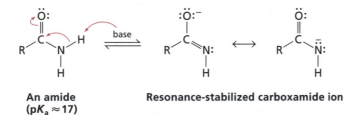

An amide
($pK_a \approx 17$)   **Resonance-stabilized carboxamide ion**

In fact, the pK_a value of an amide is not much different from that of water, so hydroxide ion deprotonates an amide to a certain extent. Recall that an amide is the acid derivative most difficult to hydrolyze under basic conditions (Section 21.5a). Besides the presence of the strong-base leaving group, *the resonance-stabilized amide ion is less susceptible to reaction with the nucleophilic OH⁻ ion than the carbonyl group of other acid derivatives.*

$$CH_3-\overset{O}{\overset{\|}{C}}-NH_2 \quad \overset{OH^-}{\rightleftharpoons} \quad CH_3-\overset{O}{\overset{\|}{C}}\overset{}{\underset{NH}{|}} - \quad + \quad H_2O$$

Resonance stabilization decreases the electrophilicity of the carbon atom.

Because this acid–base reaction is an equilibrium process, however, there is enough of the neutral carboxamide in solution that nucleophilic addition to the carbonyl group *does* occur. Therefore, slow hydrolysis of an amide bond will take place under basic conditions.

EXERCISE 22.3

N,N-Dimethylbutanamide is hydrolyzed faster in aqueous sodium hydroxide solution than is butanamide itself. Why?

22.1d A KETONE IS MORE ACIDIC THAN AN ALKANE

With an understanding of the acid properties of carboxylic acids and their amide derivatives, consider now the topic that will occupy our attention for much of this chapter: the acidity of C–H bonds adjacent to a carbonyl group. You might expect that a methyl group adjacent to a carbonyl group is similar to the OH group in a carboxylic acid or the NH group in a carboxylic acid amide. Indeed, addition of a strong base removes a proton, generating an **enolate ion,** for which we can draw resonance forms that are similar to those that we draw for the carboxylate or carboxamide ion.

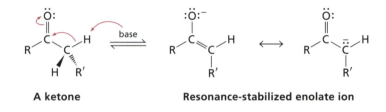

A ketone	**Resonance-stabilized enolate ion**

The proton to be removed has to be aligned with the π bond of the carbonyl group so that rehybridization of the α-carbon atom leads to delocalization of the electron pair among the three atoms.

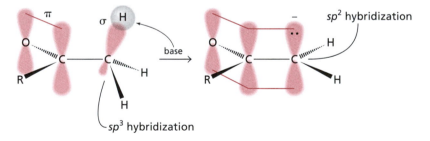

Stabilization of the negative charge by delocalization provides a way to rationalize why the pK_a value of a proton alpha to a ketone carbonyl group is ~ 20, whereas the pK_a values for C–H bonds in alkanes are closer to 50. We will see later that the magnitude of acidity varies among the different carbonyl-containing substances.

Table 23.3 summarizes the scale of pK_a values observed for the different X–H bonds. *The most important point to remember is that a carbonyl group makes protons on the neighboring atom more acidic than in the hydrocarbon analogue.*

Table 22.3 Comparison of pK_a values for carbonyl compounds and their
methylene analogues.

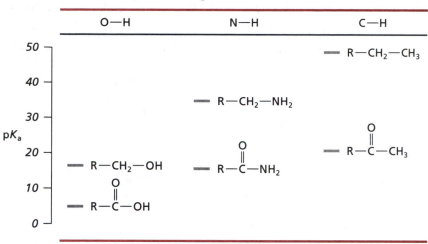

22.2   ENOLS AND ENOLATE IONS

22.2a AN ENOLATE ION IS THE CONJUGATE BASE OF AN ENOL, THE TAUTOMERIC FORM OF A CARBONYL COMPOUND

The word "enol" derives from the two suffixes *ene* and *ol*, meaning an alkene–alcohol. An **enol,** also called a *vinyl alcohol,* is normally unstable. You have encountered enols in several places already; a prominent example is the product formed by hydration of an alkyne (Section 9.2b):

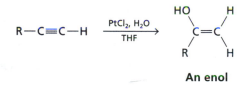

An enol

Recall, however, that this enol product does not remain as such. Instead, it undergoes a process called *tautomerism* (Section 9.2b) to form the corresponding carbonyl compound.

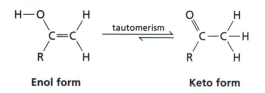

Tautomerism is driven by formation of the stronger C=O double bond and C–H bond. The keto form is thermodynamically more stable than its enol form in most instances.

EXERCISE 22.4

The tautomerism process shown below favors formation of the enol form. Why is the enol form more stable in this case?

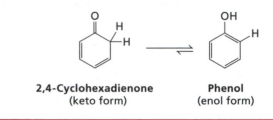

2,4-Cyclohexadienone
(keto form)

Phenol
(enol form)

An enol is the rearranged form of a carbonyl compound that has a proton on its *alpha* carbon atom. Looking at the mechanism of tautomerism, which is catalyzed either by acid or base, you can understand the acid–base relationships between carbonyls, enols, and enolate ions.

In aqueous acid, tautomerism takes place by protonation of the carbonyl oxygen atom (Section 18.1c). In Step 2, water acts as a base and removes one of the *neighboring* protons, generating the enol.

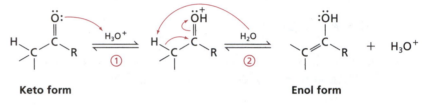

Keto form   **Enol form**

Under basic conditions, an enolate ion is formed by deprotonation of the α-carbon atom. The resulting ion removes a proton from water, generating the enol.

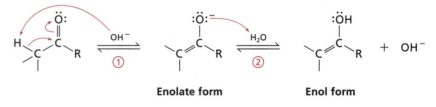

Enolate form   **Enol form**

This reaction directly above shows that *an enolate ion is the conjugate base of an enol.*

Realize, however, that aqueous conditions are not the best for making an enolate ion because only small amounts can be generated in the presence of the more acidic water molecule. Nevertheless, enolate ions can be made and do react in water. We summarize the relationships between a ketone, an enol, and an enolate ion in Figure 22.1.

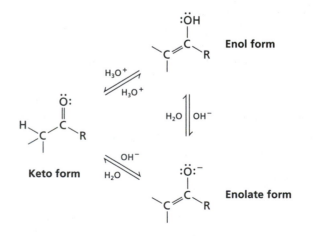

Figure 22.1

Reactions between the keto, enol, and enolate forms in acid and base solution

22.2b ISOMERIZATION REACTIONS OF CARBONYL COMPOUNDS OCCUR VIA ENOL AND ENOLATE ION DERIVATIVES

Enols and enolates are important intermediates in *isomerization reactions* (the interconversion of isomers) of aldehydes and ketones, especially in biological processes. **Carbonyl transposition**—the movement of a carbonyl group from one carbon atom to a neighboring carbon atom—constitutes one type of isomerization process that is especially important in carbohydrate metabolism. For example, in the second step of glycolysis, the energy-producing pathway for the metabolism of glucose, D-glucose-6-phosphate is converted to D-fructose-6-phosphate. The enzyme that catalyzes this transformation uses acidic groups in its active site to generate an intermediate **enediol** (a double bond with *two* OH groups attached) by acid-catalyzed tautomerism. Because the intermediate has two OH groups, tautomerism can occur in either direction. Ultimately, tautomerism regenerates the carbonyl double bond and results in isomerization of D-glucose to D-fructose (and vice versa).

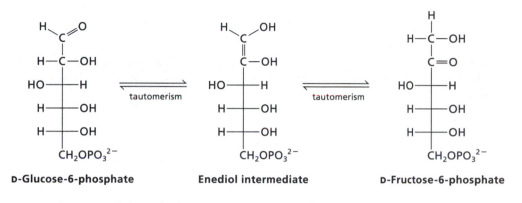

D-Glucose-6-phosphate   **Enediol intermediate**   **D-Fructose-6-phosphate**

EXERCISE 22.5

During glycolysis, the enzyme-catalyzed isomerization of glyceraldehyde-3-phosphate produces dihydroxyacetone phosphate. Propose a mechanism for this process in aqueous solution, catalyzed by acid.

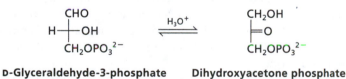

D-Glyceraldehyde-3-phosphate   **Dihydroxyacetone phosphate**

Another type of isomerization process is *racemization,* which occurs when a stereogenic center is adjacent to a carbonyl group. In the following example, an enantiomerically pure substance forms an achiral enol intermediate when treated with acid or base. Return to the keto form takes place by protonation at either face of the double bond (top or bottom in the following structures), eventually creating a racemic mixture.

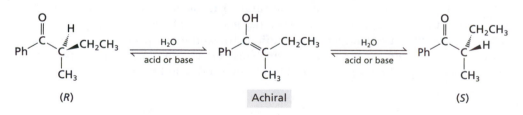

Racemization of stereogenic centers adjacent to a carbonyl group is often facile, and it can be troublesome too. Recall the case of thalidomide in which inclusion of both enantiomers in the drug formulation led to birth defects in the offspring of some women who took the substance during pregnancy (Section 16.1b). Even if the pure (*R*) form had been given to patients, the mutagenic (*S*) isomer could be formed by racemization under physiological conditions. This isomerization occurs readily because the stereogenic center is adjacent to a carbonyl group.

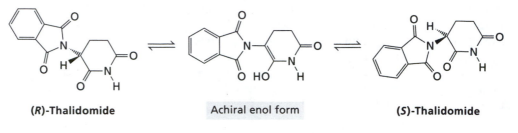

(R)-Thalidomide   Achiral enol form   **(S)-Thalidomide**

EXERCISE 22.6

Amino acids constitute another class of biomolecules that can undergo racemization. Propose a mechanism for racemization of the ethyl ester of phenylalanine under basic conditions.

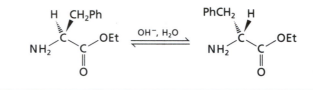

22.2c A KETONE CAN BE HALOGENATED AT THE POSITION ADJACENT TO ITS CARBONYL GROUP

If enol and enolate formation were simply reversible processes that allowed protons to be moved about within a molecule, they would have limited utility for synthetic transformations. Because an enolate ion is nucleophilic, however, it will react with electrophiles. An electrophilic process with which you are familiar is halogenation (Section 9.1d). Sure enough, an enolate ion reacts readily with chlorine, bromine, and iodine to yield α-substituted products.

When an aqueous solution of a ketone is treated with a halogen in the presence of base, the first step to occur is formation of the enolate ion. The nucleophilic carbon atom reacts with the halogen to produce the α-haloketone and halide ion.

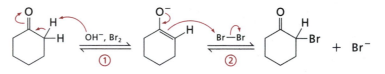

The proton attached to the newly substituted carbon atom is now more acidic than the ones in the starting material, however, because the halogen atom exerts an inductive electron-withdrawing effect (cf. Section 5.2b), so the first product will react again with base to generate a second enolate ion. This enolate ion reacts with another molecule of the halogen that is present and forms an α,α-dihalo ketone.

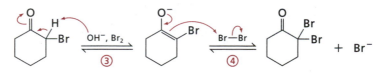

The increased reactivity of the monobromo compound toward base means that selective *mono*halogenation of a ketone is not practical under basic conditions. This reaction, however, can be used to advantage in the **haloform reaction,** which transforms a methyl ketone to a carboxylate ion with one less carbon atom. The methyl carbon atom is converted to chloroform, bromoform, or iodoform, depending on which halogen is used. For example,

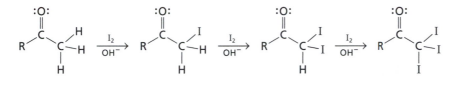

When the ketone is treated with an excess of base and halogen (iodine in the following example), the α,α,α-trisubstituted product is generated by successive steps involving enolate ion formation and reaction with the halogen molecule.

EXERCISE 22.7

Propose a mechanism for the steps shown directly above that generate the α,α,α-triiodoketone.

In the next step of the mechanism, the nucleophilic hydroxide ion *adds* to the carbonyl group. A tetrahedral intermediate is generated (Section 21.1c), which subsequently expels the trihalomethyl carbanion, a reasonable leaving group under the circumstances. (Recall that chloroform can be deprotonated using hydroxide ion; see Section 9.5b.)

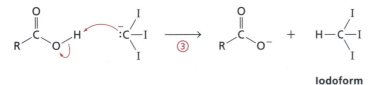

A final acid–base reaction between these initial products yields a carboxylate ion and trihalomethane, or haloform, from which the reaction derives its name. In this example, the halogenated product is iodoform.

Iodoform

The carboxylic acid itself can be obtained by adding aqueous acid during workup.

In acid solution, a ketone reacts with the halogens via its enol derivative. In fact, the acid-catalyzed process is *autocatalytic,* which means that the reaction produces its own catalyst, even if none is added initially.

The reaction begins with acid-catalyzed enol formation. The resulting enol undergoes electrophilic addition of halogen to its double bond in Step 3 by the same mechanism that you learned in Section 9.1d. The last step is an acid–base reaction that regenerates the carbonyl group, and produces more H_3O^+ than was present at the start.

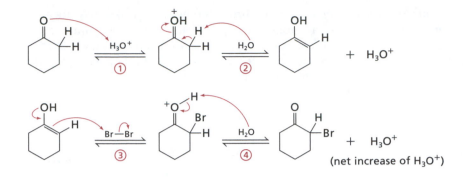

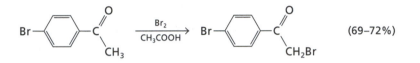

Unlike enolization under base conditions, the α-haloketone undergoes enolization *more* slowly in acid than the unsubstituted ketone does. Therefore, halogenation can be more easily controlled, and monosubstitution is possible. For example,

(69–72%)

22.2d A Strong Base Facilitates Complete Conversion of a Carbonyl Compound to Its Enolate Ion Derivative

A halogen cation is only one type of electrophile with which a nucleophilic enolate ion can react. To perform reactions with other electrophiles, stable solutions of an enolate ion must be prepared, and these solutions must be free of species that can undergo competing reactions. Because the pK_a value of a proton adjacent to a carbonyl group is ~ 20–25 (Table 5.1), a strong base is needed to remove a proton completely from the α-position of a ketone, ester, or other carbonyl-containing substance. Even though the previous sections have illustrated the preparation of enolate ions in aqueous solution, the combination of potassium hydride or LDA (Section 21.1c) in THF is more commonly used to generate a solution of a carbonyl enolate ion.

Lithium diisopropylamide is an especially useful base because it is soluble in THF, as are most lithium enolate ions. The nitrogen atom in LDA is also so hindered that it is not likely to participate in nucleophilic addition to the carbonyl group. Potassium hydride has the advantage that the only byproduct is hydrogen gas. Although its reactions can be carried out at ambient temperature, low temperatures are sometimes employed to prevent self-condensation processes that can complicate subsequent transformations.

Enolate ions formed using LDA, KH, or a related reagent are stable in THF solution, and we normally draw their structures with the negative charge on the oxygen atom. There are two reasons for this. First, oxygen is more electronegative than carbon, so a resonance form that places the charge on oxygen should contribute more toward the distribution of electron density in the ion. Second, small cations like lithium and potassium have a much higher affinity for oxygen than for carbon and are likely to be associated with an oxygen anion.

22.2e AN UNSYMMETRIC KETONE CAN FORM TWO DIFFERENT ENOLATE IONS

Cyclohexanone is symmetrical, and enolate ion formation is straightforward no matter the base employed. But a compound such as 2-methylcyclohexanone has protons that can be removed from either side of the carbonyl group, and they are in different environments. At first glance, you might expect to form approximately equal amounts of the two enolate ions when base is added. The specific reaction conditions can influence the formation of one enolate versus the other, however.

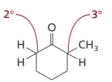

2-Methylcyclohexanone

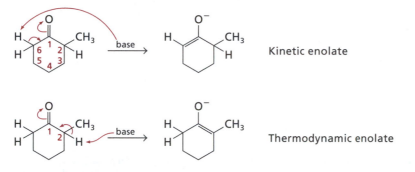

Kinetic enolate

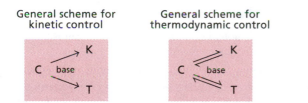

Thermodynamic enolate

The protons attached to the 6-position of 2-methylcyclohexanone are less hindered than those attached at the 2-position. They are therefore removed faster if a hindered base like LDA is used, and that product is the **kinetic enolate.**

The double bonds that are formed in the enolate ions are substituted differently in the two derivatives, so their stabilities should differ. As you learned in Section 8.1c, a tetrasubstituted double bond ($R_2C=CR_2$) is more stable than a trisubstituted one ($R_2C=CHR$), so the more highly substituted derivative is the **thermodynamic enolate** and is expected to predominate if the mixture is allowed to equilibrate.

The topic of kinetic versus thermodynamic control is one that you have seen previously (Section 10.3b). For the formation of an enolate ion from a carbonyl precursor, the processes are summarized as follows:

General scheme for   General scheme for
kinetic control   thermodynamic control

Key: C = Carbonyl compound
       K = Kinetic enolate
       T = Thermodynamic enolate

A kinetic process is favored under nonequilibrium conditions, taking advantage of the differential rates of the competing pathways. A thermodynamic process is favored when the reactions (particularly the pathway leading to the kinetic product) are equilibria, so the energy of the system eventually settles with formation of the most stable species. For enolate formation,

- The *kinetic enolate* is formed at low temperatures using a strong, hindered base.

- The *thermodynamic enolate* is formed at higher temperatures using excess carbonyl substrate.

To justify these experimental results, it is helpful to look at a reaction coordinate diagram for the two pathways. As illustrated in Figure 22.2, the free energy of activation to remove a proton from the 6-position is lower, but the enolate ion formed by abstracting the proton from the 2-position is more stable. Therefore,

- Low temperatures (commonly −78°C) provide only enough energy to cross the barrier along the pathway leading to formation of the kinetic enolate, so the thermodynamic enolate will be formed very slowly. Use of a hindered base

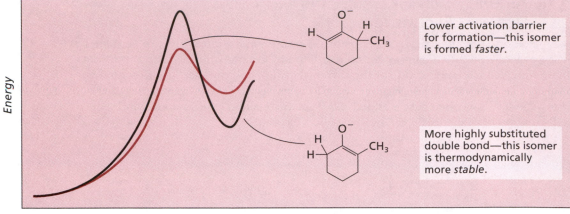

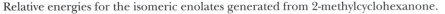

Figure 22.2
Relative energies for the isomeric enolates generated from 2-methylcyclohexanone.

ensures that the more accessible protons will be removed faster (2° vs. 3° for 2-methylcyclohexanone), and use of a strong base means that little of the carbonyl precursor will remain when the enolate ion has been formed.

• Higher temperatures mean that sufficient energy is available for the reaction to occur via the pathway with the higher free energy of activation (the one yielding the thermodynamic enolate), as well as to cross the barrier in each direction of the pathway that forms the kinetic enolate. But even if the energy is available, there has to be a way to interconvert the enolate ions. The use of excess ketone ensures that a mechanism exists for equilibration to take place:

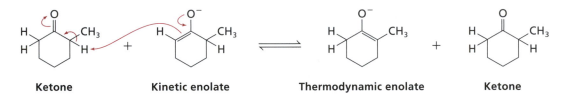

So how do these data relate to the information on enolate ions presented in Section 22.2d? First, they confirm that LDA is an excellent base for making kinetic enolate ions from unsymmetrical ketones: *LDA is both a hindered and strong base.* At −78°C formation of the kinetic enolate is favored. Potassium hydride is a strong base, but hydride ion is small, so it can deprotonate a more hindered carbon atom. Performing the reaction at room temperature will lead to formation of the thermodynamic product. The following equations illustrate these processes. Note that when preparing the thermodynamic enolate ion, some kinetic enolate is also likely to be present; the exact proportion depends on the stability of each species. For example, in the second equation below, the ratio is nearly 1:1.

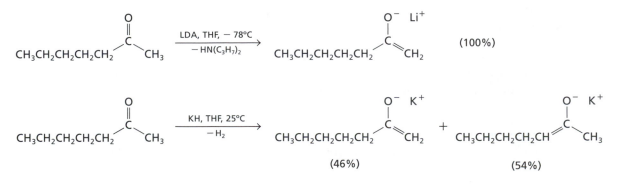

The lithium salt of the thermodynamic enolate ion is sometimes desirable if you want to take advantage of its solubility properties. In those cases, the kinetic enolate ion can be made at low temperature and allowed to warm to room temperature with a slight excess of the starting ketone. For example,

Aldehydes can also form enolate ions. Their formation does not suffer from the regiochemical problem associated with making a ketone enolate because there is only one α-position in an aldehyde. The carbonyl group of an aldehyde is unhindered, however, so an aldehyde enolate often undergoes self-condensation reactions (Section 23.1a). For this reason, preparation of aldehyde enolate ions is not generally practical.

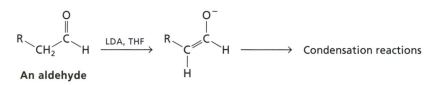

EXERCISE 22.8

Write equations showing how you would prepare the kinetic and thermodynamic enolates of the following ketones.

a.   b.

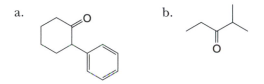

22.3   REACTIONS OF ENOLATE IONS

22.3a   A KETONE ENOLATE REACTS WITH AN ALKYL HALIDE VIA AN S$_N$2 PROCESS

A ketone enolate ion is a nucleophile, so it can react with the electrophilic carbon atom of an alkyl halide, which leads to formation of an alkylated ketone. The mechanism of the alkylation step shows the electrons moving from oxygen, regenerating the carbonyl group, while the electrons in the double bond of the enolate ion react with the electrophilic carbon atom of alkyl halide, displacing the leaving group. For example, the enolate derivative of cyclohexanone, made from its reaction with LDA in THF, reacts with methyl iodide to yield 2-methylcyclohexanone.

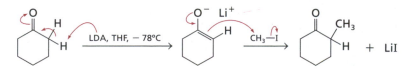

This alkylation reaction is a typical S$_N$2 process (Section 6.3a), and occurs readily when a primary alkyl halide, tosylate, or mesylate is employed as the electrophilic partner. With a secondary or tertiary substrate, elimination is a major side reaction (Section 8.3a) because the enolate ion is a strong base.

Note that alkylation occurs at the carbon atom α to the carbonyl group and not at the oxygen atom. An enolate ion has two resonance forms, each of which has an octet on each non-hydrogen atom. The carbon center is more basic (Section 5.2b), hence more nucleophilic (Section 6.1c) than the oxygen atom.

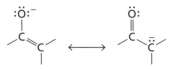

Unfortunately, the overall alkylation process is not as simple as indicated above. A major problem is polyalkylation when the alkyl group is small. The product of alkylation is a neutral ketone, *which itself has acidic protons*. Therefore, the enolate ion that is already in solution can react with the initial product by an acid–base process to form a new enolate ion and regenerate the starting ketone.

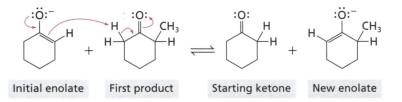

Initial enolate   First product   Starting ketone   New enolate

Now, this second enolate can react with the alkyl halide to form a dialkyl product.

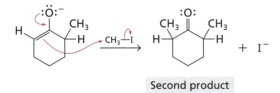

Second product

This dialkyl compound also has acidic protons, so the process can be repeated, giving rise to a mixture of compounds when the reaction is done.

Even though polyalkylation is a complication, and high yields of a pure, single product may be difficult to obtain, there are still many examples in which alkylation occurs reasonably well. Some examples follow. We can assume prudently that in any synthetic scheme in which an enolate ion is treated with an alkyl halide to prepare an alkylated product, more than one compound may be obtained.

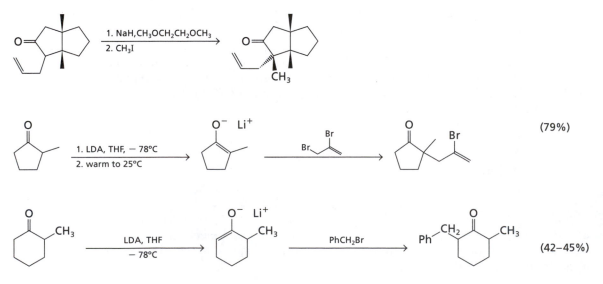

EXAMPLE 22.1

Show how you would prepare the following compound using an alkylation procedure.

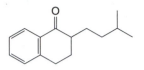

Begin with a retrosynthesis by breaking the bond between the carbon atoms that are α and β to the carbonyl group. In cyclic compounds, this disconnection is usually made outside of the ring.

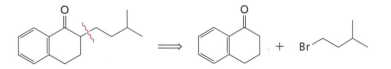

With the starting compounds identified, the synthesis starts with formation of the enolate ion and is followed by reaction with the alkyl halide.

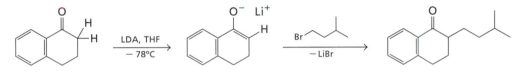

EXERCISE 22.9

Show how you would prepare the following compounds using an alkylation procedure. Start with organic compounds having eight or fewer carbons atoms.

a.   b.

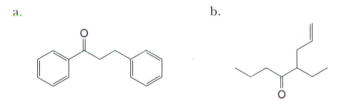

Alkylation of a ketone enolate is not the only way to prepare substituted ketones. Reactive alkyl halides (methyl, allylic, and benzylic) and α–halo carbonyl compounds react with enamines (Section 20.3b), and the three-step procedure of enamine formation, alkylation, and hydrolysis provides another route to alkyl derivatives.

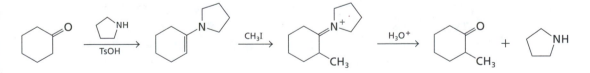

22.3b AN ENOLATE ION CAN BE USED TO ATTACH THE PHENYLSELENIUM GROUP, WHICH IS USED TO PREPARE α,β-UNSATURATED CARBONYL COMPOUNDS

Alkyl halides are not the only electrophilic reagents that react with enolate ions. The phenylselenium ion, supplied as PhSeCl, reacts readily with enolate ions and gives high yields of monosubstituted products by the following pathway:

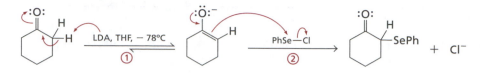

The synthetic utility of this procedure derives from the value of the phenylselenide product, which will eliminate phenylselenic acid after oxidation to its phenylselenoxide derivative.

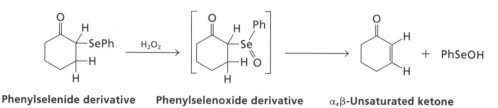

Phenylselenide derivative   **Phenylselenoxide derivative**   **α,β-Unsaturated ketone**

Elimination of PhSeOH occurs spontaneously at around room temperature to create a double bond in conjugation with the carbonyl group, and the phenylselenoxide intermediate is not actually isolated (as denoted by brackets in the preceding equation). The mechanism of this elimination process is portrayed with concerted movement of electrons via a cyclic transition state, leading to removal of a hydrogen atom from the carbon atom beta to the carbonyl group.

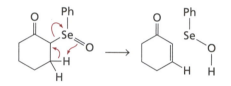

    The synthetic applications of this reaction are many, and this three-step procedure is one of the best ways to prepare α,β-unsaturated ketones (Section 24.1a). For example,

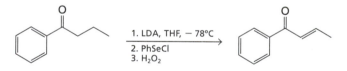

Table 22.4 summarizes the reactions that ketone enols and enolates undergo.

EXERCISE 22.10

Draw the structures of the products formed at each step of the preceding reaction sequence.

EXERCISE 22.11

Phenylsulfuryl chloride (PhSCl) and diphenyl disulfide (PhSSPh) react with an enolate ion the same way that PhSeCl reacts. Propose a mechanism for each step of the following sequence:

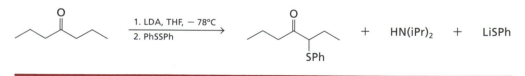

Table 22.4 A summary of the reactions between ketone enols or enolates
and electrophiles

Electrophile	Conditions	Summary (see text for details)
H^+ (HX)	Aqueous acid or base	Enolization; tautomerism; racemization of a chiral ketone
	Preformed enolate[a]	Protonation of an enolate ion regenerates the starting ketone. Racemization occurs if formation of the enolate destroys a chiral center.
Cl^+, Br^+, I^+ (X_2)	Aqueous acid	Monosubstitution at the *alpha* position is readily controlled.
	Aqueous base	Polysubstitution at the *alpha* position occurs; methyl ketones are cleaved to produce the carboxylate salt and haloform (CHX_3).
R^+ (RX)	Preformed enolate[a]	Primary alkyl halides react by an S_N2 process; secondary and tertiary alkyl halides give elimination products. Polyalkylation of the ketone can be a complication. Equilibration between kinetic and thermodynamic enolates can also give a mixture of products.
PhSe$^+$ (PhSeCl)	Preformed enolate[a]	Monosubstitution at the *alpha* position is readily controlled. Equilibration between kinetic and thermodynamic enolates is usually not a problem.

[a]Preformed enolate refers to the product of the ketone with a reagent like LDA in THF at −78°C.

22.3c THE ENOLATE DERIVATIVE OF AN ESTER IS A POTENT NUCLEOPHILE

A ketone is not the only carbonyl compound from which enolate ions can be made. Esters form enolate derivatives and nitriles form carbanion derivatives with similar properties. The pK_a value for the α-protons of a typical ester (p$K_a \approx 24$–25) is four to five orders of magnitude higher than that of a ketone (p$K_a \approx 20$), an effect that derives from resonance delocalization of the electrons between the two oxygen atoms of the ester. Because these electrons are already delocalized *before* deprotonation, the electron pair is less stabilized in an ester enolate than it is in a ketone enolate ion.

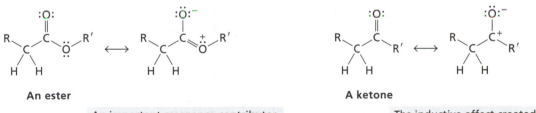

An ester

An important resonance contributor puts a positive charge on the oxygen atom rather than on the carbonyl carbon atom.

A ketone

The inductive effect created by a positive charge on the carbon atom makes the protons more acidic than those of the ester.

There is still significant stabilization relative to a hydrocarbon, however, so the acidity of an ester is much greater than that of an alkane.

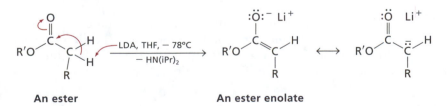

An ester   **An ester enolate**

A carbanion derived from a nitrile is also resonance stabilized, and the pK_a value of a nitrile is between 25 and 30.

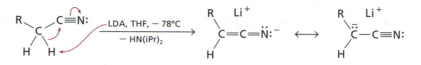

Unlike ketones, which sometimes have protons attached to both *alpha* carbon atoms, esters and nitriles have only one *alpha* carbon atom, so only one enolate ion exists for each. Furthermore, condensation reactions (Chapter 23) do not normally interfere if a strong base like LDA is used to make an ester enolate. The only ester that undergoes significant self-condensation with LDA is ethyl acetate, so *tert*-butyl acetate is often employed if a two-carbon ester is required in a synthetic procedure.

The following example illustrates the alkylation of an ester enolate. First, the enolate ion is prepared (low temperature is used to minimize possible side reactions).

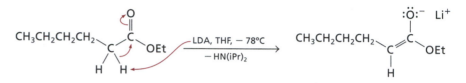

Once formed, the ester enolate ion (or nitrile carbanion) reacts with an alkyl halide by the typical S_N2 pathway and yields the alkylated product.

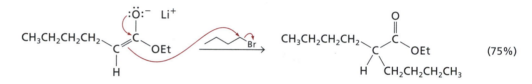

As expected for any substitution reaction that uses a strongly basic nucleophile, secondary and tertiary substrates undergo elimination. With a primary alkyl halide or epoxide, the reactions proceed in a predictable manner.

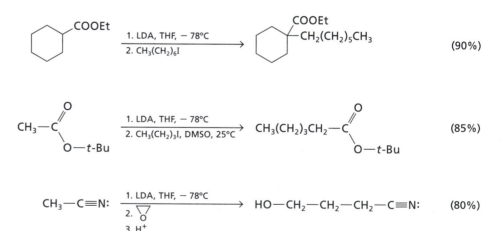

An ester enolate ion is a particularly useful nucleophile because the resulting product can be converted to many other types of molecules such as carboxylic acids (Sections 21.4c and 21.4d), amides (Section 21.4b), and alcohols (Sections 21.7c and 21.8a). Example 22.2 illustrates a specific application.

EXAMPLE 22.2

Propose a route to synthesize the following molecule that makes use of an enolate ion and starts with organic compounds with seven or fewer atoms:

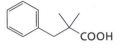

Begin by outlining the retrosynthesis. Remember first that a carboxylic acid is equivalent to an ester. To make the carbon skeleton of the ester, consider breaking the bond between carbon atoms that are α and β to the carbonyl group.

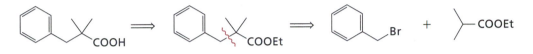

With the starting compounds identified (both have seven or fewer carbon atoms), write the steps of the synthesis; start with formation of the enolate ion and follow with the reaction between the enolate ion and the alkyl halide. Hydrolysis of the ester is required to generate the carboxylic acid product.

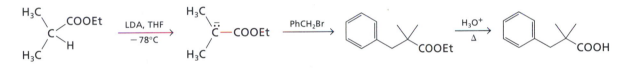

EXERCISE 22.12

A lactone can also be alkylated after formation of its enolate. What is the expected species after each step in the following scheme? Given that an enolate ion is alkylated on its less-hindered face, what is the stereochemistry of the final product? (Models may be useful here.)

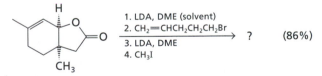

EXERCISE 22.13

Propose a route to synthesize each of the following compounds using an enolate ion and any organic compounds with seven or fewer atoms:

a.   b.

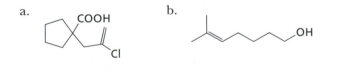

22.4   DICARBONYL COMPOUNDS

22.4a   THE PRESENCE OF TWO ELECTRON-WITHDRAWING GROUPS INCREASES THE ACIDITY OF THE ALPHA HYDROGEN ATOMS SUBSTANTIALLY

Having looked at the properties of enolate ions formed from ketones and esters, we now consider the properties of β-dicarbonyl compounds, which have *two* carbonyl groups bonded to a methylene or methine group. The most common of these are diethyl malonate and ethyl acetoacetate, which are also referred to as **active methylene compounds** because they are so easily deprotonated. Ethyl cyanoacetate, which has a carbonyl group and a nitrile group flanking a methylene group, is equally reactive.

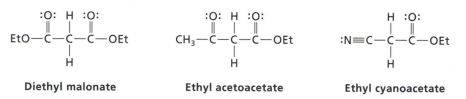

Diethyl malonate   **Ethyl acetoacetate**   **Ethyl cyanoacetate**

Molecules of this type exist partially as enols in which a proton forms a bridge between the carbonyl oxygen atoms.

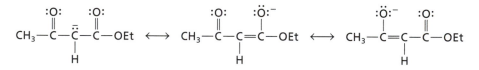

Enol form   **Dicarbonyl form**   **Enol form**

These molecules have pK_a values between 8 and 14, and their enolate derivatives are stabilized by extensive delocalization of the electron pair. For example, ethyl acetoacetate reacts with sodium ethoxide in ethanol to form the sodium salt of the enolate ion. We can draw three good resonance structures for this anion.

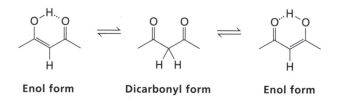

A composite representation shows that the charge is delocalized over the five atoms comprising the two carbonyl groups and the intervening methine carbon atom.

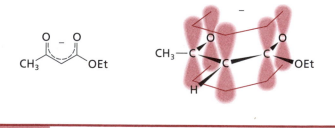

EXERCISE 22.14

Draw resonance structures for the carbanions formed from diethyl malonate and ethyl cyanoacetate.

22.4b AN ACTIVE METHYLENE COMPOUND IS READILY ALKYLATED

One advantage of using an active methylene compound is that its enolate ion must form between the two carbonyl groups, eliminating any complication that results from the involvement of kinetic and thermodynamic factors. The carbanion between the two carbonyl groups is both the kinetic and thermodynamic product.

Making and using enolates of β-dicarbonyl compounds also require less stringent control of the reaction conditions. A base much weaker than LDA is sufficient to produce the carbanion, so an alkoxide ion is often employed. The corresponding alcohol is a good solvent for the reaction, which can be conducted at room temperature or above. Because the acid–base reaction between the alkoxide and the active methylene compound is so facile, there is usually little alkoxide ion left in solution to act as a competing nucleophile.

When the carbanion is treated with an alkyl halide, an S_N2 reaction occurs to generate the alkylated product. The carbanion of an active methylene compound is a weaker base than a ketone or ester enolate, so elimination reactions are less commonplace, except with tertiary alkyl halides.

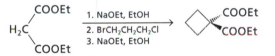

A potential side reaction is dialkylation. Formation of a dialkyl compound occurs when the monoalkyl product donates a proton to the carbanion that is still in solution, which generates a second, competing nucleophile. Dialkylation can be made to predominate if excess base is added to the active methylene compound in the presence of excess alkyl halide. For example, when an α,ω-dihaloalkane (a molecule with a halogen atom at both ends of a carbon-atom chain) is the alkylating agent and excess base is used, the product is a cycloalkane, formed by dialkylation. As is typical for ring-forming reactions, three- to six-membered rings are the easiest to prepare. Attempts to synthesize larger rings are often unsuccessful.

The overall mechanism proceeds in straightforward fashion and consists of two acid–base reactions and two S_N2 steps.

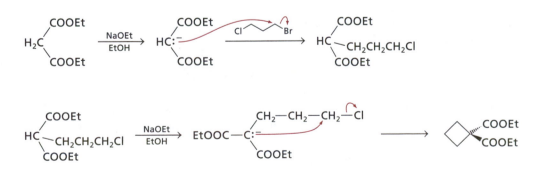

EXERCISE 22.15

Show how you would prepare each of the following compounds by making use of a β-dicarbonyl compound and an organohalide:

a.

b.

22.4c DECARBOXYLATION OCCURS READILY WHEN A CARBOXYLIC ACID GROUP IS BETA TO ANOTHER CARBONYL GROUP

The use of an active methylene compound solves many regiochemical problems in synthesis because of the specificity that can be ensured in its reactions with an alkyl halide. If the "extra" carbonyl group is not needed in the product, however, then starting with a dicarbonyl substrate just to obtain specificity would seem to make little sense. For example, to prepare 1-phenyl-3-pentanone (in the screened box, below), you might consider alkylating an enolate ion derived from 2-butanone. This scheme has all of the disadvantages inherent in making a specific enolate derivative of an unsymmetrical ketone. Employing a β-keto ester solves the regiochemistry problem, but then the product has an unwanted ester group.

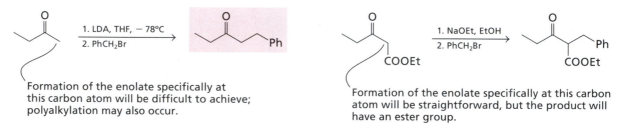

Formation of the enolate specifically at this carbon atom will be difficult to achieve; polyalkylation may also occur.

Formation of the enolate specifically at this carbon atom will be straightforward, but the product will have an ester group.

Fortunately, the ester group can be easily removed from a β-keto ester, simply by heating the compound with aqueous acid.

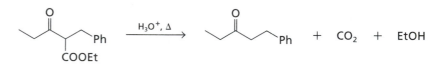

In step 1, hydrolysis of the ester group occurs as described in Section 21.4b.

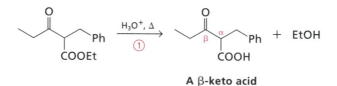

A β-keto acid

The β-keto acid formed by this hydrolysis reaction undergoes the facile loss of carbon dioxide by way of a six-membered transition state that creates an enol derivative. Tautomerism yields the ketone.

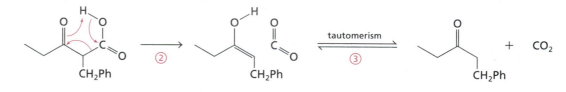

This same procedure can be performed starting with diethyl malonate. In this instance, a carboxylic acid is the final product. (The numbered steps in the following scheme corresponding to the same steps in the foregoing example.)

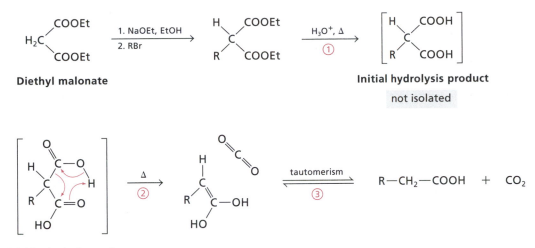

Diethyl malonate   **Initial hydrolysis product**
                                                                        not isolated

Initial hydrolysis product

Table 22.5 correlates the structures of active methylene compounds and their deprotonated forms (carbanions) with the simpler carbanions derived from acetone, 2-butanone, and ethyl acetate. The compilation is useful when you are confronted with a retrosynthesis in which a disconnection can be made between the carbon atoms that are α and β to a carbonyl group. Example 22.3 illustrates such a strategy.

Table 22.5   Active methylene compounds and their enolate equivalents.

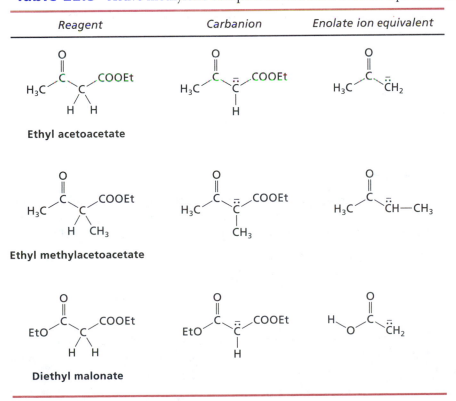

Reagent	Carbanion	Enolate ion equivalent
Ethyl acetoacetate		
Ethyl methylacetoacetate		
Diethyl malonate		

EXAMPLE 22.3

Propose a synthesis of the compound shown below from any organic compounds with
or seven fewer carbon atoms. Make use of an active methylene compound as a starting
material.

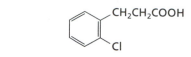

First, consider disconnections of the carbon skeleton. Breaking the bond between the
two methylene groups reveals structures of logical nucleophilic and electrophilic reactants that make use of an enolate ion as the nucleophile.

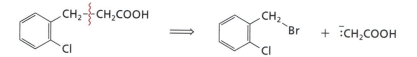

The enolate derivative ⁻:CH₂COOH requires the use of diethyl malonate (Table 22.5)
as the starting carbonyl compound. Formation of its enolate derivative, followed by
alkylation and decarboxylation, will yield the desired product.

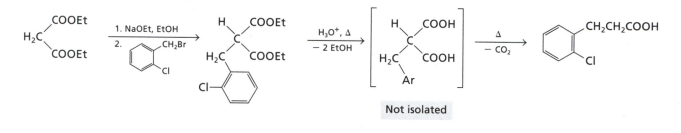

EXAMPLE 22.4

Starting with a suitable dicarbonyl precursor and any alkyl halide, show how you would
prepare the following compound.

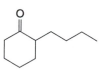

In the retrosynthetic analysis, we simply add an ester group to the substituted position
α to the carbonyl group of the product. Then we detach the alkyl substituent from that
same carbon atom, revealing the dicarbonyl precursor (in its enolate form).

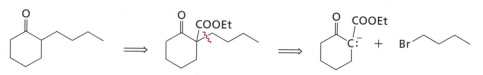

The synthesis is carried out by alkylating at the carbon atom between the two carbonyl
groups. Hydrolysis and decarboxylation occur upon heating the keto ester with aqueous
acid.

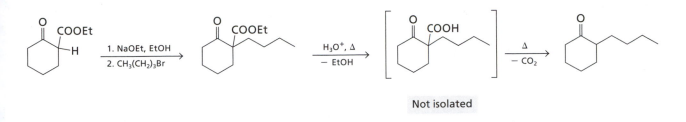

Not isolated

EXERCISE 22.16

Show how you would prepare each of the following compounds by making use of a dicarbonyl precursor and any other necessary compounds and reagents:

a.   b.

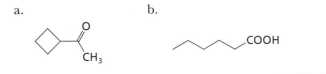

22.4d DECARBOXYLATION OF β-KETO ACIDS IS AN IMPORTANT METABOLIC TRANSFORMATION IN BIOCHEMISTRY

In the laboratory, taking advantage of the decarboxylation reaction of a β-keto acid provides a convenient way to circumvent selectivity problems associated with simpler enolate ions. In biological systems, these same decarboxylation processes are crucial to the success of several metabolic transformations.

Along the pathway that defines the citric acid cycle, the alcohol group of (2R,3S)-isocitrate is oxidized by NAD⁺. The next detectable intermediate is α-ketoglutarate, but another species is formed first: a β-keto carboxylic acid.

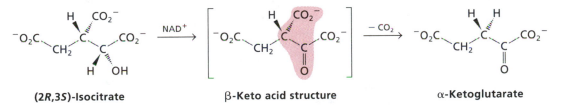

(2R,3S)-Isocitrate   **β-Keto acid structure**   **α-Ketoglutarate**

The mechanism for this enzyme-catalyzed reaction is not much different from the corresponding reaction that takes place in solution. Loss of carbon dioxide produces an enolate ion, which is stabilized by interacting with a manganese(II) ion at the enzyme active site. In solution, a β-keto carboxylic acid is protonated to form a neutral enol that subsequently undergoes tautomerism (Section 22.4c). In this enzyme active site, the metal ion associates with the original ketone group to accept the electrons donated by the carboxy group as it is converted to CO_2.

Enolate

Stabilized by Mn(II) ion

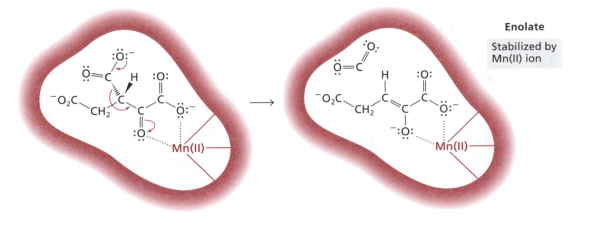

Subsequently, the enolate ion does react with a proton to generate the ketone product, which dissociates from the enzyme active site.

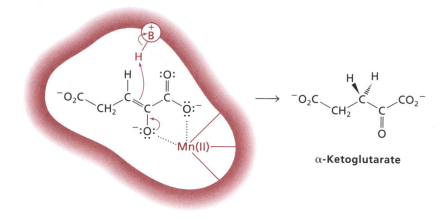

α-Ketoglutarate

22.5   REACTIONS OF AMIDE AND IMIDE IONS

22.5a   AMIDE IONS CAN BE ALKYLATED

Carboxylic amides are weak acids (Section 22.1c), so their conjugate bases—carboxamide ions—are strong bases, about the same order of basicity as hydroxide and alkoxide ions. A carboxamide ion reacts in ways similar to those seen for a ketone enolate ion; the difference is that a nitrogen atom, rather than a carbon atom, is the nucleophilic atom.

An amide is prepared by reaction of an amine with an acid chloride (Section 21.3a). Deprotonation is accomplished using a base such as NaH in an aprotic solvent like DMF. Benzamide derivatives are used most frequently to make a nucleophilic amide ion because there is no proton *alpha* to the carbonyl group besides the one on nitrogen.

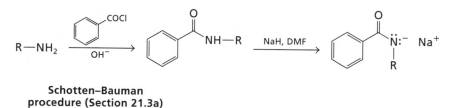

Schotten–Bauman procedure (Section 21.3a)

This carboxamide ion can be used subsequently in reactions with alkyl halides. Furthermore, the alkylated product can be hydrolyzed to form an amine. Use of acid in the hydrolysis step requires the addition of OH⁻ in order to obtain the neutral amine product.

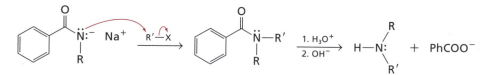

An analogous procedure can be carried out starting with an amine and benzenesulfonyl chloride. The anion of the benzenesulfonamide derivative is less basic than the amide ion, so elimination side products are less likely to be formed. Hydrolysis of a sulfonamide is conducted using aqueous sulfuric acid, and addition of OH⁻ is required to obtain the neutral amine product.

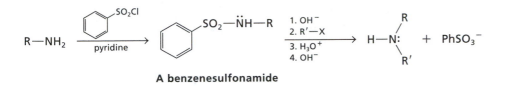

A benzenesulfonamide

EXERCISE 22.17

Show how you would prepare pure samples of butylamine, dibutylamine, and tributy-lamine from 1-bromobutane and any necessary reagents. Recall that amides can be reduced to form amines (Section 21.8b).

22.5b THE HOFMANN REARRANGEMENT TAKES PLACE DURING HALOGENATION OF A PRIMARY AMIDE

Just as a ketone can be halogenated under basic conditions (Section 22.2c), halogenation of an amide is accomplished using base and bromine (chlorine can also be employed). The nitrogen atom is first deprotonated by the base, and the nucleophilic amide ion reacts in Step 2 with Br_2, displacing bromide ion and forming the corresponding *N*-bromoamide derivative.

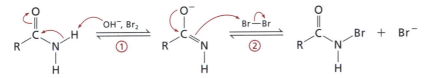

If the reaction begins with a primary amide (as it does in this example), then the electronegativity of bromine makes the second proton more acidic than those in the starting material. Deprotonation with base yields the *N*-bromoamide ion. Rather than undergoing a second reaction with bromine, this ion collapses to regenerate the carbonyl group by rearrangement: The R group migrates from carbon to nitrogen. This step occurs with the displacement of bromide ion and formation of an isocyanate molecule.

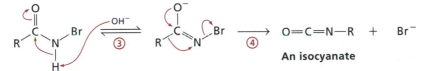

An isocyanate

Isocyanates have an electrophilic carbon atom, so they react readily with nucleophiles. Under the reaction conditions used for halogenation of a primary amide, hydroxide ion is present; it can add to the carbonyl portion of the isocyanate group to form a carbamic acid after protonation.

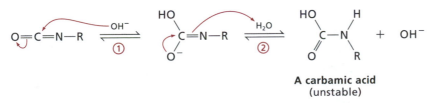

A carbamic acid
(unstable)

Carbamic acid derivatives are normally unstable and decompose to form an amine and carbon dioxide.

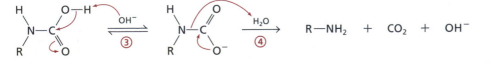

The overall process that comprises the foregoing steps can be represented by the equation shown below, which is the **Hofmann rearrangement.**

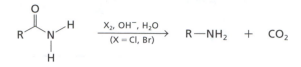

The scope of the Hofmann rearrangement is quite broad, encompassing alkyl, aryl, and heterocyclic primary amides as substrates. If the R group attached to the carbon atom is chiral, it migrates with retention of configuration.

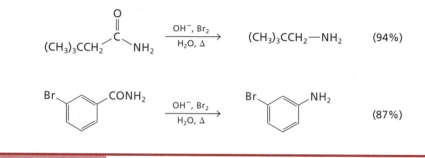

EXERCISE 22.18

If an alcohol and alkoxide ion are used in the Hofmann reaction, a carbamate ester is obtained. These compounds are stable, but they can be hydrolyzed later to produce the carbamic acid, which will decarboxylate to form the amine. Propose a reasonable mechanism for the following transformation:

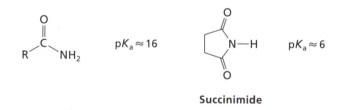

22.5c   THE GABRIEL REACTION MAKES USE OF ALKYLATION OF IMIDES

Just as an active methylene compound is more acidic than a ketone or ester (Section 22.4a), imides, which have an NH group attached to two carbonyl groups, are more acidic than amides. The pK_a value for the NH proton of a typical imide is ~ 6.

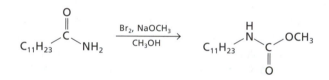

Succinimide

Resonance structures that can be drawn for the anion of succinimide are similar to those of the active methylene compounds (Section 22.4a) in which the negative charge is spread over five atoms.

In Section 22.5a, you learned that a carboxamide ion can be alkylated. An imide ion likewise can act as a nucleophile. A good way to make a primary amine is the **Gabriel reaction,** which employs the phthalimide ion as a nucleophile.

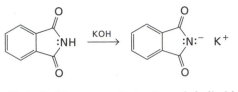

Phthalimide   Potassium phthalimide

The negative charge on the nitrogen atom makes the phthalimide ion a good nucle-ophile, and this ion reacts with an alkyl halide by the S_N2 pathway to form an *N*-alkylphthalimide derivative. After isolation and purification, the *N*-alkylphthalimide compound is hydrolyzed to liberate the primary amine and the conjugate base of phthalic acid.

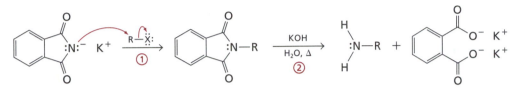

EXERCISE 22.19

Hydrazine in ethanol is often used to liberate the amine from an *N*-alkyl-phthalimide derivative. Propose a reasonable mechanism for the following reaction.

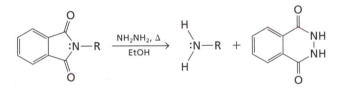

CHAPTER SUMMARY

Section 22.1   Acidity of carbonyl compounds

- The acidity of a carboxylic acid can be attributed to resonance stabilization of its conjugate base, the carboxylate ion.
- Substitution in the carbon framework of a carboxylic acid, whether aliphatic or aromatic, affects its acid strength as a result of inductive effects related to the electron-donating or electron-withdrawing nature of the substituent.
- The presence of a carbonyl group makes a carboxylic acid amide more acidic than its analogous amine, and a ketone more acidic than its corresponding hydrocarbon.

Section 22.2   Enols and enolate ions

- The carbanion generated by deprotonating a ketone is called an enolate ion.
- Enolate ion formation can also be considered to be the result of deprotonation of an enol, also called a vinyl alcohol.
- An enol is the tautomeric product of a carbonyl compound.
- Enols and enolate ions are important intermediates in isomerization reactions of aldehydes and ketones.
- Carbonyl transposition is an essential process in the isomerization reactions of carbohydrates, and it occurs with involvement of an enediol.
- Racemization of a chiral center adjacent to a carbonyl group can occur via the enol or enolate derivative, both of which are achiral.

- An enolate ion reacts with chlorine, bromine, or iodine, leading to replacement by a halogen atom of a proton that is bonded to the carbon atom adjacent to the carbonyl group.

- A methyl ketone can undergo three successive halogen substitution reactions. In the presence of strong base, the carbon–carbon bond is then cleaved, a transformation called the haloform reaction.

- Halogenation α to a ketone carbonyl group can take place under acid conditions, and substitution can be stopped after a single proton has been replaced.

- To prepare an enolate ion in high yield, a ketone is treated at low temperature with a strong base like LDA (lithium diisopropylamide) or KH (potassium hydride) in an inert solvent such as THF.

- Many unsymmetrical ketones can yield two different enolate ions. The one formed faster is called the kinetic enolate, and the more stable one (the one with the more highly substituted double bond) is the thermodynamic enolate.

- Strong base and low temperature favor formation of the kinetic enolate.

- The presence of excess ketone and ambient temperature favors formation of the thermodynamic enolate.

Section 22.3   Reactions of enolate ions

- An enolate ion acts as a nucleophile in the S_N2 reaction. Primary alkyl halides work well, but secondary and tertiary halides undergo elimination because an enolate ion is a strong base.

- An enolate ion reacts with PhSeCl to bond a phenylselenium group α to the carbonyl group. Treating this phenylselenide product with H_2O_2 leads to elimination of the PhSe group and formation of an α,β-unsaturated ketone.

- An aldehyde enolate undergoes self-condensation, so it is difficult to alkylate an aldehyde directly at the carbon atom α to its carbonyl group.

- Esters and nitriles form carbanions that can be alkylated at the carbon atom α to the carbonyl or nitrile group

Section 22.4   Dicarbonyl compounds

- Molecules with two carbonyl groups flanking a methylene or methine group are called active methylene (or methine) compounds because the carbon atom between the carbonyl groups is readily deprotonated.

- The most commonly used active methylene compounds in synthesis are ethyl acetoacetate and diethyl malonate.

- An alkoxide ion in alcohol is able to deprotonate an active methylene compound.

- The enolate ion derived from an active methylene compound is readily alkylated.

- A ketone, ester, or carboxylic acid with a COOH group attached to the carbon atom α to its carbonyl group undergoes facile decarboxylation upon heating.

- Decarboxylation of a β-keto acid or β-diacid is an important process in several metabolic transformations.

Section 22.5   Reactions of amide and imide ions

- The amide ion is nucleophilic and will participate in S_N2 reactions with alkyl halides. Sulfonamides react in the same fashion.

- In aqueous base, a primary amide undergoes halogenation at nitrogen, and the intermediate subsequently rearranges to form an isocyanate. Hydrolysis and decarboxylation occur in the aqueous medium to produce a primary amine.

- The conjugate base of phthalimide is alkylated at its nitrogen atom. Hydrolysis generates a primary amine, a transformation called the Gabriel reaction.

Section 22.1d
enolate ion

Section 22.2a
enol

Section 22.2b
carbonyl transposition
enediol

Section 22.2c
haloform reaction

Section 22.2e
kinetic enolate
thermodynamic enolate

Section 22.4a
active methylene
compound

Section 22.5b
Hofmann rearrangement

Section 22. 5c
Gabriel reaction

REACTION SUMMARY

Section 22.1e

Lithium diisopropylamide (LDA) is a strong base used to make enolate ions. It is made from the reaction of diisopropylamine with butyllithium in THF.

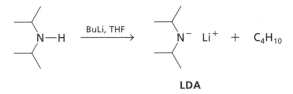

LDA

Section 22.2a

Enol formation, which is an example of tautomerism, occurs when a proton *alpha* to a carbonyl group moves to the carbonyl oxygen atom, concomitant with isomerization of the double bond; the process is catalyzed either by acid or base

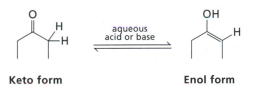

Keto form   **Enol form**

Section 22.2b

Carbohydrates undergo positional isomerism via an enediol intermediate, leading to carbonyl transposition.

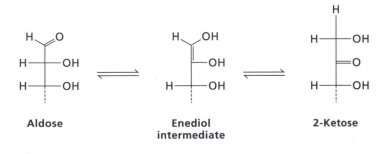

Aldose   **Enediol**   **2-Ketose**
                   intermediate

A chiral center α to a carbonyl group undergoes isomerization (racemization or epimerization) via an enol intermediate.

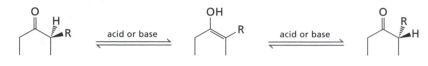

Section 22.2c

A ketone undergoes halogen substitution under basic conditions at the position α to the carbonyl group

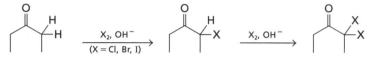

The haloform reaction. The three methyl protons of a methyl ketone undergo substitution by halogen atoms, followed by an addition–elimination step. The products are a carboxylic acid salt with one less carbon atom and trihalomethane.

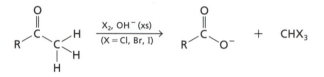

A ketone undergoes halogen substitution under acidic conditions at the position α to the carbonyl group. Halogenation can be stopped after a single substitution step.

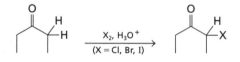

Sections 22.2d and 22.2e

The enolate ion derivative of a ketone is formed by deprotonation with LDA in THF or KH in THF. Unsymmetrical ketones can form two different enolate ion derivatives, referred to as the kinetic (faster formed) and thermodynamic (more stable) enolate.

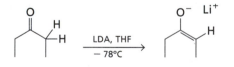

Section 22.3a

Ketone enolate ions react with alkyl halides by the S_N2 pathway. Primary alkyl halides and sulfonate esters are the best substrates; 2° and 3° substrates undergo elimination.

Section 22.3b

Ketone enolate ions react with phenylselenium chloride. Oxidation with hydrogen peroxide leads to elimination of PhSeOH and formation of α,β-unsaturated ketones.

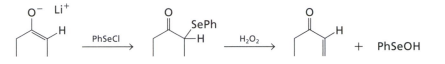

Section 22.3d

Ester enolate ions and nitrile carbanions undergo alkylation.

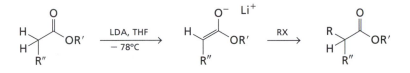

Section 22.4b

The enolate ions derived from active methylene compounds are readily alkylated under milder conditions than those required for ketone or ester enolate ions.

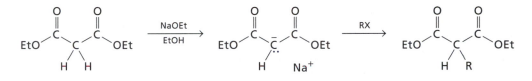

Section 22.4c

β-Keto acids or β-diacids undergo decarboxylation upon heating.

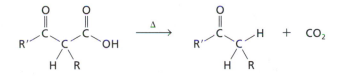

Section 22.5a

Amide and sulfonamide ions are readily alkylated.

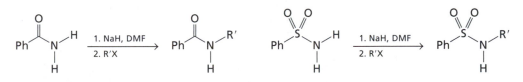

Section 22.5b

The Hofmann rearrangement. A primary amide reacts with excess halogen and hydroxide ion to produce an amine with loss of the carbonyl group. The reaction proceeds via rearrangement to form an isocyanate.

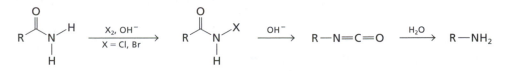

Section 22.5c

The Gabriel reaction. An alkyl halide undergoes substitution by phthalimide ion; hydrolysis liberates the corresponding amine.

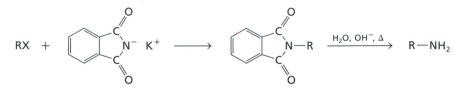

ADDITIONAL EXERCISES

22.20. Draw the structure of the conjugate base derived from each of the following compounds:

a. Methyl phenylacetate

b. Ethyl 4-methylpentanoate

c. 3-Cyano-2-butanone

d. 1,3-Cyclopentanedione

22.21. Draw the structure of the enol form of each of the following compounds:

a. 2,2-Dimethylcyclohexanone

b. Pentanal

c. Isopropyl acetoacetate

d. 1,3-Cyclohexanedione

22.22. Draw at least two good resonance structures for each of the following anions, showing all unshared electrons:

a.

b.

c. $CH_3OSO_2{}^-$

d.

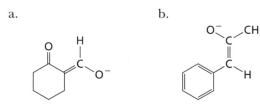

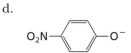

22.23. Circle the most acidic proton(s) in each of the following compounds and give its approximate pK_a value:

a.

b.

c.

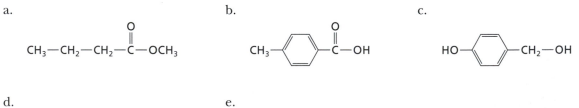

d.

e.

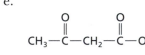

22.24. Ethyl acetoacetate reacts with 2 equivalents of LDA in THF, producing a dianion. Draw resonance structures for this species and explain why it is able to form.

22.25. Arrange the members of each set of compounds in order of decreasing acid strengths (review the material presented in Section 5.2c).

a.

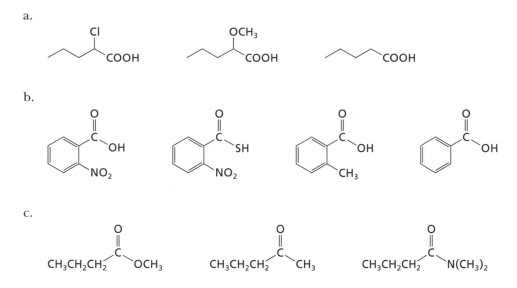

b.

c.

22.26. For the carbanion species listed in the following table, prepare a summary chart in the format given below.

Type of carbanion	General structure of starting material required	Approximate pK_a of starting material	Base and solvent required	General structure of the product after reaction with an alkyl halide, RX
Ketone enolate				
Ester enolate				
Nitrile carbanion				
β-Keto ester enolate				
β-Diester enolate				

22.27. A conjugated α,β,-unsaturated ketone is normally more stable than the corresponding β,γ isomer. In the compound below, the β,γ-unsaturated ketone has the more highly substituted double bond, so a measurable amount exists at equilibrium. Propose a mechanism for the following isomerization process, which proceeds via the enol form:

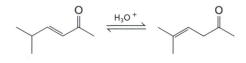

22.28. Draw the structure(s) of the major product(s) expected from each of the following reactions. Indicate the stereochemistry of the product as appropriate. Relative stereochemistry should be shown using wedges and dashed lines. If a racemic mixture will be formed, draw the structure of one enantiomer and write the word "racemic", or draw both enantiomeric structures. If diastereomers are formed, draw each structure; label meso compounds as such. If no reaction occurs, write N.R.

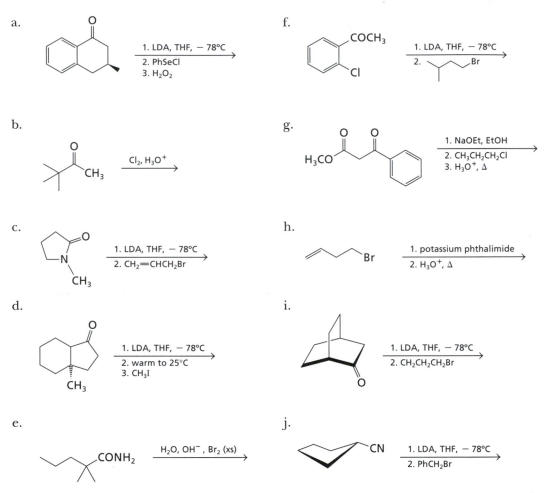

a.

1. LDA, THF, − 78°C
2. PhSeCl
3. H₂O₂

f.

1. LDA, THF, − 78°C
2.

b.

Cl₂, H₃O⁺

g.

1. NaOEt, EtOH
2. CH₃CH₂CH₂Cl
3. H₃O⁺, Δ

c.

1. LDA, THF, − 78°C
2. CH₂=CHCH₂Br

h.

1. potassium phthalimide
2. H₃O⁺, Δ

d.

1. LDA, THF, − 78°C
2. warm to 25°C
3. CH₃I

i.

1. LDA, THF, − 78°C
2. CH₂CH₂CH₂Br

e.

H₂O, OH⁻, Br₂ (xs)

j.

1. LDA, THF, − 78°C
2. PhCH₂Br

22.29. In performing each of the following reactions, indicate the principal IR absorptions that you would look for in the reactant and product to ensure that the reaction was successful. Be as specific as possible: list the position (cm⁻¹), intensity (weak, medium, or strong), and shape (sharp or broad) of the diagnostic absorption band(s).

a.

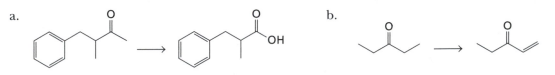

b.

22.30. For the reactions shown in Exercise 22.29, repeat the process for the ¹H and broad-band ¹³C NMR spectra. List, as applicable, the chemical shifts, integrated intensity values, splitting patterns, and DEPT results.

22.31. What reagents would you use to carry out each reaction in Exercise 22.29?

22.32. As you know, the sequence of reactions in synthetic operations is crucial. The following transformations differ only in the order that the steps are done. What is the product of each sequence? (MCPBA is a peracid.)

22.33. Show how you would prepare each of the following compounds from 1-pentanol and any reagents and solvents. More than one step may be needed.

a.   b.   c.   d.

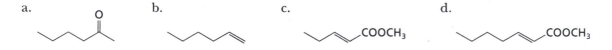

22.34. Starting with diethyl malonate or ethyl acetoacetate and any other organic compounds with seven or fewer carbon atoms, propose a synthesis for each of the following compounds:

a.   b.   c.

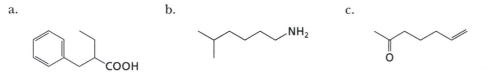

22.35. Show how you would prepare each of the following compounds in enantiomerically pure form using the given starting material and any other organic compounds that have one or two carbon atoms. You may use any reagents and solvents needed.

a.   b.

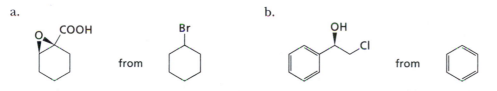

22.36. The *Curtius rearrangement* is similar to the Hofmann rearrangement, and heating an acyl azide produces an isocyanate. Propose a mechanism for the following reaction:

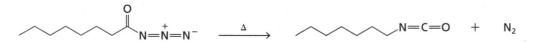

22.37. An acyl azide is made from an acid chloride and azide ion. Propose a mechanism for the following reaction:

$$C_{11}H_{23}-\overset{O}{\overset{\|}{C}}-Cl \xrightarrow[\text{DMF}]{NaN_3} C_{11}H_{23}-\overset{O}{\overset{\|}{C}}-N_3$$

22.38.   Identify each of the following compounds from its ^{1}H NMR spectra. All of the *J*-values = 7 Hz.

a. $C_{13}H_{16}O_3$

b. $C_6H_{10}O_3$

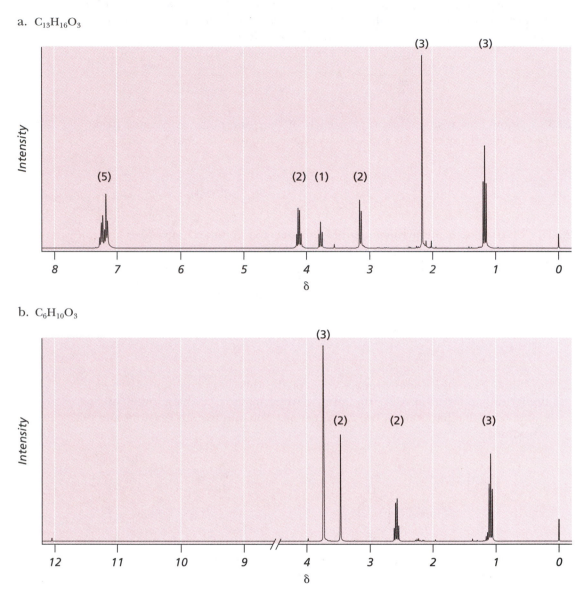

22.39.   When 2-bromocyclohexanone is treated with base, a ring contraction takes place instead of the expected E2 reaction. Propose a mechanism for this transformation, which is called the *Favorskii rearrangement.* (*Hint:* Compound **C** is an intermediate.)

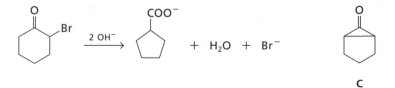

22.40. Compound **A**, $C_5H_{11}NO$, reacts with aqueous sodium hydroxide and bromine to produce **B**. The IR spectrum of **A** is shown below. When **B** is treated with dimethyldioxirane, compound **C** is formed. The IR and ^1H NMR spectra for compound **C** are shown below, too. Draw structures for **A**, **B**, and **C**.

Compound **A**

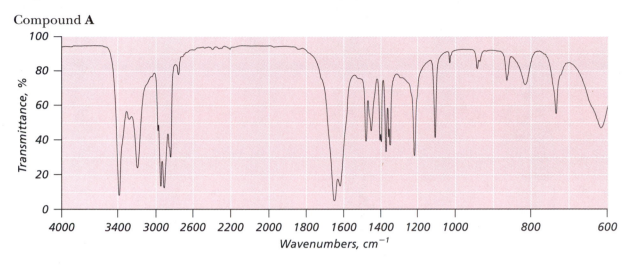

Compound **C**

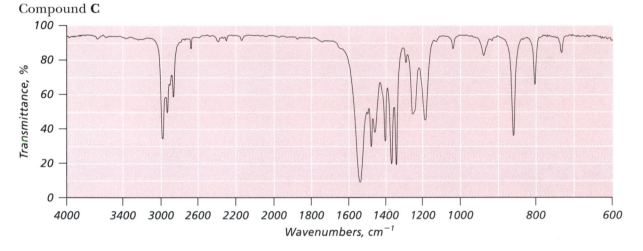

Compound **C**

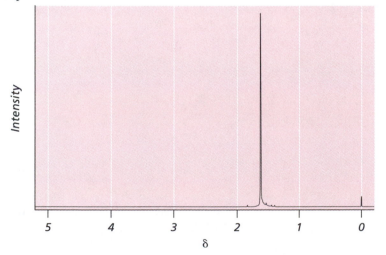

THE NUCLEOPHILIC ADDITION REACTIONS OF ENOLATE IONS

The previous chapter (Chapter 22) described the properties of enolate ions and how they can function as nucleophiles in substitution reactions. Given that a carbonyl carbon atom is electrophilic, it should not surprise you to learn that enolate ions can add to a C=O double bond, just like Grignard reagents, for example. Addition of an enolate ion to a carbonyl group forms a new carbon–carbon bond.

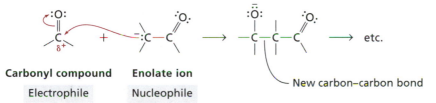

Carbonyl compound   **Enolate ion**

Electrophile   Nucleophile

New carbon–carbon bond

The addition reaction between an enol or enolate ion and a carbonyl compound can be classified in one of two ways.

1. The *aldol reaction* comprises the addition of an enol or enolate ion to the carbonyl group of an aldehyde or ketone.

2. The *Claisen condensation* is the addition–elimination reaction between an enolate ion and the carbonyl group of a carboxylic acid derivative, most frequently an ester or acid chloride.

An entire chapter is devoted to this seemingly specialized topic because these two processes play a crucial role in the making and breaking of C–C bonds during many chemical reactions of biosynthesis and metabolism.

23.1   THE ALDOL REACTION

23.1a THE SIMPLEST ALDOL REACTIONS INVOLVE THE SELF-CONDENSATION OF ALDEHYDE MOLECULES

The aldol reaction was named in 1872 by Charles Wurtz, who prepared 3-hydroxybutanal, which has the trivial name *aldol,* via self-condensation of acetaldehyde. This reaction can be catalyzed either by acid or by base.

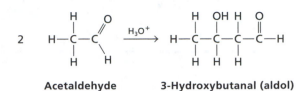

Acetaldehyde   **3-Hydroxybutanal (aldol)**

The base-catalyzed aldol reaction is the more commonly used, and its mechanism begins with formation of a resonance-stabilized enolate ion.

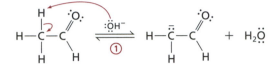

The pK_a of the protons alpha to the carbonyl group of the aldehyde has a value of ~ 20, so in an aqueous solution, only a small amount of enolate ion exists. Because a large amount of the aldehyde is still present at equilibrium, the enolate ion has a suitable concentration of electrophilic centers with which to react. Addition occurs to form a new carbon–carbon bond.

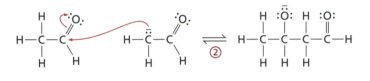

The alkoxide ion formed by this addition step removes a proton from water in the final step, which regenerates hydroxide ion and produces the β-hydroxyaldehyde.

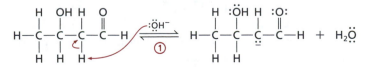

23.1b Dehydration of the Aldol Product Occurs by an E1cb Mechanism

The product of the aldol reaction also has protons alpha to the carbonyl group. If hydroxide ion removes a proton from this α-carbon atom, an enolate ion is formed. Two reactions are possible:

1. Condensation with one or more molecules of aldehyde to form trimers, oligomers, and so on.

2. Elimination of water—*dehydration*—to form an α,β-unsaturated aldehyde.

Although trimers and other oligomers can form, this reaction is of little interest or significance. Elimination of water is common, however, so it is worthwhile to look at the mechanism by which dehydration occurs.

If hydroxide ion removes a proton *alpha* to the carbonyl group (Step 1, below), the electron pair on carbon moves to displace hydroxide ion (Step 2) and create a C=C double bond. This elimination process is driven by formation of a π bond that is conjugated with the carbonyl group, the stability of which offsets the energy required to displace and solvate the highly basic leaving group (hydroxide ion).

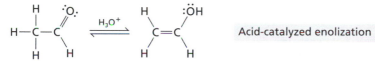

Elimination reactions of this type differ from those presented in Chapter 8 because the carbonyl group facilitates formation of a carbanion *prior* to displacement of the leaving group. Recall that in the E1 reaction, the bond between carbon and the leaving group is the first one broken. In the E2 reaction, the bond to the leaving group is broken as the proton is being removed (i.e., the process is concerted). The dehydration reaction comprising steps 1 and 2 shown above, in which the proton is removed *first,* is referred to as the **E1cb mechanism,** which stands for elimination, unimolecular, conjugate base. The unimolecular nature of this mechanism means that its rate depends only on the concentration of the hydroxy aldehyde compound and not on the amount of base that is present.

EXERCISE 23.1

Draw structures for the two products—the β-hydroxyaldehyde and the α,β-unsaturated aldehyde—formed by the base-catalyzed aldol reaction of propanal. Show the mechanism of each step by depicting electron movement with curved arrows.

23.1c THE ACID-CATALYZED ALDOL REACTION OCCURS VIA AN ENOL

The aldol reaction can also take place under acidic conditions, but an enolate ion does not exist in acid solution, so the nucleophile is the enol (Section 22.2a).

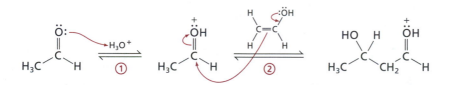

Because enols are very weak nucleophiles, the aldehyde carbonyl group (the electrophilic reactant) must be activated, which is accomplished in Step 1 by protonation (Section 18.1c). In Step 2, the enol adds to this activated carbonyl group, which leads to carbon–carbon bond formation.

Deprotonation completes the sequence and yields the β-hydroxy aldehyde product (Step 3).

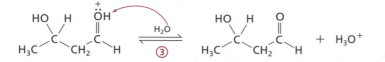

This initial product can undergo additional acid–base reactions, which lead to formation of an α,β-unsaturated aldehyde. The dehydration mechanism is E2, not E1cb, because no conjugate base is formed. Instead, the OH group is protonated to form a good leaving group (Step 4), and a concerted elimination process (E2) forms the unsaturated aldehyde (Step 5).

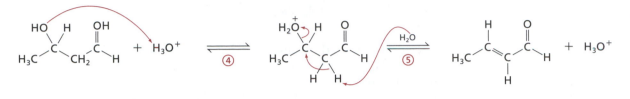

EXERCISE 23.2

Repeat Exercise 23.1 for the acid-catalyzed aldol reaction of propanal.

23.1d THE RETROALDOL REACTION BREAKS A CARBON–CARBON BOND

Whether the aldol reaction is catalyzed by acid or base, it is reversible, so a **retroaldol reaction** (recall that "retro" means reverse; Section 10.4b) can occur, breaking a C–C bond. In the first step of the base-catalyzed retroaldol reaction, hydroxide ion removes the proton from the OH group.

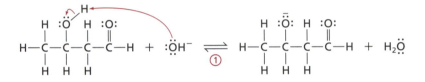

An electron pair on the oxygen atom then moves to regenerate the carbonyl group, forming a molecule of aldehyde and an equivalent of the enolate ion.

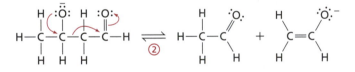

To complete the process, the enolate ion is protonated, which regenerates hydroxide ion.

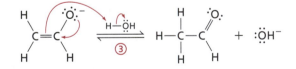

The facility of the retroaldol reaction depends on steric effects at the positions α and β to the carbonyl group. Any β-hydroxy aldehyde is a reasonable candidate for retroaldol cleavage, however.

EXERCISE 23.3

Propose a mechanism for the retroaldol reaction of 3-hydroxybutanal in aqueous acid.

23.1e KETONES CAN UNDERGO THE ALDOL REACTION

With its enolizable and reactive carbonyl group, a ketone can also undergo the aldol reaction, a transformation that was discovered 34 years before the actual "aldol" reaction was named. For ketones, however, steric effects frequently make the retroaldol process more favorable, meaning that the equilibrium favors the starting ketone.

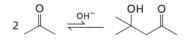

These equilibrium conditions unfavorable to the aldol reaction can be overcome by choosing conditions that lead to elimination of water and formation of the α,β-unsaturated ketone. Acidic conditions are therefore more appropriate for carrying out the aldol reaction of ketone substrates. For example, acetone undergoes an aldol reaction to form 4-methyl-3-pentene-2-one; this transformation is catalyzed by an insoluble acidic polymer, abbreviated H–(P).

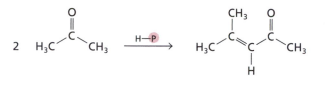

Acetone   **4-Methyl-3-pentene-2-one**

EXERCISE 23.4

Propose a mechanism for the acid-catalyzed aldol reaction of acetone to form 4-methyl-3-pentene-2-one.

23.2   CROSSED-ALDOL REACTIONS

23.2a THE CROSSED-ALDOL REACTION IS PRACTICAL WHEN ONE COMPONENT LACKS A PROTON ALPHA TO THE CARBONYL GROUP

The condensation reaction between two different molecules of an aldehyde or ketone in a protic solvent such as water or alcohol constitutes the **crossed-aldol reaction.** Such processes are normally impractical because of the different combinations with which the two molecules can condense with each other. If one of the reactants has no proton adjacent to the carbonyl group, however, then it cannot form an enolate ion. Therefore, the crossed-aldol reaction between an enolizable aldehyde or ketone and a non-enolizable molecule is practical. The enolizable reactant is usually added slowly to a solution of the non-enolizable component in base. Ketones, which do not react rapidly with themselves under typical base-catalyzed aldol conditions (Section 23.1e), are particularly useful reactants in combination with a nonenolizable aldehyde.

A common substrate for the crossed-aldol reaction is an aromatic aldehyde, which has no α-proton. Furthermore, dehydration of the initial condensation product is rapid, which leads to formation of the α,β-unsaturated ketone and prevents the retroaldol reaction from taking place.

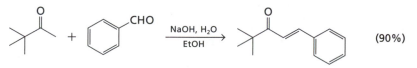

The enolate ion, which forms in the first step of a crossed-aldol reaction, adds to the carbonyl group of the aldehyde (Step 2) to generate the β-hydroxy ketone after protonation (Step 3). Dehydration occurs by the E1cb mechanism, and the (E) isomer (Section 4.1b) usually predominates.

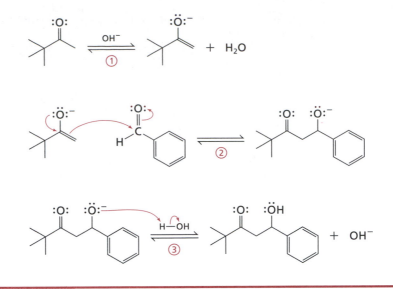

EXERCISE 23.5

Continue the steps in the preceding mechanism, showing how formation of the α,β-unsaturated ketone occurs.

If the ketone can form only a single enolate ion, then only one crossed-aldol product is formed.

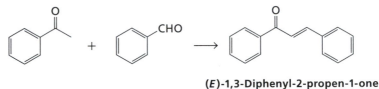

**(E)-1,3-Diphenyl-2-propen-1-one
(chalcone)**

But an unsymmetrical ketone that has protons at both α-positions can form two products. The major product is the one that derives from the kinetic enolate (Section 22.2e). The rationale for this result makes use of the fact that the product formed from the thermodynamic enolate undergoes a retroaldol reaction faster than it can eliminate water. For the product formed from the kinetic enolate, dehydration is faster, which shifts the equilibrium toward that product.

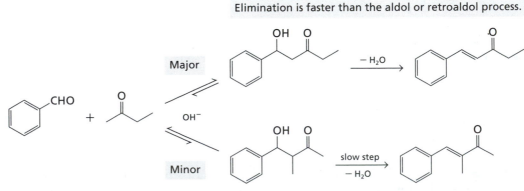

EXAMPLE 23.1

Draw the structure of the major product expected from the following crossed-aldol reaction:

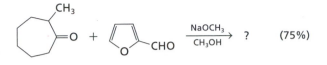

This is an example of a crossed-aldol reaction between a ketone with α-H atoms and an aldehyde with no α-hydrogen atom. The ketone is not symmetrical, and two enolate ions can form when it is treated with base:

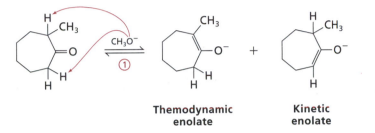

Themodynamic enolate   **Kinetic enolate**

A crossed-aldol reaction that starts with an unsymmetrical ketone will yield a major product by reaction of the kinetic enolate. Therefore, the mechanism will take place by addition of the kinetic enolate ion to the carbonyl group of the aldehyde; protonation of the resulting alkoxide ion by the solvent will produce the β-hydroxy ketone shown.

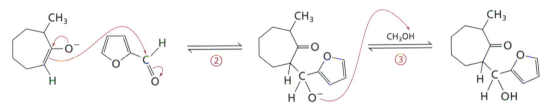

A β-hydroxy ketone in base will undergo dehydration by the E1cb mechanism; this process is particularly favorable when the aldehyde is aromatic.

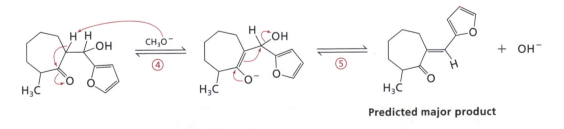

Predicted major product

EXERCISE 23.6

Draw the structure of the major product expected from the following crossed-aldol reaction:

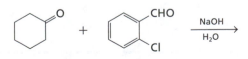

23.2b DIRECTED-ALDOL REACTIONS MAKE USE OF SPECIFIC ENOLATE DERIVATIVES

If the aldol reaction is carried out in an aprotic solvent by treating a preformed enolate derivative with an aldehyde or ketone, then a **directed-aldol reaction** occurs. (Enol derivatives with boron- or silicon-containing groups can also be used but are not covered here.) The simplest type of directed-aldol reaction begins with formation of an enolate ion by deprotonation of a carbonyl compound with LDA (Section 21.1c) in THF. In the following example, an ester enolate ion is prepared, and then it is treated with a ketone in Step 2. For an aldol reaction that makes use of a preformed enolate ion, an acid workup step is required (Step 3) to protonate the alkoxide ion formed in the addition step.

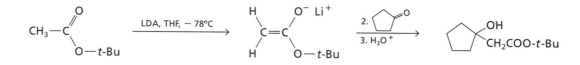

The mechanisms in this three-step procedure are straightforward, comprising (1) an acid–base reaction, (2) nucleophilic addition to a carbonyl group, and (3) an acid–base reaction.

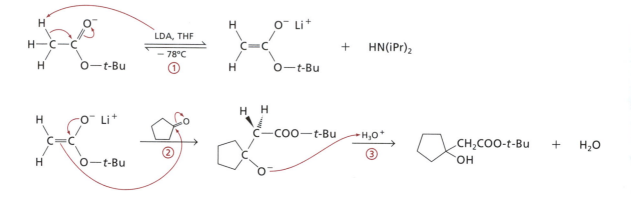

A carbanion derived from a nitrile likewise undergoes nucleophilic addition to the carbonyl group of an aldehyde or ketone.

$$CH_3CN \quad \xrightarrow[\substack{2. \\ 3.\ H_3O^+}]{1.\ BuLi,\ THF,\ -80°C} \quad \text{HO CH}_2\text{CN} \quad (68\%)$$

EXERCISE 23.7

A carbanion derived from a tertiary carboxamide also works in directed-aldol reactions. Draw the structure of the major product expected from the following reaction:

$$CH_3\text{—}\overset{\displaystyle O}{\overset{\displaystyle \|}{C}}\text{—}N(CH_3)_2 \quad \xrightarrow[\substack{3.\ H_3O^+}]{\substack{1.\ \text{LDA, Pentane} \\ 2.\ \text{Cyclohexanone}}}$$

23.2c THE METABOLISM OF GLUCOSE MAKES USE OF A RETROALDOL REACTION

The aldol reaction is remarkable because it permits the construction of C–C bonds under mild conditions in aqueous solution. Little wonder that aldol reactions are used both to form and cleave C–C bonds in biochemical systems.

The key enzymes for biochemical aldol reactions are appropriately called *aldolase* and *transaldolase*. These enzymes are responsible for making and breaking C–C bonds in reactions of carbohydrates.

Let us look at the actual enzyme-catalyzed retroaldol reaction that occurs during **glycolysis,** the pathway of glucose metabolism. The pathway starts with the phosphorylation of D-glucose at the 6-position.

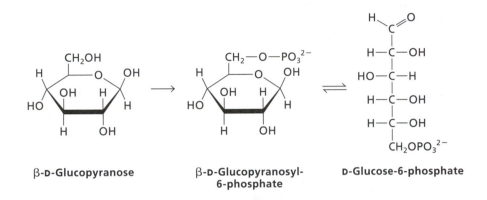

β-D-Glucopyranose   β-D-Glucopyranosyl-6-phosphate   D-Glucose-6-phosphate

D-Glucose-6-phosphate is subsequently converted to D-fructose-6-phosphate through formation of an enediol intermediate (Section 22.2b). A second phosphorylation step produces D-fructose-1,6-bisphosphate (1,6-FBP).

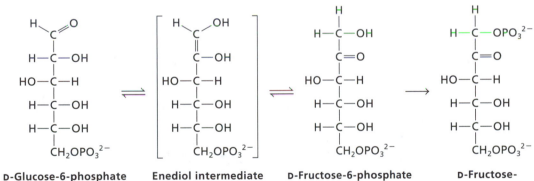

D-Glucose-6-phosphate   Enediol intermediate   D-Fructose-6-phosphate   D-Fructose-1,6-bisphosphate

The retroaldol reaction of 1,6-FBP takes place next, and this transformation yields D-glyceraldehyde-3-phosphate and dihydroxyacetone phosphate—that is, an aldehyde and a ketone—as products. The structure of 1,6-FBP below shows several atoms in color to emphasize the β-hydroxy ketone structure needed for the retroaldol reaction. Carbon atoms 1–3 of 1,6-FBP are found in the ketone product, and carbon atoms 4–6 constitute the aldehyde product.

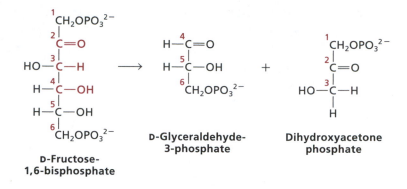

D-Fructose-1,6-bisphosphate **D-Glyceraldehyde-3-phosphate** **Dihydroxyacetone phosphate**

If the retroaldol reaction of D-fructose-1,6-bisphosphate were to occur in its carbonyl form, as illustrated in the following equation, an enolate derivative of dihydroxyacetone phosphate would be formed. This species is a strong base, which might be detrimental to an organism.

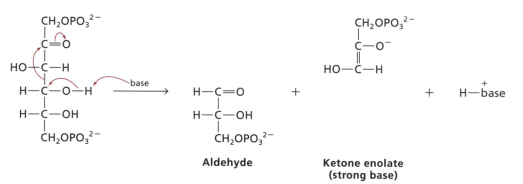

Aldehyde **Ketone enolate (strong base)**

As a way to bypass formation of an enolate ion, the enzyme *aldolase* uses a lysine amino group in its active site to first form an imine bond (Section 20.1a) with the carbonyl group of D-fructose-1,6-bisphosphate.

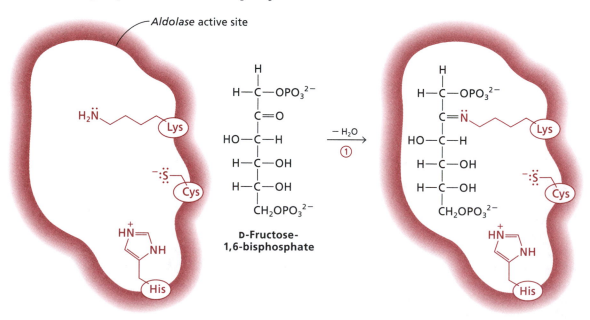

Because the nitrogen atom of an imine is a base, this intermediate actually exists as a 50:50 mixture of forms with and without a proton on the nitrogen atom (Section 20.2c). The effect of protonating this imine bond in the *aldolase*–fructose complex is to make the imine group more electrophilic, which in turn facilitates the retroaldol reaction. The process of exploiting an imine bond to create an electrophilic center—a strategy used in amino acid metabolism, too (Section 20.2c)—is called **Schiff base catalysis.**

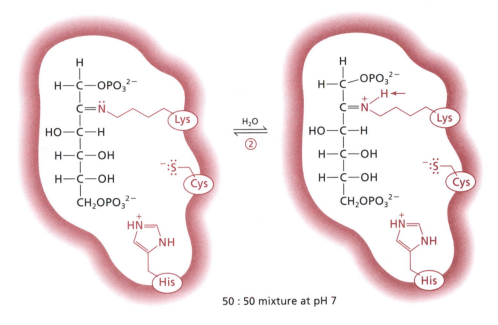

50 : 50 mixture at pH 7

The retroaldol reaction begins when the thiolate ion of a cysteine residue in the active site removes a proton from the OH group *beta* to the protonated imine group. The electrons move toward the positive charge on the nitrogen atom through the covalent bonds of the fructose derivative, thereby in one step,

- Creating the carbonyl group of glyceraldehyde.
- Breaking the C3–C4 bond.
- Forming a new π bond between C2 and C3.
- Breaking the imine π bond.

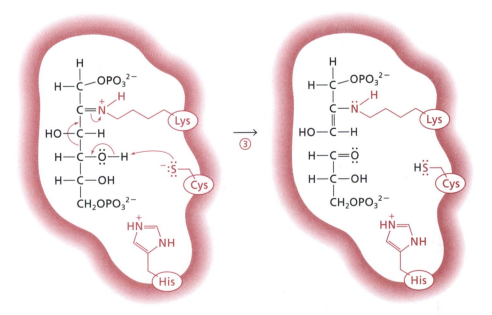

At this stage, D-glyceraldehyde-3-phosphate is released from the *aldolase* active site, leaving an enamine that is covalently attached at the active site. Recall that the carbon atom *alpha* to the carbon–nitrogen bond of an enamine is weakly nucleophilic and basic (Section 20.3b), so proton transfer occurs in the manner shown in the following equation:

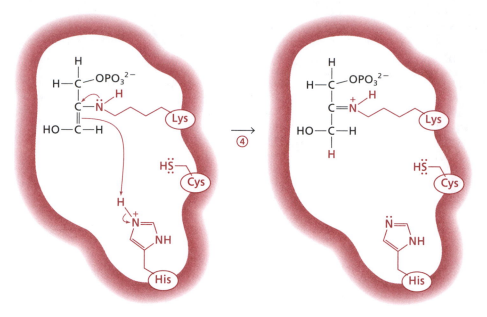

Finally, the resulting iminium group undergoes hydrolysis (Section 20.1c) to yield dihydroxyacetone phosphate, which dissociates from the enzyme active site and completes this metabolic transformation. (Proton-transfer steps subsequently return the active site to its starting form.)

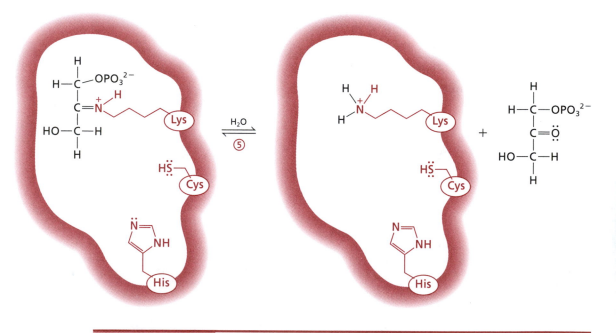

EXERCISE 23.8

Show the mechanism for the steps of the hydrolysis reaction in the preceding equation.

EXERCISE 23.9

Another way that glucose is metabolized is called the Entner–Doudoroff pathway. D-Glucose-6-phosphate is oxidized twice to form D-2-keto-3-deoxy-6-phosphogluconate. This substance is subsequently cleaved to produce pyruvate and D-glyceraldehyde-3-phosphate. Propose a mechanism for the enzyme-catalyzed transformation, which occurs via Schiff-base catalysis.

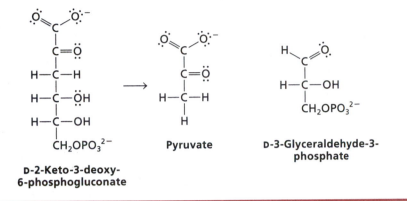

D-2-Keto-3-deoxy-6-phosphogluconate

Pyruvate

D-3-Glyceraldehyde-3-phosphate

23.2d   ALDOL AND RETROALDOL REACTIONS ARE USED TO INTERCOVERT CARBOHYDRATES

The *aldolase*-catalyzed cleavage of D-fructose-1,6-bisphosphate is reversible, so this same enzyme can also be used to *make* carbon–carbon bonds via the aldol mechanism (with Schiff-base catalysis). The enzyme *transaldolase*, which is related to *aldolase*, catalyzes carbon–carbon bond synthesis and cleavage reactions as well. *Transaldolase* operates during metabolism as one enzyme in the later phases of the pentose phosphate pathway, a route that is used to make D-fructose while extracting energy from a variety of C_5 sugars.

The starting point for the reactions catalyzed by *transaldolase* is D-sedoheptulose-7-phosphate (S7P). *Transaldolase* forms a Schiff base with the carbonyl group of S7P, and the imine nitrogen atom becomes protonated.

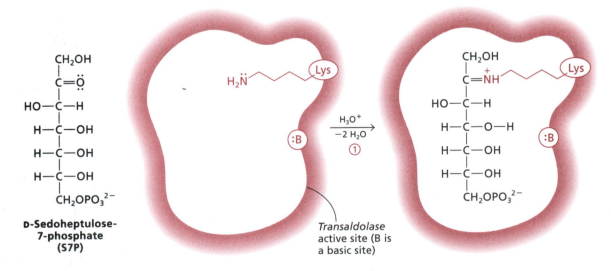

D-Sedoheptulose-7-phosphate (S7P)

Transaldolase active site (B is a basic site)

Next, a retroaldol reaction occurs, just as you saw for the reaction catalyzed by *aldolase*. This step generates D-erythrose-4-phosphate (E4P) and an enamine derivative of dihydroxyacetone.

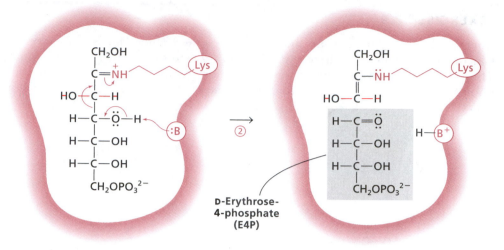

D-Erythrose-4-phosphate (E4P)

The E4P is displaced by a molecule of D-glyceraldehyde-3-phosphate, and instead of being protonated, the enamine, which is a nucleophile, reacts by addition to the carbonyl group of D-glyceraldehyde-3-phosphate. *This step is an aldol reaction.* The nucleophile is the resonance form of the enamine, which is fundamentally the imine derivative of an enolate ion.

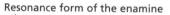

Resonance form of the enamine

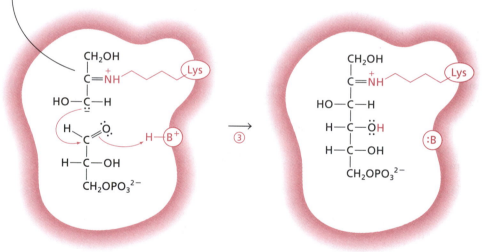

In the last step of this transformation, the product, which is D-fructose-6-phosphate, is liberated by hydrolysis.

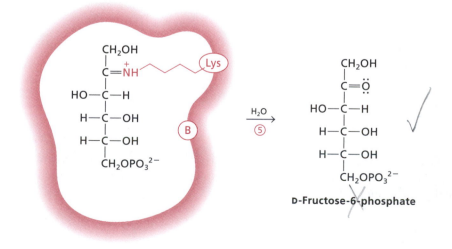

D-Fructose-6-phosphate

Overall, *transaldolase* catalyzes the redistribution of carbon atoms from a C_7 and a C_3 sugar to produce a C_4 and a C_6 sugar.

Sedoheptulose (C_7) → Enamine intermediate (C_3) + Erythrose (C_4)

enamine intermediate (C_3) + Glyceraldehyde (C_3) → Fructose (C_6)

The reaction to produce D-fructose-6-phosphate from dihydroxyacetone and D-glyceraldehyde-3-phosphate is both stereo- and regiospecific. This specificity results from combining the enamine and a carbonyl component *within the active site of the enzyme*. In the absence of an enzyme, these two reactants could form several products.

EXERCISE 23.10

Suppose that the *transaldolase*-catalyzed reaction were regiospecific but not stereospecific. Draw the structures of the four sugars (as their phosphate derivatives) that would be obtained from the process shown below.

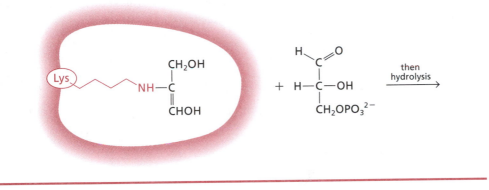

23.2e   CITRATE IS SYNTHESIZED BY A CROSSED-ALDOL REACTION

In living systems, the aldol reaction is not just used to manipulate carbohydrate structures. The biosynthesis of citrate, a key reactant in the citric acid cycle, results from the condensation reaction between an enolate derivative of Acetyl-CoA and oxaloacetate, according to the following equation:

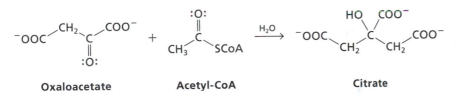

Oxaloacetate   **Acetyl-CoA**   **Citrate**

The thioester group is a reactive carboxylic acid derivative (Section 21.3c), but it is important in another respect, especially for Acetyl-CoA: *A thioester group makes the adjacent protons acidic enough to permit enolate ion formation under physiological conditions.* The protons *alpha* to the carbonyl group are acidic because of the resonance effect of the C=O double bond as well as the inductive effect of the sulfur atom.

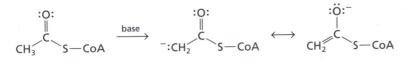

Citrate synthase is the enzyme that catalyzes the aldol reaction between Acetyl-CoA and oxaloacetate. The enzyme has been characterized by X-ray crystallography, and it employs a histidine imidazole ring as the base to generate the enolate ion.

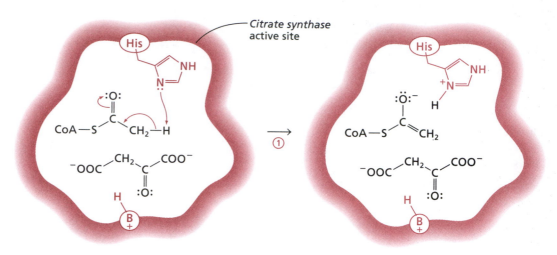

The enolate ion subsequently adds to the carbonyl group of the ketone, producing cityl CoA. Hydrolysis of this thioester in a separate step yields citrate.

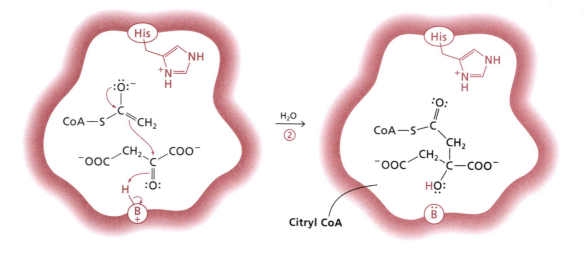

Citryl CoA

23.3   THE CLAISEN CONDENSATION

23.3a   ESTERS UNDERGO THE CLAISEN CONDENSATION IN A MANNER SIMILAR TO THE ALDOL REACTION

The other type of condensation reaction that occurs between enolate ions and carbonyl compounds is the *Claisen condensation,* as noted in the introductory section of this chapter. An ester can undergo self-condensation to form a β-keto ester after aqueous acid workup. For example, ethyl acetate forms ethyl acetoactetate via the Claisen condensation.

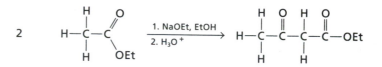

The mechanism of the Claisen condensation is not unlike that of the aldol reaction. In the first step, the enolate ion is formed. The protons *alpha* to an ester carbonyl are less acidic than those *alpha* to an aldehyde carbonyl group, so a stronger base is needed for the Claisen condensation. Sodium ethoxide in ethanol is a suitable combination.

The enolate ion subsequently adds to the carbonyl group of a second ester molecule to form a tetrahedral intermediate (Section 21.1c).

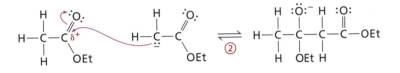

The tetrahedral intermediate collapses to regenerate the carbonyl group, displacing an equivalent of alkoxide ion.

The reaction looks like it should stop here, but this product has two carbonyl groups flanking the central methylene group, which makes the methylene protons more acidic than those of either the solvent or the starting ester molecule. Therefore, an acid–base reaction occurs to form the highly stabilized carbanion (Section 22.4a).

Unlike the aldol reaction, which needs only a catalytic amount of base, the Claisen condensation requires a full equivalent of base in order to proceed to completion because of this acid–base reaction that occurs at the position between the two carbonyl groups. The formation of this ionic species also requires that a separate acid workup step is needed to yield the final, neutral product.

The Claisen condensation of ethyl acetate has a special name: the *acetoacetic ester condensation*. This name is sometimes applied as well when other esters are used.

23.3b THE CLAISEN REACTION IS FAVORED WHEN THE PRODUCT CAN BE DEPROTONATED

All the steps of the Claisen condensation are equilibria, so the process is reversible. The final acid–base reaction, however, makes the **retro-Claisen condensation**—collapse of the product to regenerate the starting ester—difficult unless this acid–base step cannot occur. For example, if ethyl isobutyrate is treated with ethoxide ion, enolate ion formation can take place, as well as the subsequent addition–elimination steps.

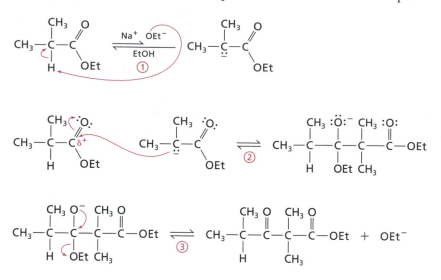

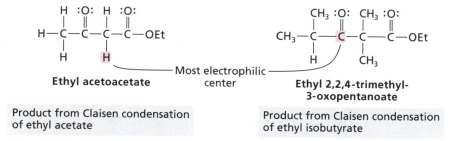

The β-keto ester formed by this mechanism does not have a proton between the carbonyl groups, however, so it cannot undergo an acid–base reaction as ethyl acetoacetate does. Instead, the most electrophilic center is now the ketone carbon atom.

Ethyl acetoacetate

Product from Claisen condensation of ethyl acetate

— Most electrophilic center —

Ethyl 2,2,4-trimethyl-3-oxopentanoate

Product from Claisen condensation of ethyl isobutyrate

As a result, ethoxide ion reacts differently with the two compounds shown above. With ethyl acetoacetate, an acid–base reaction occurs to form a stabilized carbanion, so the carbonyl groups are not susceptible to addition, and a retro-Claisen condensation does not occur. On the other hand, ethoxide ion *adds* to the ketone carbonyl group of ethyl 2,2,4-trimethyl-3-oxopentanoate, initiating the retro process and regenerating ethyl isobutyrate. Thus, *if enolate ion formation cannot take place, then a retro-Claisen reaction occurs.*

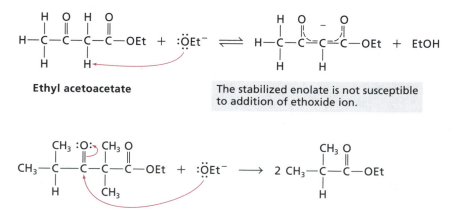

Ethyl acetoacetate

The stabilized enolate is not susceptible to addition of ethoxide ion.

Ethyl 2,2,4-trimethyl-3-oxopentanoate

To prepare a β–keto ester that lacks protons between the carbonyl groups, we first prepare the enolate derivative of the portion of the product that includes the ester group and the central carbon atom.

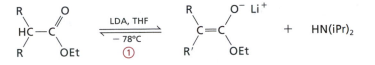

That ion is treated with the acid chloride that corresponds to the ketone portion of the product. Because chloride ion is a good leaving group and a weak base nucleophile, this equilibrium lies to the right, so a retro reaction does not occur.

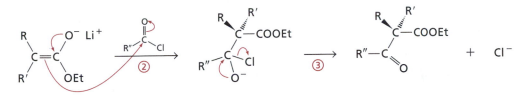

EXERCISE 23.11

What are the major products expected from the following reaction sequence?

$$CH_3-\overset{O}{\underset{\|}{C}}-CH_2-\overset{O}{\underset{\|}{C}}-OEt \quad \xrightarrow[\text{2. NaOEt, EtOH, }\Delta]{\text{1. 2 equiv NaH, THF, 2 equiv CH}_3\text{I}}$$

23.3c THE DIECKMANN CYCLIZATION IS AN INTRAMOLECULAR VERSION OF THE CLAISEN CONDENSATION

A diester can undergo an intramolecular Claisen condensation called the **Dieckmann cyclization,** which is an excellent method to make cyclic ketones and their derivatives.

The mechanism is precisely the same as that of the Claisen condensation, comprising enolate ion formation, addition–elimination, and an acid–base reaction.

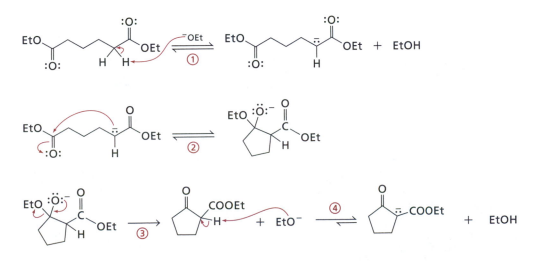

As before, aqueous acid is added during workup to generate the neutral product, a cyclic β-keto ester.

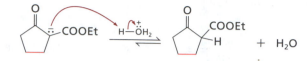

The Dieckmann reaction is especially useful for making five- and six-membered rings. Recall, too, that a β-keto ester readily undergoes hydrolysis and decarboxylation (Section 22.4c), so cyclic ketones are readily prepared by this sequence. For example,

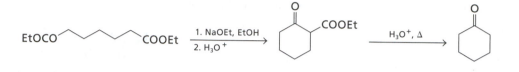

EXERCISE 23.12

Provide the mechanistic details for the steps in the preceding sequence.

23.4 CROSSED-CLAISEN CONDENSATIONS

23.4a THE CROSSED-CLAISEN CONDENSATION COUPLES DIFFERENT ESTERS

Just as with an aldol reaction, it is possible to carry out a **crossed-Claisen condensation,** which is the reaction between an enolate ion and an ester (or acid chloride). The reaction works best when the ester has no proton *alpha* to the carbonyl group, so benzoate, formate, and oxalate esters are excellent substrates.

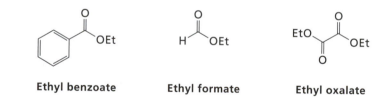

Ethyl benzoate   **Ethyl formate**   **Ethyl oxalate**

These compounds react with an enolate ion, as shown below for the reaction between ethyl octadecanoate and ethyl oxalate.

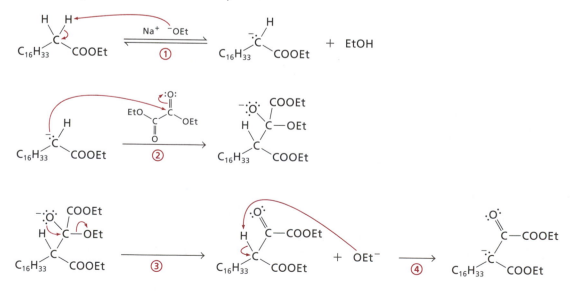

As before, aqueous acid is required for the workup stage of this procedure.

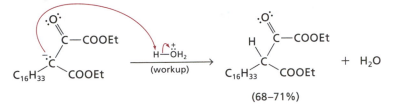

The reactions of unsymmetrical *diesters* constitute a special class of crossed-Claisen reactions. A diester in which one carbonyl group lacks protons at its α-carbon atom can form a single product via the Dieckmann cyclization.

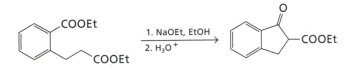

EXERCISE 23.13

Propose a reasonable mechanism for the preceding reaction scheme.

A crossed-Claisen condensation can also be done by treating a preformed enolate ion with an acid chloride, as shown previously (Section 23.3b). Both ketone and ester enolate ions work well. The reaction between a ketone enolate and an acid chloride produces a β-diketone. For example, the enolate ion derived from 3,3-dimethyl-2-butanone reacts with propanoyl chloride to form 2,2-dimethyl-3,5-heptanedione.

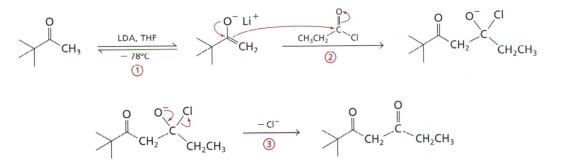

When the product has a proton attached to the carbon atom between the carbonyl groups, as in this example, the product undergoes an acid–base reaction with the enolate ion. Therefore 2 equiv of the enolate ion are needed, and workup with aqueous acid is also required.

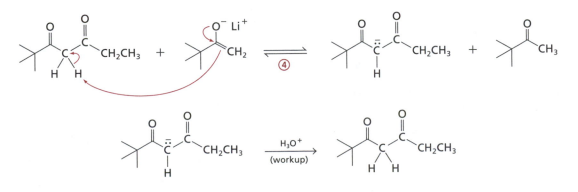

EXAMPLE 23.2

Draw the structure of the major product expected from the following crossed-Claisen condensation.

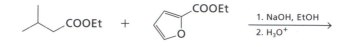

This example is of a crossed-Claisen condensation between two esters, one with α-hydrogen atoms and one with no α-hydrogen atom. The reaction begins with formation of the enolate ion of the ester having α-hydrogen atoms, and then that enolate ion adds to the carbonyl group of the ester with no α-H atom. The tetrahedral intermediate collapses to regenerate the carbonyl group of the product. (In the mechanism, there are two additional acid–base reactions, but these need not be shown to predict the structure of the product, which is set by Step 3.) (Note: In the following scheme, iPr is the isopropyl group.)

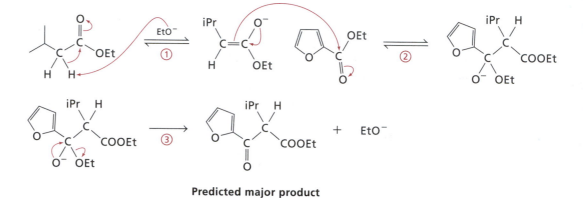

Predicted major product

EXERCISE 23.14

Draw the structure of the major product expected from the crossed-Claisen reaction shown here.

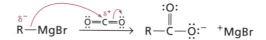

23.4b CARBOXYLATION IS A VARIANT OF THE CROSSED-CLAISEN CONDENSATION

Carboxylation occurs when an organic molecule reacts with carbon dioxide to form a carboxylic acid. You have already learned one type of carboxylation reaction, namely, the one that takes place with Grignard reagents (Section 15.2b).

An enolate ion reacts in the same way as a Grignard reagent, with one crucial difference: The product, which is a β-carboxy carbonyl compound, can readily undergo *decarboxylation* (Section 22.4c), so enolate carboxylation reactions are reversible.

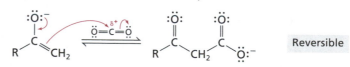

An enolate ion

There are several ways to circumvent the decarboxylation reaction. The simplest involves preparing the corresponding ester derivative, which can be done by a crossed-Claisen condensation with ethyl chloroformate or diethyl carbonate, which are derivatives of carbonic acid, the hydrated form of carbon dioxide.

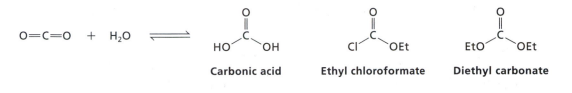

Carbonic acid   **Ethyl chloroformate**   **Diethyl carbonate**

A preformed enolate ion, when treated with either of these compounds, undergoes addition–elimination to form the β-keto ester. For example,

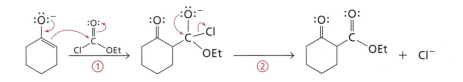

Careful saponification of this ester with acid workup at zero°C can be performed to prepare the carboxylic acid, if desired.

EXERCISE 23.15

What is the expected product of the sequence shown below? (Step 3 is drastic enough to cause both hydrolysis and decarboxylation.)

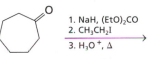

1. NaH, (EtO)$_2$CO
2. CH$_3$CH$_2$I
3. H$_3$O$^+$, Δ

23.4c CARBOXYLATION IN BIOLOGICAL SYSTEMS MAKES USE OF THE COENZYME BIOTIN

Carboxylation processes that occur in Nature utilize carbon dioxide, which is the most abundant source of carbon in the biosphere. The coenzyme **biotin,** which is attached to carboxylase enzymes by an amide bond, serves to make CO$_2$ available for reactions with enolate ion species.

Arrows for this!

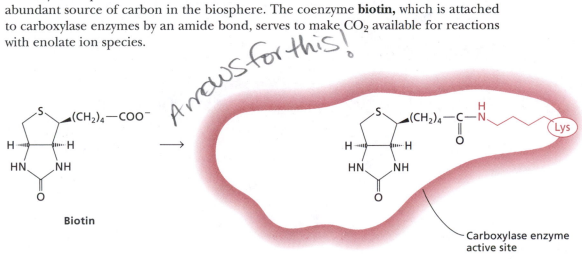

Biotin

Carboxylase enzyme active site

The carbon dioxide is bound to one of the biotin nitrogen atoms as a carbamic acid derivative (Section 20.5b)

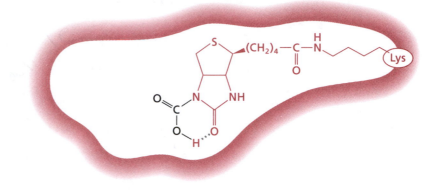

As a specific example of a biological carboxylation process, consider the biosynthesis of oxaloacetate from pyruvate.

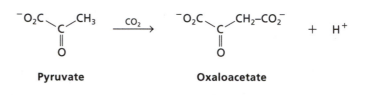

After formation of an enolate ion derivative of pyruvate, addition to the carbonyl group of the biotin–CO_2 adduct occurs.

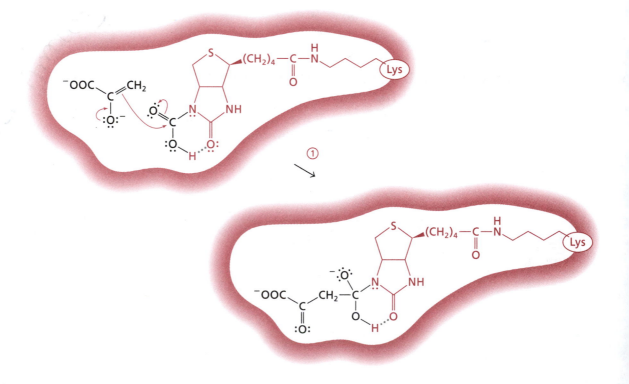

The tetrahedral intermediate collapses to form oxaloacetate and the enol form of biotin, which reacts with carbon dioxide to regenerate the adduct.

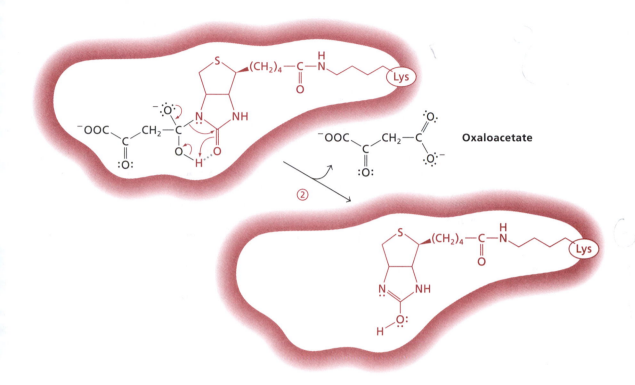

Oxaloacetate

23.4d THE RETRO-CLAISEN CONDENSATION IS IMPORTANT FOR DEGRADING FATTY ACIDS IN LIVING SYSTEMS

Just as the aldol reaction is the key to C–C bond breaking and forming reactions in carbohydrates, the Claisen condensation is the means by which fatty acids are made and broken down.

Fats are triesters of glycerol, $HOCH_2–CH(OH)–CH_2OH$ (Section 21.4d). They are ideal substances for storing chemical energy because of their preponderance of C–H and C–C bonds that during aerobic metabolism are converted to carbon dioxide and water.

Fat metabolism starts with hydrolysis of the ester bonds to generate the individual fatty acid molecules, which are carboxylic acids with long hydrocarbon chains. These are phosphorylated and then converted to CoA thioester derivatives according to the following scheme:

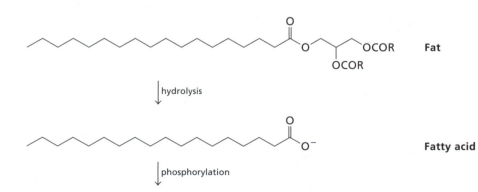

Fat

hydrolysis

Fatty acid

phosphorylation

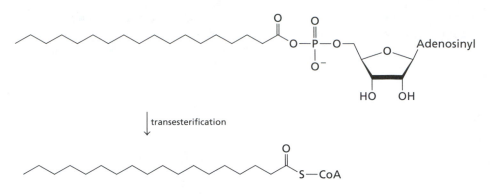

Once the thioester has been generated, a fatty acid is ready to be degraded. The protons *alpha* to the carbonyl group are slightly acidic, so one can be removed to form an enolate ion derivative. Subsequently, one of the β-hydrogen atoms is transferred to the coenzyme FAD (flavin adenine dinucleotide), which is a hydrogen-atom acceptor. Without worrying about the details of this process, notice that the removal of two hydrogen atoms creates a C=C double bond that is conjugated with the carbonyl group.

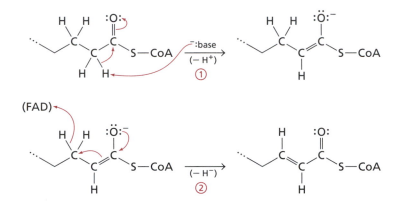

In the next step, water adds to the double bond (Section 24.2a) to form a β-hydroxy thioester. This alcohol group is oxidized by NAD⁺ (Section 11.4e) to form a β-keto thioester.

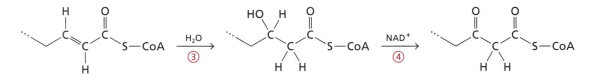

The structure of a β-keto thioester is analogous to that of a β-keto ester, which is the product of the Claisen condensation, so it is reasonable that a β-keto thioester can undergo a retro-Claisen reaction as well. The enzyme that catalyses such a reaction is *thiolase*, the name of which provides a clue about the mechanism of the transformation: A thiolate group of a cysteine residue in the active site adds to the ketone carbonyl group of the β-keto thioester.

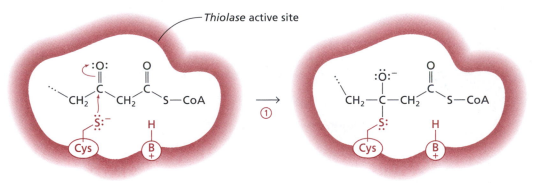

Once the tetrahedral intermediate forms, it collapses to regenerate the carbonyl group and expel an enolate ion.

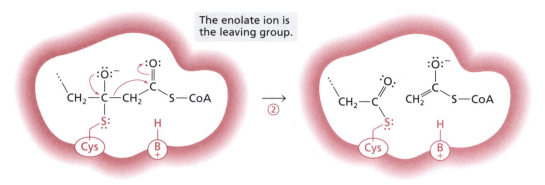

The enolate ion is protonated by an acidic group in the active site and forms acetyl-CoA.

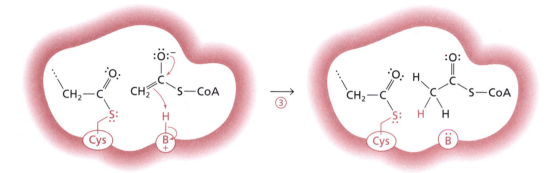

Next, acetyl-CoA dissociates from the active site to enter the citric acid cycle. A molecule of CoA enters the active site, and transesterification takes place between the cysteine thioester and the thiol group of CoA.

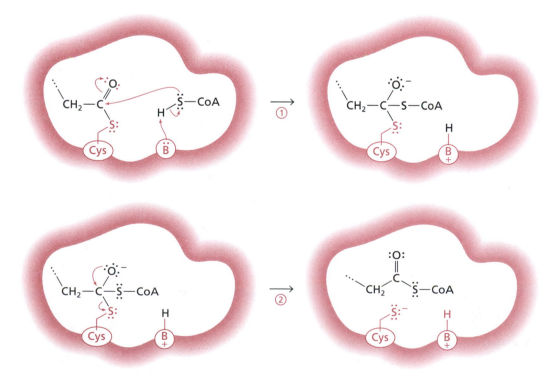

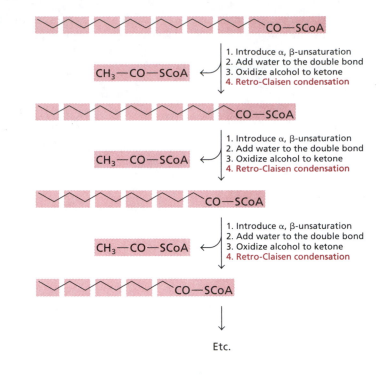

Figure 23.1
Cycles of β-cleavage that define the pathway of fatty acid degradation.

This new acyl-CoA dissociates from the active site, two carbon atoms shorter. Notice that this acyl-CoA molecule has the same general structure as the fatty acid–CoA molecule that entered the metabolic pathway. The process can therefore continue until all of the original carbon atoms have been converted into acetyl-CoA molecules, as summarized in Figure 23.1.

23.4e THE BIOSYNTHESIS OF FATTY ACIDS UTILIZES THE CLAISEN CONDENSATION

The pathway for the biosynthesis of fatty acids is nearly the reverse of the degradation process. A Claisen condensation is the key carbon–carbon bond-forming reaction in the biosynthetic pathway. The primary difference is the way in which the enolate ion is formed.

The building blocks are molecules of acetyl-CoA, but the acetyl group is first carboxylated to form malonyl-CoA. This malonyl group is then transferred to a protein abbreviated ACP (acyl carrier protein), which delivers the malonyl group to the enzyme in which the carbon skeleton of the fatty acid will be assembled.

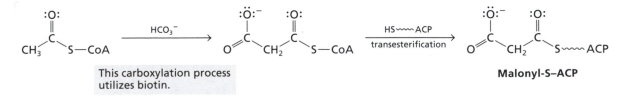

This carboxylation process utilizes biotin.

Malonyl-S–ACP

In the first round, the malonyl group reacts via a Claisen condensation with an acetyl group within the enzyme active site. Decarboxylation of the malonyl group generates the enolate ion needed to undergo addition to the acetyl-cysteine thioester group.

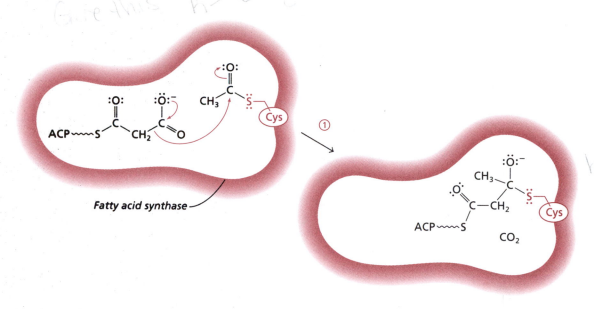

Elimination regenerates the carbonyl group to form acetoacetyl-S–ACP.

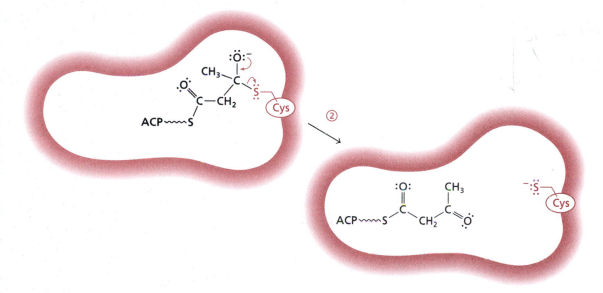

Next, the ketone group is reduced with NADPH (Section 18.3c), and dehydration occurs by an E1cb mechanism (Section 23.1b) to form the α,β-unsaturated thioester. Reduction of the C=C double bond yields butanoyl-S–ACP.

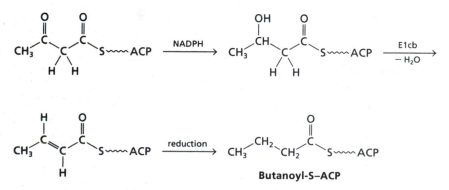

Butanoyl-S–ACP

The butanoyl group is transferred to the cysteine residue in the active site of *fatty acid synthase*, and another malonyl-S–ACP condenses with the C_4 piece to make the C_6

acyl-S–ACP. The process continues until the fatty acid skeleton (usually 16–18 carbon atoms) is built.

EXERCISE 23.16

Outline a scheme for the biosynthesis of fatty acids similar to that shown in Figure 23.1.

CHAPTER SUMMARY

Section 23.1   The aldol reaction

- Enolate ions undergo addition to the C=O double bond of carbonyl-containing compounds like aldehydes, ketones, and esters.
- The aldol reaction is the addition of an enolate ion to the carbonyl group of an aldehyde or ketone.
- The product of the aldol reaction is a β-hydroxy aldehyde or ketone. Elimination of water can occur from the aldol product to form an α,β-unsaturated aldehyde or ketone. The mechanism of this elimination reaction is designated E1cb.
- The aldol reaction is also catalyzed by acid, in which case the enol acts as a nucleophile, and the acid helps activate the carbonyl group toward addition.
- The retroaldol reaction is the reverse of the aldol reaction, and it leads to cleavage of a carbon–carbon bond.

Section 23.2   Crossed-aldol reactions

- The crossed-aldol reaction occurs between two different molecules of aldehydes and/or ketones. The best examples of the crossed-aldol reaction occur between a ketone that can form an enolate and an aldehyde that cannot form an enolate.
- Directed-aldol reactions are aldol reactions in which a specific enolate derivative reacts with an aldehyde or ketone in an aprotic solvent.
- Biological processes that metabolize and synthesize carbohydrates make use of the aldol reaction.
- The retroaldol reaction that metabolizes sugars makes use of a Schiff base derivative, which bypasses the formation of an enolate ion in the enzyme active site.
- The aldol reaction is also used to make citrate, a key molecule in the citric acid cycle.

Section 23.3   The Claisen condensation

- The Claisen condensation is the addition of an enolate ion to the carbonyl group of an ester or acid chloride.
- The Claisen condensation carried out under equilibrium conditions involves condensation of esters in the presence of a strong base such as alkoxide ion.
- The Claisen condensation goes to completion because a stabilized enolate ion is the initial product. An acid workup step is required to liberate the final product, a β-keto ester.
- The retro-Claisen condensation occurs if the β-keto ester cannot form an enolate derivative.
- The product of the Claisen condensation can undergo hydrolysis to form a β-keto acid, which subsequently undergoes decarboxylation.
- The Dieckmann condensation is an intramolecular version of the Claisen condensation.

Section 23.4 Crossed-Claisen condensations

- The crossed-Claisen condensation occurs between two different molecules of esters. The best examples of the crossed reaction occur between an ester that can form an enolate and one that cannot.

- An acid chloride is sometimes used as one component in the crossed-Claisen condensation, especially if the product does not have protons attached to the carbon atom between the carbonyl groups.

- Carboxylation is a variant of the crossed-Claisen condensation, but the product can also decarboxylate, so carbonate esters are used instead of CO_2.

- Carboxylation in biological systems utilizes the coenzyme biotin.

- Biological processes that are involved in the metabolism and biosynthesis of fatty acids make use of the Claisen condensation.

- Biosynthesis of fatty acids utilizes a Claisen condensation in which the enolate is made by decarboxylation of a β-keto acid.

KEY TERMS

Introduction
aldol reaction
Claisen condensation

Section 23.1c
E1cb mechanism

Section 23.1d
retroaldol reaction

Section 23.2a
crossed-aldol reaction

Section 23.2b
directed-aldol reaction

Section 23.2c
glycolysis
Schiff-base catalysis

Section 23.3b
retro-Claisen
   condensation

Section 23.3c
Dieckmann cyclization

Section 23.4a
crossed-Claisen
   condensation

Section 23.4b
carboxylation

Section 23.4c
biotin

REACTION SUMMARY

Section 23.1a

The aldol reaction. Aldehydes and ketones undergo self-condensation when treated with aqueous base.

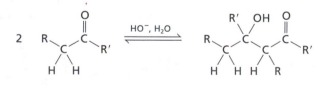

Section 23.1b

The products of an aldol reaction—a β-hydroxy aldehyde or ketone—undergo dehydration when treated with base; the mechanism is designated E1cb.

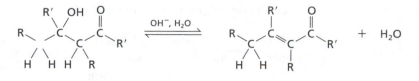

Section 23.1c

The aldol reaction can be catalyzed by acid; dehydration can occur under these conditions too.

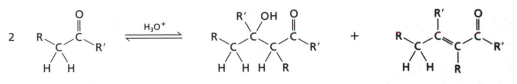

Section 23.1d

A β-hydroxy aldehyde or ketone (product of an aldol reaction) can undergo the retroaldol reaction to regenerate two molecules of aldehyde or ketone.

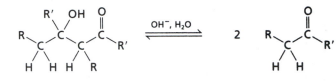

Section 23.2a

The crossed-aldol reaction. This transformation works best when one reactant lacks protons on the carbon atom adjacent to the carbonyl group.

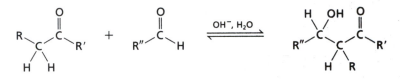

Section 23.2b

The directed-aldol reaction. An enolate ion is formed before adding the second carbonyl reactant.

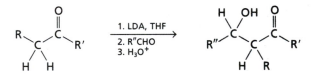

Section 23.3a

The Claisen condensation. Esters undergo self-condensation when treated with alcoholic base. An aqueous acid workup step is required to form the neutral product.

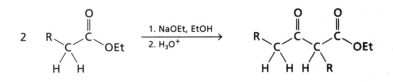

Section 23.3c

The Dieckmann cyclization. This transformation works best for the formation of five- and six-membered rings, but other size rings can be formed as well.

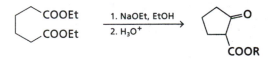

Section 23.4a

The crossed-Claisen condensation. This transformation works best when one component lacks protons on the carbon atom adjacent to the carbonyl group.

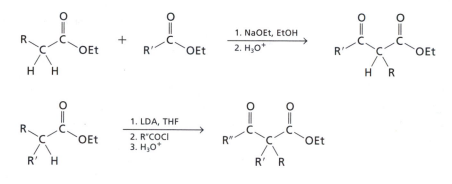

Section 23.4b

Carbonyl compounds react via their enolate derivatives with ethyl chloroformate to form β-dicarbonyl compounds with at least one ester group.

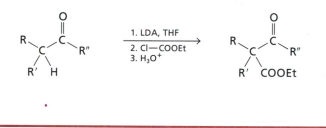

23.17.   Draw the structure for each of the following substances:

   a.   The enolate ion of *tert*-butyl acetate

   b.   Methyl 3-keto-2methylbutanoate

   c.   (2*R*, 3*S*)-2,3-Dihydroxypentanoic acid

   d.   3-^{14}C-dihydroxyacetone phosphate

23.18.   Draw the structure of the major product expected from each of the following aldol and related reactions:

a.   b.

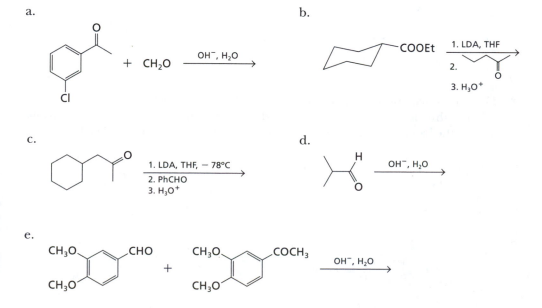

23.19.   The *Henry reaction* is a variant of the aldol reaction in which the conjugate base of a nitroalkane is the nucleophile and an amine base is used. Propose a mechanism for the following reaction:

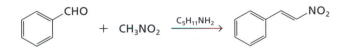

23.20.   The *Knoevenagel reaction* is a variant of the aldol reaction in which the conjugate base of an active methylene compound is the nucleophile and an amine base is used. Propose a reasonable mechanism for the following reaction:

23.21.   Draw the structure of the major product expected from each of the following Claisen condensations and related reactions:

a.

$$CH_3CH_2COOEt \xrightarrow[\text{2. } H_3O^+]{\text{1. } OEt^-, EtOH}$$

b.

$$CH_3COOt\text{-Bu} \xrightarrow[\substack{\text{2. } PhCH_2COCl \\ \text{3. } H_3O^+}]{\text{1. LDA, THF, } -78°C}$$

c.

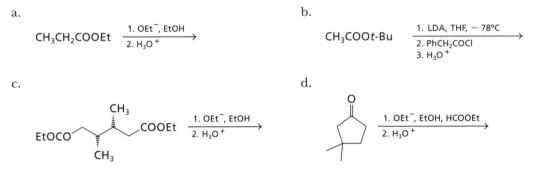

d.

23.22.   The *Thorpe reaction* is the nitrile analogue of the aldol reaction. Propose a reasonable mechanism for the following transformation: The first step is an intramolecular Thorpe reaction. There are three different reactions represented by the second arrow: nitrile hydrolysis, enamine hydrolysis, and decarboxylation.

23.23.   The *Perkin reaction* is a variant of the aldol reaction in which the conjugate base of an anhydride is the nucleophile, and the salt of the corresponding carboxylic acid is the base. Propose a reasonable mechanism for the following transformation:

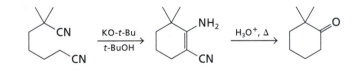

23.24.   The *Darzens reaction* is a crossed-aldol reaction between an aldehyde and the enolate derivative of an α-halo carboxylic ester, followed by an intramolecular S_N2 reaction. Propose a reasonable mechanism for the following transformation:

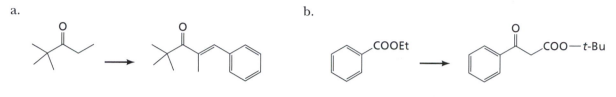

23.25.   In performing each of the following reactions, indicate the principal IR absorptions that you would look for in the reactant and product to ensure that the reaction was successful. Be as specific as possible: list the position (cm^{-1}), intensity (weak, medium, or strong), and shape (sharp or broad) of the diagnostic absorption band(s).

a.   b.

23.26.   For the reactions shown in Exercise 23.25, repeat the process for the ^{1}H and broad-band ^{13}C NMR spectra. List, as applicable, the chemical shifts, integrated intensity values, splitting patterns, and DEPT results.

23.27.   What reagents would you use to carry out each reaction in Exercise 23.25.

23.28.   How would you prepare each of the following substances via a Dieckmann reaction, starting with any acyclic organic compounds with six or fewer carbon atoms? You may use any needed reagents and solvents; several steps may be necessary.

a.   b.

23.29.   Identify each of the following compounds from its ^{1}H NMR spectrum. Each compound has a strong carbonyl absorption in the IR spectrum.

a. $C_6H_{10}O$: all *J*-values = 7 Hz.

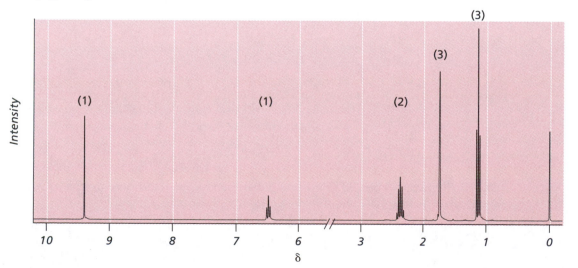

b. $C_8H_{12}O$

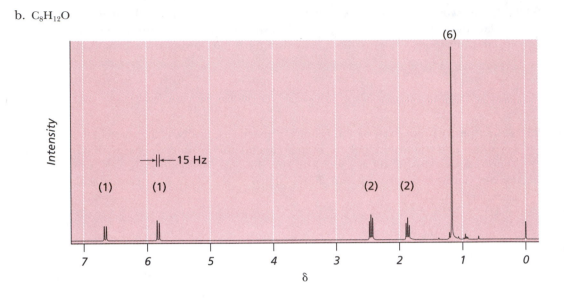

23.30. D-Glucose is converted during glycolysis to two, three-carbon sugars via D-fructose-1,6-bisphosphate according to the following scheme:

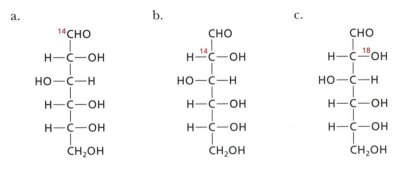

One way to study metabolic pathways is to utilize a compound with a labeled atom (C, N, or O). After adding the labeled compound to a mixture of enzymes needed to carry out the metabolic reactions, you isolate the product and determine at which position the labeled atom appears. For each of the following isotopically labeled D-glucose derivatives, indicate where the labeled atom will appear in D-glyceraldehyde-3-phosphate and dihydroxyacetone phosphate. Notice that these two products are interconverted by an enzyme that creates an enediol intermediate (Section 22.2b). If the labeled atom does not appear in either carbohydrate product, indicate where is goes and at which step.

a.

$$^{14}CHO$$
$$H-C-OH$$
$$HO-C-H$$
$$H-C-OH$$
$$H-C-OH$$
$$CH_2OH$$

b.

$$CHO$$
$$H-{}^{14}C-OH$$
$$HO-C-H$$
$$H-C-OH$$
$$H-C-OH$$
$$CH_2OH$$

c.

$$CHO$$
$$H-C-{}^{18}OH$$
$$HO-C-H$$
$$H-C-OH$$
$$H-C-OH$$
$$CH_2OH$$

23.31. Repeat Exercise 23.30 assuming that the enzyme that interconverts D-glyceraldehyde-3-phosphate and dihydroxyacetone phosphate is missing.

CONJUGATE ADDITION REACTIONS OF UNSATURATED CARBONYL COMPOUNDS

24.1 α,β-UNSATURATED CARBONYL COMPOUNDS

24.2 CONJUGATE ADDITION REACTIONS

24.3 CONJUGATE ADDITION REACTIONS OF CARBANIONS

24.4 REDUCTION REACTIONS

CHAPTER SUMMARY

The previous six chapters have described chemical transformations of carbonyl-containing compounds in which nucleophilic addition reactions are key steps. The fundamental mechanisms of these processes, which include variations such as addition–substitution and addition–elimination reactions, enable us to understand transformations that are pervasive throughout the fields of chemical synthesis and biochemistry.

This chapter describes reactions that can occur when a C=C double bond is conjugated (Section 10.1b) with a carbonyl group. Just as conjugated dienes often differ in their reactions with electrophiles when compared to the corresponding reactions of simple alkenes (Chapters 10 and 9, respectively), α,β-unsaturated carbonyl compounds can react with nucleophiles in ways that differ from their simpler aldehyde, ketone, and ester analogues.

In particular, a nucleophile can add either to the carbon–oxygen *or* to the carbon–carbon π bond of an α,β-unsaturated carbonyl compound. This addition to the terminal carbon atom, called **conjugate addition** or **1,4-addition,** occurs by the concerted movement of electrons through the carbon–carbon π bond toward the oxygen atom of the carbonyl group.

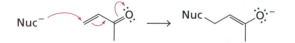

Addition of a nucleophile to the alkene portion of an α,β-unsaturated carbonyl molecule takes place because the terminal carbon atom of the conjugated π system is electrophilic, as one of its resonance forms reveals.

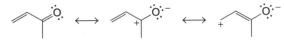

The conjugate addition reactions of enolate ions (Section 24.3) are particularly valuable because they contribute versatile methods for preparing carbon–carbon bonds to the synthetic repertoire.

24.1   α,β-UNSATURATED CARBONYL COMPOUNDS

24.1a   α,β-UNSATURATED CARBONYL COMPOUNDS ARE READILY PREPARED

You have already learned several ways to prepare compounds that have a C=C double bond conjugated with a carbonyl-containing functional group: the Horner–Emmons modification of the Wittig reaction (Section 20.4d), elimination of a phenylselenium group via oxidation (Section 22.3b), and the aldol reaction (Section 23.1b). The Horner–Emmons and aldol reactions are valuable ways to create a carbon skeleton from smaller fragments; use of the phenylselenium group is practical when the carbon framework already exists.

Examples of these transformations that are used to prepare α,β-unsaturated carbonyl compounds are summarized in Figure 24.1. Included in this compilation is a method that will be described in Section 24.1b: the *Mannich reaction.*

The Horner–Emmons reaction (Section 20.4d)

Elimination of the PhSe group (Section 22.3b)

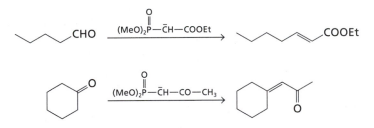

The aldol reaction (E1cb) (Section 23.1b)

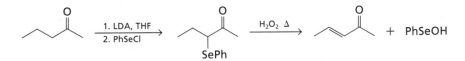

The Mannich reaction (Section 24.1b)

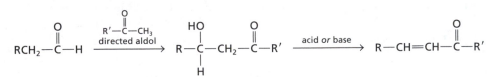

Figure 24.1
Synthetic methods commonly used to prepare α,β-unsaturated carbonyl compounds.

EXERCISE 24.1

By using the synthetic methods that you already know—the Horner–Emmons reaction, the aldol reaction, and PhSeOH elimination—show how you would prepare each of the following compounds from the given starting material. You may use any other organic compounds, reagents, and solvents.

a.   b.

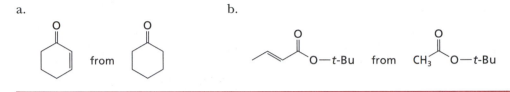

24.1b The Mannich Reaction Is Used to Make α,β-Unsaturated Ketones

The **Mannich reaction** occurs when the enol form of a ketone reacts with an iminium cation, the most common of which is $R_2N^+{=}CH_2$. These ions are formed in situ by the reaction between aqueous formaldehyde, a secondary amine, and an acid catalyst.

The first step in the Mannich reaction is the same as that for enamine formation (Section 20.3a), namely, nucleophilic addition of the amine nitrogen atom to the carbonyl group of formaldehyde.

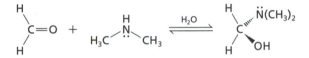

EXERCISE 24.2

Using curved arrows, illustrate the mechanism for the reaction shown directly above.

Because formaldehyde has no β-proton, elimination of water occurs under the acidic conditions of the reaction to form a dialkyiminium ion.

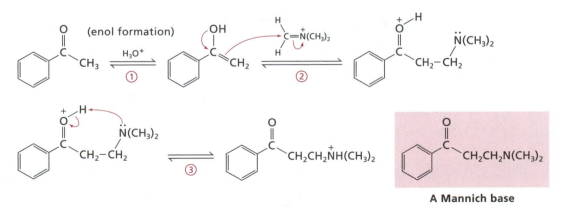

An iminium ion is electrophilic, so it can react with nucleophiles. For example, a ketone enol, which is produced in acidic solution, reacts with this iminium ion to form a protonated derivative of the β-dimethylamino ketone. (The neutral derivative of this product, shown in the following scheme in the screened box, is known as a **Mannich base.**)

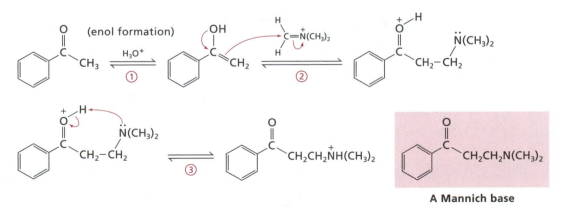

A Mannich base

The most convenient way to carry out a Mannich reaction utilizes the hydrochloride salt of the dialkylamine, which provides an acid catalyst for the reaction.

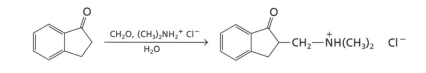

(66–75%)

EXAMPLE 24.1

Propose a reasonable mechanism for the following Mannich reaction, in which a ketone reacts with dimethylamine hydrochloride and formaldehyde.

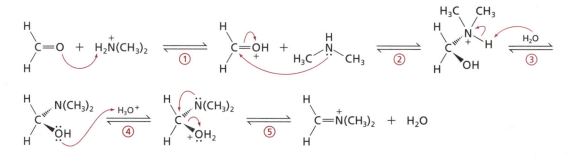

The Mannich reaction takes place in two stages. In the first stage, the reactive iminium ion is formed by the acid-catalyzed addition of dimethylamine to formaldehdye.

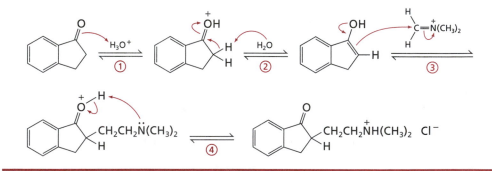

In the second stage of the transformation, the ketone undergoes enol formation by protonation–deprotonation, and then the enol reacts with the iminium ion to form the product.

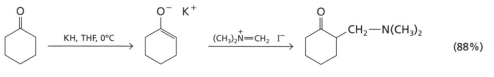

Instead of combining an amine and formaldehyde to make the iminium ion, we can use the commercially available iminium species, *Eschenmoser's salt.* This compound is particularly useful if nonaqueous conditions will be employed, for example, when a ketone enolate is first made using KH or LDA in THF (Section 22.2d). No acid is present under these conditions, so the Mannich base itself is formed.

$$H_3C\underset{H_3C}{\overset{+}{\underset{|}{}N}}=CH_2 \;\; I^-$$

Eschenmoser's salt
(commercially
available)

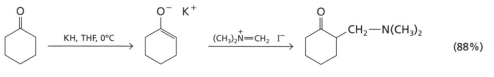

(88%)

A Mannich base

Once made, Mannich bases and their conjugate acids undergo thermal elimination of the amine to form the unsaturated ketone, called an **enone.**

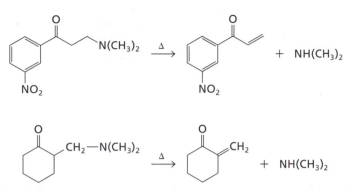

If mild reaction conditions are desired to make the α,β-unsaturated carbonyl compound, the nitrogen atom is first alkylated with methyl iodide, which makes the amine into a better leaving group; elimination is then carried out under E1cb conditions (Section 23.1b). For example, a method to prepare α-methylene lactones—a functional group found in many antitumor drugs—is performed in four steps:

1. Formation of the enolate ion.
2. Alkylation of an enolate ion with Eschenmoser's salt.
3. Methylation of the amine nitrogen atom.
4. Elimination of trimethylamine by treatment with a weak base (HCO_3^-).

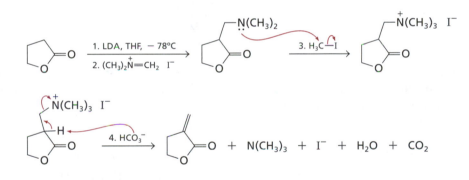

EXERCISE 24.3

Draw the product expected when the following molecule is treated with LDA and then Eschenmoser's salt. Write a mechanism for each step of the transformation.

24.2   CONJUGATE ADDITION REACTIONS

24.2a THE CONJUGATE ADDITION OF WATER FOLLOWS THE REVERSE OF THE E1CB MECHANISM

The general mechanism of conjugate addition processes can be illustrated with a simple but important reaction: hydration. This reaction—the conjugate addition of water—is a key step in several metabolic pathways. Conjugate addition is also called *1,4-addition* to differentiate it from direct addition, which is *1,2-addition* (Section 18.1b). Counting from the oxygen atom, which is designated atom No. 1, conjugate addition takes place when the nucleophile reacts at atom No. 4. Because the molecule that adds often has the form HX (X = the nucleophile), tautomerism occurs after conjugate addition has taken place, so it looks as if HX has added only to the C=C double bond. For example, when HX = H–OH (hydration), the conjugate addition process occurs overall as follows:

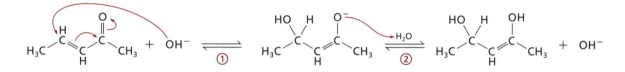

The product appears to have been formed by addition of H and OH to the alkene portion of the starting unsaturated ketone; however, addition actually occurs at atoms 1 and 4, and then tautomerism of the enol occurs to form the product.

In basic solution, hydroxide ion is the nucleophile that reacts with the α,β-unsaturated carbonyl compound, and it reacts at the end of the conjugated system (Step 1). The electrons move through the carbon framework and onto the oxygen atom, producing an enolate ion. This enolate ion is protonated by the solvent, water (Step 2),

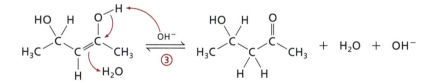

The enol tautomerizes to form the product, a β-hydroxy ketone (Step 3). If you look at the E1cb mechanism (Section 23.1b), you will see it is simply the reverse of the base-catalyzed conjugate addition of water.

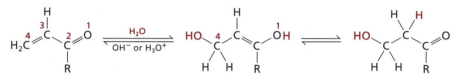

Hydration can also be catalyzed by acid, and the first step is protonation of the carbonyl oxygen atom, which activates the carbonyl group (Section 18.1c). A molecule of water then adds to the terminus of the conjugated system in Step 2. Deprotonation (Step 3) is followed by tautomerism (Step 4) to form the product.

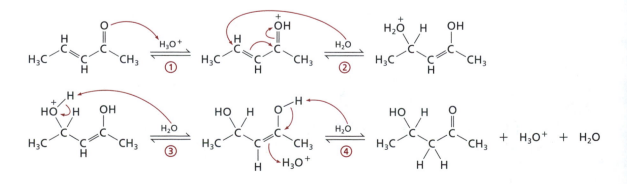

EXERCISE 24.4

When a heteroatom substituent is attached to the β-carbon atom of an α,β-unsaturated ketone, the acid-catalyzed hydration reaction yields the enol form of a β-diketone. Propose a reasonable mechanism for the following transformation.

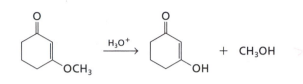

24.2b THE CONJUGATE ADDITION OF WATER DEFINES KEY STEPS IN SEVERAL METABOLIC PATHWAYS

The addition of water to an unsaturated carbonyl compound (conjugate addition of water) is an important reaction in biochemistry. In the citric acid cycle, this conjugate addition reaction is crucial for the isomerization of citrate to isocitrate, which proceeds via aconitate, an α,β-unsaturated diacid. The enzyme that catalyses this reaction is *aconitase*, which first catalyzes the E1cb dehydration of citrate to aconitate via specific deprotonation of the citrate molecule.

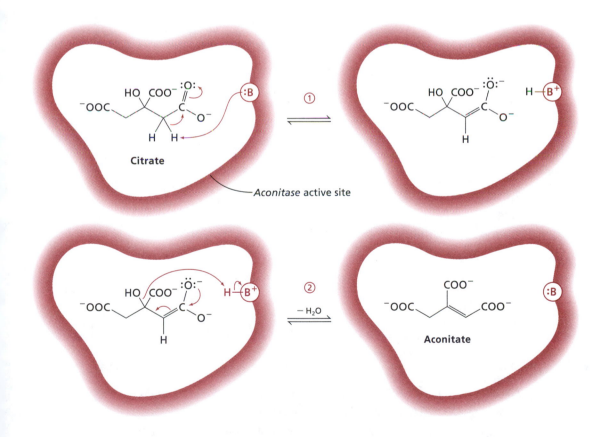

In the next stage, water undergoes conjugate addition to the double bond of aconitate. This transformation is enantiospecific, producing only (2R,3S)-isocitrate. Notice that the C=C double bond is conjugated with two different carbonyl groups, so the water molecule can add with either regiochemistry. In one direction, citrate is regenerated; in the other orientation, (2R,3S)-isocitrate is formed, as shown below.

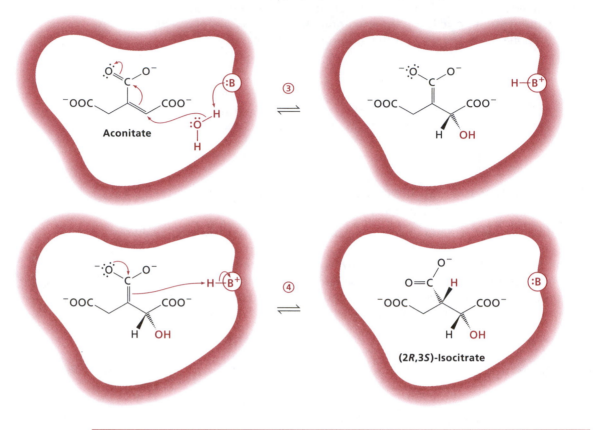

EXERCISE 24.5

Show the steps in the mechanism for the regeneration of citrate from aconitate.

In fatty acid metabolism, the conjugate addition of water is also a key step during the four-step sequence in each round of carbon-chain degradation, which was presented in Figure 23.1. After the first step—oxidation of the thioester to form an α,β-unsaturated thioester—water undergoes conjugate addition to the double bond:

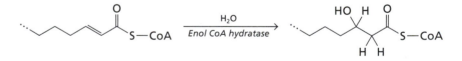

The resulting alcohol is subsequently oxidized to the β-keto thioester, and then a retro-Claisen reaction takes place (Section 23.4d).

24.2c MANY COMPOUNDS OF THE FORM H–X UNDERGO CONJUGATE ADDITION REACTIONS

The conjugate addition of water is the prototype for a series of reactions in which a molecule H–X (X = CN, N_3, NR_2, OR, SR. Cl, Br, and I) adds to the conjugated π bonds. The mechanisms of addition are identical to those observed for water, and most proceed successfully under basic conditions.

When HCN reacts with an unsaturated carbonyl compound in the presence of base, which generates a small amount of cyanide ion, conjugate addition occurs, followed by protonation and tautomerism. An acid–base reaction occurs between HX and hydroxide ion at the end of the reaction.

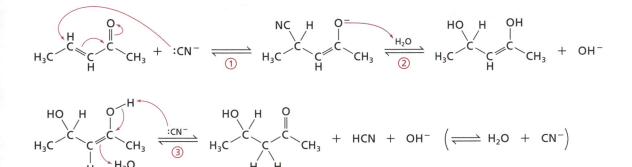

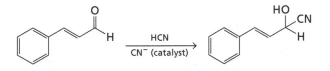

Recall that cyanide also adds to the carbonyl group of aldehydes under basic conditions (Section 18.2a). With α,β-unsaturated aldehydes, this direct addition reaction often predominates because the aldehyde carbonyl group is unhindered.

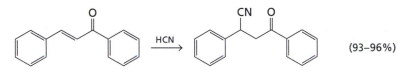

With α,β-unsaturated ketones and esters, the 1,4-addition pathway is the dominant one.

The versatility of the nitrile group makes conjugate addition of HCN an important synthetic reaction. Hydrolysis (Section 21.6) and reduction of the CN group (Section 11.2c) leads to formation of acids and amines, respectively.

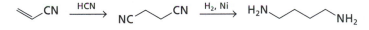

Ethyl acrylate

Acrylonitrile

Amines undergo conjugate addition reactions because they are good nucleophiles. The product of Step 1 is a zwitterion, which is relatively acidic ($pK_a \approx 10$), so the proton is transferred from nitrogen to oxygen. Tautomerism yields the final product. Notice that the overall process is the reverse of elimination from a Mannich base (Section 24.1b).

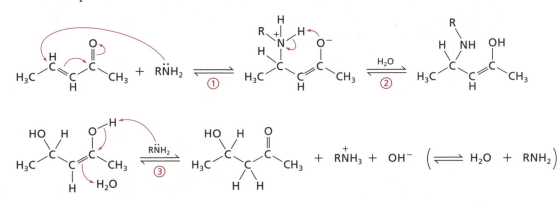

EXERCISE 24.6

Show how you would prepare each of the following compounds from cyclohexanone and any other reagents. Several steps may be necessary.

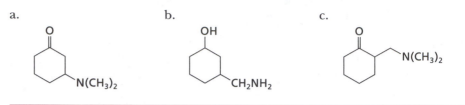

a.   b.   c.

Alcohols and thiols undergo conjugate addition with either acid or base catalysis, but the base-catalyzed reaction is more common and usually gives higher yields. You have already seen one example of thiol conjugate addition in the mechanism by which calicheamicin functions (Section 12.4d). Two synthetic examples are shown below.

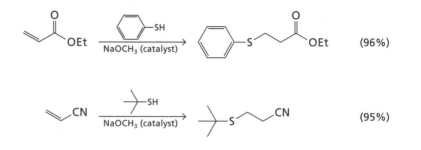

EXERCISE 24.7

Propose a reasonable mechanism for the reaction between thiophenol (C_6H_5SH) and ethyl acrylate, shown directly above.

24.2d EPOXIDES ARE FORMED BY CONJUGATE ADDITION OF THE HYDROPEROXIDE ION TO ENONES

An addend that differs from those described in Section 24.2c is the hydroperoxide ion. The initial conjugate addition step occurs in the way that is typical for nucleophiles.

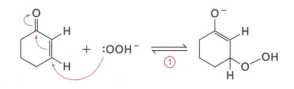

However, the intermediate enolate ion then reacts as a nucleophile toward the *oxygen atom* of the hydroperoxo group, displacing hydroxide ion as a leaving group. The product is an epoxide.

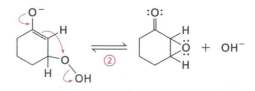

The contrast between the epoxidation reactions of enones and alkenes is significant. Recall that alkenes form epoxides when treated with a peracid: an alkene π bond is nucleophilic, and the oxygen atom is an electrophile (Section 11.3e). For an enone, the hydroperoxide ion is a nucleophile, which means the carbon–carbon π bond is electrophilic.

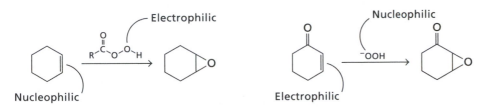

Further support for the notion that the carbon–carbon π bond of an enone is not nucleophilic comes from looking at competitive reactions between a peracid and double bonds in the same molecule. Only the nonconjugated double bond undergoes epoxidation.

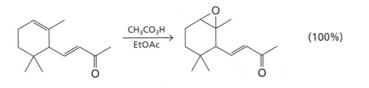

EXAMPLE 24.2

How would you prepare the epoxide shown here starting with any compound that has seven carbon atoms or fewer?

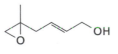

The most direct method for making an epoxide starts with an alkene, and the retrosynthetic analysis for this example reveals the diene shown below. Next, assess the reactivity profile of the double bonds—are they nucleophilic or electrophilic? In this diene, both are nucleophilic.

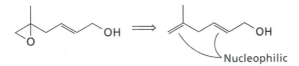

Next, consider how each double bond is substituted: As noted in Section 11.3e, more highly substituted alkenes react faster with peracids. Both of the double bonds is this molecule are disubstituted (Section 8.1c), so it would be difficult to epoxidize one and not the other. Therefore, the next retrosynthetic step is used to modify the molecule to differentiate the alkene groups. If the alcohol functional group were an aldehyde, then the adjacent double bond would be electrophilic, hence unreactive toward the peracid.

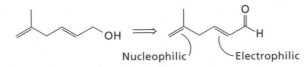

The aldehyde shown above has seven carbon atoms, so there is no need for additional retrosynthetic analysis. The synthesis therefore comprises two steps: epoxidation with a peracid, which reacts only with the nucleophilic double bond; and reduction of the aldehyde to yield the alcohol. Sodium borohydride is mild enough that it will not open the epoxide ring.

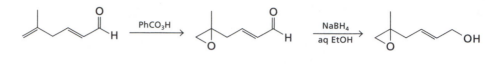

EXERCISE 24.8

Show how you would prepare the epoxide shown below, starting with any compounds with six or fewer carbon atoms.

24.3 CONJUGATE ADDITION REACTIONS OF CARBANIONS

24.3a STABILIZED ENOLATE IONS UNDERGO A CONJUGATE ADDITION PROCESS CALLED THE MICHAEL REACTION

The 1,2-addition of an enolate ion to a carbonyl group is the basis for the aldol and Claisen condensation reactions, two versatile methods to make carbon–carbon bonds (Chapter 23). When a carbanion is derived from a β-dicarbonyl compound, however, the equilibrium for the addition step lies toward the starting compounds.

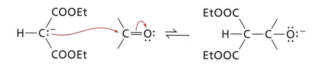

Stabilized enolate ions will react with an α,β-unsaturated carbonyl compound by *conjugate addition*, however. This transformation is called the **Michael reaction.** For example, 3-butene-2-one reacts with diethyl malonate in the presence of sodium ethoxide in ethanol.

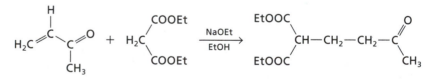

When an active methylene compound (Section 22.4a) is added to ethanol in which a small amount of ethoxide salt is dissolved, a carbanion is formed in Step 1 by an acid–base reaction. This carbanion subsequently undergoes 1,4-addition with the α,β-unsaturated carbonyl compound in Step 2.

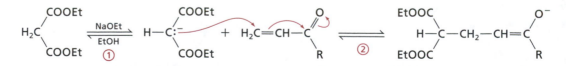

The enolate ion formed by the addition step subsequently removes a proton from the starting active methylene compound to form the product and regenerate the carbanion, which reinitiates Step 2.

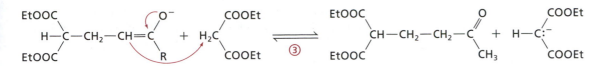

The following examples illustrate the range of reactions that occur by this process. The unsaturated component, called a **Michael acceptor**, is not limited to carbonyl compounds: Acrylonitrile ($CH_2=CH-CN$, propenenitrile) is an excellent Michael acceptor. The reactants that form the carbanion nucleophiles usually have pK_a values less than ~ 14. These reactants include β-diesters, β-diketones, β-keto esters, β-cyano esters, and nitro compounds.

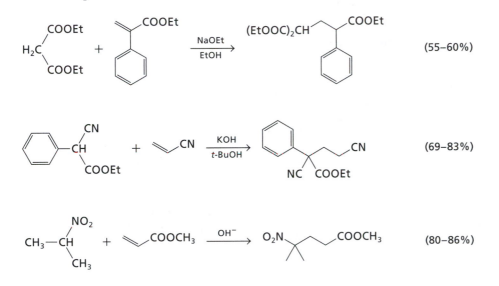

24.3b UNSTABILIZED ENOLATE IONS CAN ALSO UNDERGO THE MICHAEL REACTION

In a protic solvent such as an alcohol, the enolate ion reactant in the Michael reaction must derive from a molecule with a pK_a value less than ~ 14. If an aprotic solvent such as THF is used, then any enolate ion can be used.

The enolate ion is formed using a strong base such as LDA (Section 22.2d). Addition of a Michael acceptor results in conjugate addition and formation of a second enolate ion.

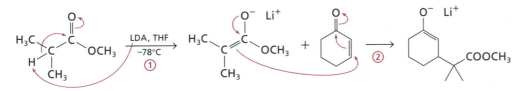

Aqueous acid workup leads to protonation of the enolate ion and formation of the product.

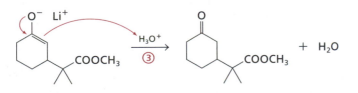

This procedure works well in many instances, giving high yields of 1,4-addition products. Aldol and Claisen condensation products can also be formed, which is a potential complication, so highly reactive Michael acceptors such as cyanoacrylates, which comprise the active ingredient in super glue, are sometimes used to circumvent this problem.

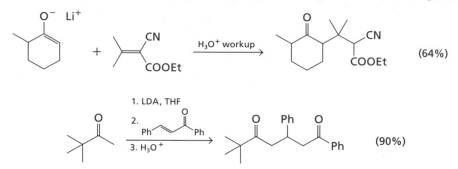

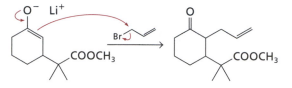

Conjugate addition of a nonstabilized enolate ion has a benefit, however, that the Michael reaction of a stabilized enolate ion lacks. After conjugate addition occurs to form the second enolate ion, workup can be performed by adding an alkyl halide instead of aqueous acid.

The product of this procedure, called **tandem addition–alkylation,** has two substituents more than the starting material. This strategy is represented schematically by the following equation, comprising two steps: reaction with a nucleophile followed by reaction with an electrophile.

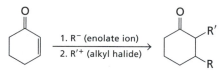

EXAMPLE 24.3

Show how you would prepare the product shown here from compounds with seven or fewer carbon atoms. One of the steps should be a Michael reaction.

To prepare a ketone using a Michael reaction, plan the retrosynthesis by breaking a bond to the carbon atom β to the carbonyl group: The fragment that is disconnected is attached as its enolate ion. Any group α to the carbonyl group can be attached using an alkyl halide in the second phase of the procedure. For this compound, the disconnections are made as follows:

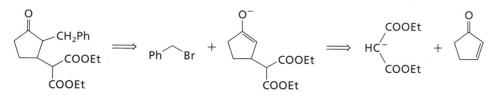

The synthesis follows the reverse of these steps and includes formation of the enolate ion. If an alkylation step is included in the synthesis, as it is here, then an aprotic solvent should be used.

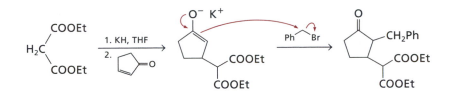

EXERCISE 24.9

Show how you would prepare each of the following products from compounds with eight or fewer carbon atoms. One of the steps should be a Michael reaction.

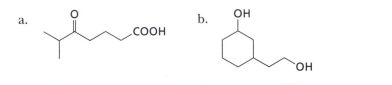

24.3c THE ROBINSON ANNULATION PROCESS MAKES USE OF THE MICHAEL AND ALDOL REACTIONS

A particularly useful variant of the Michael reaction is the **Robinson annulation,** which is used to prepare cyclohexenone derivatives.

The first step is formation of an enolate ion, which undergoes a Michael reaction with methyl vinyl ketone (3-buten-2-one). (Other unsaturated ketones can also be used in this procedure.)

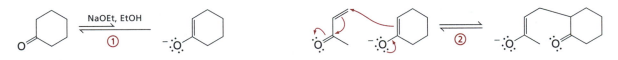

The enolate ion formed in this step undergoes two acid–base reactions to form its isomeric enolate ion.

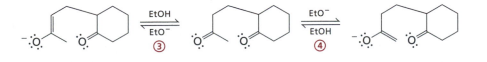

EXERCISE 24.10

Show the details of the mechanism by which the two enolate ions in the preceding scheme are interconverted; use curved arrows to denote the movement of electrons.

This enolate ion adds to the other ketone group in a crossed-aldol reaction, which creates the six-membered ring. The alkoxide ion that forms is subsequently protonated and undergoes elimination (E1cb mechanism) to form the cyclohexenone product.

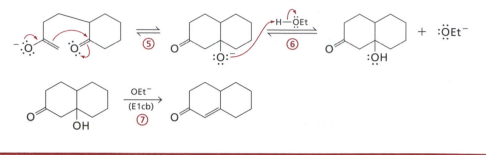

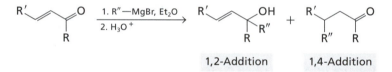

EXERCISE 24.11

Write the steps for the mechanism of the E1cb reaction shown directly above.

24.3d   ORGANOMETALLIC REAGENTS VARY IN THEIR REACTIONS WITH α,β-UNSATURATED CARBONYL COMPOUNDS

Enolate ions are carbanions that are stabilized by delocalization with a carbonyl group, and they generally undergo conjugate addition with α,β-unsaturated carbonyl compounds. Unstabilized carbanions—those in organometallic compounds—can react quite differently; in fact, Grignard reagents undergo mostly 1,2-addition with α,β-unsaturated carbonyl compounds unless the carbonyl group is very hindered. Organolithium compounds are even more prone to yield only 1,2-addition products because they are less sensitive to steric effects.

In contrast to the reactions of the highly reactive RMgX and RLi reagents with α,β-unsaturated carbonyl compounds, organocuprates react almost exclusively by conjugate addition pathways. These reactions include both the Gilman and the higher order cuprate reagents (Section 15.3a).

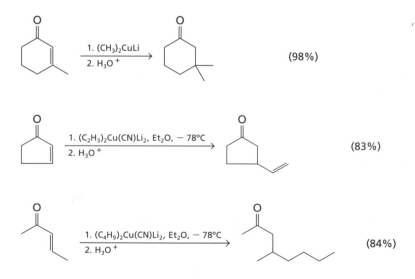

Neither type of organocuprate reagent—Gilman or higher order—reacts well with unsaturated esters, however. Organocuprate–boron trifluoride adducts, RCn (BF$_3$) are preferred for such reactions.

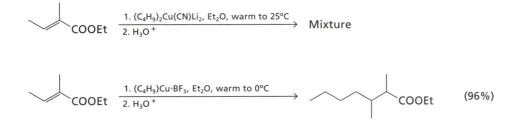

EXAMPLE 24.4

Using an organometallic reagent and any organic compounds with seven or fewer carbon atoms, show how you would prepare the following compound:

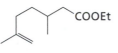

For a carbonyl compound, the retrosynthesis can normally be planned by breaking a bond to the carbon atom β to the carbonyl group: The fragment that is disconnected is attached during synthesis by using an organocuprate reagent, which in turn is made from an organolithium compound. For this compound, the disconnection is made as follows:

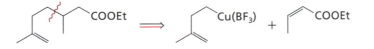

To perform the conjugate addition reaction between an organocuprate reagent and an ester, the cuprate reagent is made with 1 equiv of copper ion, followed by addition of BF$_3$.

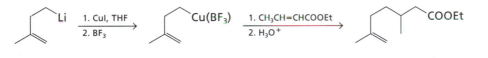

Using an organometallic reagent and any organic compounds with seven or fewer carbon atoms, show how you would prepare the compounds below:

a.   b.

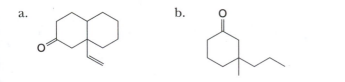

24.3e CONJUGATE ADDITION BY ORGANOCUPRATE REAGENTS CAN BE COMBINED WITH ENOLATE ION ALKYLATION

Just like conjugate addition reactions of enolate ions in aprotic solvents, conjugate addition reactions of organocuprates produce enolate ions. Therefore, workup can be carried out by adding an alkyl halide instead of aqueous acid, and the product will have two additional substituents. This procedure is another example of *tandem addition–alkylation* (Section 24.3b).

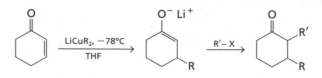

The two organic groups that are introduced by this strategy often have a transrelationship to each other because that stereochemistry minimizes steric interactions between the two groups. If the cis-isomer is formed, then epimerization via the enolate ion is often possible (Section 22.2b). In the following example, even a weak base such as acetate ion is sufficient to epimerize the carbon atom adjacent to the carbonyl group, producing the thermodynamically favored trans product.

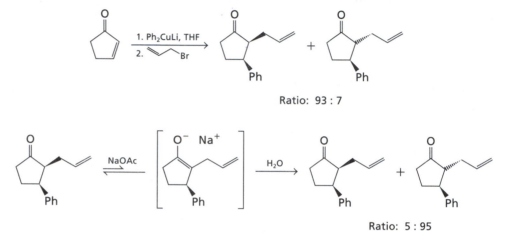

Ratio: 93 : 7

Ratio: 5 : 95

Furthermore, if PhSeCl is added to the product from the conjugate addition step with an organocuprate reagent, then the conjugated system can be regenerated by oxidation. Tandem addition–phenylselenation takes place in Stage 1 of the following reaction scheme. Oxidation of that product using hydrogen peroxide in Stage 2 leads to formation of the enone.

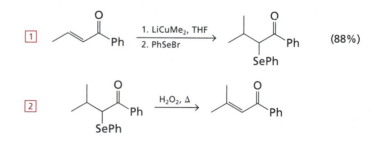

This series of transformations can be exploited so that *two* groups are added to the β-carbon atom of an enone, as summarized by the following scheme:

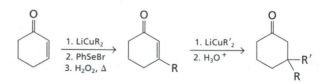

EXERCISE 24.13

Show how you would prepare the following compounds from cyclohexenone and any other organic or organometallic compounds with six or fewer carbon atoms.

a.

b.

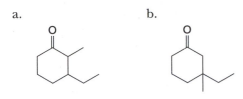

24.3f THE TANDEM ADDITION–ALKYLATION PROCEDURE HAS A BIOCHEMICAL COUNTERPART

The use of an organocuprate reagent to form a specific enolate ion is valuable because the enolate ion can be further elaborated by alkylation. The nonaqueous conditions required for the tandem addition–alkylation procedure might lead you to conclude that no counterpart exists in biological systems, yet Nature *does* make use of the tandem addition–alkylation strategy using different nucleophiles for the initial conjugate addition step. A specific example of this process comprises the biosynthesis of thymidine, which is one of the building blocks of *deoxyribonucleic acid* (DNA), the molecule used to store and transmit genetic information in organisms.

Thymidine is a *nucleotide,* a derivative of ribose with a phosphate group at the 5′-position and a heterocycle attached via a glycosidic bond (Section 17.4) to the 1′-position (the prime marks denote substitution in the sugar ring). During biosynthesis, 2-deoxyuridine monophosphate is methylated to form thymidine monophosphate. The enzyme *thymidylate synthase* catalyzes this reaction, which also requires the coenzyme N^5,N^{10}-methylenetetrahydrofolate (MeTHF).

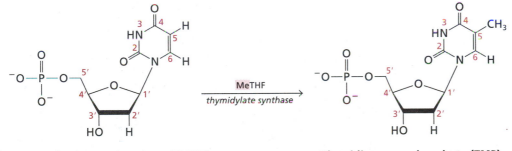

2′-Deoxyuridine monophosphate (dUMP)
Not found in RNA or in DNA

Thymidine monophosphate (TMP)
A component of DNA;
not found in RNA

The methyl group at position 5 in TMP comes from a methylene group and a hydrogen atom (highlighted in the following structure) of the coenzyme.

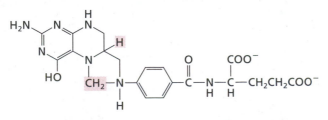

N^5,N^{10}-**Methylenetetrahydrofolate (MeTHF)**

The active site of the enzyme has two reactive centers: a cysteine thiolate group and an acidic site. It also has binding sites for the substrate and coenzyme. In the first step, the thiolate ion undergoes conjugate addition to the heterocycle. This step creates an enolate ion.

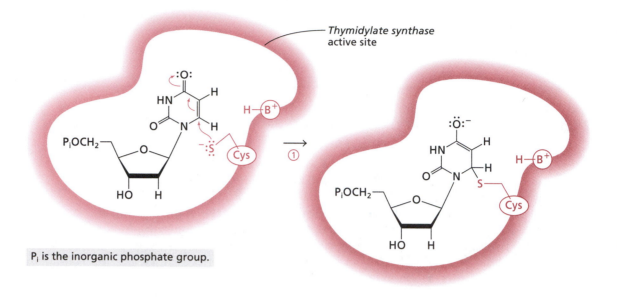

P$_i$ is the inorganic phosphate group.

In the second step—the alkylation step—the enolate ion reacts with the methylene group of MeTHF. The methylene carbon atom is electrophilic because it is attached to two nitrogen atoms. During reaction at this methylene group, the N^{10} atom functions as the leaving group, acquiring a proton as its bond to the CH$_2$ group is broken.

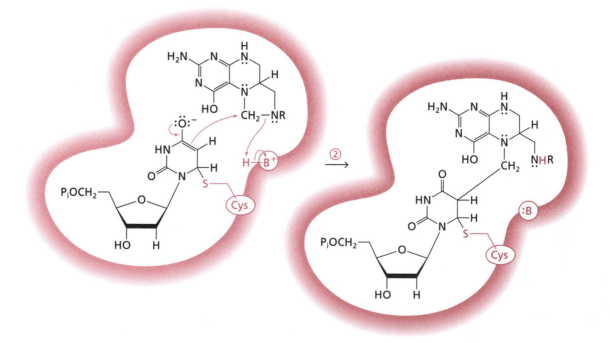

The next step creates an α-methylene carbonyl compound by elimination.

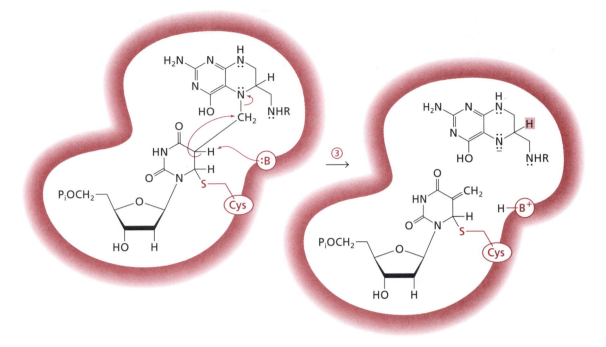

The final stage of the overall transformation, which is not shown, involves a conjugate reduction reaction in which the hydrogen atom (highlighted in the structure at the right, above) is transferred to the methylene group to create the methyl group. This example serves to illustrate that tandem addition–alkylation is not limited to laboratory reactions but rather is involved in critical biochemical transformations as well.

24.4   REDUCTION REACTIONS

24.4a   METAL HYDRIDE REAGENTS CAN BE USED FOR THE SELECTIVE REDUCTION OF UNSATURATED CARBONYL COMPOUNDS

Hydride ion and its equivalent forms (e.g., H_2 and a metal catalyst) can reduce either or both of the double bonds in a conjugated π-system. If conjugate addition occurs first, then further reduction of the carbonyl group can follow, and a mixture of products is often obtained under such conditions.

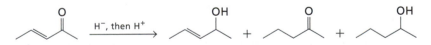

Sodium borohydride and lithium aluminum hydride react *preferentially* with the carbonyl group of an α,β–unsaturated aldehyde or ketone, but the reduction of both π bonds is not uncommon. When exclusive 1,2-addition is desired, the combination of cerium (III) chloride and $NaBH_4$ ensures that the allylic alcohol is the major product. The cerium ion most likely coordinates with the oxygen atom, activating the carbonyl group for addition and directing hydride reduction from a coordinated borohydride group, as shown here:

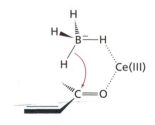

The following examples illustrate the contrast during the course of enone reduction by sodium borohydride in the absence and presence of cerium chloride:

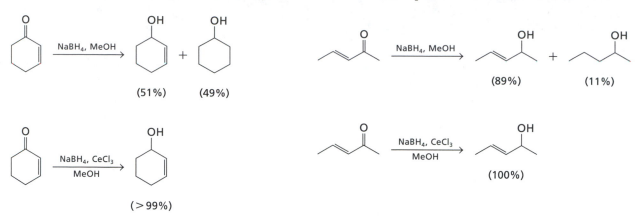

To reduce the carbonyl group of an unsaturated ester, AlH_3 or diisobutylaluminum hydride (Dibal–H) promote clean reduction to yield the allylic alcohol.

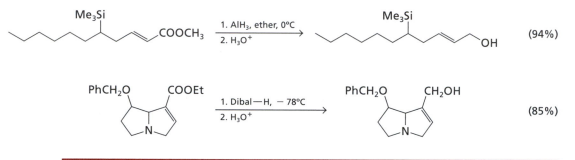

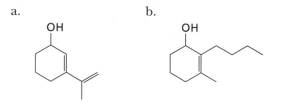

EXERCISE 24.14

Show how you would prepare each of the following compounds starting with cyclohexenone and any other organic compound with six or fewer carbon atoms.

a.

b.

24.4b CATALYTIC HYDROGENATION CAN BE USED TO REDUCE ENONES

For many α,β-unsaturated carbonyl compounds, reduction of the alkene portion of the molecule is straightforward and makes use of heterogeneous catalytic hydrogenation. An example is shown below. Only recently has it become practical to reduce the carbonyl group of α,β-unsaturated ketones catalytically without affecting the C=C double bond.

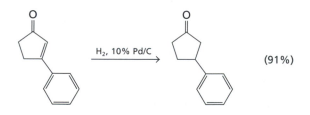

Homogeneous ruthenium catalysts are especially useful for reducing only the carbonyl portion of α,β-unsaturated ketones, and their chiral analogues exhibit remarkable levels of enantioselectivity as well. To perform the reduction without regard for the stereochemistry of the alcohol product, one employs catalysts of the form $RuCl_2(Ar_3P)_2(en)$, where "en" is $H_2NCH_2CH_2NH_2$ (ethylenediamine). A base such as $KOC(CH_3)_3$ (K_2CO_3, if the compound is sensitive to strong base) is added to a 2-propanol solution of the ketone and catalyst, and the mixture is allowed to react under several atmospheres of H_2 pressure. Both chemical yields (> 96%) and degree of chemoselectivity (C=O > C=C) are high.

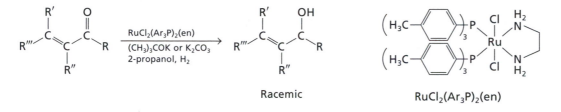

Racemic $RuCl_2(Ar_3P)_2(en)$

To prepare a chiral product, the same Noyori catalysts that were described in Section 18.3c are used. In reactions that are characterized by high chemical yields (>96%) and nearly complete chemoselectivity, these chiral catalysts reduce α,β-unsaturated ketones to allylic alcohols with a large degree of enantioselectivity as well (nearly 100%, ee). The following examples show the range of allylic alcohols that can be made with these catalysts.

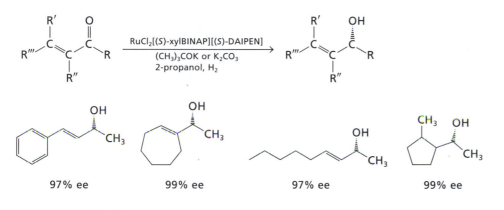

| 97% ee | 99% ee | 97% ee | 99% ee |

EXERCISE 24.15

Show how you would prepare each of the following compounds from any organic compounds that have six or fewer carbon atoms. You may use any reagents or solvents also.

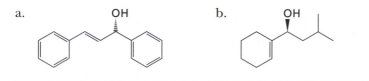

a.   b.

24.4c SOLVATED ELECTRONS CAN BE USED TO REDUCE THE C=C BOND OF ENONES

A method for the conjugate reduction of α,β-unsaturated ketones proceeds by addition of electrons from a metal like lithium dissolved in liquid ammonia—*a dissolving metal reduction*. You have learned previously that this procedure is used to make cis-alkenes from alkynes (Section 12.2c). Likewise, enones readily add an electron to produce radical-anion intermediates.

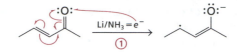

Transfer of a second electron, followed by two protonation steps, gives the ketone product.

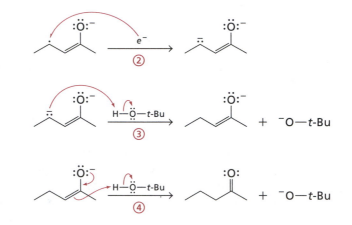

As an acid, liquid ammonia is too weak to protonate the enolate ion in Step 4, so a proton donor such as alcohol (*t*-BuOH or EtOH) is usually added to the reaction mixture. A cosolvent of ether or THF is also added to dissolve the organic reactants and products.

The stereochemistry of the product depends on the structure of the intermediate just prior to the second protonation step. If a single stereocenter is created, a racemic mixture is formed.

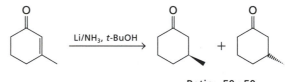

Ratio: 50 : 50

On the other hand, if the starting enone is chiral, then the products are diastereomers, and the stereochemistry depends on the experimental conditions. A well-studied class of compounds comprises the $\Delta^{1,9}$-2-octalones, which normally give a product that has a trans-ring junction.

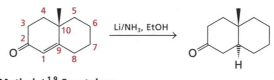

10-Methyl-$\Delta^{1,9}$-2-octalone

Δ means a double bond is present.

24.4d SOLVATED ELECTRONS CAN BE USED TO FORM SPECIFIC ENOLATE ION DERIVATIVES OF ENONES

If you look at the mechanism for the enone reduction reaction described in Section 24.4c, you will see that the final step (Step 4) is an acid–base reaction that results in protonation of the enolate ion. An acid with a pK_a value < 20 is required for this step, and alcohols or water are often used for this purpose. If no proton source other than liquid ammonia is supplied, then the reduction process stops at the enolate ion stage. (In some cases, 1 equiv of water or an alcohol is added to provide a proton for Step 3 of the scheme shown in Section 24.4c.) If the enolate ion is not protonated, it can be alkylated with methyl, primary, allylic, and benzylic halides. It is sometimes necessary to remove the ammonia, which as a nucleophile can compete with the enolate ion for the alkylating agent.

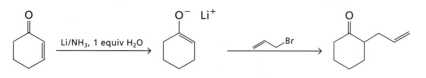

The dissolving metal reduction of an enone is one of the best ways to generate a specific enolate ion regioselectively. For example, in the following case, deprotonation would yield a mixture of enolate ions because a methylene group flanks each side of the carbonyl group (Section 22.2e).

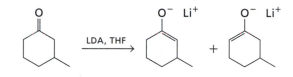

A dissolving metal reduction, on the other hand, produces the enolate ion on the side of the ketone that has the double bond, and that is the side of the carbonyl group at which alkylation also occurs. The following two examples illustrate the specific alkylation of an enolate ion produced by reduction of an enone.

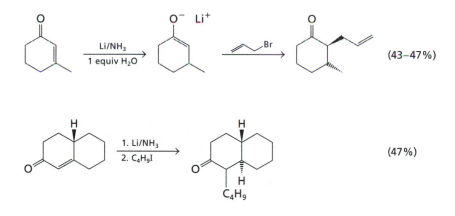

(43–47%)

(47%)

EXERCISE 24.16

What is the major product expected from each of the following sequences?

a.   b.

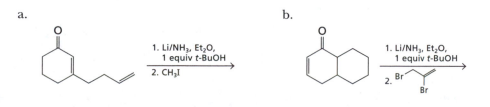

CHAPTER SUMMARY

Section 24.1 α,β-Unsaturated carbonyl compounds

- Three general methods to prepare an α,β-unsaturated carbonyl compound include the Horner–Emmons reaction, phenylselenation followed by oxidation and elimination, and the aldol reaction followed by dehydration.

- The Mannich reaction generates an α,β-unsaturated ketone by elimination of an amine from the product of the reaction between an enolizable ketone and an iminium salt of formaldehyde.

Section 24.2 Conjugate addition reactions

- Conjugate addition, also called 1,4-addition, is the reaction of a nucleophile at the carbon atom β to the carbonyl group of an α,β-unsaturated carbonyl compound.

- If a nucleophile that undergoes conjugate addition is added to the reaction mixture as its conjugate acid (H–Nuc), the product of the addition step will have a proton attached to the oxygen atom of the carbonyl group; tautomerism regenerates the carbonyl group.

- Water undergoes conjugate addition to an α,β-unsaturated carbonyl compound.

- Conjugate addition of water is a key step in several important biochemical transformations including fatty acid metabolism and isomerization of citrate during turnover of the citric acid cycle.

- Alcohols, thiols, amines, and HCN participate in conjugate addition reactions with α,β-unsaturated carbonyl compounds. These reactions are catalyzed by either acid or base.

- Conjugate addition of the hydroperoxide ion is used to make the epoxide derivative of an α,β-unsaturated ketone.

Section 24.3 Conjugate addition reactions of carbanions

- Active methylene compounds participate in conjugate addition reactions under basic conditions; the general process is called the Michael reaction.

- Enolate derivatives of ketones and esters participate in the Michael reaction if the reaction is performed in an aprotic solvent.

- The product of a Michael reaction between an α,β-unsaturated ketone and a preformed enolate ion is a new enolate ion that can subsequently be alkylated. This procedure is called tandem addition–alkylation.

- The Robinson annulation reaction is a two-step process that generates a cyclohexenone derivative; the two steps comprise conjugate addition followed by an intramolecular aldol reaction.

- Grignard and organolithium compounds normally undergo direct addition (1,2-addition) to α,β-unsaturated carbonyl compounds.

- Organocuprate reagents generally undergo conjugate addition when they react with α,β-unsaturated carbonyl compounds.

- The enolate product that is formed by conjugate addition of an organocuprate to an α,β-unsaturated ketone can be alkylated; this transformation is another example of tandem addition–alkylation.

- The enolate product formed by conjugate addition of an organocuprate to an α,β-unsaturated ketone can be treated with PhSeCl; oxidation of the resulting PhSe derivative regenerates the α,β-unsaturated carbonyl functionality.

- A biochemical example of tandem addition–alkylation is found in the biosynthesis of thymidine, a building block of DNA.

Section 24.4 Reduction reactions

- The combination of $NaBH_4$ and $CeCl_3$ is used to reduce an α,β-unsaturated ketone to the corresponding allylic alcohol.
- Aluminum hydrides (other than $LiAlH_4$) are used to reduce an α,β-unsaturated ester to the corresponding allylic alcohol.
- Ruthenium catalysts are used to reduce α,β-unsaturated ketones with H_2 to form the corresponding allylic alcohols. Chiral ruthenium catalysts lead to enantioselective reduction of the carbonyl group.
- Lithium in liquid ammonia, with excess protons from a source such as *tert*-butyl alcohol, reduces α,β-unsaturated ketones to the corresponding saturated ketones.
- Lithium in liquid ammonia *without* an added proton source (or with 1 equiv of a proton source) generates an enolate ion from an α,β-unsaturated ketone. The enolate ion can be alkylated at the position α to the carbonyl group.

KEY TERMS

Introduction
conjugate addition
1,4-addition

Section 24.1b
Mannich reaction
Mannich base
enone

Section 24.3a
Michael reaction
Michael acceptor

Section 24.3b
tandem
   addition–alkylation

Section 24.3c
Robinson annulation

REACTION SUMMARY

Section 24.1a

General methods to prepare α,β-unsaturated carbonyl compounds are summarized in Figure 24.1

Section 24.1b

The Mannich reaction. A ketone reacts with the salt of an amine and formaldehyde to form an adduct that upon heating undergoes elimination to form the corresponding α,β-unsaturated ketone.

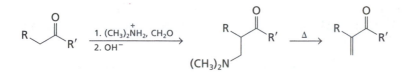

Section 24.2a

α,β-Unsaturated carbonyl compounds react with water by conjugate addition pathways. The reaction can be catalyzed by acid or base.

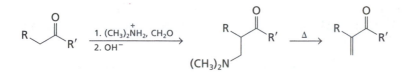

Section 24.2c

α,β-Unsaturated carbonyl compounds react with alcohols, thiols, amines, and HCN by conjugate addition pathways. The reaction is normally catalyzed by base.

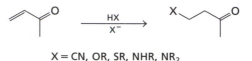

X = CN, OR, SR, NHR, NR₂

Section 24.2d

α,β-Unsaturated ketones react with the hydroperoxide ion to form epoxides at the carbon–carbon double-bond portion of the molecule.

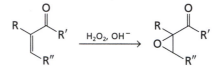

Section 24.3a

The Michael reaction: α,β-unsaturated carbonyl compounds undergo conjugate addition reactions in alcohol solvents with enolate ions derived from β-dicarbonyl compounds.

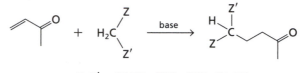

Z, Z' = COOEt, CHO, COR, CN, NO₂

Section 24.3b

α,β-Unsaturated carbonyl compounds undergo conjugate addition reactions with non-stabilized enolate ions in an aprotic solvent.

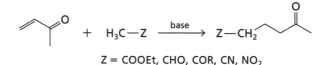

Z = COOEt, CHO, COR, CN, NO₂

Section 24.3c

The Robinson annulation. An α,β-unsaturated ketone reacts with methyl vinyl ketone by a combination of Michael and aldol reactions to form a cyclohexenone derivative.

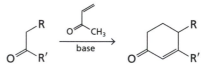

Section 24.3d

α,β-Unsaturated carbonyl compounds undergo direct (1,2-) addition with Grignard and organolithium reagents.

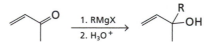

α,β-Unsaturated carbonyl compounds undergo conjugate addition reactions with organ-ocuprate reagents.

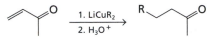

Section 24.3e

The enolate ion intermediate formed by conjugate addition of an organocuprate reagent can be alkylated or treated with phenylselenyl chloride.

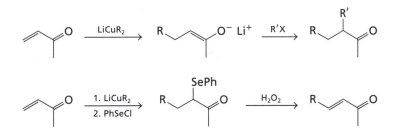

Section 24.4a

The carbonyl portion of α,β-unsaturated carbonyl compounds can be reduced with certain hydride reagents.

Section 24.4b

Hydrogenation of α,β-unsaturated carbonyl compounds. Chiral Noyori catalysts lead to enantioselective reduction of the carbonyl group.

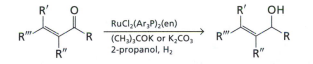

Section 24.4c

Treating an α,β-unsaturated carbonyl compound with lithium in liquid ammonia and *tert*-butyl alcohol (added as a proton source) leads to reduction of the C=C double bond.

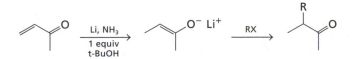

Section 24.4d

The enolate ion formed via dissolving metal reductions can be alkylated.

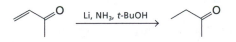

ADDITIONAL EXERCISES

24.17.   Draw the structure that corresponds to each of the following substances:

a. Methyl 2-phenylselenylbutanoate

b. (S)-3-Mehyl-2-isopropylidenecyclopentanone

c. 2,3-Epoxycyclohexanone

d. 2-Dimethylaminomethylcycloheptanone

24.18.   Provide a systematic name for each of the following molecules:

a. 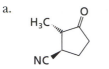   b.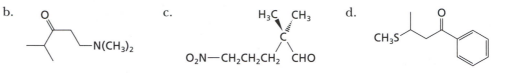

24.19.   Propose a reasonable mechanism for the following transformation, which begins with conjugate addition of water.

24.20.   A chemist decided to prepare compound **A** by carrying out a base-catalyzed conjugate addition reaction between water and 2-isopropylidenecyclohexanone. Upon workup, he detects the odor of acetone and isolates compound **B** instead. What is the structure of **B**? Propose a mechanism to explain what happened.

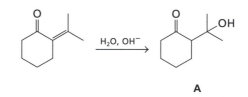

A

24.21.   The synthesis of diethyl *tert*-butylmalonate cannot be accomplished by the following route. Explain. Instead it is made by conjugate addition of CH₃MgBr to the 2-isopropylidene derivative of diethyl malonate. Show how you would make this unsaturated compound from diethyl malonate and any other reagents and solvents (*Hint:* See Exercise 23.20).

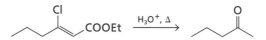

24.22.   Starting with compounds that have six or fewer carbon atoms, show how you would synthesize each of the following via a Michael reaction. Several steps may be necessary.

a.   b.

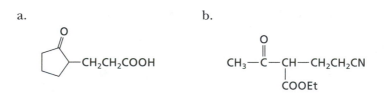

24.23. The following transformation proceeds by a combination of Michael and crossed-aldol reactions. Propose a mechanism for this process, called the *Baylis–Hillman reaction.*

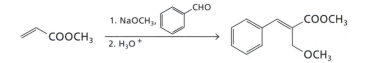

24.24. Conjugate addition of cyanide ion to an enone can be accomplished by using diethylaluminum cyanide. Aluminum is a very oxophilic (oxygen loving) element, so it acts as a Lewis acid toward oxygen. Propose a reasonable mechanism for the first part of the following reaction sequence.

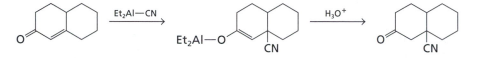

24.25. Draw the structure(s) of the major product(s) expected from each of the following reactions. Indicate the stereochemistry of the product as appropriate. Relative stereochemistry should be shown using wedges and dashed lines. If a racemic mixture will be formed, draw the structure of one enantiomer and write the word "racemic", or draw both enantiomeric structures. If diastereomers are formed, draw each structure; label meso compounds as such. If no reaction occurs, write N.R.

a.

b.

c.

d.

e.

f.

g.

h.

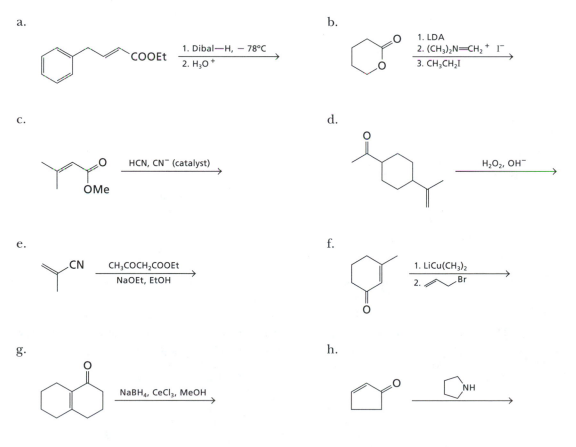

24.26. In the conversion of citrate to aconitate (Section 24.2b), the *pro-R* hydrogen atom of the *pro-R* carboxymethyl group (Exercise 16.27) is removed along with the OH group.

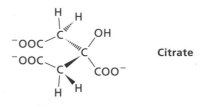

Citrate

a. Assign *pro-R* and *pro-S* to each of the two carboxymethyl groups, assuming that one of the hydrogen atoms of one of the methylene groups is replaced by OH.

b. Assign *pro-R* and *pro-S* to each of the hydrogen atoms of each methylene group, assuming replacement by OH.

24.27. In many organisms, aerobic metabolism makes use of a sequence of reactions called the *citric acid cycle,* summarized below. Draw the structure for each compound in this scheme. Propose a mechanism for each numbered step based on the descriptive phrase given.

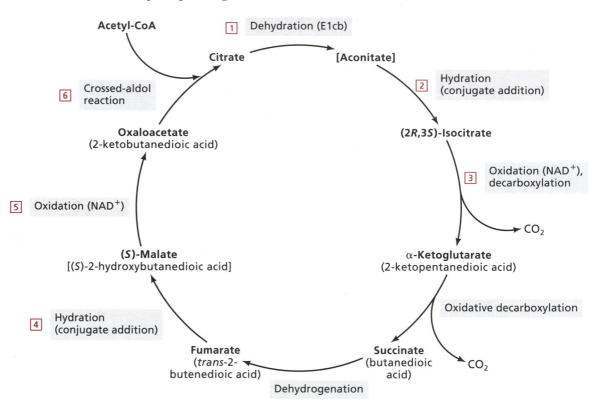

24.28. Starting with any compounds that have six or fewer carbon atoms, show how you would synthesize each of the following via the Robinson annulation procedure with methyl vinyl ketone as one reactant. Several steps may be necessary.

a.   b.   c.

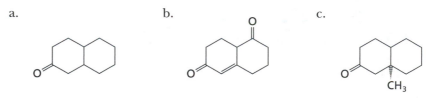

24.29. In performing each of the following reactions, indicate what you would look for in the ^{1}H and broad-band decoupled ^{13}C NMR spectra to be certain that the reaction occurred and was successful. List, as applicable, the chemical shifts, integrated intensity values, splitting patterns, and DEPT results.

a.

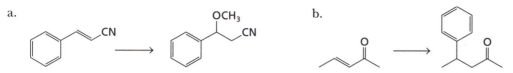

b.

24.30. Draw the structure of the product expected from each of the following sequences:

a.   b.

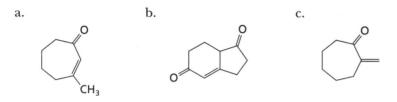

24.31. Starting with cycloheptanone and any compounds that have five or fewer carbon atoms, show how you would synthesize each of the following compounds. Several steps may be necessary.

a.   b.   c.

24.32. Compound **B** is the product of a Michael reaction between **A**, C_5H_8O, and diethylmalonate that has been subsequently heated with 10% aqueous hydrochloric acid for several hours. Compound **B** can also be obtained by ozonolysis of **C**, C_7H_{12}, followed by oxidative workup. The ^{1}H NMR spectrum of **B** has no feature that appears a singlet besides the signal for the carboxylic acid proton. If **C** is treated in sequence with (1) borane—THF; (2) hydrogen peroxide and sodium hydroxide; (3) chromium oxide in acetic acid; (4) and *m*-chloroperoxybenzoic acid (MCPBA), then compound **D**, $C_7H_{12}O_2$, is obtained. Compound **D** can also be made from **B** by stirring the latter with sodium borohydride in aqueous ethanol followed by workup with aqueous sulfuric acid. Draw structures for compounds **A–D**.

24.33. A radical can undergo conjugate addition to an unsaturated carbonyl compound. Propose a mechanism for the following cyclization reaction. (Review Chapter 12 to recall how such reactions are initiated and terminated.)

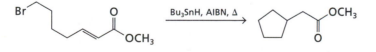

24.34.   Identify each of the following compounds from their ^{1}H NMR spectra. These compounds each have a strong carbonyl absorption in their IR spectra.

a. C_4H_6O

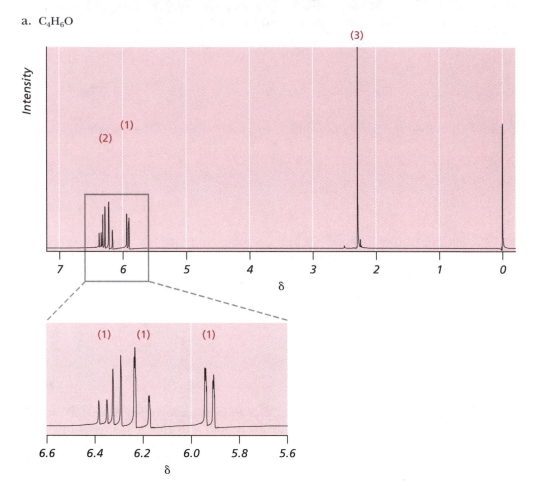

b. $C_5H_8O_2$

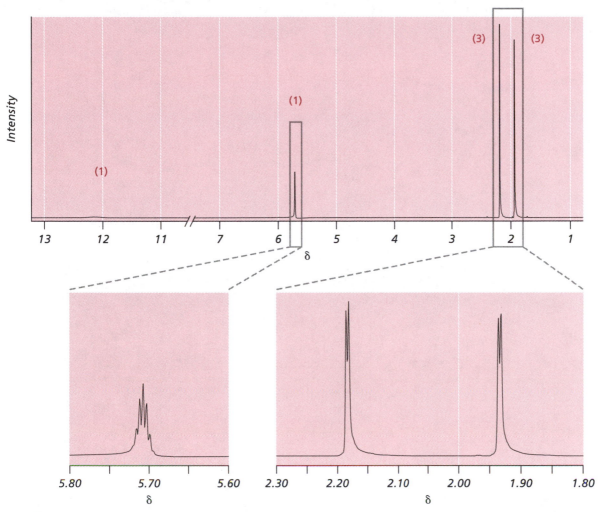

THE CHEMISTRY OF POLYCYCLIC AND HETEROCYCLIC ARENES

25.1 POLYCYCLIC AROMATIC COMPOUNDS

25.2 PYRIDINE AND RELATED HETEROCYCLES

25.3 PYRROLE AND RELATED HETEROCYCLES

25.4 AZOLES

CHAPTER SUMMARY

The chemistry of aromatic compounds, with an emphasis on benzene and its derivatives, was the subject of Chapter 17. Molecules with multiple rings (*polycyclic compounds*) and those that contain heteroatoms (*heterocycles*) can also be aromatic, and these were mentioned briefly in connection with Hückel's rule (Section 17.1e). This chapter presents more detailed descriptions of the reaction chemistry of polycyclic and heterocyclic aromatic compounds.

The properties of aromatic heterocycles can be quite different from their carbocyclic analogues. Aromatic nitrogen heterocycles in particular react in many instances more like carbonyl compounds than as benzene derivatives. Because of the vast assortment of aromatic heterocycles that exist, only a handful can be studied in any detail if you want to come away from this presentation with any useful knowledge. For that reason, we will look at the structures and reactions of heterocycles that are relevant to those found in biochemical systems.

25.1 POLYCYCLIC AROMATIC COMPOUNDS

25.1a FUSED POLYCYCLIC AROMATIC COMPOUNDS HAVE LESS THAN THE EXPECTED RESONANCE STABILIZATION ENERGY

Substances that have fused rings with delocalized π electrons are aromatic according to the $4n + 2$ rule (Section 17.1e). Naphthalene is the simplest such compound, and it has 10 π electrons in the two rings. Anthracene and phenanthrene each have 14 π electrons among three rings. Note in their structures that the carbon atoms without attached hydrogen atoms are not numbered.

Naphthalene

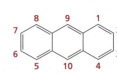

Anthracene

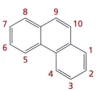

Phenanthrene

The π electrons in naphthalene are delocalized within the MOs, just as in benzene.

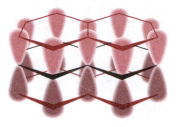

Delocalization of the π electrons can readily occur in naphthalene because the *p* orbitals are aligned.

When more than one ring is present, however, the resonance energy is not as great as might be expected. For example, the resonance energy for naphthalene is not equal to twice that of benzene's.

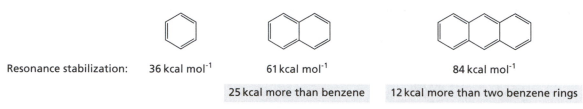

Resonance stabilization: 36 kcal mol⁻¹   61 kcal mol⁻¹   84 kcal mol⁻¹

25 kcal more than benzene   12 kcal more than two benzene rings

The reason for this inequality is a result of the unequal dispersion of electrons among all positions. For naphthalene, two of the three resonance forms have a double bond between carbon atoms 1 and 2. Likewise, two of the three forms have only a single bond between C2 and C3. The observed bond lengths in naphthalene reflect the unequal electron densities predicted by these resonance forms.

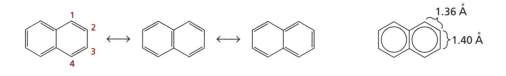

EXERCISE 25.1

Draw the important resonance forms for anthracene and phenanthrene. Predict the length of the C9–C10 bond in phenanthrene relative to the lengths of its other bonds.

25.1b NAPHTHALENE UNDERGOES ELECTROPHILIC AROMATIC SUBSTITUTION

Like benzene, naphthalene undergoes electrophilic aromatic substitution. However, electrophiles react more rapidly with naphthalene than with benzene, because the loss of resonance energy is only 25 kcal mol⁻¹ for the former but 36 kcal mol⁻¹ for benzene itself.

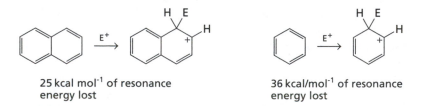

25 kcal mol⁻¹ of resonance   36 kcal/mol⁻¹ of resonance
energy lost   energy lost

The 1-position in naphthalene (also called the α position in the older literature) is more reactive than the 2-position (also called the β position) because the intermediate that is formed is more highly stabilized by resonance.

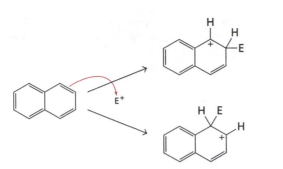

More highly stabilized
by resonance

In fact, seven structures can be drawn for the intermediate in which the electrophile is attached at C1, and four of these have an intact benzene ring. These four are expected to be the dominant resonance contributors. For the 2-substituted intermediate, we cannot draw as many good resonance contributors, suggesting that this carbocation is not as well stabilized.

These four structures are the major resonance contributors because they still have an intact benzene ring.

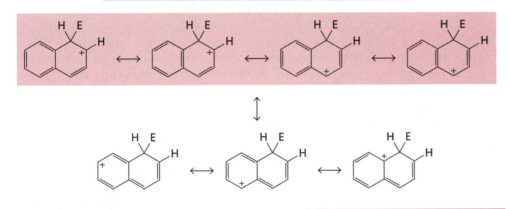

EXERCISE 25.2

Draw the expected resonance structures for the intermediate 2-substituted cation derived from naphthalene. How many of these structures have an intact benzene ring?

Stabilization of the cation intermediate by resonance means that 1-substituted naphthalenes are readily prepared by electrophilic substitution reactions.

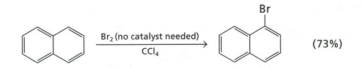

Steric hindrance can become important, however, when the electrophile is larger than a single atom. For example, nitration of naphthalene gives 1-nitro- and 2-nitro-naphthalene in a ratio of ~ 10:1. The presence of the 2-isomer is a consequence of steric hindrance between the substituents at C1 and C8.

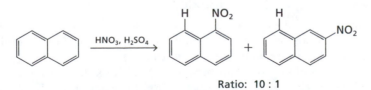

Ratio: 10 : 1

During sulfonation, which is a reversible process (see Exercise 17.8), the effect of steric crowding is even more notable. In fact, the 2-isomer is the major product if the reaction is conducted at high temperatures.

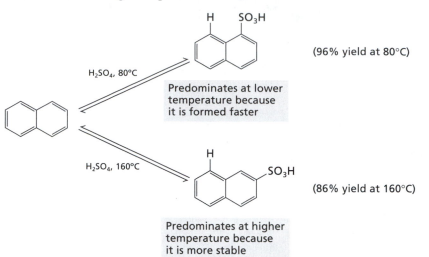

(96% yield at 80°C)

Predominates at lower temperature because it is formed faster

Predominates at higher temperature because it is more stable

(86% yield at 160°C)

This reaction illustrates yet another case of thermodynamic versus kinetic control (Section 10.3b). At the lower temperature of 80°C, the 1-isomer is formed faster because the intermediate cation is better stabilized. At higher temperatures, the more stable 2-isomer is the major product, because its formation minimizes the unfavorable steric interactions between positions C1 and C8.

Additional substitution reactions of naphthalene derivatives depend on the nature of the substituents already present. For example, 1-nitronaphthalene yields 1,5- and 1,8-dinitronaphthalene upon treatment with nitric and sulfuric acids. The ring with the nitro group is deactivated, so the adjacent ring is the one that reacts.

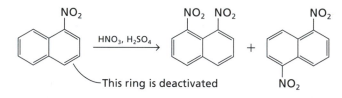

The substitution process becomes even more complicated when more than one substituent is present.

EXERCISE 25.3

What are the major products expected from the reaction shown here?

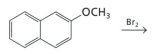

25.1c BI- AND TRICYCLIC ARENES ARE MORE REACTIVE THAN BENZENE TOWARD ADDITION AND OXIDATION

Resonance stabilization in a polycyclic arene is not overall as good as for benzene itself, consequently, their reactions with electrophiles are more facile. For example, benzene itself is inert toward most oxidants (see Section 17.4a).

In contrast, the middle ring of anthracene is readily oxidized upon treatment with chromium oxide in acetic acid. Remember that anthracene has only 12 kcal mol^{-1} of resonance energy more than that of two benzene molecules (Section 25.1a). There-

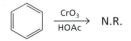

Benzene

fore, it makes sense that reactions will take place in the middle ring of anthracene, leading to the formation of two, isolated benzene rings.

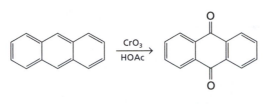

Anthracene   **9,10-Anthraquinone**

EXERCISE 25.4

Draw the structure of the product obtained by treating phenanthrene with CrO_3 and HOAc.

Anthracene reacts with reagents other than oxidants, and its reactions parallel those of 1,3-dienes (Section 10.3). For example, anthracene undergoes 1,4-addition with bromine, forming 9,10-dibromo-9,10-dihydroanthracene. When heated, this molecule *will* lose HBr to form 9-bromoanthracene because a small amount of resonance energy is gained by rearomatizing.

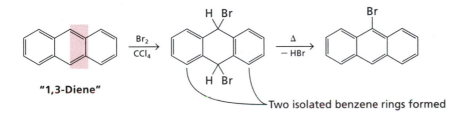

By comparing the reactions of bromine with a variety of aromatic hydrocarbons, you can see that the larger the π system, the more reactive and "alkene-like" the arene will be. The information needed to assess how a polycyclic arene will react is the number of resonance structures that retain "benzene-like" rings. A compound with more than three rings will not necessarily be *more* prone toward addition processes than anthracene.

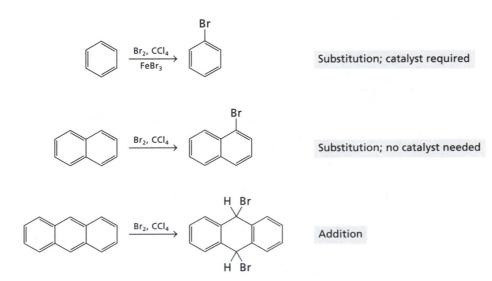

What product(s) would you expect to obtain from treating phenanthrene with bromine?

25.1d C$_{60}$ IS A CLOSED SHELL OF CARBON ATOMS THAT HAS BENZENE-LIKE RINGS

If you think about even larger polycyclic aromatic compounds, you can readily conceive of molecules with even more extended π systems. As early as the 1960s, theoretical studies suggested that a large π system might tend to curve back onto itself and form spherical arrays of carbon atoms. In 1985, H. W. Kroto, R. E. Smalley, R. F. Curl, and co-workers discovered that evaporating graphite under reduced pressure in an electric arc produced a substance that dissolved in benzene to form a wine-red solution. The molecular weight of this species, as well as other measured physical properties, suggested that the molecule had 60 carbon atoms, but no other elements.

The proposed structure consisted of a truncated icosahedron in which there are 60 vertices. Placing a carbon atom at each vertex forms 32 faces, 12 of which are pentagonal and 20 hexagonal. Each carbon atom has sp^2 hybridization, so each six-membered ring has three π bonds through which delocalization occurs over the entire molecule. Because its structure resembles the geodesic domes constructed by designer Buckminster Fuller, the C$_{60}$ molecule was called **buckminsterfullerene.**

In 1990, Wolfgang Krätschmer and co-workers described a method to make large quantities of C$_{60}$, and subsequent work led to the isolation of several derivatives. Because the system of rings closes back on itself, there is no need for hydrogen atoms to complete the valence shell of each carbon atom. Each carbon atom forms two C–C single bonds and a C=C double bond. No two five-membered rings are fused together, and each six-membered ring is fused to an alternating array of six- and five-membered rings.

It is clear that C$_{60}$ cannot react by electrophilic aromatic substitution even though it appears aromatic, because there are no hydrogen atoms to replace. On the other hand, there are plenty of π bonds, and many addition reactions that are observed with alkenes occur readily with C$_{60}$.

Buckminsterfullerene

25.2   PYRIDINE AND RELATED HETEROCYCLES

25.2a PYRIDINE, A HETEROCYCLIC ANALOGUE OF BENZENE, CAN FUNCTION AS A BASE AND A NUCLEOPHILE

Aromatic heterocycles, which are cyclic compounds having at least one heteroatom in their rings, have delocalized π systems with $4n + 2$ electrons (Section 17.1e). Pyridine is a six-membered ring compound that at first glance resembles benzene, except that a CH group has been replaced by N. Pyridine and benzene have in common similar Lewis structures, which comprise three alternating single and double bonds, and two prominent resonance forms. Each non-hydrogen atom has the expected octet of electrons.

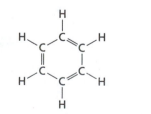

Benzene

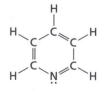

Pyridine

The valence bond representation for pyridine also looks the same as the one drawn for benzene. Each atom has a *p* orbital perpendicular to the plane of the ring, and each contributes a single electron to the π system.

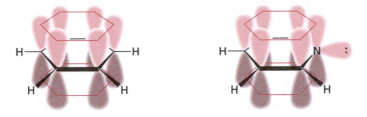

The unshared electron pair on nitrogen is in the plane of the ring and lies along the same line as a C–H bond in benzene. This feature is important for pyridine because *the nitrogen atom can function as a Lewis base or as a nucleophile without disrupting the aromatic π system.* For example, pyridine reacts as a base with the hydronium ion to form the pyridinium ion:

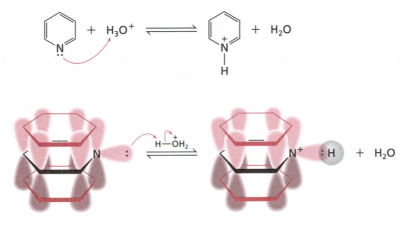

Pyridine is commonly used as a solvent to react with acidic byproducts in chemical reactions. For example, when an alcohol is treated with *p*-toluenesulfonyl chloride to make a tosylate derivative (Section 7.1b), pyridine not only dissolves the reactants, it also removes the HCl that is produced. This byproduct is pyridinium chloride, also called pyridine hydrochloride.

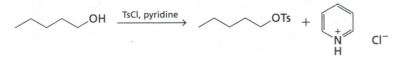

Pyridine is a weak base compared with an aliphatic or alicyclic amine (Section 5.2d). The pK_a value of the pyridinium ion is ~ 5.2, which is near that of acetic acid (pK_a = 4.75). Therefore, the basicity of pyridine is about the same as that of the acetate ion.

The basicity of pyridine also manifests itself by its reactions with metal ions, many of which are Lewis acids. Recall that chromium(VI) oxide reacts with pyridine (Py) to form $CrO_3 \cdot 2$ Py, a reagent used to oxidize primary alcohols to aldehydes and 2° alcohols to ketones (Section 11.4c).

Because of its basicity, pyridine also reacts as a nucleophile. With alkyl halides, for example, the corresponding alkylpyridinium halide salt can be isolated, as shown by the reaction here, in which methyl iodide reacts with pyridine to form methylpyridinium iodide.

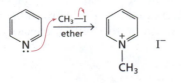

Methylpyridinium iodide

The aromatic ring of pyridine, like that of benzene, is inert toward oxidizing agents because oxidation of the ring would destroy its aromatic character. An alkyl group attached to the ring can be converted to a carboxylic acid group in the same way that an alkylbenzene is oxidized to a benzoic acid derivative (Section 17.4a).

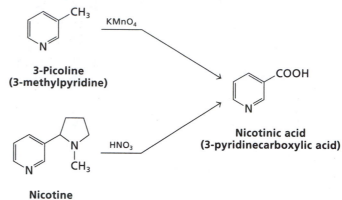

3-Picoline
(3-methylpyridine)

Nicotine
(3-[2-(*N*-methylpyrrolinyl)]pyridine)

Nicotinic acid
(3-pyridinecarboxylic acid)

The nitrogen atom of pyridine can also be oxidized by reaction with a peracid, just as tertiary amines are oxidized (Section 11.5). The product from this reaction with pyridine is pyridine-*N*-oxide.

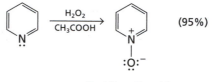

(95%)

Pyridine-*N*-oxide

25.2b PYRIDINE UNDERGOES ELECTROPHILIC SUBSTITUTION WITH DIFFICULTY

The principal *difference* between pyridine and benzene is the difficulty with which pyridine undergoes electrophilic aromatic substitution (Section 17.2a). If you look at the resonance forms of pyridine, you see several in which a negative charge resides on nitrogen, leaving a positive charge at positions 2, 4, and 6 in the ring. For this reason, we usually draw pyridine with double bonds instead of using a structure that has a circle inside the hexagon.

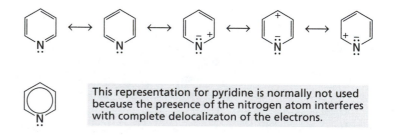

This representation for pyridine is normally not used because the presence of the nitrogen atom interferes with complete delocalizaton of the electrons.

Pyridine is about as reactive as nitrobenzene in electrophilic aromatic substitution reactions, and the *protonated* pyridine ring is estimated to be less reactive than pyridine itself by a factor of ~ 10^{12}. Pyridine can undergo electrophilic substitution under harsh, forcing conditions, however, as shown in the following example:

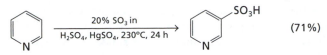

(71%)

Substitution normally occurs in the 3-position, which is the least electron-deficient atom, according to the resonance forms that can be drawn. When an electrophile reacts at the 3-position, the positive charge in the intermediate is delocalized on three carbon atoms, not on nitrogen.

Electrophile at C3

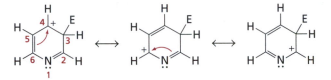

EXAMPLE 25.1

Draw the resonance forms for the intermediate that would form when an electrophile reacts at C2 of the pyridine ring.

First, draw the structure with the electrophile attached at the appropriate carbon atom.

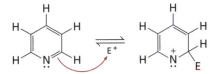

Then, move the electrons around the ring to generate the other resonance forms.

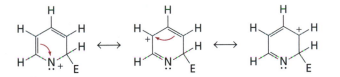

EXERCISE 25.6

Draw the resonance forms for the intermediate that would form when an electrophile reacts at C4 of the pyridine ring.

Derivatives of pyridine that have an electron-donating group on the ring are activated toward electrophilic substitution, so reactions can occur under milder conditions. For example, 2-aminopyridine undergoes bromination in acetic acid at 20°C to produce 2-amino-5-bromopyridine. Reaction occurs at the expected carbon atom (which is the 3-position if you ignore the nomenclature rules and simply number the ring clockwise from N) and para to the amino group, an o,p-director (Section 17.3c). (2-Amino-3-bromopyridine is also formed as a minor product.)

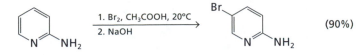

(90%)

When an o,p-directing and activating group is in the 3-position of the pyridine ring, then the activating strength of the substituent is crucial (see Fig. 17.10). An amino group (and its derivatives) can overshadow the normal reactivity patterns of a pyridine ring, but an alkyl group cannot. The following examples illustrate this difference:

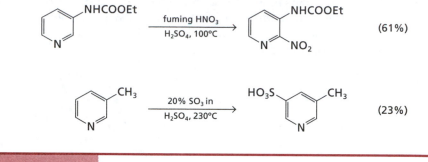

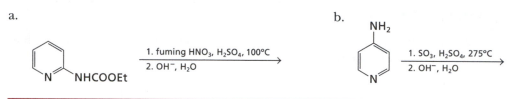

EXERCISE 25.7

What is the expected major product in each of the following schemes?

a.

b.

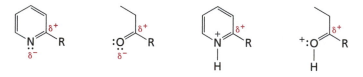

25.2c THE REACTIONS OF PYRIDINE MIRROR THOSE OF CARBONYL COMPOUNDS

While pyridine reacts slowly with electrophiles in aromatic substitution reactions, it is surprisingly reactive toward *nucleophiles*. In fact, derivatives of pyridine react in many ways like ketones. The double bond between the nitrogen atom and C2 is polarized in the same way as a carbonyl group; protonation of the nitrogen atom therefore corresponds to proton activation of a carbonyl group (Section 18.1c).

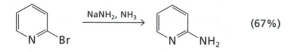

Table 25.1 summarizes a comparison of several reactions presented in Chapters 18–24 with those of pyridine and its derivatives. The next two sections will discuss some of these reactions in more detail.

25.2d PYRIDINE UNDERGOES NUCLEOPHILIC ADDITION REACTIONS

A charged nucleophile reacts with 2-chloro- or 2-bromopyridine and replaces the halogen atom. This reaction corresponds to the reaction between an acid chloride and a nucleophile, which occurs by addition–elimination (Section 21.3a). For example, sodium amide reacts with 2-bromopyridine to form 2-aminopyridine.

Table 25.1 A comparison between reactions of carbonyl compounds and pyridine derivatives.

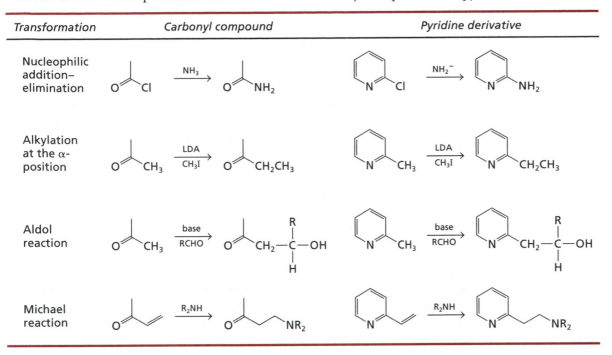

Transformation	Carbonyl compound	Pyridine derivative
Nucleophilic addition–elimination		
Alkylation at the α-position		
Aldol reaction		
Michael reaction		

Like any addition–elimination reaction of a carboxylic acid derivative, the first step is addition to the double bond (carbon–oxygen for an acid chloride; carbon–nitrogen for pyridine) and formation of a tetrahedral intermediate. In the case of pyridine, this intermediate is stabilized by resonance:

Elimination of bromide ion regenerates the double bond and yields the substitution product.

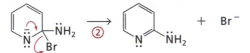

EXERCISE 25.8

4-Halopyridine derivatives also react with nucleophiles. Propose a mechanism for the following reaction:

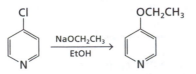

A more surprising example of nucleophilic addition–elimination takes place with pyridine itself.

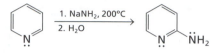

This reaction occurs in the same way as the one starting with 2-bromopyridine, except that hydride ion is eliminated! The hydride ion is not usually considered to be a good leaving group, but regeneration of the aromatic system drives this reaction to completion for pyridine.

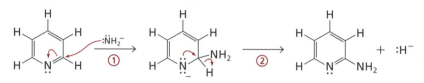

Because hydride ion is a powerful base, an acid–base reaction follows the addition–elimination steps. The hydride ion reacts with the amino group to produce hydrogen, H_2, which makes the reaction irreversible. Workup in Step 4 is the addition of water to protonate the basic amide ion.

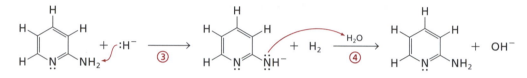

In a similar vein, organolithium compounds and Grignard reagents add to the 2-position of pyridine. For example, treatment of pyridine with phenyllithium gives 2-phenylpyridine after elimination of lithium hydride. The LiH byproduct is destroyed during aqueous workup.

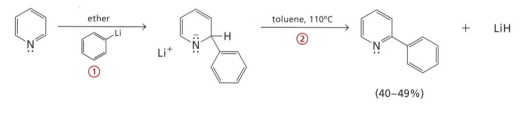

(40–49%)

EXERCISE 25.9

Pyridine-*N*-oxide reacts with organometallic compounds under milder conditions than pyridine itself. Propose a mechanism for the following transformation, which produces a neutral pyridine derivative:

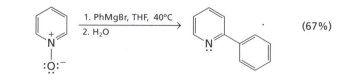

(67%)

Nucleophiles such as Grignard reagents can also react with pyridine at the 4-position, a process that is equivalent to the conjugate addition reactions observed for α,β-unsaturated ketones (Section 24.3d).

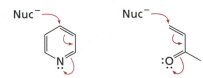

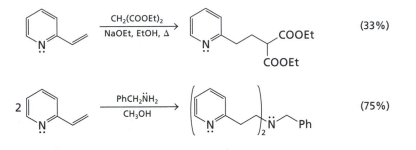

Reaction of nucleophiles at the 4 position of pyridine is equivalent to conjugate addition with an α,β-unsaturated ketone.

Conjugate addition reactions are not limited to the ring. If a vinyl group is attached at the 2- or 4-position, then weaker nucleophiles such as stabilized enolate ions, amines, and thiols can participate in conjugate addition processes (Sections 24.2c and 24.3a).

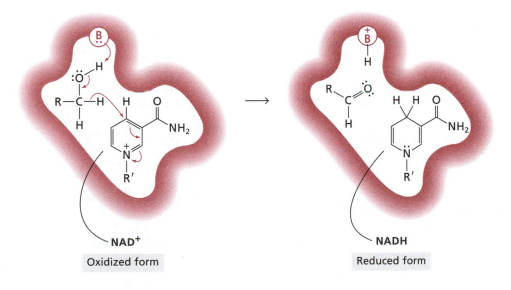

EXERCISE 25.10

Propose a mechanism for the Michael reaction (Section 24.3a) of 2-vinylpyridine with the anion of diethyl malonate.

Conjugate addition of a nucleophile to the pyridine ring is actually not a new reaction for you: Recall that a pyridine derivative comprises the functional unit of the coenzyme NAD^+, which is used for the oxidation of alcohols in biological systems (Section 11.4e). In biochemical alcohol oxidation reactions, what amounts to a hydride ion adds to the pyridinium ring at the 4-position.

25.2e ALKYLPYRIDINE DERIVATIVES CAN FORM CARBANIONS

Like any carbonyl compound that has α-protons, a pyridine derivative with an alkyl group at the 2-position can be deprotonated, forming a carbanion that is stabilized by resonance. The pK_a value of the proton attached to a carbon atom *alpha* to the ring is ~ 20, nearly the same as that of a proton *alpha* to the carbonyl group of a ketone.

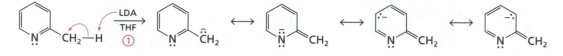

Subsequent addition of a methyl or primary alkyl halide results in substitution of the halide ion by the nucleophilic carbon atom.

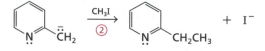

The carbanion that is formed *alpha* to the pyridine ring can also participate in addition to a carbonyl group—the aldol reaction (Section 23.1a). With aldehydes and ketones, alcohols are formed after workup.

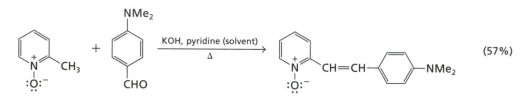

Charged derivatives of the alkylpyridine are even more acidic, and a base such as hydroxide ion is sufficient to generate the corresponding carbanion. In the following example, which is a crossed-aldol reaction, water is eliminated from the alcohol product by an E1cb mechanism (Section 23.1b).

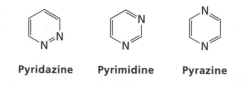

EXERCISE 25.11

Propose a mechanism for the preceding crossed-aldol reaction sequence. Draw resonance structures to explain why the methyl group of 2-methylpyridine-*N*-oxide is readily deprotonated by hydroxide ion.

25.2f DIAZINES DISPLAY THE SAME REACTIVITY PATTERNS AS PYRIDINE

The diazines, which comprise pyridazine, pyrimidine, and pyrazine, are heterocycles with two nitrogen atoms in a six-membered ring. Pyrimidine is the most interesting member of this group because its ring system is found in three of the bases that constitute the nucleic acids (Chapter 28).

Pyridazine   Pyrimidine   Pyrazine

The presence of the second nitrogen atom in the six-membered ring either diminishes or enhances the reactivity of diazine relative to that of pyridine. These compounds are weaker bases than pyridine, and they are essentially inert in electrophilic aromatic substitution reactions. They are more reactive toward nucleophiles, however.

In a pyrimidine ring, the carbon atom flanked by the two nitrogen atoms is particularly reactive toward nucleophiles. Substitution reactions at this carbon center can be up to 1 million times faster than those at a carbon atom adjacent to the nitrogen atom of pyridine.

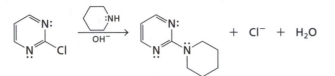

Other transformations are also facilitated at the 2-position of pyrimidine. For example, the following crossed-aldol reaction occurs with a mild Lewis acid catalyst:

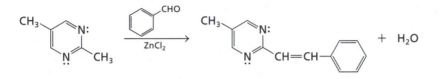

EXERCISE 25.12

Propose a mechanism for the following reaction that occurs between the deprotonated pyrazine derivative and the illustrated ester, which is an example of a crossed-Claisen condensation reaction.

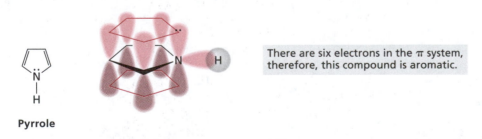

25.3 PYRROLE AND RELATED HETEROCYCLES

25.3a PYRROLE IS AN UNEXPECTEDLY WEAK BASE AND HAS SUBSTANTIAL AROMATIC CHARACTER

Pyrrole is an aromatic nitrogen heterocycle with a five-membered ring. Each atom has sp^2 hybridization and the valence bond representation shows that the p orbitals form a cyclic π system. Unlike the unshared electron pair on the nitrogen atom in pyridine, the electron pair on the nitrogen atom of pyrrole is *included* in the π system, giving a total of six electrons, which makes pyrrole aromatic.

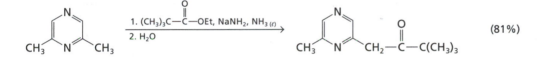

There are six electrons in the π system, therefore, this compound is aromatic.

Pyrrole

A significant consequence of pyrrole's structure is that the electron pair associated with the nitrogen atom is *not* available to act as a base or nucleophile, as it is in pyridine. Protonation would destroy the aromatic system, which explains why the pK_a value of pyrrole is estimated to be about –4, which makes it a stronger acid than H_3O^+. Protonation actually occurs on one of the carbon atoms in the ring, which still disrupts its aromaticity.

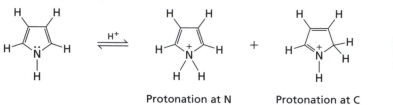

Protonation at N   Protonation at C   Formation of neither product is favorable because protonation disrupts the aromaticity of the ring.

The resonance forms that can be drawn for pyrrole place a negative charge on each carbon atom of the ring, which suggests that the pyrrole ring should react readily with electrophiles, as in fact, it does.

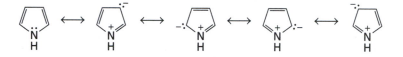

Pyrrole is an important heterocycle in biochemistry, serving as an integral part of many metal-binding groups in proteins, notably the porphyrins. Porphine, the parent compound from which porphyrins are derived by substitution of the pyrrole rings, has four pyrrole units linked by CH groups; it is aromatic with 18 electrons in the planar, conjugated π system (shown in color, below). Note that two of the double bonds of the porphine ring are not included in the electron count. In applying Hückel's rule, the atoms to be included must be part of an uninterrupted cyclic system (Section 17.1b).

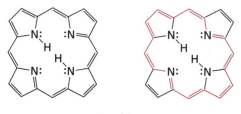

Porphine

The iron derivative of protoporphyrin IX, commonly called **heme,** is the functional unit in hemoglobin and myoglobin, which are used by mammals to transport and store dioxygen, respectively. Chlorophyll has a magnesium ion bonded to a related macrocycle called chlorin, in which one double bond of one of the pyrrole rings has been reduced. This more saturated compound is still aromatic.

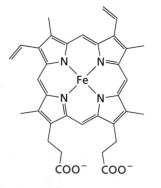

Heme
(Iron protoporphyrin IX)

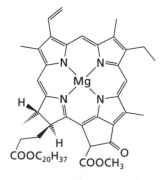

Chlorophyll *a*

Derivatives of pyrrole can be prepared by several routes, but the simplest employs the reaction of an amine with a 1,4-diketone. This procedure relies on imine bond formation (Section 18.1a), followed by nucleophilic addition of the nitrogen atom to the second carbonyl group, elimination of water, and deprotonation.

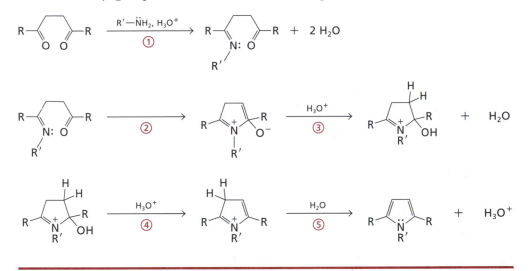

EXAMPLE 25.2

With curved arrows, illustrate the electron movement in Step 1 of the preceding synthesis scheme for pyrrole.

As for any polar reaction, nucleophiles react with electrophiles. The amine is nucleophilic, so it reacts with one of the carbonyl carbon atoms, which are electrophilic. After the addition reaction in Step 1a, a proton-transfer step occurs (Step 1b) to form the hydroxy amine.

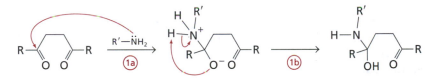

The final product of Step 1 is an imine, which requires that a molecule of water is lost from the hydroxy amine formed in Step 1b. Protonation of the OH group generates a good leaving group (Step 1c), and water acts as a base in Step 1d to generate the C=N double bond.

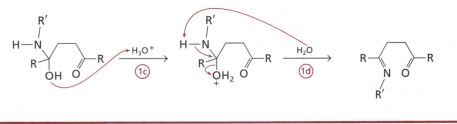

EXERCISE 25.13

With curved arrows, illustrate the electron movements that occur in Steps 2–5 of the preceding scheme for the synthesis of pyrrole.

Furan and thiophene are structurally similar to pyrrole. Furan has already been mentioned in the context of carbohydrate nomenclature (Section 19.3b), and THF, its saturated analogue, is an inert solvent used for reactions involving organometallic compounds. For furan and thiophene, the oxygen or sulfur atom contributes one electron pair to the aromatic π system, while the other electron pair is perpendicular to the π system, directed away from the ring.

Furan   **Thiophene**   (X = O or S)

EXERCISE 25.14

Draw resonance structures for furan and thiophene to show how charges can be distributed within the ring.

25.3b PYRROLE UNDERGOES FACILE ELECTROPHILIC AROMATIC SUBSTITUTION REACTIONS

Unlike pyridine, pyrrole undergoes electrophilic substitution reactions readily, even more readily than highly activated benzene derivatives. For example, pyrrole is acylated by its reaction with acetic anhydride; no Lewis acid catalyst is needed, which is unlike the case for most Friedel–Crafts reactions (Section 17.2e).

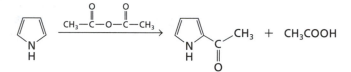

To understand why substitution occurs at the 2-position, we need only look at resonance structures for the intermediates that are formed when an electrophile reacts at either of the two positions, C2 and C3. The reaction at C2 produces a more stabilized cation intermediate.

Electrophile at C2   **Electrophile at C3**

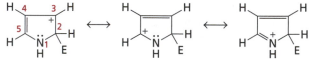   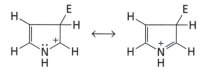

As with any aromatic substitution process, regeneration of the aromatic system is accomplished by deprotonation of the cationic intermediate.

Reactions of pyrrole are complicated by the instability of its ring toward mineral acids, which often leads to polymer formation. For this reason, mildly acidic or neutral electrophiles are required. Nitration can be done by using acetyl nitrate, $CH_3CO_2NO_2$, formed by mixing acetic anhydride with nitric acid. The major product is 2-nitropyrrole, as expected.

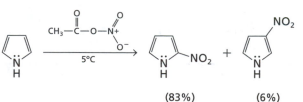

(83%)   (6%)

25.3c FURAN AND THIOPHENE ALSO UNDERGO FACILE ELECTROPHILIC AROMATIC SUBSTITUTION REACTIONS

Among the group of heterocycles comprising furan, thiophene, and pyrrole, furan is the least "aromatic". From thermochemical measurements, the resonance stabilization energy for furan is estimated to be ~ 11 kcal mol^{-1}, compared with 36 kcal mol^{-1} for benzene (Section 17.1a). Recall that anthracene has a similarly low value for its central ring (Section 25.1a), and as a consequence, it often reacts like a diene. Furan follows this same pattern, and undergoes both addition and substitution reactions. For example, furan reacts with acetic anhydride with BF$_3$ as a catalyst to form 2-acetylfuran.

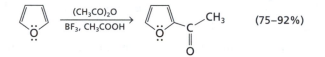

With acetyl nitrate, however, addition occurs to produce a dihydrofuran product.

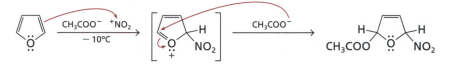

Addition of a mild base such as pyridine removes a proton and restores the aromatic π system with concomitant elimination of acetate ion.

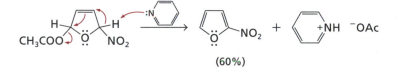

Thiophene is neither as sensitive to acid as pyrrole and furan, nor is it quite as reactive. It is still more reactive than benzene, however, and substitution products are readily obtained. The following equations illustrate nitration, Friedel–Crafts acylation, and sulfonation:

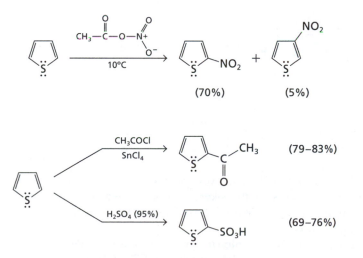

Interestingly, thiophene reacts with iodine, which is normally considered to be a weak electrophile. Benzene is used as the solvent for this reaction, illustrating the reactivity difference displayed by these two aromatic compounds.

EXERCISE 25.15

What is the product expected from each of the following reaction steps?

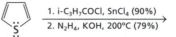

25.3d INDOLE IS A FUSED-RING PYRROLE DERIVATIVE THAT CONSTITUTES THE SIDE CHAIN OF THE AMINO ACID TRYPTOPHAN

Pyrrole is important in biology not just because it is part of the porphyrin ring. It is also part of indole, a heterocycle with a benzene ring fused to the C2–C3 bond of pyrrole that makes up the side chain of the amino acid tryptophan.

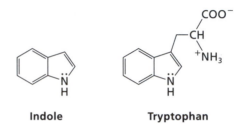

Indole   **Tryptophan**

The influence of the benzene ring on the reactivity of the five-membered ring portion of indole is significant, and its resonance structures, shown below, help to explain why. (The first four structures are the major contributors.)

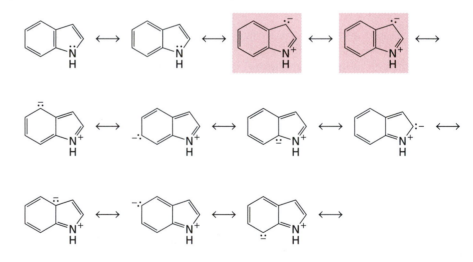

A negative charge can be placed at every position in both rings, but only the two highlighted charge-separated structures keep the benzene π system intact. From these two structures, you might expect that the 3-position of the indole ring will be the most reactive, and indeed it is.

It is significant that the indole ring of tryptophan is attached to the side chain at its 3-position. This fact suggests that an electrophilic substitution process might be involved in its biosynthesis. In fact, tryptophan is constructed by coupling of the indole ring with an electrophilic center of an unsaturated carboxylic acid derived from serine, which is first attached to PLP (Section 20.2c).

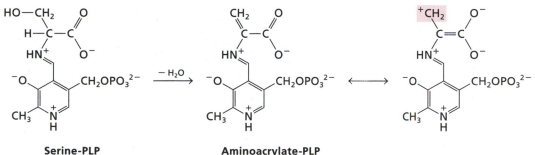

Serine-PLP   **Aminoacrylate-PLP**

The carbocation highlighted in the previous equation is the electrophile that reacts with the indole ring to form tryptophan, after hydrolysis of the imine bond to PLP.

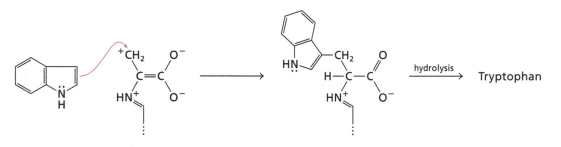

EXERCISE 25.16

Propose a reasonable mechanism for the electrophilic substitution reaction in the preceding equation.

25.4   AZOLES

25.4a   AZOLES ARE AROMATIC HETEROCYCLES THAT HAVE TWO HETEROATOMS IN A FIVE-MEMBERED RING

Indole is not the only heterocycle found in the side chain of an amino acid. **Azoles** are heterocycles that have two heteroatoms, one of which is nitrogen, in a five-membered ring. The heteroatoms can occupy positions 1 and 2 or 1 and 3 in the ring; the 1,3-azoles are the more important in biological systems. Imidazole, for example, is in the side chain of the amino acid histidine.

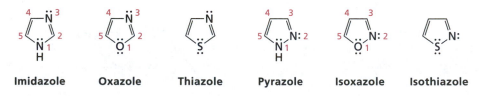

Imidazole   Oxazole   Thiazole   Pyrazole   Isoxazole   Isothiazole

The valence bond representation for imidazole (right) shows that one nitrogen atom is like that in pyrrole, contributing its electron pair to the aromatic π system. The other nitrogen atom has its electron pair in an orbital directed away from the ring, as in pyridine. This electron pair can act as a base because protonation does not disrupt the compound's aromaticity. All of the azoles are bases, too, and are readily protonated.

Imidazole is more basic than pyridine because of symmetry in the π system of the conjugate acid, which creates energetically equivalent resonance contributors, hence greater stabilization. Imidazole is *much* more basic than pyrrole, in which the electron pair on nitrogen is part of the π system.

The electron pair on this nitrogen atom is part of the π system.

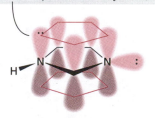

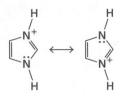

> The two resonance forms for a protonated imidazole ring have the same energy.

The basicity of imidazole is crucial in biochemistry because it means that at pH 7, imidazole can exist as a mixture with its protonated form. For this reason, histidine is frequently found at the active site of enzymes *where it can act either as an acid* (protonated form) *or as a base* (deprotonated form).

EXERCISE 25.17

Draw valence bond representations for thiazole, oxazole, and pyrazole.

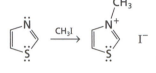

25.4b AZOLES REACTS READILY AS NUCLEOPHILES

The basicity of the azole ring means that the nitrogen atom has nucleophilic properties. All of the azoles react with alkylating agents to form isolable salts.

Imidazole can be converted into an even more potent nucleophile by treating it with a strong base such as NaH in a polar aprotic solvent like DMF. This reaction removes the proton from nitrogen. Like its conjugate acid, the resulting conjugate base of imidazole is resonance stabilized. Nucleophilic displacement of a leaving group from an alkyl halide yields the corresponding *N*-alkylimidazole.

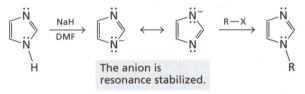

> The anion is resonance stabilized.

If the imidazole ring is already substituted, then isomeric products will be formed.

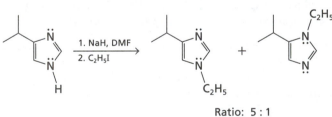

Ratio: 5 : 1

A more interesting acid–base reaction occurs with *N*-alkyl- or *N*-arylimidazoles. When one of these compounds is treated with butyllithium or LDA in THF, then the proton at C2 is removed, and an organolithium compound is formed. This organometallic compound reacts with carbonyl compounds to form alcohols. Thiazoles react in the same way.

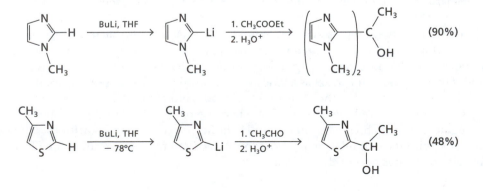

25.4c AN ALKYLATED THIAZOLE ION CONSTITUTES THE REACTIVE PORTION OF THIAMINE AND ITS DERIVATIVES

Imidazole, which is a reactive group in the active site of many enzymes, is not the only important 1,3-azole in biochemistry. Thiazole, which is a portion of *thiamine* (vitamin B_1), is also crucial to the success of certain metabolic processes. The pyrophosphate derivative of thiamine, abbreviated TPP, is a coenzyme for the decarboxylation of α-keto acids and for aldol reactions.

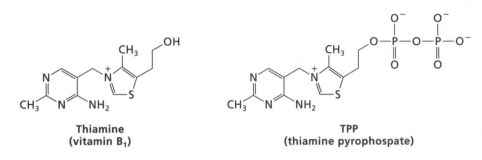

| Thiamine (vitamin B₁) | TPP (thiamine pyrophospate) |

Both processes that involve TPP take advantage of carbocation stabilization by the thiazole ring. This stabilization comes about because the ring carries a positive charge that can offset the negative charge of a carbanion center. When the thiazole nitrogen atom is alkylated, the pK_a value of the C2 proton is ~ 20, about the same as the protons *alpha* to the carbonyl group of a ketone.

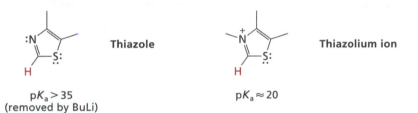

| Thiazole | Thiazolium ion |
| $pK_a > 35$ (removed by BuLi) | $pK_a \approx 20$ |

25.4d THIAMINE PYROPHOSPHATE IS A COENZYME USED IN THE DECARBOXYLATION OF A-KETO ACIDS

The decarboxylation reactions of α-keto acids are catalyzed by TPP. The conversion of pyruvate to acetaldehyde can be used to illustrate the steps of this mechanism.

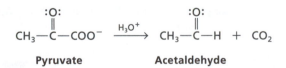

Pyruvate, a three-carbon α-keto acid, is made by oxidation of glyceraldehyde-3-phosphate, the product of glycolysis (Section 23.2c). It undergoes decarboxylation to form acetaldehyde, which can be reduced to ethanol by the enzyme *alcohol dehydrogenase* (Section 18.3c), using NADH as a coenzyme.

$$CH_3-\overset{\overset{\displaystyle :O:}{\|}}{C}-H \xrightarrow[\text{\textit{Alcohol dehydrogenase}}]{\text{NADH}} CH_3-CH_2-\ddot{O}H$$

This decarboxylation–reduction process is more commonly known as *fermentation*, which is the chemistry behind the making of alcoholic beverages. If the carbon dioxide is not allowed to escape, then the solution becomes carbonated, producing beer or champagne, depending on the source of the glucose.

The decarboxylation reaction of pyruvate begins with removal of the proton at the 2-position of TPP within the active site of *pyruvate decarboxylase.*

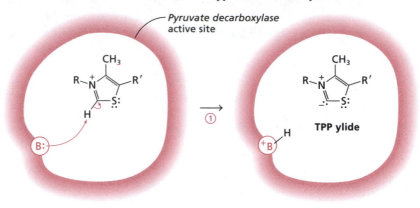

The ylide derivative of TPP is stabilized by resonance, as follows:

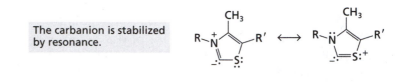

The carbanion is stabilized by resonance.

Once formed, the carbanion adds to the ketone carbonyl group of pyruvate, and the oxygen atom is protonated by an acid group to form the alcohol.

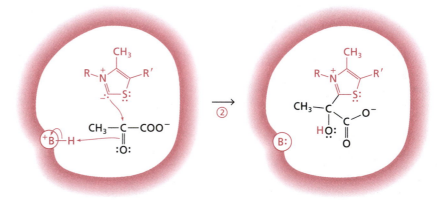

Decarboxylation can now take place, with the thiazolium ring acting as an acceptor for electrons from the carboxylate group.

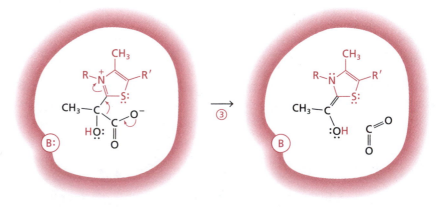

The enol intermediate formed by decarboxylation is similar to an enamine. Recall that enamines have a resonance form that places a negative charge on the carbon atom *beta* to the nitrogen atom (Section 20.3b). Therefore, protonation of the thiazoliumenamine intermediate converts it to the 2-(1-hydroxyethyl)thiazolium ion, as shown in Step 4.

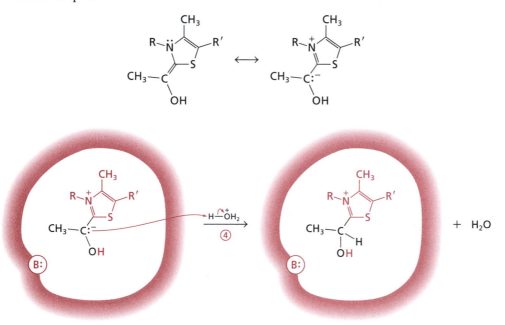

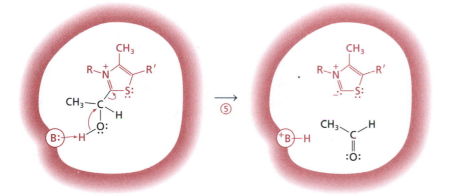

In the penultimate step, a base in the active site of the enzyme removes a proton from the hydroxyl group, regenerating both the carbonyl group and the TPP carbanion, which functions here as a leaving group. (You have seen carbanions as leaving groups before in the retroaldol and retro-Claisen reactions; Sections 23.1d and 23.4d.)

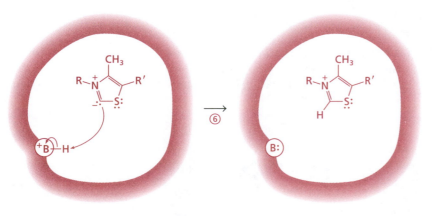

The final step in the overall transformation is protonation of the TPP ylide to regenerate the TPP coenzyme.

25.4e THIAMINE PYROPHOSPHATE STABILIZES CARBANIONS THAT ARE INVOLVED IN ALDOL REACTIONS OF CARBOHYDRATES

Besides aiding decarboxylation reactions, the stabilized TPP ylide can also interconvert carbohydrates via aldol reactions. We will look only at one carbon–carbon bond-forming reaction to see the general type of aldol reaction that TPP catalyzes.

One reactant in this crossed-aldol reaction is a thiazolium-enamine molecule, which is made by a *retroaldol* reaction from xylulose-5-phosphate. This enamine has a structure similar to the enamine derivative formed during the decarboxylation reaction of pyruvate, differing only by the presence of an additional OH group.

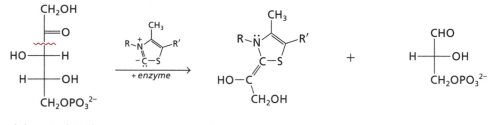

D-Xylulose-5-phosphate   **Thiazolium-enamine**   **D-Glyceraldehyde-3-phosphate**

This enamine reacts by nucleophilic addition to the aldehyde group of D-ribose-5-phosphate (R5P), as shown in Step 1. This reaction forms a new C–C bond.

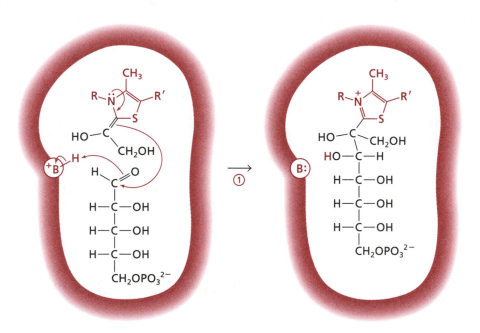

In Step 2, regeneration of a carbonyl group at the 2-position of this aldol product yields S7P, which is subsequently cleaved by *transaldolase* (Section 23.2d). Notice that the thiazolium ion acts again as a leaving group, just as it did in the decarboxylation reaction shown in Section 25.4d.

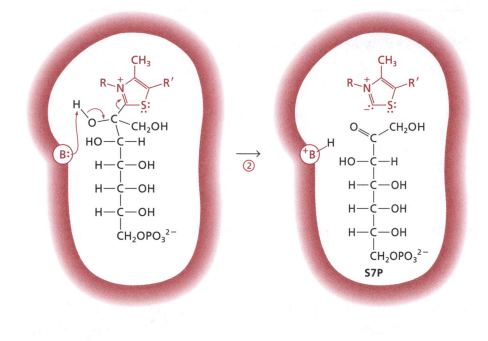

EXERCISE 25.18

In the laboratory, thiamine, in the presence of strong base, catalyzes the following reaction without the need of an enzyme. Propose a mechanism for this transformation known as the benzoin condensation. This reaction is unusual in that the first intermediate, which is formed by nucleophilic addition to the carbonyl group, undergoes an *intra*molecular acid–base reaction to create a carbanion. Nucleophilic addition to a second molecule of aldehyde is followed by an acid–base reaction, then elimination, which regenerates thiamine.

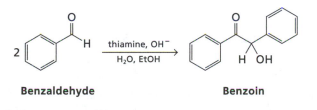

Benzaldehyde   **Benzoin**

EXERCISE 25.19

The condensation of benzaldehyde to form benzoin can also be catalyzed by cyanide ion. Propose a reasonable mechanism for the following transformation:

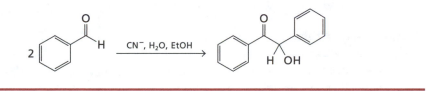

CHAPTER SUMMARY

Section 25.1   Polycyclic aromatic compounds

- Polycyclic aromatic hydrocarbons have one or more aromatic rings fused onto a benzene ring. The $4n + 2$ π electron rule still applies, but the resonance energy is less per ring than that of benzene.

- Polycyclic arenes react more readily than benzene toward substitution. Addition to the π bonds becomes increasingly facile for these compounds, a property related inversely to how many resonance forms retain benzene-like structures.

- Buckminsterfullerene (C_{60}) is a fullerene, a closed shell of carbon atoms. Fullerenes have aromatic properties, but many reagents that also react with alkenes add to the π bonds of C_{60}.

Section 25.2   Pyridine and related heterocycles

- Pyridine is an analogue of benzene in which a CH group of benzene has been replaced by a nitrogen atom with its unshared pair of electrons.

- Pyridine is a Lewis base, and it reacts as a nucleophile at its nitrogen atom.

- The nitrogen atom of pyridine can be oxidized to form pyridine-*N*-oxide.

- The pyridine ring is not easily oxidized, but alkyl substituents attached to the ring can be converted to carboxylic acid groups.

- Pyridine undergoes electrophilic aromatic substitution reactions, but the reactions are slow. Substitution occurs primarily at the 3-position.

- The double bond between nitrogen and the adjacent carbon atom in pyridine reacts much like a carbonyl group.

- Similarities between the reactions exhibited by pyridine and those of carbonyl compounds are summarized in Table 25.1 and include addition–elimination, alkylation, aldol, and conjugate addition reactions.

- Diazines are analogues of benzene that have two nitrogen atoms in the ring. Their reactions are like those of pyridine.

Section 25.3   Pyrrole and related heterocycles

- Pyrrole is a five-membered aromatic heterocycle. Pyrrole is an unexpectedly weak base because protonation would destroy its aromaticity.

- Pyrrole is highly reactive toward electrophilic aromatic substitution processes, which occurs predominantly at the carbon atom adjacent to nitrogen.

- Furan and thiophene are analogues of pyrrole in which the nitrogen atom has been replaced by oxygen or sulfur, respectively. They also undergo electrophilic aromatic substitution reactions, but furan is susceptible to addition as well.

- Indole is a bicyclic compound that has a pyrrole ring fused with a benzene ring. It is found in the side chain of the amino acid tryptophan.

Section 25.4   Azoles

- Azoles are five-membered ring compounds with two heteroatoms in the ring, one of which is nitrogen. Imidazole has two nitrogen atoms and is found in the side chain of the amino acid histidine.

- Azoles are bases and nucleophiles. For 1,3-azoles, the hydrogen atom at C2 can be removed using a strong base.

- Thiazole has a nitrogen and a sulfur atom in the five-membered azole ring. An alkylated thiazole ring constitutes part of the thiamine (vitamin B_1) skeleton.

- Thiamine, as its pyrophosphate derivative (TPP), is an important coenzyme that stabilizes a carbanion center as an ylide.

- The compound TPP is a coenzyme in the decarboxylation reactions of α–keto acids.
- The compound TPP stabilizes carbanions derived from carbohydrates via retroaldol reactions. These carbanions react with the carbonyl group of other carbohydrates to form new carbon–carbon bonds via reactions catalyzed by the enzyme *transketolase*.

Section 25.1d
buckminsterfullerene

Section 25.3a
heme

Section 25.4a
azoles

REACTION SUMMARY

Section 25.1b

Electrophilic substitution reactions of naphthalene: Substitution at the 1-position predominates, except at higher temperatures.

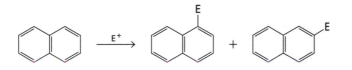

Section 25.1c

Reactions of anthracene: Addition and oxidation are more facile than for benzene.

Section 25.2a

Pyridine reacts as a base toward acids, HX, and as a nucleophile toward alkyl halides.

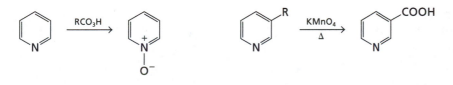

The nitrogen atom of pyridine can be oxidized to form pyridine-*N*-oxide. Alkyl groups attached to a pyridine ring are oxidized to COOH groups.

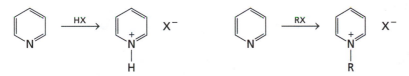

Section 25.2b

Electrophilic aromatic substitution reactions of pyridine occur at the 3-position and require harsh conditions.

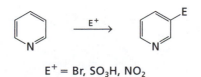

$E^+ = Br, SO_3H, NO_2$

Section 25.2d

Pyridine and 2-halopyridine derivatives undergo addition–elimination reactions when treated with nucleophiles.

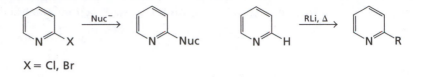

X = Cl, Br

Pyridine and its 2-vinyl derivatives can undergo conjugate addition reactions.

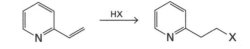

Section 25.2e

The α-carbon atom of 2-alkyl pyridine derivatives can be deprotonated with strong base; the resulting carbanion reacts with electrophiles (RX and R₂C=O).

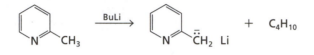

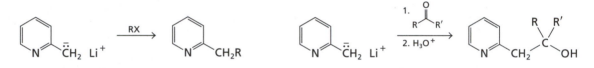

Section 25.2f

Diazines undergo reactions of the same types seen for pyridine.

Section 25.3a

Pyrrole derivatives can be made from diketones and amines.

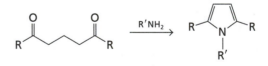

Section 25.3b

Pyrrole and its derivatives readily undergo electrophilic aromatic substitution reactions.

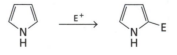

E⁺ = Br, RCO, NO₂

Section 25.3d

Thiophene and furan undergo electrophilic substitution reactions in the same manner seen for pyrrole.

Section 25.4a

Azoles react as bases or nucleophiles at their nitrogen atoms.

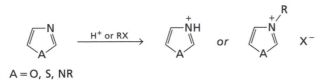

$A = O, S, NR$

1,3-Azoles can be deprotonated at nitrogen or at the intervening carbon atom, depending on the exact structure and reaction conditions.

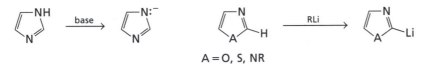

$A = O, S, NR$

25.20.   Draw the structure for each of the following substances:

a. 2-Chloro-5-methylpyridine
b. (*S*)-3-Methyltetrahydrothiophene
c. 3-Thiacyclobutane-1-one
d. 3-Hydroxypyridine-*N*-oxide
e. Methyl 3-pyrrolecarboxylate
f. 2-Mercapto-4-nitroimidazole
g. 3-Acetyl-5-*tert*-butyl-1-methylpyrazole

25.21.   Provide an acceptable name for each of the following compounds:

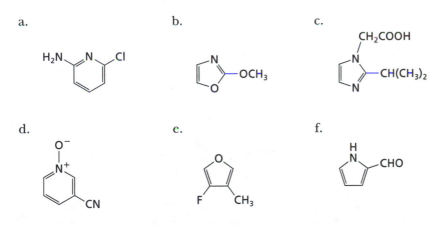

25.22.   In each of the following substances, identify the parent heterocycle(s) that are present:

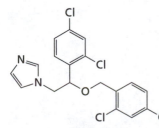

Miconazole

Antifungal

Propiram

Analgesic

Isolan

Insecticide

25.23. The compound 2-hydroxypyridine is also called 2-pyridone, which is a tautomer of the hydroxy compound. Draw the structure of this tautomer, including the unshared electrons.

2-Hydroxypyridine
2-Pyridone

Draw resonance structures for the anion derived from deprotonating this compound with base. What two products are expected to form upon treating this anion with methyl iodide?

25.24. Because pyridine and many of its derivatives are poor substrates for electrophilic substitution reactions, the *N*-oxide derivatives are sometimes used as starting materials. For example, nitration of pyridine-*N*-oxide yields the 4-nitro product. There are two notable aspects of this reaction: (1) the reaction conditions are milder than for pyridine itself; and (2) substitution occurs at the 4-position, not at the 3-position.

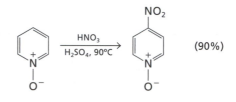

(90%)

Explain these differences by commenting on the resonance structures that can be drawn for the starting compounds and for the carbocation intermediates produced during the electrophilic substitution process.

25.25. Pyridine-*N*-oxide can be activated further by alkylating the oxygen atom. Propose a mechanism for each step of the following sequence. Notice that the products no longer have the oxygen atom attached to nitrogen.

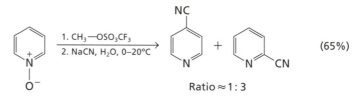

(65%)

Ratio ≈ 1 : 3

25.26. Shown below is the ^{1}H NMR spectrum of pyridine. Account for the spin–spin splitting patterns.

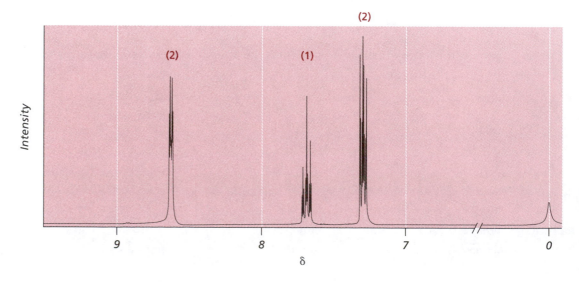

25.27. What is the major product expected from treating 3-methylthiophene with each of the following sets of reagents. If no reaction occurs, write N.R.

a. Bromine, acetic acid
b. Benzoyl chloride, $SnCl_4$
c. H_2 (1 atm), Pd/C, 25°C
d. Concentrated sulfuric acid
e. HNO_3, acetic anhydride

25.28. Draw the structure of the major product expected from each of the following reactions. Show stereochemistry where appropriate. If no reaction occurs, write N.R.

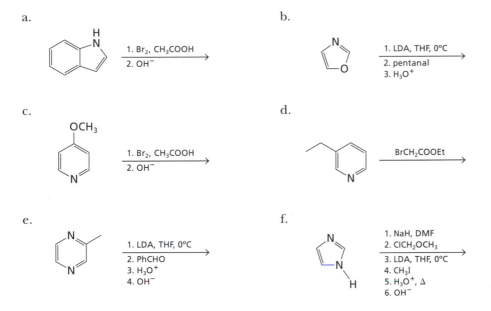

a.

1. Br₂, CH₃COOH
2. OH⁻

b.

1. LDA, THF, 0°C
2. pentanal
3. H₃O⁺

c.

OCH₃

1. Br₂, CH₃COOH
2. OH⁻

d.

BrCH₂COOEt

e.

1. LDA, THF, 0°C
2. PhCHO
3. H₃O⁺
4. OH⁻

f.

1. NaH, DMF
2. ClCH₂OCH₃
3. LDA, THF, 0°C
4. CH₃I
5. H₃O⁺, Δ
6. OH⁻

25.29. Hismanal is an antihistamine that contains a benzimidazole ring.

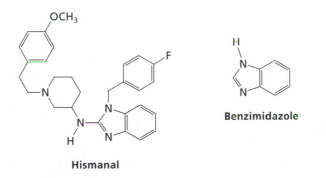

Hismanal

Benzimidazole

a. Draw a valence bond representation for benzimidazole.

b. Benzimidazole is alkylated at nitrogen upon treatment with NaH in DMF followed by addition of 1-bromopropane. Draw the structure of the product and an equation for the overall transformation.

c. Propose a reasonable mechanism for the synthesis of a 2-alkylbenzimidazole molecule from *o*-phenylenediamine and a carboxylic acid according to the following equation:

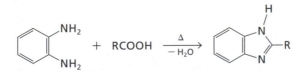

25.30. Oxazoles can be made by heating an α-chloroketone molecule with formamide, $HCONH_2$. Propose a mechanism for the following transformation:

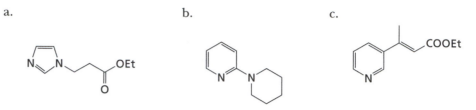

25.31. Show how you would prepare each of the following compounds, starting with any organic compounds that have six or fewer carbon atoms. Several steps may be necessary.

a.   b.   c.

25.32 Compound **A**, $C_5H_4O_2$, is obtained as a product from the hydrolysis of oat hulls, and its ^1H NMR spectrum is shown below. When **A** is treated with 1,3-propanedithiol and a Lewis acid catalyst, followed by Raney nickel, compound **B** is obtained, which has a singlet (integrated intensity of three protons) in the NMR spectrum at ~ δ 2.4. Heating **B** with aqueous acid produces cyclopent-2-ene-1-one as one of the products. Draw a structure for **A** and for **B**, and write equations for the transformations that are taking place.

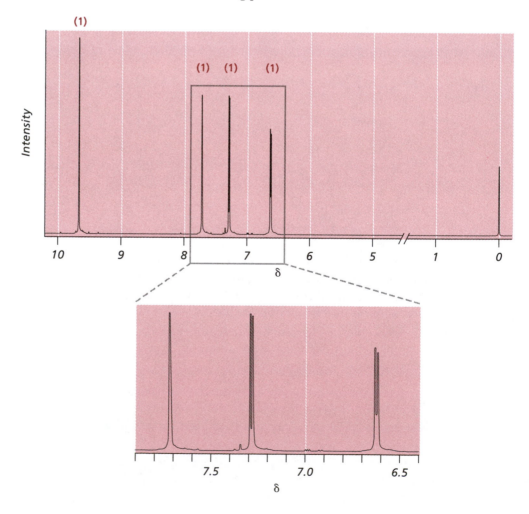

POLYMERS AND POLYMERIZATION

Much of the reaction chemistry of organic molecules presented in this text has focused attention on compounds with molecular weights in the range of 10^2–10^3 amu. Even descriptions of biochemical processes have been centered on transformations that occur when a small-molecule substrate undergoes an enzyme-catalyzed reaction. The final three chapters of this text present chemical aspects of *macromolecules,* that is, substances with molecular weights in the 10^4–10^6 amu range. These macromolecules include synthetic polymers and plastics, which are described in this chapter, proteins (Chapter 27), and nucleic acids (Chapter 28).

In this chapter, you will learn how polymers are made from relatively simple building blocks. The chapter begins with a description of some physical properties of these materials. What follows is the reaction chemistry that is used to create synthetic polymers. Some of the rudimentary reaction mechanisms have been presented earlier, but they will be reviewed here.

26.1 POLYMERS AND THEIR PROPERTIES

26.1a MANY POLYMERS CONSIST OF A STRING OF IDENTICAL REPEATING UNITS

The word **polymer** comes from the Greek for "many parts". The corresponding "single parts" are therefore **monomers.** The structure of a polymeric molecule often consists of long chains of the monomeric building blocks, called the **repeat unit.** We represent the structure of a polymer by enclosing the structure of the repeat unit within a set of parentheses, and we place the subscript n outside the parentheses to indicate that many such units constitute the molecule.

Propylene   **Polypropylene**
(monomer)   (polymer)

The value of n is usually not given, but we could write an actual numeral to specify a polymer's average degree of polymerization (if we were to determine what that average value is).

The name of a polymer is normally crafted by specifying the monomer used in its preparation. The polymer's name carries the prefix poly-, and parentheses are used if the monomer name comprises more than one word. Notice in the example of polyethylene that the structure is written as $-(CH_2CH_2)_n-$ even though the repeating unit could be written as $-(CH_2)_n-$. Generally, we use the structure of the monomer as the repeat unit even if the polymer's structure can be condensed further.

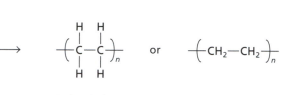

Vinylidene dichloride   **Poly (vinylidene dichloride)**

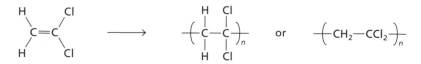

Ethylene   **Polyethylene**

EXERCISE 26.1

What monomer is used to prepare each of the following polymers? Name each polymer.

a.

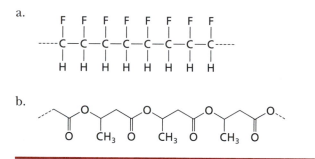

b.

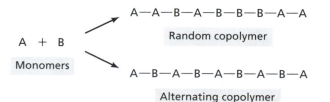

26.1b COPOLYMERS ARE FORMED WITH USE OF MORE THAN ONE MONOMER

When a single monomer is used to form a polymer, the product is called a **homopolymer.** Polyethylene, polypropylene, and poly(vinylidene dichloride) are examples of homopolymers.

A **copolymer** is formed when two or more monomers are incorporated into the growing polymer chain. Three common types are the **random copolymer,** which, as the name implies, has no regular pattern to its sequence; the **alternating copolymer,** which has a regular repeat of the two monomers in its sequence; and **segmented copolymers.**

A—A—B—A—B—B—B—A—A

Random copolymer

A + B

Monomers

A—B—A—B—A—B—A—B—A

Alternating copolymer

An example of an alternating copolymer is Saran, manufactured from vinyl chloride and vinylidene dichloride. In practice, alternation does not occur perfectly, so defects in the structure of this type of copolymer are usually present.

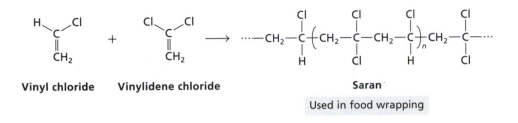

Vinyl chloride   **Vinylidene chloride**   Saran

Used in food wrapping

Segmented copolymers are normally classified in two categories: **block copolymers** and **graft copolymers.** Block copolymers have long segments that are homopolymeric; these segments are made by initiating polymerization with one monomer, then adding a large excess of a second monomer. As the concentration of the second monomer decreases, the first monomer is added again. This procedure is repeated as needed.

$$nA \xrightarrow{\text{Initiator, Init*}} \text{Init—A—A—A—A—A—A*} \xrightarrow[\text{a time interval}]{\text{add B after}}$$

$$\text{Init—A—A—A—A—A—A—B—B—B—B—B*} \xrightarrow[\text{second time interval}]{\text{add A after a}}$$

$$\text{Init—A—A—A—A—A—A—B—B—B—B—B—A—A—A—A—A—A—A*} \xrightarrow[\text{a time interval}]{\text{add B after}} \text{etc.}$$

The asterisk represents a reactive group (carbocation, radical, etc.) that can keep the polymerization process going.

Graft copolymers are made by creating reactive sites, usually free radicals (see Section 26.2c), at random sites along a homopolymer chain. These sites are used to initiate polymerization with a second monomer. A common way of generating radical sites in the middle of a hydrocarbon chain is by irradiation with gamma rays.

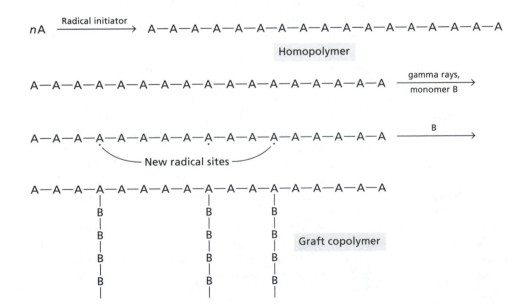

26.1c PHYSICAL PROPERTIES OF POLYMERS ARE INFLUENCED BY SIZE, MASS, AND MORPHOLOGY

Polymers, like all molecules, have mass and shape. Many of the polymers illustrated so far have been shown as linear chains. As described later, branching of these chains occurs in certain instances. In Section 26.1b, the structure of a graft copolymer was shown to have a comblike structure. Other interactions between chains, especially formation of hydrogen bonds, can also occur.

The mass of a polymer cannot be determined in the same way as that of a small molecule. A single chain of a polymer can certainly be represented by a molecular formula, but polymerization processes generally lead to formation of a mixture of polymer molecules and a range of molecular weight values. Two common ways to specify the mass of a polymer are given by the formulas in Eqs. 26.1 and 26.2.

Number average molecular weight (M_n)

$$M_n = \frac{\sum\limits_j (N_j)(M_j)}{\sum\limits_j (N_j)} \tag{26.1}$$

Weight average molecular weight (M_w)

$$M_w = \frac{\sum\limits_j (N_j)(M_j)^2}{\sum\limits_j (N_j)(M_j)} \tag{26.2}$$

In both cases, N_i is the number of chains with a particular molecular weight, M_i. The symbol i is simply an index that specifies that we will include all the components of the mixture. As an example, if we have a polymer mixture having three chains with MW = 1000, two chains with MW = 1200, and five chains with MW = 1400, the values of M_n and M_w are calculated as follows:

$$M_n = \frac{3(1000) + 2(1200) + 5(1400)}{3 + 2 + 5} = 1240 \qquad M_w = \frac{3(1000)^2 + 2(1200)^2 + 5(1400)^2}{3(1000) + 2(1200) + 5(1400)} = 1265$$

Notice that the value of M_w is greater than that of M_n. This result is always obtained because of the greater contribution, in the calculation, by polymers with larger molecular weights.

The ratio M_w/M_n is a useful quantity called the *polydispersity index*. This value is a measure of how broad the range of molecular weights is. If $M_w/M_n = 1.0$, then the polymer is said to be *monodisperse*, which means that all chains have the same molecular weight. Biopolymers such as proteins and nucleic acids that are formed in living organisms are often monodisperse (e.g., all polypeptide chains of the protein myoglobin have the same composition, length, and mass), but synthetic polymers are not monodisperse unless extra steps are taken to separate the product into its constituents.

The properties of polymers vary by more than the molecular weight. Composition and structure (linear, branched, cross-linked, etc.) play roles in defining the physical state of a polymer. Many polymers are soluble in organic solvents, from which the polymer can be cast as a film by evaporation. Polymers with cross-links, a topic to be discussed shortly, are normally insoluble. **Plastics** are defined as polymers that can be molded when hot and then retain their shape when cooled. **Thermoplastics** are those that can be melted by heating to form a fluid that is readily molded. A **thermoset plastic,** on the other hand, can be molded only when it is first prepared. After cooling, it hardens and cannot be melted again. An **elastomer** is a type of plastic that can be

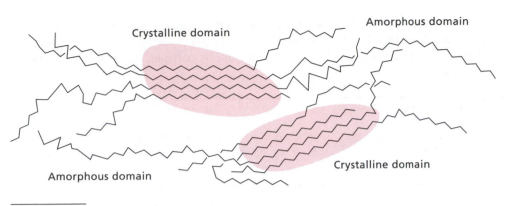

Figure 26.1
The generalized structure of a solid polymer illustrating amorphous and crystalline regions.

stretched without breaking, and it returns to its original shape and size when the stretching force is released.

Polymers tend to form crystalline regions within the larger mass of material, especially when the chains can form regular, ordered structures or where intermolecular forces such as hydrogen bonds can form. Figure 26.1 shows a schematic view of a polymer in the solid state. *Crystalline domains* differ from those regions with no ordered structure, the *amorphous domains*. Crystallinity tends to increase the strength and stiffness of a polymer. For example, high-density (or linear) polyethylene (HDPE) comprises straight chains of the polymer, which pack together tightly and are held together by London forces.

This packing of chains correlates with a high proportion of crystalline domains, which produces an opaque material used to make items such as bottle caps and computer cases. Low-density polyethylene (LDPE), by contrast, has more amorphous domains, giving a material that is transparent and more flexible. Food storage bags and squeezable plastic bottles are made from LDPE.

26.1d SOME POLYMERS HAVE STEREOGENIC CENTERS

Making a macromolecule does not change the fundamental structural features of bonded carbon atoms. In particular, a carbon atom with four different groups is chiral, whether that atom is in a small molecule or in a polymer with a molecular weight in the millions. Therefore, polymerization of an alkene that is unsymmetrical with respect to the C=C bond leads to formation of a material with new stereogenic centers.

The stereochemical outcome for polymerization of a terminal alkene is classified in one of three ways: **isotactic, syndiotactic,** and **atactic,** which are illustrated below.

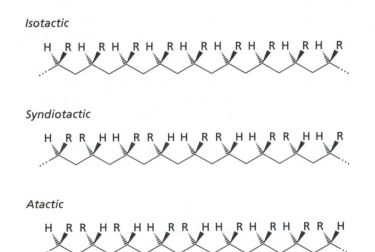

An isotactic polymer has all of its substituents on the same side of a chain that is viewed in its zigzag conformation. A syndiotactic polymer exists with its substituents alternating from side to side along such a chain. In an atactic polymer, the substituents are oriented randomly.

EXERCISE 26.2

For the short sections of polymer shown in the preceding scheme, assign absolute configurations (R and S) to each asymmetric carbon atom of the isotactic and syndiotactic polymers when R = CH_3.

26.2 CHAIN-GROWTH POLYMERIZATION

26.2a ADDITION POLYMERS ARE FORMED FROM ALKENE MONOMERS

The most common reaction that an alkene undergoes is addition: electrophilic addition (Section 9.1a), concerted addition (Section 9.4a), radical addition (Section 12.3), and conjugate (nucleophilic) addition (Section 24.2). When addition occurs sequentially within a collection of alkene molecules, then a polymer forms. This process is called **chain-growth polymerization** or **addition polymerization** because the polymer chain is constructed by a series of addition reactions between a reactive group at the end of the growing chain and molecules of the monomer.

26.2b POLYMERIZATION OCCURS VIA CARBOCATION ADDITION TO AN ALKENE IN THE MARKOVNIKOV ORIENTATION

In Section 9.3b, you learned that a carbocation is able to add to the π bond of an alkene, with Markovnikov regiochemistry; that is, addition occurs such that the new carbocation is the more stable of the two carbocations possible. An acid such as $H[BF_3(OH)]$, which is a strong acid with a poorly nucleophilic anion, catalyzes the polymerization reactions of electron-rich alkenes (Section 9.3b). For example, isobutylene forms poly(isobutylene):

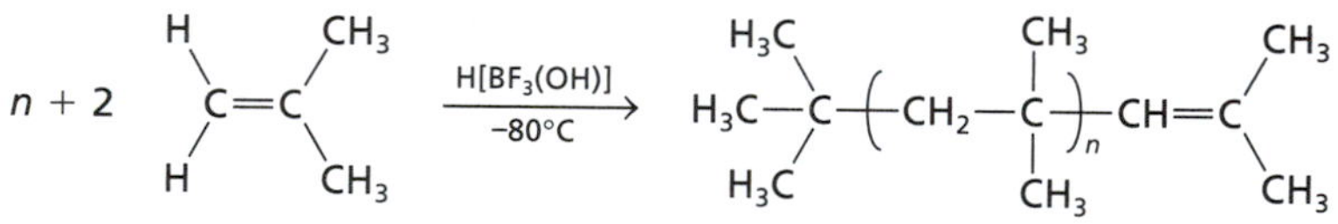

The product comprises a mixture of polyisobutylene molecules having a range of molecular weight values, as described in Section 26.1c.

Cationic polymerization, which is the general term for the process illustrated here for isobutylene, requires that stabilized carbocations are produced at each step. Therefore, electron-withdrawing groups like halogen atoms cannot be attached to the alkene carbon atoms of the monomer. Besides isobutylene, only vinyl ethers, styrene, and vinylbenzene derivatives are normally polymerized via cation intermediates.

Biological systems, however, can readily generate and stabilize carbocations so you might expect cationic polymerization to be involved in the formation of naturally occurring polymers. Natural rubber is formally a polymer of isoprene, but its formation occurs via cationic polymerization of isopentenyl pyrophosphate. The mechanism for this polymerization process follows the same pathway described previously for the biosynthesis of squalene (Section 9.3c).

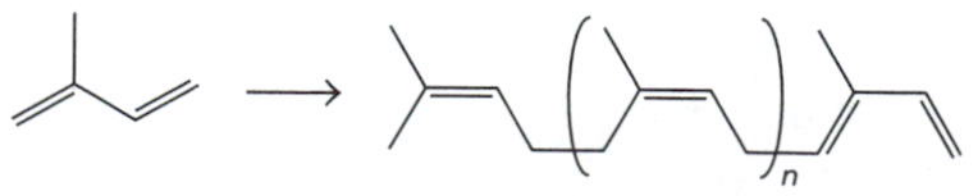

Isoprene   **Natural rubber**

Butyl rubber is a copolymer made by adding ~ 10% isoprene to the normal mixture of isobuylene and H[BF$_3$(OH)]. Isoprene occasionally intercepts a carbocation in the growing polymer chain. Draw the structure of the intermediate carbocation that is generated when this step occurs. Review the steps in the mechanism of alkene polymerization illustrated in Section 9.3b.

26.2c ADDITION OF AN ALKYL RADICAL TO AN ALKENE PROVIDES A GENERAL METHOD FOR THE PREPARATION OF POLYMERS

You learned in Chapter 12 that radical species add to the π bond of an alkene. Sequential addition of radicals to alkene molecules is called **radical polymerization** (Section 12.3b). To review the mechanism briefly, *initiation* is normally carried out using a compound like benzoyl peroxide, which decomposes upon heating to form benzoyloxy radicals and phenyl radicals (Section 12.3b). If AIBN is used instead, the initiating species is the dimethylcyanomethyl radical.

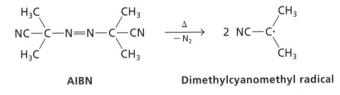

AIBN	**Dimethylcyanomethyl radical**

Any of these radicals [PhCOO·, Ph·, (CH$_3$)$_2$(CN)C·] will add to the π bond of an alkene to generate an alkyl radical, and successive addition steps with additional alkene molecules will eventually generate a polymer.

For example, styrene reacts with the benzoyloxy radical to form polystyrene. The first step creates the alkyl radial. The regiochemistry of this step is such that the more stable of the two possible radicals is formed.

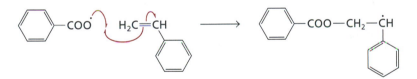

The carbon-centered radical in turn adds to a second molecule of styrene, producing yet another radical, and the process is repeated over and over (*propagation*) to form the polymer. *Termination* occurs whenever any two radical species combine (Section 12.1a), or when the polymer chain loses a hydrogen atom to form a double bond.

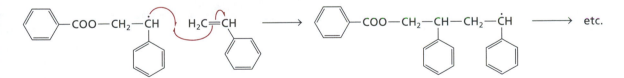

Notice that the end of the chain where polymerization began retains the initiating group. The longer the polymer chain becomes, the less effect this "end group" has on the polymer's properties. For specialized applications, the end group can be modified or removed.

EXERCISE 26.4

Methyl methacrylate is used to make poly(methyl methacrylate), a clear plastic, best known as Plexiglass. Show the first three steps of the polymerization process that makes use of AIBN to initiate the reaction.

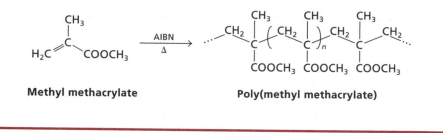

Methyl methacrylate   **Poly(methyl methacrylate)**

Ethylene is commonly polymerized using free radical initiation (Section 12.3b), shown in the following equation with AIBN as the initiator and a polymer product that stopped propagating by loss of a hydrogen atom.

$$n + 1 \quad H_2C{=}CH_2 \xrightarrow{\text{AIBN, } \Delta} (CH_3)_2C(CN){-}\!\!\left(\!CH_2{-}CH_2\!\right)_{\!n}\!\!{-}CH{=}CH_2$$

The actual structure of the polyethylene product in this equation is not as simple as the structural formula foregoing would indicate, however. When ethylene is polymerized by radical methods, the resulting polymer has many chains that branch-off from the principal carbon chain. These **branched polymers** form if the radical site of a growing polymer chain abstracts a hydrogen atom during the polymerization process.

During radical polymerization of ethylene, two types of branched polymers are formed. The first type results from what is called *short-chain branching*, which occurs via a six-membered ring transition state and leaves a chain of four carbon atoms dangling at random points of the chain.

Short-chain branching

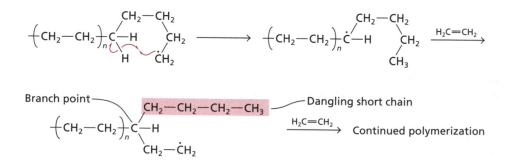

Alternatively, the radical of the growing chain can abstract a hydrogen atom from another chain. A new polymer subsequently begins to grow in the middle of an existing chain, and leads to a process called *long-chain branching*.

Long-chain branching

$\left(CH_2-CH_2\right)_n CH_2-CH_2-CH_2-CH_2-\overset{\bullet}{C}H_2$ + $\left(CH_2-CH_2\right)_n CH_2-CH_2\left(CH_2-CH_2\right)_m$

Growing chain   Polymer already formed

↓

$\left(CH_2-CH_2\right)_n CH_2-CH_2-CH_2-CH_2-CH_3$ + $\left(CH_2-CH_2\right)_n \overset{\bullet}{C}H-CH_2\left(CH_2-CH_2\right)_m$

Terminated chain   New radical in the middle of a polymer

$\downarrow H_2C{=}CH_2$

$\left(CH_2-CH_2\right)_n \underset{\underset{\underset{\overset{|}{\overset{\bullet}{C}H_2}}{\overset{|}{CH_2}}}{\overset{|}{CH}}}{}-CH_2\left(CH_2-CH_2\right)_m$

$\downarrow H_2C{=}CH_2$

Continued polymerization

Formation of branched chains usually leads to a preponderance of amorphous domains (Section 26.1c) in the resulting polymer, which leads to more flexible and lower melting plastics. Branched polyethylene is called LDPE. When there is no branching, the product is called HDPE, a material that is usually made by the Ziegler–Natta methods described in Section 26.2d. Linear polyethylene is much stronger than branched polyethylene, but branched polyethylene is cheaper and easier to make.

Low-density polyethylene, which is made by radical initiation methods, is commonly used for soft plastic bags for food storage. High-density polyethylene is found in bottles used to package products that have a short shelf-life such as milk. Polyethylene, being a hydrocarbon, is chemically nonreactive, which allows it to be used for making containers that are used for household and industrial chemicals.

You are probably familiar with these materials even if you have not studied polymer chemistry previously. The recycling symbols that are stamped into plastic items indicate the polymer that was used in their manufacture, and these are summarized in Table 26.1. You have probably seen numbers 1 and 2 most often because these polymers account for ~ 95% of all plastics used to make bottles. Polyethylene terephthalate (**1**) is a copolymer that is the relatively hard plastic used in the manufacture of soft drink bottles. (The preparation of PET is described in Section 26.3b.) High-density polyethylene (**2**) is a softer plastic that is made into gallon milk cartons, for example.

26.2d TRANSITION METAL COMPLEXES CAN BE USED TO POLYMERIZE ALKENES

Radical and carbocation species are commonly used to initiate polymerization processes, but polymers having a defined stereochemistry cannot be made by those methods because the growing chain is achiral at the reactive carbon atoms. As you have seen repeatedly, stereochemical control is easier to attain when a reaction takes place in a concerted fashion. For polymerization, a concerted process occurs when a transition metal complex catalyzes the reaction, because migratory insertion (Section 15.3c), the

Table 26.1 A summary of the numbers, abbreviations, and structures of recyclable plastics.[a]

Number	Abbreviation	Polymer name	Polymer formula
1	PET	Polyethylene terephthalate	$\left(\text{OCH}_2\text{CH}_2\text{O}-\overset{\text{O}}{\overset{\|}{\text{C}}}-\bigcirc-\overset{\text{O}}{\overset{\|}{\text{C}}}\right)_n$
2	HDPE	High-density polyethylene	$\left(\text{CH}_2-\text{CH}_2\right)_n$
3	PVC	Poly(vinyl chloride)	$\left(\text{CH}_2-\underset{\text{Cl}}{\text{CH}}\right)_n$
4	LDPE	Low-density polyethylene	$\left(\text{CH}_2-\text{CH}_2\right)_n$
5	PP	Polypropylene	$\left(\text{CH}_2-\underset{\text{CH}_3}{\text{CH}}\right)_n$
6	PS	Polystyrene	$\left(\text{CH}_2-\underset{\text{C}_6\text{H}_5}{\text{CH}}\right)_n$

[a]The symbol for PET is shown as an example.

key carbon–carbon bond-forming step, occurs between two species bonded to a metal ion.

$$n \ \overset{\text{H}}{\underset{\text{H}}{\text{C}}}=\overset{\text{H}}{\underset{\text{R}}{\text{C}} \xrightarrow[\text{metal catalyst}]{\text{transition}} \ \text{X}\left(\overset{\text{H}}{\underset{\text{H}}{\text{C}}}-\overset{\text{H}}{\underset{\text{R}}{\text{C}}}\right)_n$$

A common type of transition metal catalyzed process is **Ziegler–Natta polymerization,** named for Karl Ziegler and Giuilio Natta, the two scientists who discovered and developed this chemistry, which earned them the 1955 Nobel Prize in chemistry. The simplest Ziegler–Natta catalysts are made by combining a transition metal salt and a Lewis acid. For example, titanium(III) chloride and diethylaluminum chloride react to form a complex that subsequently dissociates to produce an alkyl(chloro)titanium species, the exact structure of which is unknown. We will use $\text{Ti}(\text{C}_2\text{H}_5)\text{Cl}_2$ as the formula of the initiating species.

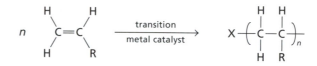

An alkene, for example, ethylene, reacts with the titanium ion in the same way that an alkene reacts with Wilkinson's catalyst (Section 15.3e), that is, by binding to an open coordination site.

Migratory insertion of the ethyl group to the alkene produces a butyl(titanium) intermediate, concomitant with C–C bond formation.

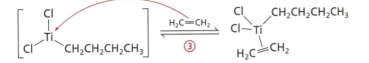

This migration step creates a vacant coordination site at the metal center, allowing another molecule of ethylene to bind to the titanium ion.

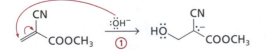

The next migratory insertion step generates a hexyl(titanium)dichloro species.

This process of migration, which opens a coordination site, followed by binding of ethylene, leads to polymer formation. Termination of the process occurs when a hydrogen atom is transferred from the solvent or another polymer chain to the metal ion. This transfer step is followed by reductive elimination of an alkane, which is no longer able to undergo addition to form a longer chain. Molecular hydrogen is sometimes added to terminate the reaction.

Presently, the catalyst used to prepare HDPE (Section 26.2c) by the Ziegler–Natta method is a chromium salt supported on silica, but the sequence of steps outlined above still applies. Heterogeneous catalysts formed by adsorbing or attaching metal complexes to silica are especially useful in controlling the stereochemistry of Ziegler–Natta polymerization of *substituted* ethylene derivatives ($RCH=CH_2$). High degrees of isotactic structure can be obtained under certain conditions, and > 95% isotacticity is obtainable. Radical and cationic polymerization processes generate atactic polymers.

26.2e ANIONIC POLYMERIZATION OFTEN MAKES USE OF CONJUGATE ADDITION

Addition of a carbanion to an α,β-unsaturated carbonyl compound produces another carbanion (Section 24.3), so polymerization can occur if the conditions are right. In fact, **anionic polymerization,** the sequential addition of carbanions to alkenes to form polymers, occurs with a variety of Michael acceptors (Section 24.3a), some of which follow:

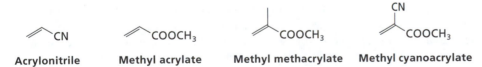

Acrylonitrile   **Methyl acrylate**   **Methyl methacrylate**   **Methyl cyanoacrylate**

The initiator for anionic polymerization is a nucleophile that can generate an enolate ion by addition to a Michael acceptor. A good example is methyl cyanoacrylate, which constitutes Super Glue. If hydroxide ion undergoes conjugate addition to methyl cyanoacrylate, a carbanion intermediate (enolate ion) is generated.

In the normal Michael reaction, this carbanion would be protonated. But conjugate addition to another double bond can occur in the absence of other electrophiles, and repeated addition steps lead to polymer formation.

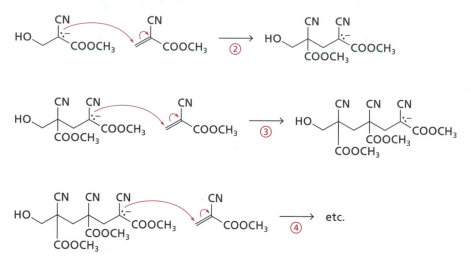

Anionic polymerization is not used as often to make specific polymers because other addition methods are easier to control. For an application like adhesion, however, the composition is not important (as long as it holds!), so anionic conjugate addition constitutes an efficient and readily initiated process.

EXERCISE 26.5

A substantial portion of skin is protein. Indicate which amino acid side chains might act as nucleophiles to initiate the anionic polymerization of methyl cyanoacrylate, accounting for the warning on products like Super Glue that caution against gluing one's fingers together. See Table 27.1 for structures of the common amino acids.

26.3   STEP-GROWTH POLYMERIZATION

26.3a CONDENSATION POLYMERS ARE FORMED BY REACTIONS BETWEEN DIFUNCTIONAL MOLECULES

So far, you have seen how polymers are made by repeated addition to the π bonds of alkenes by a radical, a carbocation, or an alkyl group attached to a metal ion. Polymers can be made by other methods, however, and **condensation polymers,** also known as **step-growth polymers,** are made by reactions between pairs of functional groups. Polyesters, polyamides, polyurethanes, and related materials are commonly made by step-growth polymerization. Step-growth polymerization differs from chain-growth polymerization in that each end of each intermediate formed during the polymerization process can react. During the chain-growth process, a reactive intermediate is present only at one end of the growing chain.

Wallace Carothers, a DuPont chemist who is credited with the discovery and development of nylon, was a pioneer in the study of step-growth polymerization. He proposed a correlation between the number average degree of polymerization, X_n (the average number of monomer units in the polymer chain) and the extent of reaction, p. This relationship has come to be known as the **Carothers equation** (Eq. 26.3).

$$\bar{X}_n = \frac{1}{(1-p)} \tag{26.3}$$

Table 26.2   A comparison between $\bar{X}_n$ and p for a typical step-growth polymerization process.

p	50%	90%	99%	99.9%	99.99%
$\bar{X}_n$	2	10	100	1000	10,000

A summary of the results from applying the Carothers equation is shown in Table 26.2, and the take-home lesson is this: *The process must reach an extremely high conversion level if a reasonable length of polymer is to be obtained.*

Even when the reaction is 99% complete, the average polymer in the mixture is only ~ 100 units long, which is often too short to have much utility. Let us look at what makes step-growth polymerization inefficient until the reaction is nearly complete.

Consider the reaction between a diacid and a diol, catalyzed by HX, which produces a polyester. The first step produces a monoester with elimination of a molecule of water, and this first product contains an acid and an alcohol functional group.

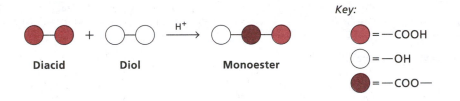

Key:

⬤ = —COOH
◯ = —OH
⬤ = —COO—

What we would like to see happen is a reaction between two molecules of this monoester product to form a triester. A polyester is the eventual product of this pathway.

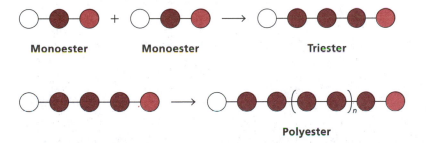

The actual sequence of events is directed by the concentrations of the components within the reaction mixture. After formation of the monoester, much greater amounts of diacid and diol are present, so the likelihood that two monoesters will find each other is low. As a result, intermediates that form in early stages of the reaction have the *same* functional group at each end, so they must diffuse through the reaction mixture until they collide with a molecule that has the other functional partner.

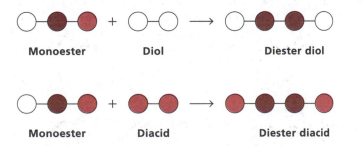

Notice that if functional groups are *different* at each end of the molecule, then *every* other molecule in the mixture has a functional partner, so productive collisions occur much more often. As the reaction progresses, the starting diacid and diol concentrations finally become low enough that the larger pieces begin to condense with each other more readily, and toward the end of the reaction (>99% conversion), appreciable amounts of polymer are finally produced.

26.3b POLYESTERS ARE MADE BY TRANSESTERIFICATION

Polyesters are prepared by condensation processes, but they are not often made by the reaction between a diacid and diol, described in Section 26.3a, because direct esterification reactions rarely exceed 98% conversion. Polyesters made in that way would have < 100 units in each chain. However, transesterification (Section 21.4e) provides a viable alternative route. For example, PET, which is used to make recyclable soft drink bottles, is prepared by the reaction between dimethyl terephthalate and ethylene glycol.

The first stage of the process is a transesterification step that replaces methanol with ethylene glycol units. The methanol can be recovered and recycled, aspects of the process that constitute important economic and environmental considerations on a large scale. Each year 50 billion tons of PET are made in the United States alone.

Dimethyl terephthalate

Then, the mixture is heated even further, to ~ 200°C, in the presence of an antimony oxide catalyst. The OH groups of the ethylene glycol units participate in a second transesterification step that creates the polyester and regenerates half of the ethylene glycol, which is also recycled.

The mechanism of transesterification was presented in Section 21.4e: The OH group of the hydroxyethyl ester acts as a nucleophile to generate a tetrahedral intermediate, which collapses to regenerate the carbonyl group. This process is repeated thousands of times to generate the polymeric product.

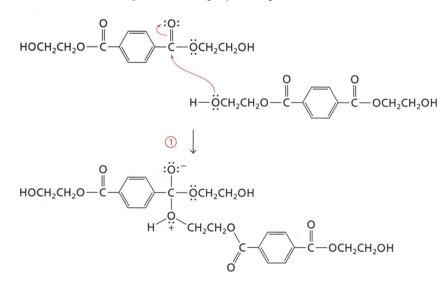

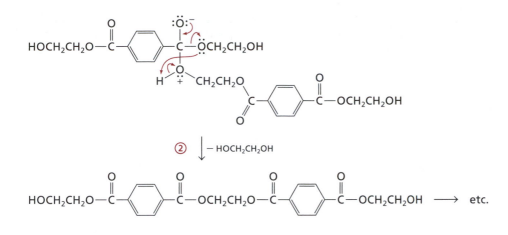

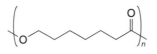

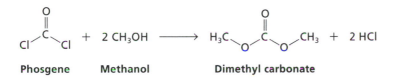

EXERCISE 26.6

What monomer would you use to prepare the polyester shown here? Propose a reasonable mechanism for the first three coupling steps.

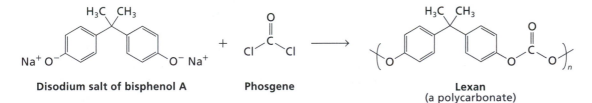

Polycarbonates are similar in structure to polyesters. A carbonate is the diester derivative of carbonic acid, but these esters are normally made by the reaction between an alcohol and phosgene, not the acid itself, which in unstable.

When a diol reacts with phosgene, then a polycarbonate is formed. A familiar polycarbonate is Lexan, which is both transparent and resistant to shattering when struck, so it is used to make materials such as safety glass and helmets for football players and cyclists.

EXERCISE 26.7

Propose a reasonable mechanism for the preparation of dimethyl carbonate from phosgene and methanol.

26.3c POLYAMIDES ARE MADE FROM DIAMINES AND DICARBOXYLIC ACIDS OR FROM AMINO ACIDS

The amide group is an integral part of proteins, and we will look more closely at the chemistry of amino acids and polypeptides in Chapter 27. The amide group is also present in polyamides that find application in the manufacture of textiles and other materials. Nylon is probably the best-known polyamide, and it is prepared in much the same way as polyesters. If a diamine is treated with a dicarboxylic acid under conditions that lead to expulsion of water (usually high temperatures), a step-growth process occurs to form the polyamide.

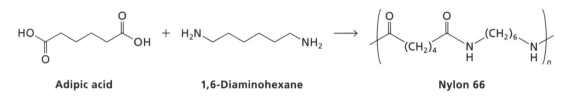

For example, adipic acid and 1,6-diaminohexane react to form Nylon 66.

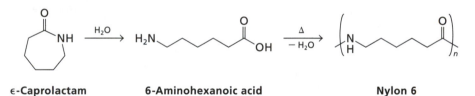

Adipic acid   **1,6-Diaminohexane**   **Nylon 66**

As you might expect, a diester or diacid chloride can also be used in polyamide synthesis, but these acid derivatives are much more expensive than the diacid itself. The use of an acid chloride also requires the addition of base to neutralize the HCl that is generated.

Another way to prepare a polyamide is to start with an amino acid. For example, 6-aminohexanoic acid reacts with itself to form Nylon 6. The starting material is made by hydrolysis of ε-caprolactam, which is made by the Beckman rearrangement of cyclohexanone oxime (Section 20.1d). Once the nylon chain starts growing, its amino group can open the caprolactam ring directly, a process called *ring-opening polymerization*. Nylon 6 and Nylon 66 have very similar properties because the amide groups are spaced similarly in each polymer.

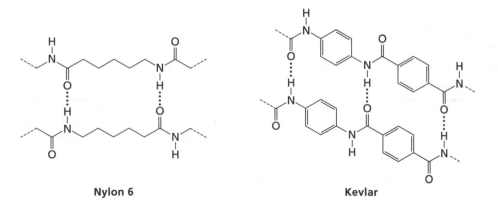

ε-Caprolactam   **6-Aminohexanoic acid**   **Nylon 6**

Polyamides often display structural regularities that result from formation of hydrogen bonds between chains. These same types of hydrogen bonds are created within the interior of many proteins and account for their structures (Section 27.5b). The N–H portion of an amide group is a good hydrogen-bond donor, and the carbonyl oxygen atom functions as a hydrogen-bond acceptor (Section 2.8b). Nylon 6 and Kevlar are two examples of polyamides that have regular structures stabilized by hydrogen bonds.

Nylon 6   **Kevlar**

26.3d MOLECULES WITH OTHER FUNCTIONAL GROUPS CAN BE USED AS REACTANTS IN STEP-GROWTH POLYMERIZATION PROCESSES

Many of the reactions you have learned so far in this course can be used to create polymers. Instead of performing a reaction between two compounds, each of which has one functional group, you could choose compounds that each have two (or more) functional groups. For example, the nucleophilic substitution reaction between an alkyl halide and an alkoxide ion (the Williamson ether synthesis—Section 7.2b) might be one to consider:

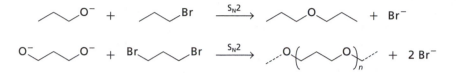

Practical considerations, however, often limit the types of reactions that are actually used to make polymers. In the case of nucleophilic substitution reactions, solvents needed to dissolve the bis(alkoxide) ion may not dissolve the polymeric product well. So as the reaction progresses, the polymer may begin to precipitate before it reaches a reasonable length. There are ways to circumvent these limitations, such as using two immiscible solvents, one to dissolve the salt, and one to dissolve the growing polymer chain. This latter strategy is actually used to make Lexan (Section 26.3a).

Nevertheless, there are several functional groups besides esters and amides that are suitable for polymer formation. Epoxides are reactive functional groups that can be opened under a variety of conditions with different reagents (Section 7.2e). For example, ethylene oxide is opened in water to form ethylene glycol.

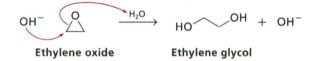

Ethylene oxide   **Ethylene glycol**

Poly(ethylene oxide) can be made by base-induced ring opening of ethylene oxide in the *absence* of water.

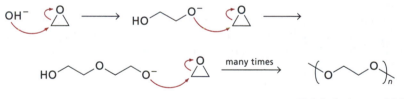

Poly(ethylene oxide)

Epoxide resins, as found in cured two-component epoxy glues, are made in a similar manner by the reaction between a bis(epoxide) and a diamine, which leads to opening of the strained three-member rings.

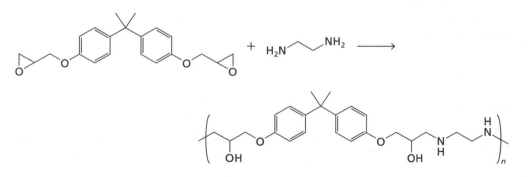

Epichlorohydrin is a difunctional molecule that reacts with 2 equiv of alcohol to form a diether.

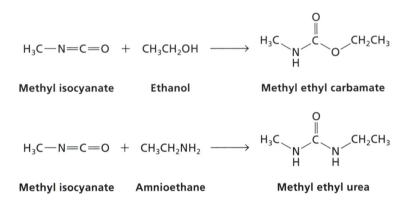

In the first step, the epoxide ring is opened by a molecule of alcohol, which generates a secondary alcohol. This secondary OH group reacts with the adjacent carbon atom by displacing the chloride ion and forming an epoxide ring. The second molecule of alcohol (methanol in this exercise) subsequently reacts to open the second epoxide ring. Show the steps of the overall mechanism just described (*cf* Exercise 16.23).

Epichlorohydrin can react in a similar way with diols to form polymers. Draw the structure of the polymer formed from the reaction between epichlorohydrin and 1,3-dihydroxypropane.

The **isocyanate** group is another highly reactive functional group that finds application in polymer synthesis. Isocyanates react with alcohols to form **carbamates,** which are also called **urethanes,** and with amines to form derivatives of **urea.**

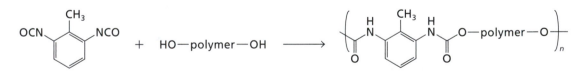

Methyl isocyanate   Ethanol   Methyl ethyl carbamate

Methyl isocyanate   Amnioethane   Methyl ethyl urea

Polyurethanes are prepared from reactions between diisocyanates and diols. Often, the diol is a polymer with a hydroxy group at each end of the chain. The length of the polymeric diol affects the specific properties of the resulting polyurethane.

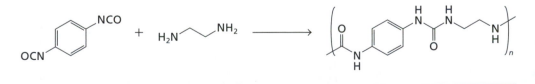

Propose a reasonable mechanism for the reactions leading to urea formation in the preparation of the following poly(urea):

26.4 MODIFICATION OF POLYMERS

26.4a CROSS-LINKS MAKE POLYMERS MORE RIGID AND LESS SOLUBLE

Polystyrene is formed by the radical-initiated polymerization of styrene, as described in Section 26.2c. Because it is soluble in a variety of organic solvents, polystyrene objects can be made in practically any shape; with injection of an inert gas during the evaporation or casting process, styrofoam is generated.

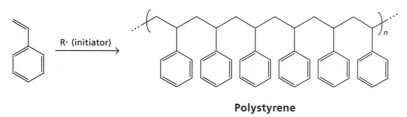

Polystyrene

Soluble in many organic solvents

When a mixture of styrene (98%) and *p*-divinylbenzene (2%) is allowed to polymerize, an insoluble *cross-linked* copolymer is formed. This material is much more rigid than the material without cross-links. Recall that proteins such as collagen are also made more rigid by forming crosslinks (Section 20.2b). Cross-linked polystyrene (cps) is often made as an emulsion in a polar solvent, which creates small spheres as it forms. These beads, as they are called, find applications as supports for reagents (Section 26.4b) and for solid-phase peptide synthesis (Section 27.4c).

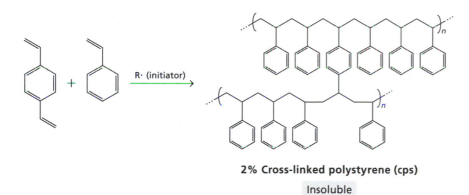

2% Cross-linked polystyrene (cps)

Insoluble

Cps has channels and voids within the polymer matrix that allow solvents and reagents to diffuse throughout its interior. Many solvents cause swelling of the 2% cross-linked polymer, increasing its flexibility and enhancing diffusion processes. With use of larger amounts of divinylbenzene, cross-linking becomes extensive enough to inhibit flexibility.

26.4b FUNCTIONALIZED POLYMERS CAN BE USED TO ATTACH OR SCAVENGE REAGENTS

The phenyl rings in cps react like any alkylbenzene derivative, so they can undergo substitution by electrophilic reagents. Friedel–Crafts alkylation (Section 17.2e) makes use of chloromethyl methyl ether and a Lewis acid catalyst such as $SnCl_4$. The electrophile is the $SnCl_5^-$ salt of $CH_3OCH_2^+$, and substitution occurs randomly throughout the polymer.

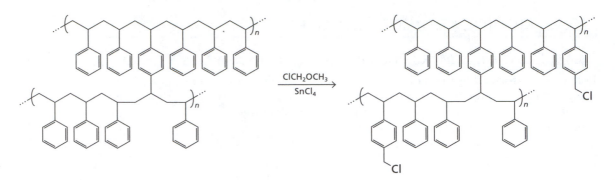

After attachment of chloromethyl groups to the phenyl rings of cross-linked polystyrene, it is possible to carry out substitution reactions at the benzylic positions. With use of appropriate reagents, a variety of functional groups can be attached to the insoluble polymer.

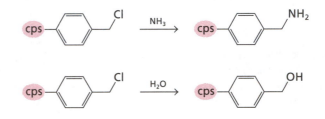

Another reaction that is generally successful with cross-linked polystyrene is electrophilic bromination.

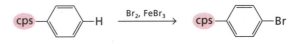

Once the ring has been brominated, then lithium–halogen exchange generates an organometallic intermediate that can be used to attach other functional groups to the rings. As with all of operations carried out with an insoluble polymer, purification is a matter of washing away unreacted reagents or unwanted side products. The functional group remains attached to the polymer and is ready for subsequent reactions (S_8 = elemental sulfur).

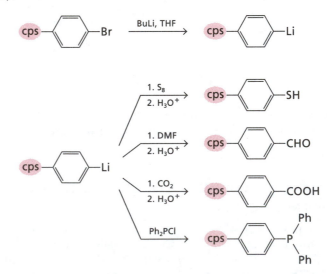

When a diphenylphosphino group is attached to cps (shown directly above), it can be used to perform a Wittig reaction as a "polymer-supported reagent". In this procedure, the phosphorus atom is alkylated in the usual way, and a base (butyllithium in this example) is added to form the ylide (Section 20.4a). Addition of a ketone leads to formation of the corresponding alkene. By using the polymer-supported Wittig reagent, the phosphine oxide byproduct remains attached to the polymer, which is insoluble. To isolate the desired alkene product, one has only to filter the reaction mixture and evaporate the solvent.

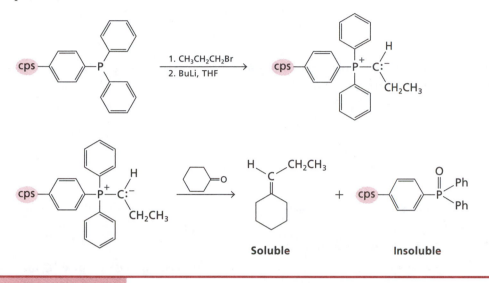

EXERCISE 26.10

Starting with the polymer-supported thiol illustrated here, show how you could create a reagent to carry out epoxidation reactions of aldehydes and ketones via a sulfur ylide (Section 20.4e).

Polymers with functional groups can also be used to scavenge reagents from a reaction mixture. For example, if you were alkylating an amine with benzyl bromide, $PhCH_2Br$, you might add excess $PhCH_2Br$ to ensure complete alkylation. Both the desired product and the alkylating agent are soluble in organic solvents, however, so purification might prove to be tedious.

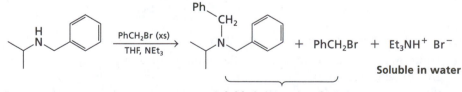

With subsequent addition of a thiol-functionalized polymer, you would be able to remove the excess alkyl halide by alkylation of the sulfur atom, an excellent nucleophile, thereby ensuring that further alkylation of the amine did not occur during its purification. The insolubility of the polymer makes workup straightforward.

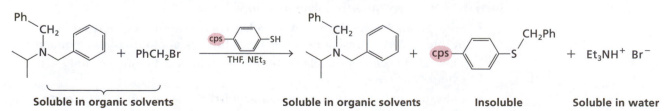

EXERCISE 26.11

You carry out hydrolysis of the following imine and wish to isolate the amine product:

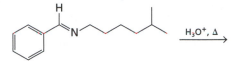

$$\xrightarrow{\text{H}_3\text{O}^+,\ \Delta}$$

Show how you would make use of aminomethylpolystryrene [(cps)–CH$_2$NH$_2$, shown here] to act as a scavenger of the byproduct.

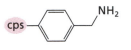

CHAPTER SUMMARY

Section 26.1   Polymers and their properties

- Polymers are high molecular weight compounds composed of repeating units of a small molecule called a monomer.
- A polymer is often named with the prefix "poly-" plus the name of the monomer.
- Copolymers are made from more than one monomer, and they can be classified as random, alternating, or segmented; segmented polymers comprise block and graft copolymers.
- The molecular weight of a polymer is an average of the masses of individual chains that constitute the bulk material.
- A polymer in which all of the chains have identical lengths is called monodisperse.
- Plastics are polymeric materials that can be molded when hot and then retain their shape when cooled.
- Polymers tend to have portions that are crystalline and some that are amorphous. Those with more crystalline than amorphous domains are often stronger and stiffer.
- Polymers with chiral carbon atoms can be classified by their types of stereochemistry, which are isotactic, syndiotactic, or atactic.

Section 26.2   Chain-growth polymerization

- Addition polymers are made from alkene monomers by a process called chain-growth polymerization.
- Chain-growth polymers are made by addition of a carbocation, radical, or carbanion at the end of a growing polymer chain to the π bonds of alkene monomers.
- Chain-growth polymerization can also be catalyzed by transition metal compounds, via migratory insertion of an alkyl group to the π bond of the alkene, a process called Ziegler–Natta polymerization.

Section 26.3   Step-growth polymerization

- Step-growth polymerization, also called condensation polymerization, occurs by intermolecular reactions between two functional groups of the monomers.
- The Carothers equation relates the average number of monomer units with the extent of reaction.
- Polyesters, polyamides, polyurethanes, polycarbonates, and polyethers are common materials made by the step-growth method.

Section 26.4   Modification of polymers

- Polymer chains can be connected by bonds called cross-links, which generally make the polymer insoluble and more rigid.

- Cross-linked polymers are useful materials to which chemical reagents can be attached. Separation of excess reagent from a soluble product is easier when the reagent is attached to an insoluble polymer; filtration suffices to separate the two.

KEY TERMS

Section 26.1a
polymer
monomer
repeat unit

Section 26.1b
homopolymer
copolymer
random copolymer
alternating copolymer
segmented copolymer
block copolymer
graft copolymer

Section 26.1c
plastic
thermoplastic
thermoset plastic
elastomer

Section 26.1d
isotactic
syndiotactic
atactic

Section 26.2a
chain growth
   polymerization
addition polymerization

Section 26.2b
cationic polymerization

Section 26.2c
radical polymerization
branched polymers

Section 26.2d
Ziegler–Natta
   polymerization

Section 26.2e
anionic polymerization

Section 26.3a
condensation polymer
step-growth polymer
Carothers equation

Section 26.3d
isocyanate
carbamate
urethane
urea

REACTION SUMMARY

Section 26.2b

Cationic polymerization; a carbocation adds with Markovnikov regiochemisty to an electron-rich alkene (styrene derivatives, vinyl ethers, and some aliphatic alkenes).

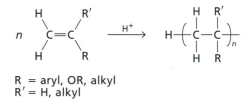

R = aryl, OR, alkyl
R′ = H, alkyl

Section 26.2c

Radical polymerization: A radical adds to a double bond to form the more stable of the two possible radicals.

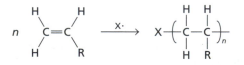

Section 26.2d

Ziegler–Natta polymerization: A transition metal complex catalyzes the addition of a metal-bonded alkyl group to a coordinated alkene.

Section 26.2e

Anionic polymerization: Conjugate addition to an α,β-unsaturated carbonyl compound or nitrile leads to formation of a carbanion, which adds to another molecule of the α,β-unsaturated carbonyl compound, perpetuating a chain reaction.

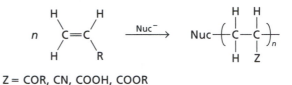

Z = COR, CN, COOH, COOR

Section 26.3b

Diols and diesters or hydroxyesters undergo transesterification reactions to form polyesters.

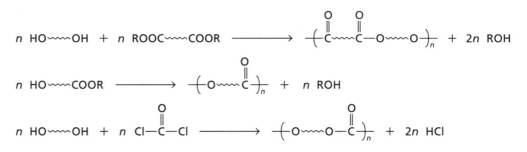

Section 26.3c

Diamines react with diacids, or amino acids react with themselves, to form polyamides.

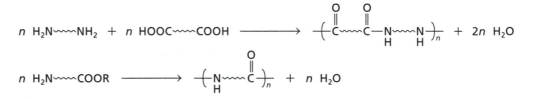

Section 26.3d

Ring-opening reactions of epoxides by nucleophiles leads to formation of polyethers and epoxy resins.

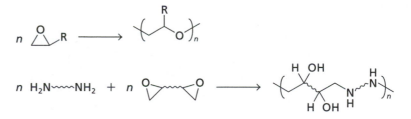

Carbamates (urethanes) are prepared by the reactions between isocyanates and alcohols.

$$R-N=C=O \ + \ R'OH \ \longrightarrow$$

Urea derivatives are prepared by the reactions between isocyanates and amines.

$$R-N=C=O \ + \ R'NH_2 \ \longrightarrow$$

Diols react with diisocyanates to form polyurethanes.

$$n \ HO\text{\sim\sim}OH \ + \ n \ OCN\text{\sim\sim}NCO \ \longrightarrow$$

26.12. Draw a portion of the structure for each of the following (random) copolymers:

    a. Viton, used in gaskets: from hexafluoropropene and 2,2-difluoroethylene.

    b. Nitrile rubber, used in gasoline hoses: from 1,3-butadiene and 2-propene-nitrile (acrylonitrile).

    c. Styrene–butadiene rubber (SBR), used in tires: from styrene and 1,3-butadiene.

26.13. Draw the structure of the polymer you would expect to generate by the acid-initiated polymerization of each of the following monomers. Show the first three steps of the polymerization process, with H^+ as the initiator.

    a. Styrene, C_6H_5–CH=CH$_2$

    b. methyl vinyl ether, CH_3–O–CH=CH$_2$

26.14. With free radical initiation, tetrafluoroethylene is converted to the polymer Teflon. Propose a mechanism for the first three steps of this process.

26.15. A chemist inserted a pipette contaminated with sodium hydroxide solution into a bottle of β-propiolactone. A month later, a co-worker took the bottle from the shelf and discovered that it was full of a viscous material, which she correctly concluded was a polymer. What was the structure of the polymeric material? Predict the appearance of the IR and ^1H NMR spectra of this substance.

26.16. Bisphenol A can be prepared from phenol and acetone under acidic conditions. From your knowledge of electrophilic aromatic substitution, as well as addition reactions of aldehydes and ketones, propose a reasonable mechanism for this transformation.

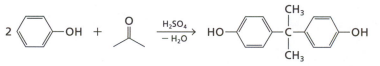

Bisphenol A

26.17. One way to make Lexan is by a transesterification process between diphenyl carbonate and bisphenol A (Exercise 26.16). Show the mechanism by which two bisphenol A molecules are linked in the first stage of this polymerization process.

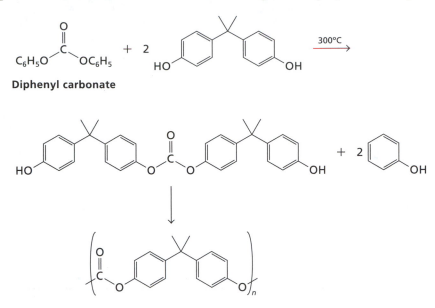

26.18. Urea undergoes a reaction with itself, forming a heterocycle called *melamine*. Formaldehyde reacts with melamine to form a three-dimensional array that is used to manufacture Melmac dishes and Formica. Propose a mechanism for the formation of melamine from urea, catalyzed by acid.

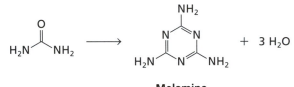

Melamine

Propose a mechanism for the formation of the polymer illustrated below, which is formed via the tris(hydroxymethyl) derivative of melamine.

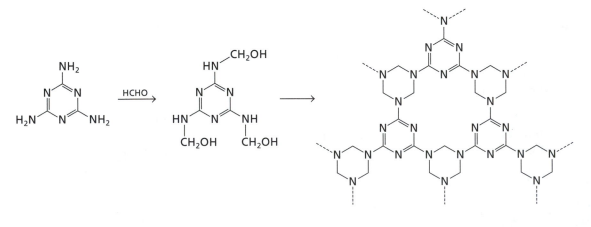

26.19. Qiana is a silky polyamide fabric that has the following structure.

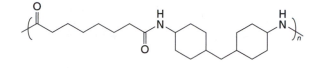

What are the monomers used in its manufacture? Illustrate how this polymer can form intermolecular hydrogen bonds to create a three-dimensional network.

26.20. Urea and formaldehyde condense to form a polymer with the following structure. Propose a reasonable mechanism for its formation, which occurs via *N*-hydroxymethylurea. A possible mechanism follows a pathway analogous to that of the Mannich reaction (Section 24.1b).

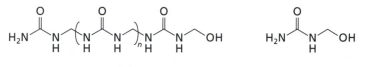

***N*-Hydroxymethylurea**

26.21. Draw the structure of poly(vinyl alcohol). This polymer cannot be made by polymerization of vinyl alcohol because the latter is unstable. Explain why vinyl alcohol does not exist as a stable molecule. Poly(vinyl alcohol) is made instead by hydrolysis of poly(vinyl acetate), a constituent of chewing gum. Draw the structure of poly(vinyl acetate) and show the mechanism for the first three steps of polymerization via radical initiation.

26.22. Natural rubber (Section 26.2b) is normally a sticky substance. Charles Goodyear discovered that heating natural rubber with sulfur, a process called *vulcanization*, produces a harder, more durable substance that he subsequently used in the manufacture of bicycle tires. Vulcanization is a free radical reaction, initiated by heat, whereby adjacent polymer chains are cross-linked by disulfide bonds. Schematically, this process can be illustrated as follows:

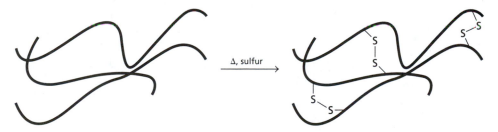

From your knowledge about the stability of free radicals (Section 12.1d), predict at which position(s) these cross-links are likely to form in the rubber chains.

26.23. Another method for the synthesis of polycarbonates makes use of nucleophilic aromatic substitution (Section 17.4e) with the carbonate ion as the nucleophile. Propose a reasonable mechanism for the following transformation:

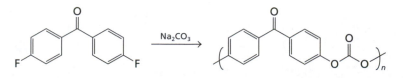

26.24. Neoprene is made by Ziegler–Natta polymerization of 2-chloro-1,3-butadiene. Draw a portion of this polymer assuming that the double bonds in neoprene have the (*Z*) configuration.

26.25. Sulfonation (Section 17.2d) of cross-linked polystyrene can occur on some of the rings of the polymer. Illustrate a portion of the polymer with the appropriate functional groups attached. Such a polymer could be used as a reagent or scavenger in organic reactions: Give some specific examples of reactions in which you might use this polymeric reagent.

26.26. How might the following polymer (Section 26.4b) be used to make a "heterogenized homogeneous catalyst", that is, an insoluble version of the soluble Wilkinson hydrogenation catalyst (Section 15.3e)?

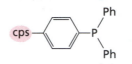

26.27. A mixture of divinylbenzene, styrene, and 4-vinylpyridine can be copolymerized to form an insoluble polymer that has pyridine groups throughout the matrix. What reagents might be attached to this polymer to carry out oxidation processes? (*Hint:* See Section 11.4c.)

26.28. Show how you could make a polymer-supported reagent to carry out the Swern oxidation (Section 11.4b).

chapter 27

AMINO ACIDS, PEPTIDES, AND PROTEINS

Proteins have many roles in the functions of living systems: they provide the structures that compose and support bones, skin, and muscle; they constitute the framework of enzymes, which are catalysts for chemical reactions in organisms; they act as stoichiometric "reagents" involved in the transport and storage of small molecules and ions; they regulate and control growth and differentiation; and they recognize foreign substances. The fact that they are composed mostly of the same 20 building blocks, the α-amino acids, is intriguing.

This chapter describes the chemistry of these two types of molecules: the α-amino acids and proteins. For both topics, the presentation focuses on the structural features and preparative methods of these substances. By necessity, the discussion can only provide a brief introduction to these subjects, but the fundamental aspects that are covered here will provide a foundation for further studies of biochemistry and molecular biology.

27.1   AMINO ACIDS

27.1a   THE α-AMINO ACIDS ARE THE MOST IMPORTANT IN LIVING SYSTEMS

Technically, any organic molecule that has an amino group and a carboxylic acid group is an amino acid. The term amino acid, however, is usually applied to the α-amino acids that constitute proteins. A compilation of the structures and names of the 20 genetically coded amino acids is presented in Figure 27.1.

The term "genetically coded amino acid" deserves comment. In living organisms, most proteins are made by enzymes that interpret the genetic code, the sequence of DNA molecules in the genes that dictate an organism's characteristics. The code is first transcribed to yield individual ribonucleic acid (RNA) molecules. Then, during translation, RNA molecules bonded to amino acids in the cell deliver them, one by one in the encoded order, to a growing protein chain. This scheme, known as "the central dogma of molecular biology", specifies the flow of information from genes to proteins.

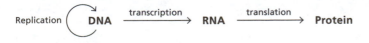

Replication (DNA $\xrightarrow{\text{transcription}}$ RNA $\xrightarrow{\text{translation}}$ **Protein**

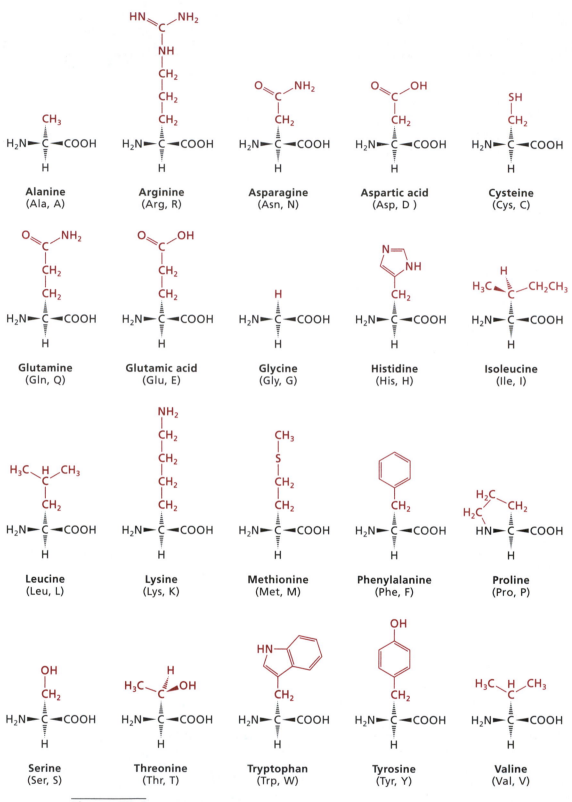

Figure 27.1

Structures, names, and abbreviations for the genetically coded amino acids.

The **genetic code** is carried by groups of three nucleic acid units called **codons.** Sixty-four possible codons can be derived from triplet combinations of the four nucleic acids that constitute RNA, so a maximum of 64 amino acids could be specified, or encoded, by nucleic acid sequences (the structures of the nucleic acids are discussed in Chapter 28). For a variety of reasons, however, the genetic code, as these correlations between codons and amino acids are known, is degenerate, meaning that more than one codon normally stands for the identity of each amino acid. As a result, only 20 amino acids correlate with the codon combinations, and these are the ones given in Figure 27.1.

Some proteins have amino acids other than the 20 associated with gene sequences. In most cases, these other structures are modified forms of one of the genetically coded amino acids. Reactions such as oxidation, carboxylation, or substitution serve to alter the original functional groups, and several of these modified amino acids are shown in Table 27.1. Because these reactions take place after translation, the amino acids are said to arise from post-translational modification.

The α-amino acids that are incorporated into proteins by transcription and translation generally have the L configuration. Some proteins in bacteria are made from D-amino acids. Recall that D and L refer to the absolute configuration of a molecule as its structure relates to that of either D- or L-glyceraldehyde (Section 4.2c). For amino acids, the use of D and L is valuable because these designations reflect the harmony

Table 27.1 The structures of some amino acids made by post-translational modification.

Amino acid	Post-translational modification
γ-Carboxyglutamic acid	Formed by carboxylation of glutamic acid. Found in thrombin, which is used to regulate blood clotting.
ε-N,N,N-Trimethyllysine	Formed by methyllation of lysine. Found in cytochrome *c*, a protein involved with the transfer of electrons.
5-Hydroxylysine (Hyl)	Formed by oxidation of lysine. Found in collagen, a protein that strengthens skin and bones.
4-Hydroxyproline (Hyp)	Formed by oxidation of proline. Found in collagen.

among the molecules' structures (i.e., the geometry of the α-carbon atom is the same in each). If you specify the stereochemistry using the designations (S) and (R), you will see that most of the L-amino acids have the (S) configuration. Cysteine, however, has the (R) configuration because the presence of sulfur raises the priority of the side chain above that of the carboxylic acid group. Notice that the arrangement of the groups is the same for cysteine as it is for the other amino acids.

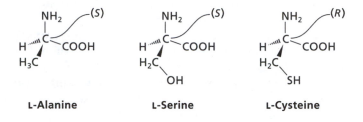

| L-Alanine | L-Serine | L-Cysteine |

Glycine is the only *achiral* amino acid that is genetically coded; two of the amino acids have two chiral carbon atoms.

EXERCISE 27.1

Which of the genetically coded amino acids have two chiral centers? Assign the absolute configuration to each chiral carbon atom in those compounds, which are shown in Figure 27.1.

27.1b AN AMINO ACID EXISTS AS A ZWITTERION AT NEUTRAL pH

In many nonpolar organic solvents, an amino acid exists in its neutral, uncharged form. Thus we draw the structures as in Figure 27.1. In *aqueous solution*, however, an equilibrium exists between the neutral form and an ionized form known as a *zwitterion*, which has two charged sites but is overall neutral. This form is created by an acid–base reaction between the acidic COOH and the basic NH$_2$ groups, and it is stabilized by the polar solvent.

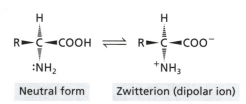

| Neutral form | Zwitterion (dipolar ion) |

If the pH is lowered to a value of 1 or less, the zwitterion form will become protonated at the carboxylate site to generate the carboxylic acid. Note that the amine remains protonated, as expected for an amine at low pH values. On the other hand, at pH 12 or higher, the ammonium site of the zwitterion will be deprotonated to form the amino group. The carboxylate ion remains deprotonated.

pH 1   pH 12

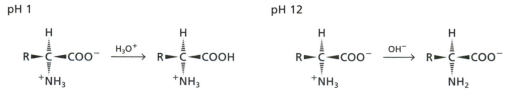

When the side chain of an amino acid has a functional group that can also function as an acid or base, its ionization state will also be affected by the pH of the solution. To predict what its form will be, you can apply the **Henderson–Hasselbalch equation** (Eq. 27.1) to calculate the degree of ionization of an acid in solution. In this equation, HA is the acid form of a group, and A^- is its conjugate base form.

$$pH = pK_a + \log \frac{[A^-]}{[HA]} \tag{27.1}$$

Using the pK_a values for acidic groups in an amino acid, which are shown in Table 27.2, you can calculate how much of the acid and base forms exist as a function of pH. Performing these calculations reveals the following qualitative results:

- If the pK_a value of a functional group is less than the pH of the solution, then the group exists primarily in its deprotonated (base) form.

- If the pK_a value of a functional group is greater than the pH of the solution, then the group exists primarily in its protonated (acid) form.

- If the pK_a value of a functional group is equal to the pH of the solution, then the acid and base forms exist in a 1:1 ratio.

For example, histidine is found at many enzyme active sites (Section 25.4a) because it can exist both as acid and base forms near physiological pH values. The pK_a value of the protonated imidazole ring is 6.0, which means that at pH 7, both imidazole (base) and imidazolium (acid) forms are present (see Exercise 27.2a).

Table 27.2 The pK_a values for the protic groups in amino acids.

Amino acid	pK_a Value		Side-chain structure (acid form)	Side-chain pK_a value
	COOH Group	NH_3^+ Group		
Ala	2.3	9.7		
Arg	2.2	9.0	$-CH_2CH_2CH_2NHC(\overset{+}{=}NH_2)NH_2$	12.5
Asn	2.0	8.8		
Asp	2.1	9.8	$-CH_2COOH$	3.9
Cys	1.9	10.5	$-CH_2SH$	8.4
Glu	2.2	9.7	$-CH_2CH_2COOH$	4.3
Gln	2.2	9.2		
Gly	2.3	9.6		
His	1.8	9.2	$-CH_2$ imidazole	6.0
Ile	2.4	9.7		
Leu	2.6	9.6		
Lys	2.2	9.0	$-CH_2CH_2CH_2CH_2\overset{+}{N}H_3$	10.8
Met	2.3	9.2		
Phe	2.2	9.2		
Pro	2.0	10.6		
Ser	2.2	9.2		
Thr	2.6	9.1		
Trp	2.4	9.4		
Tyr	2.2	9.1	$-CH_2-\text{C}_6\text{H}_4-OH$	10.0
Val	2.3	9.6		

EXAMPLE 27.1

Using the data in Table 27.2, draw the principal ionic form of cysteine that is expected to exist in solution at pH 8.2.

The ionization states of the acidic and basic groups of an amino acid are affected by the pH of the solution in which the compound is dissolved. To determine the forms of these ionized states, draw the structure of the amino acid and identify the groups with acid–base forms: cysteine has COOH, NH_2, and SH groups, and the pK_a values of their acid forms (from Table 27.2) are indicated in color. Even though this exercise asks for the form at pH 8.2, it is worthwhile to consider the ionization state of each group at extreme pH values (1 and 12) for comparison.

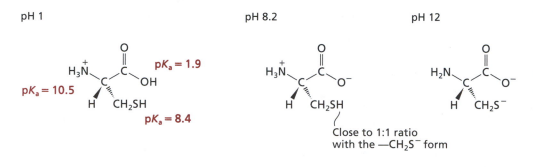

At pH 1, all of the ionizable groups are protonated. The carboxylic acid and thiol groups are neutral when protonated; the ammonium ion has a +1 charge. At pH 12, all of these groups are deprotonated. A deprotonated amino group is neutral, whereas the others are anions.

At the intermediate value of pH 8.2, groups with pK_a values < pH will be deprotonated, and those with pK_a values > pH will be protonated. Because the pK_a value for the thiol group is close to pH 8.2, the thiol and thiolate forms will exist in a ratio that is nearly 1:1, with slightly more of the thiol form present.

EXERCISE 27.2

By using the data in Table 27.2, draw the principal ionic form of the indicated amino acid that is expected to exist in solution at the given pH value.

a. Histidine at pH 7   b. Aspartic acid at pH 5   c. Tyrosine at pH 8

27.1c AMINO ACIDS CAN BE CLASSIFIED BY THE PROPERTIES OF THEIR SIDE CHAINS

Side-chain ionization is important for those amino acids that have acidic or basic groups, but the accumulated properties of all of the side chains within a protein will influence the apparently infinite number of structures that the protein can adopt. Many side-chain functional groups can form hydrogen bonds, even if they do not ionize. These hydrogen bonds can stabilize structures or orient substrates in the active site of an enzyme. Other groups are hydrophobic and fill interior spaces or exclude water molecules from the interior of a folded protein. Some classification categories are summarized in Table 27.3.

Table 27.3 A classification scheme for amino acids according to their side-chain structures.

Classification	Amino acid
Nonpolar	Ala, Val, Ile, Leu
Aromatic	Phe, Trp, Tyr
Cyclic	Pro
Acidic	Arg, Glu
Basic	Arg, His, Lys
Polar (neutral)	Asn, Cys, Gln, Gly, Met, Ser, Thr

The designations given Table 27.3 are meant to reflect the principal feature of each side chain. Some amino acids can be placed into more than one category. For example, histidine has a heterocycle (imidazole, Section 25.4a) in its side chain and thus could be designated "aromatic" as well as basic. Similarly, proline is nonpolar as well as cyclic, and the phenol group of tyrosine and the thiol group of cysteine are weak acids.

EXERCISE 27.3

The NH group in the indole ring of tryptophan can act as a hydrogen-bond donor but not as a hydrogen-bond acceptor (Section 2.8b), even though the nitrogen atom has an unshared electron pair. Why? (*Hint:* See Section 25.3).

Proline is the only cyclic amino acid, and its NH group is a secondary amine, whereas the other 19 amino acids have primary amino groups. This feature has important implications when proline forms a peptide bond because its nitrogen atom no longer has a proton attached, so it cannot be a hydrogen-bond donor (Section 2.8b).

EXERCISE 27.4

Classify each of the following modified amino acids according to the categories shown in Table 27.3. List all that apply.

a. 4-Hydroxyproline   b. 5-Hydroxylysine   c. Allysine
   (Table 27.1)   (Table 27.1)   (Section 20.2b)

27.2   CHEMICAL SYNTHESIS OF AMINO ACIDS

27.2a THE STRECKER REACTION PROVIDES A GENERAL ROUTE TO PREPARE AMINO ACIDS

The properties of the individual amino acids play a large part in defining the structures of the proteins into which they are incorporated. If you want to make proteins from their amino acid building blocks, you need a way to prepare these starting materials. Many amino acids are isolated as products from the hydrolysis of proteins, which limits the number of available substances to those shown in Figure 27.1, plus several other minor derivatives.

If you want *unnatural* amino acids, then you must be able to make them from other substances. Examples of desirable analogues include amino acids with unusual side chains; those in which an atom is isotopically labeled with ^{13}C, ^{15}N, and so on;

those that have atoms like boron in the framework; and D-isomers of the natural amino acids.

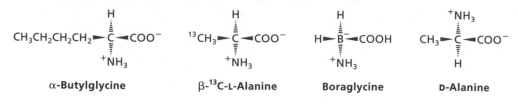

α-Butylglycine   **β-¹³C-L-Alanine**   **Boraglycine**   **D-Alanine**

One general method for preparing amino acids is the **Strecker reaction.** It begins with an addition process in which an aldehyde group is converted to an α-amino nitrile with ammonium cyanide. The nitrile group is subsequently hydrolyzed to form the carboxylic acid group.

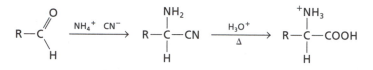

The mechanism of this transformation is straightforward and incorporates reactions with which you are already familiar. Ammonium cyanide exists in aqueous solution in equilibrium with ammonia and hydrogen cyanide. In the first step, ammonia reacts with the aldehyde carbonyl group to form an imine (Section 20.1a).

$$NH_4^+ \ CN^- \rightleftharpoons NH_3 \ + \ HCN$$

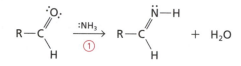

An imine is basic, so it reacts with HCN by an acid–base reaction, which creates cyanide ion, a good nucleophile, and the amino carbocation, a good electrophile. These two species react to form an α-amino nitrile. As already noted above, the acid-catalyzed hydrolysis of the nitrile group yields the carboxylic acid product, which is the amino acid.

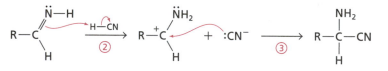

EXERCISE 27.5

Draw the structure of the aldehyde you would use to prepare each of the following amino acids via the Strecker reaction. Do any of these aldehydes present a problem with regard to its stability or reactivity?

a. Valine   b. Serine   c. Tyrosine

27.2b AMINO ACIDS CAN BE PREPARED BY SUBSTITUTION REACTIONS

Another common strategy for preparing amino acids makes use of nucleophilic substitution reactions between nitrogen-containing nucleophiles and α-halo carboxylic acids. Ammonia is one nucleophile that can be used, although polyalkylation is an undesirable side reaction (Section 6.3d).

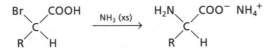

The Gabriel reaction (Section 22.5c) provides a cleaner way to introduce the amino group because the nitrogen atom of the phthalimide ion can react with only 1 equiv of the α-halo carboxylic acid.

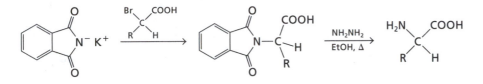

The use of azide ion followed by hydrogenation (Section 11.2c) offers another alternative.

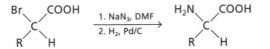

EXERCISE 27.6

Show how you would prepare each of the following compounds by a route in which the amino group is introduced by a nucleophilic substitution procedure:

a. Isoleucine   b. *p*-Methoxyphenylalanine

27.2c CHIRAL AMINO ACIDS CAN BE OBTAINED BY RESOLUTION METHODS

The methods described in Section 27.2b yield racemic products. To obtain an amino acid as a single enantiomer, the racemic material can be resolved. Most commonly, resolution can be accomplished by making a salt with a chiral base, and then separating the diastereomeric salts by crystallization (Section 16.2a).

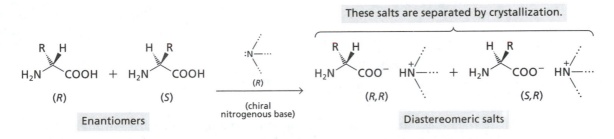

The pure enantiomer is obtained by treating the purified salt with aqueous mineral acid and separating the amino acid from the protonated resolving agent.

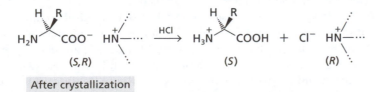

Another way to obtain an enantiomerically pure amino acid is to carry out an enzyme-catalyzed kinetic resolution (Section 16.3d). Amino acids and their derivatives are natural substrates for many transformations that are catalyzed by enzymes. For example, *carboxypeptidase* is an enzyme that catalyzes the cleavage of an amide bond at the C-terminus of a protein. It therefore catalyzes the hydrolysis of the acetyl group from the acetamide derivative of an amino acid. The natural (*S*) isomer, unlike the (*R*) isomer, is converted to the amino acid by this enzyme.

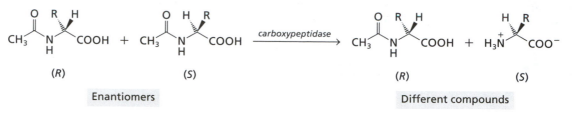

The two products are readily separated because they have different chemical properties. Other enzymes called lipases and hydratases behave in a similar way; the only drawback is that 50% of the material does not react, so it has to be converted to a usable product by another method.

27.3   ASYMMETRIC SYNTHESIS OF AMINO ACIDS

27.3a   ASYMMETRIC SYNTHESES CAN BE USED TO MAKE ENANTIOMERICALLY PURE AMINO ACIDS

Resolution of enantiomeric amino acids might be practical in specific cases, but an asymmetric synthesis often provides enantiomerically pure material in fewer steps. Of the many general routes that have been devised, three common strategies are shown below:

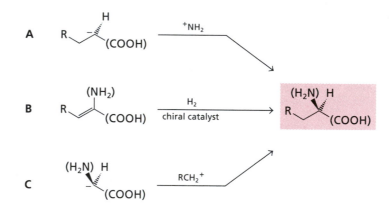

These equations represent *formal* reactions, illustrating generalized starting materials and reagents. Parentheses in these formal reactions indicate functional groups that are protected. Protected derivatives are often preferable for the synthesis of amino acids because the same protecting groups are needed to make peptides. The protecting groups can always be removed if the amino acid itself is desired.

Routes A and C in the preceding scheme make use of a **chiral auxiliary,** which is a covalently bound fragment that directs the reaction so that one diastereomer is produced preferentially. After the chiral center is established, the auxiliary is removed in a subsequent step, giving the pure enantiomeric product. In the best cases, the chiral auxiliary is recovered and recycled, minimizing waste. Route B (above) uses a chiral catalyst to induce enantioselectivity. Catalyzed reactions are preferable for asymmetric syntheses because extra steps are not required to remove a chiral auxiliary.

27.3b PROTECTING GROUPS ARE OFTEN USED FOR REACTIONS OF AMINO ACIDS

Many of the functional groups in amino acids either react with each other or are incompatible with reagents that are used in their syntheses, so protecting groups are routinely used (Section 15.5b). The classic protecting groups for the amino function are the *tert*-butyloxycarbonyl (Boc) moiety and the benzyloxycarbonyl (Cbz) unit. These groups are incorporated by reaction of the corresponding anhydride or acid chloride derivative, respectively.

Boc protecting group

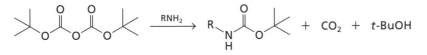

Boc group

Cbz protecting group

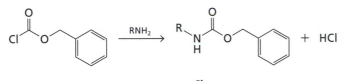

Cbz group

Deprotection of these derivatives is also straightforward, a required feature of a good protecting group. Trifluoroacetic acid readily removes the Boc group by protonating the carbonyl oxygen atom. The intermediate cation undergoes dissociation, generating the stabilized *tert*-butyl carbocation and the unstable carbamic acid.

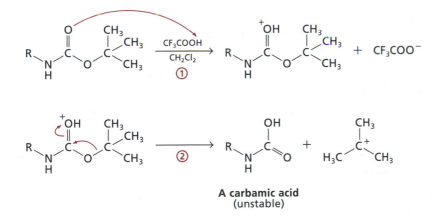

A carbamic acid
(unstable)

The *tert*-butyl cation loses a proton from one of its methyl groups to form isobutylene

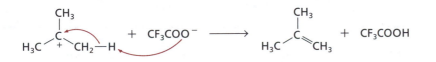

The carbamic acid loses carbon dioxide (Section 22.5b) to form the amine. The amine is readily purified because both byproducts—isobutylene and carbon dioxide—are gases.

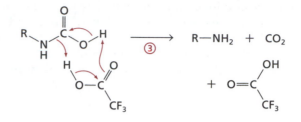

Removal of the Cbz group is also facile. Hydrogenolysis (Section 12.2a) cleaves the benzyl group, producing toluene. Again, the unstable carbamic acid loses carbon dioxide to form the amine.

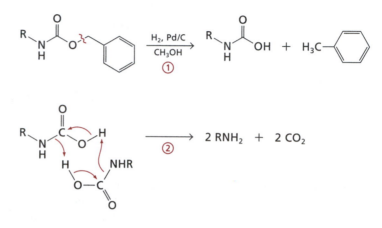

Other protecting groups are also available for amines, and some are removed under milder conditions than the ones shown here.

The carboxylic acid group is often protected as its ester. A *tert*-butyl ester is cleaved with acid, a benzyl ester can be removed by hydrogenolysis, and a methyl ester is hydrolyzed under basic conditions with LiOH in aqueous THF.

EXERCISE 27.7

Write equations showing how to prepare the *tert*-butyl and benzyl ester derivatives of carboxylic acids.

EXERCISE 27.8

Propose a mechanism for the following reaction, deprotection of a *tert*-butyl ester using CF_3COOH:

$$R-\overset{\overset{O}{\|}}{C}-O-C(CH_3)_3 \xrightarrow[CH_2Cl_2]{CF_3COOH} R-\overset{\overset{O}{\|}}{C}-OH \ + \ H_2C{=}C(CH_3)_2$$

Most of the time, it is desirable to protect the amino and carboxylic acid groups with groups that are removable under *different* conditions. For example, the *tert*-butyl ester and the Boc group are not used in the same compound because they are both cleaved with CF_3COOH. You should also realize that any functional group in the side chain of an amino acid will likely need protection too. Protection and deprotection of those groups have to be compatible with the other functional groups in the molecule.

27.3c AMINO ACIDS CAN BE PREPARED FROM THE REACTION BETWEEN THE GLYCINE ENOLATE ION AND AN ALKYLATING AGENT

One strategy for making an amino acid that uses a chiral auxiliary starts with an enantiopure amino alcohol having the diphenylethane skeleton. Treating bromoacetic acid with (1R, 2S)-1,2-diphenyl-2-aminoethanol yields a six-membered ring heterocycle. The amino group displaces the bromide ion, and then the acid and alcohol groups react to form the ester linkage (a lactone). The nitrogen atom of this heterocycle can be protected as its Boc derivative in the usual way.

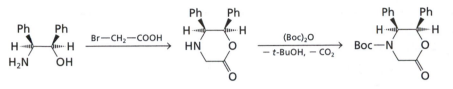

When this heterocycle is treated with LDA in THF at –78°C, an enolate ion is generated. This species corresponds to the enolate form of a protected glycine derivative.

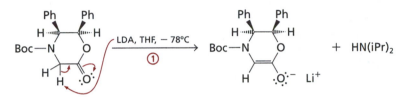

As a consequence of steric effects, this enolate ion undergoes alkylation from underneath the ring, effectively producing one stereoisomer.

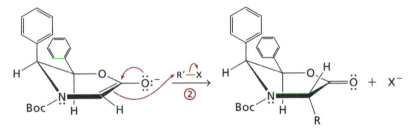

The benzylic C–N and C–O bonds in this alkylated species are then cleaved using lithium in liquid ammonia or by hydrogenolysis.

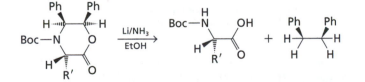

The products are the enantiomerically pure Boc-protected amino acid and diphenylethane. The chiral auxiliary is destroyed in this last step, so it cannot be reused.

A limitation of this method is that only reactive alkylating agents can be used, so R′ in the above equations must be methyl, allylic, benzylic, or R′COCH$_2$. The alkylated heterocycle can be alkylated on the bottom face a second time using methyl iodide, and this procedure is an effective way to make α-methyl amino acids.

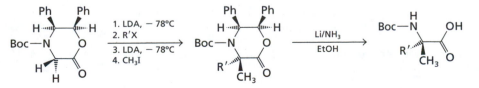

EXERCISE 27.9

Using the reactions just presented, show how you would prepare the following amino acids with the illustrated stereochemistry:

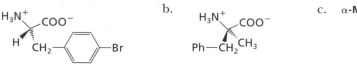

a.

b.

c. α-Methyl L-aspartic acid

27.3d HOMOGENEOUS CATALYSTS PROMOTE ENANTIOSELECTIVE HYDROGENATION

William S. Knowles was one of the first chemists to investigate enantioselective hydrogenation reactions (Section 16.4b), discovering along the way a useful method to prepare L-DOPA, an amino acid used in the treatment of Parkinson's disease.

More recent work from the laboratories at the DuPont Company uncovered a new class of catalysts for enantioselective reduction reactions that work even better than Knowles's catalysts. These so-called "DuPHOS" ligands are used to make rhodium complexes that also have a 1,5-cyclooctadiene (cod) group bonded to the Rh(I) ion: The π bonds of the diene bind to two coordination positions of the metal ion.

MeDuPHOS	R = CH_3—
EtDuPHOS	R = CH_3CH_2—
PrDuPHOS	R = $CH_3CH_2CH_2$—

The counterion is the triflate ion, $CF_3SO_3^-$, abbreviated ^-OTf.

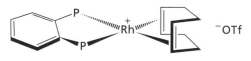

[Rh(DuPHOS*)cod](OTf)

(The other groups attached to the phosphorus atoms have been omitted for clarity.)

In methanol solution, the cod group is replaced from Rh(DuPHOS*)(cod)$^+$ by two molecules of methanol, producing Rh(DuPHOS*)(CH$_3$OH)$_2^+$, which is highly reactive toward alkenes having other functional groups. Esters of 2-(acetamido)acrylic acid, which are known as *enamides*, undergo hydrogenation to form protected amino acids, often with > 99% ee.

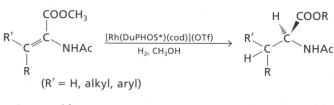

(R' = H, alkyl, aryl)

An enamide

An amino acid (protected form)

Boc-protected enamides are also good substrates, and most other substituents do not interfere with the hydrogenation reaction. Even substituents on the β-carbon atom have little effect, a problem that affected earlier catalysts. For example,

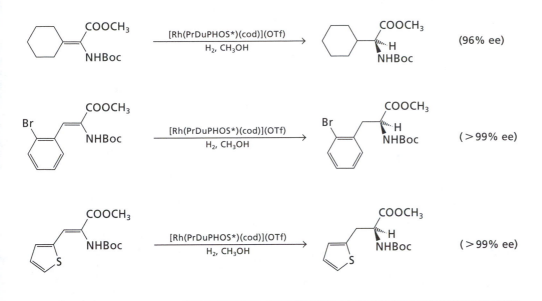

<hr>

EXERCISE 27.10

The DuPHOS catalysts can be used to hydrogenate related substrates also. What is the major product expected from each of the following transformations?

a.   b.

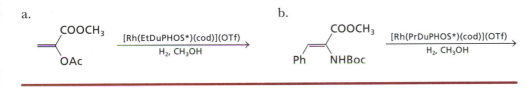

<hr>

27.4   PEPTIDE SYNTHESIS AND ANALYSIS

27.4a   A PEPTIDE BOND IS AN AMIDE FUNCTIONAL GROUP

We now turn our attention to the chemical synthesis of polypeptides and proteins from the amino acids. First, recall that polyamides can be made from amino acids by self-condensation (Section 26.3c). This process, which is used to prepare nylon 6, proceeds with the concomitant removal of water and formation of an amide bond between the carboxyl and amino groups of each molecule.

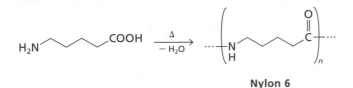

Nylon 6

The condensation reaction between α-amino acids has an additional limitation, however. Besides forming polymers, α-amino acids can form dimeric compounds, namely the six-membered ring diketopiperazines. As a result of this ring-forming reaction, polymer formation is actually quite difficult for the α-amino acids.

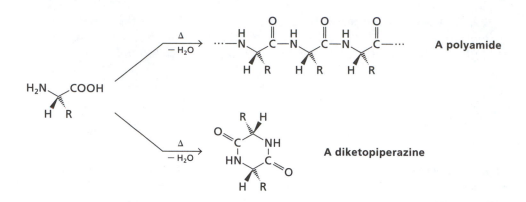

A polyamide

A diketopiperazine

When making a protein from α-amino acids, you face another constraint: It matters what *sequence* the amino acids have in the product. To accomplish control of which amino acids are linked together, a reactive acid derivative must be formed—it is not enough simply to remove water to make amide linkages. Acid chlorides are commonly used to make amides in the laboratory (Section 21.3a), but amino acids are often sensitive to acids, so thionyl chloride is not a very useful reagent for this purpose because HCl is a product of the reactions between $SOCl_2$ and carboxylic acids. Protecting groups for the functional groups and the use of mild coupling methods are required for making peptides.

27.4b A PEPTIDE BOND IS PREPARED BY THE COUPLING REACTION BETWEEN PROTECTED AMINO ACIDS

The stepwise coupling of amino acids to form polypeptides and proteins (Section 5.5b) can be carried out in the laboratory using a procedure that brings an amino group together with an activated carboxylic acid derivative. The result is an iterative process that depends only on how many amino acids are to be linked. The preparation of a dipeptide—the coupling of two amino acids (Section 5.5b)—can be represented by the following three-step scheme.

1 H_2N—(aa$_1$)—COOH $\xrightarrow{\text{protect acid}}$ H_2N—(aa$_1$)—COOR

The carboxylic acid group of an amino acid is protected.

2 H_2N—(aa$_2$)—COOH $\xrightarrow{\text{protect amine}}$ BocNH—(aa$_2$)—COOH $\xrightarrow{\text{activate acid}}$ BocNH—(aa$_2$)—C(=O)—X

The amino group of a second amino acid is protected; then an activated carboxylic acid derivative is made.

3 BocNH—(aa$_2$)—C(=O)—X + H_2N—(aa$_1$)—COOR $\longrightarrow$ BocNH—(aa$_2$)—C(=O)—NH—(aa$_1$)—COOR

The two components are mixed, and the peptide bond is formed.

When another amino acid is to be added to the amino end of this dipeptide to form a tripeptide, then deprotection of the dipeptide's amino group is performed, and a second coupling step is carried out.

4 BocNH—(aa₂)—C(=O)—NH—(aa₁)—COOR → (deprotect amine) → H₂N—(aa₂)—C(=O)—NH—(aa₁)—COOR

The protecting group on the amine function is removed.

5 BocNH—(aa₃)—C(=O)—X + H₂N—(aa₂)—C(=O)—NH—(aa₁)—COOR ⟶

Another amine-protected, carboxylic acid-activated amino acid is added...

BocNH—(aa₃)—C(=O)—NH—(aa₂)—C(=O)—NH—(aa₁)—COOR

...and a second peptide bond is formed.

This application of deprotection and coupling (Steps 4 and 5, above) can be repeated as many times as you desire to construct longer and longer polypeptides. It is important to realize that side-chain groups may also require protection so that they will not react with the activated acid derivative.

By convention, a peptide sequence is written as a string of the one- or three-letter abbreviations for the constituent amino acids (see Example 5.8). The first amino acid in a sequence (reading from left to right) is the N-terminus (so it has the amino group, H₂N), and the last one defines the C-terminus (COOH group). In between, amide groups link the amino acids together.

Let us now look at the details of the reactions needed to prepare the dipeptide Ser-Ala, which has the following structure:

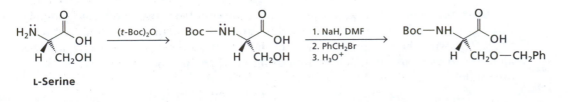

Ser–Ala

Alanine, which will be the C-terminus, is protected as its methyl ester. Its side chain needs no protecting group. The hydroxide ion in the second step of the following scheme removes HCl to make certain the amino group is not protonated:

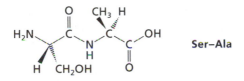

L-Alanine

Serine, which will be the N-terminus, is protected at its amino group with the Boc group; the alcohol group in the side chain is converted to its benzyl ether derivative using an S_N2 reaction (the Williamson ether synthesis—Section 7.2b).

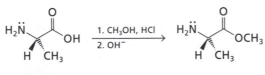

L-Serine

With the two building blocks in their protected forms, peptide bond formation can be carried out: The mixture of the two components is treated with dicyclohexylcarbodiimide (DCC).

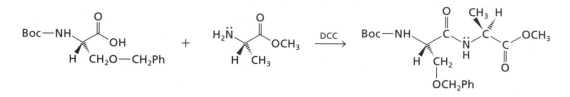

DCC, or any carbodiimide, reacts with carboxylic acids to form derivatives that are similar in structure to anhydrides (Section 21.3b).

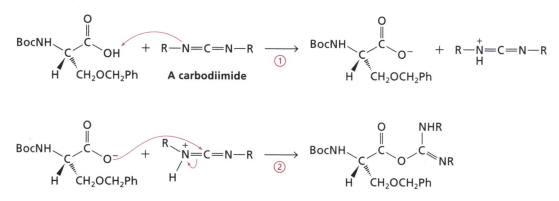

This acyl-activated intermediate then reacts with the amino group of alanine methyl ester to generate the peptide bond via an addition–elimination reaction: The amino group adds to the carbonyl group to form a tetrahedral intermediate, and regeneration of the carbonyl group displaces dicyclohexylurea (DCU) as a leaving group. The conditions are mild, the yield is high, and the reactants can all be combined at the same time in the reaction flask.

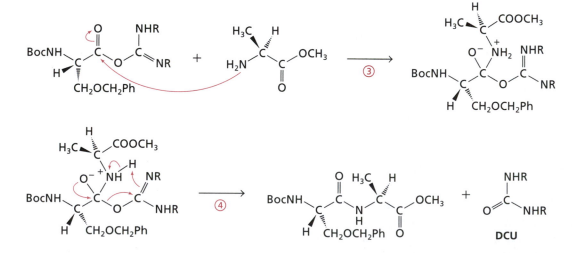

Once the peptide bond has been formed, the product must be deprotected. The three groups are differentially protected, so any one of them can be removed selectively. Removing the Boc group unveils the amino group, which can be used if another amino acid were to be attached. Removing all three protecting groups yields the dipeptide.

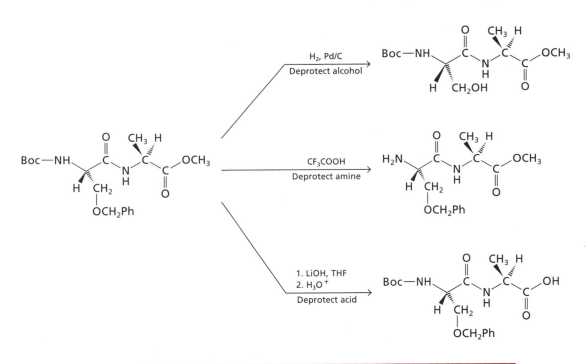

EXERCISE 27.11

Show how you would prepare the tripeptide Phe-Ser-Ala starting with the fully protected dipeptide in the preceding sheme and a suitably protected derivative of phenylalanine.

27.4c PEPTIDE SYNTHESIS HAS BEEN AUTOMATED

The general strategy for the synthesis of a polypeptide from protected amino acids consists primarily of alternating coupling and deprotection steps, so automated methods for preparing polypeptides have evolved during the past 30 years. The technique known as *solid-phase peptide synthesis* earned R. Bruce Merrifield the 1986 Nobel Prize in chemistry for his pioneering work to make automated peptide synthesis a reality.

A protected amino acid is attached to chloromethylated cps (Section 26.4b) using nucleophilic substitution of the chloride ion by the carboxylate ion.

1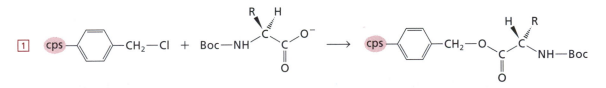

After this first amino acid is attached, it is deprotected by treatment with trifluoroacetic acid.

2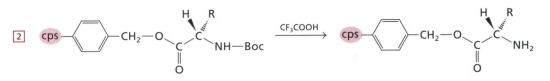

Next, a second amino acid is added, along with DCC. The amide bond forms just as it did in solution, except that the dipeptide is attached to the insoluble resin. Dicyclohexylurea is removed by washing.

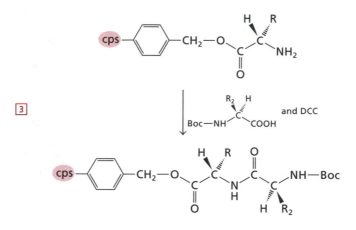

3

Deprotection and condensation with a third amino acid produces a tripeptide, and these cycles of deprotection and coupling can be repeated until the polypeptide reaches the desired length.

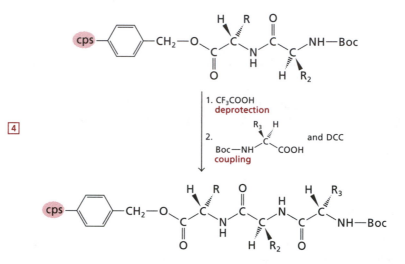

4

When all of the amino acids have been coupled, the benzylic ester that exists between the polystyrene and the polypeptide is hydrolyzed by treatment with HF. The polypeptide, which is soluble, is then purified by the normal methods.

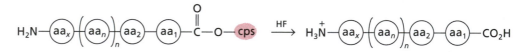

Peptide-synthesizing instruments work by injecting one reagent into the reaction vessel that contains the polymer, followed by shaking of the mixture for a specified amount of time. The reaction vessel has a fritted glass disk that permits the solvents and reagents to be drained away from the insoluble polystyrene beads. Clean solvent is injected to ensure that any excess reagent is dissolved, and that solution is also drained from the polymer. The alternating processes of injecting reagents, amino acids, and solvents continues until the polypeptide is of the specified length and sequence.

Although most synthetic polypeptides are made by this automated process, there is a significant limitation to performing reactions on an insoluble substrate. If a coupling procedure does not occur completely, then some of the growing chains will have the wrong sequence when the next amino acid is attached. This problem becomes more pronounced as the polypeptide becomes longer. Much work has been done to optimize the efficiency of the reactions, and their yields now routinely approach better than 99%.

27.4d THE AMINO ACID SEQUENCE OF A PROTEIN CAN BE DEDUCED BY A SERIES OF REACTIONS CALLED THE EDMAN DEGRADATION

The laboratory preparation of polypeptides and proteins has become routine enough that essentially any conceivable product can be made. Once isolated, however, the protein must be characterized to ensure that its structure is correct. The sequence of the amino acids in a protein is called its **primary structure,** and it is important because of its relationship to the folded protein's overall conformation. Frequently, a protein folds so as to place hydrophobic residues in the interior and hydrophilic groups on the exterior where they can interact with an aqueous environment.

Complete hydrolysis of a protein will reveal how many molecules of each amino acid are present. Acidic conditions lead to hydrolysis of the side-chain amide groups of glutamine and asparagine as well, so the presence of Gln and Asn residues cannot be determined for a native protein by the complete hydrolysis technique. Furthermore, Trp also decomposes in acid.

To determine the sequence of the amino acids in a particular protein, one needs to use reactions that are more selective than hydrolysis. One common method is the **Edman degradation,** which sequentially removes one amino acid at a time from the N-terminus of the polypeptide. The Edman reagent is phenylisothiocyanate, which undergoes addition with the amino group at the N-terminus.

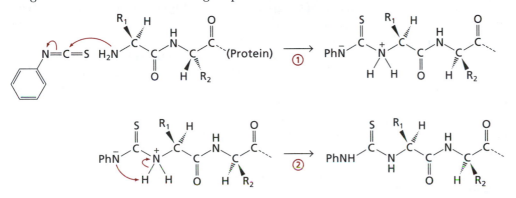

In a second operation, anhydrous HF is added, which protonates the carbonyl oxygen atom and activates the carbonyl carbon atom for reaction with the nucleophilic sulfur atom of the thiourea group.

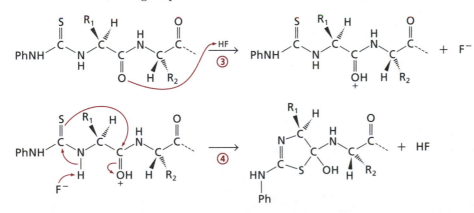

Regeneration of the carbonyl group in the next step eliminates the N-terminus of a new polypeptide that has one less residue.

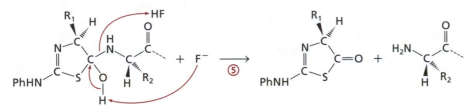

The heterocycle produced in Step 5 is subsequently extracted into aqueous acid, where it rearranges to the *N*-phenylthiohydantoin, shown below. The identity of the amino acid at the N-terminus is deduced by comparing the *N*-phenylthiohydantoin that has been produced with known samples prepared from the common amino acids.

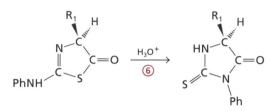

EXERCISE 27.12

Propose a mechanism for the rearrangement reaction shown directly above.

Once the polypeptide has been shortened by one residue, a second reaction can take place because a new amino group is present to react with the reagent, so the procedure can be repeated. Analysis of the specific *N*-phenylthiohydantoin molecules produced with each round of Edman degradation equals the sequence of the protein. As you might expect, this method has been automated.

If the rate of reaction between each amino acid and phenylisothiocyanate were the same, then sequencing would be highly accurate. Because the amide bonds between different amino acids react at different rates, however, the proteins in the reaction mixture can have different lengths after several rounds of the Edman reaction. As a result, the practicality of the Edman process is limited to proteins with 25–50 residues. For large proteins, sequencing is performed on fragments that are formed by cleaving the parent protein with enzymes.

EXERCISE 27.13

A polypeptide is injected into an amino acid analyzer that operates by sequential Edman reactions. The *major* products detected in order ($r_n = n$th round) have the following structures. What is the sequence of the first four residues at the N-terminus of the polypeptide?

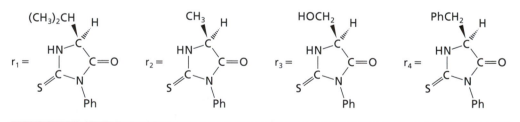

27.5   PROTEIN STRUCTURES

27.5a   PEPTIDE BONDS DISPLAY RESTRICTED ROTATION

Before looking at the higher orders of protein structure, consider the properties of the peptide bond that connects the amino acids. You might ask why Nature chose the amide group as the backbone of proteins—why not an ester or ether functional group, for example? The stability of each type of bond is a consideration. An amide is the least reactive of the carboxylic acid derivatives, yet it can be hydrolyzed readily by proteolytic enzymes that are able to stabilize the tetrahedral intermediate (Section 21.5b). Esters are too easily hydrolyzed and ethers are too unreactive to be readily metabolized.

Another feature of the amide group, which is crucial to the definition and maintenance of protein structures, is its geometry. While many conformations can exist for an amide group, the carbon–nitrogen bond has partial double-bond character.

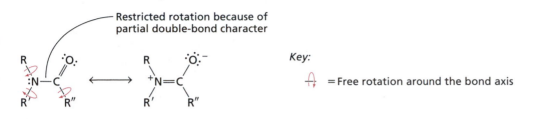

As a result, six of the atoms lie in the same plane. The carbon–nitrogen bond length of an amide is ~ 1.32 Å, which lies between that of a carbon–nitrogen single (1.47 Å) and double (1.28 Å) bond.

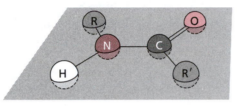

The geometric constraint imposed by the presence of a partial double bond can be demonstrated by ^{1}H NMR spectroscopy for a simple compound such as DMF (Fig, 27.2). The restricted rotation about the C–N bond makes the methyl groups nonequivalent—one is cis to the oxygen atom, and one is trans. The methyl groups experience different magnetic fields as a result, so they display different chemical shifts.

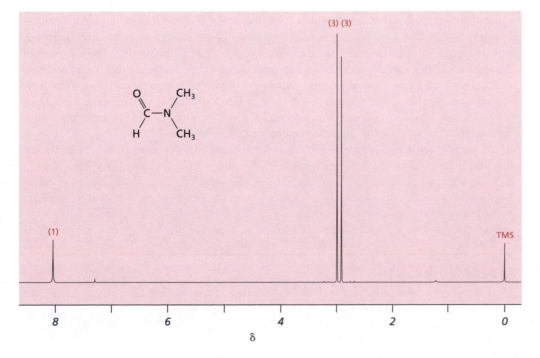

Figure 27.2
The 300-MHz ^{1}H NMR spectrum of DMF showing the two signals for the nonequivalent methyl groups at δ ~2.8.

In a protein, the restricted rotation of the amide bonds influences the ways that the polypeptide backbone can fold upon itself. In turn, this overall conformation has a direct influence on a protein's function in biological systems. Proteins are not rigid substances, however. The presence of single bonds between the carbonyl carbon atom and its *alpha*-carbon atom and between the nitrogen atom and its *alpha*-carbon atom means that many conformations are accessible in proteins. One exception occurs when the protein sequence contains proline, the only genetically coded amino acid that is cyclic. The protein's conformation is constrained by the five-membered ring linking the amino group with the *alpha*-carbon atom. This rigidity in proline's structure is an important reason so many proline residues are found in collagen (Section 20.2b), a protein that contributes to the strength and durability of skin and bones.

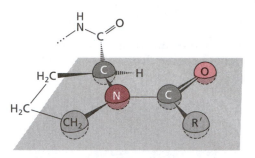

The structure of a proline residue in the middle of a protein, showing the rigid, planar arrangement of atoms.

27.5b SECONDARY AND TERTIARY STRUCTURES INFLUENCE THE OVERALL THREE-DIMENSIONAL FORM OF A PROTEIN

The sequence of amino acids in a protein defines its primary structure. The **secondary structure** of a protein refers to the *regular* conformations within the polypeptide backbone. The most common secondary structural elements are the **α-helix** and the **β-sheet.** Both are stabilized by formation of hydrogen bonds between the carbonyl oxygen atom of one amino acid and the amide N–H of another amino acid that is not adjacent in the sequence.

For the α-helix, hydrogen bonds form between residues that are separated by three intervening amino acids, as illustrated in Figure 27.3. The α-helix is right handed, being composed of L-amino acids, and its formation, although stabilized by hydrogen bonds, probably results from folding to optimize attractive dispersion forces (Section 2.8a) among the various side chains by removing them from the aqueous environment. Once the hydrophobic groups are placed within the protein's interior away from water, many of the amide groups are displaced from the aqueous milieu as well. Their highly polar nature warrants the formation of hydrogen-bond interactions, and these interactions are satisfied by the establishment of the α-helix.

A β-sheet is formed between *separated* segments within the interior of a protein. Unlike the α-helix, there is no regular spacing of residues that dictates the formation of a β-sheet. Instead, the protein chains in these regions exist as extended sequences called β-strands; when these strands are aligned with neighboring ones, they form hydrogen bonds that define the sheet structures. Figure 27.4 illustrates how adjacent strands of a polypeptide within a β-sheet are stabilized by hydrogen bonds.

EXERCISE 27.14

Draw a portion of an antiparallel β-sheet and show the hydrogen bonds.

a.   b.   c.

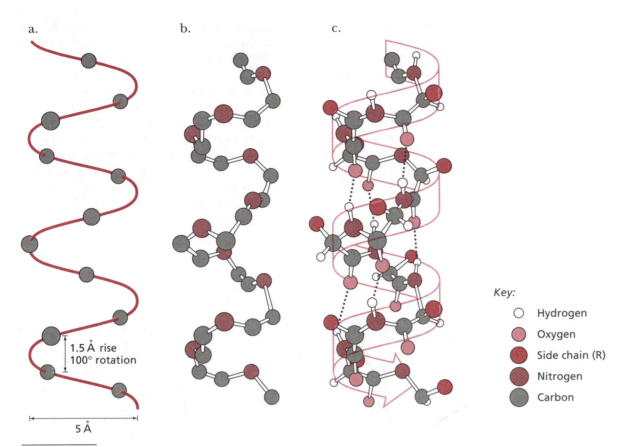

1.5 Å rise
100° rotation

5 Å

Key:

○ Hydrogen
● Oxygen
● Side chain (R)
● Nitrogen
● Carbon

Figure 27.3
A model of a right-handed α-helix, showing (*a*) the positions of the α-carbon atoms of each amino acid; (*b*) the carbon and nitrogen atoms of the backbone; and (*c*) the entire helix. Hydrogen bonds in (*c*) are shown as dotted lines.

A **β-turn** is another element of secondary structure, and β-turns are found at places where a polypeptide strand folds back on itself to form a β-sheet. This feature, also called a *reverse turn,* is illustrated in Figure 27.5, which shows the hydrogen bonds that stabilize this assembly of four amino acids.

Secondary structures can be disrupted by outside influences that break the stabilizing hydrogen bonds. This process, called **denaturation,** leads to formation of a **random coil conformation,** which has no defined secondary structure. Denaturation is

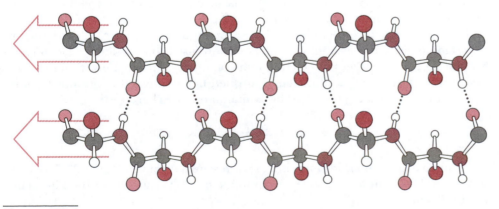

Figure 27.4
The structure of a β-sheet between two different strands of a polypeptide. This structure is a parallel β-sheet because the sequences run in the same direction. Antiparallel β-sheets also exist.

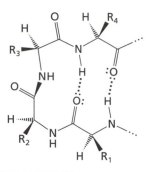

Figure 27.5
The structure of a β-turn.

valuable for sequencing work, when the polypeptide chain needs to be completely exposed to reagents in solution. The side chains of amino acids can disrupt secondary structures as well, if they can form strong hydrogen bonds that break the ones stabilizing an α-helix or a β-sheet.

Ribbon diagrams provide a particularly convenient way to represent the features of secondary structure. Where chains are linked and spread out, these are the flat ribbons that represent β-sheets. The α-helix is represented by a coil of the backbone. Figure 27.6 illustrates several protein structures using ribbon diagrams and shows that some proteins have no sheet structure, only helical domains. Others are just the opposite.

The **tertiary structure** of a protein is its overall shape, created by folding of the secondary structure onto itself as shown for the three proteins in Figure 27.6. Common terms used to refer to this level of structure are *globular*—for example, myoglobin (Fig. 5.6)—or *fibrous*—collagen (Fig. 20.2).

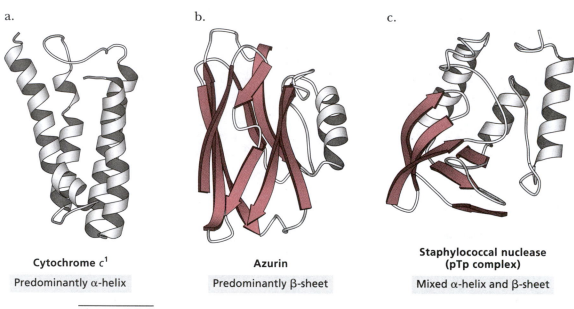

a.

b.

c.

Cytochrome c[1]

Predominantly α-helix

Azurin

Predominantly β-sheet

Staphylococcal nuclease (pTp complex)

Mixed α-helix and β-sheet

Figure 27.6
Ribbon diagrams for some proteins, showing α-helix, β-sheet, and random coil regions.

Proteins composed of several polypetide chains, which are subunits, can also have a **quaternery structure**. Hemoglobin is the classic example of a multisubunit protein, consisting of two pairs of identical chains.

Once folded, proteins can be stabilized by the formation of covalent bonds between amino acids. For example, imine bonds are used to cross-link collagen chains (Section 20.2b). A covalent interaction that commonly links regions within the *same* polypeptide chain is the disulfide bond, formed between two thiol groups from cysteine residues. A disulfide is formed by oxidation of the SH groups.

$$R\text{—}SH \xrightarrow{[O]} R\text{—}S\text{—}S\text{—}R$$

The sequence of amino acids in insulin, shown schematically in Figure 27.7, is a standard example of a protein stabilized by disulfide bonds because it has both intrachain and interchain links.

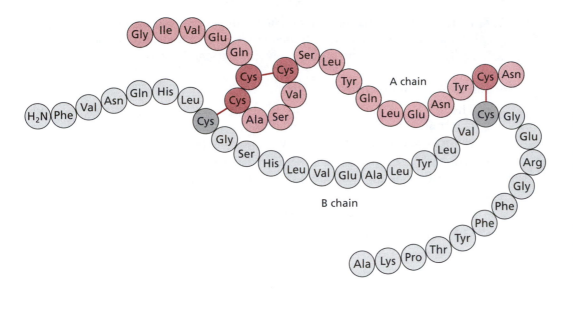

Figure 27.7
The primary structure of insulin, showing the interchain and intrachain disulfide bonds as dark lines. The A and B chains are held together by two disulfide bonds, and the A chain conformation is stabilized by a single disulfide bond.

Section 27.1   Amino acids

- The α-amino acids are the building blocks of proteins.
- Twenty of the amino acids found in proteins are encoded by DNA sequences. Other amino acids are formed after a protein has been synthesized in the cell.
- An amino acid generally exists in aqueous solution as a zwitterion, which results from an acid–base reaction between its carboxylic acid and amino groups.
- The ratio of an acid to its conjugate base for a functional group in the side chain of an α-amino acid can be calculated from the Henderson–Hasselbalch equation.
- The amino acids in proteins can be classified by the properties of their side chains. Classifications include: nonpolar, aromatic, cyclic, acidic, basic, and polar.

Section 27.2   Chemical synthesis of amino acids

- Two general methods used to prepare racemic α-amino acids are the Strecker reaction, which takes place between ammonium cyanide and an aldehyde, and substitution of an α-halo carboxylic acid with a nitrogen-containing nucleophile.
- Enantiomerically pure α-amino acids are obtained by resolution of racemic mixtures via formation of diastereomeric salts or by asymmetric synthesis.

Section 27.3.   Asymmetric synthesis of amino acids

- Glycine enolates and enantioselective hydrogenation of enamides are two routes that can be used in the asymmetric syntheses of α-amino acids.
- Protecting groups are commonly used in reactions of α-amino acids.
- The Boc or Cbz groups are useful amine protecting groups; a carboxylic acid group in an amino acid is frequently protected as its ester.

Section 27.4   Peptide synthesis and analysis

- Polypeptide synthesis is an iterative procedure that couples two amino acids in their protected forms. Dicyclohexylcarbodiimide, DCC, is a commonly used reagent to form the amide functional group that defines the peptide bond.

- Peptide synthesis has been automated with use of insoluble polymeric resins to which an amino acid and, subsequently, a growing polypeptide chain is attached. This technique is called solid-phase peptide synthesis.

Section 27.5   Protein structures

- The primary structure of a protein is its amino acid sequence, which is readily determined by the Edman degradation, a reaction scheme that removes one amino acid at a time from the N-terminus of a polypeptide.

- Proteins also display secondary and tertiary structure, which constitute localized conformations and the overall shape, respectively.

- Secondary structures result in part from restricted rotation around peptide bonds because of partial double-bond character in an amide group.

- The common types of secondary structure are the α-helix, β-sheet, and β-turn.

- After a protein folds into its natural shape, secondary structural motifs are stabilized by formation of hydrogen bonds, which take the place of the hydrogen bonds that an amide functional group would form with water molecules in the absence of folding.

- A protein can be unfolded by certain reagents, a process called denaturation.

- Proteins that have multiple subunits can also have quaternary structure, which refers to the interaction between different subunits.

- The tertiary structure of some proteins can be stabilized by the covalent bonds of imine or disulfide functional groups.

KEY TERMS

Section 27.1a
α-amino acids
codon
genetic code

Section 27.1b
Henderson–Hasselbalch
   equation

Section 27.2a
Strecker reaction

Section 27.3a
chiral auxiliary

Section 27.4d
primary structure
Edman degradation

Section 27.5b
secondary structure
α-helix
β-sheet
β-turn
denaturation
random coil conformation
ribbon diagram
tertiary structure
quaternary structure

REACTION SUMMARY

Section 27.2a

The Strecker reaction: An aldehyde reacts with ammonium cyanide via an imine intermediate to form an amino nitrile. The nitrile group is hydrolyzed to form a carboxylic acid.

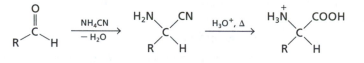

Section 27.2b

Amino acids can be prepared by methods involving substitution of a halide ion by a nitrogen-containing nucleophile.

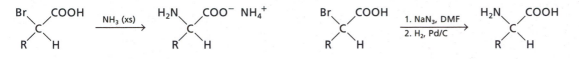

The Gabriel reaction involves substitution of a halide ion by phthalimide ion followed by removal of phthaloyl group.

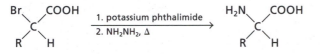

Section 27.3b

Protection and deprotection of amino groups: Boc and cbz are common protecting groups for amino acids.

Boc:

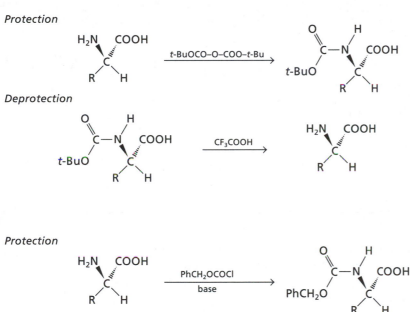

Cbz:

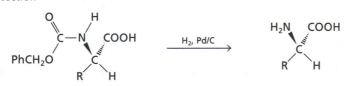

Section 27.3c

The enantioselective synthesis of amino acids can be accomplished by alkylation of a chiral glycine enolate ion equivalent.

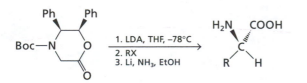

Section 27.3d

The enantioselective synthesis of amino acids can be carried out by hydrogenation of enamides with use of a chiral catalyst.

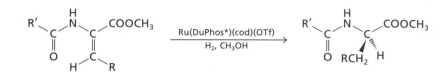

Section 27.4b

Peptides are prepared by coupling reactions between protected amino acid derivatives.

Section 27.4d

The Edman degradation: The sequence of amino acids in a polypeptide is determined by a series of reactions between the polypeptide and phenylisothiocyanate.

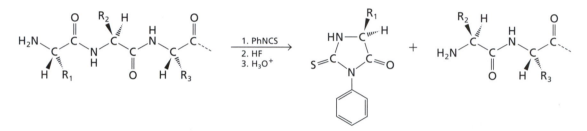

ADDITIONAL EXERCISES

27.15. Without looking at Figure 27.1, indicate the identity of X in the following general structure of the following common amino acids.

a. Glutamic acid   b. Serine

c. Cysteine   d. Aspartic acid

e. Asparagine   f. Phenylalanine

g. Alanine   h. Leucine

i. Tryptophan   j. Tyrosine

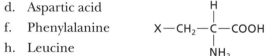

27.16. Repeat Exercise 27.15 for the following names and general structure:

a. Glutamic acid

b. Glutamine

c. Methionine

27.17.   Repeat Exercise 27.15 for the following names and general structure:

a. Threonine

b. Valine

c. Isoleucine

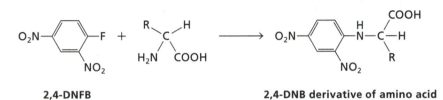

27.18.   Draw the structures expected to predominate for each of the following amino acids in water at (1) pH 1, (2) pH 7, and (3) pH 11.

a. Tyrosine   b. Histidine   c. Serine   d. Glutamic acid

27.19.   Draw the structure of each of the following peptides showing the charge on each functional group in water at pH 7. Remember that amide groups are neutral, so they are neither protonated nor deprotonated in the pH range 1–12.

a. Ala-Glu-Val   b. Phe-Tyr-Lys   c. Leu-His-Asn-Ser

27.20.   Nucleophilic aromatic substitution (Section 17.4e) has been utilized as a way to identify amino acids. The amino acid is treated with 2,4-difluoronitrobenzene (2,4-DNFB), and substitution generates a highly colored derivative that can be isolated and identified by comparison of its melting point with those of known compounds.

| 2,4-DNFB | 2,4-DNB derivative of amino acid |

Propose a reasonable mechanism for the reaction between the amino group of L-isoleucine and 2,4-DNFB.

27.21.   The reaction between an amino acid and 2.4-DNFB (Exercise 27.20) provides a method for identifying the amino acid at the N-terminus of the polypeptide chain. Draw the products expected from the following reaction sequence:

$$\text{Val—Asn—Phe—Glu—Ile—Gly—Gly—Ala} \xrightarrow[\text{2. } H_3O^+, \Delta]{\text{1. 2,4-DNFB}}$$

27.22.   After treating the following pentapeptide with 2,4-DNFB, followed by complete hydrolysis, a chemist isolates two different 2,4-DNFB derivatized amino acids.

$$H_2N\text{-Ile-Ala-Phe-Lys-Ser-COOH}$$

a. Draw the structure, including the 2,4-DNFB group, if present, for each of the five amino acids obtained after the hydrolysis step.

b. How can you be sure which amino acid was at the N-terminus of the starting polypeptide?

27.23. Draw the structure of the major product expected from each of the following reactions. Show stereochemistry where appropriate. If no reaction occurs, write N.R.

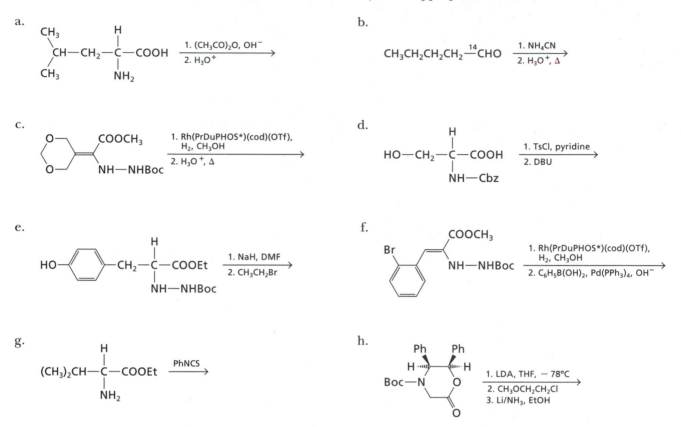

a.

CH₃
|
CH—CH₂—C—COOH   1. (CH₃CO)₂O, OH⁻
| | ───────────────→
CH₃   NH₂   2. H₃O⁺
     H

b.

CH₃CH₂CH₂CH₂—¹⁴CHO   1. NH₄CN
                       ─────────→
                       2. H₃O⁺, Δ

c.

COOCH₃   1. Rh(PrDuPHOS*)(cod)(OTf),
                H₂, CH₃OH
          ───────────────────────→
NH—NHBoc   2. H₃O⁺, Δ

d.

      H
      |
HO—CH₂—C—COOH   1. TsCl, pyridine
      | ───────────────→
      NH—Cbz   2. DBU

e.

         H
         |
HO—⬡—CH₂—C—COOEt   1. NaH, DMF
         | ───────────→
         NH—NHBoc   2. CH₃CH₂Br

f.

      COOCH₃   1. Rh(PrDuPHOS*)(cod)(OTf),
Br   H₂, CH₃OH
  ⬡   ───────────────────────────→
      NH—NHBoc   2. C₆H₅B(OH)₂, Pd(PPh₃)₄, OH⁻

g.

         H
         |
(CH₃)₂CH—C—COOEt   PhNCS
         | ──────→
         NH₂

h.

   Ph   Ph
H ⫶⫶ ⫶⫶ H   1. LDA, THF, −78°C
Boc—N   O   ──────────────────→
               2. CH₃OCH₂CH₂Cl
     ⬠   3. Li/NH₃, EtOH
     O

27.24. Show how you would prepare each of the following amino acids in an enantiomerically pure form, starting with any carboxylic acid:

a.

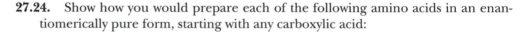

Br—⬡—CH₂—CH(COOH)(NH₂)

b.

oxetane—CH₂—CH(COOH)(NH₂)

27.25. Nutra-Sweet is the brand name for aspartame, an artificial sweetener that is the methyl ester of the dipeptide Asp-Phe. Draw the structure of aspartame at pH 7.6.

27.26. Show how you would prepare the tripeptide Ala-Gly-Ile in the laboratory, assuming you do not have an automated peptide synthesizer.

27.27. Poly(L-lysine) exists in a random coil conformation at pH 7. Adding OH⁻ raises the pH to a value of 10, and the material adopts an α-helical conformation. Explain.

27.28. γ-Carboxyglutamic acid (Table 27.1) was not discovered until 1974. What inherent structural feature accounts for the fact that it escaped detection for so long?

27.29. Among the amino acids with nitrogen in the side chain, arginine is the most basic. In fact, arginine is rarely found in its deprotonated form in proteins. The side-chain group in arginine is the guanidinium ion, which is highly stabilized. Draw resonance structures that illustrate this stabilization of the positive charge.

NUCLEIC ACIDS AND MOLECULAR RECOGNITION

The nucleic acids DNA and RNA are used to store and transmit genetic information, respectively. The nucleic acids are organic molecules, albeit very large ones, and contain carbon, hydrogen, nitrogen, oxygen, and phosphorus.

The first part of this chapter describes the structures and reactions of nucleic acids, as well as the nucleosides and nucleotides from which they are constructed, drawing on the knowledge that you have about the chemistry of carbohydrates (Chapter 19) and aromatic heterocycles (Chapter 25). The functional groups that link nucleosides to form RNA and DNA are esters of phosphoric acid, or **phosphoesters,** the reactions of which will be compared with the chemistry of carboxylic acid esters. In the second part of the chapter, you will learn how some properties of the nucleic acid structures can be applied conceptually to the structures of synthetic compounds.

In 1953, James Watson and Frances Crick unraveled the remarkable structure of DNA, for which they received the Nobel Prize in physiology or medicine in 1962. Since then, organic chemists have attempted to prepare molecules that are able to behave in the same extraordinary way, by interactions between complementary structures. Molecules used for recognition purposes must exploit noncovalent interactions instead of covalent bonds, and several such systems are described in this chapter.

28.1 NUCLEOSIDES AND NUCLEOTIDES

28.1a NOMENCLATURE OF NUCLEIC ACID BUILDING BLOCKS

Nucleosides—the building blocks of RNA and DNA—are constructed from aromatic heterocycles, called bases, attached at the anomeric carbon atom of either D-ribose or 2-deoxy-D-ribose in the β-position (Section 19.3b). Although an IUPAC name can be given to each heterocycle and nucleoside, common names adopted long ago are used almost exclusively for these molecules and are summarized in Figure 28.1.

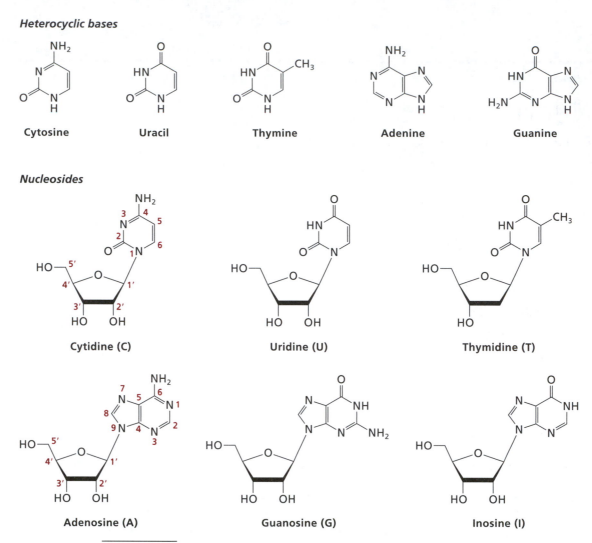

Figure 28.1

The structures of the heterocycles and the corresponding nucleosides that constitute the nucleic acids. The nucleosides A, C, G, and U are components of RNA. The 2′-deoxyy derivatives dA, dC, and dG (d means deoxy) as well as T are components of DNA. Inosine is found only in *t*-RNA.

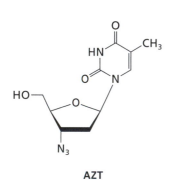

AZT

Because of the growing awareness about the relationship between the nucleic acids and the genetic basis of certain diseases, new derivatives of the nucleosides are being made continually, and naming these substances becomes a matter of specifying the identities and positions of substituents using the nucleoside name as the root.

Substitution on the heterocyclic portion is denoted by the numbers that are shown on the structures in Figure 28.1. Substitution in the carbohydrate ring is specified using a numeral with a prime mark. As an example, 3′-azido-3′-deoxythymidine (AZT), the first drug prescribed for the treatment of acquired immune deficiency syndrome (AIDS), is thymidine with an azide substituent in place of the OH group normally attached at the 3′-position. Its name—3′-azido-3′-deoxythymidine—specifies the *absence* of the OH group as well as the inclusion of the azide substituent. The prefix *deoxy* is therefore common in the names of nucleoside derivatives.

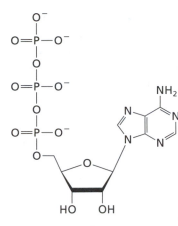

Adenosine 5′-triphosphate
(ATP)

Adenosine 5′-monophosphate
(AMP)

Thymidine 5′-monophosphate

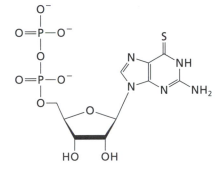

2′-Deoxycytidine 5′-diphosphate   Uridine 5′-monophosphate   6-Thioguanosine 5′-diphosphate

Figure 28.2
The structures of some natural and unnatural nucleotides.

Nucleotides are derivatives of nucleosides that have a phosphate group at the 5′ position. The phosphate group, which can have one, two, or three phosphorus atoms along with the appropriate number of oxygen atoms, is included as a suffix. Examples of some nucleotide names are given in Figure 28.2.

EXERCISE 28.1

Draw the structure for each of the following molecules:

a. 8-Mercaptoadenine
b. 2-Fluoroadenosine
c. 2′,3′-dideoxycytidine

d. 2′-Deoxyadenosine-5′-diphosphate
e. 5-Trifluoromethyluracil

28.1b NUCLEIC ACID BASES ARE DERIVATIVES OF PYRIMIDINE AND PURINE

The heterocycles shown in Figure 28.1 have either one or two rings. Cytosine, uracil, and thymine are six-membered ring compounds with nitrogen atoms at positions 1 and 3 and bear a resemblance to pyrimidine, one of the isomeric diazines (Section 25.2f). Consequently, these heterocycles are called pyrimidine bases, and cytidine, uridine, and thymidine are referred to as **pyrimidine nucleosides.**

Purine, a compound with a diazine ring fused to a 1,3-azole, is the heterocycle that bears a resemblance to the bases in adenosine, guanosine, and inosine, which are called **purine nucleosides.** The purine nucleus is also found in natural products such as caffeine (a component of tea and coffee), theobromine (a constituent of cacao beans), and uric acid (the end product of nitrogen metabolism in some animals).

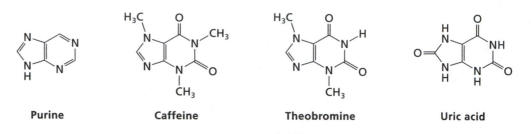

Purine   **Caffeine**   **Theobromine**   **Uric acid**

Purine is aromatic, but it does not undergo reactions such as electrophilic aromatic substitution because its high nitrogen content makes the ring prone to addition reactions. For example, adenine undergoes deamination by way of addition–elimination; the double bond between C6 and N1 reacts as an imine group does (Section 20.1).

Addition

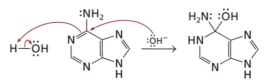

Elimination

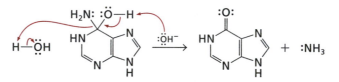

Purine derivatives are made by condensation reactions between amines and carbonyl precursors, which form the carbon–nitrogen bonds. However, the combination of ammonia and hydrogen cyanide is sufficient for the preparation of adenine in the laboratory, a point that has been made to support the contention that adenine was created in the prebiotic environment of earth, when HCN may have been abundant.

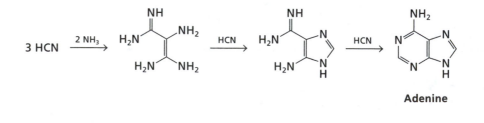

Adenine

EXERCISE 28.2

Draw valence bond representations for purine and pyrimidine, like those illustrated for pyridine and imidazole (Sections 25.2a and 25.4a). Which unshared electron pairs are part of the π system, and which are perpendicular to the π bonds?

28.1c HYDROXY-SUBSTITUTED HETEROCYCLES EXIST AS TAUTOMERS

In his book *The Double Helix,* James Watson describes how he was initially led astray by making assumptions about the structures of the bases that constitute DNA:

> My scheme [that each base paired with itself] was torn to shreds by the following noon. Against me was the awkward chemical fact that I had chosen the wrong tautomeric forms of guanine and thymine . . .
>
> I no sooner got to the office and began explaining my scheme than the American crystallographer Jerry Donohue protested that the idea would not work. The tautomeric forms I had copied out of Davidson's book were, in Jerry's opinion, incorrectly assigned. My immediate retort that several other texts also pictured guanine and thymine in the enol form cut no ice with Jerry. Happily, he let out that for years organic chemists had been arbitrarily favoring particular tautomeric forms over their alternatives on only the flimsiest of grounds. In fact, organic chemistry textbooks were littered with pictures of highly improbable tautomeric forms. The guanine picture I was thrusting toward his face was almost certainly bogus. All his chemical intuition told him that it would occur in the keto form. He was just as sure that thymine was also wrongly assigned an enol configuration. Again, he strongly favored the keto alternative.

Why did early textbooks suggest that the enol forms of the heterocyclic bases are the predominant ones? Most likely it had to do with the stability of phenol (an enol) compared with its keto form. Hydroxy-substituted heterocycles might be expected to behave in the same way, but their keto forms are actually favored. A simple molecule that can be used to illustrate these trends is 2-hydroxypyridine, which is disfavored by a ratio of 1:300 compared with its tautomer 2-pyridone.

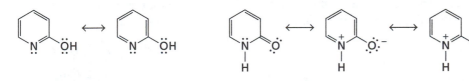

Phenol   **2-Pyridone**

One could argue that 2-hydroxypyridine should be favored because it is aromatic. But 2-pyridone is also aromatic—the difference is that some resonance forms of 2-pyridone have charged atoms. All of the atoms in these resonance forms satisfy the octet rule, however, and the negative charge resides on the more electronegative oxygen atom. Furthermore, these structures have six electrons in planar rings with conjugated π bonds.

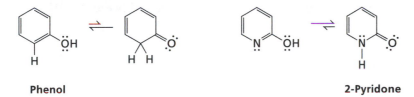

2-Hydroxypyridine   **2-Pyridone**

Similar tautomeric forms are observed for the pyrimidine and purine derivatives found in the nucleosides. When a carbonyl group is bonded to two nitrogen atoms, then additional tautomers exist, and the keto form is more predominant. The keto and enol forms of these heterocycles are designated the *lactam* and *lactim* forms, respectively.

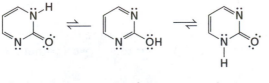

Lactam   **Lactim**   **Lactam**

The importance of the lactam forms relates to the situation that initially fooled Watson: *The hydrogen bonding patterns of the heterocyclic bases are related to the positions at which the hydrogen atoms are attached.* We will return to this topic shortly.

EXERCISE 28.3

Draw the tautomeric forms for uric acid (Section 28.1b), uracil (Fig. 28.1), and guanine (Fig. 28.1).

28.1d A GLYCOSIDIC BOND IS FORMED DURING NUCLEOSIDE SYNTHESIS

A nucleoside is made by forming a bond between the nitrogen atom of the heterocycle base and the anomeric carbon atom of D-ribofuranose or D-2-deoxyribofuranose. This link, called the **glycosidic bond,** is similar to the bond formed between an alcohol and a carbohydrate when an acetal derivative—a glycoside—is made (Section 19.4a).

There are many ways to prepare nucleosides in the laboratory, but one example will suffice to illustrate the general process. Unlike acetals, nucleosides cannot be made using a strong protic acid because the proton will react with the heterocycle and make it a cation. Instead, the anomeric acetate derivative of the carbohydrate (with the OH groups protected as esters) is converted to a cation intermediate by treatment with $SnCl_4$, a Lewis acid. The heterocycle itself, which is protected at its oxygen atoms with trimethylsilyl, $Si(CH_3)_3$, groups, does not react appreciably with tin(IV) chloride.

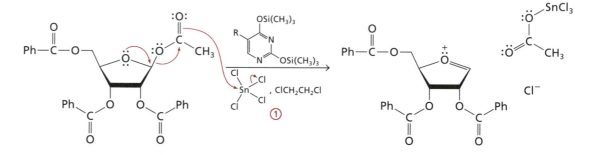

Once formed, the cation intermediate is intercepted by the nitrogen atom of the heterocycle. This step occurs at the top face of the ring because it is less hindered than the bottom (as drawn in the following scheme).

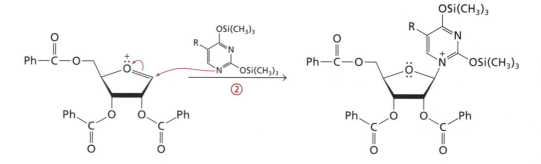

Chloride ion subsequently (or perhaps at the same time the nitrogen atom of the heterocycle is reacting with the cation) reacts to remove a trimethylsilyl group and to unmask the carbonyl group.

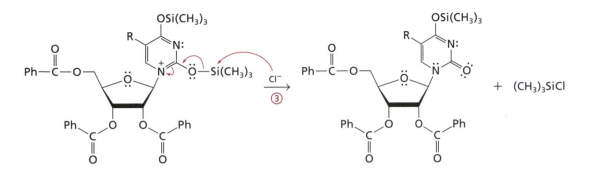

The other trimethylsilyl group is cleaved later (not shown here) to form the other carbonyl group.

28.1e PHOSPHATE ESTER FORMATION IS THE FOUNDATION FOR THE BIOSYNTHESIS OF THE NUCLEOTIDES AND NUCLEIC ACIDS

The formation of the glycosidic bond during *biosynthesis* actually occurs with a phosphorylated ribose derivative, so nucleotides, rather than nucleosides, are formed directly from the carbohydrate and heterocycle. Differences exist between the detailed reactions involved in the biosynthesis and laboratory synthesis of the nucleotides, of course, but the key step is the same: *A cation intermediate is trapped by a nitrogen-containing nucleophile to form the glycosidic bond.*

In the first stage of nucleotide biosynthesis, D-ribofuranose is converted to its 5-phosphate derivative, and then the anomeric OH group reacts to form a pyrophosphate ($P_2O_7^{3-}$) derivative (the nomenclature used here follows that in standard biochemistry texts).

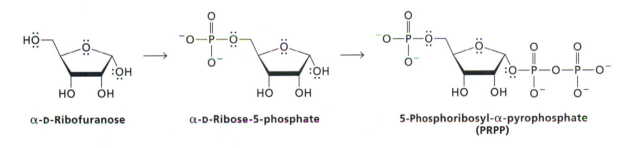

| α-D-Ribofuranose | α-D-Ribose-5-phosphate | 5-Phosphoribosyl-α-pyrophosphate (PRPP) |

The phosphorylation steps shown in the preceding scheme are common reactions in nucleic acid chemistry. The products are phosphate esters, which are formally the condensation products of phosphoric acid with alcohols.

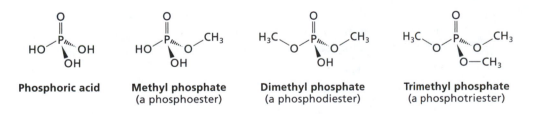

| Phosphoric acid | Methyl phosphate (a phosphoester) | Dimethyl phosphate (a phosphodiester) | Trimethyl phosphate (a phosphotriester) |

Just as carboxylic acids can be converted to esters via the reaction between alcohols and derivatives such as acid chlorides and anhydrides (Section 21.3), phosphate esters are made from alcohols and phosphoric acid anhydrides. In biochemistry,

ATP is the source of phosphate groups. The three phosphate groups constitute a bis(anhydride).

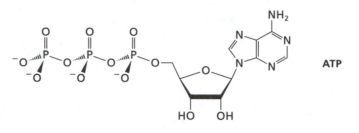

ATP

The mechanism for phosphoester formation is similar to that for ester formation. The alcohol or its conjugage base (nucleophile) reacts at the electrophilic phosphorus atom (Ribo-A in the following equations is an abbreviation for the ribose and adenosine rings of ATP). Instead of forming a tetrahedral intermediate, which is the species formed by addition of a nucleophile to a carbonyl group (Section 21.1c), a phosphoric acid derivative reacts by forming a trigonal-bipyramidal intermediate (the geometry of the phosphorus atom is trigonal bipyramidal).

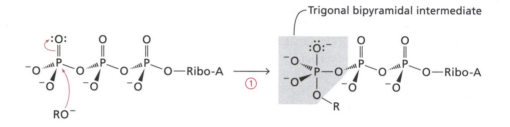

Next, just as a tetrahedral intermediate collapses to regenerate a carbonyl group (Section 21.1c), the trigonal-bipyramidal phosphorus intermediate of phosphorylation collapses to regenerate the phosphorus–oxygen double bond. Adenosine diphosphate (ADP) is the leaving group in this process.

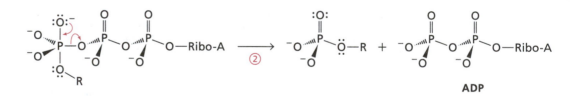

EXERCISE 28.4

Propose a mechanism for the synthesis of a diphosphate ester, also called a pyrophosphate ester, a specific example of which is given by the following equation. Adenosine monophosphate (AMP) is the leaving group. This transformation is used in the activation of the anomeric OH group of ribofuranose, which is involved in the biosynthesis of PRPP.

ROH + ATP ⟶

Returning to the route by which nucleotides are biosynthesized, PRPP undergoes ionization by displacement of the pyrophosphate leaving group by an electron pair on the oxygen atom in the ribofuranose ring.

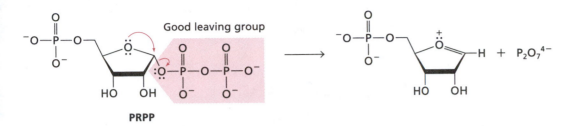

PRPP

This intermediate is subsequently trapped by a nitrogen-containing nucleophile, yielding the nucleotide precursor. For the pyrimidine nucleotides, the nucleophile is a heterocycle called *orotate*. For the purine nucleotides, the nucleophile is ammonia, generated by decomposition of the amide side-chain group of glutamine. Further reactions are required to generate the actual heterocycles found in the nucleic acids.

Pyrimidine nucleotides

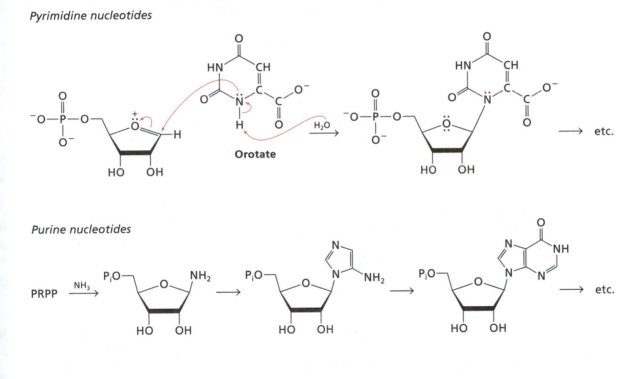

Purine nucleotides

28.2   NUCLEIC ACID STRUCTURES

28.2a   THE PRIMARY STRUCTURE OF A NUCLEIC ACID IS ITS SEQUENCE

The terms describing the structures of the nucleic acids parallel those used for proteins. Thus, the primary structure of a nucleic acid is its sequence of nucleosides. By convention, RNA and DNA sequences are specified starting at the 5′ end.

To illustrate how such sequences are denoted, consider the trinucleotide structure shown here.

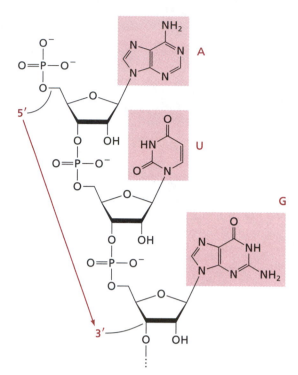

This sequence is one of RNA because the 2′ position of the sugar ring has an OH group. The phosphate groups are the same between each pair of nucleosides in a nucleic acid, so the sequence can be specified simply by listing the identities of the nucleosides/bases, starting at the 5′ end. This sequence is written AUG.

For DNA molecules, the sequences are written in exactly the same way even though the nucleoside units are derivatives of 2′-deoxyribose. To differentiate a sequence of DNA from RNA, you look for the letter T (thymine), which is found only in DNA, or for the letter U (uracil), which is present only in RNA. For example,

GCGATAGCGATCAGGATCAGG   must be a sequence of DNA because T
                               (but not U) is present

CCGAUAGCGAUUAGGACUAGA   must be a sequence of RNA because U
                               (but not T) is present

Remember that the printed sequence of a nucleic acid runs in the direction from 5′ to 3′, just as a given protein sequence runs from the N-terminus to the C-terminus. The order of the nucleotides is crucial, so, for example, the trinucleotide ACG is different from CGA, just as the tripeptide Ala-Ser-Gly is different from Ser-Gly-Ala.

EXERCISE 28.5

Draw the full structure of the teranucleotide ATGC, showing all of the atoms.

28.2b NUCLEIC ACID BIOSYNTHESIS IS AN ENZYME-CATALYZED PROCESS

Nucleic acids are prepared by coupling nucleotide triphosphate precursors via enzyme-catalyzed processes. The enzyme *DNA polymerase* replicates DNA, and *RNA polymerase* is the enzyme that transcribes DNA sequences to build RNA molecules. Both polymerases require a template—a strand of DNA—in addition to the nucleotide triphosphate building blocks. For DNA replication, a small piece of DNA called a primer is bonded at a complementary sequence via hydrogen-bond formation (Section 28.2c), and this sequence becomes part of the new DNA chain, as shown below. In the first step, a complementary nucleotide triphosphate binds to the template strand through hydrogen bonds.

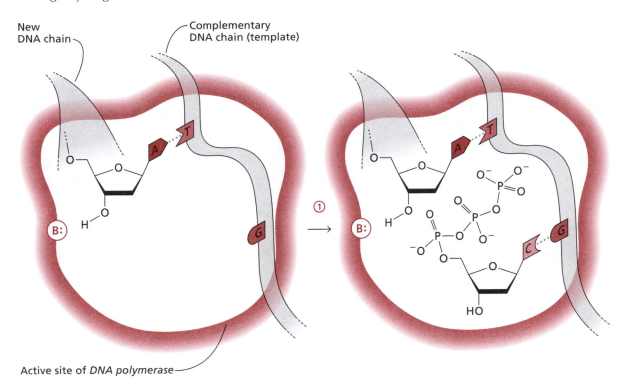

Nucleophilic reaction of the 3′-OH oxygen atom at the electrophilic phosphorus atom then takes place, which generates a trigonal-bipyramidal intermediate.

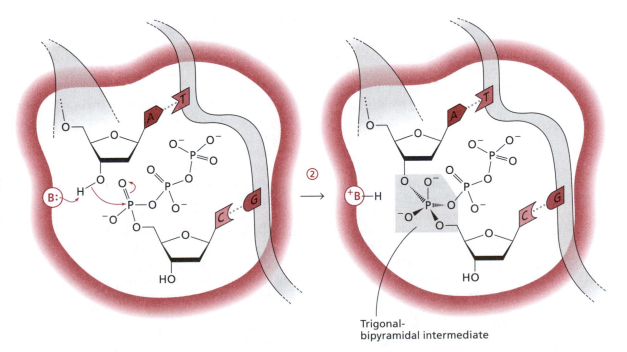

Collapse of the five-coordinate phosphate ion produces the new phosphoester bond and expels pyrophosphate ion as the leaving group.

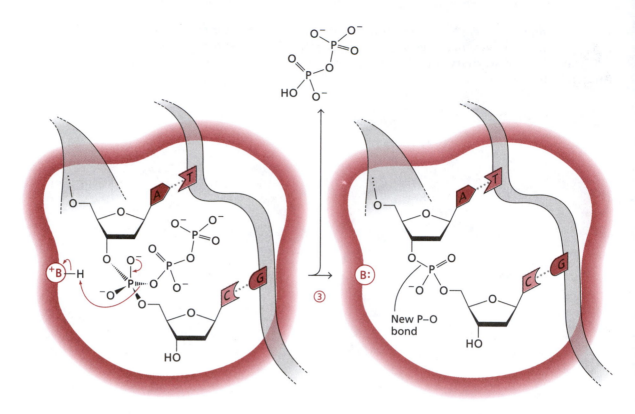

28.2c IN ITS MOST STABLE FORM, DNA EXISTS AS A DOUBLE HELIX

The assumption was made in Section 28.2b that you are already somewhat familiar with the double-helical form of DNA, in which hydrogen bonds form between complementary bases across the helix axis. Other types of structures exist for the nucleic acids, so it is worthwhile to look at the structural details more closely.

The common double helix is called **B-DNA,** which was the first regular structure deduced (Figure 28.3). The winding of the chains about the helix axis creates two crevices along the exterior of the DNA molecule, called the major and minor grooves. These regions allow small molecules to interact with the base pairs without the necessity of unwinding the two strands of the helix.

An important feature of the double-stranded helical structure is its preponderance of negative charges, which result from the presence of the anionic phosphate groups lining the edges of the helix. These anionic groups can effectively repel other anions such as hydroxide ion, preventing dissociation of the helix and hydrolysis of the phosphodiester bonds. Species with positive charges are attracted to the exterior of the DNA molecule, however. Fortunately, potent electrophiles are not common in aqueous solution because they would readily react with water, which is itself a good nucleophile.

The greatest influences on the formation of the DNA double helix are the attractive dispersion and dipole–dipole forces (Section 2.8a) between the faces of the heterocyclic rings. These phenomena are often referred to as pi-stacking interactions, which result from the dipoles or induced dipoles of the heterocycles' π systems. Once the heterocycles interact to form stacked arrays of their rings along the helix axis, hydrogen bonds can form *across* the helix between adjacent, antiparallel strands of DNA, as shown in Figure 28.4. An important fact that led to the solution of the DNA structure was an earlier observation that the ratios of adenine-to-thymine and cytosine-to-guanine were 1:1. When Watson realized that the N–H and O=C groups would form hydrogen bonds with each other across the helix axis, the significance of the equimolar ratios became apparent: *Hydrogen bonds only occur between a pyrimidine and a purine base,* that is, between A and T or C and G. These interactions lead to equidistant positions

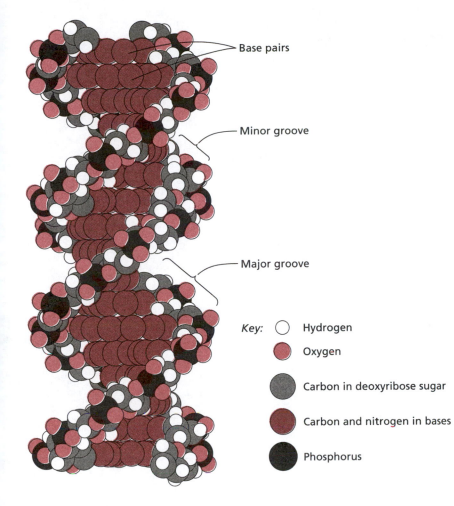

Base pairs

Minor groove

Major groove

Key: ○ Hydrogen

⬤ Oxygen

⬤ Carbon in deoxyribose sugar

⬤ Carbon and nitrogen in bases

● Phosphorus

Figure 28.3
The structure of B-DNA.

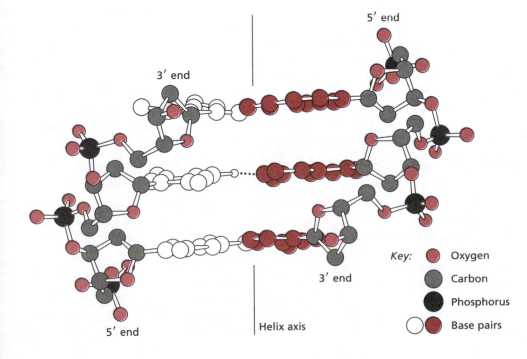

5′ end

3′ end

3′ end

5′ end

Helix axis

Key: ⬤ Oxygen
        ⬤ Carbon
        ● Phosphorus
        ○⬤ Base pairs

Figure 28.4
A close-up view of three base pairs of B-DNA. The strands are antiparallel, meaning that the directions 5′ → 3′ are spatially opposite. Hydrogen atoms have been omitted for clarity.

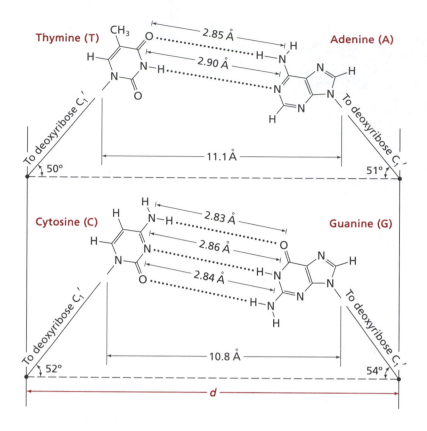

Figure 28.5
Hydrogen bonds between complementary bases in DNA. The distance d between $C_{1'}$ on adjacent strands (shown by the black dots) is the same for each pyrimidine-purine pair.

between the anomeric carbon atoms of the deoxyribofuranoside units across the diameter of the helix (Fig. 28.5) and to an even, ordered structure.

These precise base-pairing combinations provide the anchor for our understanding of DNA and RNA biosynthesis. What is arguably the most eloquent understatement in all of science is the sentence in Watson and Crick's original *Nature* publication that reads: "It has not escaped our notice that the specific pairing we have postulated immediately suggests a possible copying mechanism for the genetic material."

As important as hydrogen bonds are for the recognition processes involved in replication, the hydrogen bonds that can form with the edges of the bases within the major and minor groove can be just as significant for other phenomena such as gene regulation. The standard base pairing illustrated in Figure 28.5 leaves several hydrogen atoms and heteroatoms with no association to other groups in the nucleic acid itself. These groups line the "floor" and sides of the grooves on the exterior of the helix, and small molecules such as drugs and larger species such as proteins can interact with the nucleic acids by forming hydrogen bonds with these available donors and acceptors.

28.2d DNA CAN HAVE A VARIETY OF STRUCTURES THAT ARE INFLUENCED BY THE CONFORMATIONS OF THE BASES AND SUGAR RINGS

Just as proteins display a variety of secondary structural elements, the secondary structures of nucleic acids can vary too. When fibers of B-DNA are dried, they assume a conformation called A-DNA, in which the helix is compressed, and the base pairs tilt 19° with respect to the helix axis. The A-DNA structure, illustrated in Figure 28.6, still maintains the normal A-T and C-G base pairs, but the conformation of the deoxyribose rings change, as shown in Figure 28.7. In B-DNA, the five-membered sugar rings are puckered so that $C_{2'}$ lies above the plane of the other four atoms in the ring; this conformation is called $C_{2'}$-*endo*. (Recall from Section 3.2a that five-membered rings exist in a form that looks like a partially opened envelop.) In A-DNA, $C_{3'}$ is the atom that lies

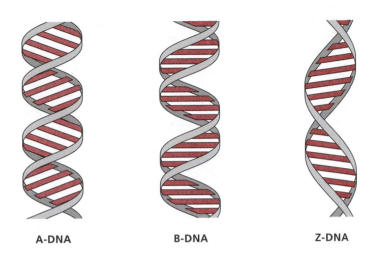

A-DNA   **B-DNA**   **Z-DNA**

Figure 28.6
Illustrations of the helix
structures in B-, A-, and
Z-DNA.

above the plane of the carbohydrate ring; its conformation is C$_3'$–*endo*. The changes in conformation from C$_2'$–*endo* to C$_3'$–*endo* leads to the tilt of the base pairs relative to the helix axis.

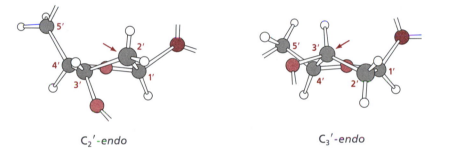

C$_2'$-*endo*   C$_3'$-*endo*

Figure 28.7
Sugar puckering in B-, and A-
DNA. The terms C$_2'$-*endo* and
C$_3'$-*endo* denote which atoms
of the deoxyribose ring lie out
of plane. The consequences of
puckering affect other conformations in the DNA backbone
by twisting of the phosphate
and glycosidic bonds.

A third type of secondary structure is Z-DNA, which is also illustrated in Figure 28.6. It is radically different from A- and B-DNA because it exists as a *left-handed* helix. The conformations that lead to formation of Z-DNA are illustrated in Figure 28.8, and they result from the orientation of the base about the glycosidic bond. Normally, the glycosidic bond is anti, but in Z-DNA, the purine bases adopt a syn conformation.

The properties of the three types of secondary structure are summarized in Table 28.1.

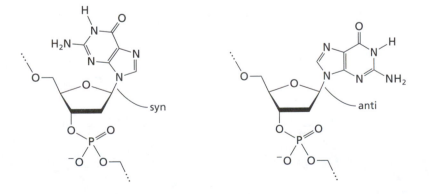

syn   anti

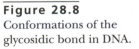

Figure 28.8
Conformations of the
glycosidic bond in DNA.

Table 28.1 A comparison of the properties of A-, B-, and Z-DNA structures.

Feature	Helix type		
	A	B	Z
Shape	Broadest	Intermediate	Most elongated
Rise per base pair	2.3 Å	3.4 Å	3.8 Å
Helix diameter	25.5 Å	23.7 Å	18.4 Å
Screw sense	Right-handed	Right-handed	Left-handed
Glycosidic bond	anti	anti	anti for C, T syn for G
Base pairs per turn of helix	11	10.4	12
Pitch per turn of helix	24.6 Å	33.2 Å	45.6 Å
Tilt of base pairs from normal to helix axis	19°	1°	9°
Major groove	Narrow and deep	Wide and deep	Flat
Minor groove	Broad and shallow	Narrow and deep	Narrow and deep

The secondary structures of RNA are even more diverse than those observed for DNA. A major difference is the scarcity of double helix structures in RNA. The OH group in the 2′-position of ribose prevents formation of the B-type helix, but double strands can be generated if the RNA molecule adopts an A-type helix. The structure of a typical t-RNA molecule is shown in Figure 28.9. Some double-stranded portions are seen, but they do not have the regular, ordered structure of the DNA double helix; many unpaired regions also exist.

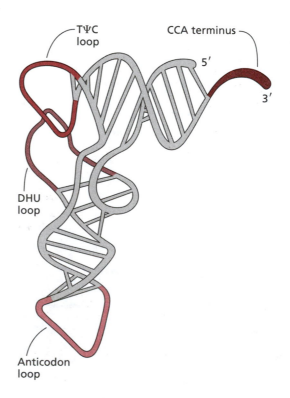

Figure 28.9
A model for the three-dimensional structure of yeast phenylalanine t-RNA.

28.3   MOLECULAR RECOGNITION

The notion that synthetic systems could be used to "recognize" other molecules was strengthened by the elucidation of DNAs structure; but as early as 1890, Emil Fischer likened an enzyme's active site and its substrate to a lock and key, a concept that dominated the field of enzymology for decades. Still, it is fair to say that the properties of catalysis, recognition, and transport were long thought to be unique to biomolecules, so the paradigm shift that brought synthetic systems to the fore did not take place until the developments in molecular biology occurred during the 1950s and 1960s.

For organic chemists, the general topic of molecular recognition has followed many avenues of exploration. Often, molecular recognition is described as *host–guest chemistry*, a term coined in 1974 by Donald Cram, who shared the 1987 Nobel Prize in chemistry on this topic. Research in the field of host–guest chemistry focuses on understanding how a **receptor** (the host) can interact with another molecule (the guest) in the absence of covalent bonding.

The host molecule is generally larger than the guest, or at least larger than the portion of the guest molecule or ion that is bound, which is called the *epitope*. The epitope is complementary to the host's binding site with regard to charge and steric requirements.

When a host and a guest molecule interact, they generate a *complex*. You have seen many times in this text that enzymes are excellent hosts; they have an active site that binds a substrate, often with high specificity. The DNA molecule is not normally considered to be a host–guest complex, but the noncovalent interactions that lead to replication are the same as those found in host–guest complexes (Section 2.8a). In such complexes, the receptors use these three types of noncovalent interactions:

- *Electrostatic forces:* ion–ion, ion–dipole, and electron donor–acceptor.
- *Dispersion forces:* induced dipole-induced dipole, π stacking.
- *Hydrogen bonds:* –O–H ⋯ O–, –O–H ⋯ N–, –N–H ⋯ O–, etc.

In the next sections, you will be introduced to several examples of synthetic molecules that have been prepared to exploit noncovalent interactions in order to study recognition processes.

28.4   CROWN ETHERS AND CRYPTANDS

28.4a   CROWN ETHERS ARE PREPARED BY SUBSTITUTION REACTIONS

The **crown ethers,** also called *coronands*, were among the first artificial recognition systems studied. The name derives from the shapes of these molecules in a particular conformation, which resembles a crown.

18-Crown-6

The simplest and most common crown ethers have CH_2CH_2 groups linking oxygen atoms, and these are named with two numbers, the larger one designating the total number of atoms in the ring, and the smaller one indicating how many oxygen atoms are present.

Crown ethers are made by a variant of the Williamson ether synthesis (Section 7.2b) in which an alkoxide ion displaces a tosylate or halide ion by an S_N2 mechanism.

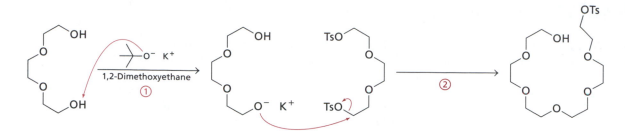

Additional base deprotonates the second alcohol group, and the cation brings together the two ends of the macrocycle. This *template effect,* as it is called, is an important part of crown ether synthesis, accounting for the high yield of ring formation.

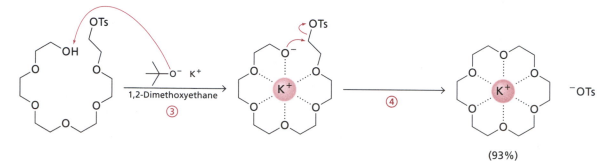

(93%)

The general procedure illustrated above for the synthesis of 18-crown-6 can be used to make many similar compounds with nitrogen, oxygen, and sulfur atoms in the ring. To prepare *azacrowns,* molecules with nitrogen atoms in the ring, protecting groups or alternate methods are sometimes required. A common alternate route is to make an amide link, and reduce the carbonyl group with LiAlH$_4$ (Section 21.8b). Some structures and names of representative coronands are shown below.

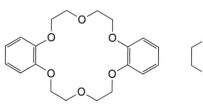

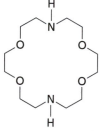

Dibenzo-18-crown-6   **1,10-Diaza-4,7,13,16-tetra-**   **1,4-Dithia-7,10,13-tri-**   **1,4,7-Triazacyclononane**
                           oxacyclooctadecane   **oxacyclopentadecane**

EXERCISE 28.6

Show how you would prepare dibenzo-18-crown-6 from catechol and bis(2-hydroxy-ethyl)ether.

28.4b   CROWN ETHERS RECOGNIZE IONS BY THEIR SIZES

Charles Pedersen, a chemist at DuPont, first prepared and studied the crown ethers in 1967, demonstrating that they bind alkali metal ions. This achievement was recognized by his sharing the 1987 Nobel Prize in chemistry with Jean-Marie Lehn and Donald Cram. It is remarkable how these compounds have been utilized since that time. One important application of crown ethers is to dissolve salts in solvents of low dielectric strength. For example, most sodium and potassium salts of simple inorganic anions are insoluble in solvents such as acetonitrile and benzene. Addition of crown ethers leads to ready dissolution.

A specific example illustrates the utility of this procedure. Acetate ion is a notoriously poor nucleophile in typical S_N2 reactions. If water is present, hydrogen bonds are formed between the acetate ion and water; if a polar, aprotic solvent is used, the salt is not soluble. With even 10 mol% of 18-crown-6, however, potassium acetate dissolves in acetonitrile, and the acetate ion acts as a potent nucleophile:

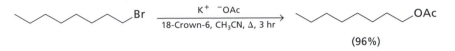

The crown ether binds the potassium ion strongly, carrying the acetate ion into solution as a "naked" (unsolvated) nucleophile. Literally hundreds of examples like this one are known.

Crown ethers are specific for the cation they bind, and this specificity is related to the size of the cavity. Thus, 18-crown-6 binds K^+ preferentially, but smaller crown ethers can bind Li^+ or Na^+ ions.

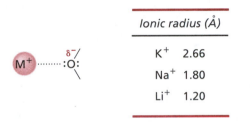

Ionic radius (Å)	
K^+	2.66
Na^+	1.80
Li^+	1.20

The binding interaction is based primarily on the electrostatic attraction between the unshared electron pairs of the oxygen atoms and the cationic metal ion.

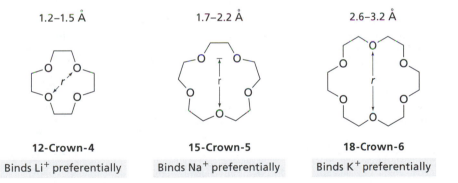

Metal ions are not the only guests that form complexes with crown ethers. Ammonium ions, which also carry a positive charge, can interact through two or three of the crown's heteroatoms via hydrogen-bond formation. Azacrowns are even better at binding ammonium ions because the nitrogen atoms are more basic than oxygen atoms. The following example shows a crown ether with a pyridine ring in the core structure:

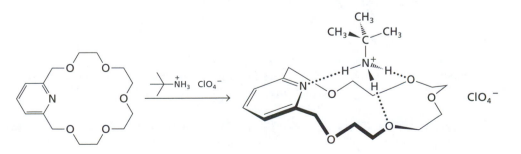

Show how you would prepare the pyridine crown ether in the preceding scheme from 2,6-lutidine (2,6-dimethylpyridine), bis(2-hydroxyethyl)ether, and chloroethanol. What is the expected order of binding toward *tert*-butylammonium perchlorate, (*tert*-butyl)methylammonium perchlorate, and (*tert*-butyl)dimethylammonium perchlorate?

28.4c CRYPTANDS AND SPHERANDS EXPLOIT THE PROPERTY OF PREORGANIZATION

The binding of a metal ion by a crown ether depends on matching the size of the cavity with the size of the metal ion, a property called *complementarity*. The magnitude of binding—shown by the value of the equilibrium constant, K_a—is related to the change in free energy of the system when the host (H) binds the guest (G) molecule. This free energy change in turn is related to changes in the enthalpy and entropy of the system.

$$H_{(solv)} + G_{(solv)} \rightleftharpoons (H \cdot G)_{(solv)} \quad K_a = \frac{[H \cdot G]}{[H][G]} \tag{28.1}$$

$$\Delta G° = \Delta H° - T\Delta S° = -RT\ln K_a = -RT\ln\left(\frac{[H \cdot G]}{[H][G]}\right) \tag{28.2}$$

One way to influence the strength of binding is to minimize the change in entropy when the guest is bound. Minimizing entropy can be accomplished by **preorganization,** in which the conformation of the host changes little as the complex forms. The *cryptands* are molecules that display the property of preorganization. They have a specifically tailored cavity to optimize binding of a metal ion by encapsulation. Jean-Marie Lehn, who shared the 1987 Nobel Prize, prepared and studied many of these substances, which are like crown ethers except that they have an additional "strap" to create the cavity. A typical cryptand host is prepared by making a diamide from a diaza crown ether; the amide groups are subsequently reduced using diborane (which reacts like LiAlH$_4$).

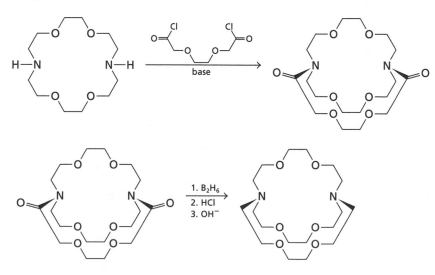

The binding of potassium ion by this cryptand is more favorable by several kilocalories per mol than the binding by more flexible crown ethers, as shown by the data presented in the following two equations:

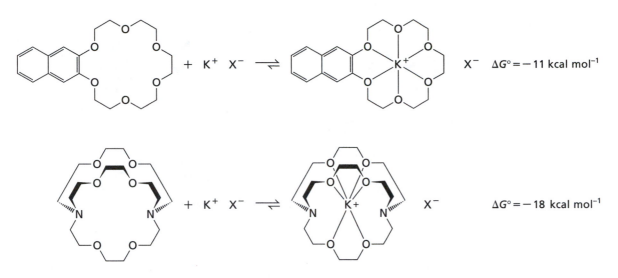

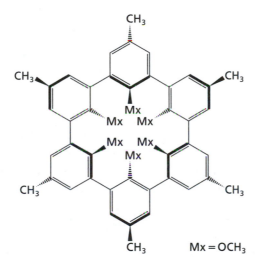

Spherands constitute another class of compounds that are preorganized to bind metal ions. These cyclic methoxybenzene derivatives, prepared and studied by Cram and co-workers, are even more rigid than the cryptands. Some spherands can bind alkali metal ions even more tightly than either crown ethers or cryptands. The compound shown below, for example, binds a lithium ion with $\Delta G° > -23$ kcal mol^{-1} but a potassium ion with $\Delta G° < -6$ kcal mol^{-1}.

A spherand with eight aromatic rings will bind a cesium ion in preference to the other alkali metal ions.

28.4d SOME ANTIBIOTICS FUNCTION AS CROWN ETHERS

Certain microorganisms make use of crown ether-like molecules called **ionophores** to transport metal ions through cell membranes, which are composed of hydrophobic molecules. Fatty acids and cholesterol are two principal constituents of cell membranes, and their hydrocarbon portions prevent polar molecules and ions from crossing their boundaries. Metal ions are usually solvated by water, which enhances their degree of lipophobicity (fear of fat) even more.

Ionophores are antibiotics that facilitate the transport of metal ions through cell membranes by binding the metal ion within their interior and presenting a hydrocarbon exterior to the membrane surroundings. Two well-studied examples are valinomycin and nonactin. Valinomycin is a cyclic trimer consisting of L-lactate, L-valine, D-hydroxyvalerate, and D-valine. The six valine carbonyl groups adopt an octahedral

a.

b.

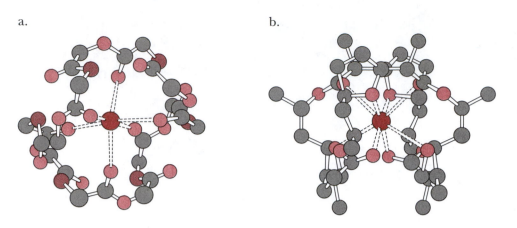

Figure 28.10
Models of the potassium complexes of (*a*) valinomycin and (*b*) nonactin. Hydrogen atoms have been omitted for clarity of presentation.

arrangement in the presence of potassium ion, creating a binding site that places the nonpolar isopropyl groups on the exterior of the molecule. Nonactin uses four carbonyl groups and four THF oxygen atoms to bind a potassium ion as an eight-coordinate complex. These potassium ion complexes are illustrated in Figure 28.10.

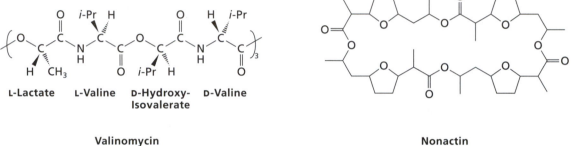

L-Lactate   L-Valine   D-Hydroxy-   D-Valine
                           Isovalerate

Valinomycin   **Nonactin**

Just as 18-crown-6 binds potassium in favor of sodium or lithium ions, valinomycin binds K⁺ more tightly than the others by a factor or > 1000. Valinomycin without a metal ion has an open and flexible conformation compared with its potassium complex. The folded form has a rigid cavity that is too large to coordinate its six carbonyl oxygen atoms to Na⁺ or Li⁺.

Besides having smaller sizes, Na⁺ and Li⁺ ions also have higher solvation energies than K⁺. So even when the binding pocket is the correct size, additional stabilization is needed to compensate for the energy that is lost when water molecules no longer bind to these ions.

28.5   CAVITY-CONTAINING MOLECULES

28.5a   CYCLODEXTRINS USE DISPERSION FORCES TO BIND GUEST MOLECULES

Cryptands and spherands bind metal ions well because they have small cavities with several electronegative atoms that can interact with metal cations by dipole–ion forces. Some host molecules are able to accommodate more structurally diverse guests because they can take advantage of weak interactions such as dispersion forces (induced dipoles, Section 2.8a) that extend over large molecular surfaces.

The **cyclodextrins** are degradation products of starch that are formed by specific enzyme-catalyzed processes. These derivatives of D-glucose have six, seven, or eight carbohydrate rings connected head-to-tail to create a cylindrical shape. The holes in these cylinders vary in size, but they are all nonpolar. In contrast, the hydroxyl groups on the sugar rings make the cyclodextrin exteriors hydrophilic, so they are soluble in water.

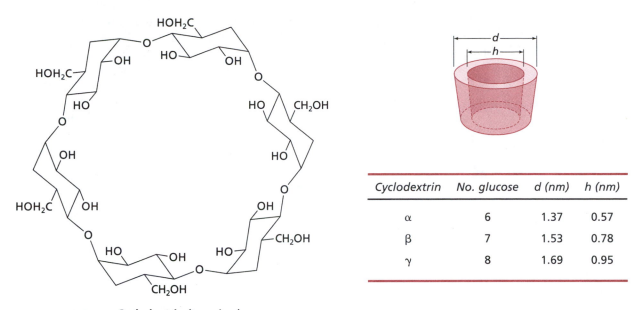

α-**Cyclodextrin** (top view)

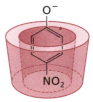

Cyclodextrin	No. glucose	d (nm)	h (nm)
α	6	1.37	0.57
β	7	1.53	0.78
γ	8	1.69	0.95

The binding of guest molecules by the cyclodextrins has received much attention and led to their current commercial use in fabric fresheners such as Febreze. The interior cavity of these hosts traps molecules that cause odors, and their water solubility means that they can be washed away, if desired. Cyclodextrins are also used in chromatography applications. For example, α-cyclodextrin binds *o*-, *m*-, and *p*-nitrophenolate ion with equilibrium constants of 200, 500, and 2439, respectively. The strong binding of the para isomer, the complex of which is shown at the right, results from dispersion forces between the benzene ring of the guest and the hydrocarbon portion of the glucose rings in the cylinder cavity. The meta and ortho isomers do not fit as well into the cavity because of their larger cross-sectional widths, so they are not bound as strongly.

28.5b CYCLOPHANES ARE OLIGOMERS OF BENZENE AND ITS DERIVATIVES

When phenol and resorcinol derivatives are made to react with aldehydes, cyclic oligomers are formed. These compounds are members of a class of substances called **cyclophanes** and are named as *calixarenes* and *calixresorcinarenes*, respectively. These names derive from their resemblance in certain conformations to an ancient Greek vase called a *calix crater*. The non-systematic nomenclature used for these bowl-shaped molecules starts with the prefix calix, followed by a number in brackets for the number of benzene rings, followed by the suffix -arene or -resorcinarene. The structures of two common cyclophanes follow:

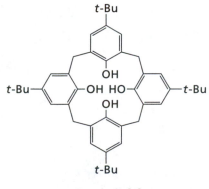

p-tert-Butylcalix[4]arene

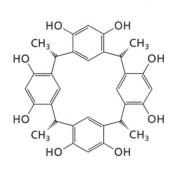

C-Methylcalix[4]resorcinarene

Calixarenes can adopt one of several conformations because rotations around the bonds linking the aromatic rings are relatively unhindered. Alkylation of the phenol oxygen atoms in calix[4]arene locks the molecule into its "cone" conformation. Calixarenes bind guest molecules such as toluene, shown below at the right, by taking advantage of dispersion forces between its aromatic rings and the hydrocarbon portions of the guest molecules. In this example, a sodium ion is also bound by the methoxy oxygen atoms on the bottom (the lower rim) of the cone-shaped host molecule.

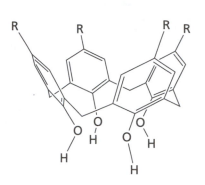

The cone formation of
p-tert-Butylcalix[4]arene
(R = *t*-Bu)

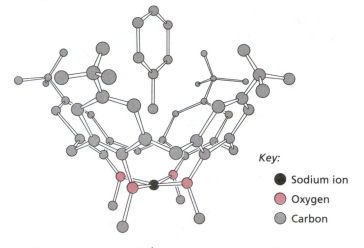

Key:
● Sodium ion
● Oxygen
● Carbon

The structure of the Na⁺ and toluene complex of
Tetra-*O*-methyl(p-tert-Butylcalix[4]arene)

Resorcinarenes exist naturally as bowl-shaped hosts because of hydrogen-bond formation between pairs of adjacent OH groups. The cavity is hydrophobic, but the OH groups can contribute hydrogen-bond acceptors and donors on the upper rim.

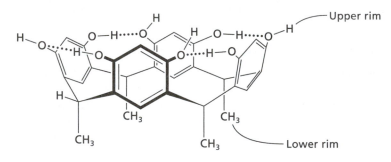

EXERCISE 28.8

Treating 2,4-dimethoxybenzyl alcohol with acid produces a trimeric product called cyclotriveratrylene, which has a shallow bowl shape. Propose a mechanism for this transformation, which starts with protonation of the benzylic OH group, dissociation of water to form a carbocation, and electrophilic aromatic substitution reactions.

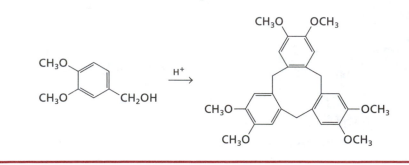

28.6 RECOGNITION USING HYDROGEN BONDS

28.6a UREA-CONTAINING HOST MOLECULES RECOGNIZE GUESTS BY HYDROGEN-BOND FORMATION

The third type of noncovalent interactions that receptors use to recognize guest molecules are hydrogen bonds (Section 28.3). Carboxylic acids are ideal guest molecules to study with respect to hydrogen-bond formation because carboxylic acids are common in biochemical systems: fatty acids, amino acids, and many metabolic intermediates contain COOH groups. Furthermore, carboxylic acids are structurally simple (compared to nucleic acids, for example) and have the ability to form hydrogen bonds in multiple ways. For example, dimers form readily in nonaqueous solutions, as illustrated here:

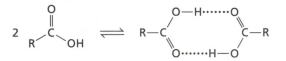

Host molecules that bind carboxylic acids often have nitrogen-containing functional groups because of their tendency to form hydrogen bonds. Derivatives of urea are neutral, and they can form two hydrogen bonds that are aligned in the same direction. Furthermore, the size of the urea functional group matches the size of a carboxylate ion, so two hydrogen bonds can be formed, as shown here.

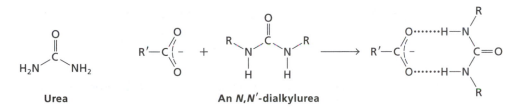

Urea   **An *N,N′*-dialkylurea**

Each oxygen atom of a carboxylate ion has two electron pairs. The one directed away from the R group is called the syn pair, and it is considered to be a stronger base than the other, designated anti.

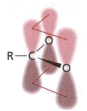

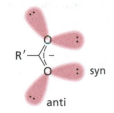

| Side view, showing the π system | Top view, showing the nonbonding electron pairs; the syn electron pairs are more basic than the anti ones. |

Carboxylate ions that are reactive within enzyme active sites employ the syn electron pair as nucleophiles, an observation that supports the notion of their higher basicity. Thus, if a carboxylate ion is to be bound by a single urea group within a synthetic host, then its two syn pairs will be engaged instead of a syn and an anti pair.

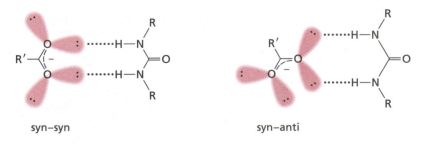

syn–syn   syn–anti

If one or two anti pairs can be utilized *in addition to* the syn pairs, however, then binding should be enhanced.

The binding capabilities of several urea-based receptors for carboxylate ions support the ideas just presented. A receptor with a single urea group binds to a carboxylate ion through the latter's syn electron pairs; the value of the equilibrium constant for this interaction is $K_a = 400$. A di(urea) receptor that binds the same carboxylate ion has $K_a = 2 \times 10^5$. Notice that the receptor with two urea groups takes advantage of interactions with the *anti* electron pairs in addition to its interactions with the *syn* pairs.

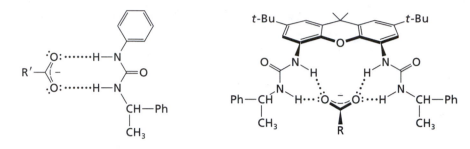

Preparing receptors with urea groups is straightforward because amines react with isocyanates in high yield. This is the same reaction used to make polyureas (Section 26.3d).

$$R-N{=}C{=}O \ + \ R'-NH_2 \ \longrightarrow \ RNH-\overset{\overset{\textstyle O}{\|}}{C}-NHR'$$

EXERCISE 28.9

Propose a reasonable mechanism for the reaction between phenylisocyanate and 1-aminobutane to form *N*-butyl-*N*-phenylurea.

The guanidinium ion, which is the side chain functional group in the amino acid arginine, is also a good steric match for the carboxylate ion, and it carries a positive charge. This feature leads to exceptionally strong interactions between guanidinum (+1) and carboxylate (–1) ions.

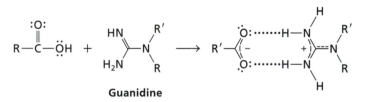

Many enzymes use the guanidinium ion of arginine to bind substrates with phosphate groups, too.

EXERCISE 28.10

Illustrate the noncovalent binding combinations of an alkylphosphate ion, $RO{-}PO_3^{2-}$, with the guanidinium ion and with urea. What types of interactions are important for each: hydrogen bonds, electrostatic attractions, dispersion forces, or some combination?

A synthetic receptor that makes use of the guanidinium ion to bind carboxylate ions is shown below. In this host compound, the guanidinium ion is part of a bicyclic ring system, and two naphthyl groups are also appended to the receptor framework. Upon treatment of the host with the salt of *p*-nitrobenzoic acid, a complex is formed that utilizes the recognition properties of hydrogen bonds in addition to an attractive electrostatic (ion-ion) interaction. Furthermore, π-stacking interactions are present between the planar π-systems of the aromatic rings, resembling those in B-DNA (Section 28.2c)

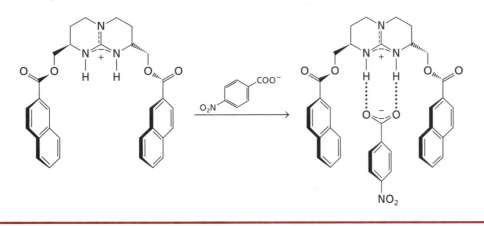

28.6b ADENINE AND OTHER NUCLEIC ACID BASES CAN BE RECOGNIZED BY SPECIFIC RECEPTORS WITH CONVERGENT FUNCTIONALITIES

As progress was made in the synthesis of hosts and receptors to bind metal ions and carboxylic acids, attention turned during the past two decades to the construction of more sophisticated host molecules for binding compounds such as nucleosides and their bases. The problem inherent in this task can be framed by considering how a molecule like adenine could be recognized and bound. An initial response might be that hydrogen bonds would suffice, just as adenosine is recognized by pairing with thymidine during DNA replication. This type of binding is called *Watson–Crick base pairing*. An alternate scheme is possible, called *Hoogsteen base pairing*, which utilizes the C6-amino group of adenine, along with N7, instead of N1.

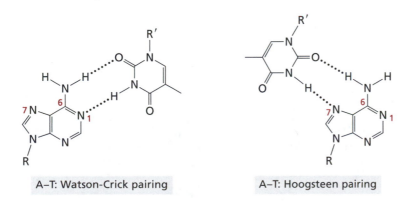

A–T: Watson-Crick pairing   A–T: Hoogsteen pairing

The formation of hydrogen bonds—whether they be of the Watson–Crick or Hoogsteen type—is not sufficient, however, for the binding of adenine by a receptor, because adenine forms perfectly good hydrogen bonds with water in aqueous solution. Simply substituting hydrogen bonds of a receptor for those of water is not so favorable that it leads to complex formation. For this reason, the concept of **convergent functionality** has become an important consideration in the design of host systems, adding to the ideas of complementarity and preorganization.

Simply stated, convergent functionality makes use of two (or more) functional groups aimed at a binding site, or cleft, within a host structure to recognize or bind more than one structural feature of the guest molecule. The di(urea) host described previously (Section 28.6a) is one example of a host with convergent functionality.

A common scaffold that has been used to attach functional groups is *all cis*-1,3-5-trimethyl-1,3,5-cyclohexanetricarboxylic acid, a compound known as *Kemp's triacid*. This molecule exists in the conformation that has all of the carboxylic acid groups in the axial positions. When heated with 1 equiv of an amine, the N-substituted imide group is produced from two of the acid groups.

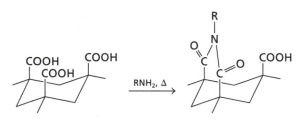

Kemp's triacid

If a rigid diamine reacts with Kemp's triacid, then the natural arrangement of functional groups will orient the two remaining carboxylic acid groups toward one another. With a small spacer such as the *m*-phenylene group, receptor **1** can be made, in which the carboxylic acid groups form hydrogen bonds as they do in solution. If the spacer is larger, as in **2,** then the two acidic groups converge on the interior of a cleft created by the scaffold structure. Functional groups and heteroatoms can be incorporated to provide additional binding groups, as in **3.**

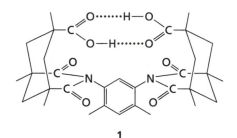

1

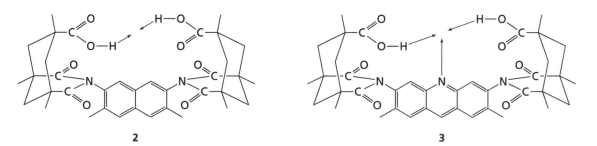

2   **3**

EXERCISE 28.11

Draw structures to show how host **3** can use its carboxylic acid groups to bind pyridine, which has one recognition point; pyrazine (two recognition points); and purine (three recognition points). Which guest will be bound most strongly? Most weakly?

So how can hosts with convergent functionality be used to bind heterocycles such as adenine? Heating Kemp's triacid with urea (which decomposes to produce ammonia) yields the imide derivative, **4.**

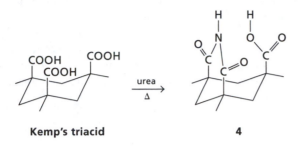

Kemp's triacid   **4**

With its N–H and carbonyl groups, an imide furnishes the same hydrogen-bond donor and acceptor pattern as thymine (Fig. 28.5), so **4** should pair with adenine to form specific hydrogen bonds. If the carboxylic acid group of **4** is subsequently converted to an amide derivative, then hosts such **5** can be made.

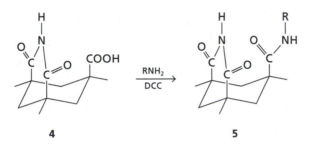

4   **5**

Analogues of host **5** have been used to assess the importance of the π-stacking interactions for recognition. A comparison among a series of hosts that provide the same hydrogen-bond patterns but different sized π systems shows that larger aromatic groups lead to enhanced binding of 9-ethyladenine (shown in color). The binding constant K_a decreases as the area of the π system shrinks.

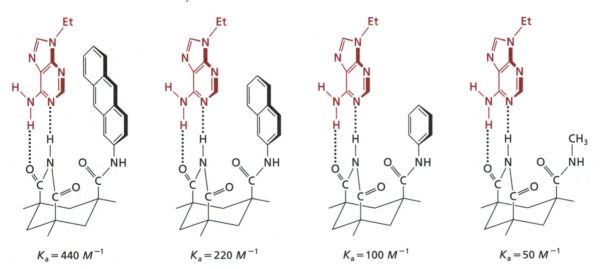

$K_a = 440\ M^{-1}$ $K_a = 220\ M^{-1}$ $K_a = 100\ M^{-1}$ $K_a = 50\ M^{-1}$

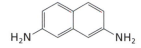

2,7-Diaminonaphthalene

If 2,7-diaminonaphthalene is condensed with compound **4**, then host **6** can be made, which has convergent imide functional groups. Employing a combination of Watson–Crick and Hoogsteen hydrogen bonds, host **6** (in black) binds 9-ethyladenine (shown in color) in chloroform solution with an association constant $K_a = 1.1 \times 10^4$, as illustrated below. The naphthalene ring is important because it adds π-stacking interactions. This receptor is able to extract 1 equiv of adenine from water into chloroform, illustrating that appropriate recognition features can overcome the normal hydrogen bonds that form between water and a water-soluble guest molecule. Other nucleotide bases like cytosine and guanine do not bind to this receptor because the hydrogen-bond patterns are not complementary.

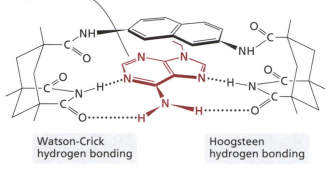

Watson-Crick
hydrogen bonding

Hoogsteen
hydrogen bonding

EXERCISE 28.12

Draw a structure to show how the following host can be used to recognize and bind thymine. Hydrogen bonds and π-stacking are both important. How would you prepare this receptor from 2,7-naphthalenediol, any ester, and any pyridine derivative?

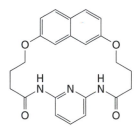

CHAPTER SUMMARY

Section 28.1   Nucleosides and nucleotides

- Nucleosides are molecules in which a heterocyclic base is attached to the anomeric carbon atom of D-ribose or 2-deoxy-D-ribose.

- Nucleotides are nucleosides that have phosphate groups attached to the OH group bonded to C5 of the carbohydrate.

- The heterocycles in nucleosides are derivatives of pyrimidine and purine, and are called adenine, cytosine, uracil, guanine, and thymine.

- For heterocycles that have an OH group attached to the carbon atom adjacent to a nitrogen atom, the keto form (a lactam) is generally more stable than the enol form (a lactim).

- A nucleoside is prepared in the laboratory by treating a heterocycle with a derivative of ribose (or 2-deoxyribose) that readily forms a cation intermediate; the mechanism is similar to that involved in acetal formation by carbohydrates, and the bond between nitrogen and the sugar is called the glycosidic bond.
- Nucleic acids are made by coupling the 3′-OH group of a nucleotide with the phosphate group attached to C5′ of another nucleotide.
- The biosynthesis of nucleic acids is an enzyme-catalyzed process that makes use of nucleotide triphosphate building blocks. An intermediate is formed that has a five-coordinate phosphorus atom, and a pyrophosphate ion is displaced from the intermediate to create the phosphodiester link.

Section 28.2   Nucleic acid structures

- The sequence of a nucleic acid is given by listing, in order, the nucleosides, starting at the end with the 5′-phosphate (or OH) group.
- The compounds DNA and RNA exist in several forms, many of which have two strands that associate by formation of hydrogen bonds between complementary pairs of heterocyclic bases. The nucleoside A forms hydrogen bonds with T (or U), and C forms hydrogen bonds with G.

Section 28.3   Molecular recognition

- Molecular recognition refers to the specific interactions between two or more molecules that result from noncovalent interactions such as hydrogen bonds and forces between ions, dipoles, and induced dipoles.
- Host–guest chemistry is aimed at understanding how one molecule (the host) binds another (the guest) to form a complex.

Section 28.4   Crown ethers and cryptands

- Crown ethers and cryptands are molecules that bind ions, using size and charge to recognize and to differentiate them.
- Crown ethers and cryptands are readily made by substitution reactions.
- Preorganization allows a host molecule to circumvent entropy changes during binding and recognition.
- Some antibiotics function in the same way as crown ethers, by binding ions according to size and charge.

Section 28.5   Cavity-containing molecules

- Cavity-containing molecules that are cyclic oligomers of benzene or carbohydrate derivatives can be used to bind nonionic guest molecules on the basis of size and dispersion forces.

Section 28.6   Recognition using hydrogen bonds

- Derivatives of urea and guanidine are used frequently in the construction of host molecules. These substances can form hydrogen bonds with guest molecules to differentiate them.
- Carboxylic acids are readily differentiated because they form strongly directional hydrogen bonds and anionic carboxylate ions.
- Nucleoside bases can be recognized on the basis of their unique hydrogen-bonding proclivities.
- Convergent functionality in a host molecule creates a site to which interacting groups, especially those that form hydrogen bonds, are directed, leading to specificity of structural recognition.

KEY TERMS

Introduction
phosphoester

Section 28.1a
nucleoside
nucleotide

Section 28.1b
pyrimidine nucleoside
purine nucleosides

Section 28.1d
glycosidic bond

Section 28.2c
B-DNA

Section 28.3
host–guest chemistry
receptor

Section 28.4a
crown ether

Section 28.4c
preorganization

Section 28.4d
ionophore

Section 28.5a
cyclodextrin

Section 28.5b
cyclophane

Section 28.6b
convergent functionality

REACTION SUMMARY

Section 28.1d

Nucleosides can be prepared from the reaction between a heterocycle and 1-*O*-acetyl-ribose derivatives with acid catalysis.

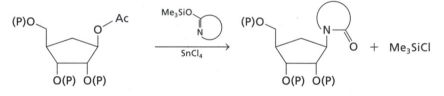

(P) = protecting group

Section 28.5a

Crown ethers and cryptands are commonly prepared by substitution reactions between alkoxide ions and alkyl dihalides.

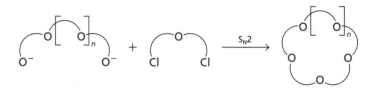

Azacrown ethers and cryptands can be made by the reduction of amide derivatives with $LiAlH_4$ or BH_3.

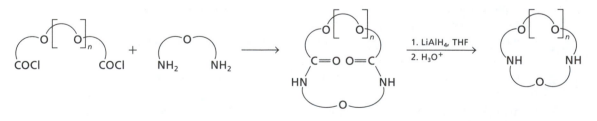

28.13. Draw a structure for each of the following compounds:

a. 6-Bromothymine   b. *N,N*-Dibenzylurea

c. 2′-Deoxycytidine-5′-monophosphate   d. 8-Fluoroadenosine-5′-triphosphate

e. 21-Crown-7   f. 2-Methoxyinosine

28.14. Give an acceptable name for each of the following molecules:

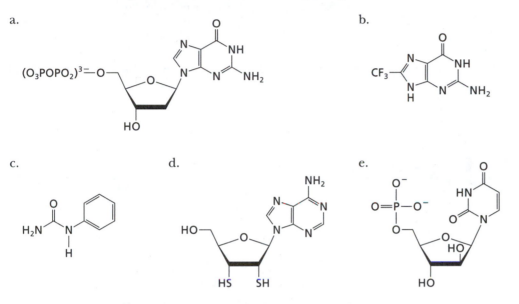

28.15. Draw the full structure for each of the following oligonucleotides, showing all of the atoms:

a. TGC   b. AUU   c. CCAT

28.16. The nucleoside inosine is a common one in RNAs because it can form complementary hydrogen bonds with A, C, and U. Illustrate the hydrogen bonds that can exist between these pairs of bases (I···A, I···C, and I···U).

28.17. Two minor bases found in RNA are pseudouridine and 1-methylinosine. Show how pseudouridine can form hydrogen bonds with adenosine.

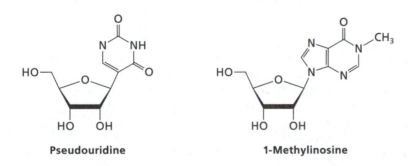

Pseudouridine   1-Methylinosine

28.18. 5-Bromouracil can be incorporated into DNA during replication. After inclusion, it can create mutations in subsequent generations because sometimes it pairs with G instead of A. The reason is related to the fact that a higher proportion of the lactim tautomer is present. Draw the structure of 5-bromouracil and its lactim tautomer. Illustrate how the lactam forms satisfactory hydrogen bonds with A but the lactim tautomer forms better hydrogen bonds with G.

28.19.   7,9-Dimethylguanine is a zwitterion that exists as a 1:1 mixture of the overall neutral and overall +1 (i.e., protonated) forms at physiological pH. As a result, the compound exists in solution as a dimeric species linked by three hydrogen bonds. Illustrate the three hydrogen bonds that can exist between the two different forms.

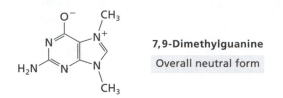

7,9-Dimethylguanine

Overall neutral form

28.20.   Show how you would prepare the following crown ethers from any achiral compounds that have six or fewer carbon atoms:

a.   b.   c.

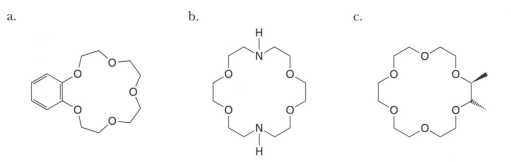

28.21.   Guanidinium carbonate condenses with dimethyl malonate to produce 4,6-dihydroxy-2-aminopyrimidine. Propose a reasonable mechanism for this reaction.

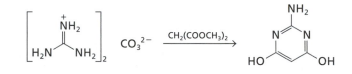

28.22.   Derivatives of guanine can be made by reactions that you have learned previously. Propose a mechanism for each step in the following sequence:

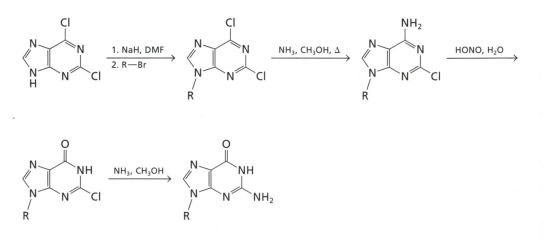

28.23.   Of the two receptors **X** and **Y**, shown below, which one is expected to bind the guest molecule 2-aminopyrimidine better? Show for each receptor how the guest molecule binds.

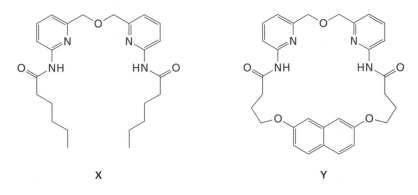

28.24.   Show how you would prepare receptor **X** in Exercise 28.23 from 2-bromo-6-methylpyridine and any other compounds, reagents, and solvents.

28.25.   Detection of creatinine can be used to diagnose conditions such as kidney failure. The following receptor, **A**, as a zwitterion, was designed to bind creatinine specifically. The host–guest complex is a different color from that of the receptor itself, and this feature provides a way to measure the concentration of the metabolite in blood serum. Draw the three possible zwitterionic forms of receptor **A**. Which zwitterion can bind creatinine by forming three hydrogen bonds, two to the creatinine amino group and one to its imine nitrogen atom?

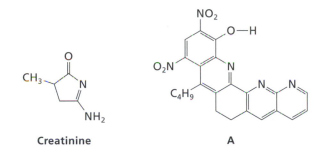

Creatinine   **A**

28.26.   The following two receptors bind dihydrogen phosphate ion in preference to hydrogen sulfate, nitrate, or perchlorate ions. Illustrate how these receptors use four hydrogen bonds to recognize the dihydrogen phosphate ion.

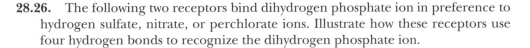

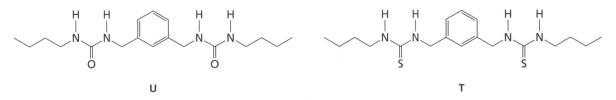

U   T

28.27.   Show how you would prepare receptor **U** in Exercise 28.26 from *m*-xylene and any other compounds with five or fewer carbon atoms. You may use any other reagents and solvents.

28.28. Host **A** binds acetate ion ~ 10 times better than *N,N*-dimethylurea (1,3-DMU). Illustrate how compound **A** and 1,3-DMU each bind to acetate ion.

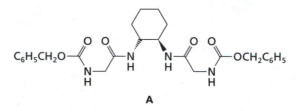

A

SUMMARY OF PREPARATIVE METHODS

This section summarizes the chemical reactions by which common functional groups can be prepared. A brief description about the reaction is given, and reference is made to the section number where the discussion of that reaction is given. Most difunctional molecules are listed under the functional group with the higher priority (Table 1.1). Many of the synthetic methods for nitrogen-, phosphorus-, and sulfur-containing molecules that are prepared by only a single method are listed in the Index. Name reactions are included in the following list in bold type.

Reaction	*Section Number*
Acetals	
Addition of alcohols to aldehydes and ketones	19.1b
Addition of alcohols to vinyl ethers	19.2b
Acid Chlorides	
Reaction of carboxylic acids with thionyl chloride	21.2c
Alcohols	
Addition of water to alkenes	9.1c
Enantioselective reduction of ketones	18.3b
Oxidative hydrolysis of organoboranes	9.4c
Hydrogenolysis of benzylic ethers	12.2a
Hydrolysis of esters	21.4c
Oxymercuration of alkenes followed by reduction with NaBH$_4$	9.1f
Reactions between organocuprate reagents and epoxides	15.3b
Reactions between Grignard reagents and epoxides	15.2c
Reactions between organometallic compounds and aldehydes or ketones	18.2c
Reactions between organometallic compounds and esters	21.7c
Reduction of aldehydes and ketones using metal hydride reagents	18.3a
Reduction of carboxylic acid derivatives using metal hydride reagents	21.8a
Ring-opening reactions of epoxides with nucleophiles	7.2e
Solvolysis of alkyl halides or alkyl sulfonate esters using water	6.2a
Substitution reactions of alkyl halides using hydroxide ion	6.3a
Aldehydes	
Alkylation of enamines followed by hydrolysis	20.3b
Oxidative hydrolysis or alkenylboranes	9.4c
Hydrolysis of acetals	19.1d
Hydrolysis of compounds with C=N bonds (imines, hydrazones, oximes)	20.1c
Hydrolysis of vinyl ethers	19.1c
Hydrolysis of thioacetals	19.2c
Oxidation of diols using NaIO$_4$	11.3d
Oxidation of alkenes using OsO$_4$ and NaIO$_4$ (**Lemeiux–Johnson cleavage**)	11.3d
Oxidation of primary alcohols with DMSO and oxalyl chloride (**Swern oxidation**)	11.4b

Reaction	Section Number
Aldehydes (*cont.*)	
Oxidation of primary alcohols with metal oxides	11.4c
Oxidation of primary alcohols with high-valent iodine compounds	11.4d
Ozonolysis of alkenes followed by reductive workup	11.3b
Reactions between organometallic compounds and formamide	21.7b
Reduction of acid chlorides with metal hydride reagents	21.8c
Alicyclic compounds	
Alkylation of an active methylene compound with a dihaloalkane	22.4b
Cationic cyclization reactions of alkenes	9.3d
Conjugate addition reactions followed by the aldol reaction (**Robinson annulation**)	24.3c
Cyclopropanation with carbenoid reagents	9.5a
Cycloaddition reactions between dienes and alkenes (**Diels–Alder reaction**)	10.4b
Free radical cyclization of an unsaturated alkyl halide	12.3c
Hydrogenation of aromatic compounds	17.1a
Intramolecular diester condensation (**Dieckmann cyclization**)	23.3c
Alkanes	
Desulfurization of sulfides and thiols using Raney nickel	12.2b
Hydrogenation of alkenes	11.2a
Hydrogenation of alkynes	11.2b
Hydrogenolysis of benzylic alcohols, ethers, and halides	12.2a
Protonolysis of organoboranes using carboxylic acids	9.4d
Protonolysis of organometallic compounds	15.2b
Reactions between organocuprates and organohalides	15.3b
Reduction of aldehydes and ketones (**Clemmensen reduction**)	18.3e
Reduction of aldehydes and ketones (**Wolff–Kishner reaction**)	20.1e
Reduction of aldehydes and ketones via thioacetal derivatives	19.2c
Reduction of alkyl halides using tributyltin hydride	12.2d
Alkenes	
Coupling of haloalkenes with organoboranes (**Suzuki coupling**)	15.3f
Dimerization reactions of alkenes	9.3a
Elimination of HX from alkyl halides (dehydrohalogenation)	8.2a
Elimination of water from alcohols (dehydration)	8.1a
Hydrogenation of alkynes using a poisoned catalyst	11.2b
Protonation of alkenylboranes using a carboxylic acid	9.4d
Reactions of organocuprates with haloalkenes	15.3b
Reactions of phosphorus ylides with aldehydes and ketones (**Wittig reaction**)	20.4b
Reduction of haloalkenes with tributyltin hydride	12.2d
Reduction of alkynes using Li or Na in liquid ammonia (dissolving metal reduction)	12.2c
Alkyl halides	
Cleavage of ethers using HBr, HI, or iodotrimethylsilane	7.2d
Electrophilic addition of HX to alkene π bonds	9.1b
Free radical addition of HBr to alkene π bonds using peroxides	12.3a
Free radical bromination of alkanes using NBS	12.1f
Free radical bromination and chlorination of alkanes using X_2	12.1e
Substitution of alcohol OH groups using HX	7.1a
Substitution of alcohol OH groups using PBr_3 or $SOCl_2$	7.1c
Aryl halides	
Halogenation of benzene and its derivatives	17.2b
Reaction of aryl diazonium ions with CuX and KX	17.4c
Alkynes	
Alkylation of alkyne carbanions with alkyl halides	15.2f
Elimination of HX from dihaloalkanes or haloalkenes (dehydrohalogenation)	8.2e

Reaction	*Section Number*
Allylic alcohols	
Reduction of α,β-unsaturated carbonyl compounds using metal hydride reagents	24.4a
Reduction of α,β-unsaturated carbonyl compounds via catalytic hydrogenation	24.4b
Amides (carboxamides)	
Alkylation of carboxamides	22.5a
Partial hydrolysis of nitriles	21.6
Reaction of amines with acid chlorides	21.3a
Reaction of amines with anhydrides	21.3b
Reaction of amines with esters	21.4b
Rearrangements of oximes (**Beckmann rearrangement**)	20.1d
Amines	
Alkylation of ammonia or amines with alkyl halides	6.3d
Aminoalkylation of carbonyl compounds (**Mannich reaction**)	24.1b
Hydrolysis reactions of amides	21.5a
Hydrolysis reactions of enamines	20.3b
Hydrolysis reactions of imines	20.1c
Hydrolysis of phthalimide derivatives (**Gabriel reaction**)	22.5c
Reaction of carboxamides with Br_2 and OH (**Hofmann rearrangement**)	22.5b
Reduction of alkyl azides	11.2c
Reduction of amides	21.8b
Reduction of imines	20.2a
Reduction of nitriles	11.2c
Reduction of nitro compounds	11.2c
Reduction of oximes	20.2a
Reductive amination of aldehydes and ketones with $NaBH_3CN$	20.2a
Amino acids	
Hydrogenation of enamides (enantioselective)	27.3d
Reaction between aldehydes and ammonium cyanide (**Strecker synthesis**)	27.2a
Arenes	
Alkylation of arenes (**Friedel–Crafts alkylation**)	17.2e
Coupling of aryl halides with organoboranes (**Suzuki reaction**)	15.3f
Coupling of aryl halides with organocuprate reagents	15.3b
Hydrogenolysis of benzylic alcohols, ethers, and halides	12.2a
Reduction of diazonium ions using hypophosphorus acid	17.4c
Carboxylic acids	
Alkylation of malonic ester followed by hydrolysis and decarboxylation	22.4b
Cleavage of methyl ketones with halogen and base (**Haloform reaction**)	22.2c
Decarboxylation of β-diacids	22.4c
Hydrogenation of α,β-unsaturated carboxylic acids (enantioselective)	16.4b
Hydrolysis of esters	21.4c
Hydrolysis of carboxamides	21.5a
Hydrolysis of nitriles	21.6
Oxidation of alkyl benzene derivatives	17.4a
Oxidation of primary alcohols	11.4c
Oxidation of aldehydes	18.4a
Ozonolysis of alkenes followed by oxidative workup	11.3b
Reaction of organometallic compounds with carbon dioxide	15.2b
β-Dicarbonyl compounds	
Alkylation of active methylene compounds	22.4b
Carboxylation of enolate ions	23.4b
Self-condensation of esters (**Claisen condensation**)	23.3a
Condensation reactions of esters with enolate ions (**crossed-Claisen condensation**)	23.4a
Conjugate addition reactions using active methylene compounds (**Michael reaction**)	24.3a

Reaction	Section Number
gem-Dihaloalkanes	
Addition of HX to alkynes	9.2a
Cyclopropanation with dichlorocarbene	9.5b
vic-Dihaloalkanes	
Addition of X_2 to alkenes	9.1d
Addition of X_2 to conjugated dienes	10.3a
1,2-Diols	
Reaction of alkenes with osmium tetroxide	11.3c
Reaction of alkenes with osmium tetroxide (enantioselective)	16.4e
Ring-opening reactions of epoxides using water	7.2e
Epoxides	
Addition–elimination reactions of sulfur ylides with aldehydes or ketones	20.4e
Reaction of alkenes with ketones and persulfate ion (enantioselective)	16.4d
Reaction of alkenes with NaOCl and Mn(III) compounds (enantioselective)	16.4d
Reaction of alkenes with peracids	11.3e
Reaction of vic-halohydrins with base	7.2c
Reaction of α,β-unsaturated ketones with hydrogen peroxide and base	24.2d
Esters	
Alkylation of esters via their enolate derivatives	22.3d
Conjugate addition of ester enolates with α,β-unsaturated carbonyl compounds	24.3b
Conjugate addition reactions of organocuprates with α,β-unsaturated esters	24.3d
Oxidation of ketones with peracids (**Baeyer–Villiger reaction**)	18.4b
Reaction of alcohols with carboxylic acids under **Mitsunobu reaction** conditions	7.1d
Reaction of carboxylic acids with alcohols (esterification)	21.2a
Reaction of carboxylic acids with diazomethane	21.4a
Reaction of esters with alcohols (transesterification)	21.4e
Reaction of acid chlorides with alcohols	21.3a
Ethers	
Dehydration of primary alcohols	8.2d
Substitution reactions between alkyl halides and alcohols (S_N1 reaction)	6.2a
Substitution reactions of alkyl halides with alkoxide ions (**Williamson synthesis**)	7.2b
Substitution reactions between nitrohalobenzenes and alkoxide ions	17.4e
Haloalkenes	
Addition of HX to alkynes	9.2a
Elimination of HX from gem- or vic-dihaloalkanes	8.2e
vic-Halohydrins	
Addition of X_2 and water to alkenes	9.1e
Reaction between epoxides and HX	7.2e
Ketones	
Acylation of benzene and its derivatives (**Friedel–Crafts acylation**)	17.2e
Addition of water to alkynes	9.2b
Alkylation of enamines followed by hydrolysis	20.3b
Alkylation of ketone enolate ions	22.3a
Alkylation of ketone enolate ions formed by conjugate reduction of enones	24.4d
Alkylation of ethyl acetoacetate followed by hydrolysis and decarboxylation	22.4b
Conjugate addition reactions of ketone enolates	24.3b
Conjugate addition reactions of nucleophiles with α,β-unsaturated ketones	24.2c
Conjugate addition reactions of organocuprates with α,β-unsaturated ketones	24.3d
Decarboxylation of β-keto acids	22.4c
Hydrolysis of acetals	19.1d
Hydrolysis of compounds with C=N bonds (imines, hydrazones, oximes)	20.1c
Hydrolysis of vinyl ethers	19.1c
Hydrolysis of thioacetals	19.2c

Reaction	Section Number
Ketones (cont.)	
Oxidation of diols using $NaIO_4$	11.3d
Oxidation of alkenes using OsO_4 and $NaIO_4$ (**Lemeiux–Johnson cleavage**)	11.3d
Oxidation of secondary alcohols with DMSO and oxalyl chloride (**Swern oxidation**)	11.4b
Oxidation of secondary alcohols with metal oxides	11.4c
Oxidation of secondary alcohols with high-valent iodine compounds	11.4d
Ozonolysis of alkenes followed by reductive or oxidative workup	11.3b
Reaction between organometallic compounds and acid chlorides	21.7a
Reaction between organometallic compounds and tertiary amides	21.7b
Reduction of α,β-unsaturated ketones	24.4c
Nitriles	
Addition of HCN to aldehydes and ketones (cyanohydrin formation)	18.2a
Conjugate addition reactions with acrylonitrile	24.2c
Conjugate addition of HCN to activated alkenes	24.2c
Dehydration of primary amides	21.6
Dehydration of oximes	20.1d
Substitution reactions between cyanide ion and alkyl halides	6.1c
Reaction of aryl diazonium compounds with CuCN and KCN	17.4c
Nitro compounds	
Nitration of benzene and its derivatives	17.2c
Oxidation of primary amines	11.5
Organoboranes	
Addition of borane and its derivatives to alkenes	9.4a
Addition of borane and its derivatives to alkynes	9.4b
Enantioselective hydroboration	16.4c
Organometallic compounds	
Acid–base reactions between terminal alkynes and RMgX or RLi	15.2f
Reaction of organolithium compounds with copper(I) salts	15.3a
Reaction of organohalides with lithium metal	15.2e
Reaction of organohalides with magnesium metal (**Grignard reagents**)	15.2a
Phenols	
Cleavage of phenol ethers using HBr, HI, or iodotrimethylsilane	7.2d
Reactions of aryl diazonium ions with water	17.4c
Sulfur and selenium compounds	
Phenylselenation of enolate ions	22.3b
Mitsunobu reaction between alcohols and SH containing compounds	7.1d
Substitution reactions between alkyl halides and alkylthiolate ions	7.3a
α,β-Unsaturated carbonyl compounds	
Co-condensation reactions of aldehydes and/or ketones (**aldol reaction**)	23.1a
Condensation reactions of ketones with aldehydes (**crossed aldol reaction**)	23.2a
Condensation reactions of enolates with aldehydes or ketones (**directed aldol reaction**)	23.2b
Elimination of PhSeOH from α-phenylselenyl carbonyl compounds	22.3b
Elimination of PhSeOH from compounds formed via conjugate addition procedures	24.3e
Elimination of amines from Mannich bases	24.1b
Horner–Emmons reaction of phosphonate ylides with aldehydes or ketones	20.4d

Definitions of the boldfaced terms appearing in the text are listed here followed in parenthesis by the section number in which the term first appears. The names of the common functional groups, specific named reagents, and name reactions (see Section 5.1a) have not been included here, but can be found in the Index. Terms that begin with a numeral, symbol, or Greek letter are listed according to the first letter of the first English word. For example, "1,2-addition" is listed under A before the other terms beginning with the word "addition," "4*n* + 2 rule" is listed under R as "rule," and "π anti-bond" is listed under A before other terms beginning with "antibond."

A

acetal   A functional group that has two OR groups attached to a single carbon atom (19.1b).

acetyl coenzyme A   A molecule used in biochemical systems to carry and transfer the acetyl group (21.3c).

achiral   A term used to describe a substance that is superimposable on its mirror image (4.2a).

active methylene compound   A compound that has a CH_2 group bonded to two electron-withdrawing groups (22.4a).

active site   The portion of an enzyme in which catalysis of a chemical reaction occurs (5.5c).

acyl group   A substituent of the form R–C=O that is attached via its carbonyl carbon atom (21.1b).

1,2-addition   A process in which addition occurs to a single π bond of a conjugated system (10.3a).

1,4-addition   *See* **conjugate addition** (10.3a).

addition polymerization   A process of forming polymers in which the reactive intermediate adds in each step to the π bond of a monomer molecule (26.2a).

addition reaction   A type of chemical transformation in which a π bond is broken and new sigma bonds are formed (5.1b).

addition–elimination reaction   A process that occurs by addition to a π bond followed by the loss of a group to regenerate a π bond (21.1c).

aldaric acid   A carbohydrate derivative in which the first and last carbon atoms in the chain are carboxylic acid groups (19.5b).

alditol   A carbohydrate derivative in which each carbon atom is attached to an OH group (19.5a).

aldonic acid   A carbohydrate derivative in which the original aldehyde group has been oxidized to form a carboxylic acid group (19.5b).

aldose   A carbohydrate that has the aldehyde group (19.3a).

alicyclic   A term indicating that a molecule has one or more rings, but lacks the conjugated π bonds that characterize an aromatic compound (1.3a).

aliphatic compounds   Molecules having hydrocarbon chains that usually contain portions of carbon atoms with four single bonds (1.3a).

alkane   An aliphatic hydrocarbon molecule that has only single bonds between carbon atoms (1.3a).

alkene   An aliphatic hydrocarbon molecule that has at least one C=C double bond (1.3b).

alkenyl carbon atom   The carbon atom in an alkene that forms a double bond with the adjacent carbon atom; also called a **vinyl carbon atom** (1.4c).

alkoxy group   A substituent that consists of an alkyl group attached to an oxygen atom (1.4b).

alkyl group   A substituent formed by removing an H atom from an alkane (1.4b).

alkyl sulfonate ester   A molecule of the form ROSO$_2$R' in which R' is an alkyl group and R is another organic group (7.1b).

alkyne   An aliphatic hydrocarbon molecule that has at least one C≡C triple bond (1.3b).

alkynyl carbon atom   The carbon atom in an alkyne that forms a triple bond with the adjacent carbon atom (1.4c).

alternating copolymer   A polymer derived from two different monomers that has the regular sequence —ABABABAB— (26.1b).

anionic polymerization   A process of forming polymers that is initiated by a species bearing a negative charge (26.2e).

annulenes   Cyclic molecules with alternating single and double bonds (17.1c).

anomers   Stereoisomers of the cyclic forms of a carbohydrate that differ in the configuration at the hemiacetal or acetal carbon atom (19.3b).

anomeric carbon atom   The hemiacetal or acetal carbon atom of a carbohydrate in its cyclic form (19.3b).

anti-**addition**   The reaction stereochemistry associated with adding two incoming groups to opposite faces of a C=C double bond; also called *trans*-**addition** (9.1d).

antiaromatic   An planar annulene that has 4, 8, 12 . . . (4n) π electrons (17.1c).

σ antibond   The mathematical cancellation of wave functions for orbitals oriented along the line between two nuclei (10.2a).

π antibond   The mathematical cancellation of wave functions for p orbitals that would correspond to a π bond (10.2c).

antibonding molecular orbital   The wave function formed by subtracting one atomic orbital wave function from another (10.2a).

antiperiplanar conformation   The staggered conformation of a compound in which the largest substituents attached to adjacent carbon atoms are as far apart as possible (3.1c).

Ar   The abbreviation for a substituent that is derived from an aromatic compound (1.3c).

arene   A synonym for an aromatic compound (1.3a).

aromatic compound   A planar annulene that has 2, 6, 10 . . . (4n + 2) π electrons (1.3a; 17.1b).

aromaticity   The property of certain annulenes that confers a special stability to the molecule (17.1b).

aryl carbon atom   Any carbon atom in the ring of an aromatic compound that contributes a p orbital to form the π system (1.4c).

aryne   A reactive derivative of an aromatic compound that has a triple bond within the ring (17.4f).

asymmetric carbon atom   *See* **chiral carbon atom** (4.2a).

asymmetric reaction   A reaction that produces an enantio-enriched product from an achiral starting material, usually with involvement of a chiral reagent or catalyst (16.1c).

atactic   A polymer with chiral carbon atoms that are randomly oriented along the polymer chain (26.1d).

atom economy   A term referring to the amount of the reactant atoms that appear in the product molecule(s) (15.5a).

axial bonds   The bonds in a cyclohexane derivative that point straight up or down when the ring is drawn in its chair form (3.2b).

azoles   Five-membered ring heterocycles that contain two heteroatoms, one of which is nitrogen (25.4a).

B

Baeyer strain   The energy stored in three- and four-membered rings that results from compressing the tetrahedral bond angles below their normal ranges (3.2a).

base peak   The peak in a mass spectrum with the highest relative intensity value (14.2a).

B-DNA The common double-stranded form of DNA (28.2c).

benzyne A reactive intermediate, C_6H_4, that has a triple bond between two adjacent carbon atoms of the benzene ring (17.4f).

bimolecular A term describing the kinetic profile of a reaction in which the rate is dependent on the concentration of two reactants (6.1d).

biotin A coenzyme involved in carboxylation reactions (23.4c).

block copolymer A polymer made of two or more monomers in which the repeating units appear in grouped sequences: —AAAAAAABBBBBBBAAAABBBBB— (26.1b).

boat conformation A term used to describe the conformation of cyclohexane and its derivatives in which two opposite vertices of the six-membered ring are bent toward each other (3.2c).

bond dissociation energy The energy required for a bond to undergo homolysis (12.1b).

bonding molecular orbital The wave function formed by the additive combination of one atomic orbital wave function with another (10.2a).

branched polymers A polymer in which the backbone chain has other carbon atom chains attached (26.2c).

bridged bicyclic compound A molecule with two rings in which each ring shares at least three atoms (3.4c).

bridgehead carbon atom The atoms in a bridged bicyclic compound that are connected by the carbon atom chains creating the rings (3.4c).

broad-band decoupling A technique used to record ^{13}C NMR spectra in which a second radio frequency is used to saturate the nuclei of the protons; the resulting spectrum has only singlet peaks that correspond to each carbon atom type (13.5c).

Brønsted–Lowry acid A substance that is a proton donor (5.2a).

Brønsted–Lowry base A substance that is a proton acceptor (5.2a).

Buckminsterfullerene A closed sphere module composed only of carbon atoms; this term usually refers specifically to the substance C_{60} (25.1d).

C

Cahn–Ingold–Prelog convention The system by which atoms and groups are assigned priorities that are subsequently used to assign the configurations of alkene double bonds and stereogenic centers (4.1b).

carbamate A functional group with the general formula RNH-CO-OR' (26.3d).

carbanion A negatively charged species having a carbon atom with three attached groups and an unshared pair of electrons (2.5e).

carbene A neutral species having a carbon atom with two attached groups and an unshared pair of electrons (2.5e).

carbocation A positively charged species having a carbon atom with three attached groups and no other electrons on the carbon atom (2.5e).

carbohydrate Any of a number of polyhydroxy aldehydes or ketones with the general formula $C_n(H_2O)_n$ (19.3a).

carbonyl group The molecular fragment comprising the C=O double bond (1.3c).

carbonyl transposition A process that results in the movement of a carbonyl group from one position to the adjacent position (22.2b).

carboxylation A chemical process in which carbon dioxide is incorporated into a molecule's structure (23.4b).

catalyst A substance that increases the rate of a chemical reaction, but is not consumed during the process (5.1a).

cationic polymerization A process of forming polymers that is initiated by a species bearing a positive charge (26.2b).

chain growth polymerization *See* **addition polymerization** (26.2a).

chair conformation A term used to describe the lowest energy structure of cyclohexane and its derivatives (3.2b).

chemical shift In NMR spectroscopy, the frequency of electromagnetic radiation that causes resonance of the nuclear spin with the precession frequency created by an external magnetic field. This quantity is expressed as an arithmetic difference with respect to the frequency of the protons or carbon atoms of tetramethylsilane (13.1c).

chemoselectivity The discrimination displayed by a specific reagent in its reactions with different functional groups (11.2a).

chiral The property of a substance that makes it not superimposable on its mirror image (4.2a).

chiral auxiliary A compound used to react with either a reagent or substrate to make it chiral, enabling an enantioselective reaction to occur (27.3a).

chiral carbon atom A carbon atom with four different groups attached—also called an **asymmetric carbon atom** or a **stereogenic carbon atom** (4.2a).

chiral pool The collection of chiral substances occurring in Nature that provide starting materials for the preparation of chiral molecules without resorting to resolution or enantioselective reactions (16.1a).

cis isomer The structural form of an alicyclic compound in which two specified groups are on the same side of ring (3.3d). Also, the structural form of a C=C double bond in which two protons—each attached to the separate carbon atoms—are on the same side of the bond connecting the carbon atoms (4.1a).

***cis*-addition** The reaction stereochemistry associated with adding two incoming groups to the same face of a C=C double bond; also called *syn*-addition (9.4b).

citric acid cycle The series of chemical reactions that take place in many living organisms during aerobic metabolism; also called the **Krebs cycle** (11.4e).

codon The triplet sequences of DNA that encode the amino acid sequences of proteins (27.1a).

coenzyme An organic molecule that is required along with an enzyme to carry out a biochemical transformation (11.4e).

coenzyme A The coenzyme employed in biochemical systems to transfer, biosynthesize, and metabolize acyl groups (21.3c).

concerted A term describing chemical processes in which bonds are made and broken simultaneously as reactants are converted to products (6.3a).

condensation polymer A polymer formed by reactions between two functional groups, which is accompanied by the expulsion of another molecule, usually water (26.3a).

condensed structure A shorthand notation for structural formulas that makes use of lines, angles, and polygons to indicate the carbon atom positions (1.2a).

configuration The three-dimensional arrangement of groups attached to a stereogenic center (4.2a).

conformation The instantaneous orientation of the atoms in a molecule that results from rotations about carbon–carbon σ bonds (3.1a).

conformer Any specific spatial arrangement at a given instant for a molecule that can exist in multiple orientations related by rotations about C–C σ bonds (3.1a).

conjugate acid The acid, HA, that corresponds to a given base, A⁻ (5.2a).

conjugate addition A process whereby a species reacts at the carbon atom on the end of a four-atom unit with the general structure C=C–C=O or C=C–C≡N; also called **1,4-additon** or **Michael addition** (Introduction to Chapter 23).

conjugate base The base, A⁻, that corresponds to a given acid, HA (5.2a).

conjugated double bond A structural unit comprising the C=C double bond separated by a single bond from another π-bonded pair of atoms (10.1b).

constitutional isomers Compounds with the same molecular formulas that differ in the connectivities between their atoms (1.4a).

convergent functionality Two or more functional groups that are oriented toward a common point in space, usually a cavity or cleft within a molecule's structure (28.6b).

copolymer A polymer that is formed from the reactions between at least two different molecules (26.1b).

coupling constant, J The value in hertz (Hz) that indicates the magnitude of the proton spin interactions between neighboring protons in an NMR spectrum (13.2a).

cross-link A group that connects two different molecular chains within the overall structure of a protein or polymeric material (20.2b).

crown ether A cyclic polyether molecule with a shape resembling a crown when it is viewed from the side (28.4a).

C-terminus The end of a protein chain that has the unlinked –COOH group (5.5b).

cumulated double bond A structural unit comprising the C=C double bond attached directly to another π-bonded pair of atoms (10.1b).

cyanohydrin A molecule (or functional group) having an OH and a CN group attached to the same carbon atom (18.2a).

[4+2] cycloaddition A reaction between a reactant with four electrons and one with two electrons that results in formation of a five- or six-membered ring (10.4b).

cyclodextrin A derivative of D-glucose derived from starch that has six, seven, or eight carbohydrate rings connected head-to-tail to create a cylindrical shape (28.5a).

cyclophane Cyclic molecules comprising two or more aromatic rings linked by sp^3-hybridized carbon atoms or chains (28.5b).

cyclopropanation The process whereby an alkene is converted to a three-membered cycloalkane (9.5a).

D

dehydration The process whereby water is removed from a molecule (8.1a).

dehydrohalogenation The process whereby HX (X = F, Cl, Br, or I) is removed from a molecule (8.2a).

delocalized A term indicating that electrons are distributed through π bonds among three or more atoms (2.7a).

denaturation The process by which the higher levels of protein structures are disrupted (27.5b).

DEPT ^{13}C NMR spectra DEPT stands for distortionless enhancement by polarization transfer, and these types of ^{13}C NMR spectra are used to obtain data about the presence of CH_3, CH_2, and CH groups in a molecule (13.5d).

deshielded The phenomenon in NMR spectroscopy indicating that the external magnetic field is not moderated by the electrons surrounding a particular nucleus as much as this same field might be offset by the electrons surrounding other nuclei in the molecule (13.1c).

deuterium exchange A process in which an acidic proton, usually one bonded to an oxygen atom, is replaced with deuterium by treating a compound with D_2O; used to detect OH protons by looking at which absorption disappears in a proton NMR spectrum when D_2O is added to the analyte (13.3).

dextrorotatory The property of a compound that rotates plane-polarized light in the clockwise direction (4.2b).

diastereomer One of any pair of stereoisomers that are not enantiomers; the plural (diastereomers) refers to any two compounds that are stereoisomers and not enantiomers (4.3a).

diastereotopic A term for two like atoms or groups indicating that the replacement of each of the groups separately will generate diastereomers (16.3b).

1,3-diaxial interaction The through-space contact that occurs between axial groups attached to alternate carbon atoms in the chair form of cyclohexane derivatives (3.3a)

diazonium compound A molecule having the N_2^+ group attached to a carbon atom (17.4b).

dienophile An alkene, usually with an electron-withdrawing substituent, that reacts readily with a conjugated diene in the Diels–Alder reaction (10.4b).

dihedral angle   When sighting along the bond between two atoms, the angle formed between two substituents attached to adjacent atoms; also called the **torsional angle** (3.1b).

dipole   The polarity difference that results from the unequal sharing of the electrons that constitute a bond (2.2a).

downfield   A term indicating a feature or region in an NMR spectrum that lies to the left of or has a larger chemical shift value than that of another feature or region (13.1c).

E

eclipsed conformation   A term describing the relationships between groups on adjacent saturated carbon atoms in which the groups are aligned (3.1b).

elastomer   A type of plastic that stretches without breaking and returns to its original shape and size when the stretching force is released (26.1c).

electronegativity   The property of an atom to attract electrons to itself (2.2a).

electrophile   Any species that is electron deficient (5.3b).

electrophilic addition   A mode of addition to π bonds that begins with the reaction of the π electrons with an electrophile (9.1a).

electrophilic aromatic substitution   The process by which aromatic compounds undergo substitution in which the first step is the reaction between the ring and an electrophile, and the second step is regeneration of the aromatic π system (17.2a).

electrostatic forces   The attractive or repulsive influences that occur between species with full or partial positive and negative charges (2.8a).

β-elimination   A term describing the process in which groups are removed from adjacent carton atoms to form a π bond (8.2a).

α-elimination   A term describing the process in which two groups are removed or lost from the same carton atom to form a carbene species (9.5b).

elimination reaction   A type of chemical transformation that leads to formation of a π bond (5.1b).

enamine   A molecule that has an amino group attached to a C=C double bond (20.3a).

enantiomer   A stereoisomer that is nonsuperimposable on its mirror image (4.2a).

enantiomeric excess, ee   A numeric value that quantifies the efficiency of an enantioselective reaction and is defined as the proportion of the major enantiomer minus that of the minor enantiomer, expressed as a percentage (16.3c).

enantiomeric resolution   A process used to obtain a pure enantiomer from a racemic mixture that makes use of separation after forming diastereomeric derivatives (16.1c).

enantioselective   A term used to describe a reaction that produces an excess of one enantiomer relative to formation of its mirror image (9.1d).

enantiospecific   A term used to describe a reaction that produces one enantiomer exclusively (9.1d).

enantiotopic   A term for two like atoms or groups indicating that the replacement of each of the groups separately will generate enantiomers (16.3b).

endergonic   A term describing the free energy of a process in which the products are less stable than the reactants (5.4a).

endocyclic   A term meaning that a group or bond is inside a ring (8.1c).

enediol   A substructure with an OH group attached to each carbon atom of a C=C double bond (22.2b).

enol ether   A type of molecule having an OR group attached to a C=C double bond (19.1c).

enol   A type of molecule having an OH group attached to a C=C double bond (22.2a).

enolate ion   The conjugate base of an enol (22.1d).

enone   A type of molecule having a ketone functional group conjugated with a C=C double bond (24.1b).

enzyme   Normally a protein molecule that catalyzes a chemical reaction by binding the reactant molecules and stabilizing the transition state of the reaction (5.5a).

epimerization   A process in which the configuration of a single chiral center is inverted (19.3c).

epimers   Stereoisomers with two or more stereogenic centers that differ in their configurations at only a single center (19.3c).

equatorial bonds   The bonds in the chair forms of cyclohexane derivatives that point away from the center of the ring and lie roughly in the plane of the ring (3.2b).

esterification   The process of forming esters from carboxylic acids and alcohols (21.2a).

exergonic   A term describing the free energy of a process in which the products are more stable than the reactants (5.4a).

exocyclic   A term meaning that a group or bond is outside of a ring (8.1c).

F

fingerprint region   The portion of an IR spectrum that allows one to confirm the suspected structure of a molecule by the unambiguous match with absorption bands of known compounds (14.3b).

Fischer projection   The structural representation of an aliphatic compound in which the molecule's three-dimensional structure appears to have been flattened onto the plane of a page (4.4).

formal charge   The electronic charge calculated for an atom that accounts for the fact that its valence electrons have been redistributed in order to give neighboring atoms an octet of electrons (2.1b).

fragmentation   A process that results in the breaking of bonds when a molecule is subjected to the removal of electrons in the vacuum chamber of a mass spectrometer (14.2a).

free energy of activation   The amount of energy that is required in order for reactants to come together to form products in a chemical process (5.4a).

functional group equivalents   In a retrosynthetic analysis, the alternate functional groups that can be readily converted to the actual functional group that appears in the product of a synthetic procedure (15.4b).

functional group region   The portion of an IR spectrum in which generic absorptions associated with each functional group appear (14.3b).

functional group   A group of atoms, other than those connected by C–C and C–H single bonds, that define the reactive portions of a molecule (1.1b).

functional isomers   Molecules with identical formulas that differ by the identities of the functional groups that are present (1.4a).

furanose   The five-membered cyclic hemiacetal form of a carbohydrate formed by the addition of an OH group to the sugar's carbonyl group (19.3b).

furanoside   The five-membered cyclic form of a carbohydrate acetal (glycoside) formed by the addition of an OH group to the sugar's carbonyl group followed by substitution of the anomeric OH group with another heteroatom-containing substituent (19.4a).

fused bicyclic compound   A compound with two rings that share a common bond (3.4c).

G

gauche conformation   The orientation between groups on adjacent carbon atoms in which the dihedral angle between them is 60° (3.1c).

geminal   A term denoting that substituents are attached to the same atom (8.2e).

geminal coupling   In proton NMR spectroscopy, the nuclear spin–spin interactions that occur between protons attached to the same carbon atom (13.2a).

geminal diol   A molecule in which a carbon atom bears two OH group; also called a **hydrate** (18.2b).

genetic code   The correlation between DNA codons and the amino acids that each triplet sequence of nucleotides represents (27.1a).

geometric isomer   One of a pair of molecules with the identical molecular formula that differs from the other member of the pair by the spatial relationship of substituents attached to a C=C double bond or alicyclic ring (4.1a).

Gillman reagent   An organocopper compound with the general formula LiR_2Cu (15.3a).

glycolysis   The metabolic process by which carbohydrates are broken down into smaller molecules to provide energy for an organism (23.2c).

glycoside   The acetal derivative of a carbohydrate (19.4a).

glycosidic bond   The exocyclic bond between the anomeric carbon atom of a cyclic carbohydrate derivative and an OR group or heterocycle (28.1d).

graft copolymer   A polymer derived from different monomers in which chains generated from one monomer have been attached to the chain prepared from the other monomer (26.1b).

Grignard reagent   An organomagnesium compound with the general formula RMgX, where X is a halogen atom (15.2a).

H

halonium ion   A reactive intermediate comprising a three-membered ring with one halogen and two carbon atoms that bears a positive charge on the halogen atom (9.1d).

Hammond postulate   A theory suggesting that the structure of a transition state should resemble the species (reactant, intermediate, or product) to which it is closest in energy (6.2c).

Haworth projection   A three-dimensional representation of cyclic carbohydrate structures in which the viewer is looking from slightly above at the front edge of a five- or six-membered ring (19.3b).

α-helix   A regular conformation within protein structures formed by coiling of the polypeptide chain (27.5b).

heme   The iron derivative of protoporphyrin IX that functions to bind molecular oxygen in proteins such as myoglobin and hemoglobin (25.3a).

hemiacetal   A functional group in which an OH and an OR group are attached to the same carbon atom (19.1a).

hemiaminal   A functional group in which an OH and an amino group are attached to the same carbon atom (20.1a).

Henderson–Hasselbalch equation   A mathematical relationship that relates the degree of protonation with the relative acid strength of a functional group and the pH of the solution in which the particular molecule is dissolved (27.1b).

heteroatom   A term referring to any element besides carbon and hydrogen (1.1b).

heterocycles   A term referring to molecules that exist as rings and contain at least one heteroatom (17.1e).

heterolysis   A bond-breaking process that occurs by movement of the bonding electron pair to one of the atoms and results in the formation of a nucleophile and an electrophile (5.1c).

higher order cuprate   An organocuprate reagent of the general formula $Li_2Cu(CN)R_2$, which is made by treating CuCN with an organolithium compound (15.3a).

high-resolution mass spectrometry   An instrumental technique used to separate and detect ions that differ in mass by only thousandths or ten-thousandths of an amu (14.2b).

homolysis   A bond-breaking process that occurs by movement of one electron to each of the bonded atoms and results in the formation of two free radicals (5.1c).

homopolymer   A polymer that is prepared from a single monomer (26.1b).

homotropic   A term referring to three atoms or groups attached to a given carbon atom; the replacement of each of the atoms or groups separately will generate the same molecule (16.3b).

host–guest chemistry   The research field that focuses on understanding how a receptor can interact with another molecule in the absence of covalent bond formation (28.3).

Hückel's rule   The statement that a molecule is aromatic when it has a planar, uninterrupted, and cyclic π system that contains $4n + 2$ electrons, where $n = 0, 1, 2, 3, 4$, and so on. (17.1b).

hybrid orbital   A quantum mechanical concept that explains molecular shapes by the formation of bonds created from atomic orbitals that have been mixed with other atomic orbitals of similar energies (2.5a).

hydrate   *See* **geminal diol** (18.2b).

hydration   A reaction in which H_2O adds to a π bond (9.1c).

hydrazone   A molecule with the C=N–NH$_2$ substructure, which is formed by the reaction between hydrazine and an aldehyde or ketone (20.1b).

hydroboration   A reaction in which a boron atom and a hydrogen atom add to a π bond (9.4a).

hydrocarbon   Any molecule that contains only C and H (1.2a).

hydrogen-bond acceptor   The component in a hydrogen bond that provides the unshared electron pair to attract the hydrogen atom (2.8b).

hydrogen-bond donor   The group, usually OH or NH, in a hydrogen bond that provides the hydrogen atom (2.8b).

hydrogen bond   A non-covalent interaction that results when an O–H or N–H group is close enough to a heteroatom, usually O or N, that an attraction exists between the unshared electron pair on the heteroatom and the nucleus of the hydrogen atom (2.8b).

hydrogenolysis   A process that uses H_2 to cleave a sigma bond (12.2a).

hyperconjugation   A phenomenon stating that electron donation from adjacent σ bonds to the vacant p orbital of a carbocation stabilizes the electron-deficient carbon center (6.2b).

I

imine   A derivative of an aldehyde or ketone in which the oxygen atom of the original carbonyl group has been replaced by the RN= group (20.1a).

inductive effect   A term referring to electron-donating or -withdrawing influences that operate through sigma bonds and result from the electronegativity differences between bonded atoms (5.2c).

initiation   The first step of a radical chain reaction that generates the free radicals (12.1a).

integrated intensity values   In a proton NMR spectrum, the ratios of the relative numbers of protons that create the absorption signals (13.1d).

intermolecular   A term referring to any influences that operate between molecules (2.8b).

intramolecular   A term referring to any influences that operate within a molecule (2.8).

inversion of configuration   A process in which the configuration of an atom is transformed into its mirror-image structure (6.2f).

ionophore   Any type of polyether molecule that is used to transport metal ions through cell membranes (28.4d).

isocyanate   A functional group with the –N=C=O structure (26.3d).

isolated double bond   A C=C double bond that is neither conjugated nor cumulated with another C=C double bond (10.1b).

isomers   A term referring to any set of molecules with the same molecular formulas and differing in the connectivities or spatial arrangements of the constituent atoms (1.4a).

isotactic   A term for the stereochemistry of the stereogenic centers in a polymer in which all of the substituents are on the same side of a chain viewed in its zigzag conformation (26.1d).

isotope peaks   The normally smaller lines in a mass spectrum that result from the masses of the minor isotopes of the constituent atoms (14.2a).

IUPAC rules   The system devised by the International Union of Pure and Applied Chemistry that specifies the names of chemical substances (1.3a).

J

J **value**   *See* **coupling constant,** *J* (13.2a).

K

ketose   A carbohydrate that contains the ketone functional group (19.3a).

kinetic enolate   Of the two enolate ions that can be formed from an unsymmetrical ketone, the enolate ion that is formed more rapidly (22.2f).

kinetic resolution   A process in which only one enantiomer of a racemic mixture reacts preferentially with a chiral catalyst or reagent (16.3d).

L

lactone   A cyclic ester, which is formed when a molecule contains both carboxylic acid and alcohol groups (21.2b).

Le Chatelier's principle   The postulate stating that the position of an equilibrium can be shifted by varying the concentrations of either reactants or products (5.4b).

leaving group   A group that is replaced by another in a polar substitution reaction (6.1a).

levorotatory   The property of a compound that rotates plane-polarized light in the counterclockwise direction (4.2b).

Lewis acid   A substance that is an electron pair acceptor (5.2g).

Lewis base   A substance that is an electron pair donor (5.2g).

Lewis structure   Line structures that also include all of the unshared electron pairs (2.1a).

localized   A term describing the situation in which electrons are associated with a specific atom or are shared only by a pair of adjacent atoms (2.3a).

M

magnetic anisotropy   A term used to describe the effects caused by localized magnetic fields that are created by the bonding electrons in a molecule, especially those in π bonds (13.1c).

magnetically equivalent   A term referring to atoms or groups that experience identical influences when placed in an external magnetic field (13.1b).

Mannich base   A type of molecule formed by the reaction between a ketone, an aldehyde, and a secondary amine in which the amino group is bonded β to the ketone carbonyl group (24.1b).

Markovnikov addition   A term that describes the orientation of addition between alkenes and polar reagents in which the more electrophilic portion of the reagent becomes attached to the less highly substituted carbon atom of the original C=C double bond (9.1b).

Markovnikov's rule   The statement that explains how polar reagents are oriented when they undergo addition to an alkene (9.1b).

McLafferty rearrangement   A type of fragmentation process in mass spectrometry involving a cyclic transition state (14.2d).

mechanism   A term referring either to the set of molecular events that results in the observed conversion of reactants to products or to the process by which one rationalizes the movement of electrons during the conversion of reactants to products (5.3a).

meso **compound**   A molecule with two or more stereogenic centers that has an internal mirror plane such that the overall molecule is achiral (4.3b).

mesylate   An abbreviated name for the methanesulfonate group, CH_3SO_3— (7.1b).

meta   A term denoting the relationship between groups attached to a 1,3-disubstituted benzene ring (17.1f).

Michael acceptor   A type of molecule that undergoes conjugate addition reactions (24.3a).

migratory insertion   A type of addition reaction that takes place with organotransition metal complexes in which one group attached to a metal ion adds to an unsaturated molecule attached to the same metal ion (15.3c).

molecular ion   The species in mass spectrometry that corresponds to a molecule that has lost a single electron (14.2a).

molecular orbital theory   A simplification of quantum theory that rationalizes bond formation by specifying the electronic structure of a group of atoms as a single entity (10.2a).

molecular sieves   An aluminosilicate material containing small channels that only water or other small molecules can enter (19.1d).

monomer   The smallest repeating unit of a polymer; the molecule from which a polymer is made (26.1a).

multiplicity   A term used to describe the appearance of absorption peaks in a proton NMR spectrum that result from spin coupling between neighboring protons (13.2a).

N

$n + 1$ rule   A statement of the relationship between the number (n) of neighboring protons that a specific proton experiences and the number of peaks that are observed in the NMR spectrum for that specified proton (13.2a).

Newman projection   A term used to describe the representation of a C–C bond and its attached substituents with the view oriented to look along that bond from one end (3.1b).

nitronium ion   An electrophilic species with the formula NO_2^+ (17.2c).

node   The portion of an atomic or molecular orbital in which the probability of finding an electron is mathematically zero (2.4a).

N-terminus   The end of a protein chain that has the unlinked $-NH_2$ group (5.5b).

nucleophile   A species with at least one unshared pair of electrons that tends to react with electron-deficient species (5.3b).

nucleoside   A building block of RNA and DNA constructed from aromatic heterocycles, called bases, attached at the anomeric carbon atom of either D-ribose or D-2-dexoyribose (28.1a).

nucleotide   Derivatives of nucleosides with a phosphate group attached to C5 of the D-ribose or D-2-dexoyribose ring (28.1a).

O

off-resonance decoupling   A technique in ^{13}C NMR spectroscopy that makes use of a second radio frequency to provide information about the number of protons attached to a given carbon atom (13.5d).

olefin   A synonym for an alkene (Introduction to Chapter 8)

optically active   A term referring to molecules that rotate the angle of plane-polarized light passing through their solutions (4.2b).

organoborane   A derivative of borane, BH_3, in which a carbon-containing group substitutes for one or more of the hydrogen atoms (9.4a).

organolithium compound   An compound with the general formula RLi (15.2e).

ortho coupling   A term referring to the magnetic interaction that occurs between benzene protons attached to adjacent positions of the ring (17.1g).

ortho effect   The influence that causes more pronounced chemical shifts in a proton NMR spectrum for the protons attached to a benzene ring adjacent to a substituent (17.3g).

ortho   A term denoting the relationship between groups attached to a 1,2-disubstituted benzene ring (17.1f).

oxazaborolidine   A class of reagents used to carry out the enantioselective reduction reactions of ketones to form alcohols (18.3b).

oxidation level   A term referring to the number of heteroatoms attached to a carbon atom (11.1a).

oxidation reaction   A chemical process in which hydrogen atoms are removed or heteroatoms are added to an organic molecule during the course of the reaction (5.1b).

oxidative addition   A process for organotransition metal complexes in which the reaction of a molecule with a metal ion leads to an increase in the metal ion's coordination number as its oxidation state also increases (15.3c).

oxime   A functional group that has the C=N–OH substructure (20.1b).

oxymercuration   The process that results in addition of the $Hg(O_2CCH_3)$ and OH groups to an alkene π bond (9.1f).

ozonide   The product formed after treating an alkene with ozone (11.3b).

P

para   A term denoting the relationship between groups attached to a 1,4-disubstituted benzene ring (17.1f).

peptide bond   The carbonyl–nitrogen atom bond in the amide functional group formed from the reaction between two amino acids (5.5b).

pericyclic reaction   A reaction process that occurs by the movement of electrons within a cyclic transition state (5.1c).

phosphoester   A derivative of phosphoric acid $[O=P(OH)_3]$ in which at least one OH group has been replaced by an OR group (Introduction to Chapter 28)

pi (π) bond   The interaction that occurs between aligned p orbitals on neighboring atoms (2.5c).

pK_a value   A quantity that provides a measure of how strong an acid is; mathematically, it is equal to $-\log(K_a)$, where K_a is the equilibrium constant of the reaction between the acidic compound and water (5.2a).

plane-polarized light   Electromagnetic radiation that has been passed through a filter that removes the waves from all but a single plane (4.2b).

plastic   A material that can be molded when hot, yet retains its shape after cooling (26.1c).

poisoned catalyst   A dispersed metal-containing material that has been treated with other chemical compounds to decrease its normal reactivity, usually toward the hydrogenation reactions of π bonds (11.2b).

polar aprotic solvent   A term referring to the liquid molecules used to dissolve reactants in a chemical reaction that have polar bonds yet lack acidic (OH and NH) groups (6.3c).

polar reaction   The classification of a chemical reaction in which the bond to be broken is polarized so that one atom assumes a positive, or partial positive, charge, and the other assumes a negative, or partial negative, charge (5.1c).

polarizable   A term referring to an atom or molecule in which the distribution of its electrons is readily distorted or deformed by outside influences (2.2a).

polymer   A molecule formed by linking together a large number of smaller molecules (26.1a).

polypeptide   A molecule that has been constructed by linking many amino acids together via formation of amide functional groups (5.5b).

positional isomers   Compounds with the same molecular formulas that differ in the position at which a particular substituent is attached in each (1.4a).

preorganization   A property of receptor molecules in which their conformations change little as a guest molecule is bound (28.4c).

primary carbon atom   A term referring to a carbon atom that is attached to only one other carbon atom (1.4a).

primary structure   A term referring to the sequence of amino acids in a protein (27.4d) or nucleotides in a nucleic acid (28.2a).

pro-(R)　　The atom, group, or π-bond face in a prochiral molecule that will react to form the enantiomer having the (R) configuration at the new chiral center (16.3b).

pro-(S)　　The atom, group, or π-bond face in a prochiral molecule that will react to form the enantiomer having the (S) configuration at the new chiral center (16.3b).

prochiral　　A term referring to a molecule that can be converted to a chiral substance with a single substitution or addition reaction (16.3b).

product favored　　A term referring to a chemical process in which the products have a lower energy state than the reactants (5.4a).

products　　The molecules in a chemical reaction that are formed from other substances, which are called reactants (5.1a).

propagation　　The steps in a free-radical process that perpetuate the chain reaction (12.1a).

protecting groups　　A molecular fragment added to a functional group to keep the functional group from reacting with certain reagents (15.4a).

proton-transfer reaction　　A chemical process that results in the movement of H^+ from one atom to another (5.1b).

protonolysis　　A chemical process that results in the breaking of a bond by the reaction of a molecule with H^+ (9.4d).

purine nucleosides　　A term that refers to the nucleosides adenosine, guanosine, and inosine (28.1b).

pyranose　　The six-membered cyclic hemiacetal form of a carbohydrate formed by the addition of an OH group to the sugar's carbonyl group (19.3b).

pyranoside　　The six-membered cyclic form of a carbohydrate acetal (glycoside) formed by the addition of an OH group to the sugar's carbonyl group followed by substitution of the anomeric OH group with another heteroatom-containing substituent (19.4a).

pyrimidine nucleoside　　A term that refers to the nucleosides cytidine, uridine, and thymidine (28.1b).

Q

quaternary carbon atom　　A carbon atom that is attached to four other carbon atoms (1.4a).

quaternary structure　　The highest level of protein structure, which occurs when a protein consists of multiple subunits that can interact with one another (27.5b).

R

R　　The abbreviation for a substituent that is derived from an aliphatic compound (1.3c).

racemate　　A term referring to an equal mixture of enantiomers (4.2b).

racemic mixture　　*See* **racemate** (4.2b).

racemization　　A process that leads to the formation of an equal mixture of enantiomers from a single chiral substance (6.2f).

radical polymerization　　A process of forming polymers that is initiated by a species bearing an unpaired electron (26.2c).

radical reaction　　A term used to describe chemical processes that proceed via intermediates with unpaired electrons (5.1c).

radical　　Any species that has a single unshared electron (2.5e).

random coil conformation　　The spatial arrangement of amino acids in a protein with no regular structure (27.5b).

random copolymer　　A polymer derived from two different monomers in which no particular sequence exists (26.1b).

rate-determining step　　The slowest process in a multistep chemical reaction (5.3a).

reactant favored　　A chemical process in which the reactants have a lower energy state than the products (5.4a).

reactants   The molecules in a chemical reaction that undergo conversion to other substances, which are called products (5.1a).

reaction coordinate   A term referring to the progression of changes in molecular structures as reactants are converted to products (5.4a).

reagent   The reactants that are of secondary importance in a chemical process; the principal reactant is the *substrate* (5.1a).

rearrangement reaction   A chemical process that changes the connectivities among the atoms in the carbon skeleton of the substrate molecule (5.1b).

receptor   A synonym for a host molecule that is involved in molecular recognition processes (28.3).

reducing sugar   A term for a carbohydrate that reacts with an oxidizing agent to form a carboxylic acid group at the C1 position (19.5b).

reduction reaction   A chemical process in which hydrogen atoms are added or heteroatoms are removed from an organic molecule during the course of the reaction (5.1b).

reductive elimination   A process for organotransition metal complexes in which reaction leads to formation of an organic molecule along with a decrease in the metal ion's coordination number and a decrease of the metal ion's oxidation state (15.3c).

regiochemistry   A term referring to the orientation by which a reagent adds to a π bond (9.1a).

regioselective   A term referring to a chemical reaction in which one orientation for addition of a reagent predominates with respect to the opposite orientation (9.1b).

regiospecific   A term referring to an reaction that occurs with a single orientation for addition of a reagent (9.1b).

repeat unit   The smallest molecular fragment replicated over and over to form a polymer's sequence (26.1a).

residue   A synonym for a general amino acid in a protein structure (5.5b).

resonance effects   The influences on chemical reactivity in a molecule that are caused by the delocalization of electrons through π bonds (5.2b).

resonance energy   The stabilizing influence in an aromatic compound that results from the delocalization of electrons through the π system (17.1a).

resonance hybrid   The electronic structure of a molecule that is the composite of the individual resonance structures (2.3a).

resonance structure   A structure with localized electrons that represents one possible electron distribution for a molecule that has delocalized electrons (2.3a).

retention of configuration   A term used in referring to reactions in which the configuration of a product is the same as that of the reactant (6.2f).

retro   The reverse direction of a chemical reaction (10.4b).

retrosynthesis   A planning process for chemical synthesis that is carried out in reverse, by considering what reactants are needed to make a particular product (15.4a).

ribbon diagram   A representation of protein structures that shows the overall structure of the backbone, but not the individual atoms (27.5b).

ring flip   A change in the conformation of cyclohexane derivatives in which the equatorial substituents become axial and vice versa (3.2c).

4n+2 rule   *See* **Hückel's rule** (17.1b).

S

saccharide   A synonym for a carbohydrate (19.3a).

saponification   The base-induced hydrolysis reaction of an ester to form the carboxylate ion and an alcohol molecule (21.4d).

sawhorse projection   A perspective side-view drawing of an aliphatic compound showing a specific C–C bond with its six substituents (3.1b).

Saytzeff product   In an elimination reaction that can form more than one product, the product with the most stable double bond (8.1b).

scalemic mixture   A term referring to an unequal proportion of enantiomers (16.3c).

Schiff base   A synonym for the imine functional group (20.1a).

Schiff-base catalysis   A reaction in which its rate is increased by the formation of an imine intermediate (23.2c).

secondary carbon atom   A carbon atom that is attached to two other carbon atoms (1.4a).

secondary structure   The regular conformations created by the spatial arrangements of the amino acids in a protein (27.5b).

segmented copolymer   A polymer derived from two different monomers in which there are separate extended sequences of each monomer (26.1b).

serine proteases   Enzymes that utilize a serine side chain to cleave peptide bonds (21.5b).

β-sheet   A regular conformation within protein structures consisting of extended strands of polypeptide chains (27.5b).

shielded   The phenomenon in NMR spectroscopy indicating that the external magnetic field is moderated more by the electrons surrounding a particular nucleus than this same field is offset by the electrons surrounding other nuclei in the molecule (13.1c).

sigma (σ) bond   The overlap of orbitals between two atoms that lie along the line passing through each nucleus (2.4b).

sites of unsaturation   A term referring to the number of π bonds and rings in a molecule (14.1c).

skeletal isomers   Molecules with the same molecular formulas differing in the connectivities between their carbon atoms (1.4a).

solvolysis   A type of nucleophilic substitution reaction in which the nucleophile is the solvent (6.2a).

solvomercuration   A type of electrophilic addition reaction in which the electrophile is a mercury ion and the nucleophile is the solvent, usually water or an alcohol molecule (9.1f).

spin–spin splitting   In ^{1}H NMR spectra, the pattern of peaks that results from magnetic coupling between the protons on the adjacent atoms (13.2a).

spirocyclic compound   A molecule in which two rings share one common atom (3.4c).

staggered conformation   A term describing the relationships between groups on adjacent saturated carbon atoms in which the groups are offset from each other as far as possible (3.1b).

step-growth polymer   *See* **condensation polymer** (26.3a).

stereogenic carbon atom   *See* **chiral carbon atom** (4.2a).

stereoisomers   Molecules with the same chemical formulas that differ by the spatial arrangements of their atoms (4.1a).

stereoselective   A term used to classify chemical reactions that generate one stereoisomer preferentially with respect to another (9.1d).

stereospecific   A term indicating that a single stereoisomer is formed in a chemical reaction (9.1d).

steric effect   The influence through-space of electrostatic repulsions that parallel the sizes of interacting groups (3.1c).

substitution reaction   A chemical process in which an atom or group in a molecule is replaced by a different one (5.1b).

substrate   The non-enzyme reactant in a biochemical process (5.5c), or the principal reactant in a chemical reaction (5.1a).

syn-addition   *See* **cis-addition** (9.4b).

syndiotactic   A term for the stereochemistry of the stereogenic centers in a polymer in which the substituents alternate from top to bottom of a chain viewed in a planar zigzag conformation (26.1d).

syn-periplanar conformation   An eclipsed conformation in which the two largest groups on adjacent atoms are aligned (3.1c).

synthetic tree   In a retrosynthesis, the scheme of possible reactions that might be used to prepare a particular molecule from different starting materials (15.4a).

T

tandem addition–alkylation   A process that makes use of a conjugate addition reaction followed by treatment with an alkyl halide (24.3b).

tautomerism   A process involving a pair of acid–base reactions that interchange the positions of a proton and an adjacent π bond (9.2b).

tautomers   Isomers that are interconverted by the concomitant movement of a proton and an adjacent π bond (9.3b).

termination   Any processes in a radical chain reaction that eliminates free radicals (12.1a).

tertiary carbon atom   A carbon atom that is attached to three other carbon atoms (1.4a).

tertiary structure   The level of protein structure that describes its overall three-dimensional shape (27.5b).

tetrahedral intermediate   The species in addition–elimination reactions of carboxylic acid derivatives that has an sp^3-hybridized carbon atom after addition to the carbonyl group has occurred (21.1c).

thermodynamic control   A term describing reactions that can yield two products and denoting that formation of the more stable product predominates (8.1b).

thermodynamic enolate   Of the two enolate ions that can be formed from an unsymmetrical ketone, the enolate ion that is more stable (22.2f).

thermoplastic   A polymer that can be melted by heating to form a fluid that is readily molded (26.1c).

thermoset plastic   A polymer that can be molded only when it is first formed as a molten solid; once cooled, the polymer cannot be melted and molded again (26.1c).

thioacetal   A functional group that has two –SR groups attached to a single carbon atom (19.2c).

thiolate ion   The conjugate base of a thiol (7.3a).

torsional angle   *See* **dihedral angle** (3.1b).

torsional strain   The energy stored in cyclic compounds as the result of eclipsed interactions between neighboring atoms (3.22).

tosylate   An abbreviated name for the *p*-toluenesulfonate group, $CH_3C_6H_4SO_3$- (7.1d).

trans isomer   The structural form of an alicyclic compound in which two specified groups are on opposite sides of the ring (3.3d). Also, the structural form of a C=C double bond in which two protons—each attached to the separate carbon atoms—are on opposite sides of the bond connecting the carbon atoms (4.1a).

trans-addition   *See anti*-addition (9.1d).

transannular interactions   Effects between groups that occur across a ring (3.4b).

transesterification   A process in which one ester is converted to another by reaction with a different alcohol molecule (21.4e).

transition state   The highest energy point of a reaction step (5.4a).

triflate   An abbreviated name for the trifluoromethanesulfonate group, CF_3SO_3- (7.1b).

β-turn   A regular conformation within protein structures that permits a polypeptide chain to fold onto itself to form a β- sheet (27.5b).

U

unimolecular   A term used to describe the kinetic profile of a reaction in which the rate is dependent on the concentration of only one reactant (6.1d).

upfield   A term indicating a feature or region in an NMR spectrum that lies to the right of or has a smaller chemical shift value than another feature or region (13.1c).

urea   A functional group that has two amino groups attached to a carbonyl group (26.3d).

urethane   *See* **carbamate** (26.3d).

V

valence bond theory   A simplification of quantum mechanics that defines bond formation as the result of overlap between atomic or hybrid orbitals that occupy the same region of space (2.4b).

vicinal   A term denoting that substituents are attached to adjacent atoms (8.2e).

vicinal coupling   In proton NMR spectroscopy, the nuclear spin–spin interactions that occur between protons attached to adjacent carbon atoms (13.2a).

vinyl carbon atom   *See* **alkenyl carbon atom** (1.4c).

vinyl ether   *See* **enol ether** (19.1c).

VSEPR model   A concept used to rationalize molecular shapes and based on the idea that electron pair repulsions lead to geometries that represent the lowest energy structures (2.5a).

W

Wagner–Meerwein rearrangement   A term for reactions involving carbocation intermediates that lead to structural changes in the carbon skeletons of aliphatic molecules (6.2e).

Walden inversion   The change of configuration from (R) to (S) or vice versa that takes place during the S_N2 reactions of chiral substrates (6.3b).

wave function   The solutions of quantum mechanical wave equations that define the shapes of atomic and molecular orbitals (2.4a).

wavenumbers   The quantity in an infrared spectrum that defines the frequency of an absorption band (14.3a).

Y

ylide   A species with a carbanion center bonded directly to a heteroatom bearing a positive charge (11.4b).

A classification of nucleophiles according to their basicity and nucleophilicity properties (Table 6.3).

Weak Base (pK_a conjugate acid < 5)		Moderate Base (pK_a conjugate acid ~5–12)		Strong Base (pK_a conjugate acid >15)	
Poor nucleophile	*Good nucleophile*	*Poor nucleophile*	*Good nucleophile*	*Poor nucleophile*	*Good nucleophile*
HSO_4^-	H_2O	$RCOO^-$	N_3^-	$(CH_3)_3CO^-$	OH^-
$ROSO_2^-$	1° ROH	R_3N	CN^-	LDA	RO^-
$H_2PO_4^-$	RCOOH		ArS^-		H^-
F^-	H_2S		RS^-		NH_2^-
ROH	RSH		NH_3		
	R_3P		RNH_2		
	Cl^-		R_2NH		
	Br^-		ArO^-		
	I^-				

Some electrophiles and nucleophiles commonly encountered in chemical reactions (Table 5.7).

Substances with electrophilic centers			Substances with nucleophilic centers		
General form	*Example*	*Species*	*General form*	*Example*	*Species*
H_3O^+	H_2SO_4 in H_2O	Proton (acid)	X^-	$:\ddot{C}l:^-$	Chloride ion
				$H\ddot{O}:^-$	Hydroxide ion
$\overset{\delta^+}{C}\!-\!\overset{\delta^-}{X}$	CH_3-Br	Alkyl halide		$CH_3\ddot{O}:^-$	Methoxide ion
	CH_3-OH	Alcohol		$^-:CN:$	Cyanide ion
$:\overset{\delta^-}{\ddot{O}}$ ‖ $\overset{\delta^+}{C}$	CH_3-CHO	Aldehyde	$\overset{\delta^-}{C}\!-\!\overset{\delta^+}{M}$	CH_3-Li	Organolithium compound
	$CH_3-CO-CH_3$	Ketone			
	$CH_3-CO-Cl$	Acid chloride		CH_3-MgBr	Grignard reagent
	$CH_3-CO-OCH_3$	Ester			
B	BH_3	Borane	N	$:N(CH_3)_3$	Amine
	BF_3	Boron trifluoride			
$:\ddot{O}:^-$ S^+	$Cl-SO-Cl$	Thionyl chloride	P	$:P(CH_3)_3$	Phosphine
			$C\!=\!C$	$H_2C=CH_2$	π Bond
$\overset{\delta^-}{X}$ $\overset{\delta^+}{P}$	PCl_3	Phosphorus trichloride	$-C\!\equiv\!C-$	$HC\equiv CH$	π Bond